Indium
Nitride
and
Related
Alloys

Indium Nitride and Related Alloys

Edited by
T. D. Veal
C. F. McConville
W. J. Schaff

CRC Press
Taylor & Francis Group
Boca Raton London New York

CRC Press is an imprint of the
Taylor & Francis Group, an **informa** business

CRC Press
Taylor & Francis Group
6000 Broken Sound Parkway NW, Suite 300
Boca Raton, FL 33487-2742

First issued in paperback 2017

ISBN 13: 978-1-138-11672-6 (pbk)
ISBN 13: 978-1-4200-7809-1 (hbk)

Library of Congress Cataloging-in-Publication Data

Indium nitride and related alloys / editors, T.D. Veal, C.F. McConville, and W.J. Schaff.
    p. cm.
    Includes bibliographical references and index.
    ISBN 978-1-4200-7809-1 (hard back : alk. paper)
    1. Indium. 2. Indium alloys. 3. Nitrides. 4. Semiconductors--Materials. I. Veal, T. D. (Tim D.) II. McConville, C. F. (Chris F.) III. Schaff, William Joseph, 1956- IV. Title.

    TA480.I53I54 2010
    620.1'89--dc22
                                                                                    2009018061

Contents

3   Polarity-dependent epitaxy control of InN, InGaN and InAlN   83
X. Q. Wang and A. Yoshikawa

4   InN in brief: Conductivity and chemical trends   121
P. D. C. King, T. D. Veal, and C. F. McConville

Preface

Following the discovery between 2000 and 2002 that the band gap of indium nitride (InN) is close to 0.7 eV rather than the previously accepted value of ~1.9 eV, intensive and rapidly progressing research has focused on determining the physical properties of this semiconductor material. Since a consensus has emerged about the narrow band gap nature of InN, it seems to be an appropriate time to summarize the current state of knowledge of the properties of InN and thereby move toward developing device applications of what is perhaps the last of the 'common' III-V semiconductors to be exploited.

The advancement of our knowledge of InN and In-rich alloys was dramatically accelerated by the progress made during a series of three International InN Workshops co-chaired by William J. Schaff, K. Scott Butcher, and José Fernando Chubaci. Many of the contributors met for the first time at these events and created a remarkable informal network of collaborations. The meetings were sponsored by: U. S. Air Force Office of Scientific Research (AFOSR), Asian Office of Aerospace Research and Development, Australian Nuclear Science and Technology Organisation, Macquarie University, European Office of Aerospace Research and Development, U. S. Office of Naval Research (ONR), ONR Global. Dr. Colin Wood of ONR initiated this Workshop series with the encouragement of Professor Trevor Tansley of Macquarie University in 2003 and they were successfully coordinated by AFOSR staff. The results of the collaborations initiated and developed at these workshops form a significant part of the content of the following chapters.

The purpose of this book is to provide a comprehensive account of the advances made in the growth, characterization, and understanding of InN and related alloys. It is intended to be both an invaluable resource to established InN researchers and an excellent starting point for those new to the field. Hopefully, it gives the reader a clear description of both important results and current understanding that will inform future work.

Chapters 1–3 concentrate on different aspects of epitaxial growth of InN and related alloys. Chapter 4 gives a brief introduction to the electrical conductivity behavior of InN and places it within the context of the chemical trends of III-V semiconductors. The transport properties of InN are comprehensively described in Chapter 5, while optical properties of InN and related alloys are reviewed in Chapters 6 and 7. After the results of bulk band structure calculations are outlined in Chapter 8, optical properties are further investigated in Chapter 9 by spectroscopic ellipsometry. Chapters 10 and 11 review experimental and theoretical approaches to studying defects and doping, while experimental and theoretical aspects of InN surfaces are described in Chapters 12 and 13, respectively. Chapter 14 details microscopy studies of the structure of InN and InGaN and Chapter 15 considers the potential of Mn- or Cr-doped InN as a dilute magnetic semiconductor. Chapter 16 explores the growth and properties of InN-based low dimensional structures. Finally, recent progress on InN nanocolumn research is presented in Chapter 17.

We are very grateful to all the authors who have contributed chapters. Thanks to Phil King at Warwick for typesetting several of the chapters and providing LATEX assistance. We would also like to thank all the staff at CRC Press/ Taylor & Francis for their help and patience with this project. Tim Veal would like to thank Ann-Marie Wyatt for her invaluable help and support. He also acknowledges financial support from the University of Warwick's Contract Research Staff Study Leave Programme and an Engineering and Physical Sciences Research Council (EPSRC), UK, Career Acceleration Fellowship. Chris McConville would like to thank his wife Hayley for her continual love and support.

Someone once said at an InN meeting that we wrote the book on InN at Cornell. This was not true at the time. However, once we had developed direct-write patterning using a focused thermal beam during molecular-beam epitaxy, as shown in Fig. 1, we could then say, "At Cornell, we wrote 'The Book' on InN". And now, with the excellent contributions from many of the leaders in the field of InN research, we can say in a different way that we have written the book on InN!

FIGURE 1
Scanning electron microscope image of InN after direct-write patterning during molecular-beam epitaxy using an *in situ* focused thermal beam.

Tim Veal and Chris McConville
University of Warwick

William J. Schaff
Cornell University

List of contributors

J. W. Ager III
Materials Sciences Division
Lawrence Berkeley National Laboratory
Berkeley, California 94720
United States of America

O. Ambacher
Fraunhofer Inst. for Applied Solid State Physics
Tullastr. 72, 79108 Freiburg
Germany

T. Araki
Department of Photonics
Ritsumeikan University
1-1-1 Noji-higashi, Kusatsu 525-8577
Japan

F. Bechstedt
Institut für Festkörpertheorie und -optik
Friedrich-Schiller-Universität and
European Theoretical Spectroscopy Facility
Max-Wien-Platz 1, D-07743 Jena
Germany

E. Calleja
ISOM-Departamento Ingeniería Electrónica
ETSI Telecomunicación
Universidad Politécnica de Madrid
Ciudad Universitaria, 28040 Madrid
Spain

J. M. Calleja
Department Física de Materiales
Universidad Autónoma de Madrid
E-28049 Madrid
Spain

S. B. Che
Graduate School of Electrical and
Electronics Engineering
Chiba University
1-33 Yayoi-cho, Inage-ku, Chiba 263-8522
Japan

V. Cimalla
Fraunhofer Inst. for Applied Solid State Physics
Tullastr. 72, 79108 Freiburg
Germany

V. Yu. Davydov
Ioffe Physico-Technical Institute
Russian Academy of Science
Polytechnicheskaya 26
194021 St. Petersburg
Russia

S. M. Durbin
Department of Electrical and
Computer Engineering
The MacDiarmid Institute for
Advanced Materials and Nanotechnology
University of Canterbury
Te Whare Wānanga o Waitaha
Christchurch 8140
New Zealand

G. Ecke
Institut für Mikro- und Nanotechnologien
Technische Universität Ilmenau
Gustav-Kirchhoff Str. 7, 98693 Ilmenau
Germany

F. Fuchs
Institut für Festkörpertheorie und -optik
Friedrich-Schiller-Universität and
European Theoretical Spectroscopy Facility
Max-Wien-Platz 1, D-07743 Jena
Germany

J. Furthmüller
Institut für Festkörpertheorie und -optik
Friedrich-Schiller-Universität and
European Theoretical Spectroscopy Facility
Max-Wien-Platz 1, D-07743 Jena
Germany

E. Gallardo
Department Física de Materiales
Universidad Autónoma de Madrid
E-28049 Madrid
Spain

C. S. Gallinat
Materials Department
University of California
Santa Barbara, California 93106-5050
United States of America

R. Goldhahn
Institute of Physics and Institute of Micro- and
Nanotechnologies
Technical University of Ilmenau
PF 100565, 98684 Ilmenau
Germany

J. Grandal
ISOM-Departamento Ingeniería Electrónica
ETSI Telecomunicación
Universidad Politécnica de Madrid
Ciudad Universitaria, 28040 Madrid
Spain

A. Janotti
Materials Department
University of California
Santa Barbara, California 93106-5050
United States of America

R. E. Jones
Materials Sciences Division
Lawrence Berkeley National Laboratory and
Department of Materials Science and Engineering
University of California
Berkeley, California 94720
United States of America

P. D. C. King
Department of Physics
University of Warwick
Coventry, CV4 7AL
United Kingdom

A. A. Klochikhin
Ioffe Physico-Technical Institute
Russian Academy of Science
Polytechnicheskaya 26
194021 St. Petersburg
Russia

G. Koblmüller
Materials Department
University of California
Santa Barbara, California 93106-5050
United States of America

S. Lazić
Department Física de Materiales
Universidad Autónoma de Madrid
E-28049 Madrid
Spain

V. Lebedev
Fraunhofer Inst. for Applied Solid State Physics
Tullastr. 72, 79108 Freiburg
Germany

Z. Liliental-Weber
Materials Sciences Division
Lawrence Berkeley National Laboratory
Berkeley, California 94720
United States of America

E. Luna
Paul-Drude-Institut für Festkörperelektronik
Hausvogteiplatz 5-7, 10117 Berlin
Germany

C. F. McConville
Department of Physics
University of Warwick
Coventry, CV4 7AL
United Kingdom

N. Miller
Materials Sciences Division
Lawrence Berkeley National Laboratory and
Department of Materials Science and Engineering
University of California
Berkeley, California 94720
United States of America

T. H. Myers
Materials Science and Engineering Program
Texas State University
601 University Drive
San Marcos, TX 78666
United States of America

Y. Nanishi
Department of Photonics
Ritsumeikan University
1-1-1 Noji-higashi, Kusatsu 525-8577
Japan

M. Niebelschütz
Institut für Mikro- und Nanotechnologien
Technische Universität Ilmenau
Gustav-Kirchhoff Str. 7, 98693 Ilmenau
Germany

V. M. Polyakov
Institut für Mikro- und Nanotechnologien
Technische Universität Ilmenau
Gustav-Kirchhoff Str. 7, 98693 Ilmenau
Germany

M. Röppischer
Institute for Analytical Sciences
Department Berlin
Albert-Einstein-Str. 9, 12489 Berlin
Germany

M. A. Sánchez-García
ISOM-Departamento Ingeniería Electrónica
ETSI Telecomunicación
Universidad Politécnica de Madrid
Ciudad Universitaria
28040 Madrid
Spain

W. J. Schaff
Department of Electrical and Computer
Engineering
Cornell University
Ithaca, New York 14853
United States of America

P. Schley
Institute of Physics and Institute of Micro- and
Nanotechnologies
Technical University of Ilmenau
PF 100565, 98684 Ilmenau
Germany

F. Schwierz
Institut für Mikro- und Nanotechnologien
Technische Universität Ilmenau
Gustav-Kirchhoff Str. 7, 98693 Ilmenau
Germany

J. S. Speck
Materials Department
University of California
Santa Barbara, California 93106-5050
United States of America

A. Trampert
Paul-Drude-Institut für Festkörperelektronik
Hausvogteiplatz 5-7, 10117 Berlin
Germany

C. G. Van de Walle
Materials Department
University of California
Santa Barbara, California 93106-5050
United States of America

T. D. Veal
Department of Physics
University of Warwick
Coventry, CV4 7AL
United Kingdom

W. Walukiewicz
Materials Sciences Division
Lawrence Berkeley National Laboratory
Berkeley, California 94720
United States of America

X. Q. Wang
Department of Electronics and
Mechanical Engineering
Chiba University
1-33 Yayoi-cho, Inage-ku, Chiba 263-8522
Japan

J. Wu
Materials Sciences Division
Lawrence Berkeley National Laboratory and
Department of Materials Science and Engineering
University of California
Berkeley, California 94720
United States of America

T. Yamaguchi
Department of Photonics
Ritsumeikan University
1-1-1 Noji-higashi, Kusatsu 525-8577
Japan

J. W. L. Yim
Materials Sciences Division
Lawrence Berkeley National Laboratory and
Department of Materials Science and Engineering
University of California
Berkeley, California 94720
United States of America

A. Yoshikawa
Department of Electronics and
Mechanical Engineering
Chiba University
1-33 Yayoi-cho, Inage-ku, Chiba 263-8522
Japan

K. M. Yu
Materials Sciences Division
Lawrence Berkeley National Laboratory
Berkeley, California 94720
United States of America

1

Molecular-beam epitaxy of InN

Y. Nanishi, T. Araki, and T. Yamaguchi

Department of Photonics, Ritsumeikan University, 1-1-1 Noji-higashi, Kusatsu 525-8577, Japan

1.1 Introduction

For a long time, difficulty in obtaining high-quality InN due to the low dissociation temperature of InN and the extremely high equilibrium vapor pressure of nitrogen has hindered the understanding of fundamental physical properties of InN and their applications. During the past few years, properties of InN grown by radio frequency plasma-assisted molecular beam epitaxy (RF-MBE) have been significantly improved [1.1]. Metalorganic vapor phase epitaxy (MOVPE) has an inherent disadvantage because it must satisfy the conditions for NH_3 pyrolysis and prevention of InN dissociation, which impose conflicting temperature requirements. In contrast, RF-MBE has an essential advantage over MOVPE for obtaining high-quality InN. In this growth method, excited nitrogen radicals can be generated separately in a plasma source, enabling us to select growth temperature without considering the requirements of NH_3 pyrolysis. Due to this inherent advantage, the quality of InN grown by RF-MBE has improved very quickly over a relatively short period of time [1.1–1.5] and a room temperature electron mobility over 2100 cm^2/Vs and a residual carrier concentration close to 3×10^{17} cm^{-3} were obtained [1.1].

Using single crystalline InN films obtained by MBE and MOVPE, studies using photoluminescence (PL) and optical absorption have been performed. The fundamental band gap of single crystalline InN was argued to be around 0.8 eV [1.6–1.11] instead of 1.9 eV which was determined from polycrystalline InN. Now, a narrow band gap of around 0.65 eV is widely recognized after several years of discussion. This new finding expanded the field of applications of group III nitrides, with the alloy systems spanning from 6.2 eV of AlN to 0.65 eV of InN. However, many scientific and technical issues hinder applications of InN-based material systems to actual devices. These include very high concentrations of residual donors, high densities of dislocations due to large lattice mismatch, surface accumulation of carriers, *p*-type doping and formation of high-quality heterointerface. To overcome these critical issues, further developments in MBE growth of InN-based materials are strongly required. In this chapter, we will review comprehensive studies of MBE growth of InN and related alloys.

1.2   MBE growth of InN on c-plane (0001) sapphire

RF-MBE growth of InN is performed in the MBE growth chamber equipped with a high-speed turbo-molecular pump (1500 l/s), which can be evacuated to 1×10^{-10} Torr. This MBE system is also equipped with a SVTA radical cell (model 2.75) for nitrogen source supply. The substrates used were mainly c-plane (0001) sapphire substrates. The backside was coated with 1 μm thick molybdenum. They were set on an In-free molybdenum holder after cleaning by organic solvents. High-purity (6N) In and Ga are used as group III materials and are evaporated from standard effusion cells. High-purity N_2 gas (6N) is supplied to the RF plasma cell after passing through a SAES Getters purifier (model Monotorr Phase II 3000) to remove impurities including H_2O, H_2, CO, CO_2 gases. InN growth was performed under 1×10^{-4} Torr when 2 standard cubic centimeters per minute (sccm) of nitrogen gas was supplied. Growth temperatures were carefully monitored by pyrometer during each growth run.

In this section, basic growth processes to obtain high-quality single-crystal InN films on c-plane (0001) sapphire are reviewed; these include the substrate nitridation process, two-step growth processes, the precise control of the V/III ratio and growth temperature. Then, we describe how to further improve crystal quality of InN films by novel growth techniques.

1.2.1   Nitridation of c-plane (0001) sapphire substrate

A nitridation process of c-plane (0001) sapphire at 550°C for 1 hour prior to growth was found to be essential to obtain single-crystalline InN on sapphire [1.12]. Profiles of x-ray diffraction (XRD) pole figure of InN grown without and with the nitridation process are compared in Fig. 1.1. Without this process, InN tends to grow with both $[10\bar{1}0]_{InN} \| [11\bar{2}0]_{sapph}$ and $[11\bar{2}0]_{InN} \| [11\bar{2}0]_{sapph}$ epitaxial relationships, resulting in the formation of poly-crystalline InN with both epitaxial relationships as shown in Fig. 1.1(a) [1.13, 1.14]. This is due to the fact that lattice mismatch values between InN and sapphire are very close to each other in these cases as summarized in Fig. 1.2. This relationship is in contrast to GaN or AlN growth on sapphire, where lattice mismatch for the case of $[10\bar{1}0]_{GaN,AlN} \| [11\bar{2}0]_{sapph}$ is much smaller than that of $[11\bar{2}0]_{GaN,AlN} \| [11\bar{2}0]_{sapph}$. With the nitridation process, on the other hand, the epitaxial relationship can be uniquely determined to be $[10\bar{1}0]_{InN} \| [11\bar{2}0]_{sapph}$ as shown in Fig. 1.1(b). This is probably due to the formation of an AlN or AlON layer with $[10\bar{1}0]_{AlN} \| [11\bar{2}0]_{sapph}$, which results in the $[10\bar{1}0]_{InN} \| [10\bar{1}0]_{AlN} \| [11\bar{2}0]_{sapph}$ relationship.

(a) (b)

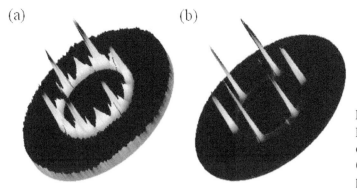

FIGURE 1.1
Profiles of XRD pole figure of InN grown (a) without and (b) with the nitridation of sapphire process.

	$[11\bar{2}0]_{III-N}//[11\bar{2}0]_{Sap}$	$[11\bar{2}0]_{III-N}//[10\bar{1}0]_{Sap}$
Atomic arrangement (e.g. InN)	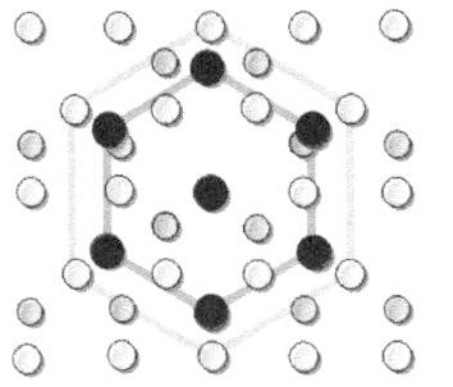	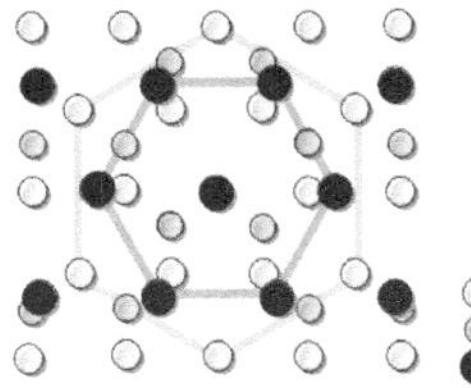
AlN a=3.112Å	-34.6%	+13.3%
GaN a=3.189Å	-33.0%	+16.1%
InN a=3.548Å	-25.4%	+29.2%

FIGURE 1.2
Atomic arrangements and lattice mismatch values for III-nitride semiconductor/sapphire heterointerfaces.

1.2.2 Low-temperature InN buffer layer

It is essential to obtain atomically flat surfaces to grow high-quality InN films. The application of the two-step growth method which utilizes low-temperature buffer growth at around 550°C prior to main epitaxial growth at around 1000°C is known to be very effective to grow flat and high-quality GaN films on sapphire substrates by MOVPE [1.15]. In this study, InN buffer layers with around 30 nm in thickness were grown at 300°C for 10 min. Consecutive annealing at 550°C was performed [1.16]. Figure 1.3 shows the RHEED patterns and surface morphologies of an InN buffer layer grown at 300°C (a) before and (b) after annealing at 550°C. This figure indicates that the surface morphology was improved by annealing at 550°C. The crystal quality of the buffer layer grown at 300°C is poor due to shortage of In migration on the surface. However, the crystallinity of InN film was remarkably improved by consecutive annealing as well as improvement of surface morphology. Improvement in crystal quality is evident from the RBS results before and after annealing at 550°C [1.4].

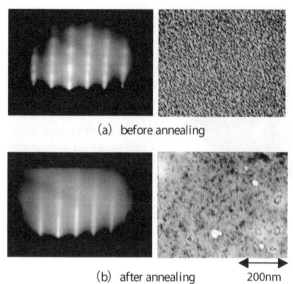

(a) before annealing

(b) after annealing   200nm

FIGURE 1.3
RHEED patterns (e$\|$[11$\bar{2}$0]) and sur-face morphology by SEM observation of InN buffer layer; (a) before and (b) after annealing at 550°C.

1.2.3   V/III ratio and growth temperature

The control of the V/III ratio is by far the most important issue to obtain high-quality InN by MBE [1.17]. At temperatures below the dissociation temperature of InN (around 550°C), the vapor pressure of N over InN is much higher than In vapor pressure from In metal [1.18]. To suppress the dissociation of InN films, one should supply a nitrogen pressure higher than the thermo-equilibrium pressure throughout the growth process. Once the beam equivalent pressure of In becomes higher than the equivalent nitrogen pressure, In droplets should form on the surface. They cannot be evaporated from the surface at the selected growth temperature under the dissociation temperature. This implies that InN should be grown under N-rich conditions. On the other hand, it is well known that high-quality GaN can be grown under slightly Ga-rich conditions due to the enhanced migration of Ga. This should be the case also for InN growth by MBE. Considering all these limitations for the InN MBE growth process, one can predict that the V/III ratio should be controlled as closely as possible to one from the N-rich side. Growth temperature should be selected at the highest possible temperature without noticeable dissociation of InN film in order to guarantee enough migration of In. Keeping the growth temperature, beam equivalent pressure of In, and N_2 flow rate constant at 550°C, 3.2×10^{-7} Torr, and 2.0 sccm, the effective V/III ratio on the growing surface was carefully controlled by changing the RF-plasma power between 230 and 250 W. Cross-sectional scanning electron microscopy (SEM) images of these InN films are shown in Fig. 1.4 [1.19]. A high density of In droplets is observed for the sample grown under an RF-plasma power of 230 W, apparently due to a lack of active nitrogen. The pit size of InN grown at 240 W is much smaller compared with that of the 250 W sample. Presumably, the 240 W sample was grown under the condition closer to surface stoichiometry and the 250 W sample should have been grown under deeper N-rich conditions. From the results of a ω-2θ XRD scan of InN for the plasma power of 240 W and 250 W, only two diffraction peaks corresponding to the (0002) diffraction peak from wurtzite InN and the (0006) diffraction peak from sapphire are observed from the 250 W sample. A very small trace of a peak corresponding to In metal was observed, however, from the 240 W sample, probably due to slightly In-rich conditions. Thus, precise control of the V/III ratio is vital to obtain high-quality InN by RF-MBE.

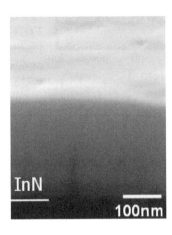

FIGURE 1.4
SEM images of the surface morphology of InN grown at plasma powers of (a) 230 W, (b) 240 W and (c) 250 W.

Optimization of the growth temperature is another critical issue to obtain high-quality InN [1.9]. InN films were grown at various growth temperatures between 460°C and 550°C and corresponding Raman spectra were compared. Raman spectra for InN films grown at 550°C, 530°C, 500°C and 460°C are shown in Fig. 1.5. As the growth temperature increases, the full width at half maximum (FWHM) of the E_2(high)-phonon-mode decreases. A FWHM as narrow as 3.7 cm^{-1} was obtained for 550°C growth. These results indicated that the crystalline quality of InN was improved by increasing the growth temperature within the dissociation limit of InN. The dependence of InN PL spectra on growth temperature again indicates that the crystalline quality is sensitive to growth temperature and improves as the growth temperature increases below the dissociation limit of InN. Based on these comprehensive studies, the typical growth diagram shown in Fig. 1.6 was fixed as a standard growth process for high-quality InN by RF-MBE.

1.2.4 Optimization of low-temperature nitridation process

As described in section 1.2.1, nitridation of sapphire substrates prior to growth improves the crystalline quality of InN because: (i) the formation of AlN on the sap-

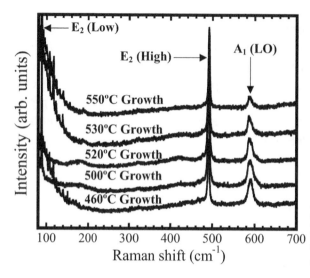

FIGURE 1.5

Raman spectra for InN films grown at 550°C, 530°C, 520°C, 500°C and 460°C.

phire surface reduces the in-plane lattice mismatch between InN and the underlying layer from ~25% for $[11\bar{2}0]_{InN} \| [11\bar{2}0]_{sapph}$ to ~13% for $[10\bar{1}0]_{InN} \| [10\bar{1}0]_{AlN}$; and (ii) AlN formation suppresses the generation of a multi-domain InN structure. However, the correlation between nitridation conditions and InN crystalline quality has not been investigated in detail. In this section, we review the effects of nitridation of sapphire substrates on InN crystalline quality by varying the nitridation conditions [1.20].

Nitridation of the sapphire substrates was performed at 300–800°C for 1–3 hours without using the ion deflector to remove nitrogen ions. The LT-InN buffer layers were deposited at 300°C for 10 min. After deposition of the buffer layer, the InN films were grown at 530°C for 1 hour. The sapphire surface was studied by *in situ* RHEED and XPS before and after nitridation under varying conditions. RHEED patterns revealed that higher temperature and longer nitridation times resulted in a shorter distance between RHEED streaks. This indicates that the substrate surface changed from sapphire to $AlO_{1-x}N_x$ or AlN. Figure 1.7 shows the XPS results. In Fig. 1.7(a), the integrated N $1s$ peak intensity versus nitridation time for a fixed nitridation temperature of 300°C is shown. The intensity increased linearly for the first 2 hours and began to saturate at ~800 counts/s thereafter. Heinlein *et al.* reported that integrated N $1s$ peak intensity of nitridated sapphire as a function of nitridation time has two distinct regimes: linear increase and saturation [1.21]. This agrees with the results of the present study. Heinlein *et al.* concluded that the saturation regime signals the completion of the formation of 1 monolayer (ML) of AlN on the sapphire surface. Figure 1.7(b) shows the integrated N $1s$ peak intensity for the nitridated sapphire versus nitridation temperature for a fixed nitridation time of 2 hours. The intensity increased with nitridation temperature. The intensities were higher than ~800 counts/s. These results imply that more than 1 ML of AlN was formed on the sapphire surface at higher nitridation temperatures.

Figure 1.8 shows the dependence of XRC-FWHM values for (0002), $(10\bar{1}2)$ and

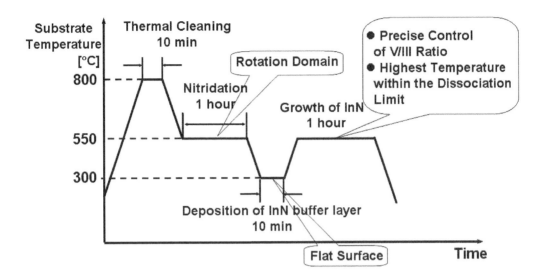

FIGURE 1.6
Typical growth diagram for high-quality InN by RF-MBE.

$(10\bar{1}0)$ InN on nitridation conditions, for InN films grown on LT-InN buffer layers. XRC-FWHM values for $(10\bar{1}0)$ InN were theoretically estimated from those for (0002) and $(10\bar{1}2)$ InN [1.22]. The dependence on nitridation temperature and time are shown in Figs. 1.8(a) and (b), respectively, where sapphire nitridation was carried out for a fixed time of 2 hours for the former and at a fixed temperature of 300°C for the latter. It can be seen that XRC-FWHM values for $(10\bar{1}0)$ InN stay small and almost constant at approximately 80 arcmin, independent of nitridation time and temperature. We conclude that the LT-InN buffer layers are effective in suppressing twisting in InN. On the other hand, the XRC-FWHM values for (0002) InN vary dramatically with nitridation temperature and time. Higher-temperature nitridation, which is thought to yield a thicker AlN layer on the sapphire surface, resulted in larger tilt distributions in InN, as shown in Fig. 1.8(a). This is attributed to the rougher sapphire surface that results from the higher nitridation temperatures. However, longer nitridation times at 300°C produced smaller tilt distributions in InN, as shown in Fig. 1.8(b). The tilt distributions in InN were presumably influenced by both surface roughness and coverage of the sapphire by the nitrided layers. The XRC-FWHM value of 54 arcsec for (0002) InN, the lowest value reported to date, was achieved by using the LT-InN buffer layer and sapphire nitridation at 300°C for 3 hours.

1.2.5 Regrowth on micro-faceted InN template

In section 1.2.4, we found that low-temperature nitridation of the sapphire substrate is effective in improving the tilt distribution of InN films. We have already achieved InN films of very small tilt distribution. The FWHMs of x-ray rocking curves

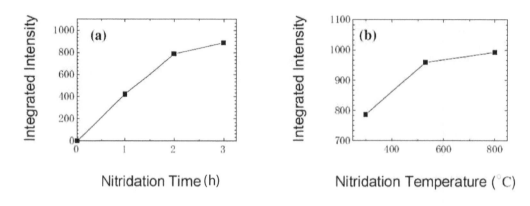

FIGURE 1.7
Integrated N 1s peak intensity for nitridated sapphire versus (a) nitridation time and (b) nitridation temperature. Nitridation was carried out (a) at a fixed temperature of 300°C and (b) for a fixed time of 2 hours.

for the (0002) reflection from these InN films were reproducibly as narrow as 1 arcmin [1.20]. However, the InN films still exhibit large twist distribution. Simply growing InN films of over 1 μm thick is a possible way to achieve high-quality InN films of small tilt and twist distributions [1.4, 1.23]. However, it is difficult to obtain a relatively flat surface in such thick InN films due to the difficulty of precisely controlling and maintaining the optimum growth conditions [1.1]. In the case of GaN grown by MOVPE, it has been quite recently reported that the twist distribution could be improved by using micro-faceted GaN templates [1.24]. In this section we demonstrate successful improvement of the twist distribution in InN films regrown on micro-faceted InN templates formed by KOH wet etching [1.25].

InN films were grown on micro-faceted and flat surface N-polar InN templates by RF-MBE. These InN templates were fabricated on the same piece of (0001) sapphire. Prior to growth, the sapphire substrate was nitrided at 300°C for 2 hours. After a low-temperature InN buffer layer was deposited at 300°C for 10 min, an InN film was grown at 550°C for 1 hour. Then the sample was taken out of the MBE chamber. After half of the sample was masked by wax, the sample was etched by a 10 mol/l KOH solution for 2 hours at room temperature (RT). After the wax was removed, surface morphologies of the sample were observed by SEM. The etched area had a rough surface with hexagonal InN pyramids with {1011} facets [1.25]. On the other hand, the area that had been covered by wax kept a flat surface. After these templates fabricated on the same piece were treated in HCl solution for 30 sec to remove the native oxides from their surface, they were transferred back into the MBE chamber. InN with a thickness of 375 nm was then grown at 550°C for 1 hour. The resultant total thickness of InN/micro-faceted InN and InN/flat InN template is 650 and 750 nm, respectively.

In order to evaluate the crystallographic quality of the InN films, XRCs for the

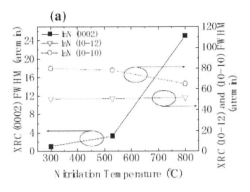 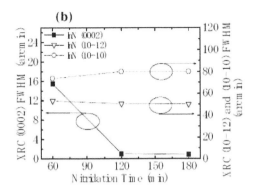

FIGURE 1.8

Dependencies of XRC-FWHM values for the (0002), (10$\bar{1}$2) and (10$\bar{1}$0) InN on nitridation conditions, for InN films grown on LT-InN buffer layers. The nitridation was carried out: (a) for a fixed time of 2 hours, and (b) at a fixed temperature of 300°C.

(0002), (10$\bar{1}$2) and (30$\bar{3}$2) InN reflections were measured. Additionally, the XRC-FWHM values for the (10$\bar{1}$0) InN reflection were theoretically estimated from those for the (0002), (10$\bar{1}$2) and (30$\bar{3}$2) InN reflections [1.22]. These results are shown in Table 1.1. XRC-FWHM values for the (0002)/(10$\bar{1}$0) InN reflection were 1.13/36.5 and 1.19/50.6 arcmin for the InN films grown on the micro-faceted templates and those on flat surface InN templates, respectively. The FWHM values for the (0002) reflection, which correspond to the tilt distributions, were almost the same for both types of sample, keeping their excellently narrow original values of the underlying templates. As for the FWHM values for the (10$\bar{1}$0) reflection, which corresponds to the twist distributions, the InN grown on the micro-faceted template showed a considerably smaller FWHM value than that grown on the flat InN template. From these results, it was revealed that regrowth of InN on the micro-faceted InN template is an effective way to improve the twist distribution without any deterioration of the excellent tilt distribution.

TABLE 1.1

XRC-FWHM values of InN grown on micro-faceted and flat InN templates.

structure	XRD rocking curve FWHM (arcmin)			
	(0002)	(10$\bar{1}$2)	(30$\bar{3}$2)	(10$\bar{1}$0)
InN/micro-faceted InN template	1.13	25.2	34.5	36.5
InN/flat InN template	1.19	35.3	45.7	50.6

Dislocations in the InN films were observed by TEM. Figures 1.9(a) and (b) shows the cross-sectional TEM images of InN films grown on micro-faceted InN template with g = 0002 and g = 1$\bar{1}$00. It was found that both InN templates and regrown layers had dramatically low screw dislocation density. These results are in good agreement with their excellently small tilt distributions determined from the small XRC-FWHM values for the (0002) InN reflection. On the other hand, a high density of dislocations with edge component was clearly observed in micro-faceted InN templates. However, the InN layer regrown on the micro-faceted InN template had much lower edge dislocation density than that grown on the flat InN template. It is found that edge-type dislocations tend to bend toward the top of hexagonal pyramids at the regrowth interface. Then, edge-type dislocations in the InN template assembled on the top of the hexagonal pyramid and propagated into the regrown InN film. As a result, apart from at the top of hexagonal pyramids, the regrowth area exhibited relatively low dislocation density due to the bending of the underlying dislocation. This bending behavior of edge-type dislocations at the regrowth interface resulted in a reduction of the dislocation density in the film surface region by around one order of magnitude. We also found that the PL intensity of the InN grown on the micro-faceted InN template was four times greater than the intensity from the InN layer grown on the flat InN template. Figure 1.10 shows a panchromatic cross-sectional cathodoluminescence (CL) image of this sample under an electron beam acceleration voltage of 3 kV [1.26]. The region between the dashed lines corresponds to the InN film. The upper area is the InN regrowth region and the lower area is the InN template. We observed strong CL intensity from the upper regions, where the threading dislocation (TD) density was reduced by the regrowth method. On the other hand, we observed only very weak CL intensity from the InN template, where the dislocation density was quite high. These results also indicate that TDs act as a nonradiative recombination center, resulting in CL dark regions.

1.2.6 Growth of InN-based nanostructures

Growth of InN nanostructures is expected to be one of the effective ways to obtain high-quality InN crystal without generation of threading dislocations. In this section, the successful fabrication of position-controlled InN nanodots [1.27] and nanocolumns [1.28, 1.29] on a hole-patterned GaN template prepared by a focused ion beam (FIB) is reviewed.

Ga-polar GaN epitaxial layers grown by MOVPE on (0001) sapphire were used as templates. Rectangular patterns of $\sim$500$\times$500 holes were prepared by the FIB technique using Ga$^+$ ions on the GaN template. Figure 1.11 shows images of the patterned GaN template with a typical rectangular pattern of holes, taken by SEM and atomic force microscopy (AFM). The patterned area was $\sim$200$\times$200 μm^2. The hole diameter and depth in the periodic structures were $\sim$100 nm and $\sim$10 nm, respectively. The hole pitch was varied from 210 to 300 nm. On these patterned substrates, InN dots were grown by electron cyclotron resonance (ECR)-MBE under N-rich conditions to suppress In droplet formation. The growth temperature of InN dots was changed from 390°C to 410°C. To enhance the migration of In atoms, InN dot growth was carried out with the following four-step source-supply sequence: In supply for 30 s, source-supply interruption for 10 s, nitrogen plasma

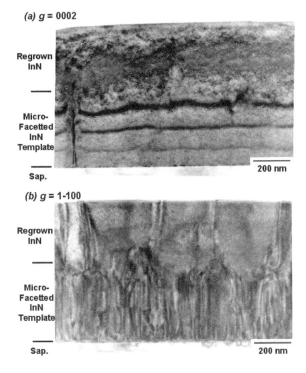

(a) g = 0002

Regrown InN

Micro-Facetted InN Template

Sap.

200 nm

(b) g = 1-100

Regrown InN

Micro-Facetted InN Template

Sap.

200 nm

FIGURE 1.9

Cross-sectional TEM images of the InN/micro-faceted InN template with (a) g = 0002 and (b) g = $1\bar{1}00$.

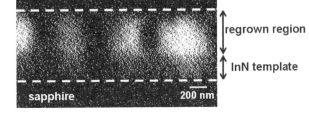

regrown region

InN template

sapphire

200 nm

FIGURE 1.10

Cross-sectional panchromatic CL image of InN regrown on a micro-faceted InN template under an electron-beam acceleration voltage of 3 kV.

supply for 20 s, and simultaneous supply of In and nitrogen plasma for 60 s. After the growth, the InN dots were annealed for 10 minutes at a temperature approximately 10°C lower than the growth temperature. We investigated the hole pitch dependence of the size of InN nanodots.

Figure 1.12 shows typical AFM images of InN dots grown with the use of three differently patterned substrates (hole pitch: 300 nm, 230 nm, 210 nm). Here, the growth was carried out at 390°C, and the subsequent annealing was performed at 380°C for 10 minutes. In case of the 300 nm hole pitch (Fig. 1.12(a)), we confirmed that InN dots varied greatly in size. Furthermore, InN dots were also observed even in the flat region between the holes. These results suggest that the surface migration length of In atoms is shorter than 300 nm, and thus In atoms could not uniformly enter every hole. According to the AFM measurement, the average diameter of InN dots was ∼120 nm, and the average height was ∼25 nm. On the other hand, when the hole pitch was 230 nm and 210 nm, we confirmed that InN dots were formed only in holes, as shown in Fig. 1.12(b), (c), and Fig. 1.13. We believe

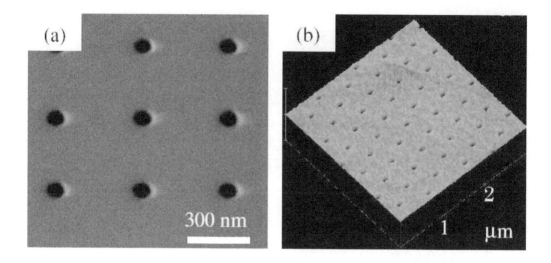

FIGURE 1.11
(a) SEM and (b) AFM images of the hole-patterned GaN template prepared by FIB.

that, by narrowing the hole pitch, the distance between the holes became shorter than or closer to the migration length of In atoms, and thus In atoms were able to uniformly enter every hole, making it possible to nucleate InN dots only at holes by the subsequent nitrogen plasma supply. In particular, InN dots grown with the narrowest hole pitch (210 nm) demonstrated the highest size uniformity among the three patterns. In case of the 230 nm hole pitch, the average diameter of InN dots was ∼120 nm, and the average height was ∼23 nm. In case of the 210 nm hole pitch, the average diameter was ∼100 nm, and the average height was ∼20 nm. These results also revealed that when the hole pitch decreased (that is, the hole density in the patterned area increased), the dot size decreased.

Next, we examined the growth temperature dependence of InN dot size. SEM

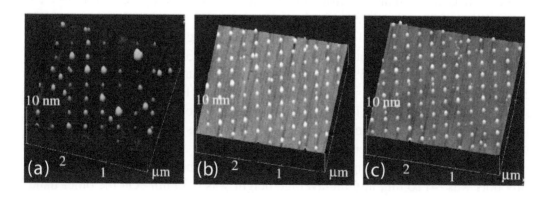

FIGURE 1.12
AFM images of InN dots grown by varying the hole pitch: (a) 300 nm; (b) 230 nm; and (c) 210 nm.

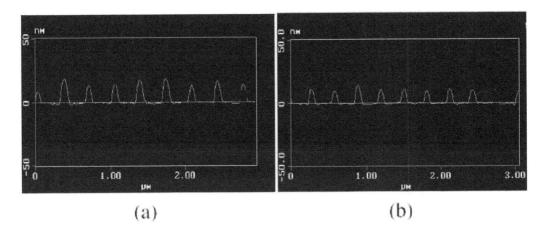

FIGURE 1.13

Cross-sectional line scan profiles of AFM images of InN dots grown with (a) 230 nm and (b) 210 nm hole pitch.

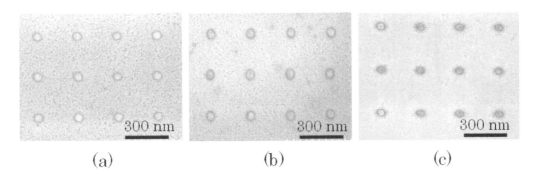

FIGURE 1.14

SEM images of InN dots grown at different temperatures: (a) 390°C; (b) 400°C; and (c) 410°C.

images of InN dots grown at temperatures of 390°C, 400°C, and 410°C are shown in Fig. 1.14. The hole pitch used in this experiment was 210 nm. When the growth temperature was 390°C (Fig. 1.14(a)) and 400°C (Fig. 1.14(b)), as well as InN dot formation in the holes, a thin InN layer was grown even on the flat region between the holes. However, the InN layer was not observed on the flat region when the growth was carried out at 410°C. We can explain this difference by taking the polarity of the InN layer into account. The InN layer grown on the flat region should have In polarity because of the use of a Ga polar GaN template grown by MOVPE. Several papers reported that a suitable growth temperature (the highest possible temperature below the dissociation limit) for In-polar InN was approximately 100°C lower than that for N-polar InN [1.10, 1.30, 1.31]. Therefore, it is considered that the proper growth temperature for In-polar InN should be around 400°C since the typical growth temperature for N-polar InN on sapphire is around

500°C using our ECR-MBE system [1.32]. Thus, we estimate that InN could not be grown on the flat region of the GaN template at 410°C. An additional possibility is that a thin layer of InN barely deposited on the flat region at this temperature was decomposed during the subsequent annealing at 400°C. The results presented here indicate that the proper selection of growth temperature and annealing temperature is essential for the perfect selective growth of InN dots. The dot size (both the diameter and the height) tended to decrease with increasing growth temperature probably because the sticking coefficients of source atoms decreased.

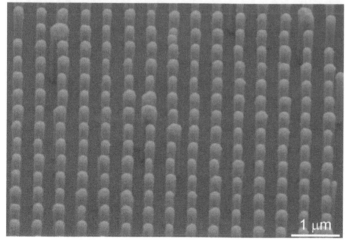

1 μm

FIGURE 1.15
'Bird's eye view' image of patterned InN nanocolumns.

By applying the fabrication technique of position-controlled InN nanodots, we have successfully grown patterned InN nanocolumns on the hole-patterned GaN templates. We also used ECR-MBE to grow InN nanocolumns. Square arrays of holes were prepared on the MOVPE-grown GaN templates by FIB using Ga$^+$ ions. InN growth was carried out at 450°C for 2 hours on the hole-patterned GaN templates. With optimized growth conditions and hole pattern on the GaN template, we obtained patterned InN nanocolumns with well-ordered height and diameter as shown in Fig. 1.15.

We performed structural characterization of InN nanocolumns by TEM [1.29]. Figure 1.16 clearly shows that InN nanocolumns with diameter of $\sim$300 nm were uniformly aligned on the GaN template. A selected area electron diffraction (SAED) pattern obtained from the interface between the InN nanocolumns and the GaN template indicated that the InN nanocolumns were grown on the GaN template with the same in-plane orientation. The a-axis lattice constant of the InN nanocolumns measured from the experimental SAED pattern was 3.527±0.013Å, which was almost the same as that of unstrained InN. Therefore, fully relaxed InN nanocolumns were epitaxially grown on the GaN template.

Then, we focused on threading dislocations generated at the interface between the InN nanocolumns and the GaN template, which are introduced in order to re-

FIGURE 1.16
Cross-sectional TEM image of InN nanocolumns grown on a GaN template. Reprinted with permission from S. Harui, H. Tamiya, T. Akagi, H. Miyake, K. Hiramatsu, T. Araki, and Y. Nanishi, Japanese Journal of Applied Physics, 47 (2008) 5330. Copyright 2008, The Japan Society of Applied Physics.

lax the strain. We observed bright and dark field images of InN nanocolumns by TEM with various electron diffraction vectors. Figure 1.17 shows cross-sectional TEM images of InN nanocolumns grown on the GaN template with (a) g = 0002 (screw component) and (b) g = 11$\bar{2}$0 (edge component). As shown in Fig. 1.17, neither screw- nor edge-type dislocations were detected in the majority of the InN nanocolumns. As discussed in more detail in Chapter 17, Grandal *et al.* reported that misfit dislocations were periodically generated at the InN nanocolumn/AlN interface. However, these dislocations were only present at the InN nanocolumn/AlN interface with no threading arms passing into the InN nanocolumns. Therefore, InN nanocolumns grow free of extended defects [1.33]. Calleja *et al.* also reported that the large lattice mismatch between GaN and Si was accommodated at the GaN nanocolumn/Si substrate hetero-interface by misfit dislocations that ended at the nanocolumn free surface [1.34]. By taking these results into account, there should be a certain number of misfit dislocations at the InN nanocolumn/GaN template interface corresponding to their lattice mismatch. We speculate that, if only a single InN nucleus is formed in each hole, the misfit dislocations are abruptly bent at the interface since the InN nucleus grows laterally along the GaN surface until they become a certain diameter or might make loops with one another. As a result, generally no threading dislocations are detected in the InN nanocolumns. On the other hand, in a few of the InN nanocolumns, threading dislocations are generated from the interface. This might be because multiple nuclei are formed in holes, which coalesce at the initial stage of the growth, and then threading dislocations can be generated at the boundaries between nuclei. Therefore, it is important to optimize the growth conditions as well as the depth and diameter of holes in order to obtain InN nanocolumns free of threading dislocations. Using convergent beam electron diffraction (CBED), we determined the polarity of InN nanocolumns to be In-polarity, which was consistent with that of the underlying GaN template grown by MOVPE.

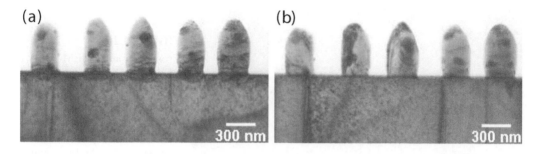

FIGURE 1.17

Cross-sectional TEM images of InN nanocolumns grown on a GaN template: (a) screw component with g=0002; and (b) edge component with g=11$\bar{2}$0. Reprinted with permission from S. Harui, H. Tamiya, T. Akagi, H. Miyake, K. Hiramatsu, T. Araki, and Y. Nanishi, Japanese Journal of Applied Physics, 47 (2008) 5330. Copyright 2008, The Japan Society of Applied Physics.

1.3   Effect of substrate polarity

Considering the potential applications of InN, both for high-speed electronic devices, possibly utilizing the piezoelectric effect, and for opto-electronic devices utilizing quantum wells, it is essential to control the polarity of the InN crystal structure. Unlike the case of GaN [1.35–1.37], there are only a small number of reports on the polarity of InN [1.10, 1.30, 1.38]. It is not even clearly understood how the polarity of a substrate affects the InN growth mechanisms, and the quality, as well as polarity, of the grown InN layers. We have been making a thorough investigation of these issues using co-axial impact collision ion scattering spectroscopy (CAICISS) [1.38], CBED [1.39], and by growing InN directly on the (0001) Ga-face and the (000$\bar{1}$) N-face of free-standing GaN substrates [1.30]. We have also found that the polarity of InN can be controlled by growing InN on the (0001) Si-face and the (000$\bar{1}$) C-face of 6H-SiC substrates [1.30]. In this section, the results we have obtained so far are reviewed.

1.3.1   Growth of InN on (0001) Ga-face and (000$\bar{1}$) N-face free-standing GaN substrates

In order to investigate how the polarity of InN affects its growth behavior, growth experiments were carried out by using (0001) Ga-face and (000$\bar{1}$) N-face free-standing GaN substrates, on which InN layers were grown with In-polarity and N-polarity, respectively. The free-standing GaN substrates were prepared by hydride vapor phase epitaxy (HVPE) with subsequent mechanical polishing of both sides. The thickness of the GaN substrates was approximately 300 μm. Prior to growth, the free-standing GaN substrates were thermally cleaned at 800°C in vacuum for 10 min. Then, InN layers were grown directly on the substrates in the temperature range of 450–550°C. The thickness of the InN layers was typically 300 nm. At a growth temperature of 550°C it was not possible to obtain an InN layer on the Ga-face of the GaN substrates, while it was possible for growth on the N-face. However, by decreasing the growth temperature to 450°C, an InN layer was obtained on the Ga-face of the GaN substrates. Figure 1.18 shows x-ray diffraction 2θ-ω scan profiles of these samples. With the expected diffraction peak corresponding to (0002) GaN, the diffraction peak corresponding to (0002) InN is clearly observed for the samples grown on the N-face GaN substrate at 550°C (Fig. 1.18(c)) and on the Ga-face GaN substrate at 450°C (Fig. 1.18(b)). For the sample grown on the Ga-face GaN substrate at 550°C, however, the diffraction from In ($2\theta \sim 33°$) is clearly observed, and the diffraction from InN (0002) is almost invisible (Fig. 1.18(a)).

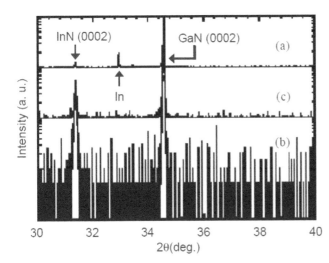

FIGURE 1.18
XRD 2θ-ω profiles of the samples grown on (a) Ga-face GaN at 550°C, (b) Ga-face GaN at 450°C and (c) N-face GaN at 550°C.

These XRD characteristics are consistent with SEM images of these samples as shown in Fig. 1.19. The sample grown on the Ga-face GaN substrate at 550°C (Fig. 1.19(a)) contains many In droplets, while the other samples show a smooth surface for the InN layers. The grain size of the InN layer grown on the N-face GaN substrate at 550°C (Fig. 1.19(c)) is larger than that of the InN layer grown on the Ga-face GaN substrate at 450°C (Fig. 1.19(b)), which can be attributed to the difference in the growth temperatures. The FWHM of (0002) XRCs was 37 and 58 arcmin for the InN layers grown on the N-face (at 550°C) and the Ga-face (at 450°C) GaN substrates, respectively. The above results confirm that InN with N-polarity can be grown at higher temperature than InN with In-polarity. The higher growth temperature for InN layers with N-polarity (550°C) is also consistent with that for the high-quality InN layers grown on the (0001) sapphire substrates, for which the InN layers were revealed to have N-polarity.

The difference in growth temperature (450°C for In-polarity and 550°C for N-

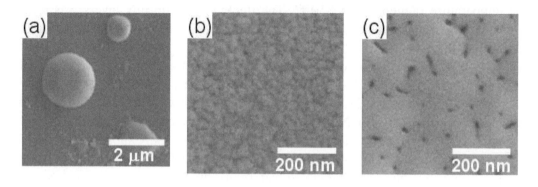

FIGURE 1.19

Surface SEM images of samples grown on (a) Ga-face GaN at 550°C, (b) Ga-face GaN at 450°C and (c) N-face GaN at 550°C.

polarity) can be explained by the different bonding configuration of the surface N atoms with the underlying In atoms for the two growth processes with In-polarity and with N-polarity, as shown in Fig. 1.20. In the case where InN is growing in In-polarity, the surface N atoms are bound to one underlying In atom. On the other hand, when InN is growing in N-polarity, the surface N atoms are bound to three underlying In atoms. It is therefore reasonable to conclude that the surface N atoms desorb more easily from the surface of InN with In-polarity than from the surface of InN with N-polarity. Due to the enhanced desorption of the surface N atoms, the effective V/III ratio at the surface must be lower for the growth of InN with In-polarity, resulting in the formation of In droplets at 550°C (Fig. 1.19(a)). Consequently, the successful growth of InN with In-polarity is accomplished at a decreased temperature of 450°C, where the desorption of the surface N atoms is suppressed.

1.3.2 Growth of InN on (0001) Si-face and (000$\bar{1}$) C-face 6H-SiC substrates

InN layers were grown also on (0001) Si-face and (000$\bar{1}$) C-face 6H-SiC substrates, with the aim of controlling the polarity of InN. Prior to growth, the substrates were thermally cleaned at 850°C in vacuum for 10 min. Then, an AlN nucleation layer was grown at 800°C. Subsequently, InN layers were grown in the temperature range of 450–550°C. The thickness of the AlN layer and that of the InN layer were typically 30 and 250 nm, respectively. This growth experiment is based on our previous experimental results for which the GaN layers grown on (0001) Si-face and (000$\bar{1}$) C-face 6H-SiC substrates with the AlN nucleation layer had Ga-polarity and N-polarity, respectively, as confirmed by etching the GaN surface using KOH solution. Thus, polarity control is also expected for InN growth using the 6H-SiC substrates, that is, we expect InN layers with In-polarity and with N-polarity on Si-face and C-face SiC substrates, respectively.

Figure 1.21 shows XRD 2θ-ω profiles for the InN samples grown at 550°C on

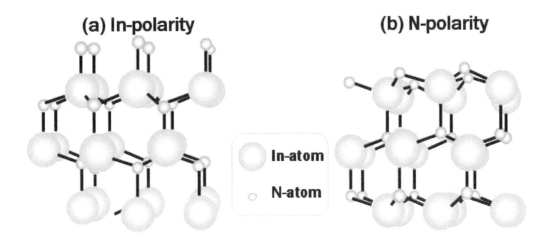

FIGURE 1.20
Bonding configuration between In atoms and N atoms in wurtzite InN: (a) In-polarity and (b) N-polarity.

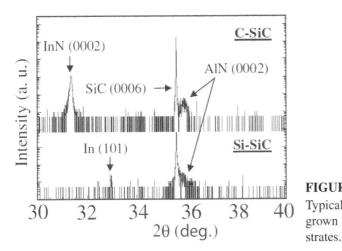

FIGURE 1.21

Typical 2θ-ω XRD profiles of InN grown on Si-face and C-face SiC substrates.

the Si-face and the C-face 6H-SiC substrates. The expected diffraction peaks corresponding to (0006) SiC and (0002) AlN can be seen in each profile. With these diffraction peaks, the peak corresponding to (0002) InN is clearly observed for the sample grown on the C-face SiC substrate, indicating that an InN layer was grown. In contrast, only the In diffraction line at $2\theta \sim 33°$ was observed for the sample grown on the Si-face SiC substrate. By decreasing the growth temperature to 450°C, an InN layer was obtained on the Si-face SiC substrate.

These XRD characteristics are consistent with SEM images of these samples as shown in Fig. 1.22. The sample grown on the Si-face SiC substrate at 550°C (Fig. 1.22(a)) contains many In droplets while the other samples show a smooth InN surface. The grain size for the InN layer grown on the C-face SiC substrate at 550°C (Fig. 1.22(b)) is larger than that of the InN layer grown on the Si-face SiC

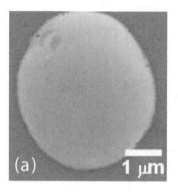 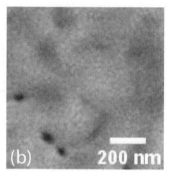 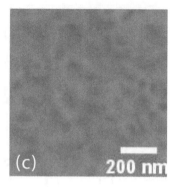

FIGURE 1.22

Surface SEM images of samples grown on (a) Si-face SiC at 550°C, (b) C-face SiC at 550°C and (c) Si-face SiC at 450°C.

substrate at 450°C (Fig. 1.22(c)), which can also be attributed to the difference in the growth temperatures. The FWHM of (0002) XRCs was 12.7 and 68.7 arcmin for the InN layers grown on the C-face (at 550°C) and the Si-face (at 450°C) SiC substrates, respectively. By comparing the above experimental results obtained using 6H-SiC substrates with those obtained using free-standing GaN substrates, it is confirmed that the InN layers grown on Si-face and C-face SiC have In-polarity and N-polarity, respectively. The polarity of the InN layers grown on the SiC substrates was further confirmed by etching the InN surface using KOH solution.

1.4   Growth of non-polar InN

Growth techniques to produce high-quality wurtzite InN (h-InN) in the direction of the c-axis and its properties have been extensively studied. However, it is well known that c-axis oriented opto-electronic devices suffer from large piezoelectric polarization fields, resulting in low electron/hole recombination efficiency. One of the most useful approaches to eliminating the piezoelectric polarization effects is to fabricate nitride-based devices along non-polar directions on, for example, *a*-plane (11$\bar{2}$0) and *m*-plane (1$\bar{1}$00). Although growth of GaN with non-polar orientation has been extensively studied using various substrates, there are few reports about growth of non-polar InN. It should be one of the key approaches to the fabrication of InN-based opto-electronic devices operating in the infrared wavelength region without the effects of piezoelectric polarization. Furthermore, the growth of non-polar InN has attracted much attention because, as discussed in detail in Chapters 12 and 13, elimination of the surface electron accumulation layer was theoretically predicted on non-polar InN surfaces [1.40]. In this section, we review the MBE growth of non-polar *a*-plane InN, *a*-plane In-rich InGaN and *m*-plane InN.

(a)

e // [100]   e // [110]

e // <100>

RHEED Simulated Pattern

(b)

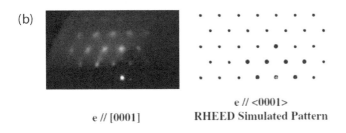

e // [0001]

e // <0001>

RHEED Simulated Pattern

FIGURE 1.23

RHEED patterns of (a) c-AlN(001) with [100] and [110] azimuths after nitridation and (b) *a*-plane (11$\bar{2}$0) InN with [0001] azimuth after InN growth.

1.4.1   Growth of *a*-plane InN on nitrided *r*-plane sapphire

Generally, non-polar *a*-plane (11$\bar{2}$0) GaN can be successfully grown on (10$\bar{1}$2) *r*-plane sapphire [1.41]. In the case of direct InN growth on *r*-plane sapphire, however, (001) cubic InN growth was reported rather than *a*-plane InN due to the small lattice mismatch between (001) cubic InN and *r*-plane sapphire [1.42, 1.43]. Inserting an *a*-plane GaN buffer layer is necessary to obtain *a*-plane InN on *r*-plane sapphire [1.44]. However, additional defects might be generated at the InN/template interface. In addition, the growth sequence is not simple since the optimized growth temperature of GaN is different from that of InN. In this section, we report the successful growth of *a*-plane (11$\bar{2}$0) InN on *r*-plane (10$\bar{1}$2) sapphire without any template by ECR-MBE [1.45].

The InN layers were grown on the *r*-plane (10$\bar{1}$2) sapphire substrates after substrate nitridation without buffer layers. The substrate nitridation of *r*-plane sapphire by ECR nitrogen plasma was performed at 430°C for 10 and 15 min prior to the InN growth. Microwave power and nitrogen gas flow rate for the nitridation were kept constant at 120 W and 10 sccm, respectively. The subsequent InN growth was carried out at 360°C for 1 hour.

Figures 1.23(a) and (b) show RHEED patterns observed after nitridation of *r*-plane sapphire and after the subsequent InN growth, respectively. As shown in Fig. 1.23(a), we confirmed by using two different azimuths that a cubic AlN layer with the [001] growth direction was obtained. After InN growth, the RHEED shown in Fig. 1.23(b) also shows a typical diffraction pattern for the [11$\bar{2}$0] oriented wurtzite structure. Therefore, these results indicate that a single crystalline *a*-plane (11$\bar{2}$0) InN layer was successfully obtained on *r*-plane (10$\bar{1}$2) sapphire.

Based on these results, we propose the following mechanism for the growth of *a*-plane (11$\bar{2}$0) InN on nitrided *r*-plane (10$\bar{1}$2) sapphire. Figure 1.24 shows the epitaxial relationship and lattice mismatch between (001) cubic AlN and (11$\bar{2}$0) hexagonal InN. RHEED observations after nitridation of r-plane sapphire confirmed that (001) cubic AlN was formed on the surface of r-plane sapphire. The lattice mismatch between (001) cubic AlN and (1120) a-plane InN ($\sim$7.0 % for [0001]InN$\|$[110]c-AlN and $\sim$0.8 % for [1100]InN$\|$[110]c-AlN) is much smaller than that between (001) c-InN and (001) c-AlN ($\sim$13.5 %) and that between a-plane GaN and a-plane InN ($\sim$11 %). Thus, a-plane (11$\bar{2}$0) InN grows directly on nitrided r-plane sapphire. This is the clear reason why a-plane InN was grown on cubic AlN formed by nitridation of r-plane sapphire. In other words, we can conclude nitridation of r-plane sapphire is very effective to grow a-plane InN.

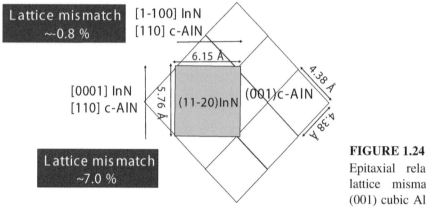

FIGURE 1.24

Epitaxial relationship and lattice mismatch between (001) cubic AlN and (11$\bar{1}$0) hexagonal InN.

Figure 1.25 shows XRD profiles of InN layers grown on r-plane sapphire with nitridation for (a) 10 and (b) 15 min. The diffraction peak of (11$\bar{2}$0) hexagonal InN at 52.5° was clearly observed in both samples. However, a weak diffraction peak of (002) cubic InN at 35.8° was also observed, together with a-plane hexagonal InN in the case of the 10 min nitridation. These results suggested that the InN film deposited on the sapphire substrate nitrided for 10 min was a mixture of (11$\bar{2}$0) hexagonal and (001) cubic InN. In order to improve the quality of a-plane InN, the formation of cubic InN inclusions should be suppressed. As the origin of cubic InN inclusions, we considered that cubic InN was directly grown on r-plane (10$\bar{1}$2) sapphire where cubic AlN was not formed due to insufficient nitridation. Therefore, uniform formation of cubic AlN by sufficient nitridation of r-plane sapphire is required to suppress the formation of cubic InN inclusions, resulting in crystal quality improvement of a-plane InN. In Fig. 1.25(b), when we increased the substrate nitridation time to 15 min, only a single diffraction peak of (11$\bar{2}$0) hexagonal InN was observed. This result suggests that sufficient nitridation (15 min) of r-plane sapphire formed a uniform cubic AlN layer, and successfully suppressed the inclusion of the cubic InN phase. Regarding the x-ray rocking curve of a-plane (11$\bar{2}$0) InN, as substrate nitridation time was varied from 10 min to 15 min, the FWHM was improved from 128 arcmin to 80 arcmin. As a result, when the substrate nitridation time was longer, the crystalline quality of *a*-plane (11$\bar{2}$0) InN was improved. We conclude that nitridation of *r*-plane (10$\bar{1}$2) sapphire by ECR nitrogen plasma is a very effective approach to grow *a*-plane InN on *r*-plane sapphire without any template.

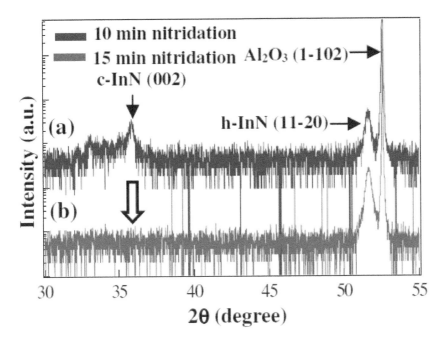

FIGURE 1.25 XRD profiles of InN layers grown on *r*-plane sapphire with nitridation for (a) 10 and (b) 15 min.

We also carried out TEM characterization of *a*-plane (11$\bar{2}$0) InN grown on nitrided *r*-plane (10$\bar{1}$2) sapphire [1.46]. Figure 1.26 shows a schematic image of the epitaxial relationship for *a*-plane InN grown on *r*-plane sapphire, which was confirmed by SAED. The epitaxial relationship between *a*-plane InN and *r*-plane sapphire is (11$\bar{2}$0)$_{InN}$||(1$\bar{1}$02)$_{sapph}$, [1$\bar{1}$00]$_{InN}$||[$\bar{1}$$\bar{1}$20]$_{sapph}$ and [0001]$_{InN}$||[$\bar{1}$101]$_{sapph}$.

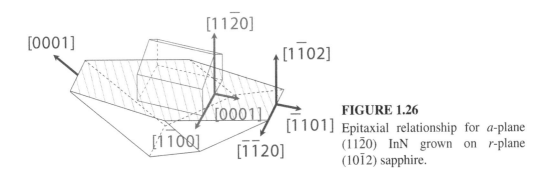

FIGURE 1.26 Epitaxial relationship for *a*-plane (11$\bar{2}$0) InN grown on *r*-plane (10$\bar{1}$2) sapphire.

A high-resolution TEM image of an *a*-plane InN/*r*-plane sapphire interface region is shown in Fig. 1.27. It seems likely that a cubic AlN layer formed by nitridation, which plays an important role for the *a*-plane InN growth on *r*-plane sapphire (the in-plane epitaxial relationship between *a*-plane InN and cubic AlN is $[0001]_{InN} \| [1\bar{1}0]_{AlN}$ and $[1\bar{1}00]_{InN} \| [110]_{AlN}$), was found at the InN/sapphire interface region. At the interface region, it was revealed that *a*-plane InN was successfully grown on *r*-plane sapphire with the cubic AlN layer by introducing a misfit dislocation almost every 11 lattice spacings. Lattice mismatch between $(11\bar{2}0)$ hexagonal InN and $(1\bar{1}02)$ sapphire is around 14.0% for $[0001]_{InN} \| [1\bar{1}01]_{sapph}$, suggesting that a misfit dislocation should be introduced every 7 lattice spacings. From these results, it is considered that lattice misfit is adequately relaxed by the cubic AlN layer at the interface region.

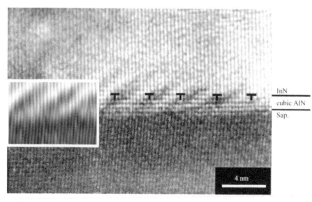

FIGURE 1.27

High-resolution TEM lattice image of the interface *a*-plane InN/cubic AlN/*r*-plane sapphire. The insert shows the Fourier filtered image at the interface region.

To characterize the type of dislocations in *a*-plane InN on *r*-plane sapphire, cross-sectional TEM observation with different diffraction vectors g was carried out. Figures 1.28(a) and (b) show TEM images of the *a*-plane InN film with g=0002 and g=11$\bar{2}$0, respectively. Generally, the Burgers vectors b of threading dislocations in InN are defined as b=(1/3)<11$\bar{2}$0>, b=±[0001], b=(1/3)<11$\bar{2}$3>. With the threading dislocation line parallel to the growth direction of [11$\bar{2}$0], pure screw dislocations in *a*-plane InN will have Burgers vectors b=(1/3)<11$\bar{2}$0>, and pure edge dislocation will have b=±[0001] [1.47]. Therefore, the TEM image with g=0002 shows dislocations edge-type component and the image with g=11$\bar{2}$0 shows screw-type component. Figure 1.28 reveals a high density of threading dislocations in the film. The density of screw-type dislocation was estimated to be 5×10^{10} cm^{-2}, which is about ten times higher than the estimated density for edge-type dislocations. In *c*-plane InN growth, the edge-type dislocation density was much higher than the screw-type dislocation density [1.39]. Since the Burgers vectors b=(1/3) <11$\bar{2}$0> is the direction with the smallest Peierls stress in the wurtzite structure, dislocations with Burgers vectors b=(1/3)<11$\bar{2}$0> (edge dislocations in *c*-plane InN and screw dislocations in *a*-plane InN) are mainly observed.

We have also demonstrated non-polar *a*-plane (11$\bar{2}$0) In-rich InGaN growth on *r*-plane (10$\bar{1}$2) sapphire with an *a*-plane (11$\bar{2}$0) InN template by RF-MBE [1.48].

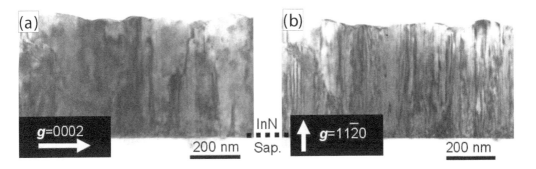

FIGURE 1.28
Cross-sectional TEM images of *a*-plane InN film with (a) g=0002 and (b) g=11$\bar{2}$0.

Figure 1.29 shows the typical XRD (2θ-ω) profile of samples. As shown in Fig. 1.29, the diffraction peak of *a*-plane (11$\bar{2}$0) hexagonal InGaN at 53.3° was observed with the expected peaks from the hexagonal *a*-plane (11$\bar{2}$0) InN and the (20$\bar{2}$4) sapphire at 51.6° and 52.6°, respectively. Moreover, the reciprocal space mapping (RSM) measurement around the asymmetric (12$\bar{3}$0) reflection showed that the In-GaN layer was fully relaxed. The In-composition of the InGaN layer was estimated to be approximately 0.71, assuming Vegard's law. The *a*-plane InGaN layer exhibited PL emission at approximately 1.1 eV. These results indicate the possibility of fabricating non-polar InGaN/InN heterostructures and quantum well structures.

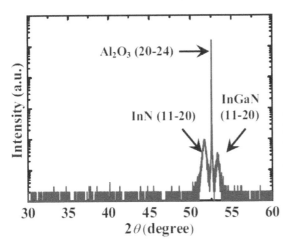

FIGURE 1.29
XRD (2θ-ω) profile of InGaN layer grown on *r*-plane sapphire with an InN template layer. Reprinted with permission from M. Noda, Y. Kumagai, S. Takado, D. Muto, H. Na, H. Naoi, T. Araki, Y. Nanishi, physica status solidi (c), 4, 2560 (2007). Copyright (2007) by Wiley-VCH Publishers, Inc.

1.4.2   Growth of *m*-plane InN on LiAlO$_2$(001)

Compared to the growth of *a*-plane InN, there are few reports on the growth of *m*-plane InN. This is probably due to fewer candidates for the substrate to grow *m*-plane InN. For example, hexagonal *m*-plane GaN, ZnO, and 6H-SiC are good candidates for growing hexagonal *m*-plane InN, but these substrates are rather expensive. For the growth of *m*-plane GaN, a LiAlO$_2$(001) substrate has been generally used due to the relatively small lattice mismatch. Here we demonstrate that this substrate is also useful for growing *m*-plane InN.

The *m*-plane InN thin film was grown on LiAlO$_2$(001) substrate, which was coated on the back by molybdenum for efficient heat absorption. Prior to growth, the substrate was thermally cleaned at 800°C for 10 min in vacuum. After thermal cleaning, an *m*-plane InN thin film was grown at 450°C for 60 min directly without a buffer layer under both In-rich and N-rich conditions. Figure 1.30 shows the surface SEM images of InN on LiAlO$_2$(001) substrates grown under In-rich (Fig. 1.30(a)) and N-rich (Fig. 1.30(b)) conditions. From Fig. 1.30(a), we observed slate-like morphology. In the sample grown under an N-rich condition, however, hexagonal facets were observed in the slate-like morphology, as shown by the white circle in Fig. 1.30(b). This result suggests that polycrystalline InN was grown on the substrate under N-rich conditions.

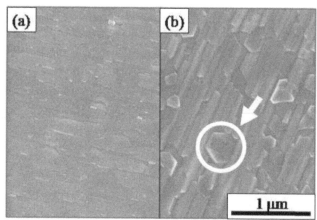

FIGURE 1.30
SEM images of *m*-plane InN grown on LiAlO$_2$(001) under (a) In-rich and (b) N-rich conditions.

Figure 1.31 shows the XRD 2θ-ω profiles for both samples. We can observe two peaks, which are from hexagonal *m*-plane (10$\bar{1}$0) InN and (002) LiAlO$_2$ at approximately 29.1° and 34.7°, respectively, for the sample grown under In-rich conditions. An additional peak from hexagonal *c*-plane InN (0002) at approximately 31.3° was detected for the sample grown under N-rich conditions probably due to reduced migration of indium atoms.

The epitaxial relationship of *m*-plane InN (for the sample grown under In-rich conditions without the polycrystallinity) was investigated using XRD φ-scan measurements. The results shown in Fig. 1.32 indicate that the epitaxial relationships are (10$\bar{1}$0)$_{InN}$∥(001)$_{LiAlO_2}$, [0001]$_{InN}$∥[010]$_{LiAlO_2}$ and [11$\bar{1}$0]$_{InN}$∥[001]$_{LiAlO_2}$. A detailed description of the epitaxial relationship is shown in Fig. 1.33. These relationships are exactly the same as the case of *m*-plane GaN on LiAlO$_2$(001).

We also characterized *m*-plane InN by TEM. Figure 1.34 shows a cross-sectional

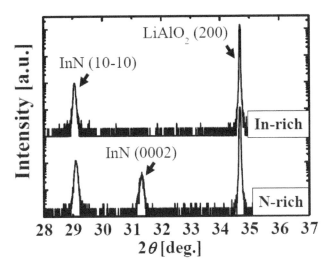

FIGURE 1.31

XRD 2θ-ω profiles of *m*-plane InN grown on LiAlO$_2$(001) under In-rich and N-rich conditions.

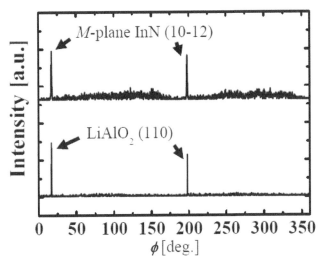

FIGURE 1.32

XRD ϕ-scan of single *m*-plane InN grown on LiAlO$_2$(001).

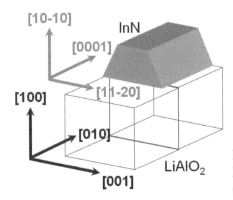

FIGURE 1.33

Epitaxial relationship for *m*-plane ($1\bar{1}00$) InN grown on LiAlO$_2$(001).

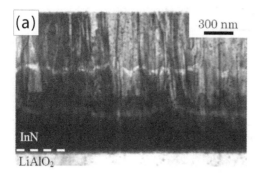
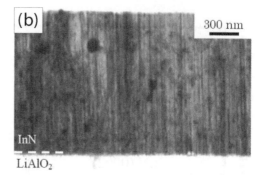

FIGURE 1.34

Cross-sectional TEM image of *m*-plane InN with incident azimuth of (a) $[0001]_{InN}$ and (b) $[11\bar{2}0]_{InN}$.

TEM image of *m*-plane InN. Figure 1.34(a) was taken with an electron incident azimuth of $[0001]_{InN}$ and $g = 1\bar{1}00$. Thus, threading dislocations observed in this image should have a screw component. As shown in Fig. 1.34, a high density of threading dislocations propagating from the interface to the surface was clearly observed. The density of threading dislocations with a screw component is estimated to be 2.0×10^{10} cm^{-2}, which is more than ten times higher than that of typical *c*-plane InN grown on (0001) sapphire by RF-MBE.

Figure 1.34(b) shows a cross-sectional TEM image of *m*-plane InN with the electron incident azimuth of $[11\bar{2}0]_{InN}$. In this image, the c-axis of InN is parallel to the InN/LiAlO$_2$ interface. We observed a high density of straight line defects propagating from the interface to the surface. These defects are basal plane stacking faults as observed in *m*-plane GaN grown on LiAlO$_2$ [1.49]. The linear stacking fault density, as derived from the TEM micrograph, is found to be about 1.6×10^5 cm^{-1}, which is almost comparable with the density of stacking faults in *m*-plane GaN grown on LiAlO$_2$ [1.50]. Such high densities of defects in *m*-plane InN are considered to be caused by the large lattice mismatch between InN and LiAlO$_2$. Proper initial growth processes, such as two-step growth, need to be developed for improved crystal quality of *m*-plane InN.

1.5   Growth of Mg-doped InN

InN films of device quality have not yet been achieved because the films suffer from a high residual carrier concentration and very high electron accumulation at the surface [1.51–1.53]. Therefore, it is still difficult to obtain and characterize *p*-type InN. Even though *p*-type doping of InN by Mg has been reported [1.54], it is difficult to precisely determine the bulk conductivity of the *p*-type layer because of a large electron accumulation at the surface. Furthermore, the effect of film polarity on Mg-doping is also very important. In the case of GaN grown by RF-MBE, it was reported that GaN grown on Ga-polarity was preferred to achieve *p*-type conductivity because the Mg incorporation rate was higher than that of N-polarity. However, it was reported that the polarity of GaN was inverted from Ga- to N-polar by high levels of Mg doping [1.55]. In the case of In-polar InN films, polarity inversion with a high density of V-shaped inversion domains occurred at Mg concentrations over 1×10^{19} cm^{-3} [1.56]. Taking into account these results, we expect that Mg doping levels sufficient for obtaining *p*-type InN could be achieved using N-polar InN growth. In this section, we report investigations of the growth properties of Mg-doped N-polar InN films grown by RF-MBE [1.57].

Figure 1.35 shows the growth time chart of Mg-doped InN used in this study. Nitridation of the sapphire substrates was carried out at 280°C for 2 hours. After a low-temperature InN buffer layer was deposited at 280°C for 10 min, an N-polar InN intermediate layer was grown at 530°C for 10 min without Mg doping. Then, undoped and Mg-doped InN layers were grown at 530°C for 1 hour under nitrogen-rich conditions to avoid the risk of In droplet formation on the surface. Mg was supplied by a conventional effusion cell, and the doping amount was varied by changing the cell temperature between 130 and 140°C. The thickness of the InN films was approximately 420 nm. The Mg concentrations of the samples were measured by secondary ion mass spectrometry (SIMS) calibrated with a Mg-ion implanted InN standard. Mg-doped InN grown with Mg cell temperatures (T_{Mg}) of 125, 130, 135, and 140°C had Mg concentrations of 2.5×10^{20}, 4.0×10^{20}, 6.3×10^{20}, and 7.3×10^{20} cm^{-3}, respectively. We found that all the samples had higher Mg concentrations than the critical value for polarity inversion of In-polar InN [1.56]. Also, we found that the Mg concentration of InN became higher with increasing Mg cell temperature.

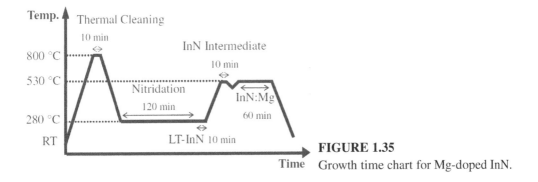

FIGURE 1.35
Growth time chart for Mg-doped InN.

Figure 1.36 shows the surface morphology of undoped and Mg-doped InN. These samples were grown under nitrogen-rich conditions to investigate the effect of Mg doping on grain size. The images show that the Mg-doped InN had smaller grain size than the undoped InN. We also found that the grain size of Mg-doped InN samples became smaller as the amount of doping was increased. These results indicate that Mg doping decreases the surface migration length of In atoms.

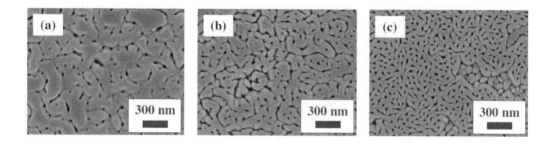

FIGURE 1.36

SEM images of the surface morphologies of InN grown on N-polar InN intermediate layers: (a) undoped InN and Mg-doped InN grown with Mg cell temperatures of (b) 130°C and (c) 140°C.

To investigate the crystal structure of undoped and Mg-doped InN, we carried out XRD (0002) 2θ-ω scans. From the undoped and Mg-doped InN with a Mg concentration of 2.5×10^{20} cm^{-3}, a single peak of (0002) hexagonal InN (h-InN) was observed, as shown in Figs. 1.37(a) and (b). On the other hand, the Mg-doped InN with Mg concentrations of 6.3×10^{20}, and 7.3×10^{20} cm^{-3} produced a weak diffraction peak of (111) cubic InN (c-InN) in addition to a strong diffraction peak of (0002) h-InN, as shown in Figs. 1.37(c) and (d). From these results, we found that a Mg concentration of around 2.5×10^{20} cm^{-3} is the critical value for prevention of c-InN inclusions. We also found that the ratio of the content of c-InN phases in the h-InN matrix increased with an increase in Mg cell temperature. We assume that the poor surface migration of In atoms, which was caused by Mg-doping, resulted in a higher content of c-InN.

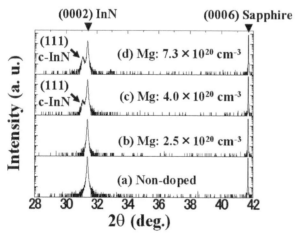

FIGURE 1.37

XRD 2θ-ω profiles: (a) undoped InN, Mg-doped InN grown with Mg cell temperatures of (b) 130°C and (c) 140°C.

All the samples exhibited *n*-type conductivity by single-field Hall effect measurement. However, as discussed in Chapter 10, the results of single-field Hall effect measurements do not necessarily accurately represent the properties of Mg-doped InN films because of the presence of the surface electron accumulation layer.

In summary, we have successfully grown highly Mg-doped InN with Mg concentrations of over 2.5×10^{20} cm^{-3} using N-polar InN. This value is higher than the critical value for the polarity inversion of In-polar InN. We found that a Mg concentration of around 2.5×10^{20} cm^{-3} is the critical value for prevention of c-InN inclusions.

1.6 Growth of In-rich InGaN

Following the finding that the true band gap of InN is ~0.65 eV rather than 1.9 eV, InN and InGaN alloys with high In-contents (In-rich InGaN) are considered attractive materials for future devices, such as temperature-insensitive high-efficiency infrared laser diodes for optical communication and high-efficiency tandem solar cells. However, just as for InN, the studies on In-rich InGaN have been hindered for a long time by the difficulty of growing high-quality crystals. The properties of InN and In-rich InGaN layers grown by both RF-MBE and MOVPE, however, have been improved considerably in recent years. Motivated by these improvements, the trial growth of a double heterostructure (DH) of an $In_{0.9}Ga_{0.1}N/In_{0.75}Ga_{0.25}N$ [1.58] and InN/GaN [1.59], single quantum well (SQW) of InN/In-rich InGaN [1.60] and InN/GaN [1.59], and multiple quantum well (MQW) of InN/In-rich InGaN [1.60, 1.61] have been performed quite recently by RF-MBE. In this section, we will review RF-MBE growth of In-rich InGaN including comprehensive data sets on the surface morphology and crystalline quality of In-rich InGaN and fabrication of InN/InGaN quantum well structures [1.60, 1.62–1.65].

1.6.1 Growth of In-rich InGaN on InN templates

In-rich $In_xGa_{1x}N$ ($0.70 = x = 0.94$) samples were grown on (0001) sapphire substrates by RF-MBE. Prior to growth, (0001) sapphire substrates were thermally cleaned at 800°C for 10 min in vacuum. Subsequently the substrates were nitrided at 550°C for 1 hour (conventional nitridation conditions). The temperature and duration of the nitridation process were later modified to 300°C and 2 hours, respectively (optimized nitridation conditions), allowing the growth of InN templates of dramatically improved crystalline quality [1.20]. After the nitridation of the substrates, LT-InN buffer layers were grown at 300°C for 10 min. InN templates were grown at 530°C to a thickness of 500 nm on the LT-InN buffer layers. In-rich $In_xGa_{1-x}N$ films ($0.71<x<1$) were grown at 550°C to a thickness of 250 nm on both LT-InN buffer layers and the InN templates. In-rich InGaN films were also grown on high-crystalline-quality InN templates that were grown on LT-InN buffer layers with the optimized nitridation process.

Figure 1.38 shows FWHMs of (0002) and ($10\bar{1}0$) XRCs of In-rich InGaN layers grown on LT-InN buffer layers and on InN templates grown on LT-InN buffer layers. Both conventional and optimized nitridation-process-based samples are described in this figure. The FWHM of the ($10\bar{1}0$) XRC for each InGaN layer was again theoretically estimated from that of the (0002), ($10\bar{1}3$), ($10\bar{1}2$), ($10\bar{1}1$), and ($30\bar{3}2$) XRCs [1.22]. It has been found that the InN template and the optimized nitridation process were both very effective in improving c-axis orientation of In-rich InGaN layers, as shown in Fig. 1.38(a). In particular, the combination of these two techniques has been found to be the most effective in improving not only the c- but also the a-axis orientations of In-rich InGaN layers. Figure 1.39 shows typical SEM images of the surface of these InGaN layers. The best results were again obtained when the In-rich InGaN layers were grown on the high-crystalline-quality InN templates grown via the optimized nitridation process.

1.6.2 Growth and characterization of InN/In$_{0.8}$Ga$_{0.2}$N quantum well structures

As shown in section 1.6.1, the combination of the optimized nitridation process and the InN template was found to be the most effective in improving both the crystalline quality and the surface morphology of In-rich InGaN layers. Based on these results, an InN/In$_{0.8}$Ga$_{0.2}$N MQW structure and SQW structures with different well widths were grown on 200-nm thick In$_{0.8}$Ga$_{0.2}$N layers grown on 400-nm thick InN templates that were grown via the optimized nitridation process. Here, the 400-nm thick InN templates were grown at 530°C, and the 200 nm In$_{0.8}$Ga$_{0.2}$N layers were grown at a slightly reduced temperature of 520°C. For the SQW structures, InN well layers with the thicknesses of 2.3, 3.4, and 5.7 nm were grown on the 200-nm In$_{0.8}$Ga$_{0.2}$N layers, and the InN well layers were capped by 11.2-nm thick In$_{0.8}$Ga$_{0.2}$N barrier layers. As for the MQW structure, 20 pairs of 5-nm thick In$_{0.8}$Ga$_{0.2}$N barrier and 5-nm thick InN well layers were grown on the 200-nm thick In$_{0.8}$Ga$_{0.2}$N layers. The InN well and In$_{0.8}$Ga$_{0.2}$N barrier layers in these SQW and MQW structures were grown at 520°C, the same temperature as that used for the growth of the 200-nm thick underlying In$_{0.8}$Ga$_{0.2}$N layers which were used as the bottom barrier layer in these QW structures. The use of this slightly reduced temperature of 520°C for the growth of InN well and In$_{0.8}$Ga$_{0.2}$N barrier layers aimed at reducing the risk of intermixing between the well and barrier layers.

Figure 1.40 shows the RSM around the asymmetric ($10\bar{1}4$) reflection for the SQW structure with an InN well width of 5.7 nm. The three peaks observed in this map are from the In$_{0.8}$Ga$_{0.2}$N barrier layers (mostly from the relaxed bottom barrier layer), the relaxed InN template, and the strained InN well layer. This map confirms that the InN well layer was grown coherently on the relaxed In$_{0.8}$Ga$_{0.2}$N bottom barrier layer. Thus the additional generation of dislocations at the heterointerface between the InN well and In$_{0.8}$Ga$_{0.2}$N barrier layers is expected to be

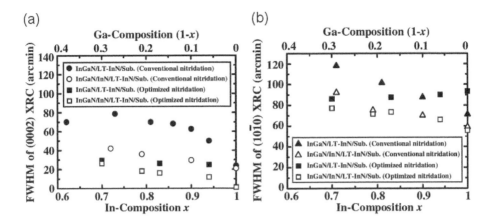

FIGURE 1.38

Compositional dependencies of FWHMs of XRCs of In-rich InGaN layers: (a) for the (0002) reflection and (b) for the (10$\bar{1}$0) reflection.

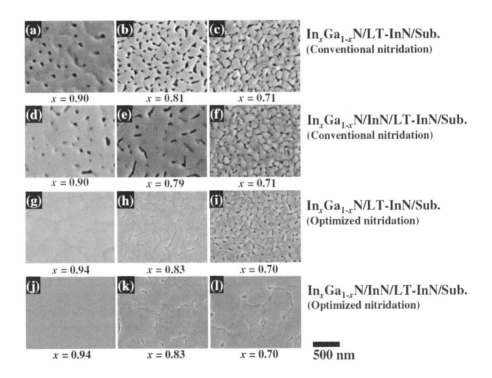

FIGURE 1.39

Surface morphologies of In-rich In$_x$Ga$_{1x}$N layers: (a), (b), and (c) grown on LT-InN buffer layers via conventional nitridation; (d), (e), and (f) grown on InN templates via conventional nitridation; (g), (h), and (i) grown on LT-InN buffer layers via optimized nitridation; and (j), (k), and (l) grown on InN templates via optimized nitridation. In-compositions of In$_x$Ga$_{1-x}$N layers are denoted by x.

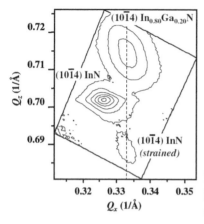

FIGURE 1.40
Reciprocal space map around the asymmetric ($10\bar{1}4$) reflection for the SQW sample with an InN well width of 5.7 nm.

suppressed not only for this SQW sample but also for the rest of the SQW samples with even narrower InN well widths (2.3 nm and 3.4 nm). This expectation was further supported by the critical thickness estimates for InN grown on $In_{0.8}Ga_{0.2}N$, using the proposed models [1.66–1.68]. Among these models, the smallest critical thickness of InN was given by the model proposed by Fischer *et al.* [1.68]. The critical thickness of InN calculated by this model was approximately 7 nm, which is larger than the InN well widths in the SQW structures (2.3, 3.4, and 5.7 nm).

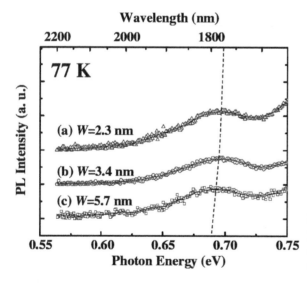

FIGURE 1.41
PL spectra at 77 K for InN template and MQW samples with the InN well widths, W, of 2.3, 3.4 and 5.7 nm.

Figure 1.41 shows 77 K PL spectra of the SQW samples in the photon-energy range of interest. These PL peaks around 0.7 eV observed on the tail of the $In_{0.8}Ga_{0.2}N$ luminescence peaks were not from InN template layers but were from the InN well layers, because the argon ion laser beam irradiated on the SQW samples was absorbed by the 200-nm $In_{0.8}Ga_{0.2}N$ bottom barrier layer and therefore could not excite the InN template layer. Noteworthy is that the luminescence peak from the InN well layers slightly shifts to the higher energy side with decreasing well width; moreover their peak energy values are all below the 0.71 eV of the bulk InN.

These PL characteristics, that is, the very weak well-width dependence of the PL peak energy and the energy values lower than those of the bulk InN, can be explained by the combined effects of the quantum size effect, the quantum confined Stark effect (QCSE) induced by large strain in the InN well layers, as evidenced in Fig. 1.41, and the band filling effect caused by high residual carrier concentration in the InN well layers. Disordering of the QW layered structures caused by the high density of threading dislocations should also be taken into consideration.

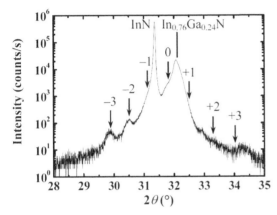

FIGURE 1.42
XRD 2θ-ω profile of the InN/ $In_{0.76}Ga_{0.24}N$ MQW sample.

Figure 1.42 shows the XRD 2θ-ω curve for (0002) reflection of the MQW structure with 3.4-nm thick InN well layers. Satellite peaks (marked by arrows) were observed, as well as the expected diffraction peaks from (0002) InN and (0002) $In_{0.76}Ga_{0.24}N$, indicating an abrupt heterointerface and uniform well and barrier layer thicknesses in the structure. The total thickness of well and barrier estimated from the period of satellite peaks was approximately 15 nm, corresponding to our structure design.

Figure 1.43 shows PL spectra from a 400-nm thick InN template and several InN/$In_{0.76}Ga_{0.24}N$ MQW structures. A PL peak at an energy of 0.705 eV was observed from the InN template. From the MQW structure, two PL peaks were observed near the photon energies of 0.73 and 1.05 eV. These peaks are assigned to the emissions from the InN QW layer and the thick $In_{0.76}Ga_{0.24}N$ barrier layers, respectively. The energy of the peak from the InN well layers is higher than that of bulk InN (InN template). In our previous report [1.63], a 5.7-nm thick InN well layer grown on $In_{0.8}Ga_{0.2}N$ was fully strained. InN well layers of our MQW structure will be fully strained because their thicknesses are not more than 5.7 nm. The formation of defects at the interface will be suppressed. The PL peaks of InN well layers would be blueshifted due to the quantum size effect (QSE) with decreasing InN well widths.

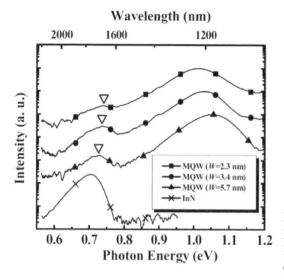

FIGURE 1.43

PL spectra at 77 K for InN template and MQW samples with the InN well widths, W, of 2.3, 3.4 and 5.7 nm.

Figure 1.44 shows the dependence of PL peak energy on InN well width. The calculation result from the Kronig-Penny model is shown in Fig. 1.44. Since large strain would also be induced in the InN well layers of the MQW structure, the QCSE should be taken into account in the peak shift. The energy shift with the decrease of well width with QCSE is larger than that of QSE in no field. However, this peak shift is smaller than our expectation. These results can be explained by the band filling effect. Residual carrier concentration of InN was typically measured to be over 10^{18} cm^{-3}. In the degenerately doped InN wells, the Fermi level is assumed to lie far above the lowest quantized energy level of the conduction band, and thus it is thought that QSE and QCSE could not be observed clearly from our InN/InGaN MQW structures.

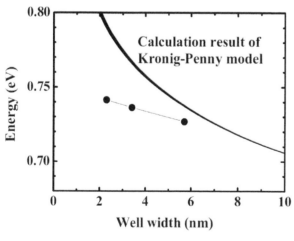

FIGURE 1.44

Dependence of PL peak energy on InN well width and the result of Kronig-Penny model calculations.

1.7 Growth of InN on Si

So far, InN films are generally grown on sapphire substrates even though Si substrates have several advantages over sapphire. For example, Si substrates are easily obtained at high quality and low cost, and Si substrates can be used in optoelectronic integrated circuits (OEICs) with well-developed Si-based devices. For InN, in addition, the lattice mismatch with the Si substrate (7.6%) is much smaller than that for the sapphire substrate (25.4%). However, growing single crystalline InN films on Si has been difficult [1.69–1.75]. In this section, we describe how to obtain single crystalline InN films on Si substrates and improve their quality using RF-MBE. The effects of substrate nitridation [1.76, 1.77] and insertion of an AlN buffer layer [1.78, 1.79] on MBE growth of InN on Si substrate are discussed.

1.7.1 Substrate nitridation

Generally, the quality of InN films grown on Si substrates is much worse than that on sapphire substrates because InN films easily become polycrystalline due to the formation of amorphous-like SiN_x on the surface of the Si substrates. Figure 1.45 shows the RHEED patterns of the Si substrates before and after substrate nitridation. Before substrate nitridation, a clear (7×7)-reconstruction RHEED pattern typical of a clean Si (111) surface was obtained (Fig. 1.45). When the Si substrate surface was exposed to the nitrogen plasma, the RHEED pattern changed from (7×7) to (1×1). After a brief substrate nitridation (3 min), the pattern became diffuse (Fig. 1.45(b)).

FIGURE 1.45

RHEED patterns of Si substrates (a) before substrate nitridation, (b) after substrate nitridation for 3 min, and (c) after substrate nitridation for 30 min.

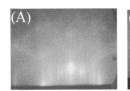

We used XPS to determine whether or not SiN_x formed on the substrate surface during the 3-min substrate nitridation. Figure 1.46(a) shows the Si $2p$ spectra before and after the nitridation. An additional peak representing Si-N bonding appeared in the substrate with nitridation. The N $1s$ spectrum for the substrate without nitridation (Fig.1.46(b)) shows no peak, whereas the peak is clearly observed for the substrate with nitridation. The RHEED and XPS results suggest that SiN_x formed on the Si substrate even when the nitridation was carried out for only 3 min, and the SiN_x would be formed on only part of the Si substrate surface. On the other hand, when the substrate nitridation lasted over 30 min (Fig. 1.45(c)), the pattern changed to a halo pattern with weak streaks, suggesting that an amorphous SiN_x layer had formed on the substrate surface.

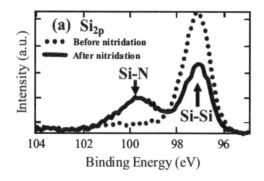 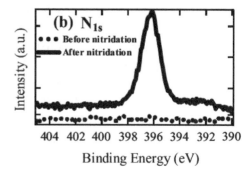

FIGURE 1.46

(a) Si $2p$ and (b) N $1s$ XPS spectra of the Si substrate surface before and after a brief period of nitridation. Reprinted with permission from T. Yamaguchi, Y. Saito, C. Morioka, K. Yorozu, T. Araki, A. Suzuki, and Y. Nanishi, physica status solidi (b), 240, 429 (2003). Copyright (2003) by Wiley-VCH Publishers, Inc.

After 3 min of nitridation of a Si(111) substrate at 800°C, a low-temperature InN buffer layer was deposited at 300°C to obtain a smooth InN surface, then growth of InN at approximately 400°C followed. Figure 1.47 shows (a) a RHEED pattern and (b) an XRD pole-figure spectra of InN grown on Si with a brief period of nitridation (3 min). A clear streak pattern appeared in RHEED and no metastable rotation domains were observed in the XRD pole figure. These results confirm that single crystalline hexagonal InN film can be grown on Si substrates with a brief substrate nitridation. The resultant epitaxial relationships between InN and Si were: $(0001)_{InN}\|(111)_{Si}$ and $[11\bar{2}0]_{InN}\|[1\bar{1}0]_{Si}$. The InN epitaxial layers with thicknesses of 200–300 nm showed FWHM values a little broader than 1 degree for (0002) XRC. On the other hand, when we carried out substrate nitridation for 1 hour, RHEED observations showed a diffuse ring pattern due to the existence of an amorphous SiN_x layer formed by the lengthy substrate nitridation. From these results, we conclude that the nitridation time is an important parameter for the growth of single crystalline InN films on Si substrates.

1.7.2   AlN buffer

The insertion of an AlN buffer layer was also attempted. The AlN buffer layer was grown on a Si substrate at approximately 800°C prior to the growth of the low-temperature InN buffer layer. Insertion of an AlN buffer layer was found to be a very effective way to improve the crystallinity of InN on a Si substrate. The best

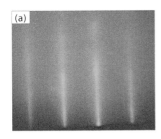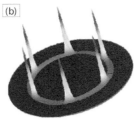

FIGURE 1.47

(a) RHEED pattern and (b) $\{11\bar{2}2\}$ XRD pole-figure of InN grown on Si with a brief period of substrate nitridation.

value of the XRC-FWHM was 31.3 arcmin for InN films with a thickness of 300 nm, which was significantly smaller than that obtained without the AlN buffer layer ($>$ 60 arcmin). Figure 1.48 shows a typical RHEED pattern and an SEM image of the InN film with the AlN buffer layer. The surface morphology of the InN layer was very smooth. In particular, a surface reconstruction of $(\sqrt{3} \times \sqrt{3})R30°$ was observed by RHEED for the InN layer grown with the AlN buffer layer, as shown in Fig. 1.49. This indicates the realization of an atomically flat surface of InN on Si substrates.

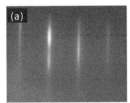

FIGURE 1.48

(a) Typical RHEED pattern and (b) SEM image of InN grown on Si with AlN buffer. Reprinted with permission from T. Yamaguchi, Y. Saito, C. Morioka, K. Yorozu, T. Araki, A. Suzuki, and Y. Nanishi, physica status solidi (b), 240, 429 (2003). Copyright (2003) by Wiley-VCH Publishers, Inc.

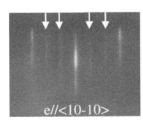

FIGURE 1.49

RHEED patterns of InN grown on Si with an AlN buffer. Reprinted with permission from T. Yamaguchi, Y. Saito, C. Morioka, K. Yorozu, T. Araki, A. Suzuki, and Y. Nanishi, physica status solidi (b), 240, 429 (2003). Copyright (2003) by Wiley-VCH Publishers, Inc.

1.7.3   MEE AlN buffer

A brief and careful nitridation process of the Si substrates or the insertion of an AlN buffer layer between InN and Si could remarkably improve the crystal quality of InN films grown. The crystal quality of InN films grown on Si(111) substrates, however, is inferior to that on (0001) sapphire substrates. We found that the insertion of an AlN buffer layer that was grown by the migration enhanced epitaxy method (MEE-AlN) was very effective for further improving the crystal quality of InN films.

We grew InN films on Si(111) substrates by RF-MBE. An AlN buffer layer was grown at 800°C to a thickness of approximately 100 nm by the MEE method without using a substrate-nitridation process. Here, the MEE growth of AlN used a basic cycle consisting of Al supply for 4 sec, source-supply interruption for 2 sec, N radical supply for 2 sec, and source-supply interruption for 2 sec. Finally, InN films were grown at 450°C to a thickness of approximately 350 nm. For comparison, we also grew samples having the same layer-stacking structure but with a normal AlN buffer layer that was grown by simultaneous supply from the sources.

The MEE-AlN buffer layers exhibited RHEED reconstruction patterns when they were brought to the temperature (450°C) for InN growth as shown in Fig. 1.50. These RHEED images indicate that the MEE-AlN buffer layers have an atomically flat surface. The normal AlN buffer layers, on the other hand, showed only (1×1) RHEED reconstruction patterns. XRC measurements revealed that the crystal quality of the MEE-AlN buffer layers was superior to that of the normal AlN buffer layers. The (0002) XRC-FWHMs of the MEE-AlN buffer layers and the normal AlN buffer layers were approximately 20 arcmin and 30 arcmin, respectively, indicating that the MEE-AlN buffer layers have a narrower tilt distribution than that of the normal AlN buffer layers. The twist distribution of these AlN buffer layers, on the other hand, could not be determined because XRC measurements for asymmetric reflections produced only a weak and noisy diffraction peak due to the insufficient thickness of the AlN layers.

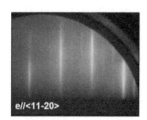
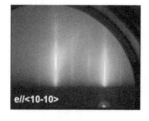

FIGURE 1.50
RHEED patterns of the MEE-AlN buffer layer.

Typical XRC-FWHM values for the InN films grown on Si(111) substrates with these AlN buffer layers are summarized in Table 1.2. FWHM data obtained from our high-crystal-quality InN films grown on (0001) sapphire substrate using a microfaceted InN template [1.25] is also included in this table. Here, the $(10\bar{1}0)$ XRC-

FWHM values of the InN films on Si (111) substrates were theoretically estimated from those of the (0002), ($10\bar{1}3$), ($10\bar{1}2$), ($10\bar{1}1$), and ($30\bar{3}2$) XRCs [1.23]. The FWHM values of (0002) and ($10\bar{1}0$) XRCs of the InN film grown on Si(111) by using the MEE-AlN buffer layer (MEE-AlN-based InN film) were 18.1 arcmin and 42.2 arcmin, respectively, and those by using the normal AlN buffer layer were 31.0 arcmin and 72.7 arcmin, respectively.

TABLE 1.2
FWHM values of symmetric and asymmetric XRCs of InN films grown on MEE-AlN/Si(111), AlN/Si(111) and micro-faceted InN templates.

Structure	XRD rocking curve FWHM (arcmin)					
	(0002)	($10\bar{1}3$)	($10\bar{1}2$)	($10\bar{1}1$)	($30\bar{3}2$)	($10\bar{1}0$)
InN/MEE-AlN/Si(111)	18.1	27.1	30.7	36.3	38.3	42.2
InN/AlN/Si(111)	31.0	46.5	54.0	61.3	69.2	72.7
InN/micro-faceted InN template	1.1	–	25.2	–	34.5	36.5

These results indicate that both the tilt and twist distributions in the MEE-AlN-based InN film are remarkably narrower than those in the normal AlN-based InN film. Moreover, the twist distribution in the MEE-AlN-based InN film is comparable with that in the high-crystal-quality InN films grown on (0001) sapphire substrates, as can be seen in Table 1.2. Although the (0002) XRC-FWHM of 18.1 arcmin obtained for the MEE-AlN-based InN film is one of the narrowest values reported for InN grown on Si, the tilt distribution in this film is still much broader than that in the InN films grown on sapphire substrates. Since the (0002) XRC-FWHM values for the MEE-AlN-based InN film (18.1 arcmin) and the normal AlN-based InN film (31.0 arcmin) are almost the same as those for the underlying MEE-AlN (20 arcmin) and normal AlN (30 arcmin) buffer layers, respectively, it seems that the tilt distribution in the present InN films on Si substrates is governed by the tilt distribution in the underlying AlN buffer layers. Thus we consider that the tilt distribution in the MEE-AlN-based InN film can be further narrowed by improving (narrowing) the tilt distribution in the underlying AlN buffer layer. The best (0002) XRC-FWHM value obtained among the MEE-AlN-based InN films is 17.5 arcmin ($\sim$1050 arcsec). Figure 1.51 shows PL spectra at 77K of the InN film grown on the Si (111) substrate with the MEE-AlN buffer layer. A PL emission peak was observed at approximately 0.75 eV. The observed PL peak energy was slightly higher than that of InN films grown on (0001) sapphire substrates, which is due to the higher residual carrier concentration in the InN grown on the Si(111) substrate. The FWHM values of PL peaks from the InN films were in the range of 85 to 100 meV.

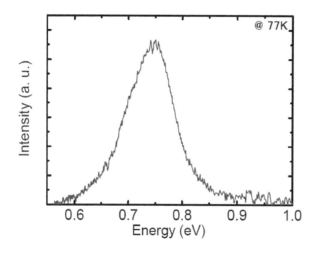

FIGURE 1.51
PL spectra of the InN film grown on Si(111) substrate with the MEE-AlN buffer layer.

1.8   Conclusion

For a long time, growth of high-quality InN has been hindered by its intrinsic problem: the low dissociation temperature and high equilibrium vapor pressure of nitrogen. During the past decade, the quality of InN has been dramatically improved by applying MBE growth. Alloy composition control over the full compositional range and heterostructure formation between InN and InGaN also became possible.

However, there are many issues to be solved before this material system is really applied to opto-electronic devices. These include: (1) very high concentration of residual donors; (2) reduction in high-density threading dislocations and defects; (3) surface electron accumulation; (4) p-type doping; and (5) formation of high-quality heterointerface. Among these issues, reduction of residual carrier concentration and elimination of surface charge accumulation are by far the most important issues. These issues make it very difficult to realize p-type doping and Schottky contact which are essential for opto-electronic device applications. By virtue of the specific nature of the InN band lineup among other semiconductor materials [1.80], almost all of the impurities, dislocations and related defects can become major sources of the high concentration of donors both in the bulk and in the surface region of InN. Accordingly, novel device structures and further development in MBE growth techniques are required.

In conclusion, we have described novel techniques for the realization of high-quality InN single crystals by MBE. Critical growth procedures to obtain high-quality InN films were investigated and (1) nitridation process of sapphire substrates, (2) low-temperature InN buffer layer, (3) precise control of V/III ratio, and (4) selection of optimum growth temperature were found to be essential. We also demonstrated that (1) optimization of nitridation process, (2) regrowth on micro-faceted InN, and (3) nanocolumn growth were very effective to further improve crystal quality of InN films. The InN regrown on the micro-faceted InN templates had considerably smaller twist distribution than that grown on the flat InN templates due to the reduction of edge dislocation density in the regrowth region. TEM studies of nanocolumns also demonstrated the remarkable effect of this technology to reduce or eliminate threading dislocations in InN. To investigate the effect of substrate polarity, InN layers were grown on a variety of substrates: (0001) sapphire substrates, (0001) Ga-face and (000$\bar{1}$) N-face free-standing GaN substrates, and (0001) Si-face and (000$\bar{1}$) C-face 6H-SiC substrates. We revealed that InN with N-polarity can be grown at higher temperatures than InN with In-polarity; this can be explained by the different bonding configuration of the surface N atoms. We have investigated growth of non-polar InN. *a*-plane (11$\bar{2}$0) InN has been successfully grown on *r*-plane (10$\bar{1}$2) sapphire substrates through the use of a substrate nitridation process. *a*-plane In-rich InGaN has also been obtained using an *a*-plane InN template. We have demonstrated that LiAlO$_2$ substrates are useful for the growth of *m*-plane (1$\bar{1}$00) InN. We have studied the growth characteristics of highly Mg-doped N-polar InN films. We have successfully grown highly Mg-doped InN with Mg concentrations over 2.5×10^{20} cm^{-3} using N-polar InN. We found that a Mg concentration of around 2.5×10^{20} cm^{-3} is the critical value for prevention of c-InN inclusions. We have demonstrated that the quality of InGaN dramatically improves by inserting high-temperature-grown InN as a template between InGaN and low-temperature grown InN buffer layer. InN/InGaN quantum well structures were successfully grown on this InN template. PL emission and its photon energy dependence on well width could be successfully observed from this material system. Single-crystalline InN films were obtained on Si substrates by a brief nitridation of the Si substrate. InN crystalline quality was improved by the insertion of an AlN buffer layer between a Si substrate and a low-temperature InN buffer layer.

Acknowledgments

The authors would like to thank Y. Saito, M. Kurouchi, F. Matsuda, D. Muto, T. Yamaguchi, Y. Kumagai, S. Watanabe, K. Nagase, T. Akagi, S. Harui, M. Noda, Y. Takagi, H. Nozawa for MBE growth and characterization experiments. The authors acknowledge J. Wu, J. W. Ager, W. Walukiewicz for optical and electrical characterization, H. Harima for Raman scattering measurements, A. Kasic for infrared spectroscopic ellipsometry, S. Pereira for RBS measurements and A. Koukitu, K. Hiramatsu, H. Miyake for supplying substrates. They also acknowledge H. Naoi, N. Teraguchi, A. Suzuki and T. Miyajima for fruitful discussions. The research work presented here has been supported by the Ministry of Education, Culture, Sports, Science and Technology, Grant-in-Aid for Scientific Research in Priority Areas "Optoelectronics Frontier by Nitride Semiconductor" No. 18069012, for Scientific Research (A) No. 18206003 and (B) No. 13450131, Academic Frontier Promotion Project and The 21st Century COE Program.

References

[1.1] H. Lu, W. J. Schaff, L. F. Eastman, J. Wu, W. Walukiewicz, D. C. Look, and R. J. Molnar, Growth of thick InN by molecular beam epitaxy, *Material Research Society Symposium Proceedings*, 743 (2003) L4.10.1–L4.10.6.

[1.2] M. Higashiwaki and T. Matsui, High-quality InN film grown on a low-temperature-grown GaN intermediate layer by plasma-assisted molecular-beam epitaxy, *Japanese Journal of Applied Physics*, 41 (2002) L540–L542.

[1.3] M. Higashiwaki and T. Matsui, Effect of low-temperature-grown GaN intermediate layer on InN growth by plasma-assisted MBE, *physica status solidi (c)*, 0 (2002) 360–363.

[1.4] Y. Nanishi, Y. Saito, and T. Yamaguchi, RF-molecular beam epitaxy growth and properties of InN and related alloys, *Japanese Journal of Applied Physics*, 42 (2003) 2549–2559.

[1.5] K. Xu, N. Hashimoto, B. Cao, T. Hata, W. Terashima, M. Yoshitani, Y. Ishitani, and A. Yoshikawa, High-quality and thick InN films grown on 2-inch sapphire substrate by molecular-beam epitaxy, *physica status solidi (c)*, 0 (2003) 2790–2793.

[1.6] V. Yu. Davydov, A. A. Klochikhin, R. P. Seisyan, V. V. Emtsev, S. V. Ivanov, F. Bechstedt, J. Furthmüller, H. Harima, A. V. Mudryi, J. Adrhold, O. Semchinova, and J. Graul, Absorption and emission of hexagonal InN. Evidence of narrow fundamental band gap, *physica status solidi (b)*, 229 (2002) R1–R3.

[1.7] J. Wu, W. Walukiewicz, K. M. Yu, J. W. Ager III, E. E. Haller, H. Lu, W. J. Schaff, Y. Saito, and Y. Nanishi, Unusual properties of the fundamental band gap of InN, *Applied Physics Letters*, 80 (2002) 3967–3969.

[1.8] T. Matsuoka, H. Okamoto, M. Nakao, H. Harima, and E. Kurimoto, Optical bandgap energy of wurtzite InN, *Applied Physics Letters*, 81 (2002) 1246–1248.

[1.9] Y. Saito, H. Harima, E. Kurimoto, T. Yamaguchi, N. Teraguchi, A. Suzuki, T. Araki, and Y. Nanishi, Growth temperature dependence of indium nitride crystalline quality grown by RF-MBE, *physica status solidi (b)*, 234 (2002) 796–800.

[1.10] K. Xu and A. Yoshikawa, Effects of film polarities on InN growth by molecular-beam epitaxy, *Applied Physics Letters*, 83 (2003) 251–253.

[1.11] M. Higashiwaki and T. Matsui, Plasma-assisted MBE growth of InN films and InAlN/InN heterostructures, *Journal of Crystal Growth*, 251 (2003) 494–498.

[1.12] T. Yamaguchi, T. Araki, Y. Saito, T. Maruyama, Y. Nanishi, N. Teraguchi, and A. Suzuki, Growth condition dependence of InN film a-axis directions on sapphire (0001) substrate, *Institute of Physics Conference Series*, 170 (2002) 765–770.

[1.13] T. Yamaguchi, Y. Saito, K. Kano, T. Araki, N. Teraguchi, A. Suzuki, and Y. Nanishi, Study of epitaxial relationship in InN growth on sapphire (0001) by RF-MBE, *physica status solidi (b)*, 228 (2001) 17–20.

[1.14] T. Yamaguchi, Y. Saito, K. Kano, T. Araki, N. Teraguchi, A. Suzuki, and Y. Nanishi, The c-axis and a-axis orientations in InN grown directly on (0001) sapphire substrate by RF-MBE, *Proceedings of the 14th International Conference on Indium Phosphide and Related Materials* (2002) pp. 643–646.

[1.15] H. Amano, N. Sawaki, I. Akasaki, and Y. Toyoda, Metalorganic vapor phase epitaxial growth of a high quality GaN film using an AlN buffer layer, *Applied Physics Letters*, 48 (1986) 353–355.

[1.16] Y. Saito, N. Teraguchi, A. Suzuki, T. Araki, and Y. Nanishi, Annealing effect of low temperature growth of InN films by RF-MBE, *Institute of Pure and Applied Physics Conference Series*, 1 (2000) 182–185.

[1.17] Y. Saito, N. Teraguchi, A. Suzuki, T. Yamaguchi, T. Araki, and Y. Nanishi, Electrical properties of InN grown by RF-MBE, *Material Research Society Symposium Proceedings*, 639 (2001) G11.18.1.

[1.18] V. Yu. Davydov, A. A. Klochikhin, S. V. Ivanov, J. Aderhold, and A. Yamamoto, Growth and properties of InN, in *Nitride semiconductors: Handbook on materials and devices*, edited by P. Ruterana, M. Albrecht, and J. Neugebauer, Wiley, New York (2003) pp. 241–294.

[1.19] Y. Saito, N. Teraguchi, A. Suzuki, T. Araki, and Y. Nanishi, Growth of high-electron-mobility InN by RF molecular beam epitaxy, *Japanese Journal of Applied Physics*, 40 (2001) L91–L93.

[1.20] D. Muto, R. Yoneda, H. Naoi, M. Kurouchi, T. Araki, and Y. Nanishi, Effects of the nitridation process of (0001) sapphire on crystalline quality of InN grown by RF-MBE, *Material Research Society Symposium Proceedings*, 831 (2005) E4.2.1.

[1.21] C. Heinlein, J. Grepstad, T. Berge, and H. Riechert, Preconditioning of c-plane sapphire for GaN epitaxy by radio frequency plasma nitridation, *Applied Physics Letters*, 71 (1997) 341–343.

[1.22] V. Srikant, J. S. Speck, and D. R. Clarke, Mosaic structure in epitaxial thin films having large lattice mismatch, *Journal of Applied Physics*, 82 (1997) 4286–4295.

[1.23] K. Xu, W. Terashima, T. Hata, N. Hashimoto, M. Yoshitani, B. Cao, Y. Ishitani, and A. Yoshikawa, Comparative study of InN growth on Ga- and N-polarity GaN templates by molecular-beam epitaxy, *physica status solidi (c)*, 0 (2003) 2814–2817.

[1.24] Y. T. Moon, Y. Fu, F. Yun, S. Dogan, M. Mikkelson, D. Johnstone, and H. Morkoç, A study of GaN regrowth on the micro-facetted GaN template formed by in-situ thermal etching, *physica status solidi (a)*, 202 (2005) 718–721.

[1.25] D. Muto, H. Naoi, T. Araki, S. Kitagawa, M. Kurouchi, H. Na, and Y. Nanishi, High-quality InN grown on KOH wet etched N-polar InN template by RF-MBE, *physica status solidi (a)*, 203 (2005) 1691–1695.

[1.26] T. Akagi, K. Kosaka, S. Harui, D. Muto, H. Naoi, T. Araki, and Y. Nanishi, Correlation between threading dislocations and nonradiative recombination centers in InN observed by IR cathodoluminescence, *Journal of Electonic Materials*, 37 (2008) 603–606.

[1.27] T. Yamaguchi, T. Araki, H. Naoi, and Y. Nanishi, Position-controlled InN nano-dot growth on patterned substrates by ECR-MBE, *Material Research Society Symposium Proceedings*, 955 (2007) 0955-I07-40.

[1.28] T. Araki, T. Yamaguchi, S. Harui, T. Tamiya, H. Miyake, K. Hiramatsu, Y. Nanishi, Oral presentation at the 7th International Conference on Nitride Semiconductors, Las Vegas, USA, 2007.

[1.29] S. Harui, H. Tamiya, T. Akagi, H. Miyake, K. Hiramatsu, T. Araki and Y. Nanishi, Transmission electron microscopy characterization of position-controlled InN nanocolumns, *Japanese Journal of Applied Physics*, 47 (2008) 5330–5332.

[1.30] F. Matsuda, Y. Saito, T. Muramatsu, T. Yamaguchi, Y. Matsuo, A. Koukitsu, T. Araki, and Y. Nanishi, Influence of substrate polarity on growth of InN films by RF-MBE, *physica status solidi (c)*, 0 (2003) 2810–2813.

[1.31] H. Naoi, F. Matsuda, T. Araki, A. Suzuki, and Y. Nanishi, The effect of substrate polarity on the growth of InN by RF-MBE, *Journal of Crystal Growth* 269 (2004) 155–161.

[1.32] T. Koizumi, J. Wada, T. Araki, H. Naoi, and Y. Nanishi, Optical emission spectroscopy during InN growth by ECRMBE, *Journal of Crystal Growth*, 275 (2004) e1073–e1075.

[1.33] J. Grandal, M. A. Sánchez-Garciá, E. Calleja, E. Luna, and A. Trampert, Accommodation mechanism of InN nanocolumns grown on Si(111) substrates by molecular beam epitaxy, *Applied Physics Letters*, 91 (2007) 021902:1–3.

[1.34] E. Calleja, J. Ristić, S. Fernández-Garrido, L. Cerutti, M. A. Sánchez-Garciá, J. Grandal, A. Trampert, U. Jahn, G. Sánchez, A. Griol, and B. Sánchez, Growth, morphology, and structural properties of group-III-nitride nanocolumns and nanodisks, *physica status solidi (b)* 244 (2007) 2816–2837.

[1.35] S. Sonoda, S. Shimizu, Y. Suzuki, K. Balakrishnan, J. Shirakashi, and H. Okumura, Characterization of polarity of plasma-assisted molecular beam epitaxial GaN{0001} film using coaxial impact collision ion scattering spectroscopy, *Japanese Journal of Applied Physics*, 39 (2000) L73–L75.

[1.36] M. Sumiya, M. Tanaka, K. Ohtsuka, S. Fuke, T. Ohnishi, I. Ohkubo, M. Yoshimoto, H. Koinuma, and M. Kawasaki, Analysis of the polar direction of GaN film growth by coaxial impact collision ion scattering spectroscopy, *Applied Physics Letters*, 75 (1999) 674–676.

[1.37] S. Fuke, H. Teshigawara, K. Kuwahara, Y. Takano, T. Ito, M. Yanagihara, K. Ohtsuka, Influences of initial nitridation and buffer layer deposition on the morphology of a (0001) GaN layer grown on sapphire substrates, *Journal of Applied Physics*, 83 (1998) 764–767.

[1.38] Y. Saito, Y. Tanabe, T. Yamaguchi, N. Teraguchi, A. Suzuki, T. Araki, and Y. Nanishi, Polarity of high-quality indium nitride grown by RF molecular beam epitaxy, *physica status solidi (b)*, 228 (2001) 13–16.

[1.39] T. Araki, S. Ueta, K. Mizuo, T. Yamaguchi, Y. Saito, Y. Nanishi, TEM characterization of InN films grown by RF-MBE, *physica status solidi (c)*, 0 (2003) 2798–2801.

[1.40] D. Segev and C. G. Van de Walle, Origins of Fermi-level pinning on GaN and InN polar and nonpolar surfaces, *Europhysics Letters*, 76 (2006) 305–311.

[1.41] B. A. Haskell, F. Wu, S. Matsuda, M. D. Craven, P. T. Fini, S. P. DenBaars, J. S. Speck, and S. Nakamura, Structural and morphological characteristics of planar ($11\bar{2}0$) a-plane gallium nitride grown by hydride vapor phase epitaxy, *Applied Physics Letters* 83 (2003) 1554–1556.

[1.42] V. Cimalla, J. Pezoldt, G. Ecke, R. Kosiba, O. Ambacher, L. Spiess, G. Teichert, H. Lu, and W. J. Schaff, Growth of cubic InN on r-plane sapphire, *Applied Physics Letters*, 83 (2003) 3468–3470.

[1.43] A. Tsuyuguchi, K. Teraki, T. Koizumi, J. Wada, T. Araki, Y. Nanishi, and H. Naoi, Cubic InN Growth on R-plane (10-12) Sapphire by ECR-MBE, *Institute of Physics Conference Series*, 184 (2005) 239–242.

[1.44] H. Lu, W. J. Schaff, L. F. Eastman, J. Wu, W. Walukiewicz, V. Cimalla, and O. Ambacher, Growth of a-plane InN on r-plane sapphire with a GaN buffer by molecular-beam epitaxy, *Applied Physics Letters*, 83 (2003) 1136–1138.

[1.45] Y. Kumagai, A. Tsuyuguchi, H. Naoi, T. Araki, and Y. Nanishi, A-plane ($11\bar{2}0$) InN growth on nitridated R-plane ($10\bar{1}2$) sapphire by ECR-MBE, *physica status solidi (b)*, 243 (2006) 1468–1471.

[1.46] S. Watanabe, Y. Kumagai, A. Tsuyuguchi, H. Naoi, T. Araki, and Y. Nanishi, Microstructure of A-plane InN grown on R-plane sapphire by ECR-MBE, *physica status solidi (c)*, 4 (2007) 2556–2559.

[1.47] M. D. Craven, S. H. Lim, F. Wu, J. S. Speck, and S. P. DenBaars, Structural characterization of nonpolar ($11\bar{2}0$) a-plane GaN thin films grown on ($1\bar{1}02$) r-plane sapphire, *Applied Physics Letters*, 81 (2002) 469–471.

[1.48] M. Noda, Y. Kumagai, S. Takado, D. Muto, H. Na, H. Naoi, T. Araki, and Y. Nanishi, Growth of A-plane ($11\bar{2}0$) In-rich InGaN on R-plane ($10\bar{1}2$) sapphire by RF-MBE, *physica status solidi (c)*, 4 (2007) 2560–2563.

[1.49] B. A. Haskell, A. Chakraborty, F. Wu, H. Sasano, P. T. Fini, S. P. Denbaars, J. S. Speck, and S. Nakamura, Microstructure and enhanced morphology of planar nonpolar m-plane GaN grown by hydride vapor phase epitaxy, *Journal of Electronic Materials*, 34 (2005) 357–360.

[1.50] J. W. Gerlach, A. Hofmann, T. Höche, F. Frost, B. Rauschenbach, and G. Benndorf, High-quality m-plane GaN thin films deposited on γ-LiAlO$_2$ by ion-beam-assisted molecular-beam epitaxy, *Applied Physics Letters*, 88 (2006) 011902:1–3.

[1.51] H. Lu, W. J. Schaff, L. F. Eastman, and C. E. Stutz, Surface charge accumulation of InN films grown by molecular-beam epitaxy, *Applied Physics Letters*, 82 (2003) 1736-1738.

[1.52] I. Mahboob, T. D. Veal, C. F. McConville, H. Lu, and W. J. Schaff, Intrinsic electron accumulation at clean InN surfaces, *Physical Review Letters*, 92 (2004) 036804:1–4.

[1.53] S. X. Li, K. M. Yu, J. Wu, R. E. Jones, W. Walukiewicz, J. W. Ager III, W. Shan, E. E. Haller, H. Lu, and W. J. Schaff, Fermi-level stabilization energy in group III nitrides, *Physical Review B*, 71 (2005) 161201:1–4.

[1.54] R. E. Jones, K. M. Yu, S. X. Li, W. Walukiewicz, J. W. Ager, E. E. Haller, H. Lu, and W. J. Schaff, Evidence for *p*-type doping of InN, *Physics Review Letters*, 96 (2006) 125505:1–4.

[1.55] L. K. Li, M. J. Jurkovic, W. I. Wang, J. M. Van Hove, and P. P. Chow, Surface polarity dependence of Mg doping in GaN grown by molecular-beam epitaxy, *Applied Physics Letters*, 76 (2000) 1740–1742.

[1.56] X. Wang, S. B. Che, Y. Ishitani, A. Yoshikawa, H. Sasaki, T. Shinagawa, and S. Yoshida, Polarity inversion in high Mg-doped In-polar InN epitaxial layers, *Applied Physics Letters*, 91 (2007) 081912:1–3.

[1.57] D. Muto, H. Naoi, S. Takado, H. Na, T. Araki, and Y. Nanishi, Mg-doped N-polar InN grown by RF-MBE, *Material Research Society Symposium Proceedings*, 955 (2007) 0955-I08-01.

[1.58] T. Ohashi, T. Kouno, M. Kawai, A. Kikuchi, and K. Kishino, Growth and characterization of InGaN double heterostructures for optical devices at 1.5-1.7 μm communication wavelengths, *physica status solidi (a)*, 201 (2004) 2850–2854.

[1.59] S. B. Che, W. Terashima, T. Ohkubo, M. Yoshitani, N. Hashimoto, K. Akasaka, Y. Ishitani, and A. Yoshikawa, InN/GaN SQW and DH structures grown by radio frequency plasma-assisted MBE, *physica status solidi (c)*, 2 (2005) 2258–2262.

[1.60] M. Kurouchi, H. Naoi, T. Araki, T. Miyajima, and Y. Nanishi, Fabrication and characterization of InN-based quantum well structures grown by radio-frequency plasma-assisted molecular-beam epitaxy, *Japanese Journal of Applied Physics*, 44 (2005) L230–L232.

[1.61] S. B. Che, W. Terashima, Y. Ishitani, A. Yoshikawa, T. Matsuda, H. Ishii, and S. Yoshida, Fine-structure N-polarity InN/InGaN multiple quantum wells grown on GaN underlayer by molecular-beam epitaxy, *Applied Physics Letters*, 86 (2005) 261903:1–3.

[1.62] M. Kurouchi, T. Araki, H. Naoi, T. Yamaguchi, A. Suzuki, and Y. Nanishi, Growth and properties of In-rich InGaN films grown on (0001) sapphire by RF-MBE, *physica status solidi (b)*, 241 (2004) 2843–2348.

[1.63] M. Kurouchi, T. Yamaguchi, H. Naoi, A. Suzuki, T. Araki, and Y. Nanishi, Growth of In-rich InGaN on InN template by radio-frequency plasma assisted molecular beam epitaxy, *Journal of Crystal Growth*, 275 (2005) e1053–e1058.

[1.64] H. Naoi, M. Kurouchi, D. Muto, S. Takado, T. Araki, and T. Miyajima, H. Na, and Y. Nanishi, Growth and properties of InN, InGaN, and InN/InGaN quantum wells, *physica status solidi (a)*, 203 (2006) 93–101.

[1.65] H. Naoi, M. Kurouchi, D. Muto, T. Araki, T. Miyajima, and Y. Nanishi, Growth of high-quality In-rich InGaN alloys by RFMBE for the fabrication of InN-based quantum well structures, *Journal of Crystal Growth*, 288 (2006) 283–288.

[1.66] J. W. Matthews and A. E. Blakeslee, Defects in epitaxial multilayers: I. Misfit dislocations, *Journal of Crystal Growth*, 27 (1974) 118–125.

[1.67] R. People and J. C. Bean, Calculation of critical layer thickness versus lattice mismatch for Ge_xSi_{1-x}/Si strained-layer heterostructures, *Applied Physics Letters*, 47 (1985) 322.

[1.68] A. Fischer, H. Kühne, and H. Richter, New approach in equilibrium theory for strained layer relaxation, *Physical Review Letters*, 73 (1994) 2712–2715.

[1.69] A. Yamamoto, M. Tsujino, M. Ohkubo, and A. Hashimoto, Nitridation effects of substrate surface on the metalorganic chemical vapor deposition growth of InN on Si and α-Al$_2$O$_3$ substrates, *Journal of Crystal Growth*, 137 (1994) 415–420.

[1.70] F. H. Yang, J. S. Hwang, K. H. Chen, Y. J. Yang, T. H. Lee, L. G. Hwa, and L. C. Chen, High growth rate deposition of oriented hexagonal InN films, *Thin Solid Films*, 405 (2002) 194–197.

[1.71] M. Yoshimoto, T. Nakano, T. Yamashita, K. Suzuki, and J. Saraie, MBE growth of InN on Si toward hole-barrier structure in Si devices, *Institute of Pure and Applied Physics Conference Series*, 1 (2000) 186–189.

[1.72] F. Agullo-Rueda, E. E. Mendez, B. Bojarczuk, and C. Guha, Raman spectroscopy of wurtzite InN films grown on Si, *Solid State Communications*, 115 (2000) 19–21.

[1.73] I. Bello, W. M. Lau, R. P. W. Lawson, and K. K. Foo, Deposition of indium nitride by low energy modulated indium and nitrogen ion beams, *Journal of Vacuum Science and Technology A*, 10 (1992) 1642–1646.

[1.74] Y. Bu, L. Ma, and M. C. Lin, Laser-assisted chemical vapor deposition of InN on Si(100), *Journal of Vacuum Science and Technology A*, 11 (1993) 2931–2937.

[1.75] T. Yodo, H. Yona, H. Ando, D. Nosei, and Y. Harada, Strong band edge luminescence from InN films grown on Si substrates by electron cyclotron resonance-assisted molecular beam epitaxy, *Applied Physics Letters*, 80 (2002) 968–970.

[1.76] T. Yamaguchi, K. Mizuo, Y. Saito, T. Araki, Y. Nanishi, and T. Miyajima, Single crystalline InN films grown on Si (111) substrates, *Institute of Physics Conference Series*, 174 (2003) 17–20.

[1.77] T. Yamaguchi, K. Mizuo, Y. Saito, T. Noguchi, T. Araki, and Y. Nanishi, Single crystalline InN films grown on Si substrates by using a brief substrate nitridation process, *Material Research Society Symposium Proceedings*, 743 (2003) 163.

[1.78] T. Yamaguchi, Y. Saito, C. Morioka, K. Yorozu, T. Araki, A. Suzuki, and Y. Nanishi, Effect of AlN buffer layer on the growth of InN epitaxial film on Si substrate, *physica status solidi (b)*, 240 (2003) 429–432.

[1.79] K. Nagase, H. Naoi, T. Araki, A. Suzuki, and Y. Nanishi, Presentation at the 3rd Asia-Pacific Workshop on Widegap Semiconductors, Jeonju, Korea, 2007.

[1.80] C. G. Van de Walle and J. Neugebauer, Universal alignment of hydrogen levels in semiconductors, insulators and solutions, *Nature*, 423 (2003) 626–628.

2

Thermal stability, surface kinetics, and MBE growth diagrams for N- and In-face InN

C. S. Gallinat, G. Koblmüller, and J. S. Speck

Materials Department, University of California, Santa Barbara, California 93106-5050, United States of America

2.1 Introduction

Although the first single-crystalline InN films grown by molecular-beam epitaxy (MBE) several years ago fueled major breakthroughs in InN research [2.1, 2.2], progress toward the current state-of-the-art InN growth has evolved slowly and has remained complicated by many challenges. For instance, the much weaker group-III-to-N atomic bond of InN (1.93 eV) in comparison to the other binaries (GaN: 2.2 eV, AlN: 2.88 eV) [2.3] gives rise to significant inherent thermal instability and large dissociation rates. On the other hand, large lattice and thermal mismatches with the most commonly available commercial substrates (SiC, sapphire, GaN) of greater than 10% provide an extra burden for producing high-quality epitaxial InN films.

Many groups reported that these two issues along with the generally low growth temperatures required for the MBE growth of InN [2.4] yield often unsatisfactory three-dimensional surfaces [2.5, 2.6], high defect densities and high background doping by unintentional impurities, causing substantial shifts in the optical bandgap and electronic transport properties of InN [2.1, 2.2]. Further problems, such as phase separation and compositional modulation, arise when alloying InN with GaN or AlN due to the large difference in optimum growth parameters of these constituents during MBE growth (see Chapter 3 for more details).

To improve the quality of InN films, recently more efforts have been dedicated to understanding the effects of InN film polarity on the surface growth kinetics, dopant and impurity incorporation and crystal quality, but also on the surface electronic properties (that is, electron accumulation). However, there is still limited knowledge about the surface growth kinetics, like adatom diffusion and the effects of metallic adlayer coverages on InN surfaces, much less than for the other binaries GaN or AlN.

In this chapter, several of these issues inherent to InN growth by plasma-assisted (PA)-MBE will be elucidated comprehensively and several routes for improving the quality of InN films will be given.

In the first part, an experimental analysis of the differences in thermal stability along the predominant orientations of *c*-plane InN, that is, In-face versus N-face InN, will be presented. These are the technologically most relevant and therefore most studied orientations, while to date almost no systematic investigations have been performed for the non-polar or semi-polar orientations. Special focus will also be directed toward the surface kinetics during growth, influenced by both the thermal decomposition and the existence of surface reconstructions and metallic In surface coverages on InN. As in PAMBE growth of GaN or AlN, it will be shown that the latter play a dominant role in producing high-quality InN films through surfactant-mediated growth at otherwise non-optimum growth conditions.

The growth mode and mechanism will be further summarized in PAMBE growth diagrams for both In-face and N-face InN, which serve as maps for identifying optimum growth conditions and for achieving the best possible physical properties of MBE-grown InN films. Finally, special attention will be given to the structural and electronic properties of In-face and N-face InN films.

2.2 Thermal decomposition of InN

2.2.1 InN film polarity

The typically reported growth directions of the wurtzite III-nitrides occur along the basal $\{0001\}$ plane formed by the stacking of bilayers consisting of alternating planes of metal and nitrogen atoms (shown in Fig. 2.1). The symmetry of the wurtzite structure prevents the stacked bilayers from having equivalent atomic arrangements on either side of the bilayer, thus changing the terminating atomic arrangement on the final $\{0001\}$ surface. This concept is referred to as crystal polarity and two different growth directions normal to the basal plane are defined by the polarity convention. The In-face orientation is defined by the crystal orientation where only single bonds from the In atoms are directed toward the surface along the c^+-axis. The opposite direction along the c^--axis, which consists of three bonds away from the In atoms (toward the N atoms) characterizes the N-face orientation. Note that in both cases the surfaces are terminated by In atoms, since a N-terminated surface may not be thermodynamically stable due to formation and desorption of N_2.

The realization of the two polarities of InN, that is, In-face and N-face, is especially sensitive to the chosen substrate material and the applied growth technique, as has been also demonstrated to a great extent for GaN. Growth of InN on Si-face [0001] SiC or Ga-face [0001] GaN substrates is known to yield commonly In-face polarity, while C-face [000$\bar{1}$] SiC or freestanding N-face [000$\bar{1}$] GaN substrates typically give N-face polarity [2.7]. For growth on sapphire, N-polarity InN

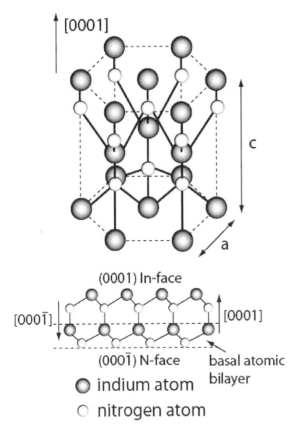

[0001] In-face

[000$\bar{1}$]

[0001]

(000$\bar{1}$) N-face   **basal atomic bilayer**

⦿ **indium atom**

○ **nitrogen atom**

FIGURE 2.1

Schematic illustration of InN wurtzite crystal structure exhibiting the polarity along the *c*-axis, In-face [0001] polarity on the top and N-face [000$\bar{1}$] polarity on the bottom side. The small and large spheres indicate N and In, respectively.

is typically achieved for either metal-organic vapor phase epitaxy [2.8] or MBE growth [2.9]. This is in contrast to GaN growth by these two techniques, where the former yields generally metal-face polarity.

Finding out which type of polarity is predominant for a grown InN film requires mostly refined electron microscopic techniques such as matching of experimental and simulated convergent beam electron diffraction (CBED) patterns [2.10]. Co-axial impact collision ion scattering spectroscopy (CAICISS) [2.11] represents another method to gain information about the InN surface polarity, although the most straightforward analysis can be obtained by wet chemical etching using KOH solution [2.7]. Using this method, the polarities of InN can be distinguished by the rough etching and the formation of hexagonal pyramids surrounded by facets of N-face InN, while In-face InN etched very smoothly [2.12]. These features observed are very similar to those reported for GaN [2.13]. Recently, also hydrogen irradiation [2.14] has been proposed to effectively determine the InN polarity, resulting in In droplet formation on N-face InN, whereas the In-face orientation was inert to this process.

Like in GaN, the InN community has also realized several differences in thermal stability between the metal-face (that is, In-face) and the N-face orientation, investigated mainly by the formation of metallic In droplets or In inclusions in the crystal [2.15–2.17] as well as the appearance of different crystal defects [2.18].

Addressing these polarity-dependent differences more fundamentally, knowledge about the actual mechanisms and pathways of thermal decomposition are needed. The following investigations concentrate on both *in situ* and *ex situ* measurements of the thermal decomposition rates of InN for the *c*-plane crystal orientations. Here, it will be clearly differentiated between thermal decomposition in N_2 ambient (that is, 10^{-5} Torr) and under standard radio-frequency (rf) PAMBE growth conditions (that is, under impinging In and N at comparable chamber pressure). Routes for suppressing thermal decomposition will also be proposed and compared with the typical decomposition behavior in other growth environments.

2.2.2 *Ex situ* analysis

A very reliable way to determine the thermal decomposition rate of InN is to measure the decrease in growth rate from the nominally expected value of thick InN films grown at temperatures where InN dissociates. This is shown below for films grown under metal-rich (that is, In-rich) growth conditions at constant N flux. Detailed explanations about precise selection of growth conditions, buffer layer growth and calibration schemes of the molecular fluxes involved can be found in the literature [2.19–2.21]. The proposed reduction in growth rate can be determined systematically by measuring, for instance, the film thickness after growth by cross-section scanning electron microscopy (SEM). This is illustrated in Fig. 2.2 for InN films grown under identical conditions (that is, constant In/N flux ratio ~ 1.3 and growth time) for a wide range of growth temperatures for both polar wurtzite orientations, N-face and In-face InN [2.21, 2.22].

The main result of the *ex situ* studies is the difference in the onset temperature for thermal decomposition between the two polar orientations. This critical temperature is defined as the temperature where the growth rate deviates from the nominally expected value (which is given by the supplied N flux under the metal-rich conditions [2.23]). While for In-face InN significant reduction in growth rate occurs at around 470°C, the equivalent onset temperature for N-face InN is ~ 570°C, allowing growth at least ~ 100°C higher without drastic reduction in growth rate.

With increasing temperature though, the InN growth rate decreases substantially, yielding no growth beyond 500°C for In-face and beyond 635°C for N-face InN. As recognized by many groups [2.24, 2.25], macroscopic In droplets form on the surface [Fig. 2.2(e)] above these temperatures, such that these temperatures mark an upper limit for the successful sublimation of InN under the given MBE growth conditions. This also suggests that the sum of excess In supplied during growth and rate of In accumulation through InN decomposition is higher than the maximum possible desorption of In from the surface [2.26]. Therefore it has been interpreted that the high InN thermal decomposition combined with the low In desorption may complicate the growth of metal-free InN films and in severe cases may even lead to growth interruption [2.24].

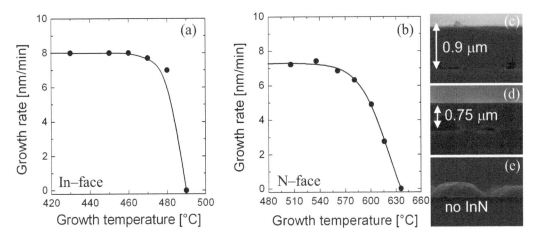

FIGURE 2.2

Growth rate evolution of (a) In-face and (b) N-face InN as a function of growth temperature for In-rich grown InN films (In/N~1.3), as determined from thickness measurements by cross-section SEM. Example SEM images of N-face InN films grown at three different temperatures are also shown for (c) 560°C, (d) 580°C and (e) 635°C. Note the drastic decrease in InN thickness with increasing temperature and the accumulation of In droplets in (e). Reprinted with permission from G. Koblmüller, C. S. Gallinat, and J. S. Speck, Journal of Applied Physics, 101 (2007) 083516. Copyright 2007, American Institute of Physics.

2.2.3   *In situ* analysis

Thermal decomposition of InN can also be determined *in situ* by various techniques (laser reflectometry, ellipsometry or reflection high energy electron diffraction (RHEED) [2.15, 2.16]). However, most experimental studies focused mainly on analyzing the decomposition rates in the N_2 ambient of the MBE environment, under conditions where no molecular fluxes are supplied to the InN surface. Dimakis *et al.* reported on the substrate temperature dependence of as-grown In-face InN films by analyzing the In droplet formation through RHEED darkening [2.24], which occurs at temperatures above ~450°C. More quantitative assessment of the decomposition rates can also be gained by quadrupole mass spectrometry (QMS), a unique method which has been recently developed for surface kinetic and desorption studies of Ga adlayers on GaN [2.27] and which also allows measurement of the In desorption during the InN dissociation process. A description of the experimental QMS setup, calibration and the data acquisition can be found in detail in the literature [2.21].

For convenience, the following experimental results are specified for the N-face InN orientation, and will be briefly compared with analogous results achieved for the In-face orientation. The N-face InN films grown for this study were produced at fairly low temperature (that is, 540°C) under slightly N-rich conditions (to prevent thermal decomposition and the formation of In droplets on the surface). After growth, the InN film was incrementally heated beyond the onset temperature of thermal decomposition (~560°C) to a final temperature of 650°C. During these temperature ramps neither nitrogen nor indium was supplied to the surface. The QMS-measured In desorption into vacuum given in growth rate equivalent units (expressed in nm/min) is shown in Fig. 2.3(a) and can be categorized by three characteristic regimes:

I A low-temperature regime, for temperatures between 560 and 595°C. In this regime, In desorption is rather low, concurrent with quite low intensity RHEED diffraction streaks (see inset). RHEED intensities of low contrast are typically associated with a surface structure consisting of a metallic adlayer and droplets on top, where the latter cause severe shadowing effects of the diffracted electron beam [2.28]. Therefore, this regime can be referred to as strong *metal In accumulation* due to thermally limited desorption of the metallic In adlayer and droplets terminating the decomposing InN surface structure.

II During further increase in temperature from 595 to 605°C (regime II), the desorption signal increased rapidly to a sharp maximum at ∼600°C. Along with this, a transition from a low intensity to a high intensity RHEED pattern was observed, indicating a significant change of the InN surface where the majority of accumulated droplets have desorbed.

III Further temperature increase causes no change in RHEED pattern, but a stepwise increase in the In desorption rate, exceeding a rate of 6 nm/min at 643°C [also shown in a closeup view in Fig. 2.3(b)]. This region (III) can be referred to a regime of *"dry" InN decomposition*, because the progressively constant high intensity RHEED pattern suggests no further metallic In accumulation on the surface.

Investigations of the temperature dependence of the In desorption within these regimes yield two different Arrhenius dependencies [Fig. 2.3(c)]: (i) at low temperatures the limited desorption process underlies a high activation energy ($E^A = 4.4$ eV); and (ii) at high temperatures above 595°C the activation energy was much lower ($E^A = 1.15$ eV). The latter describes more the true energy barrier for decomposition, in close agreement with the value determined independently from the growth rate reduction by *ex situ* cross-section SEM ($E^A = 1.2$ eV, Section 2.2.2).

Comparison of the In desorption rate (by QMS) with the growth rate reduction (by SEM) demonstrates direct proportionality between In desorption and InN decomposition rate. These results show also that the decomposition rates are nearly identical between vacuum conditions (no active nitrogen) and typical metal-rich growth conditions (with active nitrogen), as recently also found for GaN [2.29]. This invariance is expected to hold also for slight variations in In/N flux ratio from stoichiometric flux conditions, but breaks down for more N-rich growth conditions. This can be understood by viewing the meta-stable InN growth process as the competition between the forward reaction, which depends on the arrival of the active nitrogen species at the growth surface, and the reverse reaction, whose rate is limited by a kinetic barrier to decomposition. As elaborated in detail for meta-stable

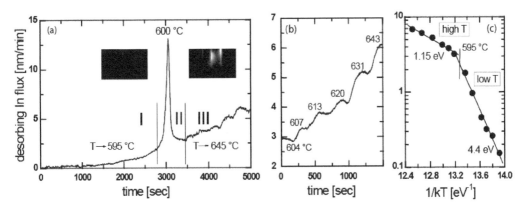

FIGURE 2.3

(a) In desorption measured by QMS during N-face InN decomposition in a temperature range between 560 and 645°C (insets show typical RHEED patterns on each side of the desorption maximum); (b) closeup view of the In desorption in the high-temperature region III; and (c) Arrhenius presentation of In desorption versus $1/kT$ yielding two activation energies of 4.4 eV (low T) and 1.15 eV (high T). Reprinted with permission from G. Koblmüller, C. S. Gallinat, and J. S. Speck, Journal of Applied Physics, 101 (2007) 083516. Copyright 2007, American Institute of Physics.

GaN growth by Newman *et al.* [2.30], this means that decomposition can be significantly suppressed if the arrival rate of active nitrogen is larger than the sublimation rate.

For the other polar orientation, In-face InN, the thermal decomposition can be simply described by the *metal In accumulation* regime (regime I). This is the only possible decomposition pathway, because at the significantly lower temperature limit for successful In-face InN growth ($< 500°C$) In desorption is generally negligible [2.22, 2.24]. For the In-face InN orientation it has even been proposed that the accumulated In droplets may catalytically enhance thermal decomposition, leading to complete growth interruption at high In/N flux ratios [2.24]. Regarding the dependence of the arrival rate of active nitrogen on the meta-stability of InN growth, it can be expected that large excess N supply may allow droplet-free growth by PAMBE at temperatures beyond the currently accepted limit of 500°C. Indeed, this would mimic the typical growth environment of ammonia (NH_3)-MBE or MOVPE, where large ammonia flows enabled growth of In-face InN even up to temperatures of 700°C [2.31].

2.3   Development of PAMBE growth diagrams

2.3.1   Metallic In accumulation and desorption

As in GaN or AlN, precise knowledge about metallic droplet formation and desorption is essential for controlling and optimizing the PAMBE growth of nitrides. To

differentiate between the contributions in In desorption arising from the decomposing InN surface with simultaneous In accumulation and from the purely metallic
bulk In state, further analysis is needed. This can be achieved by performing In
adsorption/desorption experiments on substrates different from InN, to discard the
contributions caused by InN decomposition.

In the following, In adsorption/desorption experiments are presented on adequate
substrates for both polar orientations, that is, N-face and Ga-face GaN being the
ideal candidates at hand. This can be achieved by depositing a certain amount of
In onto the GaN surfaces and measuring the subsequent desorption into vacuum
by QMS. The results are shown in Fig. 2.4 and demonstrate evidence for two distinct desorption regimes at a given temperature: a steady-state desorption regime,
followed by a monotonic decrease to zero desorption (similar to the typical ad-
sorption/desorption behavior for Ga on GaN and AlN surfaces) [2.27, 2.32]. The
steady-state desorption can be attributed to the desorption of droplets (by maintaining the adlayer in its equilibrium), while the final monotonic decrease represents
the decay of the remaining adlayer. Further details about In adlayer coverage on
InN surfaces are highlighted in Section 2.3.3.

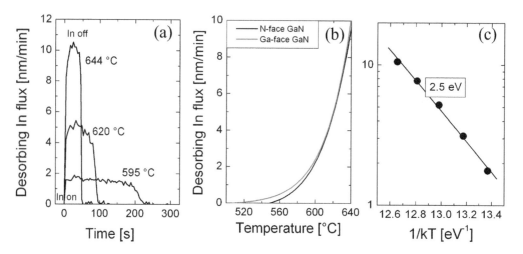

FIGURE 2.4

(a) In desorption of ≈ 40 ML In adsorbed on a N-face GaN surface at different temperatures. (b) The
steady-state desorption (after closing the In shutter) represents the maximum desorption of In (that
is, droplet boundary), yielding (c) an activation energy of 2.5 eV when plotted as a function of $1/kT$.
The In desorption behavior, temperature dependence and corresponding activation energy evaluated
on Ga-face GaN are identical. Reprinted with permission from G. Koblmüller, C. S. Gallinat, and
J. S. Speck, Journal of Applied Physics, 101 (2007) 083516. Copyright 2007, American Institute of
Physics.

According to these observations, the steady-state In desorption therefore represents the maximum desorbing In flux as the limit of In droplet accumulation. From

the plots of Figs. 2.4(b) and (c), this boundary flux can be expressed by

$$\Phi^{In(droplets)} = (6.2 \times 10^{14} \text{ nm/min}) \times \exp\left(\frac{-2.5 \text{ eV}}{kT}\right), \tag{2.1}$$

where k is the Boltzmann constant and T is the substrate temperature in degrees Kelvin. The Arrhenius-type fit to the data results in an activation energy of 2.5 eV, which agrees well with the evaporation energy of In over liquid In ($E^A = 2.49$ eV), as derived from equilibrium In vapor pressure data [2.33]. The prefactor of 6.2×10^{14} nm/min corresponds to a desorption attempt frequency of 3.95×10^{13} Hz. It is important to note that the desorption boundary for metallic In is independent of the underlying substrate, as shown for both Ga-face and N-face GaN in Fig. 2.4(b) and should also hold for InN as the host surface.

2.3.2 PAMBE growth diagrams of InN

When constructing growth diagrams for the PAMBE growth of both N-face and In-face InN, it is necessary to evaluate the net In accumulation rate on the surface as a function of substrate temperature. The net In accumulation rate can be defined as the difference of two competitive processes: (i) InN decomposition and its effective In flux from bulk InN *to* the surface; and (ii) the rate of maximum In desorption *from* the surface (that is, which is defined by the droplet boundary line in Fig. 2.4(b)). For a comprehensive analysis, two different cases will be distinguished: (a) the stationary case (where no In or N fluxes are supplied to the InN surface); and (b) the dynamic case during standard PAMBE growth conditions.

2.3.2.1 N-face InN

(a) For the stationary case, these competitive rates are illustrated in Fig. 2.5(a) for N-face InN, showing the maximum In desorption rate $\Phi^{In(droplets)}$ (solid curve) and the InN decomposition rate $\Phi^{In,dec}$ (dashed curve). Up to $\sim 610°C$, In generated on the surface from decomposing InN is apparently limited by the maximum In desorption rate, inevitably resulting in the accumulation of metallic In (a saturated adlayer and droplets). But at higher temperatures $\Phi^{In(droplets)}$ exceeds $\Phi^{In,dec}$, confirming again that InN may decompose at high temperature without the accumulation of In droplets. The net In accumulation rate on the surface is therefore given by the difference between the two curves, $\Phi^{In(droplets)} - \Phi^{In,dec}$. This situation is illustrated in the In adsorption diagram of Fig. 2.5(b).

(b) For the more complex dynamic growth case of N-face InN, the active N flux (or the In/N flux ratio) on the growth surface plays a central role. The results presented in the following are based on constant N flux of N= 7.3 nm/min and variable incident In fluxes. Considering the conservation of a 1:1 film stoichiometry of In:N atoms, InN decomposition limits the N-limited growth rate to $\Phi_N - \Phi^{In,dec}$, where $\Phi^{In,dec}$ represents again the effective In flux from bulk to surface due to decomposition. Here, $\Phi^{In,dec}$ can be extracted from the In loss, given by the growth rate

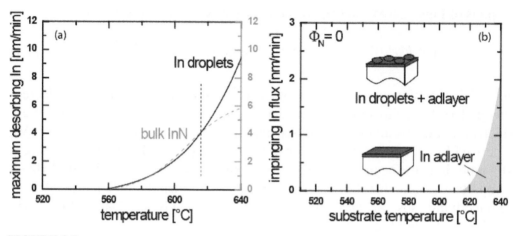

FIGURE 2.5

(a) Evaluation of the net In accumulation rate on the surface by balancing (1) the rate of thermal decomposition from bulk InN "to" the surface (dashed curve) with (2) the rate of maximum In desorption "from" the surface (solid curve) for N-face InN. The difference between the two curves yields (b), a surface structure diagram for impinging In flux versus substrate temperature (In adsorption diagram), showing a small region where an In adlayer can be formed on the surface and a large region of In droplet formation. Reprinted with permission from G. Koblmüller, C. S. Gallinat, and J. S. Speck, Journal of Applied Physics, 101 (2007) 083516. Copyright 2007, American Institute of Physics.

reduction by SEM (Fig. 2.2(b)), and as a first order approximation is independent of the incident In flux.

The combined curves for the In loss $\Phi^{In,dec}$ and the maximum In desorption $\Phi^{In(droplets)}$ (that is, droplet boundary) are illustrated in Fig. 2.6(a). Essentially, both curves follow the same trend as in the stationary (no growth) case. Calculation of the net In accumulation rate for In-rich conditions yielded again the boundary line between the In adlayer terminated surface and the surface consisting of In droplets and adlayer as shown in the growth diagram of Fig. 2.6(b). Since the growth diagram also underlies the temperature dependence of the growth rate, additional surface structures can be realized. In similarity with the possible surface structures recently evaluated for the GaN growth front [2.23, 2.34, 2.35], for N-face InN also three different growth surfaces can be classified: (i) *In droplet on top of adlayer structure* under In-rich growth conditions; (ii) *In adlayer structure* under slightly In-rich and also slightly N-rich growth conditions at high temperatures; and (iii) a *"dry" no adlayer terminated surface* under more N-rich growth conditions.

2.3.2.2 In-face InN

In the case of In-face InN, the much higher thermal decomposition compared to the N-face orientation results in a completely different growth surface diagram (Fig. 2.7). As seen in Fig. 2.7(a), the decomposition rate of InN [as evaluated by the growth rate reduction or In loss from Fig. 2.2(a)] is significant for substrate temperatures above 480°C (dashed curve). In contrast, the maximum In desorption rate

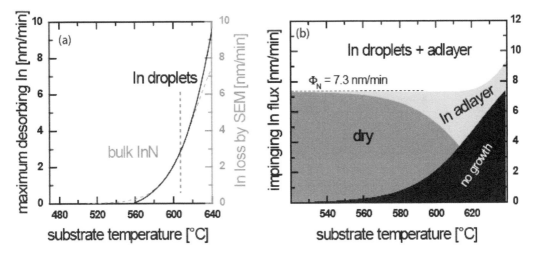

FIGURE 2.6

(a) Determination of the net In accumulation rate on the surface for the growth case of N-face InN showing (i) the rate of thermal decomposition from bulk InN, that is, the In loss from the SEM growth rate analysis (dahsed curve) and (ii) the rate of maximum In desorption "from" the surface (solid curve). Similar to Fig. 2.5, the difference between the two curves allows construction of a growth structure diagram (b) for actual growth conditions (that is, impinging In flux at constant N flux versus growth temperature). Three characteristic growth surface structures can be achieved: one dry no-adlayer terminated surface, one In adlayer stabilized surface and one consisting of In droplets on top of the adlayer. Growth is unsustainable at temperatures above ~635°C. Reprinted with permission from G. Koblmüller, C. S. Gallinat, and J. S. Speck, Journal of Applied Physics, 101 (2007) 083516. Copyright 2007, American Institute of Physics.

($\Phi^{In(droplets)}$) (solid curve) at these temperatures is negligible, and no cross-over between the two curves (as seen in Fig. 2.7(a)) occurs in the whole temperature range.

Inevitably, decomposition of In-face InN must always result in the accumulation of metallic In droplets on the surface. This is manifested in the existence of only two growth surface regimes as opposed to three observed in the case for N-face InN: (i) *In droplet on top of adlayer structure* under In-rich growth conditions and slightly N-rich growth conditions at high temperatures ($> 480°C$), and (ii) the *"dry" no-adlayer terminated surface* under more N-rich growth conditions. Most importantly, no pure *In adlayer* terminated surface structure (without In droplets) can exist under the standard PAMBE growth conditions of In-face InN.

2.3.3   In adlayers and surface reconstructions

In this section, the surface adatom reconstructions, In adlayers and their equilibrium coverages on the polar InN growth surfaces will be elucidated. For the more studied GaN and AlN surfaces, theoretical calculations have well established that metallic Ga or Al adlayers of one or two monolayers describe the energetically most favorable surface structure under standard metal-rich growth conditions in PAMBE. At the same time these adlayers increase the adatom diffusion and seem necessary to

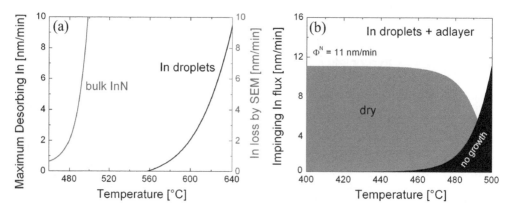

FIGURE 2.7

(a) Rate of thermal decomposition from bulk InN, that is, the In loss from the SEM growth rate analysis vs. maximum In desorption rate "from" the surface (black curve) for In-face InN. (b) Resulting growth structure diagram for typical PAMBE growth of In-face InN (with constant N flux of 10.5 nm/min). Only two characteristic growth surface structures can be realized: one dry no-adlayer terminated surface and one consisting of In droplets on top of an adlayer. Growth is unsustainable at temperatures above $\sim 500°$C. Reprinted with permission from C. S. Gallinat, G. Koblmüller, J. S. Brown, and J. S. Speck, Journal of Applied Physics, 102 (2007) 064907. Copyright 2007, American Institute of Physics.

achieve superior material quality [2.34]. Recent InGaN growth studies also point to the existence of a stable In adlayer on the surface [2.36, 2.37], but for binary InN growth very limited experimental knowledge exists about surface reconstructions, adlayer coverage, and their correlation to adatom mobility and surfactant action.

In Chapter 13, density-functional theoretical calculations highlight insights into the formation energies for the various surface reconstructions on InN surfaces as a function of the typical InN growth conditions. It will be discussed that the surface reconstructions are quite different from those common on GaN and AlN surfaces. Essentially, the formation energies for all the possible surface reconstructions are much smaller than for the other binary group-III nitrides, despite the similarity in crystal structure [2.38, 2.39]. Moreover, for the In-face InN surface, there seems no N adatom-related surface reconstruction under N-rich conditions. This is at odds with GaN or AlN, and has been attributed to a much weaker In-N bond strength in InN. Instead, a (2×2) In adatom structure appears as the most stable surface structure over the entire In/N ratio range, except for highly In-rich conditions, where the formation of a contracted In bilayer becomes more favorable. In contrast to In-face InN, GaN and AlN, on the N-face InN surface the (2×2) In adatom structure was found to be not stable over the entire In/N ratio range. The most stable form is defined by the (1×1) In adlayer structure, which consists of 1 ML of In added to the surface, making it overall metallic.

Experimentally, apart from Ref. [2.40], no determination of surface reconstructions in InN have been achieved, not even by RHEED, due to the generally high

In accumulation under the most studied In-rich growth conditions. As for the In adlayers though, post-growth x-ray photoemission spectroscopy (XPS) studies on clean InN surfaces revealed stable In coverages of 3.4 ML and 2.0 ML on In-face and N-face InN, respectively [2.41]. In each case this is 1 ML more than for Ga on GaN surfaces under metal-rich conditions in PAMBE growth.

A more direct *in situ* method to define the surface adlayer and its equilibrium coverage on a given surface can be determined by the transition boundary from a two-dimensional (2D) wetting layer (WL) to three-dimensional (3D) islands (droplets). This resembles the well-known Stranski-Krastanow growth phenomenon, where a highly strained 2D layer relaxes through the free surface of 3D islands. In the present case, this system consists of compressively strained metallic In (with In-In in-plane spacing of ~ 3.25 Å) on top of an InN substrate.

By using RHEED intensity profiling to determine the onset of In droplet formation during In adsorption experiments or by consumption of adsorbed In through exposure to nitrogen, In adlayer coverages could be determined in the MBE growth environment *in situ*. The results are shown in Fig. 2.8 for In adsorption/N consumption experiments performed at 530°C for different In coverages (0.35–7.8 ML) on a standard N-face InN surface.

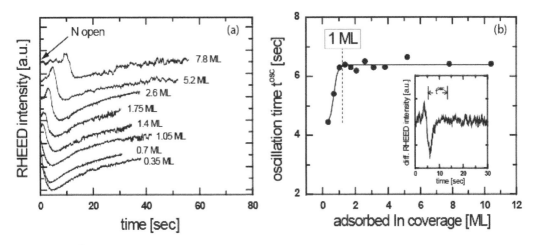

FIGURE 2.8

(a) Variation of the RHEED specular intensity during the nitrogen consumption of various different In surface coverages at 530°C on N-face InN. (b) Time period of RHEED intensity oscillation as a function of adsorbed In coverage pointing to a saturated In adlayer coverage of 1 ML. The oscillation time t^{osc} is defined by the time span between onset of oscillation and last inflection point in the differentiated RHEED intensity. Reprinted with permission from G. Koblmüller, C. S. Gallinat, and J. S. Speck, Journal of Applied Physics, 101 (2007) 083516. Copyright 2007, American Institute of Physics.

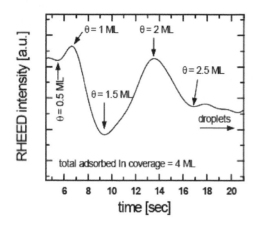

FIGURE 2.9
RHEED intensity oscillations during the adsorption of 4 ML of In onto an In-face InN surface. Each half oscillation corresponds to 0.5 ML of In adsorbing onto the surface. Reprinted with permission from C. S. Gallinat, G. Koblmüller, J. S. Brown, and J. S. Speck, Journal of Applied Physics, 102 (2007) 064907. Copyright 2007, American Institute of Physics.

For sub-ML In coverages, no full intensity oscillation was observed, while one complete intensity oscillation was seen for In coverages exceeding 1 ML. Also, delays in the onset of the oscillation are observed with higher In coverages, which could be associated with the accumulation or consumption of droplets on the surface [2.42]. Therefore it has been suggested that In coverages in excess of 1 ML accumulated as droplets, and that the maximum In adlayer coverage is therefore given by 1 ML. Investigations of the RHEED oscillation period also yield insight into the In adlayer coverage, as shown in Fig. 2.8(b). The transition point between monotonic increase in the oscillation time period for In coverages of less than 1 ML and saturated oscillation time period above 1 ML highlight once more the critical In adlayer coverage of 1 ML on the N-face InN surface.

The situation for In-face InN is described by a similar experiment performed at 420°C with representative RHEED intensity-time profile shown in Fig. 2.9 for a nominal In deposition of ∼4 ML. In contrast to the previous case, here multiple oscillations were observed, indicating a maximum In adlayer coverage of about 2.5 ML, before In droplet formation sets in (shown by the continuous decrease in RHEED intensity after the final oscillation). With these RHEED studies the maximum In adlayer coverages match more closely with the Ga adlayer coverages found on GaN surfaces, that is, ∼1 ML on the N-face, and ∼2.5 ML on the metal-face orientation.

2.3.4 Effect of In adlayer on InN surface morphologies

Since metallic adlayers on nitride surfaces influence strongly the surface growth kinetics, such as adatom diffusion, dopant and impurity incorporation, this section is dedicated to the surfactant-mediated growth behavior under adlayer terminated growth conditions. In particular, the tendency for 3D versus 2D growth behavior and the formation and suppression of detrimental surface pit defects will be closely investigated.

2.3.4.1 N-face InN

According to the PAMBE surface growth diagram of Fig. 2.6(b) for N-face InN, detailed studies of the surface morphologies for the different growth regimes have been performed. The morphologies shown in Fig. 2.10 are from nominally 1 μm thick InN layers grown on free-standing N-face GaN templates under variable In/N flux ratios and temperatures. Although N-face InN growth is commonly sustainable up to temperatures beyond 600°C, InN layers grown between 610 and 635°C directly on GaN did not exhibit the single-crystalline wurtzite phase. This result was also independent of the chosen In/N flux ratio and further agrees with the temperature limit of ∼600°C for N-face InN MBE growth determined by Xu *et al.* [2.4]. Therefore, the presented morphologies are from InN layers grown at a slightly lower temperature of 595°C with In fluxes ranging from 5 nm/min (N-rich growth) to 10 nm/min (In-rich growth) at fixed N flux of 7.3 nm/min.

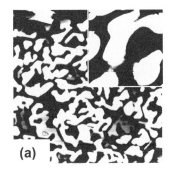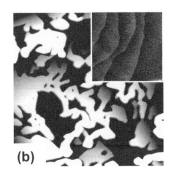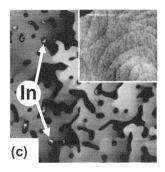

FIGURE 2.10

10×10 μm^2 and 1×1 μm^2 (insets) AFM micrographs of 1 μm thick N-face InN layers grown on N-face free-standing GaN templates under conditions of constant temperature ($T = 595$°C) but variable In fluxes of (a) 5 nm/min (dry N-rich regime), (b) 6.5 nm/min (In-adlayer N-rich regime) and (c) 10 nm/min (In-droplet regime). The height scale of the 10×10 μm^2 images is 50 nm and for the 1×1 μm^2 images is 2 nm. Reprinted with permission from G. Koblmüller, C. S. Gallinat, and J. S. Speck, Journal of Applied Physics, 101 (2007) 083516. Copyright 2007, American Institute of Physics.

It is worth noting that prior to the InN growth, a homoepitaxially grown GaN buffer layer (under Ga-rich conditions) seemed to be necessary to significantly improve both the structural and the electrical properties of InN films, by also effectively suppressing the impurity incorporation at the substrate interface [2.19].

Representative InN grown under very N-rich conditions ($\Phi_{In} = 5$ nm/min) yielded a very rough surface morphology characterized by a high density of pits and facets and a root-mean-square (rms) roughness larger than 100 nm over a 10×10 μm^2 area (Fig. 2.10(a)). Analysis of several layers grown under N-rich conditions in a temperature range of 500–600°C showed similarly rough surface structures.

Slightly smoother and more coalesced morphology with larger flat areas resulted for InN growth performed under less N-rich conditions ($\Phi_{In} = 6.5$ nm/min). Under these conditions, the surface growth diagram proposes an In adlayer (or at least fractions of it) to terminate the surface. The rms roughness decreased substantially to below 30 nm over a $10 \times 10\ \mu m^2$ and 0.15 nm over a $1 \times 1\ \mu m^2$ area, as compared to the more N-rich growth stated above. Analysis of the atomically flat plateaus revealed typical characteristics of step-flow growth with characteristic parallel steps and ~ 3 Å high monolayer steps (monolayer height, $c/2 = 2.88$ Å for bulk InN).

These characteristic step-flow morphologies persisted when growth was performed under In-rich droplet conditions ($\Phi_{In} = 10$ nm/min, Fig. 2.10(c)), where the surface is covered by In droplets on top of the adlayer. Here, the step terraces were arranged with a distinct curvature, characteristic of spiral growth hillock formation around screw-component threading dislocations [2.43]. The surface exhibits more coalesced and larger flat areas with fewer surface pits. The high density of metallic In droplets (highlighted by the arrows) resulted from the increased In accumulation under the In-rich conditions, but also from significant InN decomposition and limited In desorption at the given temperature. Excluding areas of larger In droplets, the rms roughness was reduced to less than 10 nm over a $10 \times 10\ \mu m^2$ and 0.1 nm over a $1 \times 1\ \mu m^2$ area. Similar surface morphologies and rms roughness values have been found even for much lower growth temperatures [2.20].

Further studies of the temperature dependence of the InN surface morphologies verified that the surface structure within a growth regime is reproducible and independent of the growth temperature [2.20], unlike reports by various groups [2.25]. In particular, all InN layers grown between 500 and 600°C resulted in a heavily pitted surface when grown under very N-rich conditions, while In-rich growth conditions yielded less pitted, step-flow like morphologies with spiral growth hillocks in the same temperature region. Similar to GaN growth, this points to the well known autosurfactant action of the metallic adlayer [2.34], that is, providing substantial increases in adatom surface diffusion during InN growth by the In adlayer. Under very N-rich conditions and low temperatures, the absence of the In adlayer (that is, dry N-rich surface) can be viewed as the main reason for the limited adatom mobility. Contrary to this, InN growth with an In adlayer (under slightly N-rich and In-rich conditions) is subject to enhanced surface diffusion mediated step-flow growth behavior.

2.3.4.2   In-face InN

In contrast to N-face InN, where three distinct growth regimes and surface morphologies are observed, only two surface morphologies are characteristic for In-face InN. Fig. 2.11 shows representative $5 \times 5\ \mu m^2$ AFM images for the two discernible growth regimes in In-face InN. Fig. 2.11(a) shows the surface morphology of a $1\ \mu m$ InN sample grown with excess In ($\Phi_{In} = 13.5$ nm/min and $\Phi_N = 10.5$ nm/min). The InN surface of Fig. 2.11(a) exhibited a relatively smooth, spiral hillock rich morphology typical for dislocation pinned step-flow growth [2.23, 2.43]. This is a morphology characteristic of that seen in all In-face InN films grown with excess In. As reported throughout various studies, all samples grown with excess In exhibited In droplet accumulation visible by optical microscopy. Droplets became larger and more dense as increased excess In was supplied during growth.

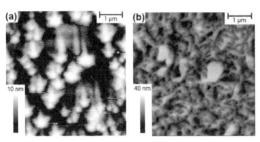

FIGURE 2.11

5×5 μm^2 AFM micrographs of In-face InN layers grown on Ga-face GaN templates under (a) In-rich conditions, and (b) N-rich conditions. The growth temperature for both layers was $T = 450°C$. Reprinted with permission from C. S. Gallinat *et al.*, Journal of Applied Physics, 102 (2007) 064907. Copyright 2007, American Institute of Physics.

As the impinging In is decreased to a flux less than the active nitrogen ($\Phi_{In} < \Phi_N$), the surface morphology roughened and the step-flow and spiral hillock termination disappeared. A representative AFM image is shown in Fig. 2.11(b) for a 1.5 μm InN film grown N-rich ($\Phi_{In} = 8.5$ nm/min and $\Phi_N = 10.5$ nm/min). This N-rich morphology was characterized by a rough, three-dimensional faceted surface similar to the morphologies reported for N-face InN and other N-rich PAMBE grown III-nitrides [2.23, 2.43]. In addition, no liquid In adlayer or droplets were observed *ex situ* in N-rich samples, suggesting a dry surface free of excess In during growth. The rms surface roughness of the N-rich films increased as Φ_{In}/Φ_N decreased.

2.4 Structural properties of InN

The constructed growth diagrams and insight into the surface kinetics during growth were then used to evaluate the structural and electrical properties of films grown under different conditions. Optimization of the structural properties and understanding the effect of dislocations on the transport properties of InN are the next steps toward integrating InN into actual devices. This part of the chapter will particularly focus on the dislocation control of In-face InN films and provide comparisons with and analysis of N-face films where appropriate.

2.4.1 Evaluation of dislocations

To study the effect of growth conditions on the structural properties of InN, films were grown at varying substrate temperatures and In-fluxes while maintaining a constant active N-limited growth rate. These films were then evaluated by x-ray diffraction (XRD). Full-width half-maximum (FWHM) values of InN ω-scan rocking curves in both on- and off-axis orientations were measured. These FWHM values were then used to assess the dislocation densities of the InN films. Plan-view and cross-sectional TEM was used to directly observe the dislocations in select InN films.

The significant lattice mismatch between InN and GaN (9.6%) leads to InN growth following the Volmer-Weber growth mode in which strained 3-dimensional InN islands first nucleate on GaN. The islands relax after roughly 2.5 ML of InN deposition according to Ng *et al.* [2.44]. These islands then grow laterally until coalescence occurs and a uniform InN film is formed. Similar to the threading dislocation structure observed in (0001) GaN growth on sapphire [2.45–2.48], three threading dislocation types arise from this island coalescence: screw-type dislocations, edge-type dislocations and dislocations with mixed screw and edge character. These dislocations can be directly observed by transmission electron microscopy (TEM), or indirectly observed by measuring x-ray rocking curves (ω-scans) full-width at half-maximum (FWHM) values. Since the x-ray measurement is non-destructive and takes significantly less time than the TEM measurement, ω-scans are the preferred method of evaluating threading dislocation densities in InN grown on GaN. Understanding the effect of these dislocations on the rocking curve FWHMs is the first step in evaluating the structural quality of the InN films.

The InN crystallites that form due to the island-like nature of the growth can be misoriented with respect to the underlying GaN in two relevant ways: (i) *tilted*, as defined by the crystallite surface normal being non-parallel to the substrate surface normal, and (ii) *twisted*, as defined by the crystallite being rotated in the basal plane relative to the substrate [2.48]. The only way to achieve tilt in the film is to have dislocations with Burger's vectors out of the basal plane, resulting in screw-component threading dislocations [2.48]. The strain fields from these dislocations distort the on-axis planes and, therefore, broaden on-axis rocking curves measurements [2.48]. Conversely, twist in the InN mosaic is achieved with dislocations exhibiting Burger's vectors in the basal plane [2.48]. Pure edge-type threading dislocations and dislocations with edge-type character fall in this category [2.46, 2.48]. The off-axis planes are disrupted by edge-type dislocations and the FWHM of off-axis rocking curves, such as $(10\bar{1}2)$ and $(20\bar{2}1)$, are broadened by the presence of such dislocations [2.46, 2.48, 2.49].

Fig. 2.12(a) shows an XRD $\omega/2\theta$ scan for a typical 1 μm InN epilayer grown on GaN. The (0002) InN peak is clearly separated from the GaN (0002) peak indicating a fully relaxed InN film (as confirmed by reciprocal space mapping of the sample). The data in Fig. 2.12(b) represents the three rocking curves discussed above used to evaluate the structural quality of InN films. The (0002) rocking curve FWHM is much narrower than either of the two off-axis FWHM. This is indicative of a dislocation structure dominated by edge-type dislocations which is analogous to the dislocation network observed in GaN growth on sapphire [2.46].

The dislocation character was more closely analyzed by cross-sectional TEM.

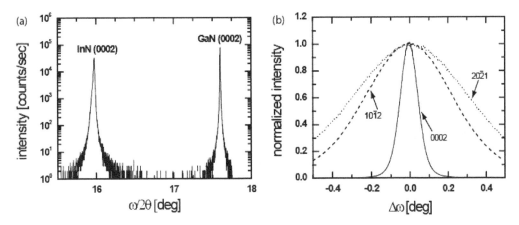

FIGURE 2.12

Representative x-ray diffraction data for a 1 μm InN sample grown. (a) The $\omega/2\theta$ scan showing a fully relaxed InN film on the underlying GaN film. (b) On- and off-axis rocking curves for the same 1 μm InN sample shown in (a).

Fig. 2.13 shows two cross-sectional images taken using the different Bragg conditions for imaging (a) pure edge-type dislocations and (b) screw-type dislocations. The TEM micrographs confirmed the XRD data that edge-type dislocations were the dominant dislocation in the InN film. The total dislocation density in this film as determined by plan view TEM was 3×10^{10} cm^{-2}. The density of dislocations with screw-type character was 3×10^8 cm^{-2}. Therefore, the overall dislocation density can be approximated by determining the density of edge-type dislocations. The dislocation densities in InN films were also calculated using the FWHM data from the on- and off-axis rocking curves. Models developed by Srikant *et al.* for crystal growth of highly lattice mismatched films [2.48] and applied specifically to GaN by Lee *et al.* [2.49] determined the calculations described here. By extrapolating the FWHM values of the increasingly off-axis rocking curves to 90° (with the on-axis rocking curve at 0° defined as the tilt angle Γ_y), according to [2.48, 2.49]

$$\Gamma = \sqrt{(\Gamma_y \cos \chi)^2 + (\Gamma_z \cos \chi)^2}, \tag{2.2}$$

where Γ is the FWHM at an angle χ, Γ_y (the tilt angle) is the on-axis (0002) rocking curve FWHM and Γ_z (the twist angle) is the extrapolated FWHM value for a rocking curve rotated 90° to the surface normal, a value for the twist angle was determined. Fig. 2.14(a) depicts the resulting curve from Eq. 2.2 for the InN sample with the corresponding plan-view TEM micrograph shown in Fig. 2.14(b).

The resulting angles can then be used to calculate the dislocation densities with screw character (using the tilt angle) and with pure edge character (using the twist angle) according to

$$\rho_s = \Gamma_y^2 / 1.88c^2 \tag{2.3}$$

$$\rho_e = \Gamma_z^2 / 1.88a^2 \tag{2.4}$$

(a)

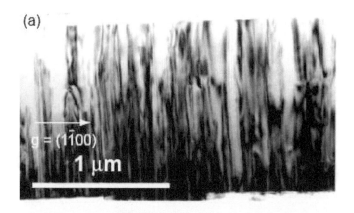

g = (1$\bar{1}$00)

1 μm

(b)

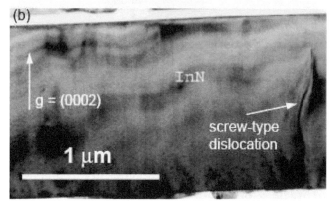

InN

g = (0002)

screw-type
dislocation

1 μm

FIGURE 2.13

Cross-sectional TEM micrographs of a 1.5 μm InN film grown on GaN. (a) $g = 1\bar{1}00$ image showing pure edge-type dislocations and (b) $g = 0002$ image showing a single screw-type dislocation. These images provide additional evidence that the dislocation structure of InN on GaN is dominated by edge-type dislocations.

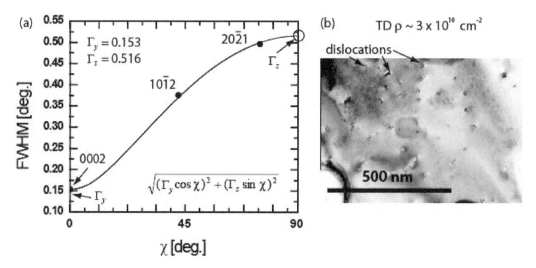

FIGURE 2.14

(a) Resulting curve of Eq. 2.2 using the FWHM values from the rocking curves shown in Fig. 2.12. A twist angle of 0.516° was determined by extrapolating the FWHM values of increasingly off-axis rocking curves. (b) Plan-view TEM of the InN sample from showing an overall threading dislocation density of $\sim 3 \times 10^{10}$ cm^{-2}.

in which ρ_s and ρ_e are the screw and edge dislocation densities, c and a are the relevant Burgers vectors (the InN c lattice constant of 5.693 Å for dislocations with screw character and the InN a lattice constant of 3.533 Å for pure edge dislocations), and 1.88 is a TEM calibration constant determined for GaN and extrapolated to InN for our purposes. These equations were derived from several classic formulae [2.50–2.52] and applied to the GaN system by Lee *et al.* [2.49]. The calculated density of dislocations having screw character for the sample shown in Fig. 2.14 is 1.1×10^9 cm^{-2} and the calculated density of pure edge-type dislocations is 3.5×10^{10} cm^{-2}, in good agreement with the observed TEM data. This calculation will be used to determine dislocation densities in InN films throughout the rest of this chapter.

Screw character threading dislocations comprised only 3% of the total density in this representative In-face InN film. This is similar to the case of GaN on sapphire which typically has a screw character dislocation population of 1% to 20% of the total dislocation density depending on the growth method and nucleation procedure [2.53].

2.4.2   Effect of growth conditions on dislocations

The systematic study of the effect of growth conditions on InN structural quality is schematically represented in Fig. 2.15(a). In-face InN films were evaluated by measuring the rocking curve FWHM of samples grown at varying substrate temperatures and Φ_{In} while maintaining a constant Φ_N. For the set of In-face InN samples, the Φ_{In} was varied such that growth occurred in the In-droplet regime, the N-rich regime, and approximately on the crossover from N-rich growth to In-droplet growth (stoichiometric). Samples grown in the In-droplet regime were all grown with an excess Φ_{In} of 2 nm/min and samples grown in the N-rich regime were grown with a deficient Φ_{In} of 2 nm/min. This precise Φ_{In} control required adjustment of the In cell temperature for each substrate temperature due to increasing InN decomposition at higher substrate temperatures.

The XRD results for In-face InN showing the FWHM of the on- and off-axis rocking curves and the calculated dislocation densities are shown in Fig. 2.16. The on-axis XRD data (Fig. 2.16(a)) and calculated screw character dislocation densities (Fig. 2.16(d)) indicate the relative insensitivity of screw-type dislocations to growth temperature across all growth regimes. This data also indicates that samples grown in the In-droplet regime have slightly lower screw-type dislocation densities than samples grown either N-rich or stoichiometric.

The off-axis XRD data (Fig. 2.16(b) and (c)) show a decrease in the FWHM at increasing growth temperatures for all three growth regimes with the In-droplet samples exhibiting the lowest values. This trend indicates a decrease in pure edge-type dislocations at higher substrate temperatures (Fig. 2.16(e)).

The observation of decreased edge-type dislocations at higher substrate temperatures can be explained by considering the effect of substrate temperature at the onset of InN deposition. The island nucleation and subsequent film coalescence

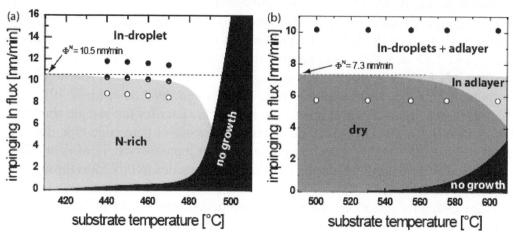

FIGURE 2.15

Schematic representations of the sample sets used to study the effect of growth conditions on the structural quality of InN. (a) In-face InN samples were grown at different substrate temperatures varying only the Φ_{In}. (b) N-face InN samples were grown at different substrate temperatures and two different Φ_{In} corresponding to growth in the In-droplet regime using the high flux (solid black circles) and either dry growth or growth with an In adlayer using the low flux (open circles).

leads to coalescence boundaries in the form of edge-type dislocations. Increased adatom mobility at elevated substrate temperatures enhances lateral island growth creating fewer coalescence boundaries (edge-type dislocations) as the islands form an epitaxial film. The lower observed edge-type dislocation density for InN films grown with excess In can also be understood by considering the behavior of island coalescence. The observed metallic In adlayer during In-droplet InN growth discussed previously was shown to increase adatom mobility leading to morphologically smoother films. The increased adatom mobility due to the presence of the In adlayer also leads to enhanced lateral island growth and fewer coalescence boundaries; therefore, films grown with excess In have fewer edge-type dislocations than films grown in the N-rich regime at the same substrate temperature.

2.4.3   Comparison to N-face InN

The N-face InN sample set (shown earlier in Fig. 2.15(b)) included samples grown with an excess Φ_{In} of 3 nm/min and samples grown with a deficient Φ_{In} of 2 nm/min. These samples were evaluated by XRD to determine FWHM of the on- and off-axis rocking curves. These FWHM values were then used to calculate dislocation densities following the procedure described above. The growth temperatures of the N-face InN were much higher than the In-face sample set in this analysis. The lowest N-face InN growth temperature considered was 500°C — a growth temperature too high for In-face InN deposition.

The resulting pure edge-type dislocation densities for both In- and N-face InN are shown in Fig. 2.17. The minority screw-type dislocations are not presented here

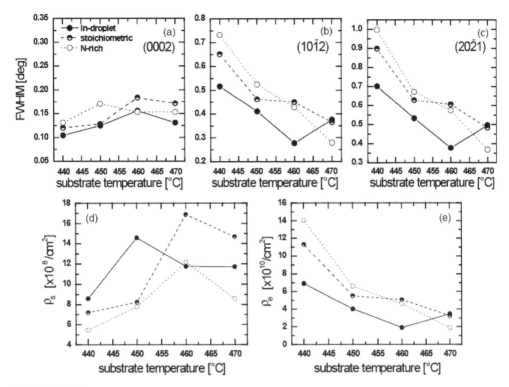

FIGURE 2.16

Measured (XRD) and calculated (dislocation densities) data for the sample set used to study the effect of growth conditions on the structural quality of InN. In-face InN samples were grown at different substrate temperatures varying only the Φ_{In} to establish In-droplet growth conditions (solid black circles), stoichiometric growth conditions (half-filled circles) and N-rich growth conditions (open circles). (a) FWHM values of the on-axis (0002) x-ray rocking curves. (b) and (c) FWHM values of the off-axis ($10\bar{1}2$) and ($20\bar{2}1$) x-ray rocking curves. (d) Calculated densities of dislocations with screw character using Eq. 2.3 and the on-axis (tilt) FWHM values from (a). (e) Calculated pure edge-type dislocation densities using Eq. 2.4 and extrapolated twist angles using the XRD data from (a–c). The y-scale for the screw dislocation density data is two orders of magnitude less than the scale for the edge dislocation data.

since they represent only a few percent of the total dislocation density. The decrease of dislocations observed with increasing growth temperature and Φ_{In} for the In-face InN sample set (Fig. 2.17(a)) was not observed for N-face InN (Fig. 2.17(b)). Increased Φ_{In} had minimal effect on the dislocation density at only the highest growth temperature.

The insensitivity of dislocation density on growth conditions for N-face InN was in contrast to the effect that growth conditions had on In-face InN. The lower growth temperatures necessary for In-face InN growth allowed for slower adatom mobility during growth than for the N-face InN samples grown at higher temperatures. Slower adatom mobility provided an environment for more numerous islands to form during the Volmer-Weber growth mode for the In-face samples than the N-face samples. The higher adatom mobility during the N-face InN growth

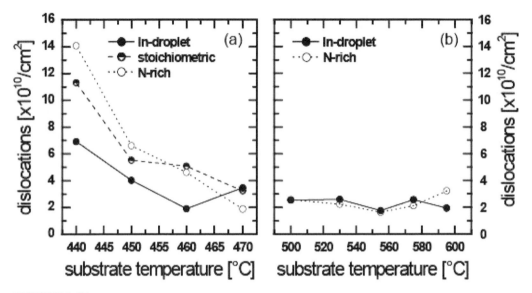

FIGURE 2.17

(a) Dependence of In-face InN edge-type dislocation density on substrate temperature and Φ_{In} (repeated from Fig. 2.16(e)). (b) Dependence of N-face InN edge-type dislocation density on temperature and growth regime. The same *y*-axis scale is used in both cases to illustrate the relative insensitivity of the dislocation density on substrate temperature for the N-face case.

also provided faster lateral island growth and quicker island coalescence than compared to In-face InN; thus, the island coalescence boundaries were fewer for N-face than In-face InN creating fewer dislocations in N-face InN for the relevant growth temperatures.

2.5   Transport properties of InN

The same sample set used to measure the dependence of the structural quality of In-face InN films on growth conditions (shown schematically in Fig. 2.15) was used to observe the dependence of the unintentionally doped electron concentration and mobility on growth conditions. Hall effect measurements at room temperature were performed on these InN films and the measured sheet density was used to calculate the carrier concentration accounting for the surface accumulation layer. A two-layer Hall model was employed to calculate the actual bulk mobility accounting for the surface electron mobility. The resulting data are the actual bulk concentration and mobility.

Fig. 2.18(a) shows the dependence of the electron concentration on growth temperature for three different Φ_{In} corresponding to In-droplet growth, stoichiometric growth, and N-rich growth. The electron concentration decreased in films grown with increasing Φ_{In} at the same substrate temperature and decreased slightly with increasing substrate temperature for InN films grown with the same Φ_{In}. All of the InN films grown with excess In had lower electron concentrations than all films grown stoichiometric or In deficient, even when comparing films grown in the In-droplet regime at the lowest substrate temperature with films grown in the N-rich regime at the highest substrate temperatures. This indicates Φ_{In} had a stronger effect than substrate temperature on electron concentration in InN films.

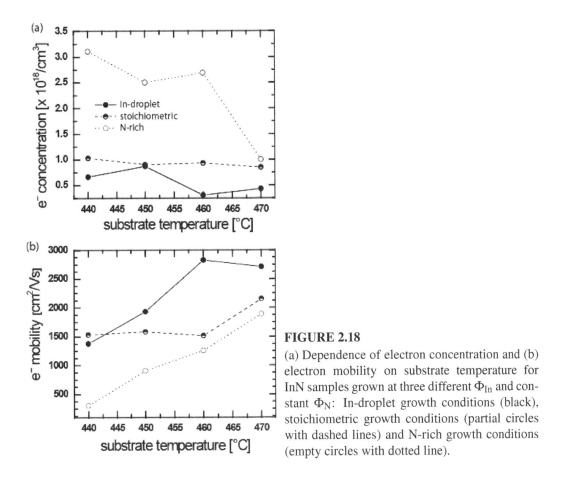

FIGURE 2.18

(a) Dependence of electron concentration and (b) electron mobility on substrate temperature for InN samples grown at three different Φ_{In} and constant Φ_{N}: In-droplet growth conditions (black), stoichiometric growth conditions (partial circles with dashed lines) and N-rich growth conditions (empty circles with dotted line).

The dependence of electron mobility on growth conditions is shown in Fig. 2.18(b). The measured electron mobilities increased with increasing Φ_{In} for samples grown at the same substrate temperature. Samples grown with the same Φ_{In} showed increasing electron mobilities with increasing substrate temperature. The InN films grown in the In-droplet regime exhibited the highest overall electron mobilities compared to all the films grown stoichiometric or N-rich, regardless of substrate temperature.

2.6   Concluding Remarks

The application of InN in actual devices requires precise control of the growth and properties of high-quality InN films. This work has shown that high-quality films can be prepared by PAMBE. Although InN exhibits high thermal decomposition at relatively low growth temperatures, films can be deposited with smooth surface morphologies, relatively low dislocation densities, and excellent electron transport properties when grown with excess In. Unfortunately, even the highest-quality samples still exhibit high unintentionally doped electron concentrations.

The highlights from this chapter begin with an increased understanding of the thermal stability of In- and N-face InN. Growth temperatures beyond 500°C were not possible for In-face films, while growth temperatures as high as 640°C were realized for N-face InN. The smoothest surface morphologies for both InN polarities were achieved in films grown with excess In, in which a metallic In-adlayer was present (2.5 ML for the In-face and ~ 1 ML for the N-face) during growth. Surface growth diagrams for both In- and N-face InN were determined by combining the observed surface morphologies, InN decomposition rates, and desorbing In-fluxes from InN and GaN surfaces. These growth diagrams have multiple purposes: to predict surface morphologies based on growth temperature and impinging In-flux, as a reference frame in which to compare physical properties of differently grown InN films, and as a general guide to other InN growers in starting to understand PAMBE growth of InN.

Utilizing this growth diagram, the dislocation density has been reduced to values as low as 2×10^{10} cm^{-2}. Similar to GaN grown on sapphire, for both In- and N-face InN films grown on GaN, pure edge-type dislocations dominated the dislocation structure and dislocations with screw-type character only accounted for 1–5% of the total dislocation density. Dislocation densities in In-face InN were shown to be growth regime and growth temperature dependent, whereas dislocations in N-face InN films showed minimal dependence on typical growth conditions. The minimum electron concentrations and maximum electron mobilities were measured for In-face InN films grown at the highest possible temperatures (prior to significant thermal decomposition) in the In-droplet growth regime.

References

[2.1] V. Y. Davydov, A. A. Klochikhin, R. P. Seisyan, V. V. Emtsev, S. V. Ivanov, F. Bechstedt, J. Furthmüller, H. Harima, A. V. Mudryi, J. Aderhold, O. Semchinova, and J. Graul, Absorption and emission of hexagonal InN. Evidence of narrow fundamental band gap, *physica status solidi (b)*, 229 (2002) R1–R3.

[2.2] J. Wu, W. Walukiewicz, K. M. Yu, J. W. Ager III, E. E. Haller, H. Lu, W. J. Schaff, Y. Saito, and Y. Nanishi, Unusual properties of the fundamental band gap of InN, *Applied Physics Letters*, 80 (2002) 3967–3969.

[2.3] J. H. Edgar, *Group-III Nitrides* (INSPEC, London, 1994).

[2.4] K. Xu and A. Yoshikawa, Effects of film polarities on InN growth by molecular-beam epitaxy, *Applied Physics Letters*, 83 (2003) 251–253.

[2.5] B. Liu, T. Kitajima, D. Chen, and S. R. Leone, Growth modes of InN (000-1) on GaN buffer layers on sapphire, *Journal of Vacuum Science and Technology A*, 23 (2005) 304–309.

[2.6] E. Dimakis, E. Iliopoulos, K. Tsagaraki, and A. Georgakilas, Physical model of InN growth on Ga-face GaN (0001) by molecular-beam epitaxy, *Applied Physics Letters*, 86 (2005) 133104:1–3.

[2.7] Y. Nanishi, Y. Saito, T. Yamaguchi, M. Hori, F. Matsuda, T. Araki, A. Suzuki, and T. Miyajima, MBE-growth, characterization and properties of InN and InGaN, *physica status solidi (a)*, 200 (2003) 202–208.

[2.8] T. Matsuoka, Y. Kobayashi, H. Takahata, T. Mitate, S. Mizuno, A. Sasaki, M. Yoshimoto, T. Ohnishi, and M. Suyima, N-polarity GaN on sapphire substrate grown by MOVPE, *physica status solidi (b)* 243 (2006) 1446–1450.

[2.9] H. Naoi, F. Matsuda, T. Araki, A. Suzuki, Y. Nanishi, The effect of substrate polarity on the growth of InN by RF-MBE, *Journal of Crystal Growth*, 269 (2004) 155–161.

[2.10] T. Mitate, S. Mizuno, H. Takahata, R. Kakegawa, T. Matsuoka, and N. Kuwano, InN polarity determination by convergent-beam electron diffraction, *Applied Physics Letters*, 86 (2005) 134103:1–3.

[2.11] Y. Saito, Y. Tanabe, T. Yamaguchi, N. Teraguchi, A. Suzuki, T. Araki, and Y. Nanishi, Polarity of High-Quality Indium Nitride Grown by RF Molecular Beam Epitaxy, *physica status solidi (b)*, 228 (2001) 13–16.

[2.12] D. Muto, T. Araki, H. Naoi, and Y. Nanishi, Polarity determination of InN by wet etching, *physica status solidi (a)*, 202 (2005) 773–776.

[2.13] A. R. Smith, R. M. Feenstra, D. W. Greve, M. S. Shin, and M. Skowronski, Determination of wurtzite GaN lattice polarity based on surface reconstruction, *Applied Physics Letters*, 72 (1998) 2114–2116.

[2.14] Y. Hayakawa, D. Muto, H. Naoi, A. Suzuki, T. Araki and Y. Nanishi, Polarity determination of InN by atomic hydrogen irradiation, *Japanese Journal of Applied Physics*, 45 (2006) L384–L386.

[2.15] M. Drago, T. Schmidtling, C. Werner, M. Pristovsek, U. W. Pohl, and W. Richter, InN growth and annealing investigations using in-situ spectroscopic ellipsometry, *Journal of Crystal Growth*, 272 (2004) 87–93.

[2.16] Y. Huang, H. Wang, Q. Sun, J. Chen, J. F. Wang, Y. T. Wang, and H. Yang, Study on the thermal stability of InN by in-situ laser reflectance system, *Journal of Crystal Growth*, 281 (2005) 310–317.

[2.17] S. V. Ivanov, T. V. Shubina, V. N. Jmerik, V. A. Vekshin, P. S. Kop'ev and B. Monemar, Plasma-assisted MBE growth and characterization of InN on sapphire, *Journal of Crystal Growth*, 269 (2004) 1–9.

[2.18] A. Laakso, J. Oila, A. Kemppinen, K. Saarinen, W. Egger, L. Liszkay, P. Sperr, H. Lu, and W. J. Schaff, Vacancy defects in epitaxial InN: identification and electrical properties, *Journal of Crystal Growth*, 269 (2004) 41–49.

[2.19] C. S. Gallinat, G. Koblmüller, J. S. Brown, S. Bernardis, J. S. Speck, G. D. Chern, E. D. Readinger, H. Shen, and M. Wraback, In-polar InN grown by plasma-assisted molecular beam epitaxy, *Applied Physics Letters*, 89 (2006) 032109:1–3.

[2.20] G. Koblmüller, C. S. Gallinat, S. Bernardis, J. S. Speck, G. D. Chern, E. D. Readinger, H. Shen, and M. Wraback, Optimization of the surface and structural quality of N-face InN grown by molecular beam epitaxy, *Applied Physics Letters*, 89 (2006) 071902:1–3.

[2.21] G. Koblmüller, C. S. Gallinat, and J. S. Speck, Surface kinetics and thermal instability of N-face InN grown by plasma-assisted molecular beam epitaxy, *Journal of Applied Physics*, 101 (2007) 083516:1–9.

[2.22] C. S. Gallinat, G. Koblmüller, J. S. Brown, and J. S. Speck, A growth diagram for plasma-assisted molecular beam epitaxy of In-face InN, *Journal of Applied Physics*, 102 (2007) 064907:1–7.

[2.23] B. Heying, R. Averbeck, L. F. Chen, E. Haus, H. Riechert, and J. S. Speck, Control of GaN surface morphologies using plasma-assisted molecular beam epitaxy, *Journal of Applied Physics*, 88 (2000) 1855–1860.

[2.24] E. Dimakis, E. Iliopoulos, K. Tsagaraki, Th. Kehagias, Ph. Komninou, and G. Georgakilas, Heteroepitaxial growth of In-face InN on GaN (0001) by plasma-assisted molecular-beam epitaxy, *Journal of Applied Physics*, 97 (2005) 113520:1–10.

[2.25] X. Wang, S. B. Che, Y. Ishitani, and A. Yoshikawa, Effect of epitaxial temperature on N-polar InN films grown by molecular beam epitaxy, *Journal of Applied Physics*, 99 (2006) 073512:1–5.

[2.26] T. Ive, O. Brandt, M. Ramsteiner, M. Giehler, H. Kostial, and K. H. Ploog, Properties of InN layers grown on 6HSiC(0001) by plasma-assisted molecular beam epitaxy, *Applied Physics Letters*, 84 (2004) 1671–1673.

[2.27] G. Koblmüller, R. Averbeck, H. Riechert, and P. Pongratz, Direct observation of different equilibrium Ga adlayer coverages and their desorption kinetics on GaN (0001) and (000$\bar{1}$) surfaces, *Physical Review B*, 69 (2004) 035325:1–9.

[2.28] G. Mula, C. Adelmann, S. Moehl, J. Oullier, and B. Daudin, Surfactant effect of gallium during molecular-beam epitaxy of GaN on AlN (0001), *Physical Review B*, 64 (2001) 195406:1–12.

[2.29] S. Fernández-Garrido, G. Koblmüller, E. Calleja, and J. S. Speck, In situ GaN decomposition analysis by quadrupole mass spectrometry and reflection high-energy electron diffraction, *Journal of Applied Physics*, 104 (2008) 033541:1–6.

[2.30] N. Newman, J. Ross, and M. Rubin, Thermodynamic and kinetic processes involved in the growth of epitaxial GaN thin films, *Applied Physics Letters*, 62 (1993) 1242–1244.

[2.31] B. Maleyre, O. Briot, and S. Ruffenach, MOVPE growth of InN films and quantum dots, *Journal of Crystal Growth*, 269 (2004) 15–21.

[2.32] J. Brown, G. Koblmüller, R. Averbeck, H. Riechert, and J. S. Speck, Quadrupole mass spectrometry desorption analysis of Ga adsorbate on AlN (0001), *Journal of Vacuum Science and Technology A*, 24 (2006) 1979–1984.

[2.33] I. Barin, *Thermodynamical data of pure substances* (VCH, Weinheim, 1993).

[2.34] G. Koblmüller, J. Brown, R. Aberbeck, H. Riechert, P. Pongratz, and J. S. Speck, Ga adlayer governed surface defect evolution of (0001)GaN films grown by plasma-assisted molecular beam epitaxy, *Japanese Journal of Applied Physics*, 44 (2005) L906–L908.

[2.35] G. Koblmüller, J. Brown, R. Aberbeck, H. Riechert, P. Pongratz, and J. S. Speck, Continuous evolution of Ga adlayer coverages during plasma-assisted molecular-beam epitaxy of (0001) GaN, *Applied Physics Letters*, 86 (2005) 041908:1–3.

[2.36] J. E. Northrup and J. Neugebauer, Indium-induced changes in GaN(0001) surface morphology, *Physical Review B*, 60 (1999) R8473–R8476.

[2.37] H. Chen, R. M. Feenstra, J. E. Northrup, T. Zywietz, and J. Neuge-bauer, Spontaneous formation of indium-rich nanostructures on InGaN(0001) surfaces, *Physical Review Letters*, 85 (2000) 1902–1905.

[2.38] C. K. Gan and D. J. Srolovitz, First-principles study of wurtzite InN (0001) and (000$\bar{1}$) surfaces, *Physical Review B*, 74 (2006) 115319:1–5.

[2.39] D. Segev and C. G. Van de Walle, Surface reconstructions on InN and GaN polar and nonpolar surfaces, *Surface Science*, 601 (2007) L15–L18.

[2.40] T. D. Veal, P. D. C. King, M. Walker, C. F. McConville, H. Lu, W. F. Schaff, In-adlayers on non-polar and polar InN surfaces: Ion scattering and photoemission studies, *Physica B*, 401–402 (2007) 351–354.

[2.41] T. D. Veal, P. D. C. King, P. H. Jefferson, L. F. J. Piper, C. F. Mc-Conville, H. Lu, W. F. Schaff, P. A. Anderson, S. M. Durbin, D. Muto, H. Naoi, and Y. Nanishi, In adlayers on *c*-plane InN surfaces: A polarity-dependent study by x-ray photoemission spectroscopy, *Physical Review B*, 76 (2007) 075313:1–8.

[2.42] C. Adelmann, J. Brault, D. Jalabert, P. Gentile, H. Mariette, G. Mula and B. Daudin, Dynamically stable gallium surface coverages during plasma-assisted molecular-beam epitaxy of (0001) GaN, *Journal of Applied Physics*, 91 (2002) 9638–9645.

[2.43] B. Heying, E. J. Tarsa, C. R. Elsass, P. Fini, S. P. DenBaars, and J. S. Speck, Dislocation mediated surface morphology of GaN, *Journal of Applied Physics*, 85 (1999) 6470–6476.

[2.44] Y. F. Ng, Y. G. Cao, M. H. Xie, X. L. Wang, and S. Y. Tong, Growth mode and strain evolution during InN growth on GaN(0001) by molecular-beam epitaxy, *Applied Physics Letters*, 81 (2002) 3960–3962.

[2.45] X. H. Wu, P. Fini, E. J. Tarsa, B. Heying, S. Keller, U. K. Mishra, S. P. DenBaars, and J. S. Speck, Dislocation generation in GaN het-eroepitaxy, *Journal of Crystal Growth*, 189–190 (1998) 231–243.

[2.46] B. Heying, X. H. Wu, S. Keller, Y. Li, D. Kapolnek, B. P. Keller, S. P. DenBaars, and J. S. Speck, Role of threading dislocation structure on the x-ray diffraction peak widths in epitaxial GaN films, *Applied Physics Letters*, 68 (1996) 643–645.

[2.47] W. Qian, M. Skowronski, M. De Graef, K. Doverspike, L. B. Rowland, and D. K. Gaskill, Microstructural characterization of α-GaN films grown on sapphire by organometallic vapor phase epitaxy, *Applied Physics Letters*, 66 (1995) 1252–1254.

[2.48] V. Srikant, J. S. Speck, and D. R. Clarke, Mosaic structure in epitaxial thin films having large lattice mismatch, *Journal of Applied Physics*, 82 (1997) 4286–4295.

[2.49] S. R. Lee, A. M. West, A. A. Allerman, K. E. Waldrip, D. M. Follstaedt, P. P. Provencio, D. D. Koleske, and C. R. Abernathy, Effect of threading dislocations on the Bragg peakwidths of GaN, AlGaN, and AlN heterolayers, *Applied Physics Letters*, 86 (2005) 241904:1–3.

[2.50] M. J. Hordon and B. L. Averbach, X-ray measurements of dislocation density in deformed copper and aluminum single crystals, *Acta Metallurgica*, 9 (1961) 237–246.

[2.51] C. G. Dunn and E. F. Kogh, Comparison of dislocation densities of primary and secondary recrystallization grains of Si-Fe, *Acta Metallurgica*, 5 (1957) 548–554.

[2.52] P. Gay, P. B. Hirsch, and A. Kelly, The estimation of dislocation densities in metals from X-ray data, *Acta Metallurgica*, 1 (1953) 315–319.

[2.53] S. K. Mathis, A. E. Romanov, L. F. Chen, G. E. Beltz, W. Pompe, and J. S. Speck, Modeling of threading dislocation reduction in growing GaN layers, *physica status solidi (a)*, 179 (2000) 125–145.

3

Polarity-dependent epitaxy control of InN, InGaN and InAlN

X. Q. Wang and A. Yoshikawa

Department of Electronics and Mechanical Engineering, Chiba University, 1-33 Yayoi-cho, Inage-ku, Chiba 263-8522, Japan

3.1   Introduction

Hexagonal III-nitrides are polar crystals along the c-direction, that is $+c$ (In-polarity for InN) and $-c$ (N-polarity). Both growth behavior and properties of III-nitrides are greatly influenced by the polarity. In general, GaN and AlN are usually grown on $+c$-polarity where the growth is under step-flow mode and atomically flat surfaces are obtained. On the other hand, GaN and AlN layers with N-polarity usually exhibit relatively rough surfaces. Similar to GaN and AlN, the growth behavior and properties of InN also greatly depend on the polarity. In particular, the maximum epitaxial temperatures for InN with different polarities are different. Figure 3.1 shows the dependence of InN growth rate on epitaxial temperature for different polarities. The growth rate steeply decreases for In- and N-polarity films at temperatures above 500 and 600°C, respectively, indicating that the maximum epitaxial temperature for N-polar InN is about 100°C higher than that for In-polar InN. Above these temperatures, InN starts to decompose and In droplets appear on the surface. Therefore, the growth process is limited by the dissociation of InN itself but not by the re-evaporation of excess In on the surface. This results in a much narrower stoichiometric growth window for InN than for GaN and AlN.

Due to the nature of the limited growth temperature, the effect of growth temperature on the growth behavior and properties of InN is more serious than for GaN and AlN. The epitaxial temperatures of In-rich InGaN and InAlN at different polarities show similar behavior to that of InN, that is, N-polar shows a higher maximum growth temperature, since they are mainly limited by the growth temperature of InN.

In this chapter, we will present a review of epitaxy control and the properties of InN, InGaN and InAlN grown by molecular beam epitaxy (MBE), paying particular attention to the control of polarity.

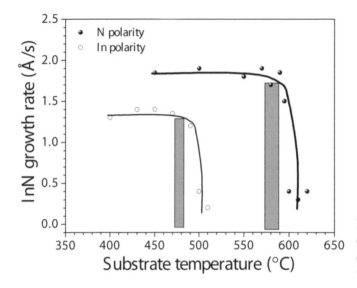

FIGURE 3.1

Growth rate of InN films with different polarities as a function of growth temperature.

3.2   Methods to determine and control the polarity of InN

There are several methods to determine the polarity of InN such as chemical etching, convergent beam electron diffraction (CBED), coaxial impact collision ion scattering spectroscopy (CAICISS) [3.1], surface reconstruction revealed from reflection high-energy electron diffraction (RHEED) and so on. CBED is used to determine the polarity using a transmission electron microscope (TEM). It can determine the polarity of a grain even on the nanoscale and thus has great advantages for analyzing domain structures of InN. The surface reconstruction revealed from RHEED has been widely used to determine the polarity of GaN and AlN, and it has also been reported that the polarity of InN could be revealed from surface reconstruction [3.2]. However, it is still more difficult to observe the surface reconstruction of InN than that of GaN due to the difficulty in epitaxy of atomically flat InN.

CAICISS is a derivative of the generic technique of ion scattering spectroscopy, in particular of low energy ion scattering spectroscopy. In a CAICISS measurement, a simple 180° backscattering geometry enables us to determine the atomic positions at both an outermost surface and slightly deeper layers. In a typical CAICISS system, a He$^+$ ion beam (about 2 keV) is chopped into pulses of 150 ns duration at a 100 kHz repetition rate. The focused He$^+$ ion beam hits the sample surface and the He$^+$ particles scattered by the atoms are detected. Since the flight times of He$^+$ ions scattered from different atoms are different, we can determine the type of atoms by their flight time. The penetration depth of He$^+$ ions is as shallow as a few atomic layers, and the detected intensity of scattered He$^+$ ions from the same atom depends on the incident angle because of shadow cone effects [3.1]. CAICISS spectra including information on the surface structure and lattice polarity can be obtained by analyzing the intensity of scattered signals as a function of incident angle. Then, the polarity can be determined by comparing the experimental CAICISS spectra to the simulated ones shown in Fig. 3.2. As shown in the figure, three peaks at 23°, 47° and 72° are observed in simulated CAICISS spectra for In-polarity while six peaks at 15°, 23°, 32°, 51°, 67° and 74° are seen for N-polarity.

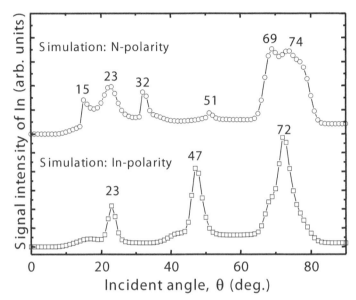

FIGURE 3.2
Simulated CAICISS spectra of InN.

Using CAICISS measurements, it was found that InN layers grown on a *c*-sapphire substrate without using nitridation, but thermally cleaned in ultra-high vacuum (UHV) at high temperatures for about 1 h, are mixed polarity, due to the Al-rich surface of sapphire substrates after thermal cleaning. On the other hand, InN layers grown on deeply nitrided sapphire are N-polarity. This result is very similar to that of GaN grown on sapphire substrates by MBE [3.3]. A thin AlN layer formed by sapphire nitridation is N-polarity. The observation of N-polarity InN on N-polarity AlN shows that the InN follows the lattice polarity of the AlN. Similar behavior has also been found in the polarity of InN grown on GaN epitaxial layers, where InN layers grown on Ga- and N-polar GaN are In- and N-polarity, respectively. Therefore, the polarity of InN is controlled by that of the underlying GaN or AlN buffer layers. Similarly, the polarity of InGaN and InAlN can also be controlled by that of underlayers, where cation and anion polarity underlayers lead to cation and anion polarity InGaN and InAlN, respectively.

Due to the chemical nature of the different polarities, the chemical stability is different, where the In-polar InN is more chemically stable and more difficult to chemically etch than the N-polar InN. Therefore, the polarity of InN can be determined by chemical etching, where aqueous solutions of KOH and NaOH are often used as the chemical etching solutions [3.2, 3.4]. Figure 3.3 shows surface morphologies of In- and N-polarity InN layers grown on Ga- and N-polarity GaN templates before and after the chemical etching, investigated by atomic force microscope (AFM). It is shown that the In-polar InN layer shows step-flow morphology. This layer was hardly etched and the step-flow structure was kept after etching for 2 h except for the appearance of a high density of small pits. Large hexagonal pits were observed after etching for 13 h, while the step-flow feature was still kept in the smooth regions. On the other hand, the as-grown N-polarity InN layer shows grain-like morphology where step-flow features can be observed within each grain. This layer was easily etched and small hexagonal islands were observed after etching for 2 h. After etching for 13 h, larger hexagonal islands appeared. In addition, the surface became much rougher for N-polar InN than for In-polar InN after etching, as indicated by the scale bar in Fig. 3.3, where the root mean square (rms) roughness was 5 and 22 nm for the In- and N-polarity InN after 13 h of etching.

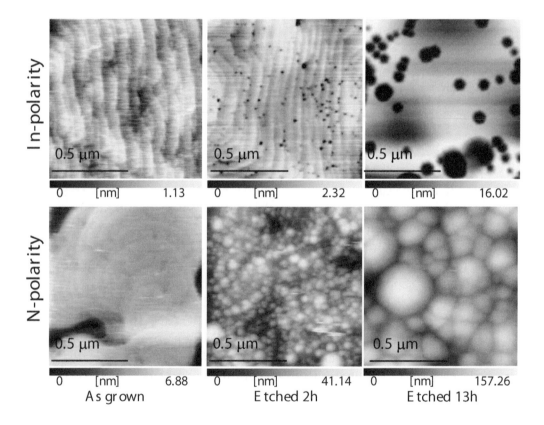

FIGURE 3.3
AFM images of as-grown, 2h-etched and 13h-etched InN layers with different polarities.

In general, the polarity of InN is controlled by selecting the polarity of the GaN and AlN underlayers, where InN films grown on cation- and anion-polar GaN or AlN are In- and N-polarity, respectively.

3.3 Epitaxy control and properties of InN

3.3.1 *In situ* investigation of growth diagram by spectroscopic ellipsometry

Spectroscopic ellipsometry (SE) is a very surface-sensitive thin film measurement technique that uses polarized light [3.5]. Besides the same advantages as other optical tools, such as non-destructiveness and non-invasiveness, the sensitivity of SE is higher than the conventional optical reflection measurement, such as laser reflectrometry. It is also an excellent technique for precise determination of the complex-valued dielectric functions of semiconductor materials. In this section, we will show that SE is very useful to check the growth stoichiometry and determine the growth regions.

Figure 3.4 shows a typical time evolution of the real part of the pseudodielectric function ($< \varepsilon_1 >$) of InN grown under In-rich, stoichiometric and N-rich conditions at incident light wavelengths of 700 and 1600 nm. For InN grown under stoichiometry, clear oscillations of $< \varepsilon_1 >$ can be observed for both wavelengths and the oscillations tend to be saturated due to the photon energy of the incident light being larger than the band gap energy of InN. The oscillations of $< \varepsilon_1 >$ for shorter wavelengths (that is, with larger photon energy) saturate more quickly. As shown in Fig. 3.4, at the same thickness, the $< \varepsilon_1 >$ of InN grown under In-rich conditions is larger than those of InN grown under stoichiometry and the oscillation of $< \varepsilon_1 >$ tends to be damped quickly, which is attributed to the effect of excess In-metal on the surface. On the other hand, the $< \varepsilon_1 >$ of InN grown under N-rich conditions is smaller than those of InN grown under stoichiometric conditions and the $< \varepsilon_1 >$ becomes smaller as growth proceeds, which is mainly ascribed to fractional depolarization of the probing light because of surface roughening.

On the basis of the effect of In beam flux (P_{In}) on the growth rate and the measured values of $< \varepsilon_1 >$ and $< \varepsilon_2 >$ at different P_{In}, it was found that the surface stoichimetry-dependent growth region/mode can be determined by the values of $< \varepsilon_1 >$ and $< \varepsilon_2 >$. Figure 3.5 shows the growth modes/regions of InN, where the guidelines for the values of $< \varepsilon_1 >$ and $< \varepsilon_2 >$ recorded during the growth (at an InN thickness of 100 nm) and at a convergent point (InN thickness of 2–8 μm) are slightly different. This may be due to the surface becoming rougher with increasing thickness. As shown in Fig. 3.5, an intermediate regime where the growth is performed under metal-rich conditions but free of droplets is too narrow to be determined under conditions with a growth rate of hundreds of nm/h and thus there are roughly two growth regimes for InN, that is, In-rich and N-rich regimes. This is

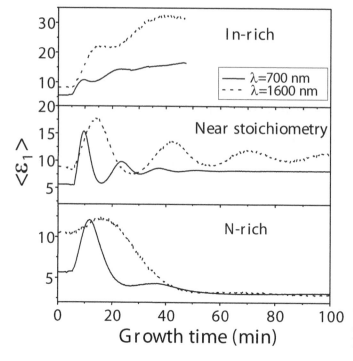

FIGURE 3.4

SE signal traces of $< \varepsilon_1 >$ for InN grown under In-rich, near stoichiometry and N-rich conditions at incident wavelengths of 700 and 1600 nm.

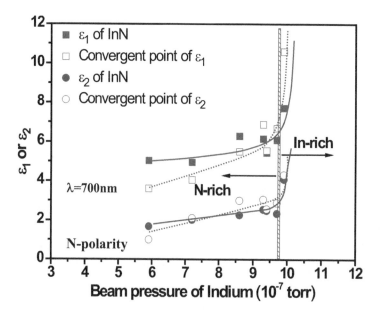

FIGURE 3.5

Growth region diagram illustrated by $< \varepsilon_1 >$ and $< \varepsilon_2 >$ at InN thickness of 100 nm and the convergent point. Solid and dotted lines show guidelines for values of $< \varepsilon_1 >$ and $< \varepsilon_2 >$ at InN thickness of 100 nm and at convergent point, respectively.

different from the three distinct growth regimes of GaN. This can be ascribed to the low growth temperature of InN, where once the In droplets appear on the surface, it is almost impossible to re-evaporate them at such a low temperature. The quite narrow or almost non-existence of an intermediate growth region leads to difficulty in epitaxy of thick InN with flat surface. SE is an important tool for *in situ* SE monitoring of the growth stoichiometry to enable the growth of thick and smooth InN layers.

3.3.2   N-polarity InN epitaxy

N-polar InN films were grown on N-polar GaN underlayers. In this case, a 500 nm thick N-polar GaN epitaxial layer was grown by MBE first and then InN was grown on it [3.6]. In this section, the influence of growth factors, such as In/N ratio and epitaxial temperature, on N-polar InN will be introduced.

3.3.2.1   Effect of In/N ratio on InN epitaxy

Surface morphologies of InN layers are greatly influenced by the In/N ratio. It was found that the surface is quite rough under N-rich conditions and the roughness becomes smaller with increasing P_{In} as stoichiometric conditions are approached. This is a normal behavior in the epitaxy of III-nitrides because of reduced surface migration of group-III adatoms under N-rich growth conditions.

Crystal quality is also greatly affected by the In/N ratio. As shown in Fig. 3.6, the full width at half-maximum (FWHM) values of x-ray diffraction (XRD) rocking curves for (002) and (102) diffractions ($FWHM_{(002)}$ and $FWHM_{(102)}$) are decreased with increasing P_{In}, showing improved crystalline quality. As for electrical properties, although electron concentrations (n_e) are not obviously affected, electron Hall mobility (μ_e) becomes higher with increasing P_{In}.

3.3.2.2   Effect of growth temperature on InN

Growth temperature (T_G) is an important factor for InN epitaxy. Surface morphologies of N-polar InN layers with a thickness of 2.3 μm grown at 440–620°C are shown in Fig. 3.7. Here, since the growth is performed under slightly N-rich conditions, the maximum growth temperature could be slightly higher, up to 620°C. The growth can be roughly divided into two regions from the surface morphologies. In region I, where T_G=540°C, dendritic growth is observed and some hexagonal grains are also seen on the surface. The branch size of the dendritic grain is increased with increasing T_G. In region II, where $T_G \leq 540$°C, spiral growth is observed. The grains show step-flow-like features with step heights of two monolayers (MLs) or multiple MLs in this region, as shown by the 1μm$\times 1\mu$m AFM images in the inset. In region II, some pits are found on the surface. The surface is rough in region I and becomes smooth in region II. The rms roughness is decreased with increasing T_G in region II. The grain size is increased with increasing T_G in both regions. The improved surface morphology of InN can be ascribed to the enhanced migration of

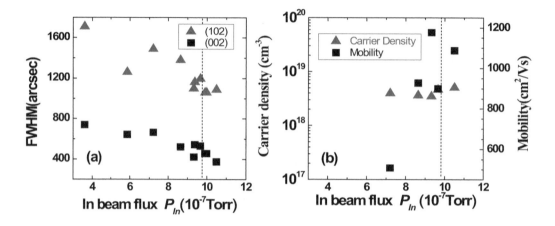

FIGURE 3.6

Structural and electrical properties of InN layers as a function of In beam flux. (a) FWHM values for (002) and (102) diffractions, (b) electron concentration and mobility. Reprinted with permission from M. Yoshitani, K. Akasaka, X. Wang, S. B. Che, Y. Ishitani, and A. Yoshikawa, Journal of Applied Physics, 99 (2006) 044913. Copyright 2006, American Institute of Physics.

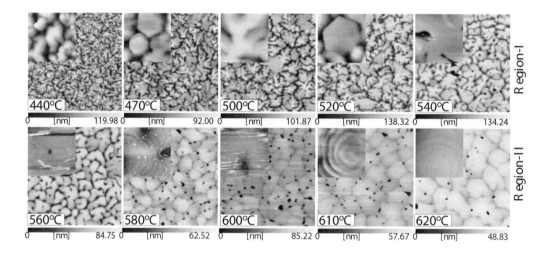

FIGURE 3.7

Surface morphologies of InN films grown at different temperatures investigated by AFM. The scanned areas are 10 μm $\times$ 10 μm and 1 μm $\times$ 1 μm (shown in the inset), respectively.

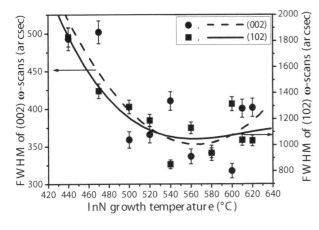

FIGURE 3.8

FWHM values of (002) ω-scans (solid circle) and (102) ω-scans (solid rectangular) as a function of InN growth temperature. The polynomial fit lines are also shown.

In adatoms on the surface due to the higher T_G.

Figure 3.8 shows FWHMs of XRD ω-scans for (002) and (102) InN as a function of growth temperature. The $FWHM_{(002)}$ are 300–500 arcsec while $FWHM_{(102)}$ are 800–1900 arcsec. The lowest $FWHM_{(102)}$ value of 800 arcsec shows comparative crystal quality of InN with that of GaN. The crystal quality becomes better with increasing T_G up to 540°C and is nearly saturated there. This coincides with the growth regions investigated from surface morphologies.

The effect of the growth temperature on n_e and μ_e is shown in Fig. 3.9. The InN films grown at 470–600°C have almost the same μ_e and n_e with values about 1400 cm^2/Vs and 3.5×10^{18} cm^{-3}, respectively. The films grown at the lower (440°C) and higher temperatures (610–620°C) show slightly smaller mobility and higher electron concentration.

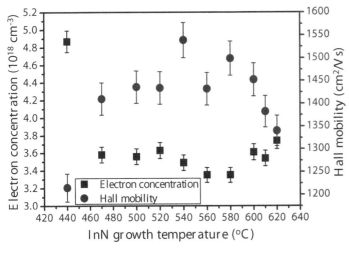

FIGURE 3.9

Room temperature residual electron concentration (solid rectangles) and Hall mobility (solid circles) of InN film as a function of InN growth temperature.

3.3.3   In-polarity InN epitaxy

In-polarity InN film is grown on Ga-polar GaN template by MBE. In this case, the Ga-polar GaN template is grown by metalorganic vapor-phase epitaxy (MOVPE) [3.7, 3.8]. In this section, the epitaxial behavior and properties of In-polar InN will be discussed.

3.3.3.1   Attempts to achieve an atomically flat surface

For In-polar InN, atomically flat surfaces are achieved. Figure 3.10 shows typical surface morphology of the In-polar InN film. An atomically flat surface with a step and terrace feature formed in step-flow growth mode is observed. From the line profile of the AFM image shown in the inset of Fig. 3.10(b), the step height is 0.28–0.29 nm, which coincides with half of the c lattice constant of InN, showing that the step height is a single monolayer. The AFM image with a 10×10 μm^2 scanned area is shown in the inset of Fig. 3.10(a). Two-dimensional hexagonal growth spirals are observed, which are influenced by threading dislocations in screw components (STDs), as will be discussed later. The surface is quite flat with rms roughnesses of 0.9, 0.4, and 0.1 nm over scanned areas of 10×10, 3×3, and 1×1 μm^2, respectively. The surface morphology of InN is similar to that of homoepitaxial GaN grown by MBE on MOVPE-grown GaN [3.9, 3.10].

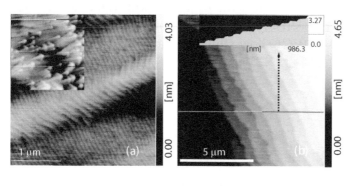

FIGURE 3.10

Surface morphologies of InN film investigated by AFM. The scanned areas are 3×3 and 1×1 μm^2 for (a) and (b), respectively. The AFM image in the scanned area of 10×10 μm^2 is shown in the inset of (a). The line profile is shown in the inset of (b).

To achieve such an atomically flat surface, four important factors are necessary. First, In-polarity is preferred. This is similar to the case of GaN and AlN, where cation-polar films usually exhibit a smoother morphology than N-polar films [3.11]. In fact, the step height is 2 or 4 ML for the smoothest surface of the N-polar InN and it is difficult to obtain an atomically flat surface with a one-monolayer step height [3.12, 3.13]. Second, a slightly In-rich growth condition is chosen. This is also similar to GaN epitaxy, where a Ga-stablized growth condition leads to a smooth morphology while growth under N-rich conditions produces a rough surface due to the higher surface migration capability of Ga atoms in the former case [3.14, 3.15]. For InN, the migration of In adatoms is also enhanced under an In-stablized condition. However, one problem for InN epitaxy is that the dissociation temperature of InN is much lower than the desorption temperature of In metal, which results in a difference between the growth behaviors of InN and GaN. In GaN epitaxy, the adsorbed Ga can re-evaporate, and a sufficiently high growth temperature can be used to obtain a steady state, at which the surface is metal-rich but free of any metal accumulation, resulting in an atomically smooth surface [3.16]. This model cannot be used for InN epitaxy because the excess In does not re-evaporate and accumulates as In droplets owing to the limited low growth temperature as we discussed above. Therefore, it is important to control the growth under a very slightly In-rich condition to minimize the formation of In droplets. Third, a GaN template with a low dislocation density should be used. As will be discussed, STDs lead to growth spirals and thus influence the flatness of the surface. In fact, InN grown on the MOVPE-grown GaN template with higher STD density exhibits a higher density of spirals. Fourth, a high growth temperature should be used in order to enhance the migration of In adatoms.

3.3.3.2 Effect of growth temperature on In-polarity InN epitaxy

Surface morphologies of the In-polar InN films are influenced by growth temperature, as shown in Fig. 3.11. It is also shown that macroscopic surface defects for these samples are growth-spiral hillocks and surface roughening is apparently accompanied by an increase in the density of spirals. The rms roughness increases from 0.9 to 2.2 nm with decreasing T_G to 400°C, and the surface morphology for the samples grown at T_G below 450°C is dominated by growth spirals. Figure 3.12 shows the counted density of growth spirals in the AFM images over an area of 25 $\times$ 25 μm^2 and the STD density as a function of T_G. The STD density shows almost the same value as the density of growth spirals at each temperature and they show the same growth temperature dependence, that is, their density quickly decreases at T_G up to 440°C and then slightly decreases at 460–500°C. This indicates that the growth spirals in InN films originate from the STDs. There are two sources of STDs in InN epilayers; one is that they originate from STDs in the GaN templates and another is that they are newly generated during epitaxy, in particular, at the hetero-interface. The higher T_G can enhance the migration capability of In adatoms and thus leads to reduction of threading dislocation density and better surface morphology.

3.3.3.3 Effect of threading dislocations on residual carrier concentration

In InN, native defects tend to act as donors and this leads to highly conducting n-type InN epilayers. Besides point defects, threading dislocations are also a candidate for the unintentional donors. Figure 3.13 shows n_e and μ_e of many In-polar InN films with thicknesses of 0.5–5 μm as a function of FWHMs of (102) InN ω-scans. These n_e values are those for the bulk, that is, the estimated contribution from surface/interface charge [3.17] of 3–5 $\times$ 10^{13} cm^{-2} has been removed [3.8].

As shown in Fig. 3.13, n_e increases from 4.0 $\times$ 10^{17} to 2.4 $\times$ 10^{18} cm^{-3} with increasing FWHM$_{(102)}$ from 950 to 2200 arcsec, indicating the increase of n_e with

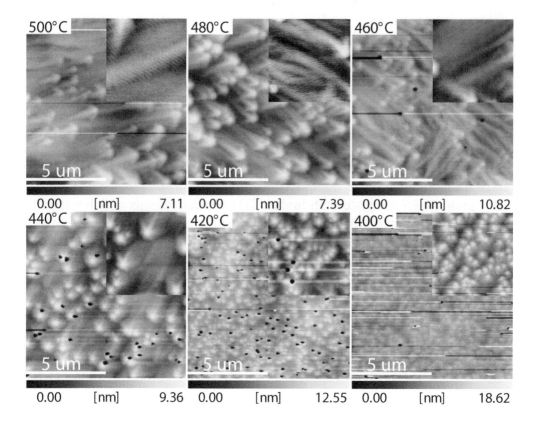

FIGURE 3.11

Surface morphologies of InN films grown at different temperatures, investigated by AFM. The scanned areas are $10 \times 10 \ \mu m^2$ and $3 \times 3 \ \mu m^2$ (shown in the inset), respectively.

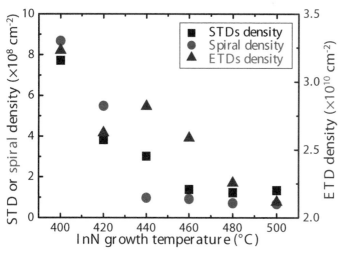

FIGURE 3.12

STD (solid rectangles) and ETD (solid triangles) densities as a function of growth temperature. The density of growth spirals is also shown (solid circles). Reprinted with permission from X. Wang, S. B. Che, Y. Ishitani, and A. Yoshikawa, Applied Physics Letters, 90 (2007) 151901. Copyright 2007, American Institute of Physics.

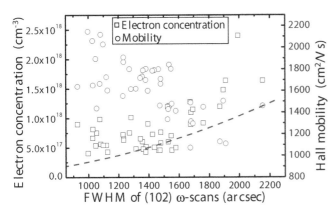

FIGURE 3.13

Electron concentrations and Hall mobilities of many InN films as a function of FWHMs of (102) ω-scans. The estimated density of dangling bonds originating from ETDs is shown by the dashed line. Reprinted with permission from X. Wang, S. B. Che, Y. Ishitani, and A. Yoshikawa, Applied Physics Letters, 90 (2007) 151901. Copyright 2007, American Institute of Physics.

increasing edge-type threading dislocation (ETD) density since the $FWHM_{(102)}$ is correlated with the ETDs. The dashed line indicates the contribution of dangling bonds to the n_e in the InN epilayers, assuming that each dangling bond at the ETDs acts as a singly ionizable donor, that is, one electron is supplied every monolayer along each ETD. It is shown that the dependence of the dashed line agrees well with that of the n_e on $FWHM_{(102)}$ and even the exact values are almost the same as the experimental data. This indicates that ETDs are probably a dominant source of n_e in high purity InN epilayers. The discrepancy between the measured n_e and the density of dangling bonds is probably caused by contributions from impurity- and point-defects-generated carrier concentrations that are not taken into account here. In addition, μ_e decreases from 2150 to 1000 cm^2/Vs with increasing $FWHM_{(102)}$ from 950 to 2200 arcsec, showing that the μe is also influenced by the ETDs.

3.4   *p*-type doping of In- and N-polarity InN

3.4.1   Properties of Mg-doping in InN

Mg-doped InN (InN:Mg) films were grown on GaN templates. A 50-nm-thick undoped InN layer and InN:Mg layer were grown in sequence under slightly In-rich growth conditions, where the T_G were 480–500°C and 600°C for In- and N-polarity, respectively [3.18]. It was found that the effect of Mg doping on the properties of InN with different polarities are very similar and thus we mainly discuss the In-polarity case here.

Figure 3.14 shows a secondary ion mass spectrometry (SIMS) profile for a multiple-InN layer-structure sample where four 390-nm-thick InN:Mg layers were grown at Mg cell temperature (T_{Mg}) varied from 200 to 275°C and each InN:Mg layer was separated by a 110-nm-thick undoped spacer-InN layer, and finally capped by a 170-nm-thick undoped InN layer. Several features of Mg-doping in InN are found: (1) the Mg-doping level in each InN:Mg layer is almost constant and the boundary at different doping levels is sharp; and (2) the Mg concentration ([Mg]) changes exponentially with T_{Mg} or it is linearly proportional to the incident Mg beam flux. This indicates that the sticking coefficient of Mg on In-polarity InN is almost unity and Mg atom diffusion is negligibly small, which are basically attributed to the growth temperature being low. The Mg concentrations in these samples estimated by using the SIMS data, calibrated using a reference Mg-ion implanted InN sample, are about 1.0×10^{18}, 5.6×10^{18}, 2.9×10^{19}, and 1.8×10^{20} cm^{-3} for $T_{Mg} =$ 200, 225, 250, and 275°C, respectively.

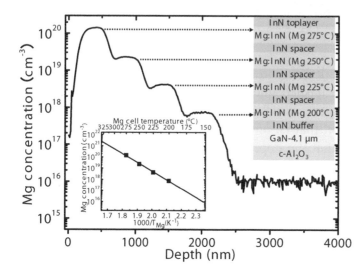

FIGURE 3.14

SIMS profile of Mg in a multiple-layer-structure InN film. The sample structure is shown in the inset at the top right and the measured [Mg] vs $1000/T_{Mg}$ is also shown in the inset at the left bottom. Reprinted with permission from X. Wang, S. B. Che, Y. Ishitani, and A. Yoshikawa, Applied Physics Letters, 90 (2007) 201913. Copyright 2007, American Institute of Physics.

In the InN:Mg samples grown under the In-polarity regime, polarity inversion occurred [3.4]. Figure 3.15 shows a TEM image of the same sample shown in the inset of Fig. 3.14. One apparent feature is the appearance of V-shaped domains in the third InN:Mg layer with [Mg] of 2.9×10^{19} cm^{-3}. CBED measurement results show that these V-shaped domains are inversion domains (IDs) which invert the In-polarity to N-polarity [3.4]. These InN IDs are similar to those IDs previously reported in InN:Mg and GaN:Mg [3.19–3.22]. Further investigation shows that the critical [Mg] value for polarity inversion in InN is about 10^{19} cm^{-3}, where the In-polarity InN:Mg changes to N-polarity when [Mg] is over 10^{19} cm^{-3}.

In Fig. 3.14, the InN:Mg layers grown with T_{Mg}=250 and 275°C were inverted to N-polarity. Thus, it seems that the polarity inversion does not influence the doping efficiency of Mg in InN. To confirm this, a N-polarity multiple-InN layer-structure sample doped with different [Mg]s has been investigated by SIMS and it was found that the [Mg] is almost the same at the same T_{Mg} for both polarities. This indicates that the sticking coefficient of Mg is almost independent of the polarity.

Surface roughness is influenced by the Mg doping. The surface is quite flat at [Mg]<10^{19} cm^{-3} and it becomes rougher and rougher with increasing [Mg] at [Mg]>10^{19} cm^{-3}. The crystal quality is also influenced by Mg doping. As

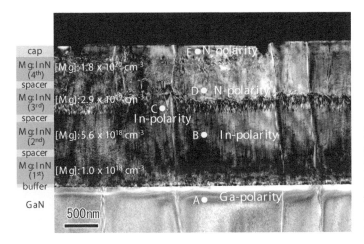

FIGURE 3.15

Cross-sectional dark-field TEM images of a multiple-layer-structure InN film recorded with g=[0002]. The sample structure is shown on the left.

shown in Fig. 3.16, the FWHMs of XRD ω-scans show that the crystal quality is first improved with increasing [Mg] at [Mg]$<10^{19}$ cm^{-3} and then deteriorates at [Mg]$>10^{19}$ cm^{-3}.

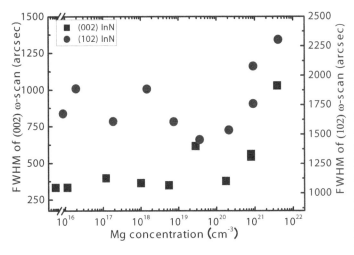

FIGURE 3.16

FWHM values of XRD ω-rocking curves for (002) and (102) InN as a function of [Mg]. Reprinted with permission from X. Wang, S. B. Che, Y. Ishitani, and A. Yoshikawa, Applied Physics Letters, 90 (2007) 201913. Copyright 2007, American Institute of Physics.

Figure 3.17 shows PL spectra and corresponding PL intensity measured at 16 K as a function of [Mg]. It is shown that both PL spectra and intensity are drastically affected by Mg-doping, in particular, the PL intensity changes by more than five orders of magnitude. At [Mg]$\sim10^{18}$ cm^{-3}, the intensity is already lower by about four orders of magnitude than that of the undoped sample, and no PL emission is detected at [Mg]=5.6-29 $\times$ 10^{18} cm^{-3}. Weak PL emission is observed again with increasing [Mg]. Further, no PL emission can be detected at [Mg]$\sim3.9 \times 10^{21}$ cm^{-3} again, which is probably because of unknown luminescence-killer processes in these poor quality samples. The exact reason for the non-luminescent nature of InN:Mg at [Mg]=5.6–29 $\times$ 10^{18} cm^{-3}, as well as *p*-type InN reported in the literature and also to be discussed in the next section is not known [3.23, 3.24], but this may be attributed to the diffusion length of minority carriers in *p*-InN and/or the introduction of complex levels by Mg-doping resulting in about five orders of magnitude higher carrier-relaxation paths.

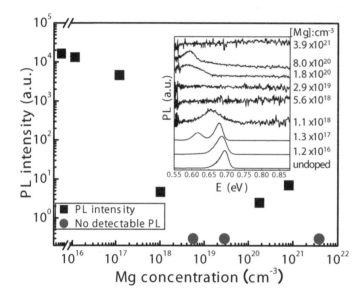

FIGURE 3.17

PL intensity and PL spectra (inset) of InN layers at 16 K as a function of [Mg]. Reprinted with permission from X. Wang, S. B. Che, Y. Ishitani, and A. Yoshikawa, Applied Physics Letters, 90 (2007) 201913. Copyright 2007, American Institute of Physics.

In the lightly Mg-doped InN samples with $[Mg]\sim 1.3 \times 10^{17}$ cm^{-3}, two emission peaks are observed at around $0.67(I_{bb})$ and $0.61(I_{fa})$ eV, respectively. Excitation power dependent PL studies of this sample are shown in Fig. 3.18(a). It is clear that the peak energies and integrated intensities of I_{bb} and I_{fa} greatly depend on the excitation power and their dependencies are summarized in Fig. 3.18(b). As shown in Fig. 3.18(b), the intensity of I_{bb} increases linearly with increasing excitation power while that of I_{fa} tends to saturation. This indicates that I_{bb} originates from the band-to-band transitions while I_{fa} comes from the free-to-acceptor transitions, where the activation energy for Mg acceptors is estimated to be about 61 meV. This value coincides with the predicted activation energies of 47–66 meV from the hydrogenic model, provided that m_h^*=0.3–0.42m_0 and ε_r=9.3ε_0 [3.25–3.27].

3.4.2 Investigation of *p*-type conduction in InN

One problem in studying *p*-type doping of InN is how to detect the *p*-type conduction. The existence of a surface electron accumulation layer [3.17] greatly influences conventional Hall effect measurements and it shows *n*-type conduction even when the bulk InN is actually *p*-type. At present, most evidence for *p*-type conduction of InN comes from electrolyte-based capacitance voltage (ECV) measurements [3.23, 3.24, 3.28].

ECV measurements are usually performed by using an electrolyte, for exam-

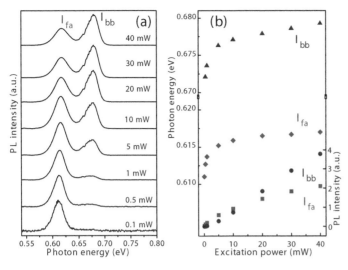

FIGURE 3.18

(a) 16 K PL spectra of partly compensated InN:Mg under different excitation powers. (b) Intensity and energy of I_{fa} and I_{bb} as a function of excitation power. Reprinted with permission from X. Wang, S. B. Che, Y. Ishitani, and A. Yoshikawa, Applied Physics Letters, 90 (2007) 201913. Copyright 2007, American Institute of Physics.

ple KOH or NaOH solutions, to form a Helmholtz-layer contact on InN. This Helmholtz layer behaves as a very thin insulator and the total structure is in principle a "metal-insulator-semiconductor (MIS) structure" provided that the layer does not break down under certain applied bias voltages (V_{bias}). Since the Helmholtz layer capacitance is very large, the apparent behavior of the MIS structure often simply looks like that of a metal-semiconductor structure for normal semiconductors. This is true when their surface Fermi levels are within the forbidden band gap and also if there is no high density of interface states or surface charges. For semiconductors having a high density of surface accumulation electrons such as InN, the total capacitance under zero V_{bias} is that only for the Helmholtz layer capacitance. When negative V_{bias} is applied to the electrolyte, however, the surface/interface electron density tends to decrease due to the compensation effect by the induced positive charges, and even such high density surface/interface charges as 3–5 $\times$ 10^{13} cm^{-2} can be depleted by carefully doing ECV measurements without inducing the breakdown of the electrolyte [3.23, 3.24, 3.28, 3.29]. This means that the surface potential and/or the Fermi level on the InN surface can be modified/and shifted into the forbidden band gap, and the measured total capacitance becomes that for the InN depletion region depending on the density of donors and/or acceptors. Further, the conduction type of InN can be determined on the basis of detecting the minimum capacitance (C_{min}) and/or the peak of C^{-2} against the V_{bias}, which depends on both the high density of surface charges and the different positions of the Fermi level. The corresponding V_{bias} at C_{min} or the peak of C^{-2} is defined as V_{peak} here and it is about 0.3–0.5 V smaller for *p*-type InN than for *n*-type InN [3.29].

Figure 3.19 shows C^{-2}-V_{bias} spectra for InN layers with different [Mg]s using a low frequency measuring signal [3.28]. As shown in the figure, for the In-polarity case, the behavior of C^{-2}-V_{bias} spectra at low [Mg]s ([Mg]$\leq$1.3 $\times$ 10^{17} cm^{-3}) is similar to that for undoped InN, indicating that these InN:Mg layers are *n*-type. For

InN:Mg samples with $1.1 \times 10^{18} \leq [Mg] \leq 2.9 \times 10^{19}$ cm^{-3}, the V_{peak} is about 0.2–0.5 V smaller than that for undoped InN, implying that these InN:Mg layers are p-type [3.29]. For samples with $[Mg] \geq 1.8 \times 10^{20}$ cm^{-3}, however, the V_{peak} tends to increase and is similar to that of undoped InN, again indicating that the InN:Mg layers change to n-type in this [Mg] range.

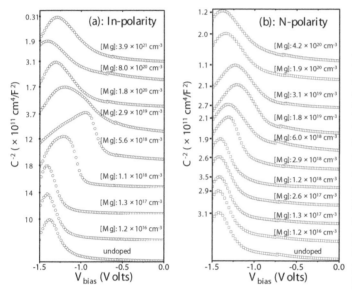

FIGURE 3.19

Mott-Schottky plots of ECV data ($C^{-2} - V$ spectra) for InN layers with (a) In- and (b) N-polarity as a function of [Mg]. The rest potential used here is 0.675 V (Ref. [3.29]). Reprinted with permission from X. Wang, S. B. Che, Y. Ishitani, and A. Yoshikawa, Applied Physics Letters, 91 (2007) 242111. Copyright 2007, American Institute of Physics.

For the N-polar InN:Mg layers shown in Fig. 3.19(b), almost the same behavior for the capacitance is observed against [Mg]s, but with a slightly different window for achieving p-type conduction, that is, showing p-type conduction for $2.9 \times 10^{18} \leq [Mg] \leq 3.1 \times 10^{19}$ cm^{-3}, while showing n-type conduction for $[Mg] \leq 1.2 \times 10^{18}$ and $\geq 1.9 \times 10^{20}$ cm^{-3}. Therefore, the achieved p-type conduction with Mg-doping is almost independent of polarity, which is basically attributed to the Mg-sticking coefficient being almost unity for both polarities. A slightly earlier appearance of p-type conduction for In-polar InN:Mg with increasing [Mg] is probably due to the residual donor density in In-polar InN being lower than that in N-polar, which has also been revealed by SIMS measurements of impurity concentrations.

Net acceptor/donor concentration (N_{ae}/N_{de}) can be estimated from the corresponding C_{min} values [3.29]. However, interface states exist on the surface of InN and contribute to the measured C_{min} and thus the directly estimated N_{ae} or N_{de} are overestimated. The effect of the interface states on ECV results can be estimated by the difference between theoretically predicted and measured capacitance [3.28]. Table 3.1 shows the measured ($C_{min-exp}$) and theoretically predicted capacitance ($C_{min-cal}$) values of several undoped InN layers. It is clear that each N_{de-ECV} value is about one order of magnitude higher than the n_{e-Hall}, and correspondingly there is a big difference in capacitance ΔC between $C_{min-exp}$ and $C_{min-cal}$, where the average ΔC values are 634 and 1320 nF/cm^2 for In- and N-polarity, respectively. The ΔC can be regarded as the contribution from the interface states, and their densities corresponding to ΔC are roughly estimated to be 4.0×10^{12} and 8.4×10^{12} cm^{-2} for In- and N-polarity. The difference between the densities is probably due to different surface properties of InN with opposite polarities.

TABLE 3.1
Electron concentration (n_{e-Hall}) in InN bulk layer, net donor concentration (N_{de-ECV}), theoretically predicted and experimental C_{min} of undoped InN layers with different polarities.

Polarity	Thickness (nm)	n_{e-Hall} (cm^{-3})	N_{de-ECV} (cm^{-3})	$C_{min-cal.}$ (nF/cm^2)	$C_{min-exp.}$ (nF/cm^2)	ΔC (nF/cm^2)
In	980	5.08×10^{17}	6.9×10^{18}	254	834	580
In	1850	5.17×10^{17}	8.9×10^{18}	256	938	682
In	1860	5.08×10^{17}	6.6×10^{18}	254	816	562
In	2340	6.01×10^{17}	1.2×10^{19}	274	1067	793
In	895	9.08×10^{17}	7.8×10^{18}	330	885	555
N	2030	1.25×10^{18}	3.8×10^{19}	382	1763	1381
N	3050	1.25×10^{18}	3.0×10^{19}	382	1650	1268
N	655	1.30×10^{18}	3.6×10^{19}	389	1794	1405
N	655	1.25×10^{18}	2.8×10^{19}	382	1603	1221

The N_{de} and N_{ae} for InN:Mg layers can be estimated by taking into account the effect of the interface states by subtracting the ΔC from measured capacitances. Here, symmetrical energy-dependence distribution of interface states centering at the midgap in the forbidden gap of InN is assumed. It means that the ΔCs obtained for *n*-type samples are used for *p*-type ones too. Figure 3.20 summarizes $N_{ae}s$ and $N_{de}s$ as a function of [Mg]. In the In-polarity case, N_{ae} increases from 7.0 $\times 10^{17}$ to 1.3×10^{19} cm^{-3} with increasing [Mg] from 1.2×10^{18} to 2.9×10^{19} cm^{-3}. For the over-doped InN at [Mg]$\geq 1.8 \times 10^{20}$ cm^{-3}, the change to *n*-type conduction is accompanied by an increase of N_{de} from 2.2×10^{18} to 2.4×10^{20} cm^{-3} with increasing [Mg] from 1.8×10^{20} to 3.9×10^{21} cm^{-3}. For InN:Mg at [Mg]$<1.2 \times 10^{18}$ cm^{-3}, compensation of residual donors has been observed and N_{de} decreases with increasing [Mg]. Here, one thing should be noticed, that is, the polarity is inverted from In- to N-polarity for the InN:Mg layers with [Mg]$>10^{19}$ cm^{-3} grown under the In-polarity regime as discussed above. Those samples are still indicated together with other In-polar samples in Fig. 3.20 though ΔC for N-polarity is used for calibration. Similar trends of $N_{ae}s$ and $N_{de}s$ as a function of [Mg] have also been observed in the N-polarity case, as shown in Fig. 3.20. Here it should be noticed that there is a relatively large error for the estimated $N_{ae}s$ and $N_{de}s$ due to the complications involved in the ECV measurements. Anyway, we can conclude that the *p*-type conduction is achieved at [Mg]$\sim 10^{18}$–3×10^{19} cm^{-3} while the InN:Mg samples in the other [Mg] regions shown in Fig. 3.19 exhibit *n*-type conduction.

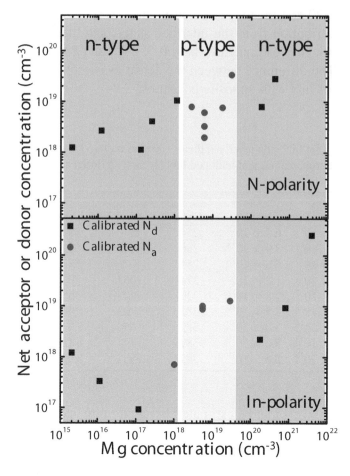

FIGURE 3.20
Conduction regions and calibrated net acceptor/donor concentration (N_{ae}s/ N_{de}s) of InN:Mg layers for In- and N-polarity as a function of [Mg].

3.4.3 Hole mobility in *p*-type InN

After confirming *p*-type conduction of InN, hole mobility (μ_h) is an important parameter. However, it is difficult to determine because *p*-type InN appears to be *n*-type by conventional Hall effect measurements, as discussed above. Meanwhile, ECV analysis cannot give any information about μ_h. Until now, only specialist techniques could be used to determine μ_h. For example, the analysis of ambipolar carrier diffusion in *n*-type InN by using the time-resolved transient grating technique has shown μ_h of 39 and 26 cm^2/Vs near the surface of InN and the interface between InN and the GaN buffer layer, respectively [3.30]. Variable magnetic field Hall effect measurements using magnetic fields of up to 14 T and quantitative mobility spectrum analysis (QMSA) have also been used to detect *p*-type conduction in InN:Mg layers, where two hole-related peaks were observed; one was around 600 cm^2/Vs and the other was broad, extending to less than 130 cm^2/Vs [3.24]. It was also shown that a hole mobility value of 50 cm^2/Vs was obtained from multiple carrier fitting (MCF) in the same work [3.24]. The problem with the QMSA method is that a very high magnetic field is necessary to identify low mobility carriers, because the resolution is limited by the squared product of mobility μ_h and maximum magnetic field B, $(\mu B)^2$, which should be at least comparable to unity or possibly much larger to get unambiguous results. Therefore, μ_h values of less than the 100 cm^2/Vs expected for InN are too low to be determined accurately, because it requires extremely strong magnetic fields. In this section, we will show a simple but effective method to estimate μ_h by analyzing their hole conduction properties, that is, the thickness dependence of total sheet conductivity. Several InN:Mg layers with different thicknesses were prepared under the same growth/doping conditions so that their N_{ae} values were almost the same at around 3–6 × 10^{18} cm^{-3}. Through a linear fit of the sheet conductivities of these layers as a function of layer thickness, the bulk conductivity of the p-type region can be extracted as the slope of the fit line assuming that the hole concentrations in those samples are almost the same as each other. An important point of this method is that the measurement of sheet conductivity is simple but accurate even for layered structure samples with different polarity carriers, because the total sheet conductivity is completely independent of carrier polarity and just a simple sum of that for each layer. Then the contribution of a high density of surface/interface electrons and that of any underlayer as well can be removed easily.

For a sample in which we have to consider parallel carrier-transport conductivity, the total sheet conductivity σ_{st} is expressed simply as a total summation of each sheet conductivity for the ith path/layer [3.31].

$$\sigma_{st} = \sum_i \sigma_i d_i \qquad (3.1)$$

where σ_i and d_i are the conductivity and thickness for the ith layer. In the present case, we take into account two components, that is, the bulk p-type layer and all other regions contributed by surface, interface and GaN underlayer, and equation 3.1 can be simplified as

$$\sigma_{st} = \sigma_b d_b + (\sigma_{ss} + \sigma_{si} + \sigma_{su}) = \sigma_b d_b + \sigma_{s-others} \qquad (3.2)$$

where σ_b and d_b are the conductivity and thickness of the bulk p-type layer, σ_{ss}, σ_{si} and σ_{su} are the sheet conductivities for surface, interface and GaN underlayer, respectively. Therefore, if we plot total sheet conductivity σ_{st} as a function of p-type layer thickness d_b, the effect of the surface electron accumulation layer can be removed and the σ_b can be extracted from the slope of the linear fit line assuming that $\sigma_{s-others}$ are the same among samples with different thicknesses. Then, the hole mobility can be calculated if we know the hole concentration p from the equation $\sigma_b = pq\mu_h$, where q is the elementary charge.

Figure 3.21 shows the dependence of total sheet conductivity of the InN:Mg layers as a function of p-type layer thickness. A data point shown by a solid square at each thickness represents an average of sheet conductivities of 5–6 samples measured over different areas (shown by open circles in Fig. 3.21), and the error bars are the standard deviation among them, which is basically due to the sample inhomogeneity. It is shown that all plotted data in Fig. 3.21 are well aligned along a linear fit line, and $\sigma_b = 8.1 \pm 0.5 \ \Omega^{-1}\text{cm}^{-1}$ can be determined for these p-type InN layers from the slope of the fit line. Through ECV measurement, it was found that the N_{ae} values are 2.0×10^{18}, 3.3×10^{18}, and $6.1 \times 10^{18} \ \text{cm}^{-3}$ for three different thick samples as summarized in Table 3.2.

TABLE 3.2
Net acceptor concentration (N_{ae}), hole concentration (p) and hole mobility (μ_h) for p-type InN layers with bulk conductivity of 8.1 ± 0.5 $\Omega^{-1}\text{cm}^{-1}$.

Sample	Thickness (μm)	N_{ae} (cm^{-3})	p (cm^{-3})	μ_h (cm^2/Vs)
A	0.655	3.3×10^{18}	2.0×10^{18}	25
B	2.29	2.0×10^{18}	1.4×10^{18}	36
C	4.34	6.1×10^{18}	3.0×10^{18}	17

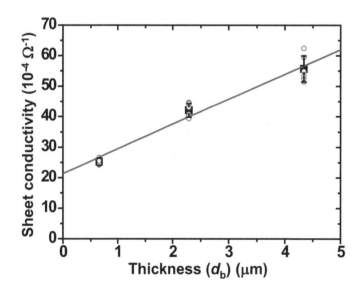

FIGURE 3.21
Sheet conductivity of bulk p-type InN:Mg layers as a function of thickness. The solid rectangle at each thickness shows an average of sheet conductivities of 5–6 samples measured at different areas which are shown by open circles.

The hole concentration p for each sample can be estimated by combining the following equations 3.3 and 3.4 [3.32]

$$p = \frac{N_{ae}}{1 + g\exp[E_a - (E_F - E_v)/k_BT]} \tag{3.3}$$

where the degeneracy factor $g = 1/4$ is used, N_{ae} is the net acceptor concentration, E_F is the Fermi level, E_v is the valence band-edge energy, E_a is the Mg acceptor activation energy, k_B is the Boltzmann's constant, and $T = 300$ K, and

$$p = 4\pi \left(\frac{2m_h^* k_B T}{h^2} \right) \int_0^\infty \frac{\eta'^{1/2} d\eta}{1 + \exp(\eta' - \eta_F')} \qquad (3.4)$$

where m_h^* is the effective mass of holes in InN, h is Planck's constant, $\eta' = (E_v - E)/k_B T$ and $\eta' = (E_v - E_F)/k_B T$.

Then using the reported values of $E_a = 61$ meV and $= 0.42 m_0$ [3.18, 3.25], hole concentrations of 2.0×10^{18}, 1.4×10^{18}, and 3.0×10^{18} cm^{-3} were obtained, as summarized in Table 3.2.

Then we could estimate corresponding μ_h values of 25, 36 and 17 cm^2/Vs, respectively, as also summarized in Table 3.2. It should be noted that although the errors for obtaining conductivity values for the *p*-type region are relatively smaller, those for estimating N_{ae} in InN by ECV are still large. Then we calculated the hole mobility also for averaged N_{ae} of 3.8×10^{18} cm^{-3} among three different thick samples. The average hole concentration was 2.2×10^{18} cm^{-3} and the obtained μ_h was 23 cm^2/Vs. These μ_h values are close to those values of 26–39 cm^2/Vs estimated by other methods [3.30]. This confirms that the obtained hole mobilities are reasonable and that this is a worthwhile approach for studying *p*-type InN through sheet conductivity analysis.

3.5 InGaN epitaxy

InGaN ternary alloys can cover a wide wavelength range from 2 μm (0.63 eV) to 360 nm (3.4 eV), and it is expected that the emission wavelength of the nitride-based light emitters can expand from blue to green, red and even near-infrared regions. To date, In$_x$Ga$_{1-x}$N films with lower In contents (about $x < 0.2$) have been extensively studied, and blue-green InGaN laser diodes (LDs) and light emitting diodes (LEDs) have been successfully fabricated. However, In$_x$Ga$_{1-x}$N films with higher In contents ($x > 0.5$) have been less studied due to the difficulty in epitaxy without any phase separation. In this section, we will describe the epitaxy and properties of In-rich InGaN layers grown by MBE.

3.5.1 Properties of InGaN with different polarities

Generally, InGaN is grown on GaN templates that are deposited by either MOVPE or MBE. In this case, the polarity of the InGaN follows that of the GaN templates, that is, InGaN grown on Ga- and N-polarity GaN are cation- and N-polarity, respectively. Since we mainly discuss InGaN with high In content in this chapter, we will describe the cation-polarity as In-polarity. Similar to InN, the growth temperatures of InGaN films depend on the polarity, where N-polar InGaN can be grown at higher temperatures. The typical growth temperatures examined here are 550°C and 450°C for N- and In-polarity, respectively.

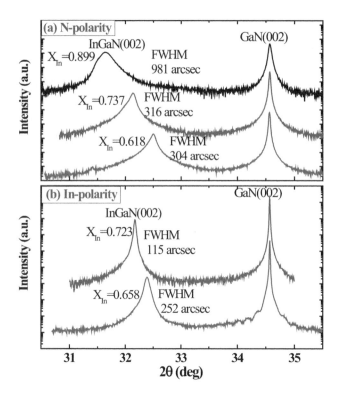

FIGURE 3.22

XRD 2θ-ω scan of In-rich In-GaN films. (a) N-polarity samples grown at 550°C and (b) In-polarity samples grown at 450°C.

The quality of the InGaN layers with different polarities was investigated by XRD, as shown in Fig. 3.22. The In content ranges from 0.6 to 0.9. Single diffraction peaks observed from the InGaN films in both polarities indicate that there is no phase separation in this composition range. Compared to the N-polar films, In-polar InGaN films show better crystalline quality: FWHM values of 2θ-ω scans for In-polar films are about 100–250 arcsec, which are much narrower than those for N-polar films. Figure 3.23 shows the FWHMs of XRD ω-rocking curves for (002) and (102) InGaN with different In contents. It is shown that the crystalline quality of In-rich InGaN films is poorer than that of InN films and the (002) and (102) diffraction peaks become broader with decreasing In content. Moreover, the diffraction peaks for the In-polarity InGaN films are narrower than those of the N-polarity films, indicating better quality for the In-polar InGaN films at present. Figure 3.24 shows surface morphologies of InGaN films with different polarities. In both polarities, the grain size is increased with increasing In content, while this tendency is clearer for the N-polarity. This may be due to the growth temperature of In-polarity InGaN being lower than for the N-polar. The surface roughness is also increased with increasing In content.

3.5.2   Effect of stoichiometry control on In-polar InGaN epitaxy

Since In-polar InGaN shows better quality, we will mainly discuss In-polar InGaN in the following. To obtain InGaN films with a smooth surface and high crystalline

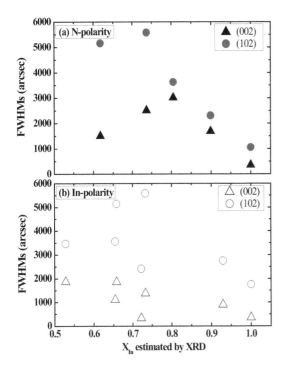

FIGURE 3.23

FWHM values of XRD ω-rocking curves for (002) and (102) diffractions. (a) N-polarity samples and (b) In-polarity samples.

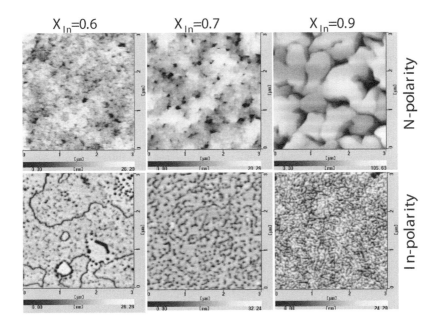

FIGURE 3.24

Surface morphologies of InGaN with different polarities and In contents.

quality, it is very important to control the surface stoichiometry [3.33]. However, it is difficult because once the Ga and/or In droplets form on the surface, they cannot re-evaporate quickly and would remain on the surface. To solve this problem, a special method called "shutter control" is used to control the surface stoichiometry in epitaxy of InGaN films. As an example, $In_{0.7}Ga_{0.3}N$ is used here because it is expected to be a suitable barrier layer for growing InN-based multiple quantum well structures. Briefly, once the In/Ga droplets are formed on the surface, active nitrogen irradiation is performed for several seconds to remove them. In this case, the surface is almost stoichiometric and free of droplets.

Figure 3.25 shows XRD 2θ-ω scans for (002) diffraction for three samples A, B, and C, that is, sample A: $In_{0.7}Ga_{0.3}N$ grown under N-rich conditions, sample B: shutter-controlled stoichiometry and sample C: metal-rich conditions. For sample C, two diffraction peaks are observed at around 32°, showing the appearance of phase separation, which results in a significant deterioration of crystalline quality. On the other hand, in samples A and B, single diffraction peaks from the $In_{0.7}Ga_{0.3}N$ films are observed and no droplets appear on their surfaces. These results suggest that growth under metal-rich conditions is one of the possible causes for phase separation.

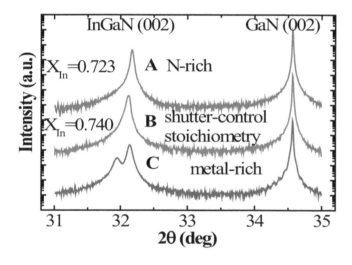

FIGURE 3.25
2θ-ω scans of $In_{0.7}Ga_{0.3}N$ films grown under N-rich conditions, shutter control stoichiometry and metal-rich conditions.

Although there is no marked difference in the 2θ-ω scans between samples A and B, a noticeable difference has been found in the ω-rocking curves for the (002) and (102) diffractions as shown in Fig. 3.26. The FWHMs for (002) and (102) diffractions are 325 and 2235 arcsec for sample B, which are smaller than those for sample A of 606 and 3852 arcsec. Moreover, sample A shows an asymmetric shape for (002) diffraction, which may be caused by the inhomogeneity of In content in the $In_{0.7}Ga_{0.3}N$ film. Anyway, a marked improvement of crystalline quality can be achieved by using the shutter-controlled growth. This quality improvement is due to the effect of enhanced surface migration by the shutter control, which is similar to the case where the MBE-grown GaN and InN show better crystalline and surface qualities under a slightly metal-rich growth condition due to the enhancement of surface migration [3.7, 3.33, 3.34].

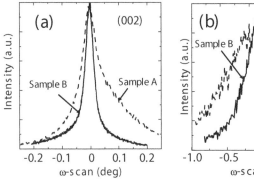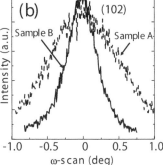

FIGURE 3.26
ω-rocking curves of samples A (dashed line) and B (solid line). (a) (002) diffraction; (b) (102) diffraction.

Growth temperature has a great effect on the crystalline quality of the $In_{0.7}Ga_{0.3}N$. The $In_{0.7}Ga_{0.3}N$ has been grown at 450, 500, 550, and 600°C. Single diffraction peaks in the 2θ-ω scans are observed from the $In_{0.7}Ga_{0.3}N$ films at T_G up to 550°C. FWHMs of the diffraction peaks of the $In_{0.7}Ga_{0.3}N$ layers grown at 500 and 550°C are 111 and 131 arcsec, respectively, showing good quality. However, epitaxial growth does not succeed at 600°C, which shows that $T_G \sim 550$°C is almost the upper limit for the In-polarity $In_{0.7}Ga_{0.3}N$ epitaxy.

Figure 3.27 shows the FWHMs of the ω-rocking curves and electrical properties of the $In_{0.7}Ga_{0.3}N$ films, grown by using the shutter control method, as a function of growth temperature. The FWHM is markedly improved with increasing T_G, particularly for that of the (102) diffraction. The best values are obtained for the sample grown at 550°C: 287 arcsec for (002) diffraction and 1714 arcsec for (102) diffraction. As for electrical properties, the sample grown at 550°C shows the best values with n_e of 2×10^{18} cm^{-3} and μ_e of 364 cm^2/Vs, respectively. In addition, the surfaces of the $In_{0.7}Ga_{0.3}N$ layers grown at 500 and 550°C exhibit lower pit densities and larger grain sizes than those grown at 450°C, suggesting that better surface morphology is obtained at higher T_G and the epitaxy at around the upper limit temperature is beneficial. This phenomenon is similar to the result of the temperature dependence of InN epitaxy [3.16].

Anyway, the precise control of V/III ratio is quite important for improving crystalline quality and this control can be realized by using the shutter control method. The growth temperature is an important factor to improve the crystalline quality of the $In_{0.7}Ga_{0.3}N$, where the samples grown at 550°C show the best quality.

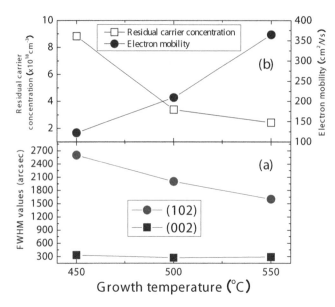

FIGURE 3.27

(a) FWHMs of ω-rocking curves for (002) and (102) diffraction and (b) electrical properties of In-GaN as a function of growth temperature.

3.5.3 Band gap energy of In-polarity InGaN alloys

Single emission peaks are usually observed from PL spectra of the In-polar InGaN layers. Figure 3.28 shows PL peak energy and the band gap energy as a function of In composition, where the measuring temperature is about 14 K. A bowing parameter of about 1.0 ± 0.2 eV can be obtained by fitting to these results. This value is smaller than the previously reported values of 1.4 and 2.5 eV [3.35]. Then the band gap of the In-polar InGaN at 14 K can be written as

$$E_g(x) = 0.675x + 3.50(1-x) - (1.0 \pm 0.2)x(1-x) \tag{3.5}$$

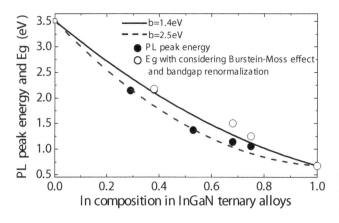

FIGURE 3.28

PL peak energy of InGaN as a function of In composition.

3.6 InAlN epitaxy

It is more difficult to grow InAlN alloys without phase separation than to grow InGaN alloys because there are very large differences in physical and chemical parameters between AlN and InN in comparison to those between GaN and InN, that is, larger differences in the lattice constants, bonding energies, and optimum growth temperatures between AlN and InN compared with those of GaN and InN [3.36, 3.37].

Similar to InGaN, the polarity of InAlN follows that of the GaN template so that it is controlled by choosing the GaN templates with different polarities. Due to the fact that the suitable growth temperature of AlN is much higher than that of InN, InAlN is grown under N-polarity at present since the maximum T_G for N-polar InN is about 100°C higher than for In-polar InN and it might be better for InAlN growth from the point of view of achieving higher T_G. In this case, a 120-nm-thick N-polarity GaN layer is grown by MBE and acts as a template for InAlN epitaxy. In this section, the effect of T_G on InAlN epitaxy and properties of InAlN will be shown.

3.6.1 Effect of growth temperature on N-polar InAlN epitaxy

Figure 3.29(a) shows the 2θ-ω XRD patterns around the (002)-plane of the InAlN ternary alloys grown at T_G values of 500, 550, 580 and 600°C. The Al composition in the InAlN alloys grown at 500 and 550°C shows the same value of 0.56. On the other hand, the Al composition tends to be smaller with further increases of T_G, that is, 0.51 at 580°C and 0.48 at 600°C. In addition, the sample grown at 600°C gives a broad diffraction peak, implying the occurrence of phase separation. This tendency is opposite to the expectation that the Al composition may shift to the Al-rich side if we consider stronger chemical bonding in AlN than in InN and the composition pulling effect by the GaN template, where the in-plane lattice constant of the InAlN epilayer approaches that of the underlying GaN layer.

In order to understand how the T_G affects the crystalline structure of the InAlN ternary alloys, the samples are characterized by the XRD reciprocal space mapping around (104) plane. Fig. 3.29(b) shows their half maximum contours and corresponding peak positions for the samples grown at 550, 580 and 600°C. The dashed line indicates the theoretical diffraction peak position for the fully relaxed InAlN alloy, that is, the line connecting the (104) diffraction peaks for the bulk AlN and InN. As shown in Fig. 3.29(b), the diffraction peak position and the shape for the sample grown at 550°C look normal, where the peak is located almost at the dashed line and the shape is symmetric around the peak. However, those for the samples grown at 580 and 600°C look irregular, that is, the peaks are apart from the dashed line and the shapes are quite asymmetrically elongated. This is probably due to the effect of enhanced phase separation in the ternary alloys grown at higher tempera-

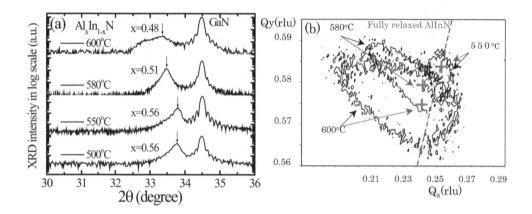

FIGURE 3.29

(a) 2θ-ω XRD patterns around (002)-plane of InAlN ternary alloys grown at different temperatures. (b) XRD reciprocal space mapping around (104)-plane for the samples grown at 550, 580 and 600°C. Half maximum contours and corresponding peak positions are indicated for each sample. The dashed line indicates the theoretical diffraction peak position for the fully relaxed InAlN alloy.

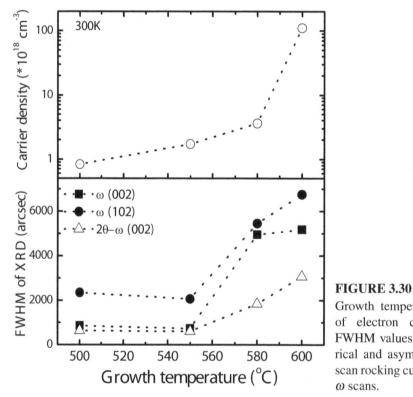

FIGURE 3.30

Growth temperature dependence of electron concentration and FWHM values for their symmetrical and asymmetrical XRD ω-scan rocking curves as well as 2θ-ω scans.

tures arising from the large immiscibility between InN and AlN.

The crystalline quality of the InAlN ternary alloys deteriorates at growth temperatures of 580 and 600°C, as shown in Fig. 3.30. Correspondingly, the electron concentration increases from 10^{18} to 10^{20} cm^{-3}. In general, the growth temperature of the InAlN ternary alloy is limited by that of InN and the composition of the ternary alloy tends to the InN-rich side with increasing growth temperature.

3.6.2 Structural properties of N-polar InAlN alloys

Figure 3.31 shows the Al-composition dependence of XRD 2θ-ω scans around the (002)-plane of In$_{1-x}$Al$_x$N ternary alloys grown at 550°C. It is shown that a single diffraction peak is observed in each In$_{1-x}$Al$_x$N alloy and the Al composition is monotonically shifted from the InN-rich to the AlN-rich side. The In$_{1-x}$Al$_x$N alloys are grown without apparent phase separation in the whole Al composition range. Figure 3.32 shows the Al composition dependence of the electron concentration and the FWHM values for several XRD characterizations for the In$_{1-x}$Al$_x$N alloys. The crystallinity tends to be poorer with increasing x up to $\sim$0.3, probably because of alloying effects between those materials having a large immiscibility gap, that is, InN and AlN. With increasing x above 0.3, however, the crystallinity tends to be improved, probably due to reduced lattice mismatch with the GaN template. The best crystalline In$_{1-x}$Al$_x$N is obtained at $x = 0.77$, which is close to the lattice matching composition ($x = 0.82$) to the GaN template. Further, as shown in Fig. 3.32, there is a strong correlation between the electron concentration and the crystallinity of the In$_{1-x}$Al$_x$N, that is, the carrier concentration tends to increase in poorer quality samples.

Surface morphologies of In$_{1-x}$Al$_x$N are shown in Fig. 3.33. With increasing Al composition, the 2D growth (InN) changes to 3D growth. When the Al composition is close to 100%, the growth tends to be 2D again. Correspondingly, the surface rms roughness becomes larger with increasing Al composition up to around 50% and then is decreased with further increasing Al composition.

3.6.3 Band gap energy of N-polar InAlN alloys

Figure 3.34 summarizes the optical band gap, PL peak energy of N-polar In$_{1-x}$Al$_x$N as a function of Al composition. The band gap energy of In$_{1-x}$Al$_x$N with $x = 0.87$ is estimated by the optical reflectance spectrum because its band gap is larger than that of the GaN template and its transmittance spectrum cannot be obtained. A bowing parameter of 4.78 ± 0.30 eV is obtained. This value is slightly larger than the reported value of 3.0 eV [3.35]. Then the band gap of N-polar In$_{1-x}$Al$_x$N at room temperature can be written as

$$E_g(x) = 6.14x + 0.64(1-x) - (4.78 \pm 0.30)x(1-x) \tag{3.6}$$

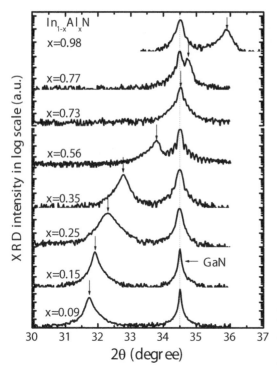

FIGURE 3.31

Al composition dependence of XRD 2θ-ω scans around the (002)-plane of InAlN ternary alloys.

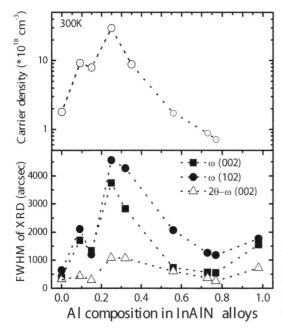

FIGURE 3.32

Al composition dependence of electron concentration at room temperature and FWHM values for their symmetrical and asymmetrical XRD ω-scan rocking curves as well as 2θ-ω scans.

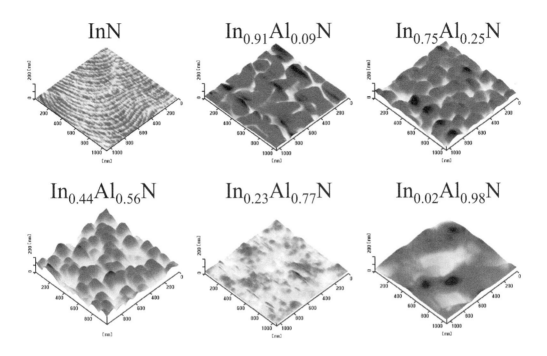

FIGURE 3.33

Surface morphologies from AFM of InAlN with different In compositions.

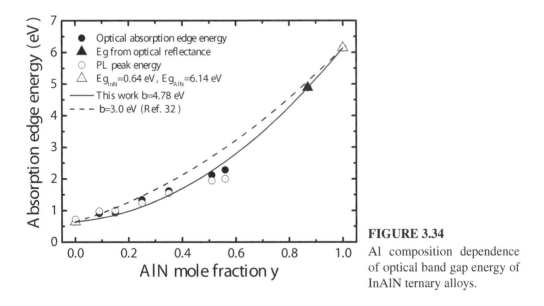

FIGURE 3.34

Al composition dependence of optical band gap energy of InAlN ternary alloys.

3.7   Conclusions

The MBE growth of In- and N-polarity InN has been reviewed. The maximum epitaxy temperature was found to be 100°C higher for N-polar growth than for In-polar growth. The determination of the polarity of InN was described, using CAICISS, CBED and also AFM of KOH- or NaOH-etched layers. *In situ* spectroscopic ellipsometry was used to monitor stoichiometry to enable thick and smooth InN layers to be grown. The effects of the growth temperature and the In/N ratio on the surface morphology, XRD peak width and residual electron concentration and mobility were investigated for In- and N-polarity InN. Mg-doping of InN was, according to ECV measurements, successful in achieving p-type conduction for $1.1 \times 10^{18} \leq [\text{Mg}] \leq 2.9 \times 10^{19}$ cm^{-3}. For [Mg] below and above this range, n-type conduction was apparent. For p-type InN, hole mobilities in the range 17–36 cm^2/Vs were determined. From PL of Mg-doped InN samples with $[\text{Mg}] < 10^{18}$ cm^{-3}, the activation energy for Mg acceptors is estimated to be 61 meV. Finally, the influence of the growth conditions and alloy composition on the structural and optical properties of InGaN and InAlN alloys was also explored.

References

[3.1] X. Wang and A. Yoshikawa, Molecular beam epitaxy growth of GaN, AlN and InN, *Progress in Crystal Growth and Characterization of Materials*, 48/49 (2004) 42–103.

[3.2] D. Muto, T. Araki, H. Naoi, F. Matsuda, and Y. Nanishi, Polarity determination of InN by wet etching, *physica status solidi (a)*, 202 (2005) 773–776.

[3.3] A. Yoshikawa and K. Xu, Polarity selection process and polarity manipulation of GaN in MOVPE and RF-MBE growth, *Thin Solid Films*, 412 (2002) 38–43.

[3.4] X. Wang, S. B. Che, Y. Ishitani, A. Yoshikawa, H. Sasaki, T. Shinagawa, and S. Yoshida, Polarity inversion in high Mg-doped In-polar InN epitaxial layers, *Applied Physics Letters*, 91 (2007) 081912:1–3.

[3.5] M. Yoshitani, K. Akasaka, X. Wang, S. B. Che, Y. Ishitani, and A. Yoshikawa, *In-situ* spectroscopic ellipsometry in plasma-assisted molecular beam epitaxy of InN under different surface stoichiometries, *Journal of Applied Physics*, 99 (2006) 044913:1–6.

[3.6] X. Wang, S. B. Che, Y. Ishitani, and A. Yoshikawa, Effect of epitaxial temperature on N-polar InN films grown by molecular beam epitaxy, *Journal of*

Applied Physics, 99 (2006) 073512:1–5.

[3.7] X. Wang, S. B. Che, Y. Ishitani, and A. Yoshikawa, Step-flow growth of In-polar InN by molecular beam epitaxy, *Japanese Journal of Applied Physics*, 45 (2006) L730–L733.

[3.8] X. Wang, S. B. Che, Y. Ishitani, and A. Yoshikawa, Threading dislocations in In-polar InN films and their effects on surface morphology and electrical properties, *Applied Physics Letters*, 90 (2007) 151901:1–3.

[3.9] B. Heying, E. J. Tarsa, C. R. Elsass, P. Fini, S. P. DenBaars, and J. S. Speck, Dislocation mediated surface morphology of GaN, *Journal of Applied Physics*, 85 (1999) 6470–6476.

[3.10] T. Ide, M. Shimizu, J. Kuo, K. Jeganathan, X. Q. Shen, and H. Okumura, Surface morphology of GaN layer grown by plasma-assisted molecular beam epitaxy on MOCVD-grown GaN template, *physica status solidi (c)*, 0 (2003) 2549–2552.

[3.11] M. Stutzmann, O. Ambacher, M. Eickhoff, U. Karrer, A. Lima Pimenta, R. Neuberger, J. Schalwig, R. Dimitrov, P. J. Schuck, and R. D. Grober, Playing with polarity, *physica status solidi (b)*, 228 (2001) 505–512.

[3.12] K. Xu and A. Yoshikawa, Effect of film polarities on InN grown by molecular beam epitaxy, *Applied Physics Letters*, 83 (2003) 251–253.

[3.13] K. Xu, W. Terashima, T. Hata, N. Hashimoto, Y. Ishitani, and A. Yoshikawa, Step-flow growth of InN on N-polarity GaN template by molecular beam epitaxy with a growth rate of 1.3 μm/h, *physica status solidi (c)*, 0 (2002) 377–381.

[3.14] R. M. Feenstra, Y. Dong, C. D. Lee, and J. E. Northrup, Recent developments in surface studies of GaN and AlN, *Journal of Vacuum Science and Technology B*, 23 (2005) 1174–1180.

[3.15] B. Heying, R. Averbeck, L. F. Chen, E. Haus, H. Riechert, and J. S. Speck, Control of surface morphology using plasma assisted molecular beam epitaxy, *Journal of Applied Physics*, 88 (2000) 1855–1860.

[3.16] T. Ive, O. Brandt, M. Ramsteiner, M. Giehler, H. Kostial, and K. H. Ploog, Properties of InN layers grown on 6H-SiC(0001) by plasma-assisted molecular beam epitaxy, *Applied Physics Letters*, 84 (2004) 1671–1673.

[3.17] I. Mahboob, T. D. Veal, C. F. McConville, H. Lu, and W. J. Schaff, Intrinsic electron accumulation at clean InN surfaces, *Physical Review Letters*, 92 (2004) 036804:1–3.

[3.18] X. Wang, S. B. Che, Y. Ishitani, and A. Yoshikawa, Growth and properties of Mg-doped In-polar InN films, *Applied Physics Letters*, 90 (2007) 201913:1–3.

[3.19] V. Ramachandran, R. M. Feenstra, W. L. Sarney, L. Salamanca-Riba,

J. E. Northrup, L. T. Romano, and D. W. Greve, Inversion of wurtzite GaN(0001) by exposure to magnesium, *Applied Physics Letters*, 75 (1999) 808–810.

[3.20] D. S. Green, E. Haus, F. Wu, L. Chen, U. K. Mishra, and J. S. Speck, Polarity control during molecular beam epitaxy growth of Mg-doped GaN, *Journal of Vacuum Science and Technology B*, 21 (2003) 1804–1811.

[3.21] S. Pezzagna, P. Vennéguès, N. Grandjean, and J. Massies, Polarity inversion of GaN(0001) by a high Mg doping, *Journal of Crystal Growth*, 269 (2004) 249–256.

[3.22] J. Jasinski, Z. Liliental-Weber, H. Lu, and W. J. Schaff, V-shaped inversion domains in InN grown on c-plane sapphire, *Applied Physics Letters*, 85 (2004) 233–235.

[3.23] R. E. Jones, K. M. Yu, S. X. Li, W. Walukiewicz, J. W. Ager, E. E. Haller, H. Lu, and W. J. Schaff, Evidence for p-type doping of InN, *Physical Review Letters*, 96 (2006) 125505:1–4.

[3.24] P. A. Anderson, C. H. Swartz, D. Carder, R. J. Reeves, S. M. Durbin, S. Chandril, and T. H. Myers, Buried p-type layers in Mg-doped InN, *Applied Physics Letters*, 89 (2006) 184104:1–3.

[3.25] A. A. Klochikhin, V. Yu. Davydov, V. V. Emtsev, A. V. Sakharov, V. A. Kapitonov, B. A. Andreev, H. Lu, and W. J. Schaff, Acceptor states in the photoluminescence spectra of n-InN, *Physical Review B*, 71 (2005) 195207:1–16.

[3.26] B. Arnaudov, T. Paskova, P. P. Paskov, B. Magnusson, E. Valcheva, B. Monemar, H. Lu, W. J. Schaff, H. Amano, and I. Akasaki, Energy position of near-band-edge emission spectra of InN epitaxial layers with different doping levels, *Physical Review B*, 69 (2004) 115216:1–5.

[3.27] R. E. Jones, H. C. M. van Genuchten, S. X. Li, L. Hsu, K. M. Yu, W. Walukiewicz, J. W. Ager III, E. E. Haller, H. Lu, and W. J. Schaff, Electron transport properties of InN, *Material Research Society Symposium Proceedings*, 892 (2006) 0892-FF06-06.

[3.28] X. Wang, S. B. Che, Y. Ishitani, and A. Yoshikawa, Systematic study of p-type doping control of InN with different Mg concentrations with both In and N polarities, *Applied Physics Letters*, 91 (2007) 242111:1–3.

[3.29] J. W. L. Yim, R. E. Jones, K. M. Yu, J. W. Ager III, W. Walukiewicz, W. J. Schaff, and J. Wu, Effects of surface states on electrical characteristics of InN and In$_{1-x}$Ga$_x$N, *Physical Review B*, 76 (2007) 041303(R):1–4.

[3.30] F. Chen, A. N. Cartwright, H. Lu, and W. J. Schaff, Hole transport and carrier lifetime in InN epilayers, *Applied Physics Letters*, 87 (2005) 212104:1–3.

[3.31] B. Arnaudov, T. Paskova, S. Evtimova, E. Valcheva, M. Heuken, and B. Monemar, Multiplayer model for Hall effect data analysis of semiconductor structures with step-changed conductivity, *Physical Review B* 67, 045314 (2003).

[3.32] D. A. Neamen, *An Introduction to Semiconductor Devices* (McGraw-Hill, 2005).

[3.33] J. Wu, W. Walukiewicz, K. M. Yu, J. W. Ager III, E. E. Haller, H. Lu, and W. J. Schaff, Small band gap bowing in $In_{1-x}Ga_xN$ alloys, *Applied Physics Letters*, 80 (2002) 4741–4743.

[3.34] C. Gallinat, G. Koblmüller, J. Brown, S. Bernardis, J. S. Speck, G. Cherns, E. Readinger, H. Shen, and M. Wraback, In-polar InN grown by plasma-assisted molecular beam epitaxy, *Applied Physics Letters*, 89 (2006) 032109:1–3.

[3.35] K. P. O'Donnell, R. W. Martin, C. Trager-Cowan, M. E. White, K. Esona, C. Deatcher, P. G. Middleton, K. Jacobs, W. van der Stricht, C. Merlet, B. Gil, A. Vantomme, and J. F. W. Mosselmans, The dependence of the optical energies on InGaN composition, *Material Science and Engineering B*, 82 (2001) 194–196.

[3.36] T. Matsuoka, Calculation of unstable mixing region in wurtzite $In_{1-x-y}Ga_xAl_yN$, *Applied Physics Letters*, 71 (1997) 105–107.

[3.37] A. Koukitu and H. Seki, Unstable region of solid composition in ternary nitride alloys grown by metalorganic vapor phase epitaxy, *Japanese Journal of Applied Physics*, 35 (1996) L1638–L1640.

4

InN in brief: Conductivity and chemical trends

P. D. C. King, T. D. Veal, and C. F. McConville

Department of Physics, University of Warwick, Coventry, CV4 7AL, United Kingdom

4.1   Introduction

In spite of an intense recent research effort, and dramatic improvements in sample quality, all nominally undoped InN films grown to date exhibit a high unintentional *n*-type conductivity. The different mechanisms contributing to this conductivity must be understood, and ultimately controlled, in order to realize the potential for InN-based electronic and optoelectronic device applications. This chapter presents a brief introduction to the *n*-type conductivity of InN. It is not intended to be a comprehensive review of all research on bulk, surface and interface electronic properties of InN — these will be discussed in significantly more detail in the subsequent chapters of this book. Instead, it aims to serve as a brief discussion of the most important aspects.

In particular, a three-region model is considered, incorporating background, surface and interface-related conduction mechanisms, which yields good agreement with experimental free-electron densities in a wide range of InN samples. Details of the bulk band structure of InN, resulting in the charge neutrality level (CNL) lying above the conduction band minimum, in contrast to the majority of other semiconductor materials, is revealed as the overriding mechanism driving its high unintentional *n*-type conductivity. Finally, as the electronic properties of InN are often considered unusual, it is instructive to consider InN's position relative to other semiconductors. In doing this, the position of the CNL in InN, and consequently its electronic properties, are reconciled within the chemical trends of common-cation and common-anion semiconductors, revealing that InN is not anomalous, but that it is rather extreme.

4.2 *n*-type conductivity in InN

Early growth of InN was generally performed by sputter deposition, resulting in polycrystalline samples with very high carrier densities ($n \sim 10^{20}$ cm^{-3}) [4.1]. The initial growth of high-quality single-crystalline films of InN by techniques such as plasma-assisted molecular beam epitaxy (PAMBE) and metal-organic vapor-phase epitaxy (MOVPE) provided a marked improvement in electrical properties [4.2, 4.3], and subsequent optimization of the growth conditions and buffer layer structures have resulted in samples with electron densities in the low $\sim 10^{17}$ cm^{-3} range and mobilities over 2000 cm^{2}V^{-1}s^{-1}, as measured by the single-field Hall effect [4.4–4.7]. While this represents a significant improvement in electrical properties, the minimum free electron densities obtained are still rather high; the origins of this require further investigation.

Analysis of single-field Hall effect results from a large number of samples grown by PAMBE on GaN and AlN revealed some interesting trends. A marked increase in carrier density was obtained with decreasing film thickness (shown for films grown on GaN buffer layers in Fig. 4.1), along with a corresponding decrease in Hall mobility [4.4, 4.5].

Due to the large lattice mismatch between InN and its buffer layers ($\sim 11\%$ for GaN, $\sim 14\%$ for AlN), the InN/buffer layer interface region is characterized by a high density of threading dislocations (TDs), acting as strain relieving mechanisms, whose density reduces with distance from the interface [4.8–4.10]. Lu *et al.* [4.4] proposed that the decrease in mobility was due to an increase in charged dislocation scattering with reducing film thickness, by analogy with GaN [4.11]. However, in GaN, the dislocations are negatively charged, and so dislocations were not originally attributed as the origin of the changes in carrier density with film thickness.

Lu *et al.* [4.12, 4.13] also considered the thickness dependence of the sheet carrier density from single-field Hall effect measurements. While a linear dependence of the sheet density on film thickness was observed, indicating an approximately constant density in the bulk of the films, an extrapolation to zero film thickness revealed a non-zero 'excess' sheet density. This was attributed to an increase in electron density near the surface and/or interface of the InN film. The presence of surface electron accumulation was confirmed by electrochemical capacitance-voltage profiling [4.13] and high-resolution electron energy loss spectroscopy [4.14], and subsequently by a number of other techniques as discussed in Chapter 12.

4.2.1 Three-region model

The work discussed above clearly demonstrates that the carrier concentration cannot be uniform with depth throughout the InN layer; consequently, the volume density determined from single-field Hall effect measurements represents a mobility-weighted average of the variations in carrier concentration throughout the InN film.

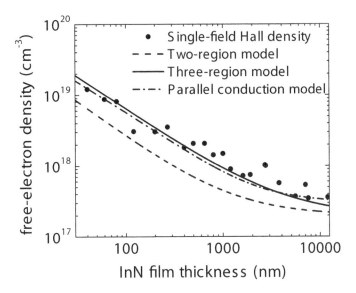

FIGURE 4.1

Free-electron density determined by single-field Hall effect measurements for a number of InN samples of varying thickness grown by PAMBE at Cornell Univeristy [4.4] on sapphire substrates incorporating GaN buffer layers. The calculated electron density assuming a two-region model (dashed line) to account for surface and bulk contributions underestimates the measured densities, reconciled by additionally including interface contributions (a three-region model (solid line), Section 4.2.1). A full parallel conduction analysis of the three-region model (dot-dashed line) supports its validity. Adapted from Refs. [4.15] and [4.16].

Here, we show that both surface and interface effects must be considered, in addition to the bulk of the semiconductor [4.15], and propose a 'three-region' model for the variation of free-electron density in InN, shown in Fig. 4.2. Each region is briefly discussed. An equivalent model has been independently proposed by Cimalla *et al.* [4.17, 4.18], as discussed in Chapter 5.

I: Bulk

The starting point for a model of conductivity in semiconductors must be a uniform background donor density due to defects and impurities throughout the material. For a thick film, a region of the sample sufficiently far from the surface and interface should represent this uniform 'bulk' contribution to the free-electron density. Indeed, for sufficiently thick films, when the surface and interface contributions to the carrier density are averaged over the entire (large) film thickness, the free-electron density measured by the single-field Hall effect should approach the background donor density. Consequently, from Fig. 4.1, the background donor density for these samples grown by PAMBE may be estimated to be in the low 10^{17} cm^{-3} region, in good agreement with the carrier density estimated for the 'bulk' region of a thick sample from multiple-field Hall effect measurements [4.19] (see Chapter 5).

A number of possibilities exist for the microscopic nature of the background

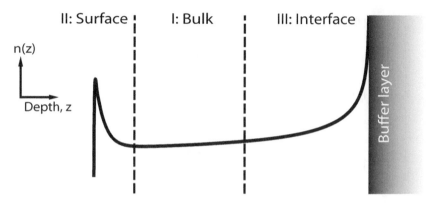

FIGURE 4.2

Schematic representation of the variation in free-carrier concentration, n, as a function of depth throughout the film, indicating I: bulk, II: surface and III: interface contributions to the total free-electron density of InN films.

donors. Theoretical calculations [4.20] indicate that the donor-type nitrogen vacancy has the lowest formation energy of the native defects, not only when the Fermi level lies within the band gap, but also when it lies significantly above the conduction band minimum (CBM). Consequently, native defects will preferentially form as donors. The formation energies for oxygen and silicon impurities acting as donors were shown to be lower than that of the nitrogen vacancy, causing them to also be suggested as a possible cause of the n-type conductivity. Hydrogen has also been theoretically [4.21, 4.22] and experimentally [4.23, 4.24] shown to be a donor even in already n-type InN. Therefore, if present in significant concentrations during the growth of InN, these impurities are also candidates for the background donors. However, secondary ion mass spectroscopy measurements of high-quality InN samples grown by PAMBE have shown that, at least in some cases, these impurities are not present in sufficient concentrations to account for all of the unintentional n-type conductivity [4.25], and its variation with film thickness, and so other mechanisms such as nitrogen vacancies and the surface and interface contributions, discussed below, must also be considered. Defects and doping in InN are discussed in detail in Chapters 10 (experiment) and 11 (theory).

II: Surface

Electron accumulation at the surface of InN results from a re-arrangement of the carriers in the near-surface region in order to screen the electric field of a number of unoccupied, and hence positively charged, donor surface states, causing the conduction and valence bands to bend downward relative to the Fermi level. Due to the large amount of downward band bending present, electron densities as high as $\sim 10^{20}$ cm^{-3} occur in the peak of the accumulation layer [4.14, 4.26]. A pronounced universality of the electron accumulation has been observed [4.26], indicating that it must always be included when considering conductivity in InN. However, the increase in electron density at the surface occurs over length scales determined by the Thomas-Fermi screening length, typically $\lesssim 10$ nm for InN samples. Therefore, while the surface electronic properties of InN are crucial, for example, in understanding the Ohmic nature of metal–InN contacts, and for 'masking' the bulk properties of *p*-type InN, the contribution of the surface free-electron density to the total (averaged) conductivity of the InN films will have only limited effect, except for very thin samples. This is confirmed by the thickness dependence of the single-field Hall effect results which are not well reproduced considering only background and surface contributions [4.15, 4.17] (Fig. 4.1, dashed line), even when the surface electrons are assumed to have a mobility as high as those in the bulk, overestimating their influence on the measured Hall density. The interface must therefore also be considered and is discussed below. The surface electronic properties of InN are discussed in detail in Chapters 12 (experiment) and 13 (theory).

III: Interface

Given the propensity for native defects and impurities to be donor-like in InN, as discussed for the bulk region, it seems highly probable that, in contrast to GaN, defects along, or impurities localized at, TDs will be donors in InN, as was previously observed for the similar metamorphic InAs/GaAs and InAs/GaP interfaces [4.27, 4.28]. Lebedev *et al.* [4.10] have shown the dislocation density to decrease exponentially with distance from the interface. Neglecting variations in mobility between the different regions, the volume free-electron density measured by the single-field Hall effect is therefore given, in the three-region model, by

$$n_{\text{Hall}} = n_{\text{back}} + \frac{N_{ss}}{d} + \frac{C}{d} \int_0^d D(x)\,\mathrm{d}x \qquad (4.1)$$

where n_{back} is the background ('bulk') volume density, N_{ss} is the sheet density of the surface electron accumulation layer, $D(x)$ is the density of dislocations (assumed to decay exponentially from the interface), C is a constant for the charge contribution per unit length of each dislocation and d is the film thickness. Using this model, the variation of free-electron density with thickness is well reproduced for InN grown on both GaN [4.15] (Fig. 4.1, solid line) and AlN [4.17, 4.18] buffer layers.

While this model is attractive in its simplicity, it does not consider the effects of varying mobility of the carriers at the surface and close to the interface, in comparison to the 'bulk' electrons. A more complete 'parallel conduction' analysis for the carrier density profile shown in Fig. 4.2, incorporating variations in mobility due to ionized impurity scattering, charged dislocation scattering and the much lower mobility of surface electrons compared to bulk electrons, is also able to reproduce the thickness dependence of the free-electron density measured by the single-field Hall effect [4.16] (Fig. 4.1, dot-dashed line), using very similar parameters as for the constant mobility three-region model presented above. Additionally, the variation of Hall mobility with film thickness is also reproduced by the parallel conduction model [4.16], supporting the importance of not only the bulk and surface, but also interface contributions to the conductivity of InN. The importance of the interface-related electron density was further confirmed by Ishitani *et al.* [4.29] who estimated an electron sheet density of $\sim 10^{13}$ cm^{-3} associated with the interface, from detailed analysis of infrared reflectivity measurements.

The 'excess' sheet density, obtained by extrapolating the single-field Hall effect sheet density of various samples to zero thickness, appears slightly larger for InN grown on AlN than on GaN buffer layers [4.12, 4.13], and for In- rather than N-polar InN [4.6, 4.7]. Additionally, Fehlberg *et al.* [4.30] observed an increase in the low-mobility peak of multiple-field Hall effect measurements with decreasing In-flux during growth. Some of these changes were originally attributed to changes in the degree of surface electron accumulation. However, an invariance of the degree of surface electron accumulation on buffer layer, polarity and growth conditions has been confirmed [4.26, 4.31, 4.32], indicating that the changes must be due to the interface rather than the surface. Due to the lower lattice mismatch, a smaller density of TDs would be expected for samples grown on GaN rather than on AlN buffer layers. Additionally, from x-ray diffraction studies, the on-axis (0002) x-ray rocking curve width was found to be slightly lower for N-polar samples than for In-polar samples, and to increase with decreasing In-flux during growth [4.7], indicating a slightly lower density of screw and mixed TDs in N-polar than In-polar samples, and for samples grown under In-rich rather than N-rich conditions. In all cases, an increase in the interface contribution can therefore be associated with an increase in the density of TDs. Further comparisons of TD density with carrier concentration in InN samples support the conclusion that TDs originating at the interface are a very important factor contributing to the high unintentional conductivity in MBE-grown InN [4.33], and extend these conclusions to samples grown by MOVPE [4.34]. The interfacial electronic properties of InN are also discussed in Chapter 5.

4.2.2   Overriding origin of the *n*-type conductivity

The position of the band extrema relative to the charge neutrality level (CNL) [4.35], also referred to as the branch point energy [4.36] or Fermi level stabilization energy [4.37], has proved very successful in explaining the surface, bulk and interface electronic properties of a wide variety of semiconductor materials [4.35–4.43]. The CNL has been experimentally [4.44] and theoretically [4.43–4.45] located approximately 1.8 eV above the valence band maximum (VBM) in InN, and consequently over 1 eV above the conduction band minimum (CBM), as shown in Fig. 4.3. This is in contrast to the majority of semiconductors, where the CNL lies within the direct band gap of the semiconductor [4.43]. This can be understood as the CNL is located close to the mid-gap energy averaged across the entire Brillouin zone; due to the particularly low Γ-point CBM of InN compared to the average conduction band edge across the rest of the Brillouin zone, shown in Fig. 4.3, the average mid-gap energy, and hence the CNL, lies well above the CBM. As the Γ-point CBM lies so low in InN, its electron affinity, defined as the CBM to vacuum level separation, is therefore one of the highest values known among the semiconductor materials.

This provides an overriding bulk band structure origin of the high *n*-type conductivity in each region of the three-region model for InN. Within the amphoteric defect model [4.37] (discussed in detail in Chapter 10), native defects preferentially form as donors if the Fermi level is below the CNL, whereas for the Fermi level above the CNL, acceptor native defects have the lower formation energy. As the CNL is so far above the CBM, donor-type native defects, such as nitrogen vacancies, will almost always be the most favorable defects to form in InN, supporting the results of the first principles calculations discussed above. This is also consistent with a stabilization of the Fermi level well into the conduction band following high energy particle irradiation [4.46]. In addition, extrinsic donors such as oxygen will not be heavily compensated by acceptor native defects until the Fermi level becomes extremely high in InN, allowing these to effectively dope this material. The donor/acceptor transition energy for hydrogen in semiconductors is also given by the CNL [4.43]. Consequently, both native defects and impurities, such as hydrogen and oxygen, will act to increase the Fermi level position in InN, contributing to its high *n*-type conductivity. These considerations also apply to impurities localized at, or native defects decorating, dislocations in the InN/buffer layer interface region. Finally, given its high position relative to the band edges, the Fermi level must pin close to, but slightly below, the CNL at the surface, resulting in the pronounced downward band bending and electron accumulation observed at the surface of InN. In summary, the electronic behavior in each region contributing to the high unintentional *n*-type conductivity in InN results from the high location of the CNL relative to the band edges in InN, which is itself a property of its bulk band structure.

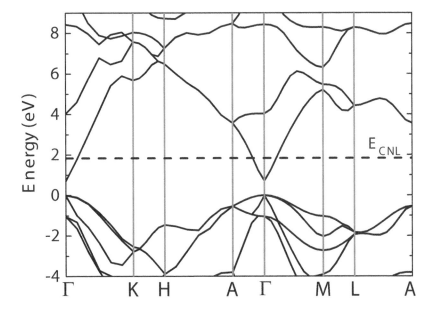

FIGURE 4.3

Calculated bulk band structure (see Chapter 8) of InN, revealing its particularly low Γ-point conduction band minimum which lies below the measured charge neutrality level position, E_{CNL}. Figure adapted from Ref. [4.44].

4.3   InN within chemical trends

It is instructive to consider whether the position of the CNL high above the CBM in InN, which gives rise to its extreme properties such as high unintentional *n*-type conductivity and surface electron accumulation, can itself be understood. First, the evolution of the valence band edge relative to the CNL in the III-N materials is considered. This is extended to a consideration of the band extrema in the common-anion (III-N) and common-cation (In-V) materials, locating InN within chemical trends [4.44].

The III-N valence band offsets (VBOs) are important not only for developing a fundamental understanding of InN and its alloys, but also for numerous technological applications involving III-N-based heterostructures. Consequently, many investigations of these quantities have been performed in recent years, both experimentally [4.47–4.55] and theoretically [4.36, 4.43, 4.45, 4.56, 4.57], although a large range of results have been obtained. This is reviewed in detail in Ref. [4.55], but can at least partly be ascribed to the use of In 4*d* and Ga 3*d* levels in the determination of the VBO by photoemission spectroscopy in a number of previous investigations. These levels, however, are very shallow in the nitrides, and so hybridize with the valence band structure, making them inappropriate for use in the VBO measurement. Recent investigations making use of alternative core-level combinations have yielded values of 0.58 ± 0.08 eV and 1.52 ± 0.17 eV for the InN/GaN and InN/AlN VBOs, respectively [4.50, 4.55], as shown in Fig. 4.4(a), in good agreement with both theoretical predictions and the relative locations of the CNL [4.44, 4.58]. From the transitivity rule, this gives the GaN/AlN VBO as ~ 0.9 eV, in agreement with previous results [4.59].

In a simple tight-binding model, the valence band edge energy is given by the bonding state of the anion and cation *p*-orbitals. Consequently, as the cation valence *p*-orbitals have very similar energies for Al, Ga and In [4.60], the VBM might be expected to be located at very similar energies in InN, GaN and AlN. However, both Ga and In have occupied shallow *d*-orbitals which, neglecting crystal-field and spin-orbit splitting, have the same symmetry representation at the Γ-point (Γ_{15}) as the N 2*p* state [4.61]. These can therefore hybridize, creating a *p–d* repulsion which pushes the VBM to higher energies. The In 4*d* levels in InN are shallower than the Ga 3*d* levels in GaN, as evident from x-ray photoemission spectra shown in Fig. 4.4(b), and so the *p–d* repulsion is stronger for InN than GaN, resulting in InN having the highest VBM position among the III-Ns. In contrast, Al has no occupied *d*-levels (Fig. 4.4(b)), and so there is no *p–d* repulsion pushing the VBM to higher energies. Therefore, the VBO is larger between AlN and (In/Ga)N than between InN and GaN.

The variation in CBM of the III-Ns can also be considered within a tight-binding model, where the conduction band edge results from the anti-bonding state of the cation *s*-orbitals and N 2*s*-orbital, shown in Fig. 4.5(b). The cation *s*-orbital energy

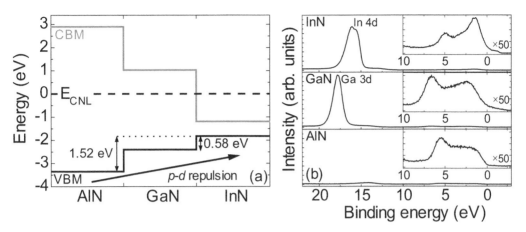

FIGURE 4.4

(a) Band lineup of the III-N semiconductors relative to the CNL determined from experimental measurements of the InN/(Ga,Al)N valence band offsets [4.50, 4.55] and the CNL position in InN [4.44] and GaN [4.58]. (b) X-ray photoemission spectra of the valence band region in the III-N materials, relative to the VBM, with the low binding energy regions shown magnified (×50) in the inset. The spectra for InN and GaN exhibit pronounced peaks due to occupied shallow cation *d*-levels, not present in AlN. A small peak due to N 2*s*–like states is seen in the AlN spectrum. These hybridize with the *d*-levels forming shoulders above and below the In 4*d* and Ga 3*d* peaks in the InN and GaN spectra, respectively.

reduces on moving from Al to Ga, but then increases on moving to In. However, the cation-anion bond length increases, and hence the strength of the *s*–*s* repulsion, which pushes the CBM to higher energies, decreases, with increasing cation atomic number. Consequently, the CBM energy does not follow the energetic ordering of the cation-*s* orbital energies; instead, the reduction in *s*–*s* repulsion causes a marked reduction in conduction band edge on moving from AlN through to InN.

Similar considerations hold for the band lineup of the common-cation semiconductors, shown in Fig. 4.5(a). A reduction of the valence band edge with decreasing anion atomic number follows the lowering anion *p*-orbital energy upon moving from Sb through to N (Fig. 4.5(b)). In addition, the spin-orbit splitting, which pushes the VBM to higher energies, also reduces with decreasing anion atomic number. From the chemical trends of the anion *s*-orbitals alone (Fig. 4.5(b)), the conduction band edge would be highest for InSb, followed by InP, InAs and finally InN. However, the reduction in cation-anion bond length on moving from InSb through to InN, acting to increase the *s*–*s* repulsion, coupled with the changing repulsion strength with In 5*s* and anion *s*-orbital energy separation, results in the conduction band edge variation observed for the common-cation compounds.

The narrow band gap of InN therefore results from the presence of shallow *d*-levels in In, leading to a relatively high-lying VBM, and the low energy of the N 2*s*-orbital coupled with the relatively large In–N bond length, a result of the large size and electronegativity mismatch between In and N, giving rise to a low-lying CBM. These factors explain why InN deviates from the so-called common-cation

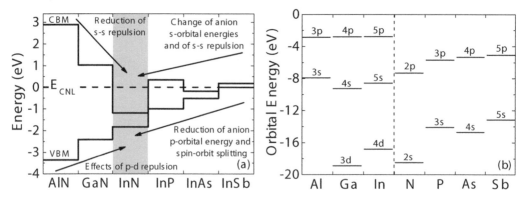

FIGURE 4.5

(a) Band extrema relative to the CNL in InN, common-cation and common-anion semiconductors, determined from experimental measurements of the CNL position [4.44, 4.58, 4.62] and valence band offsets [4.50, 4.55]. (b) Atomic orbital energies [4.60] for the constituent cations and anions, used to understand the properties of InN within chemical trends. Figure adapted from Ref. [4.44].

rule [4.60], within which the direct band gap of semiconductors generally increases with decreasing cation atomic number. Throughout the above, the CNL provides a universal energy reference. The particularly low energetic position of the CBM in InN on this energy scale results in its CBM lying below the CNL simply as a result of the chemical trends discussed here, explaining the extreme nature of InN's electronic properties.

In fact, InN is actually rather similar to InAs, where the CNL is also located above the CBM (Fig. 4.5(a)). This material also exhibits electron accumulation at the surface [4.63], and an increase in electron density associated with dislocations at the interface [4.27, 4.28]. However, due to the smaller separation between the CBM and the CNL in InAs than in InN, these effects are much less extreme than for InN. In contrast, in GaN for example, the CNL is located significantly below the CBM (Fig. 4.5(a)), and so the surface Fermi level will tend to be pinned below the bulk Fermi level for *n*-type material, leading to electron depletion at the surface [4.64], and acceptors associated with dislocations at the GaN/buffer layer interface [4.11]. However, as the CNL still lies above mid-gap in GaN, *n*-type conductivity remains more favorable than *p*-type conductivity in this material, whereas for AlN, where the CNL lies a long way from both the CBM and VBM (Fig. 4.5(a)), *n*- and *p*-type doping are both difficult [4.42]. Thus, while InN's fundamental electronic properties are certainly extreme, they are not unusual, being determined by the same overriding mechanisms as in other semiconductors.

4.4 Conclusions

A model for the high unintentional *n*-type conductivity in InN has been presented, including contributions from donor-type background impurities and native defects, carriers in a surface electron accumulation layer and native defects or impurities localized at threading dislocations resulting from the large lattice mismatch between InN and its buffer layers. This model reproduces the film-thickness dependence of the free-electron density determined from single-field Hall effect measurements. The *n*-type prevalence of each of these conduction mechanisms was understood in terms of the charge neutrality level lying well above the conduction band minimum in InN. This high position relative to the band extrema was itself explained within the chemical trends of common-cation and common-anion semiconductors, and so the fundamental properties of InN cannot be seen as anomalous; rather, they are governed by the same mechanisms as in other semiconductors. However, the large size and electronegativity mismatch between In and N ensure a rather extreme limit of these mechanisms, giving rise to many striking electronic properties in InN. These will be discussed in more detail in the subsequent chapters of this book.

Acknowledgments

We would like to thank L. F. J. Piper, W. J. Schaff, F. Fuchs, F. Bechstedt, and S. M. Durbin for useful discussions. We also thank H. Lu, W. J. Schaff, C. E. Kendrick and S. M. Durbin for providing the samples used in these investigations, and F. Fuchs, F. Bechstedt, and J. Furthmüller for providing the results of their theoretical calculations. This work was supported by the Engineering and Physical Sciences Research Council, UK, under Grant No. EP/C535553/1.

References

[4.1] A. G. Bhuiyan, A. Hashimoto, and A. Yamamoto, Indium nitride (InN): A review on growth, characterization, and properties, *Journal of Applied Physics*, 94 (2003) 2798–2808.

[4.2] H. Lu, W. J. Schaff, J. Hwang, H. Wu, W. Yeo, A. Pharkya, and L. F. Eastman, Improvement on epitaxial grown of InN by migration enhanced epitaxy, *Applied Physics Letters*, 77 (2000) 2548–2550.

[4.3] A. Yamamoto, T. Tanaka, K. Koide, and A. Hashimoto, Improved electrical properties for metalorganic vapour phase epitaxial InN films, *physica status solidi (a)*, 194 (2002) 510–514.

[4.4] H. Lu, W. J. Schaff, L. F. Eastman, J. Wu, W. Walukiewicz, D. C. Look, and R. J. Molnar, Growth of thick InN by molecular beam epitaxy, *Materials Research Society Symposium Proceedings*, 743 (2003) L4.10.1–L4.10.6.

[4.5] W. J. Schaff, H. Lu, L. F. Eastman, W. Walukiewicz, K. M. Yu, S. Keller, S. Kurtz, B. Keyes, and L. Gevilas, Electrical properties of InN grown by molecular beam epitaxy, in V. H. Ng and A. G. Baca, editors, *Fall 2004 ECS meeting proceedings, v. 2004–06 (ISBN 1-56677-419-5) State-of-the-art program on compound aemiconductors XLI-and-nitride and wide bandgap semiconductors for sensors, photonics, and electronics.*

[4.6] C. S. Gallinat, G. Koblmüller, J. S. Brown, S. Bernardis, J. S. Speck, G. D. Chern, E. D. Readinger, H. Shen, and M. Wraback, In-polar InN grown by plasma-assisted molecular beam epitaxy, *Applied Physics Letters*, 89 (2006) 032109:1–3.

[4.7] G. Koblmüller, C. S. Gallinat, S. Bernardis, J. S. Speck, G. D. Chern, E. D. Readinger, H. Shen, and M. Wraback, Optimization of the surface and structural quality of N-face InN grown by molecular beam epitaxy, *Applied Physics Letters*, 89 (2006) 071902:1–3.

[4.8] D. C. Look, H. Lu, W. J. Schaff, J. Jasinski, and Z. Liliental-Weber, Donor and acceptor concentrations in degenerate InN, *Applied Physics Letters*, 80 (2002) 258–260.

[4.9] C. J. Lu, L. A. Bendersky, H. Lu, and W. J. Schaff, Threading dislocations in epitaxial InN thin films grown on (0001) sapphire with a GaN buffer layer, *Applied Physics Letters*, 83 (2003) 2817–2819.

[4.10] V. Lebedev, V. Cimalla, J. Pezoldt, M. Himmerlich, S. Krischok, J. A. Schaefer, O. Ambacher, F. M. Morales, J. G. Lozano, and D. González, Effect of dislocations on electrical and electron transport properties of InN thin films. I. Strain relief and formation of a dislocation network, *Journal of Applied Physics*, 100 (2006) 094902:1–13.

[4.11] D. C. Look and J. R. Sizelove, Dislocation scattering in GaN, *Physical Review Letters*, 82 (1999) 1237–1240.

[4.12] H. Lu, W. J. Schaff, L. F. Eastman, and C. Wood, Study of interface properties of InN and InN-based heterostructures by molecular beam epitaxy, *Materials Research Society Symposium Proceedings*, 693 (2002) I1.5.1–I1.5.6.

[4.13] H. Lu, W. J. Schaff, L. F. Eastman, and C. E. Stutz, Surface charge accumulation of InN films grown by molecular-beam epitaxy, *Applied Physics Letters*, 82 (2003) 1736–1738.

[4.14] I. Mahboob, T. D. Veal, C. F. McConville, H. Lu, and W. J. Schaff, Intrinsic electron accumulation at clean InN surfaces, *Physical Review Letters*, 92 (2004) 036804:1–4.

[4.15] L. F. J. Piper, T. D. Veal, C. F. McConville, H. Lu, and W. J. Schaff, Origin of the *n*-type conductivity of InN: The role of positively charged dislocations, *Applied Physics Letters*, 88 (2006) 252109:1–3.

[4.16] P. D. C. King, T. D. Veal, and C. F. McConville, Unintentional conductivity of indium nitride: transport modelling and microscopic origins, *Journal of Physics: Condensed Matter*, Special Issue on Physics of III-V Nitrides, 21 (2009) 174201:1–7.

[4.17] V. Cimalla, V. Lebedev, F. M. Morales, R. Goldhahn, and O. Ambacher, Model for the thickness dependence of electron concentration in InN films, *Applied Physics Letters*, 89 (2006) 172109:1–3.

[4.18] V. Lebedev, V. Cimalla, T. Baumann, O. Ambacher, F. M. Morales, J. G. Lozano, and D. González, Effect of dislocations on electrical and electron transport properties of InN thin films. II. Density and mobility of the carriers, *Journal of Applied Physics*, 100 (2006) 094903:1–8.

[4.19] C. H. Swartz, R. P. Tompkins, N. C. Giles, T. H. Myers, H. Lu, W. J. Schaff, and L. F. Eastman, Investigation of multiple carrier effects in InN epilayers using variable magnetic field Hall measurements, *Journal of Crystal Growth*, 269 (2004) 29–34.

[4.20] C. Stampfl, C. G. Van de Walle, D. Vogel, P. Krüger, and J. Pollmann, Native defects and impurities in InN: First-principles studies using the local-density approximation and self-interaction and relaxation-corrected pseudopotentials, *Physical Review B*, 61 (2000) R7846–R7849.

[4.21] S. Limpijumnong and C. G. Van de Walle, Passivation and doping due to hydrogen in III-nitrides, *physica status solidi (b)*, 228 (2001) 303–307.

[4.22] A. Janotti and C. G. Van de Walle, Sources of unintentional conductivity in InN, *Applied Physics Letters*, 92 (2008) 032104:1–3.

[4.23] M. Losurdo, M. M. Giangregorio, G. Bruno, T.-H. Kim, S. Choi, A. S. Brown, G. Pettinari, M. Capizzi, and A. Polimeni, Behavior of hydrogen in InN investigated in real time exploiting spectroscopic ellipsometry, *Applied Physics Letters*, 91 (2007) 081917:1–3.

[4.24] G. Pettinari, F. Masia, M. Capizzi, A. Polimeni, M. Losurdo,

G. Bruno, T. H. Kim, S. Choi, A. Brown, V. Lebedev, V. Cimalla, and O. Ambacher, Experimental evidence of different hydrogen donors in n-type InN, *Physical Review B*, 77 (2008) 125207:1–6.

[4.25] J. Wu, W. Walukiewicz, S. X. Li, R. Armitage, J. C. Ho, E. R. Weber, E. E. Haller, H. Lu, W. J. Schaff, A. Barcz, and R. Jakiela, Effects of electron concentration on the optical absorption edge of InN, *Applied Physics Letters*, 84 (2004) 2805–2807.

[4.26] P. D. C. King, T. D. Veal, C. F. McConville, F. Fuchs, J. Furthmüller, F. Bechstedt, P. Schley, R. Goldhahn, J. Schörmann, D. J. As, K. Lischka, D. Muto, H. Naoi, Y. Nanishi, H. Lu, and W. J. Schaff, Universality of electron accumulation at wurtzite *c*- and *a*-plane and zincblende InN surfaces, *Applied Physics Letters*, 91 (2007) 092101:1–3.

[4.27] H. Yamaguchi, J. L. Sudijono, B. A. Joyce, T. S. Jones, C. Gatzke, and R. A. Stradling, Thickness-dependent electron accumulation in InAs thin films on GaAs(111)A: A scanning-tunneling-spectroscopy study, *Physical Review B*, 58 (1998) R4129–R4222.

[4.28] V. Gopal, E. P. Kvam, T. P. Chin, and J. M. Woodall, Evidence for misfit dislocation-related carrier accumulation at the InAs/GaP heterointerface, *Applied Physics Letters*, 72 (1998) 2319–2321.

[4.29] Y. Ishitani, X. Wang, S.-B. Che, and A. Yoshikawa, Effect of electron distribution in InN films on infrared reflectance spectrum of longitudinal optical phonon-plasmon interaction region, *Journal of Applied Physics*, 103 (2008) 053515:1–10.

[4.30] T. B. Fehlberg, C. S. Gallinat, G. A. Umana-Membreno, G. Koblmüller, B. D. Nener, J. S. Speck, and G. Parish, Effect of MBE growth conditions on multiple electron transport in InN, *Journal of Electronic Materials*, 37 (2008) 593–596.

[4.31] T. D. Veal, L. F. J. Piper, I. Mahboob, H. Lu, W. J. Schaff, and C. F. McConville, Electron accumulation at InN/AlN and InN/GaN interfaces, *physica status solidi (c)*, 2 (2005) 2246–2249.

[4.32] P. D. C. King, T. D. Veal, C. S. Gallinat, G. Koblmüller, L. R. Bailey, J. S. Speck, and C. F. McConville, Influence of growth conditions and polarity on interface-related electron density in InN, *Journal of Applied Physics* 104 (2008) 103703:1–5.

[4.33] X. Wang, S.-B. Che, Y. Ishitani, and A. Yoshikawa, Threading dislocations in In-polar InN films and their effects on surface morphology and electrical properties, *Applied Physics Letters*, 90 (2007) 151901:1–3.

[4.34] H. Wang, D. S. Jiang, L. L. Wang, X. Sun, W. B. Liu, D. G. Zhao,

J. J. Zhu, Z. S. Liu, Y. T. Wang, S. M. Zhang, and H. Yang, Investigation on the structural origin of n-type conductivity in InN films, *Journal of Physics D: Applied Physics*, 41 (2008) 135403:1–5.

[4.35] J. Tersoff, Schottky barrier heights and the continuum of gap States, *Physical Review Letters*, 52 (1984) 465–468.

[4.36] W. Mönch, Empirical tight-binding calculation of the branch-point energy of the continuum of interface-induced gap states, *Journal of Applied Physics*, 80 (1996) 5076–5082.

[4.37] W. Walukiewicz, Mechanism of Schottky barrier formation: The role of amphoteric native defects, *Journal of Vacuum Science and Technology B*, 5 (1987) 1062–1067.

[4.38] J. Tersoff, Theory of semiconductor heterojunctions: The role of quantum dipoles, *Physical Review B*, 30 (1983) 4874–4877.

[4.39] J. Tersoff, Schottky barriers and semiconductor band structures, *Physical Review B*, 32 (1985) 6968–6971.

[4.40] W. Mönch, Role of virtual gap states and defects in metal-semiconductor contacts, *Physical Review Letters*, 58 (1987) 1260–1263.

[4.41] W. Walukiewicz, Amphoteric native defects in semiconductors, *Applied Physics Letters*, 52 (1989) 2094–2096.

[4.42] W. Walukiewicz, Intrinsic limitations to the doping of wide-gap semiconductors, *Physica B*, 302–303 (2001) 123–134.

[4.43] C. G. Van de Walle and J. Neugebauer, Universal alignment of hydrogen levels in semiconductors, insulators and solutions, *Nature*, 423 (2003) 626–628.

[4.44] P. D. C. King, T. D. Veal, P. H. Jefferson, S. A. Hatfield, L. F. J. Piper, C. F. McConville, F. Fuchs, J. Furthmüller, F. Bechstedt, H. Lu, and W. J. Schaff, Determination of the branch-point energy of InN: Chemical trends in common-cation and common-anion semiconductors, *Physical Review B*, 77 (2008) 045316:1–6.

[4.45] J. Robertson and B. Falabretti, Band offsets of high K gate oxides on III-V semiconductors, *Journal of Applied Physics*, 100 (2006) 014111:1–8.

[4.46] S. X. Li, K. M. Yu, J. Wu, R. E. Jones, W. Walukiewicz, J. W. Ager III, W. Shan, E. E. Haller, H. Lu, and W. J. Schaff, Fermi-level stabilization energy in group III nitrides, *Physical Review B*, 71 (2005) 161201(R):1–4.

[4.47] G. Martin, A. Botchkarev, A. Rockett, and H. Morkoç, Valence-band discontinuities of wurtzite GaN, AlN, and InN heterojunctions measured by x-ray photoemission spectroscopy, *Applied Physics Letters*, 68 (1996) 2541–2543.

[4.48] C. F. Shih, N. C. Chen, P. H. Chang, and K. S. Liu, Band offsets of InN/GaN interface, *Japanese Journal of Applied Physics - Part 1*, 44 (2005) 7892–7895.

[4.49] C.-L. Wu, C.-H. Shen, and S. Gwo, Valence band offset of wurtzite InN/AlN heterojunction determined by photoelectron spectroscopy, *Applied Physics Letters*, 88 (2006) 032105:1–3.

[4.50] P. D. C. King, T. D. Veal, P. H. Jefferson, C. F. McConville, T. Wang, P. J. Parbrook, H. Lu, and W. J. Schaff, Valence band offset of InN/AlN heterojunctions measured by x-ray photoelectron spectroscopy, *Applied Physics Letters*, 90 (2007) 132105:1–3.

[4.51] C.-L. Wu, H.-M. Lee, C.-T. Kuo, S. Gwo, and C.-H. Hsu, Polarization-induced valence-band alignments at cation- and anion-polar InN/GaN heterojunctions, *Applied Physics Letters*, 91 (2007) 042112:1–3.

[4.52] Z. H. Mahmood, A. P. Shah, A. Kadir, M. R. Gokhale, S. Ghosh, A. Bhattacharya, and B. M. Arora, Determination of InN–GaN heterostructure band offsets from internal photoemission measurements, *Applied Physics Letters*, 91 (2007) 152108:1–3.

[4.53] K. A. Wang, C. Lian, N. Su, D. Jena, and J. Timler, Conduction band offset at the InN/GaN heterojunction, *Applied Physics Letters*, 91 (2007) 232117:1–3.

[4.54] C.-L. Wu, H.-M. Lee, C.-T. Kuo, C.-H. Chen, and S. Gwo, Cross-sectional scanning photoelectron microscopy and spectroscopy of wurtzite InN/GaN heterojunction: Measurement of "intrinsic" band lineup, *Applied Physics Letters*, 92 (2008) 162106:1–3.

[4.55] P. D. C. King, T. D. Veal, C. E. Kendrick, L. R. Bailey, S. M. Durbin, and C. F. McConville, InN/GaN valence band offset: High-resolution x-ray photoemission spectroscopy measurements, *Physical Review B*, 78 (2008) 033308:1–4.

[4.56] S.-H. Wei and A. Zunger, Valence band splittings and band offsets of AlN, GaN, and InN, *Applied Physics Letters*, 69 (1996) 2719–2721.

[4.57] C. G. Van de Walle and J. Neugebauer, Small valence-band offsets at GaN/InGaN heterojunctions, *Applied Physics Letters*, 70 (1997) 2577–2579.

[4.58] T. U. Kampen and W. Mönch, Barrier heights of GaN Schottky contacts, *Applied Surface Science*, 117–118 (1997) 388–393.

[4.59] I. Vurgaftman and J. R. Meyer, Band parameters for nitrogen-containing semiconductors, *Journal of Applied Physics*, 94 (2003) 3675–3696.

[4.60] S. H. Wei, X. L. Nie, I. G. Batyrev, and S. B. Zhang, Breakdown of the band-gap-common-cation rule: The origin of the small band gap of InN, *Physical Review B*, 67 (2003) 165209:1–4.

[4.61] S.-H. Wei and A. Zunger, Role of *d* orbitals in valence-band offsets of common-anion semiconductors, *Physical Review Letters*, 59 (1987) 144–147.

[4.62] V. N. Brudnyi, S. N. Grinyaev, and N. G. Kolin, Electronic properties of irradiated semiconductors. A model of the Fermi level pinning, *Semiconductors*, 37 (2003) 537–545.

[4.63] M. Noguchi, K. Hirakawa, and T. Ikoma, Intrinsic electron accumulation layers on reconstructed clean InAs(100) surfaces, *Physical Review Letters*, 66 (1991) 2243–2246.

[4.64] K. M. Tracy, W. J. Mecouch, R. F. Davis, and R. J. Nemanich, Preparation and characterization of atomically clean, stoichiometric surfaces of *n*- and *p*-type GaN(0001), *Journal of Applied Physics*, 94 (2003) 3163–3172.

5

Transport properties of InN

V. Cimalla, V. Lebedev, and O. Ambacher
Fraunhofer Institute for Applied Solid State Physics, Tullastr. 72, 79108 Freiburg, Germany

V. M. Polyakov, F. Schwierz, M. Niebelschütz, and G. Ecke
Institut für Mikro- und Nanotechnologien, Technische Universität Ilmenau, Gustav-Kirchhoff Str. 7, 98693 Ilmenau, Germany

T. H. Myers
Materials Science and Engineering Program, Texas State University, 601 University Drive, San Marcos, TX 78666, United States of America

W. J. Schaff
Department of Electrical and Computer Engineering, Cornell University, Ithaca, New York 14853, United States of America

5.1 Introduction

Group III-nitrides are the most advanced material system for optoelectronic devices operating in the blue and near ultraviolet. InN represents the least studied member of this family, and of importance, has a low band gap energy of about 0.7 eV (see Chapters 6–9 in this book) which expands the possible operational spectral range to the near infrared. Moreover, InN has outstanding electron transport properties such as a high electron mobility [5.1, 5.2], a large overshoot electron velocity [5.3] and the appearance of transient drift velocity oscillations [5.4], which predestine this material for high frequency devices operating up to the THz range. Unfortunately, the majority of the InN epitaxial thin films studied to date exhibit extremely high electron concentrations, which prevents the development of any kind of either electronic or optoelectronic devices.

Up to the year 2000, sputter deposition was the predominant method for growth of InN films. For those early samples, which were polycrystalline in nature, a band gap of ~ 1.9 eV [5.5], high electron mobilities up to 2700 cm^2/Vs and electron concentrations down to 5×10^{16} cm^{-3} were reported [5.6]. However, these values have not been reproduced to date for sputtered films, and more recent systematic investigations with growth by sputtering demonstrate maximum electron mobility and minimum free electron concentrations of about 300 cm^2/Vs and 10^{19} cm^{-3}, respectively.

Recent improvements in both molecular beam epitaxy (MBE) [5.7] and metalorganic chemical vapor deposition (MOCVD) [5.8] of InN thin films yielded single crystalline epitaxial layers demonstrating the fundamental band gap of < 0.7 eV and carrier densities below 10^{18} cm^{-3}. However, despite the obvious progress on the technological side, the origin of the high electron concentration in InN remains under intensive debate. Recently, several *n*-type doping mechanisms were identified for heteroepitaxial InN thin films. In this work, the current understanding of transport properties in unintentionally doped InN films is summarized based on theory and experiments and the *n*-type doping mechanisms are discussed. After this introduction, in Section 5.2, the electron transport properties of bulk wurtzite InN are evaluated by ensemble Monte Carlo simulations. Section 5.3 gives an overview of the experimental techniques used to study the electrical properties of InN as well as the results obtained. Section 5.4 discusses the different contributions to the apparent electron concentration based on experimental results. Finally, methods to improve the transport properties are briefly discussed in Section 5.5.

5.2   Electron transport in bulk wurtzite InN

Early Monte Carlo (MC) studies of electron transport in InN predicted high mobilities and high peak drift velocities [5.9–5.11]. These studies, however, assumed a conduction band based on a band gap of about 2 eV. Since the electron effective mass in the bottom valley roughly scales with the band gap energy, even higher mobility and peak drift velocity in InN could be expected for the lower band gap value ~ 0.7 eV. Moreover, the value of the band gap is related not only to the effective mass in the main conduction band valley, but has implications for the whole band structure, including satellite valleys, valley separation energies and nonparabolicities. Thus, previous MC simulation results appear to be questionable and the electron transport properties of InN should be re-examined.

This section reviews the current understanding of the electron transport in bulk wurtzite InN and InN-based heterosystems based on the corrected conduction band structure of InN with a band gap of ~ 0.7 eV. First, the electron transport characteristics of bulk wurtzite InN are compared to those of other semiconductors (GaN, GaAs and In$_{0.53}$Ga$_{0.47}$As). Then, the physical mechanisms responsible for the onset of negative differential mobility (NDM) and drift velocity overshoot in InN are discussed. Finally, a low-field mobility model for bulk wurtzite InN valid over a wide range of ionized donor concentrations is presented and compared to the available experimental results.

5.2.1 Steady-state and transient transport characteristics

For the following ensemble MC (EMC) simulations, a self-developed computer code was used. The simulations are based on an analytical nonparabolic multivalley band structure, including all relevant electron scattering mechanisms. Specifically, the lowest conduction band of wurtzite InN is approximated by the three lowest conduction band minima (Γ_1, Γ_3 and M-L valleys). Parameters such as effective masses, m^*, nonparabolicity factors for the valleys, α, and valley separation energies have been extracted from fitting the conduction bands recently calculated by the empirical pseudopotential method (EPM) [5.12]. The satellite valleys Γ_3 and M-L are assumed to be parabolic (that is, $\alpha = 0.0$), whereas the central Γ_1 valley is treated as nonparabolic in the framework of the Kane model [5.13]. Table 5.1 summarizes the band structure parameters of wurtzite InN used. Figure 5.1 compares the steady-state velocity-field characteristics of bulk wurtzite InN with those of GaN, GaAs and In$_{0.53}$Ga$_{0.47}$As. The latter is currently used as a channel material in the fastest field effect transistor (FET) structures. As clearly shown, InN demonstrates the largest peak velocity among these materials, making this material very attractive for high-frequency applications.

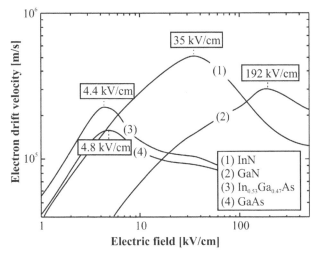

FIGURE 5.1

Steady-state drift velocity vs. electric field characteristics for different bulk semiconductors. The onsets of the negative differential mobility are marked by the dashed vertical lines. For all materials, an ionized donor concentration of 10^{17} cm^{-3} was assumed.

TABLE 5.1

Isotropic nonparabolic three-valley model of the conduction band structure of wurtzite InN. The values of the model parameters have been obtained by fitting the theoretical conduction band dispersion data given in Ref. [5.12].

Conduction band valleys	Γ_1	Γ_3	M-L
Number of equivalent valley	1	1	6
Intervalley energy separation (eV)	0.0	1.775	2.709
Effective electron mass (m^*/m_0)	0.04	0.25	1.00
Nonparabolicity factor (eV^{-1})	1.43	0.0	0.0

Next, we focus on the transient electron transport characteristics of these materials exhibiting different peak overshoot velocities, as well as longer or shorter overshoot effect durations. It is not possible to compare these semiconductors directly at the same applied electric field because the transient overshoot effect occurs at different electric fields for each material. As the drift velocity overshoot is usually related to the intervalley transfer, we can facilitate such a comparison by analyzing the drift velocity transients simulated at electric field strengths equal to twice the values where the onset of the NDM occurs (see Fig. 5.1).

Figure 5.2 compares the overshoot effects of the same four materials, that is GaN, InN, GaAs, and $In_{0.53}Ga_{0.47}As$. As indicated, InN exhibits the highest overshoot velocity over a distance of $\sim 0.3~\mu$m.

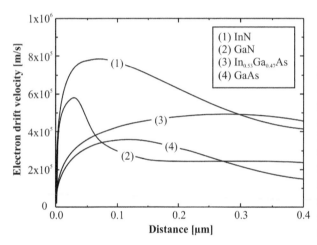

FIGURE 5.2

A comparison of the drift velocity overshoot effect among the III-nitrides, InN and GaN, and the III-V semiconductors, $In_{0.53}Ga_{0.47}As$ and GaAs. The applied electric field strengths for different semiconductors are chosen as twice the critical electric field strength at which the onset of the NDM occurs, for example 70 kV/cm for InN (see Fig. 5.1).

The NDM region of the steady-state drift velocity-electric field characteristics and the transient velocity overshoot in semiconductors are the subject of extensive investigation. The NDM effect can strongly affect the performance of electron devices in which high electric fields promote a population of the satellite valleys, whereas the drift velocity overshoot can enhance high-speed operation of nanoscaled devices. These phenomena have been investigated in many semiconductors, such as GaAs [5.14–5.17], GaN [5.18–5.21] and InN [5.3, 5.9, 5.10, 5.11, 5.22, 5.23]. As has been already discussed in the literature, there are several possible reasons for the NDM and velocity overshoot effects. Among these, intervalley electron transfer to the heavier effective mass satellite valleys in the conduction band is the conventional explanation. However, depending on the transport regime (temperature, doping, electric field strength), as well as the parameters of the band structure and of the scattering processes, other physical mechanisms can be responsible for these effects. Hauser *et al.* [5.15] have shown theoretically that nonparabolicity in the main conduction band valley alone can lead to a negative differential mobility in semiconductors where the energy separation between the central and first satellite valleys is large enough to prevent the electron intervalley transfer. Recently, several researchers have related this problem to the specific energy dependence of the electron group velocity in the central valley. Here, the nonparabolicity is characterized by the presence of an inflection point [5.24–5.26], where the derivative of the group velocity

$$\mathbf{v}_g = \frac{1}{\hbar} \frac{dE(\mathbf{k})}{d\mathbf{k}} \tag{5.1}$$

becomes zero. Attaining the maximum group velocity at this point, an electron starts to decelerate under an applied electric field even in the absence of any scattering mechanism. Consequently, if the inflection point is energetically lower than the bottom of the first satellite valley, this mechanism can be responsible for both the NDM and the overshoot effect. In contrast, a nonparabolicity of the main valley without an inflection point leads to a simple saturation of the group velocity and a velocity overshoot therefore implies the presence of scattering mechanisms.

In the following, we discuss the mechanism responsible for the onset of the negative differential mobility in InN. In the conduction band structure of InN theoretically calculated by the EPM [5.12], an inflection point of the energy dispersion $E(\mathbf{k})$ is not observed in the Γ_1 valley, at least, up to the bottom of the first upper satellite valley (1.775 eV, see Table 5.1). This suggests that the inflection point in the energy dispersion $E(\mathbf{k})$ as a possible reason for the NDM onset, mentioned above, can be neglected. Then, to clarify the mechanism responsible for NDM we intentionally discard the nonparabolicity of the central conduction band valley in InN.

Figure 5.3 shows the steady-state electron drift velocity and valley occupancies in wurtzite InN calculated as a function of the applied electric field. Here we assume the central Γ_1 valley to be (a) nonparabolic ($\alpha = 1.43$ eV^{-1} from Table 5.1) and (b) parabolic. As shown in Fig. 5.3(a), the NDM onset at $E = 5$ kV/cm is not pinned to the intervalley repopulation, which sets in only at $E = 50$ kV/cm. Instead, the peak velocity occurs at an electric field where all electrons still reside in the central valley. Even at $E = 60$ kV/cm, less than 5% of the electrons have been transferred to the satellite valleys and, correspondingly, more than 95% of electrons are still confined in the central valley. The drift velocity, however, has already dropped to 80% of the peak value. This is quite different from what is observed for semiconductors like GaAs, where the NDM is mainly caused by intervalley transfer [5.27].

In the calculation neglecting the nonparabolicity (see Fig. 5.3(b)), the NDM onset at $E = 15$ kV/cm coincides with the start of the intervalley transfer. Thus, the onset of the NDM in InN at $E = 35$ kV/cm is indeed associated with the strong nonparabolicity of the Γ_1 valley, but not with the intervalley repopulation. The influence of nonparabolicity on the drift velocity overshoot is illustrated in Fig. 5.4, which shows the transient drift velocities simulated for different nonparabolicity factors of the central valley at $E = 120$ kV/cm and room temperature. As expected, the peak velocity substantially decreases for larger nonparabolicity factors. We also observe that the shape of the velocity overshoot drastically changes with in-

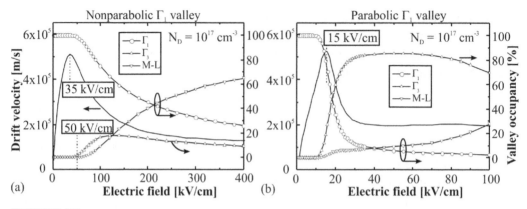

FIGURE 5.3

Steady-state drift velocity and occupancies of the central (Γ_1) and satellite (Γ_3, M-L) valleys as a function of electric field for wurtzite InN calculated at room temperature: (a) nonparabolic and (b) parabolic cases.

creasing valley nonparabolicity. For the parabolic case and for the small values of the nonparabolicity, the transient peak velocity is associated with the onset of the intervalley transfer. However, for $\alpha = 0.5$ eV^{-1} the overshoot shape flattens and the velocity peak becomes less pronounced. For the strong nonparabolicity $\alpha = 1.43$ eV^{-1} we observe a much less pronounced velocity peak occurring significantly prior to the onset of intervalley transfer.

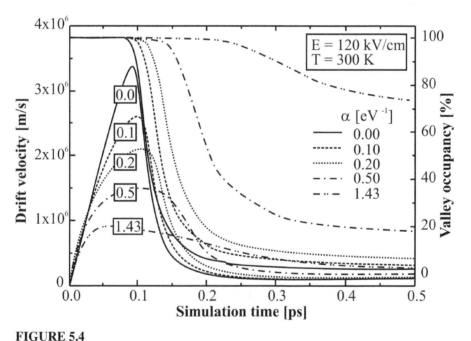

FIGURE 5.4

Transient drift velocity and valley occupancies calculated for different nonparabolicity factors of the central Γ_1 valley at $E = 120$ kV/cm and $T = 300$ K.

5.2.2 Low-field electron mobility

The low-field electron mobility has been calculated by the simple relation $\mu_0 = v_{\text{drift}}/E$ applied to the linear region of the simulated $v_{\text{drift}}(E)$ curves. As the electric fields used for the calculation of μ_0 are quite low, a long simulation time of ~ 15 ps and a large number of simulated MC particles $N_{sim} = 20000$ have been taken to improve the statistics of the MC calculation of steady-state drift velocity. The interaction of electrons with ionized impurities is incorporated in the calculation using the standard Brooks-Herring [5.28] and Conwell-Weisskopf [5.29] models.

In Fig. 5.5, the calculated low-field mobility μ_0 is shown as a function of ionized impurity concentration N_D together with the experimental data collated from the literature. It should be noted that the calculated values are drift mobilities, whereas the experimental ones are Hall mobilities.

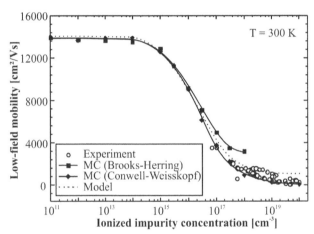

FIGURE 5.5

Calculated low-field electron mobility in wurtzite InN at room temperature using the Brooks-Herring (rectangles) and Conwell-Weisskopf (diamonds) models of ionized impurity scattering. Experimental data compiled from the literature (open circles) and the empirical model (dotted line) given by Eq. 5.2 are plotted in addition. Reprinted with permission from V. M. Polyakov and F. Schwierz, Applied Physics Letters, 88 (2006) 032101. Copyright 2006, American Institute of Physics.

The mobilities calculated using the Brooks-Herring and Conwell-Weisskopf models are very close for low impurity concentrations, but differ considerably for concentrations above 5×10^{16} cm^{-3} where the Conwell-Weisskopf model is expected to better describe the scattering. Notably, the experimental mobility data are well reproduced by the two simulated curves.

Figure 5.5 also presents the mobility (dotted line) modeled using the formula proposed by Caughey and Thomas [5.30]

$$\mu_0 = \mu_{min} + \frac{\mu_{max} - \mu_{min}}{1 + (N_D/N_{ref})^\beta} \tag{5.2}$$

with the fitting parameters: $\mu_{min} = 1030$ cm^2/Vs, $\mu_{max} = 14150$ cm^2/Vs, $N_D = 2.07 \times 10^{16}$ cm^{-3}, and $\beta = 0.6959$. As shown in Fig. 5.5, the predicted maximum room-temperature low-field electron mobility in InN is about 14000 cm^2/Vs, which is much higher than the previously proposed values of about 4000 cm^2/Vs [5.2].

Currently, In$_{0.53}$Ga$_{0.47}$As is used as a channel material for InP-based high electron mobility transistors (HEMTs) which represent the fastest FETs to date. From the results presented above, it follows that InN has: (i) a low-field mobility comparable with InP; (ii) a higher peak velocity (of $\sim 5 \times 10^7$ cm/s); and (iii) a more pronounced velocity overshoot than $In_{0.53}Ga_{0.47}As$. These superior transport properties of InN make this semiconductor very promising for radio frequency (rf) and high-speed electronics. However, simulation also clearly indicates that the high mobility can be achieved only for InN layers with carrier concentrations $<$ 10^{17} cm^{-3}, which to date has proven a challenging task. In the following sections, an overview is given of the current experimental results. These results are analyzed and models are proposed for the observed electrical properties with particular attention to the electron density.

5.3 Experimentally observed electron concentrations and mobilities

In early studies, undoped InN exhibited electron densities near 10^{20} cm^{-3} (see Fig. 5.6, for a review see Ref. [5.31]). Electron densities in the low 10^{17} cm^{-3} range were not achieved until the development of MBE techniques allowed the growth of single crystalline InN [5.7, 5.32]. Nevertheless, Hall measurements revealed that thin InN (< 100 nm) layers are typically highly degenerate with electron carrier densities above 10^{19} cm^{-3}. An important discovery was the observation of a monotonically decreasing average electron density with increasing layer thickness in high-quality epitaxial InN (Fig. 5.7) [5.33]. Additionally, an extrapolation of the sheet carrier density revealed that an excess sheet carrier density "remains" at zero layer thickness which was explained by a surface electron accumulation and electrons at the InN/buffer layer interface [5.34]. Further studies suggested the importance of interface, surface [5.35, 5.36] and piezoelectric polarization charges [5.35], along with misfit dislocations [5.37–5.40] and unintentional doping by hydrogen [5.41] or oxygen [5.42, 5.43]. As a consequence, carriers are

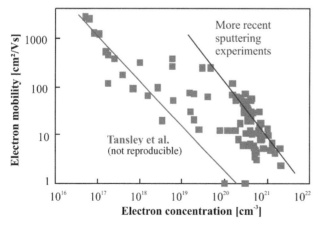

FIGURE 5.6

Electron mobility vs. concentration in sputtered InN films. The results are compiled from Refs. [5.44–5.56].

inhomogeneously distributed in InN films and any interpretation of the experimental data for the transport properties (electron density and mobility) requires use of multiple analysis techniques. Basically the following techniques have been employed to extract transport properties, with results of these measurements discussed in more detail in the following sections:

1. Single-field Hall effect measurements provide one of the cornerstone methods for evaluating the quality and reproducibility of semiconducting materials. However, this technique can be influenced by mixed transport mechanisms complicating the interpretation of the measurement.

2. Variable-magnetic-field Hall effect measurements enable analyzing carriers having different mobilities.

3. Capacitance-Voltage (C-V) depth profiling measures the distribution of the carriers versus the thickness in a thin film. Specifically for InN, this method can only be employed using special electrolyte contacts [5.34].

4. Sputter depth profiling accompanied by *in situ* resistance measurements enables the calculation of a conductivity profile.

5. Resistivity measurements are the simplest technique for electrical characterization; however, they are not able to extract carrier concentration and mobility. Such measurements are useful for nanostructures, such as nanowires, where other techniques are difficult to implement for geometrical reasons.

6. Finally, the near-surface carrier density was obtained by high-resolution electron-energy-loss spectroscopy (HREELS) [5.36, 5.57, 5.58]. This technique and its results are described in more detail in Chapter 12.

5.3.1 Single-field Hall measurements

The majority of the electrical data for InN layers have been obtained by (single-field) Hall measurements using the van der Pauw geometry. Figure 5.7 shows reported values of InN electron density and mobility as a function of the layer thickness. Despite the scatter in the data, there is an obvious general trend represented by the results of Lu and Schaff [5.7, 5.33, 5.34]: The electron mobility increases with layer thickness, while the free electron concentration decreases. Moreover, a comparison with data from InN films, which have been grown in a hydrogen-rich environment (MOCVD and remote plasma-enhanced chemical vapor deposition (RPECVD)), demonstrates the superior transport properties of the MBE-grown films. The data set of Lu and Schaff roughly represents the current experimental limit for the electron mobility and concentration that can be achieved for InN films and will be used as the basis for further analysis.

In Fig. 5.7, each measurement point represents a value for Hall density and mobility using the usual single layer approximation. This approximation is accurate

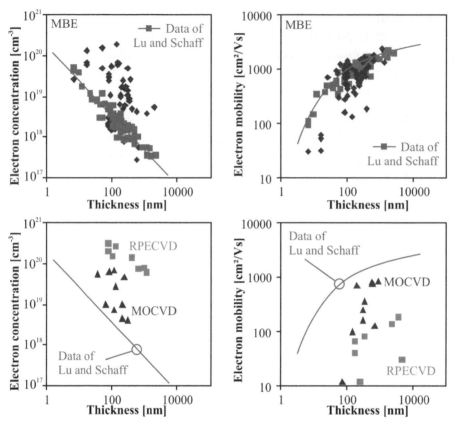

FIGURE 5.7

Free electron concentration and mobility in InN grown by MBE (data from Lu and Schaff [5.7, 5.33, 5.34] and from Refs. [5.32, 5.37, 5.59–5.73], by MOCVD [5.65, 5.74–5.77] and RPECVD [5.78]. The lines represent a fit of the data from Lu and Schaff.

only if density and mobility both have uniform distribution throughout the entire layer. Non-uniformities of electron density and mobility thickness profiles result in Hall values which are weighted averages of values from multiple layers. In a first approach to evaluate these data, layers are analyzed in differential comparisons at different thicknesses. Such differential analysis makes the approximation that a layer of incremental thickness beyond previous thinner layers is composed of two layers — a single lower layer which is a composite of all layers below, and a second top layer that is a parallel contribution to Hall conductivity. At each incremental thickness, the layers below are taken as a single layer in parallel to the incremental thickness. When the smoothed Hall data is characterized by this two layer model, the results are shown in the solid line in Fig. 5.8. It can be seen that the differential analysis leads to incremental layers with slightly higher mobility and lower electron density than determined from the single layer approximation. This treatment gives a more accurate view that layer properties change by smaller amounts with thickness than suggested by the single layer approximation. Particularly, a bulk free electron density of about $1-2 \times 10^{17}$ cm^{-3} can be estimated for very thick InN

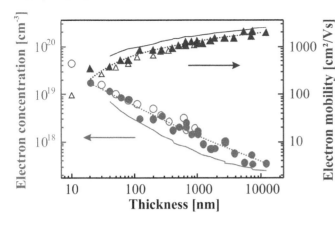

FIGURE 5.8

Electron density and mobility as a function of InN thickness. Closed symbols are grown on a GaN buffer, open are on AlN. Circles are carrier density, triangles are mobility. Dotted lines are fits to GaN characteristics to provide a smoothed function for differentiation. Solid lines are extraction of differential density and mobility [5.33].

layers.

Surface and interface properties can be estimated from the thickness dependence as well. Sheet electron densities are measured in the usual single layer approximation and plotted versus layer thickness for InN films grown on either a GaN or AlN buffer (Fig. 5.9). The sheet density change is approximately linear, which indicates that carrier density is constant for layers with small thickness differences on a particular buffer. The sheet density intercept corresponding to zero layer thickness is 2.73×10^{13} cm^{-2} for InN on GaN buffers and 4.23×10^{13} cm^{-2} for layers on AlN buffers [5.34].

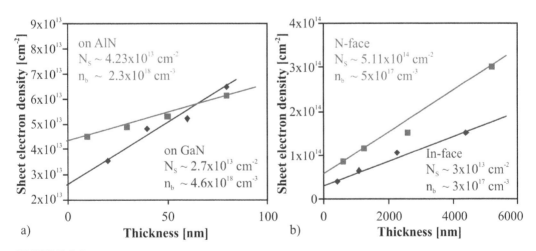

FIGURE 5.9

Electron sheet density vs. film thickness: Comparison for (a) thin InN films on AlN and GaN buffer layers [5.33, 5.34] and (b) for thick InN films with N-face [5.59] or In-face [5.32] polarity. Given parameters are the extrapolated surface/interface sheet density N_S and the extracted bulk electron density n_b.

When the intercept sheet charge density is subtracted from the single layer approximation values in Fig. 5.9, the carrier volume density is calculated to be $4.6 \times$

10^{18} cm^{-3} for InN on GaN and 2.3×10^{18} cm^{-3} on AlN, which is still much higher than the bulk value expected for thick InN layers. The extrapolation of the zero charge density using thicker films yields higher values of the surface or interface charge. The extracted bulk density is in the low 10^{17} cm^{-3} range as expected from the analysis of Fig. 5.8. Thus, different mechanisms of the conductivity in thin and in thick films have to be taken into account. Moreover, the comparison of similarly grown InN films on different substrates demonstrate the additional influence of the buffer layer and the polarity of the epitaxial InN layer on the transport properties. More reliable information about carriers with different mobilities in inhomogeneous layers is obtained by variable-field Hall measurements, which are discussed in the next section.

5.3.2   Variable-field Hall measurements

Variable-magnetic-field Hall measurements can be used to determine the influence of the carriers in a sample with different mobilities, such as occurs for multiple conduction layers, sample inhomogeneities, or a mixture of these effects [5.79, 5.80]. The utility of this technique relies on the fact that the contribution of individual carriers to both the conductivity and Hall coefficient has a field dependence on the individual carrier mobility, resulting in a mobility dispersion that can be used to identify these contributions to the variable field data. The results reported by two groups [5.81–5.83] indicate that variable-field Hall techniques are very useful for evaluating multiple carrier transport in InN epilayers, as well as distinguishing electrical conduction due to strongly *n*-type material near the interface from the higher quality, and potentially non-degenerate, material far from the interface.

InN layers with varying thickness and polarity have been investigated by us using variable-field Hall measurements [5.81, 5.82, 5.84]. These samples were grown by MBE at Cornell University [5.7, 5.34, 5.85] or the University of Canterbury [5.86, 5.87]. Resistivity and Hall measurements were carried out as a function of both temperature (4–300 K) and magnetic field (up to 12 T). The van der Pauw technique was used with soldered In contacts. At a given temperature, measurements were made at each of 22 logarithmically spaced values of the magnetic field. All contact configurations, current and field directions were measured.

The Hall and resistivity values were converted to the conductivity tensor components σ_{xx} and σ_{xy}, then quantitative mobility spectrum analysis (QMSA) and multiple-carrier fitting (MCF) were performed using standard techniques [5.79, 5.88]. It should be noted that QMSA does not depend on any *a priori* assumptions about the type and number of carriers, and can actually reveal information about potential non-uniformity, or spread, in mobility for a given carrier.

For all InN samples investigated, both MCF and QMSA analysis of the variable-field Hall measurements clearly revealed conduction is dominated by electrons with two distinctly different mobilities. Figure 5.10 shows a QMSA spectrum illustrating the observation of the two kinds of electrons in a 0.6 μm thick InN layer grown on an AlN buffer. The mobility and concentration of low mobility (< 200 cm^2/Vs)

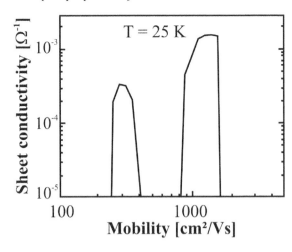

FIGURE 5.10

QMSA spectrum at 25 K for a 600 nm thick InN layer grown on a 200 nm thick AlN buffer, clearly showing both high and low mobility conduction.

carriers are often difficult to uniquely determine using the maximum magnetic fields available. Fortunately, the low mobility conductivity is a robust number. Thus, MCF can be used to remove the effect of the low mobility carriers, allowing accurate determination of the mobility and concentration values for the more mobile bulk electrons. Since Hall measurements alone cannot determine the location of the various conduction electrons within a sample, the low mobility electrons have been labelled as surface/interface electrons based on accompanying measurements pointing to the presence of significant surface and interface charges with comparable sheet concentration.

Figure 5.11(a) shows the bulk carrier concentration and mobility as a function of layer thickness as determined by MCF analysis. There is a distinct thickness dependence for the mobility and volume carrier concentration of the films. The increasing mobility and decreasing concentration imply that the film quality is non-uniform with depth, improving with distance from the substrate. The carrier concentration data can be fitted to a power law, and the "true" carrier concentration as a function of thickness inferred as illustrated in Fig. 5.11(b) (similar to what was done above for single field measurements), resulting in an inverse dependence of the carrier concentration on the square root of thickness. While there is no *a priori* reason to expect this variation with thickness, it is very likely that such a strong variation of carrier concentration does indeed exist and may be related to a changing defect structure (see Section 5.4.3). Thus one can conclude that the far-from-the-interface bulk carrier concentration (exclusive of the surface and interface-related conduction) may be significantly lower than the "average" measured, and that the first micron of the layer can completely dominate the electrical properties. Thus, the true electrical quality of thicker layers is likely masked by what occurred early during the sample growth. This model explicitly removes the effect of the low mobility surface and interface-related electrons (see Section 5.4.2) through MCF analysis.

Of particular interest, QMSA analysis also indicates a significant spread in mobility of the electron associated with the "bulk" of the sample as the layer thickness

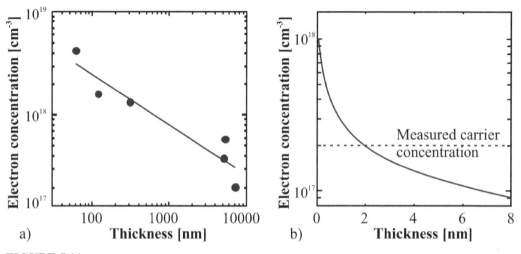

FIGURE 5.11

(a) The variation in carrier concentration of the bulk electrons with thickness at 300 K, determined from MCF analysis. The effect of low mobility surface and interface-related conduction has been removed. (b) A fit to the MCF carrier concentration vs. thickness data in (a), based on a power law. The value of the average QMSA carrier concentration of the thickest sample measured (7.5 μm) is also shown. Note that the predicted near interface carrier concentration is much larger than this average value, while the far-from-the-interface value would likely be significantly lower.

increases. One of the strengths of the QMSA technique is that it determines a spectrum of mobility values, rather than imposing a single value for mobility as in MCF. This allows the determination of potential inhomogeneities in a single sample. This is indicated in the spectra shown in Fig. 5.12 for the "bulk" electrons collected at different temperatures for a sample. Note that conduction due to the surface electrons occurs at much lower mobility, and is not shown in the plot. The QMSA analysis indicates that at room temperature some portion of the sample, most likely near the surface, has a peak mobility of about 3000 cm^2/Vs, while a different portion, most likely near the interface, has a fairly constant mobility closer to 1000 cm^2/Vs. At lower temperature the spread in mobility increases, until finally it seems to collapse in a fairly narrow "band" at 4 K. While all the carriers do not freeze out at 4 K, the strong temperature variation in the maximum mobility of the QMSA spectrum as shown in Fig. 5.12 suggests that the portion of the sample resulting in the maximum mobility may not be degenerate. Another way to interpret this is that there is a degenerate InN layer associated with the initial stages of growth similar to what has been observed in many cases for GaN [5.85]. There are also degenerate layers at the surface and interface with significantly lower mobility, but their contribution has been resolved separately by the QMSA analysis.

To extract the non-degenerate high mobility component, it was assumed that the carriers in the non-degenerate portion of the layer were frozen out during the 4 K measurement. The field dependent conductivity tensor coefficients at this temperature were then subtracted from the measurements at all other temperatures, and a single carrier fitting was performed on the remaining values, providing an

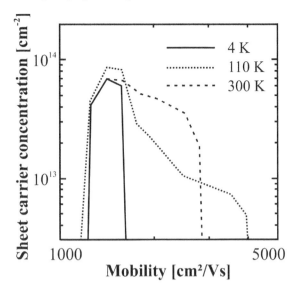

FIGURE 5.12

Variation in the bulk electron mobility spectrum for a thick InN film ($\sim 5 \ \mu$m) suggesting that different parts of the sample have different mobility dependencies on temperature.

average concentration and mobility excluding the fully degenerate carriers. The mobility obtained in this way for two thick samples grown at Cornell is plotted in Fig. 5.13(a). We have also plotted the mobility results obtained in another study [5.83] on an InN sample grown at the University of California at Santa Barbara on a MOCVD GaN template, which apparently results in a better initial layer growth. Interestingly, the room temperature mobility of both samples is about 3500 cm^2/Vs, as shown in Fig. 5.13(a), which is close to the theoretical limit for an ionized donor density of 1×10^{17} cm^{-3}, but still far lower than the predicted maximum mobility (see Fig. 5.5). Clearly, all three samples exhibit a temperature-dependent mobility.

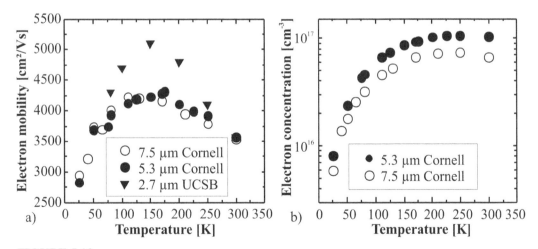

FIGURE 5.13

The temperature dependence of the transport properties of the nearly non-degenerate near surface layer: (a) electron mobility and (b) electron concentration (with the average bulk electron mobility given by a two carrier MCF analysis of "bulk" electron conduction).

The carrier concentration resulting from our subtraction of the low temperature measurement, shown in Fig. 5.13(b), indicates net electron densities in the range 5×10^{16} to 1×10^{17} cm^{-3} for the samples grown at Cornell, the lowest reported for InN. For this volume concentration, we assumed that the thickness of the non-degenerate layer was simply the thickness of the film, which is not strictly true, as an indeterminate volume of the growth is fully degenerate. Nonetheless, these carrier concentrations are still partially degenerate. At 80 K, where the carrier concentration is 4.2×10^{16} cm^{-3}, the conduction band effective density of states is a comparable 7.7×10^{16} cm^{-3}, using a density of states effective mass based on $m_{\perp}^* = 0.05 m_0$ and $m_{\parallel}^* = 0.037 m_0$ [5.89].

The temperature dependence of the electron mobility and carrier concentration for the Cornell samples were analyzed at Wright Patterson Air Force Base (WPAFB) using the full Boltzmann Transport Equation (BTE) model [5.90, 5.91]. When the full charge balance equation fit to the carrier concentration was attempted, the donor activation energy was found to be less than 1 meV [5.92], meaning the Fermi energy is likely within a few thermal energies of the conduction band at most temperatures. Hence, the use of non-degenerate statistics is a poor approximation, and a reliable activation energy is difficult to find. However, as carrier activation is clearly visible in Fig. 5.13(b), it is reasonable to conclude that the Fermi level is not in the conduction band, as it is for material near the interface/surface.

The temperature dependence of mobility shown in Fig. 5.13(b) could also be explained by either assuming a low acceptor density (near zero) with a realistic dislocation density ($\sim 8 \times 10^8$ cm^{-2}, see Section 5.4.3) or an unrealistically low dislocation density ($< 10^8$ cm^{-2}) with a "reasonable" acceptor density near 1×10^{17} cm^{-3}. Again, the sample quality, while among the best reported for InN, is in a regime that makes absolute determination of the dominant scattering mechanisms difficult.

5.3.3   C-V depth profiling

The electron accumulation at the surface of InN films complicates the formation of Schottky contacts for C-V measurements. However, a weakly rectifying contact can be formed between InN and a KOH-based electrolyte. This effect was utilized to perform C-V measurements on InN using an electrochemical profiler [5.34]. Figure 5.14 shows the C-V profile of the surface carrier concentration of three different InN samples. A gradient of carrier concentration ranging from 10^{20} to 10^{18} cm^{-3} within 6 nm in depth is observed at the InN surface. By integrating the curve of carrier concentration versus depth, a surface sheet carrier density of 1.57×10^{13} cm^{-2} is obtained, which is in the same range as the excess sheet carrier densities derived by other techniques. This observation is direct evidence of surface charge accumulation in InN films. Its value is also not affected by the bulk and interface properties. A comparison of the C-V profile of InN films grown on AlN and GaN buffer layers as well as doped samples (for example, with Be, shown in Fig. 5.14) indicates changes in the bulk transport properties; however the surface concentration of electrons is approximately the same. Similar results were recently obtained for *p*-type InN [5.84]. Consequently, the excess sheet electron density inferred from single-field Hall measurements is a combination of surface and interface effects, and interface effects can be considered as the reason for the differences of the values extrapolated from the thickness dependence of single-field Hall effect measurements in Fig. 5.9 for different buffer layers and polarities.

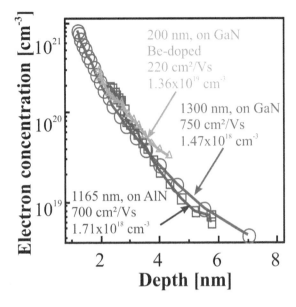

FIGURE 5.14
Electron density vs. depletion depth into the surface of InN from C-V measurements [5.33, 5.34]. Given electron concentration and mobility are the results of single-field Hall measurements.

5.3.4 Sputter depth profiling

Sputter depth profiling is based on step-wise removal of a thin surface layer by ion sputtering and simultaneous measurement of the resistance of the remaining layer. For such depth profiling, the samples were cut into narrow strips, contacted by metallic In, and loaded into an Auger electron spectroscopy (AES) chamber [5.42, 5.43]. The sputtering was performed using a low energy (1 keV) Ar^+ ion beam under gracing incidence (80°) in order to remove the InN surface layer and to decrease the cross section of the conductive InN layer on the insulating substrate. The sputtering rate was about 0.35 nm/min. The dependence of the measured film resistance on the thickness of the removed film provides information on the depth distribution of the carrier concentration. Here, sputter induced changes of the surface composition and the crystalline structure were minimized by optimized sputtering conditions. Otherwise sputtering results in highly In enriched surfaces and In droplet formation [5.93].

Figure 5.15 schematically shows the setup for these experiments. R_1 and R_2 are the resistances of the film outside the sputtered area and are constant. R_0 represents the resistance, which is changing during the sputtering. In the ideal case, assuming a constant conductivity σ, the measured resistance $R(z)$ should monotonically increase with the sputtering depth z (Fig. 5.16(a)).

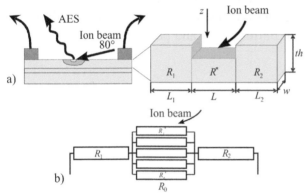

FIGURE 5.15

(a) Scheme for the sputtering of the InN stripe and (b) an equivalent circuit diagram. Reprinted with permission from V. Cimalla, G. Ecke, M. Niebelschütz, O. Ambacher, R. Goldhahn, H. Lu, and W. J. Schaff, physica status solidi (c), 2, 2254 (2005). Copyright 2005 by Wiley-VCH Publishers, Inc.

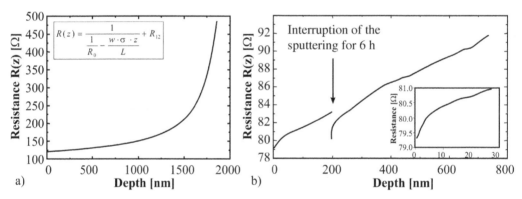

FIGURE 5.16

(a) Theoretical resistance vs. depth for a film with constant σ; insert: analytical description, (b) Measured resistance vs. depth during the grazing incidence sputtering of a 1.8 μm thick InN layer; insert: resistance evolution at the beginning of the sputtering [5.42]. Reprinted with permission from V. Cimalla, G. Ecke, M. Niebelschütz, O. Ambacher, R. Goldhahn, H. Lu, and W. J. Schaff, physica status solidi (c), 2, 2254 (2005). Copyright 2005 by Wiley-VCH Publishers, Inc.

The measured resistance during the sputtering of a thick InN layer is shown in Fig. 5.16(b). As expected, the resistance increases with decreasing layer thickness. However, two typical features disturb this almost linear trend. First, there is a strong increase in the resistance during the removal of the first 5 nm of the InN layer (shown in the insert). This strong increase of the resistivity confirms the existence of a highly conductive surface layer which can be removed by sputtering. The second feature, marked as "Interruption", appeared after stopping of the sputtering process and exposing the sample to air. After reloading in ultra high vacuum, the resistivity was observed to have significantly decreased. However, it recovered to the previous trend after again sputtering of about 5 nm of material.

Using the equivalent circuit diagram and the geometrical assumptions in Fig. 5.15, the absolute values of the differential conductivity $\sigma(z)$ were calculated by:

$$\sigma(z) = \frac{L}{w} \frac{\mathrm{d}\left(\frac{1}{R(z)-R_{12}}\right)}{\mathrm{d}z} \tag{5.3}$$

where L, w, and z are length, width and depth of the sputtered InN film, R_{12} and $R(z)$ are the resistances of the non-sputtered and the measured part, respectively. The majority carriers for the highly n-type InN layers are electrons and their concentration n can be estimated by $\sigma = e\mu n$, where μ is the mobility of the electrons. For the mobility the integrated values estimated by Hall measurements on InN layers with different thickness are used (Fig. 5.17 [5.7]).

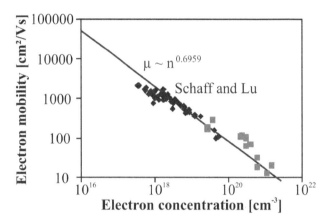

FIGURE 5.17

Electron concentration vs. mobility of MBE grown InN from single-field Hall measurements. Data are from Schaff and Lu [5.7, 5.33, 5.34] (diamonds) and selected values for high electron concentrations from [5.61, 5.94, 5.95] (squares).

From these data, the empirical relationship $\mu \sim 0.6 \times 10^{16} n^{0.6959}$ (n in cm^{-3} and μ in cm^2/Vs, the exponent was taken from the MC results fitting Eq. 5.2), was estimated and applied to calculate the carrier depth profile shown in Fig. 5.18. Obviously the sputtering interruption and/or exposure to air results in a very highly localized carrier concentration of more than 10^{20} cm^{-3} at the surface. This behavior was fully reproduced after interruption of the sputtering and exposure to air. This suggests, for air exposed samples, the microscopic origin of the surface electron might be associated with contamination from the air, most probably the presence of oxygen atoms on the surface, which act as n-type dopants creating free carriers close to the surface. AES depth profiling (Fig. 5.19) confirms the existence of oxygen in the near surface region. The correlation between the depth profiles for the concentration of oxygen and electrons suggests that, when present, oxygen may be responsible for the high conductivity of air exposed InN layers. Moreover, after removing of the uppermost oxidized layer by sputtering the electron concentration ultimately reaches a value of 4.5×10^{17} cm^{-3}, in good agreement with the best values measured on InN layers of comparable thickness (see Fig. 5.11(a), for example).

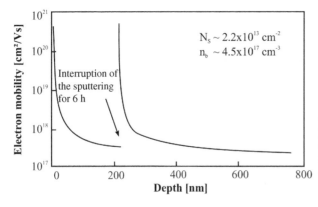

FIGURE 5.18
Depth profile of the free electron concentration calculated from the resistance evolution shown in Fig. 5.16(b).

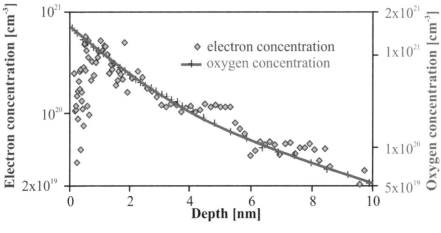

FIGURE 5.19
Depth profiles of the electron (diamonds, from Fig. 5.18) and oxygen concentration (line, AES) [5.43]. Reprinted with permission from V. Cimalla, G. Ecke, M. Niebelschütz, O. Ambacher, R. Goldhahn, H. Lu, and W. J. Schaff, physica status solidi (c), 2, 2254 (2005). Copyright 2005 by Wiley-VCH Publishers, Inc.

Finally, by integration of the excess electron density close to the surface versus depth a sheet carrier density of 2.2×10^{13} cm^{-2} is obtained. The estimated electron accumulation agrees very well with the results obtained by HREELS on clean InN surfaces [5.57, 5.58] as well as on air exposed InN by C-V depth profiling [5.34], where carrier concentrations up to around 2×10^{20} cm^{-3} close to the surface were observed.

If the InN film is completely removed by sputtering, a second peak appears indicating an increased electron concentration at the interface between InN and the AlN buffer layer (Fig. 5.20). The integration yields a sheet carrier concentration $> 5 \times 10^{12}$ cm^{-3}. This feature is substantially broader than the surface peak. The accuracy of the geometrical values (especially the sputter crater width) critically determines the accuracy of the estimated data, and a quantitative analysis can therefore only give a benchmark of the interface-related electron density.

Nevertheless, this investigation has shown that a significantly higher electron

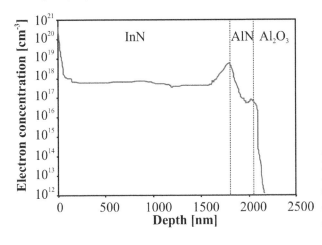

FIGURE 5.20

Depth profile of the free electron concentration calculated from the evolution of resistance during complete sputtering of an 1800 nm thick InN film.

density than in the bulk exists at both the surface of the sample and the interface with the buffer layer. This important result demonstrates that simple extrapolation of the sheet carrier density to zero thickness (Fig. 5.9) will yield the sum of both contributions.

5.3.5   Summary of the experimental observations

The above described electrical measurements clearly demonstrate that InN thin films exhibit a highly inhomogeneous carrier profile. Significantly, both surface electron accumulation and interface-related electron density play a crucial role in the transport properties in InN. High electron mobilities and low electron concentrations of 3500 cm^2/Vs and $\sim 10^{17}$ cm^{-3} have been obtained only for thick films (> 1 μm). Although the measured carrier concentration and mobility exhibited temperature dependence, the samples were still partially degenerate, precluding detailed analysis of activation energies and scattering mechanisms. In the following, the origin of the high electron concentration is discussed based on the analysis of the available data.

5.4   Contributions to the *n*-type conductivity of InN

The experimental data presented in the previous section demonstrate that InN films exhibit inhomogeneous profiles for both carrier concentration and mobility. Accordingly, multiple mechanisms have to be considered as contributing to the *n*-type conductivity in unintentionally doped InN films. Possible causes of the apparent electron concentration can be classified into three contributions: (i) a localized electron accumulation with specific sheet carrier concentration $N_{S,0}$; (ii) a homogeneous background volume concentration n_b; and (iii) an inhomogeneous carrier distribution n_{inhom} over the InN film. Consequently, the dependence of the net electron density, n, on the InN film thickness, th, will be:

$$n = n_b + \frac{1}{th} N_{S,0} + \frac{1}{th} \int_0^{th} n_{inhom} \, dz. \qquad (5.4)$$

A summary of data for the corresponding contributions published in the literature so far is given in Table 5.2.

5.4.1   Bulk concentration of electrons

In the previous section, it has been shown that thicker InN films generally have superior electrical properties. It is possible to estimate the bulk (background) contribution to the electron concentration for very thick films: $\sim 2 \times 10^{17}$ cm^{-3} from single-field Hall measurements (Fig. 5.8), $\sim 1 \times 10^{17}$ cm^{-3} in variable-field Hall measurements (Fig. 5.11(b)), and $\sim 4 \times 10^{17}$ cm^{-3} from sputtering experiments (Fig. 5.18). Several donors have been proposed to be responsible. The most commonly discussed are oxygen [5.42, 5.43, 5.59, 5.96], hydrogen [5.32, 5.41, 5.59, 5.97, 5.98, 5.99], and intrinsic point defects, such as nitrogen vacancies [5.100]. A large amount of oxygen has been found to be present in high carrier density InN films grown by sputtering [5.101] as well as by MBE [5.102]. However, elemental analysis by secondary ion mass spectroscopy (SIMS) [5.32, 5.103] and glow discharge mass spectrometry (GDMS) [5.98] revealed a background concentration of oxygen, as well as for carbon to be in the 10^{16} cm^{-3} range, that is 2–10 times smaller than the electron concentration. In contrast, hydrogen, which is a common impurity in semiconductors that can be found in almost all growth environments, was detected at significantly higher concentrations [5.32, 5.98]. Moreover, InN films grown in a hydrogen containing environment (CVD, metal organic MBE) were observed to have increased electron concentration (Fig. 5.7). Consequently, hydrogen can be considered as a strong possibility as the main doping source for the bulk electron concentration in unintentionally doped high-quality InN [5.97]. In addition, intrinsic defects like point defects and dislocations are expected to create donor states inside the conduction band. A more detailed discussion of impurities and defects in InN can be found in Chapters 10 and 11. Here, all these effects are included in the current lowest background concentration of 1×10^{17} cm^{-3} estimated by variable-magnetic-field Hall measurements [5.82, 5.83].

5.4.2   Surface electron accumulation

Lu *et al.* [5.34] first observed a strong reduction of the electron concentration with increasing film thickness and concluded a high electron accumulation at the surfaces and interfaces of about 4.3×10^{13} cm^{-2} and 2.5×10^{13} cm^{-2} for InN layers grown on AlN and GaN buffer layers, respectively (Fig. 5.9), and by C-V depth profiling in electrolyte a surface accumulation of 1.57×10^{13} cm^{-2} was estimated (Fig. 5.14). This surface electron accumulation was discussed as an intrinsic property of InN due to the extraordinary low conduction band minimum

TABLE 5.2
Summary of the published contributions to *n*-type conductivity in InN.

Sample	Location	Origin	$N_{S,0}$ (cm^{-2})	n (cm^{-3})	Technique	Ref.
InN	surface		1.57×10^{13}		C–V	[5.33, 5.34]
InN/AlN	surf/interf		4.3×10^{13}		Hall, extrapolated	[5.33, 5.34]
InN/GaN	surf/interf		2.5×10^{13}		Hall, extrapolated	[5.33, 5.34]
InN	bulk			1×10^{17}	Variable-field Hall	[5.81, 5.82, 5.83]
InN/AlN	surface	intrinsic	2.5×10^{13}		HREELS	[5.36, 5.58]
InN/GaN	surface	intrinsic	2.4×10^{13}		HREELS	[5.36, 5.57]
InN	surface	intrinsic	1.6×10^{13}		XPS	[5.104]
InN/AlN	interface	defects	1.9×10^{13}		extrapolated	[5.58]
InN/GaN	interface	defects	$< 0.5 \times 10^{13}$		extrapolated	[5.58]
InN/GaN	interface		$2 - 9 \times 10^{13}$		Hall + XPS	[5.105]
InN/GaN	interface		$\sim 1 \times 10^{13}$		IR reflectivity	[5.106]
InN/GaN,AlN	surface	oxygen	2.2×10^{13}		depth profiling	[5.42, 5.43]
InN/GaN,AlN	bulk			4.5×10^{17}	depth profiling	[5.42, 5.43]
InN/GaN	interface	defects	0.5×10^{13}		depth profiling	[5.42]
InN/AlN	interface	polarization			calculated	[5.35]
InN/GaN	interface	polarization	0.5×10^{13}		calculated	[5.35]
InN	bulk	hydrogen				[5.32, 5.41, 5.97, 5.98]
InN	bulk	nitrogen defects				[5.107]
InN	bulk	dislocations			Hall/TEM	[5.37, 5.38, 5.39, 5.40]

at the Γ-point [5.57], which allows donor-type surface states to be located inside the conduction band. As a consequence, on clean InN surfaces, a surface Fermi level of about 0.9–1.0 eV above the conduction band minimum was estimated by HREELS [5.36, 5.57, 5.58]; for a more detailed discussion see Chapters 12 and 13. The corresponding surface state density is about 2.4×10^{13} cm^{-2}. Very similar surface electron concentration and Fermi level position have been observed by x-ray photoemission spectroscopy (XPS) at In-polar, N-polar and *a*-plane InN surfaces [5.104] and on an air-exposed InN surface, extracted by *in situ* measurements of the resistance of an InN film during sputter depth profiling [5.42], see Section 5.3.4, and by ultraviolet photoelectron spectroscopy (UPS) [5.43], respectively. By a numerically self-consistent solution of the Schrödinger and Poisson equations (Fig. 5.21), a corresponding electron accumulation of about 6×10^{12} cm^{-3} was calculated [5.108].

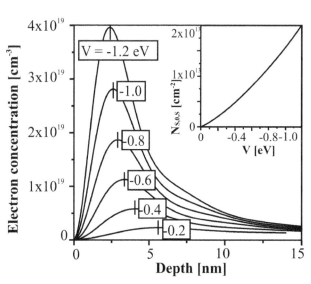

FIGURE 5.21

Electron density distribution *n* vs. depth beneath the surface calculated for different values of the surface band bending *V* (in eV). Inset: Sheet density $N_{S,0,S}$ of the accumulated carriers due to the surface band bending *V*. $N_{S,0,S}$ is the sheet carrier density obtained by integration of the profile subtracted by the value due to the background doping *n* (*th* = 200 nm) [5.108]. Reprinted with permission from V. Cimalla, V. Lebedev, Ch. Y. Wang, M. Ali, G. Ecke, V. M. Polyakov, F. Schwierz, O. Ambacher, H. Lu, and W. J. Schaff, Applied Physics Letters, 90 (2007) 152106. Copyright 2007, American Institute of Physics.

5.4.3   Influence of dislocations

As shown in Section 5.4.1, defects and impurities in InN are likely to act as donors in contrast to GaN and AlN. The primary defects penetrating through the entire film are dislocations, whose density in heteroepitaxial layers typically monotonically decreases away from the InN/buffer layer interface. Transmission electron microscopy (TEM) of a thick InN film (~ 2 μm) is shown in Fig. 5.22 [5.37, 5.38].

The dislocation density is clearly seen to decrease with increasing distance from the interface with the AlN buffer. A quantitative analysis of the areal TD density is shown in Fig. 5.22(b). Here, we do not differentiate between edge, screw and mixed

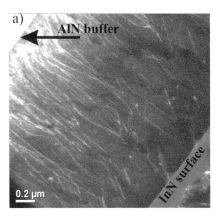
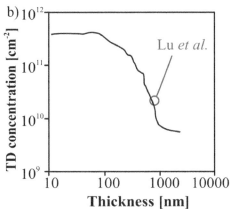

FIGURE 5.22

(a) Cross-section TEM of an InN layer on AlN, (b) calculated dislocation (TD) density versus film thickness.

dislocations. The determined TD density shows good agreement with previous TEM studies [5.112] performed on the samples of Lu and Schaff [5.7], where a TD density of 2.2×10^{10} cm^{-2} was estimated for 760 nm InN layers grown on a GaN/AlN buffer.

The influence of these defects explains qualitatively the dependence of the carrier concentration on the layer thickness. For quantification, it is assumed that each dislocation creates donors at evenly spaced intervals in the growth direction. Using the concentration profile of Fig. 5.22(a), its influence is shown in Fig. 5.23 for the TDs only ("TD") and for the sum of all of the effects discussed ("sum"). Good agreement with the measured data is obtained if the distance between two charges in the growth direction is about 1.14 nm, that is, 50% of the free bonds in dislocations are active donors. This behavior is opposite to *n*-type GaN where dislocations act as (deep) acceptors [5.113]. The underlying doping mechanisms of dislocations in InN have not been investigated and require further analysis. However, taking the similarities of the materials into account, one would expect the dominating edge dislocations [5.38] to create dangling bonds which act as donors. A recent study comparing the dislocation density obtained from x-ray diffraction measurements with the measured electron density strongly indicates that the edge-type dislocations are indeed acting as donors [5.40]. Their charge state is dependent on the Fermi level E_F relative to the Fermi level stabilization energy E_{FS} [5.39, 5.114]. For *n*-type GaN, $E_F > E_{FS}$, and thus, compensating, negatively charged vacancies (acceptors) are created. In contrast, $E_F < E_{FS}$ for InN, which stimulates the formation of positively charged nitrogen vacancies (donors), and thus a high electron concentration. The observed dependence of the electron concentration on the InN film thickness is consistent with this assumption, since the measured concentration for InN is higher than the sum of all other possible contributions.

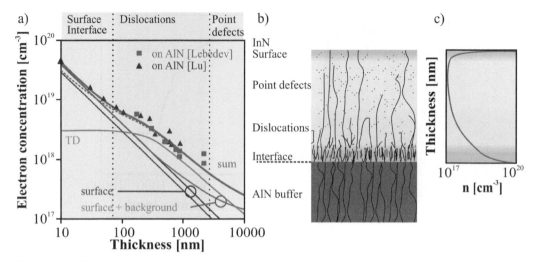

FIGURE 5.23

Summary of the electron-generating mechanisms and comparison with experimental data [5.7, 5.34, 5.38] with dependence on the InN film thickness. "surface" means the influence of a localized sheet carrier density the surface, "background" the influence of a fixed bulk concentration, "TD" the contribution of threading dislocations, and "sum" the addition of all effects. (b) Corresponding schematic layer model, and (c) qualitative free electron profile in thick InN films.

5.5 Implications for InN technology

In the previous sections it has been shown that the transport properties are affected by a number of different mechanisms, reducing the mobility and increasing the electron concentration. All of these influences have to be controlled in order to obtain the intrinsic properties of InN to realize rf electronic or long-wavelength optoelectronic devices. These are:

- Intrinsic point defects such as vacancies, controlled in MBE grown high-quality InN layers.

- Impurities such as hydrogen, oxygen and carbon have to be minimized to be less than 10^{17} cm^{-3}. This has been achieved for oxygen and carbon, while the incorporation of hydrogen requires further attention.

- The dislocation density should be considerably less than 10^{10} cm^{-2}. To date this can only be achieved for thick layers. The extremely high interface defect density requires the consideration of new lattice matched substrates or advanced epitaxial techniques, for example, lateral overgrowth methods.

- The intrinsic surface electron accumulation needs to be passivated.

- Finally, compensation can be achieved by Mg doping; however, this method is successful only for InN films with low donor densities. Particularly, the surface/interface accumulation cannot be compensated with Mg.

Currently, for most of these mechanisms no feasible strategy exists to improve the transport properties. While the bulk concentrations n_b and n_{inhom} are the subject of optimization of the epitaxial growth, surface electron accumulation is an intrinsic property of a clean InN surface; these properties are discussed in detail in Chapters 12 and 13. In this final section, the first steps toward a manipulation of the surface accumulation will be described.

First principles calculations have predicted that surface states on polar InN always pin the surface Fermi level above the conduction band minimum, while for non-polar InN in the absence of In adlayers, surfaces without electron accumulation should be possible [5.115]. Indeed, this has very recently been observed for vacuum-cleaved *a*-plane InN surfaces [5.116]. Obviously, the microscopic structure of the surface is crucial for the surface accumulation. Interestingly, during the sputtering experiments a strong reduction of the surface sheet density could be achieved (see Figs. 5.16 and 5.18); however, this surface is not stable, even in ultra-high vacuum. Nevertheless this observation indicates that there might be a surface configuration with low or no electron accumulation. Further, surface dangling bonds could be passivated by chemical modification such as sulfurization [5.117] or oxidation [5.108]. It has been shown in the previous sections that oxygen acts as a donor in InN. However, soft oxidation using ozone at room temperature produces a homogeneously oxidized surface. For such InN layers, the electron concentration is reduced and the mobility increased.

For this oxidation, the samples were exposed for 30 s to UV light ($\lambda \sim 375$ nm). Then, a gas mixture of N_2, O_2, and O_3 was flowed over the sample for 60 s. This cycle was repeated for several hours until the resistivity of the sample saturated. The oxidation process was monitored *in situ* by measuring the resistance between In contacts (Fig. 5.24).

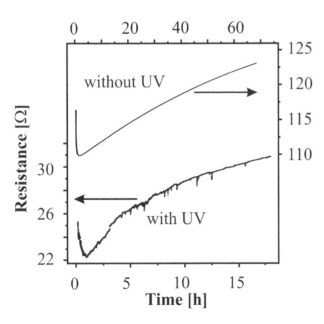

FIGURE 5.24

Resistivity of InN layers versus time during oxidation in ozone (upper curve) and the UV/ozone cycling oxidation (lower curve) of a 1000 nm thick InN film. Top and right scales correspond to the upper curve. Reprinted with permission from V. Cimalla, V. Lebedev, Ch. Y. Wang, M. Ali, G. Ecke, V. M. Polyakov, F. Schwierz, O. Ambacher, H. Lu, and W. J. Schaff, Applied Physics Letters, 90 (2007) 152106. Copyright 2007, American Institute of Physics.

Clearly, the resistivity drops during the first few UV/ozone oxidation cycles. A decreased carbon concentration and slight oxygen enrichment were found from AES. A cleaning effect by ozone as the reason for the observed decrease of resistance is unlikely, since desorption of water or polar carbon containing molecules (for example, hydrocarbons or CO_2) should result in the opposite effect on the resistance [5.118]. Rather, oxygen is diffusing into the surface without the formation of an indium oxide. Since oxygen was shown to act as donor, it leads to an enhanced free electron concentration close to the surface. Further oxidation opposes the trend and the resistance then continuously increases with the transformation of the InN surface into a thin oxidized film. When exposed to ozone without intermediate UV illumination (Fig. 5.24, upper curve), the InN surface oxidizes in a similar way; however, it requires a longer time until saturation of the resistance is observed.

The transport properties of the InN films, both as grown and after the UV/ozone oxidation, were measured for temperatures ranging from 100 to 300 K. The observed higher resistance of the InN film after the ozone oxidation is attributed to an improvement of the transport properties. At room temperature, for a 1000 nm thick InN the electron mobility increased upon oxidation from 1140 cm^2/Vs to 1230 cm^2/Vs, while the electron concentration dropped from 1.3×10^{18} cm^{-3} to 1.1×10^{18} cm^{-3}. The most striking result of these measurements is the reduction of the corresponding sheet carrier concentration of about $2 \pm 0.5 \times 10^{13}$ cm^{-2} independent of the temperature (Fig. 5.25(b)). This value is very close to the measured sheet charge density for the free electrons accumulated at the surface [5.7, 5.10] ($N_{S,0} = 1.6$–2.4×10^{13} cm^{-2} and was observed for all InN samples with a thickness greater than 700 nm (Fig. 5.25(a)). Since the oxygen only penetrates the InN surface under these conditions, a very effective passivation of the surface occurs. AES measurements revealed the existence of a closed oxide layer on top of the InN; however, this indium oxide does not correspond to cubic In_2O_3.

Two mechanisms are possible for the explanation of the effect of the ozone stimulated oxidation. First, a saturation of the In bonds on the surface by oxygen leads to a reduction of the density of surface states and consequently a reduced band bending. On the other hand, the thin oxide could exhibit bulk-like properties. The conduction band offset is high, and the native doping of the wide band gap indium oxide is *n*-type. Thus, for an InO_x/InN heterojunction an even stronger electron accumulation could be expected. Therefore it can be concluded that the ozone stimulated oxidation indeed passivates the surface; however, the exact mechanism requires further investigation.

The observed improvement of the transport properties is caused by two effects. First, if a two layer model for the electronic properties is taken into account, the film mobility μ is given by:

$$\mu = \frac{\sum_i \mu_i^2 n_i}{\sum_i \mu_i n_i},$$

(5.5)

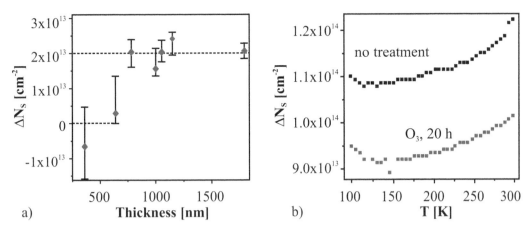

FIGURE 5.25

(a) Difference of the sheet electron concentration ΔN_S for InN films before and after ozone oxidation versus film thickness. (b) Sheet carrier concentration N_S versus temperature for a 1000 nm InN film before and after oxidation, which shows that ΔN_S is independent of the temperature. Reprinted with permission from V. Cimalla, V. Lebedev, Ch. Y. Wang, M. Ali, G. Ecke, V. M. Polyakov, F. Schwierz, O. Ambacher, H. Lu, and W. J. Schaff, Applied Physics Letters, 90 (2007) 152106. Copyright 2007, American Institute of Physics.

where μ_i and n_i are the electron mobility and concentration of the ith layer, respectively. However, reasonable values for the mobility of the accumulation layers $\mu \sim 100 - 300$ cm^2/Vs, taken from the variable-field Hall measurements (Section 5.3.2.), and a sheet carrier concentration for the surface and the interface of $N_{S,0,S} = 2.4 \times 10^{13}$ cm^{-2} and $N_{S,0,I} = 2.4 \times 10^{13}$ cm^{-2}, respectively, underestimate the mobility improvement for thicker InN layers. Thus, a second effect has to be considered.

The high degree of passivation at the surface results in decreased surface band bending, which was measured to be about -0.9 eV at the clean and the air-exposed InN surface. From a numerically self-consistent solution of the Schrödinger and Poisson equations (Fig. 5.21), for the remaining electron accumulation of about 5×10^{12} cm^{-3}, a band bending of less than -0.4 eV was determined after the ozone treatment. Figure 5.21 also demonstrates that with lower band bending the maximum of the electron concentration is shifted into the bulk from ~ 2.5 nm for the as grown InN to ~ 5.5 nm for the oxidized InN. These electrons are less affected by the surface roughness, and so the mobility should increase. For the investigated InN samples with an rms roughness ~ 1–13 nm, this effect should be more pronounced for the smoother InN (rms roughness ~ 1–3 nm) films, which was qualitatively confirmed by the experiments. Finally, it should be noted that for thin InN films with lower structural quality the improvement of the transport properties is less pronounced or even negligible (Fig. 5.25). This effect might be caused by the high density of dislocations, which are decorated by oxygen and thus negate any positive effect. In defect-free InN nanowires, the effect of the ozone oxidation was observed to be even stronger. Here the resistivity increased by up to a factor of 4.

In conclusion, a room temperature ozone-stimulated oxidation of thin InN films is proposed, which improves the transport properties and decreases the surface accumulation of free electrons in high-quality InN. The increased electron mobility is caused by the reduced number of "slow" electrons at the InN surface and the shift of the maximum electron concentration from the surface into the bulk upon the reduced surface band bending, which decreases the surface scattering. The reduction of the sheet carrier concentration after oxidation is close to the value of the intrinsic surface accumulation of free electrons in InN, which implies that the ozone treatment effectively passivates the surface. In contrast to the sputtered surface (Figs. 5.16 and 5.18), this ozone oxidized surface degrades only slightly in air (about 25% for a smooth sample after several months).

5.6 Summary

MC simulations predict that high-quality InN should exhibit excellent electron transport properties, particularly a high mobility of 14000 cm^2/Vs at room temperature and a high peak drift velocity of $\sim 5 \times 10^7$ cm/s, which makes this material promising for high-frequency devices. However, it has also been shown that these extraordinary properties require the availability of InN with carrier concentrations of 10^{16} cm^{-3} and lower. To date, this requirement has not been achieved by any growth technique used. All reported InN layers appear to be degenerate with electron concentrations of more than 10^{17} cm^{-3}. Moreover, free electrons in InN thin films are distributed in a highly inhomogeneous fashion, with different regions having different mobilities as well. In this work, several techniques were presented to analyze the carrier distribution in InN layers along with an evaluation of the obtained results. By separating electrons having different mobilities by variable-field Hall measurements and applying the QMSA technique, room temperature mobilities of 3500 cm^2/Vs and peak mobilities of 4300–5100 cm^2/Vs were obtained in samples from various laboratories, with a measured n-type carrier concentration of $\sim 1 \times 10^{17}$ cm^{-3} at room temperature. By modelling the dependence of the electron concentration on the layer thickness, different contributions to the conductivity could be identified. First, an accumulation of electrons at the surface and at the interface dominates the electronic properties of InN thin layers with $th < 300$ nm. A background electron concentration below 10^{17} cm^{-3} is required to produce non-degenerate InN. Second, the electronic properties of layers with thicknesses in the 1–2 μm range are strongly affected by edge-type threading dislocations. Finally, the background electron concentration in InN caused by impurities and/or point defects is about 1×10^{17} cm^{-3}, strongly influencing the apparent carrier concentration only for films with $th \gtrsim 10$ μm. As a consequence, for application of InN films for electronic devices, both a reduction of the density of threading dislocations and the suppression of the electron accumulation at the surface are of crucial importance. To take advantage of the predicted high electron mobility in InN, all of the discussed influences must be better controlled.

Acknowledgments

Support provided by ONR Grant N00014-02-1-0974, monitored by C. E. C. Wood, by the EU 6th Framework Program (NMP4-CT2003-505614), and by DFG Grant CI148/2.

References

[5.1] V. M. Polyakov and F. Schwierz, Low-field electron mobility in wurtzite InN, *Applied Physics Letters*, 88 (2006) 032101:1–3.

[5.2] V. W. L. Chin, T. L. Tansley, and T. Osotchan, Electron mobilities in gallium, indium, and aluminum nitrides, *Journal of Applied Physics*, 75 (1994) 7365–7372.

[5.3] V. M. Polyakov and F. Schwierz, Nonparabolicity effect on bulk transport properties in wurtzite InN, *Journal of Applied Physics*, 99 (2006) 113705:1–6.

[5.4] V. M. Polyakov and F. Schwierz, On the possibility of streaming transport due to impact ionization in wurtzite InN, *Semiconductor Science and Technology*, 21 (2006) 1651–1655.

[5.5] T. L. Tansley and C. P. Foley, Optical band gap of indium nitride, *Journal of Applied Physics*, 59 (1986) 3241–3244.

[5.6] T. L. Tansley and C. P. Foley, Electron mobility in indium nitride, *Electronics Letters*, 20 (1984) 1066–1068.

[5.7] H. Lu, W. J. Schaff, L. F. Eastmann, J. Wu, W. Walukiewicz, D. C. Look, and R. J. Molnar, Growth of thick InN by molecular beam epitaxy, *Materials Research Society Symposium Proceedings*, 743 (2003) L4.10.1–L4.10.6.

[5.8] A. Yamamoto, T. Tanaka, K. Koide, and A. Hashimoto, Improved electrical properties for metalorganic vapour phase epitaxial InN films, *physica status solidi (a)*, 194 (2002) 510–514.

[5.9] S. K. O'Leary, B. E. Foutz, M. S. Shur, U. V. Bhapkar, and L. F. Eastman, Electron transport in wurtzite indium nitride, *Journal of Applied*

Physics, 83 (1998) 826–829.

[5.10] E. Bellotti, B. K. Doshi, K. F. Brennan, J. D. Albrecht, and P. P. Ruden, Ensemble Monte Carlo study of electron transport in wurtzite InN, *Journal of Applied Physics*, 85 (1999) 916–923.

[5.11] B. E. Foutz, S. K. O'Leary, M. S. Shur, and L. F. Eastman, Transient electron transport in wurtzite GaN, InN, and AlN, *Journal of Applied Physics*, 85 (1999) 7727–7734.

[5.12] D. Fritsch, H. Schmidt, and M. Grundmann, Band dispersion relations of zinc-blende and wurtzite InN, *Physical Review B*, 69 (2004) 165204:1–5.

[5.13] E. O. Kane, Band structure of indium antimonide, *Journal of Physics and Chemistry of Solids*, 1 (1957) 249–261.

[5.14] J. G. Ruch, Electron dynamics in short channel field-effect transistors, *IEEE Transactions on Electron Devices*, 19 (1972) 652–654.

[5.15] J. R. Hauser, T. H. Glisson, and M. A. Littlejohn, Negative resistance and peak velocity in the central (000) valley of III-V semiconductors, *Solid-State Electronics*, 22 (1979) 487–493.

[5.16] J. Y.-F. Tang and K. Hess, Investigation of transient electronic transport in GaAs following high energy injection, *IEEE Transactions on Electron Devices*, 29 (1982) 1906-1911.

[5.17] K. Brennan, K. Hess, and G. J. Iafrate, Monte Carlo investigation of transient hole transport in GaAs, *Journal of Applied Physics*, 55 (1984) 3632–3635.

[5.18] J. Kolník, İ. H. Oğuzman, K. F. Brennan, R. Wang, P. P. Ruden, and Y. Wang, Electronic transport studies of bulk zincblende and wurtzite phases of GaN based on an ensemble Monte Carlo calculation including a full zone band structure, *Journal of Applied Physics*, 78 (1995) 1033–1038.

[5.19] U. V. Bhapkar and M. S. Shur, Monte Carlo calculation of velocity-field characteristics of wurtzite GaN, *Journal of Applied Physics*, 82 (1997) 1649–1655.

[5.20] J. D. Albrecht, R. P. Wang, P. P. Ruden, M. Farahmand, and K. F. Brennan, Electron transport characteristics of GaN for high temperature device modelling, *Journal of Applied Physics*, 83 (1998) 4777–4781.

[5.21] S. Yamakawa, S. Aboud, M. Saraniti, and S. M. Goodnick, Influence of the electron-phonon interaction on electron transport in wurtzite GaN, *Semiconductor Science and Technology*, 19 (2004) S475–S477.

[5.22] C. Bulutay and B. K. Ridley, Theoretical assessment of electronic transport in InN, *Superlattices and Microstructures*, 36 (2004) 465–471.

[5.23] V. M. Polyakov, F. Schwierz, D. Fritsch, and H. Schmidt, Monte Carlo study of steady-state and transient transport in wurtzite InN, *physica status solidi (c)*, 3 (2006) 598–601.

[5.24] S. Krishnamurthy, M. van Schilfgaarde, A. Sher, and A.-B. Chen, Bandstructure effect on high-field transport in GaN and GaAlN, *Applied Physics Letters*, 71 (1997) 1999–2001.

[5.25] M. Wraback, H. Shen, S. Rudin, E. Bellotti, M. Goano, J.C. Carrano, C.J. Collins, J.C. Campbell, and R.D. Dupuis, Direction-dependent band nonparabolicity effects on high-field transient electron transport in GaN, *Applied Physics Letters*, 82 (2003) 3674–3676.

[5.26] J. T. Lü and J. C. Cao, Terahertz generation and chaotic dynamics in GaN NDR diode, *Semiconductor Science and Technology*, 19 (2004) 451–456.

[5.27] N. Fitzer, A. Kuligk, R. Redmer, M. Städele, S. M. Goodnick, and W. Schattke, Full-band Monte Carlo simulations of high-field electron transport in GaAs and ZnS, *Physical Review B*, 67 (2003) 201201(R):1–4.

[5.28] H. Brooks, Scattering by ionized impurities in semiconductors, in Proceedings of the American Physical Society, *Physical Review*, 83 (1951) 868–882, p. 879.

[5.29] E. Conwell and V. P. Weisskopf, Theory of impurity scattering in semiconductors, *Physical Review*, 77 (1950) 388–390.

[5.30] D. M. Caughey and R. E. Thomas, Carrier mobilities in silicon empirically related to doping and field, *Proceedings of the IEEE*, 55 (1967) 2192–2193.

[5.31] A. G. Bhuiyan, A. Hashimoto, and A. Yamamoto, Indium nitride (InN): A review on growth, characterization, and properties, *Journal of Applied Physics*, 94 (2003) 2779–2808.

[5.32] C. S. Gallinat, G. Koblmüller, J. S. Brown, S. Bernardis, J. S. Speck, G. D. Chern, E. D. Readinger, H. Shen, and M. Wraback, In-polar InN grown by plasma-assisted molecular beam epitaxy, *Applied Physics Letters*, 89 (2006) 032109:1–3.

[5.33] W. J. Schaff, H. Lu, L. F. Eastman, W. Walukiewicz, K. M. Yu, S. Keller, S. Kurtz, B. Keyes, and L. Gevilas, in V. H. Ng and A. G. Baca, editors, *Fall 2004 ECS meeting Proceedings, v. 2004-06 (ISBN 1-56677-419-5) State-of-the-Art Program on Compound Semiconductors XLI-and-Nitride and Wide Band Gap Semiconductors for*

Sensors, Photonics, and Electronics.

[5.34] H. Lu, W. J. Schaff, L. F. Eastman, and C. E. Stutz, Surface charge accumulation of InN films grown by molecular-beam epitaxy, *Applied Physics Letters*, 82 (2003) 1736-1738.

[5.35] V. Cimalla, Ch. Förster, G. Kittler, I. Cimalla, R. Kosiba, G. Ecke, O. Ambacher, R. Goldhahn, S. Shokhovets, A. Georgakilas, H. Lu, and W. J Schaff, Correlation between strain, optical and electrical properties of InN grown by MBE, *physica status solidi (c)*, 0 (2003) 2818–2821.

[5.36] I. Mahboob, T. D. Veal, C. F. McConville, H. Lu, and W. J. Schaff, Intrinsic electron accumulation at clean InN surfaces, *Physical Review Letters*, 92 (2004) 036804:1–4.

[5.37] V. Cimalla, V. Lebedev, F. M. Morales, R. Goldhahn, and O. Ambacher, Model for the thickness dependence of electron concentration in InN films, *Applied Physics Letters*, 89 (2006) 172109:1–3.

[5.38] V. Lebedev, V. Cimalla, J. Pezoldt, M. Himmerlich, S. Krischok, J. A. Schaefer, O. Ambacher, F. M. Morales, J. G. Lozano, and D. González, Effect of dislocations on electrical and electron transport properties of InN thin films. I. Strain relief and formation of a dislocation network, *Journal of Applied Physics*, 100 (2006) 094902:1–13.

[5.39] L. F. J. Piper, T. D. Veal, C. F. McConville, H. Lu, and W. J. Schaff, Origin of the *n*-type conductivity of InN: The role of positively charged dislocations, *Applied Physics Letters*, 88 (2006) 252109:1–3.

[5.40] X. Wang, S.-B. Che, Y. Ishitani, and A. Yoshikawa, Threading dislocations in In-polar InN films and their effects on surface morphology and electrical properties, *Applied Physics Letters*, 90 (2007) 151901:1–3.

[5.41] E. A. Davis, S. F. J. Cox, R. L. Lichti, and C. G. Van de Walle, Shallow donor state of hydrogen in indium nitride, *Applied Physics Letters*, 82 (2003) 592–594.

[5.42] V. Cimalla, G. Ecke, M. Niebelschtz, O. Ambacher, R. Goldhahn, H. Lu, and W. J. Schaff, Surface conductivity of epitaxial InN, *physica status solidi (c)*, 2 (2005) 2254–2257.

[5.43] V. Cimalla, M. Niebelschtz, G. Ecke, V. Lebedev, O. Ambacher, M. Himmerlich, S. Krischok, J. A. Schaefer, H. Lu, and W. J. Schaff, Surface band bending at nominally undoped and Mg-doped InN by Auger Electron Spectroscopy, *physica status solidi (a)*, 203 (2006) 59–65.

[5.44] H. J. Hovel and J. J. Cuomo, Electrical and Optical Properties of rf-

Sputtered GaN and InN, *Applied Physics Letters*, 20 (1972) 71–73.

[5.45] Q. Guo, K. Murata, M. Nishio, and H. Ogawa, Growth of InN films on (111)GaAs substrates by reactive magnetron sputtering, *Applied Surface Science*, 169–170 (2001) 340–344.

[5.46] T. J. Kistenmacher, S. A. Ecelberger, and W. A. Bryden, Structural and electrical properties of reactively sputtered InN thin films on AlN-buffered (00.1) sapphire substrates: Dependence on buffer and film growth temperatures and thicknesses, *Journal of Applied Physics*, 74 (1993) 1684–1691.

[5.47] T. Maruyama and T. Morishita, Indium nitride thin films prepared by radio-frequency reactive sputtering, *Journal of Applied Physics*, 76 (1994) 5809–5812.

[5.48] K. Ikuta, Y. Inoue, and O. Takai, Optical and electrical properties of InN thin films grown on ZnO/α-Al_2O_3 by RF reactive magnetron sputtering, *Thin Solid Films*, 334 (1998) 49–53.

[5.49] Motlan, E. M. Goldys, and T. L. Tansley, Optical and electrical properties of InN grown by radio-frequency reactive sputtering, *Journal of Crystal Growth*, 241 (2002) 165–170.

[5.50] B. R. Natarjan, A. H. Eltoukhy, J. E. Green, and T. L. Barr, Mechanisms of reactive sputtering of indium. I: Growth of InN in mixed Ar-N_2 discharges, *Thin Solid Films*, 69 (1980) 201–216.

[5.51] N. Puychevrier and M. Menoret, Synthesis of III-V semiconductor nitrides by reactive cathodic sputtering, *Thin Solid Films*, 36 (1976) 141–145.

[5.52] S. Kosaraju, J. A. Marino, J. A. Harvey, and C. A. Wolden, The role of argon in plasma-assisted deposition of indium nitride, *Journal of Crystal Growth*, 286 (2006) 400–406.

[5.53] Q. Guo, N. Shingai, M. Nishio, and H. Ogawa, Deposition of InN thin films by radio frequency magnetron sputtering, *Journal of Crystal Growth*, 189–190 (1998) 466–470.

[5.54] W. Pan, Z. Qian, W. Shen, H. Ogawa, and Q. Guo, Capacitance characteristics in InN thin films grown by reactive sputtering on GaAs, *Japanese Journal of Applied Physics – Part 1*, 42 (2003) 5551–5556.

[5.55] Z. G. Qian, W. Z. Shen, H. Ogawa, and Q. X. Guo, Infrared reflection characteristics in InN thin films grown by magnetron sputtering for the application of plasma filters, *Journal of Applied Physics*, 92 (2002) 3683–3687.

[5.56] H. P. Zhou, W. Z. Shen, H. Ogawa, and Q. X. Guo, Temperature dependence of refractive index in InN thin films grown by reactive sputtering, *Journal of Applied Physics*, 96 (2004) 3199–3205.

[5.57] I. Mahboob, T. D. Veal, L. F. J. Piper, C. F. McConville, H. Lu, W. J. Schaff, J. Furthmüller, and F. Bechstedt, Origin of electron accumulation at wurtzite InN surfaces, *Physical Review B*, 69 (2004) 201307(R):1–4.

[5.58] T. D. Veal, L. F. J. Piper, I. Mahboob, H. Lu, W. J. Schaff, and C. F. McConville, Electron accumulation at InN/AlN and InN/GaN interfaces, *physica status solidi (c)*, 2 (2005) 2246–2249.

[5.59] G. Koblmüller, C. S. Gallinat, S. Bernardis, J. S. Speck, G. D. Chern, E. D. Readinger, H. Shen, and M. Wraback, Optimization of the surface and structural quality of N-face InN grown by molecular beam epitaxy, *Applied Physics Letters*, 89 (2006) 071902:1–3.

[5.60] V. Lebedev, Ch. Y. Wang, V. Cimalla, S. Hauguth, T. Kups, M. Ali, G. Ecke, M. Himmerlich, S. Krischok, J. A. Schaefer, O. Ambacher, V. M. Polyakov, and F. Schwierz, Effect of surface oxidation on electron transport in InN thin films, *Journal of Applied Physics*, 101 (2007) 123705:1–6.

[5.61] E. Dimakis, E. Illiopoulos, K. Tsagaraki, Th. Kehagias, Ph. Kominou, and A. Georgakilas, Heteroepitaxial growth of In-face InN on GaN (0001) by plasma-assisted molecular-beam epitaxy, *Journal of Applied Physics*, 97 (2005) 113520:1–10.

[5.62] A. Kamińska, G. Franssen, T. Suski, I. Gorczyca, N. E. Christensen, A. Svane, A. Suchocki, H. Lu, W. J. Schaff, E. Dimakis, and A. Georgakilas, Role of conduction-band filling in the dependence of InN photoluminescence on hydrostatic pressure, *Physical Review B*, 76 (2007) 075203:1–5.

[5.63] K. A. Wang, C. Lian, N. Su, D. Jena, and J. Timler, Conduction band offset at the InN/GaN heterojunction, *Applied Physics Letters*, 91 (2007) 232117:1–3.

[5.64] T. Inushima, T. Sakon, and M. Motokawa, Relationship between the optical properties and superconductivity of InN with high carrier concentration, *Journal of Crystal Growth*, 269 (2004) 173–180.

[5.65] T. Inushima, T. Takenobu, M. Motokawa, K. Koide, A. Hashimoto, A. Yamamoto, Y. Saito, T. Yamaguchi, and Y. Nanishi, Influence of growth condition on superconducting characteristics of InN on sapphire (0001), *physica status solidi (c)*, 0 (2002) 364–367.

[5.66] M. Higashiwaki and T. Matsui, High-quality InN film grown on a low-temperature-grown GaN intermediate layer by plasma-assisted molecular-beam epitaxy, *Japanese Journal of Applied Physics – Part 2*, 41 (2002) L540–L542.

[5.67] M. Higashiwaki and T. Matsui, Epitaxial growth of high-quality InN films on sapphire substrates by plasma-assisted molecular-beam epitaxy, *Journal of Crystal Growth*, 252 (2003) 128–135.

[5.68] Y. Nanishi, Y. Saito, and T. Yamaguchi, RF-molecular beam epitaxy growth and properties of InN and related alloys, *Japanese Journal of Applied Physics – Part 1*, 42 (2003) 2549–2559.

[5.69] Y. Saito, N. Teraguchi, A. Suzuki, T. Araki, and Y. Nanishi, Growth of high-electron-mobility InN by RF molecular beam epitaxy, *Japanese Journal of Applied Physics – Part 2*, 40 (2001) L91–L93.

[5.70] T. Ive, O. Brandt, M. Ramsteiner, M. Giehler, H. Kostial, and K. H. Ploog, Properties of InN layers grown on 6H-SiC(0001) by plasma-assisted molecular beam epitaxy, *Applied Physics Letters*, 84 (2004) 1671–1673.

[5.71] D.-J. Jang, G.-T. Lin, C.-L. Wu, C.-L. Hsiao, L. W. Tu, and M.-E. Lee, Energy relaxation of InN thin films, *Applied Physics Letters*, 91 (2007) 092108:1–3.

[5.72] Y. Ishitani, W. Terashima, S. B. Che, and A. Yoshikawa, Conduction and valence band edge properties of hexagonal InN characterized by optical measurements, *physica status solidi (c)*, 3 (2006) 1850–1853.

[5.73] V. Lebedev, V. Cimalla, F. M. Morales, J. G. Lozano, D. González, Ch. Maudera, and O. Ambacher, Effect of island coalescence on structural and electrical properties of InN thin films, *Journal of Crystal Growth*, 300 (2007) 50–56.

[5.74] O. Briot, B. Maleyre, S. Clur-Ruffenach, B. Gil, C. Pinquier, F. Demangeot, and J. Frandon, The value of the direct band gap of InN: a re-examination, *physica status solidi (c)*, 1 (2004) 1425–1428.

[5.75] W. Z. Shen, X. D. Pu, J. Chen, H. Ogawa, Q. X. Guo, Critical point transitions of wurtzite indium nitride, *Solid State Communications*, 137 (2006) 49–52.

[5.76] X. D. Pu, J. Chen, W. Z. Shen, H. Ogawa, and Q. X. Guo, Temperature dependence of Raman scattering in hexagonal indium nitride films, *Journal of Applied Physics*, 98 (2005) 033527:1–6.

[5.77] R. S. Qhalid Fareed, R. Jain, R. Gaska, M. S. Shur, J. Wu, W. Walukiewicz, and M. A. Khan, High quality InN/GaN heterostructures grown by migration enhanced metalorganic chemical vapor deposition, *Applied Physics Letters*, 84 (2004) 1892–1894.

[5.78] K. S. A. Butcher, M. Wintrebert-Fouquet, P. P.-T. Chen, K. E. Prince, H. Timmers, S. K. Shrestha, T. V. Shubina, S. V. Ivanov, R. Wuhrer, M. R. Phillips, and B. Monemar, Non-stoichiometry and non-homogeneity in InN, *physica status solidi (c)*, 2 (2005) 2263–2266.

[5.79] J. R. Meyer, C. A Hoffman, F. J. Bartoli, D. A. Arnold, S. Sivananthan, and J. P. Faurie, Methods for magnetotransport characterization of IR detector materials, *Semiconductor Science and Technology*, 8 (1993) 805–823.

[5.80] A. Saxler, D. C. Look, S. Elhamri, J. Sizelove, W. C. Mitchell, C. M. Sung, S. S. Park, and K. Y. Lee, High mobility in *n*-type GaN substrates, *Applied Physics Letters*, 78 (2001) 1873–1875.

[5.81] C. H. Swartz, R. P. Tomkins, T. H. Myers, H. Lu, and W. J. Schaff, Demonstration of nearly non-degenerate electron conduction in InN grown by molecular beam epitaxy, *physica status solidi (c)*, 2 (2005) 2250–2253.

[5.82] C. H. Swartz, R. P. Tompkins, N. C. Giles, T. H. Myers, H. Lu, W. J. Schaff, and L. F. Eastman, Investigation of multiple carrier effects in InN epilayers using variable magnetic field Hall measurements, *Journal of Crystal Growth*, 269 (2004) 29–34.

[5.83] T. B. Fehlberg, G. A. Umana-Membreno, B. D. Nener, G. Parish, C. S. Gallinat, G. Koblmüller, S. Rajan, S. Bernadis, and J. S. Speck, Characterisation of multiple carrier transport in indium nitride grown by molecular beam epitaxy, *Japanese Journal of Applied Physics*, 45 (2006) L1090–L1092.

[5.84] P. A. Anderson, C. H. Swartz, D. Carder, R. J. Reeves, S. M. Durbin, S. Chandril, and T. H. Myers, Buried *p*-type layers in Mg-doped InN, *Applied Physics Letters*, 89 (2006) 184104:1–3.

[5.85] H. Lu, W. J. Schaff, J. Hwang, H. Wu, W. Yeo, A. Pharkya, and L. F. Eastman, Improvement on epitaxial grown of InN by migration enhanced epitaxy, *Applied Physics Letters*, 77 (2000) 2548–2550.

[5.86] P. A. Anderson, R. J. Reeves, and S. M. Durbin, RF plasma sources for III-nitrides growth: influence of operating conditions and device geometry on active species production and InN film properties, *physica status solidi (a)*, 203 (2006) 106–111.

[5.87] R. J. Kinsey, P. A. Anderson, C. E. Kendrick, R. J. Reeves, and S. M. Durbin, Characteristics of InN thin films grown using a PAMBE technique, *Journal of Crystal Growth*, 269 (2004) 167–172.

[5.88] I. Vurgaftman, J. R. Meyer, C. A. Hoffman, D. Redfern, J. Antoszewski, L. Farone, and J. R. Lindemuth, Improved quantitative mobility spectrum analysis for Hall characterization, *Journal of Applied Physics*, 84 (1998) 4966–4973.

[5.89] T. Hofmann, V. Darakchieva, B. Monemar, H. Lu, W. J. Schaff, and M. Schubert, Optical Hall effect in hexagonal InN, *Journal of Electronic Materials*, 37 (2008) 611–615.

[5.90] D. L. Rode, Low-field electron transport, in R. K. Willardson and A. C. Beer (Eds.) *Semiconductors and semimetals*, Academic Press, New York, London 10 (1975) 1–90.

[5.91] D. C. Look and J. R. Sizelove, Predicted maximum mobility in bulk GaN, *Applied Physics Letters*, 79 (2001) 1133-1135.

[5.92] D. C. Look and J. R. Sizelove, Private communication.

[5.93] R. Kosiba, G. Ecke, V. Cimalla, L. Spieß, S. Krischok, J. A. Schaefer, O. Ambacher, and W. J. Schaff, Sputter depth profiling of InN layers, *Nuclear Instruments and Methods in Physics Research Section B*, 215 (2004) 486–494.

[5.94] W. Huang, M. Yoshimoto, K. Taguchi, H. Harima, and J. Saraie, Improved electrical properties of InN by high-temperature annealing with *in situ* capped SiN_x layers, *Japanese Journal of Applied Physics*, 43 (2004) L97–L99.

[5.95] C. R. Abernathy, S. J. Pearton, F. Ren, and P. W. Wisk, Growth of InN for ohmic contact formation by electron cyclotron resonance metalorganic molecular-beam epitaxy, *Journal of Vacuum Science and Technology B*, 11 (1993) 179–182.

[5.96] Y. Uesaka, A. Yamamoto, and A. Hashimoto, Band gap widening of MBE grown InN layers by impurity incorporation, *Journal of Crystal Growth*, 278 (2005) 402–405.

[5.97] A. Janotti and C. G. Van de Walle, Sources of unintentional conductivity in InN, *Applied Physics Letters*, 92 (2008) 032104:1–3.

[5.98] D. C. Look, H. Lu, W. J. Schaff, J. Jasinski, and Z. Liliental-Weber, Donor and acceptor concentrations in degenerate InN, *Applied Physics Letters*, 80 (2002) 258–260.

[5.99] S. Limpijumnong and C. G. Van de Walle, Passivation and doping due to hydrogen in III-Nitrides, *physica status solidi (b)*, 228 (2001) 303–307.

[5.100] J. Oila, A. Kemppinen, A. Laasko, K. Saarinen, W. Egger, L. Liszkay, P. Sperr, H. Lu, and W. J. Schaff, Influence of layer thickness on the formation of In vacancies in InN grown by molecular beam epitaxy, *Applied Physics Letters*, 84 (2004) 1486–1488.

[5.101] K. L. Westra, R. P. W. Lawson, and M. J. Brett, The effects of oxygen contamination on the properties of reactively sputtered indium nitride films, *Journal of Vacuum Science and Technology A*, 6 (1988) 1730–1732.

[5.102] P. Specht, R. Armitage, J. Ho, E. Gunawan, Q. Yang, X. Xu, C. Kisielowski, and E. R. Weber, The influence of structural properties on conductivity and luminescence of MBE grown InN, *Journal of Crystal Growth*, 269 (2004) 111–118.

[5.103] J. Wu, W. Walukiewicz, S. X. Li, R. Armitage, J. C. Ho, E. R. Weber, E. E. Haller, H. Lu, W. J. Schaff, A. Barcz, and R. Jakiela, Effects of electron concentration on the optical absorption edge of InN, *Applied Physics Letters*, 84 (2004) 2805–2807.

[5.104] P. D. C. King, T. D. Veal, C. F. McConville, F. Fuchs, J. Furthmüller, F. Bechstedt, P. Schley, R. Goldhahn, J. Schörmann, D. J. As, K. Lischka, D. Muto, H. Naoi, Y. Nanishi, H. Lu, and W. J. Schaff, Universality of electron accumulation at wurtzite *c*- and *a*-plane and zincblende InN surfaces, *Applied Physics Letters*, 91 (2007) 092101:1–3.

[5.105] P. D. C. King, T. D. Veal, C. S. Gallinat, G. Koblmüller, L. R. Bailey, J. S. Speck, and C. F. McConville, Influence of growth conditions and polarity on interface-related electron density in InN, *Journal of Applied Physics*, 104 (2008) 103703:1–5.

[5.106] Y. Ishitani, X. Wang, S.-B. Che, and A. Yoshikawa, Effect of electron distribution in InN films on infrared reflectance spectrum of longitudinal optical phonon-plasmon interaction region, *Journal of Applied Physics*, 103 (2008) 053515:1–10.

[5.107] A. V. Soldatov, A. Guda, A. Kravtsova, M. Petravic, P. N. K. Deenapanray, M. D. Fraser, Y.-W. Yang, P. A. Anderson, and S. M. Durbin, Nitrogen defect levels in InN: XANES study, *Radiation Physics and Chemistry*, 75 (2006) 1635–1637.

[5.108] V. Cimalla, V. Lebedev, Ch. Y. Wang, M. Ali, G. Ecke, V. M. Polyakov, F. Schwierz, O. Ambacher, H. Lu, and W. J. Schaff, Reduced surface electron accumulation at InN films by ozone induced oxidation, *Applied Physics Letters*, 90 (2007) 152106:1–3.

[5.109] R. P. Bhatta, B. D. Thoms, M. Alevli, and N. Dietz, Surface electron accumulation in indium nitride layers grown by high pressure chemical vapor deposition, *Surface Science*, 601 (2007) L120–L123.

[5.110] R. P. Bhatta, B. D. Thoms, A. Weerasekera, A. G. U. Perera, M. Alevli, and N. Dietz, Carrier concentration and surface electron accumulation in indium nitride layers grown by high pressure chemical vapor deposition, *Journal of Vacuum Science and Technology A*, 25 (2007) 967–970.

[5.111] T. D. Veal, I. Mahboob, L. F. J. Piper, C. F. McConville, H. Lu, and W. J. Schaff, Indium nitride: Evidence of electron accumulation, *Journal of Vacuum Science and Technology B*, 22 (2004) 2175–2178.

[5.112] C. J. Lu, L. A. Bendersky, H. Lu, and W. J. Schaff, Threading dislocations in epitaxial InN thin films grown on (0001) sapphire with a GaN buffer layer, *Applied Physics Letters*, 83 (2003) 2817–2819.

[5.113] I. Arslan, A. Bleloch, E. A. Stach, and N. D. Browning, Atomic and electronic structure of mixed and partial dislocations in GaN, *Physical Review Letters*, 94 (2005) 025504:1–4.

[5.114] W. Walukiewicz, Amphoteric native defects in semiconductors, *Applied Physics Letters*, 54 (1989) 2094–2096.

[5.115] D. Segev and C. G. Van de Walle, Origins of Fermi level pinning on GaN and InN polar and nonpolar surfaces, *Europhysics Letters*, 76 (2006) 305–311.

[5.116] C. L. Wu, H. M. Lee, C. T. Kuo, C. H. Chen, and S. Gwo, Absence of Fermi-level pinning at cleaved nonpolar InN surfaces, *Physical Review Letters*, 101 (2008) 106803:1–4.

[5.117] T. Maruyama, K. Yorozu, T. Noguchi, Y. Seki, Y. Saito, T. Araki, and Y. Nanishi, Surface treatment of GaN and InN using $(NH_4)_2S_x$, *physica status solidi (c)*, 0 (2003) 2031–2034.

[5.118] H. Lu, W. J. Schaff, and L. F. Eastman, Surface chemical modification of InN for sensor applications, *Journal of Applied Physics*, 96 (2004) 3577–3579.

6

Electronic states in InN and lattice dynamics of InN and InGaN

V. Yu. Davydov[1] **and A. A. Klochikhin**[1,2]

[1] *Ioffe Physico-Technical Institute, Russian Academy of Science, Polytechnicheskaya 26, 194021 St. Petersburg, Russia*

[2] *Nuclear Physics Institute, 188350, St. Petersburg, Russia*

6.1 Introduction

At present indium nitride attracts considerable attention from scientists. The interest is mainly stimulated by its narrow band gap (about 0.7 eV) first reported in 2002 and a strong IR luminescence, which makes InN and InGaN alloys extremely attractive candidates for application in a wide class of opto-electronic devices. Since the authors of this chapter, as part of one of the participating teams, witnessed how different teams dealing with electronic structure calculations of InN and experimental investigations of its optical properties came simultaneously and independently to the idea of revision of the InN band gap (from about 1.9 eV down to 0.7 eV), we believe that it is appropriate to describe here the history of this important event.

During a private discussion at the International Workshop on Nitride Semiconductors (IWN-2000) in Nagoya in the autumn of 2000, one of the authors (VD) mentioned to Prof. F. Bechstedt from Friedrich-Schiller-University (Jena, Germany) and Prof. H. Harima from Osaka University (Japan) that, according to the experimental results obtained at Ioffe Institute (St. Petersburg, Russia) for InN and In-rich InGaN alloys grown by plasma-assisted MBE at Ioffe Institute and metal-organic MBE at Hannover University (Germany), the true band gap of InN was of the order of 0.9 eV or even less, which was well below the commonly accepted value of 1.89 eV [6.1]. Among the main arguments was the observation of the absorption edge in the region of 0.9 eV, its high-frequency shift with increasing free carrier concentration (which was attributed to the Burstein-Moss effect [6.2]), the observation of intense IR luminescence near the absorption edge, and, lastly, a regular shift of the absorption edge and edge luminescence to higher energies with increasing Ga concentration in In-rich $In_xGa_{1-x}N$ alloys ($0.4 < x < 1$). The response of Friedhelm Bechstedt was surprising. He said that, in early 2000, his group obtained a band gap value of the order of 1 eV in *ab initio* calculations of the electronic structure of hexagonal InN. However, they had not published these results because of strong criticism from colleagues.

As a result of the discussion in Nagoya, it was decided to submit a joint paper. The manuscript, entitled "Absorption and emission of InN: A possible revision of the fundamental gap value", was ready in early summer of 2001. It reported the experimental data on luminescence and absorption of InN and In-rich $In_xGa_{1-x}N$ alloys ($0.4 < x < 1$), pointing to a narrow band gap of InN and the results of quasi-particle band structure calculations leading to the same conclusion. However, several journals rejected the paper because, as one of the referees wrote, "there is no new science in what is presented here, except perhaps some advance in crystal growth technique, which is not the point".

Meanwhile, new data on the photoluminescence and absorption for the InN samples grown by MOVPE at Fukui University (Japan) were obtained at Ioffe Institute. These data also supported the validity of the conclusions on the narrow gap nature of InN (about 0.7 eV). Finally, the paper was separated into two parts and was published as two short Rapid Research Letters by Physica Status Solidi (b) in 2002 [6.3, 6.4]. A more detailed description of the theoretical calculations of the electronic structure of InN was given by Friedhelm Bechstedt and Jürgen Furthmüller in the paper entitled "Do we know the fundamental energy gap of InN?" which was also published in 2002 [6.5].

The narrow band-gap nature of InN, established simultaneously and independently both theoretically and experimentally in 2000 and perceived at first with great skepticism, is now commonly accepted. This is probably not the only surprise InN has to offer, and we will enjoy more intriguing findings in the future.

This chapter presents results of experimental and theoretical investigations of optical properties of hexagonal InN. Interband photoluminescence and absorption spectra of InN samples with free electron concentrations in a wide range are reviewed. A model approach based on a detailed consideration of the conduction and valence bands is described. Model calculations of PL and absorption spectra have shown that the band gap of InN, in the limit of zero temperature and zero electron concentration, is about 0.665–0.670 eV. The analytical solution of the classical Thomas-Fermi equation for a planar accumulation layer of a degenerate semiconductor is also presented. This solution gives simple expressions for the band bending potential and inhomogeneous electron density, relating these functions to crystal parameters and external electrical charge. Results of experimental and theoretical studies of the lattice dynamics of hexagonal InN and InGaN alloys are reviewed. InN zone-center phonons, coupling between electron excitations and longitudinal optical phonons, and local vibrational modes in Mg-doped InN are considered in detail. The calculated phonon dispersion curves and the phonon density of states function of InN are discussed. The phonon mode behavior experimentally observed in InGaN is described, and the use of resonant Raman scattering for estimating the band gap of InGaN alloys is demonstrated.

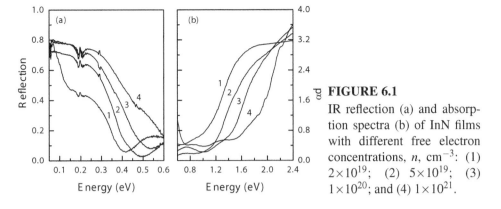

FIGURE 6.1

IR reflection (a) and absorption spectra (b) of InN films with different free electron concentrations, n, cm^{-3}: (1) 2×10^{19}; (2) 5×10^{19}; (3) 1×10^{20}; and (4) 1×10^{21}.

6.2 Interband optical spectra of heavily doped InN

6.2.1 Absorption spectra

No photoluminescence (PL) was observed in InN with very high electron concentrations, $n \sim 10^{20}$–10^{21} cm^{-3}. For this reason, the band gap of this material had previously been estimated from absorption measurements alone. The most complete study of the optical properties of InN films with high electron concentrations was performed in Ref. [6.6]. The threshold of interband absorption in films with an electron concentration of 3×10^{20} cm^{-3} was found to be 2.05 eV. Noteworthy is the unusual energy dependence of the absorption coefficient which is V-shaped, similar to curve 4 in Fig. 6.1: the film absorption is the lowest at energy $\hbar\omega \approx 1.2$–1.3 eV and steeply increases with decreasing or increasing photon energy. Such behavior of the absorption coefficient was accounted for in Ref. [6.6] in terms of two different mechanisms. The behavior of the absorption coefficient in the high-energy part of the spectrum was attributed to interband transitions. The increase in the absorption coefficient from the minimum value of 6×10^3 cm^{-1} at $\hbar\omega \approx 1.2$–1.3 eV to 2×10^4 cm^{-1} in the range of low energies at $\hbar\omega \approx 0.8$ eV was attributed to increasing absorption by free carriers. The plasma resonance revealed in the reflectance spectrum at $\hbar\omega \approx 0.6$ eV confirmed this interpretation and indicated that the transmission of a film is, indeed, strongly modified by free-carrier absorption.

A set of typical absorption and reflection spectra of heavily doped InN epilayers similar to those given in Ref. [6.6] is presented in Fig. 6.1. The spectra clearly demonstrate that the plasma reflection and interband absorption thresholds correlate well with free electron concentrations in InN.

6.2.2 Interband luminescence spectra

Interband PL was observed for the first time from InN samples grown by molecular-beam epitaxy [6.3]. It was shown that heavily doped degenerate n-type InN crystals were characterized by band-to-band PL bands which were structureless because of

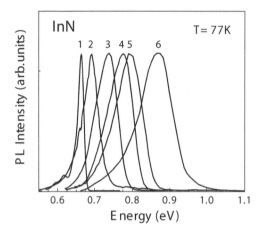

FIGURE 6.2

Shift and broadening of the InN PL band with increasing electron concentration n, cm^{-3}: (1) 3.6×10^{17}; (2) 1×10^{18}; (3) 6×10^{18}; (4) 9×10^{18}; (5) 1.1×10^{19}; and (6) 2.1×10^{19}.

screening of the Coulomb interaction followed by disappearance of excitons, shallow donor and acceptor states, and their complexes. The interband PL origin was confirmed by variations in the band shape with both electron concentration and sample temperature [6.4, 6.7]. Comparison of the PL parameters obtained for the InN crystals with the data for doped GaAs and GaN crystals demonstrated that the PL was due to interband recombination of free electrons and photoexcited holes [6.7]. As joint analysis of interband PL and absorption spectra and model calculations of the PL band shape have established, InN is a narrow-gap semiconductor with a band gap of about 0.7 eV. It has been found that the interband absorption threshold of these crystals is shifted toward higher energies [6.4] due to the band-filling Burstein-Moss effect [6.2] and can considerably exceed the real band gap. In addition, it has been established [6.7] that the band gap itself is affected by the electron concentration (see Sect. 6.2.3.) The InN band gap of about 0.7 eV was also confirmed in studies of In$_x$Ga$_{1-x}$N solid solutions with high indium concentrations [6.4]. The appearance of the interband infrared PL of InN samples with electron concentrations of $n_e \approx 1\text{--}2 \times 10^{19}$ cm^{-3} and lower was confirmed by many authors [6.8–6.12].

A set of PL spectra of heavily doped InN samples is presented in Fig. 6.2. For comparison, the PL spectra of samples with electron concentrations of about 3.6×10^{17} cm^{-3} and 1×10^{18} cm^{-3} are also presented.

The PL spectra of heavily doped samples with different electron concentrations allowed analysis of the PL band shape variations in a wide energy range [6.4, 6.7]. Our studies and the results of other authors have shown that the PL band shape depends strongly on the conduction band dispersion and the breakdown of the momentum conservation law in optical interband transitions. The nonparabolic dispersion of the InN conduction band and the momentum conservation law violation lead to a very characteristic PL band shape (see below). Another characteristic that can be deduced from the interband PL spectra of heavily doped samples is the band gap dependence on the electron concentration.

6.2.3 Effect of exchange interaction on the band gap

Studies of the photoluminescence of heavily doped InN samples have revealed that band gap (E_g) decreases with increasing carrier concentration [6.7]. This band-gap shrinkage can be attributed to the lowering of the energy of the electrically neutral system comprised of charged donors and the electron gas. The attractive and repulsive Coulomb interactions between particles in this system cancel each other out, and the theory of Fermi liquids [6.13, 6.14] should be invoked to explain the observed effect.

According to the Gell-Mann-Bruckner theory, the dependence of E_g on carrier concentration arises in a Fermi liquid because of the Hartree-Fock exchange interaction of Fermi particles [6.13, 6.14]. This dependence is to be taken into account in the analysis of the optical properties of degenerate semiconductors. The correction to the self-energy of an electron, $\Delta E_{ex}(n)$, which is associated with the electron-electron exchange interaction, depends on the electron concentration, n.

It is noteworthy that, in contrast to the Coulomb interaction, the exchange interaction is not screened or cancelled and, in the effective-mass approximation for an isotropic conduction band, this expression reduces to the known function for the ideal Fermi liquid [6.13] and gives the shift

$$\Delta E_{ex}(n) = -\frac{3}{2\pi}\frac{R_H a_B}{r_s}\left\{\frac{9\pi}{4}\right\}^{1/3} = -\left[\frac{81}{8\pi}\right]^{1/3}\frac{e^2 n^{1/3}}{2}, \tag{6.1}$$

where, R_H and a_B are the binding energy and the Bohr radius of the hydrogen atom, and r_s is the averaged inter-electron separation. For clarity, the shift can be expressed in meV, with $n_0 = 1 \times 10^{18}$ cm^{-3} used as the unit of concentration. As a result, we have

$$\Delta E_{ex}(n) = -106[n/n_0]^{1/3}. \tag{6.2}$$

In this approximation, the function of the concentration is universal and independent of other crystal parameters. Nevertheless, individual characteristics of the crystal can change $\Delta E_{ex}(n)$ because the wave functions of the electrons in the crystal differ from those of free electrons.

Using the average value of the correction for the exchange interaction to estimate this energy, we obtain

$$\Delta E_{ex}(n) = -\frac{e^2}{2}\sum_{\mathbf{p},\mathbf{p}'<p_F}\int d^3r\, d^3r' \times$$
$$\times \Psi^*_{\sigma,\mathbf{p}}(\mathbf{r})\Psi^*_{\sigma,\mathbf{p}'}(\mathbf{r})\Psi_{\sigma,\mathbf{p}}(\mathbf{r}')\Psi_{\sigma,\mathbf{p}'}(\mathbf{r}')/|\mathbf{r}-\mathbf{r}'|. \tag{6.3}$$

For example, substitution into Eq. 6.3 of the wave functions in the form

$$\Psi_{\sigma,\mathbf{p}'}(\mathbf{r}) = \frac{1}{\sqrt{V}}u_{\sigma,\mathbf{p}}(\mathbf{r})\exp(i\mathbf{pr}), \tag{6.4}$$

where $u_{\sigma,\mathbf{p}'}(\mathbf{r})$ is the Bloch factor and V is the crystal volume, shows that $\Delta E_{ex}(n)$ is proportional to the squared overlapping integrals

$$\frac{1}{v_0}\int d^3r\, u_{\sigma,\mathbf{p}}(\mathbf{r})u_{\sigma,\mathbf{p}'}(\mathbf{r}) \tag{6.5}$$

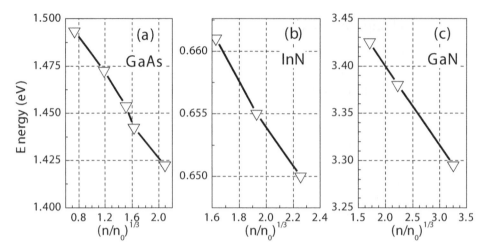

FIGURE 6.3

Dependence of E_g on electron concentration for GaAs, InN, and GaN. Concentration is expressed in units $n_0 = 1 \times 10^{18}$ cm^{-3}. Reprinted with permission from V. Yu. Davydov, A. A. Klochikhin, V. V. Emtsev, D. A. Kurdyukov, S. V. Ivanov, V. V. Vekshin, F. Bechstedt, J. Furthmüller, J. Aderhold, J. Graul, A. V. Mudryi, H. Harima, A. Hashimoto, A. Yamamoto, and E. E. Haller, physica status solidi (b), 234, 787 (2002). Copyright 2002 by Wiley-VCH Publishers, Inc.

between Bloch factors with $\mathbf{p} \neq \mathbf{p}'$. Here, v_0 is the unit lattice cell volume. In the effective-mass approximation, $\mathbf{p} = \mathbf{p}' = 0$, and the overlapping integrals are equal to unity. However, in the case of degenerate electrons, the range within which the values of p fall is rather wide, and the overlapping integrals may be, on the average, much smaller than their values in the effective-mass approximation. Therefore, Eq. 6.2 can be regarded as an estimate of the upper bound on the concentration-related shift of E_g. It follows from this estimate that the shift of the conduction band as a function of n is to be taken into account in evaluating the band gap energy. Thus, E_g in calculations is to be replaced by $E_g(n)$.

The result obtained makes it possible to analyze data on interband absorption and luminescence in heavily doped GaAs, InN, and GaN crystals [6.7]. As shown by the absorption and luminescence studies, the $E_g(n)$ dependence has the form characteristic of the Hartree-Fock exchange interaction for these three materials (see Fig. 6.3). The shift is of the same order of magnitude as the estimate of Eq. 6.2. However, the values of these factors are lower than 106 meV and different for these three materials. The highest value of about 70 meV is observed for GaAs. Extrapolation of $E_g(n)$ to a zero carrier concentration gives for this crystal $E_g = 1.5$ eV, i.e., the value close to the true band gap energy. This factor is 50 meV for the GaN crystal and about 20 meV for InN. Thus, we have a noticeable quantitative deviation from the estimate of Eq. 6.2. This fact can be explained by the fact that the overlapping integrals of Eq. 6.3 are not equal to unity if it is assumed that the deviation from the effective-mass approximation for the conductivity bands for InN and GaN exceeds that for GaAs. Linear extrapolation of $E_g(n)$ to zero concentration gives, for InN, $E_g = 0.69$ eV [6.7].

6.3 Photoluminescence and absorption spectra of InN samples with low electron concentrations

A considerable breakthrough in optical investigations of InN crystals occurred when n-InN films up to 12 μm thick, with carrier concentrations below 4×10^{17} cm^{-3} and room-temperature Hall mobilities higher than 2100 cm^2/Vs, were grown [6.15, 6.16]. At present, a good deal of experimental information concerning high-quality n-InN samples is available in Refs [6.17–6.22] and references therein.

The PL studies of degenerate n-InN samples with relatively low electron concentrations of the order of $n \sim 10^{18}$ cm^{-3} have revealed that the exponentially decreasing density-of-states tails of the valence and conduction bands formed by shallow localized electrons and holes can manifest themselves in the PL band formation. The band gap has been estimated to be about 0.670 eV, though the Urbach tails lead to some uncertainty in this value [6.23].

We consider here in detail some results obtained in Ref. [6.24] for the InN samples of improved quality. The samples were grown by MBE at Cornell University (USA). Typically, an AlN nucleation layer and a GaN buffer layer were deposited prior to InN growth [6.16]. Although the samples were not intentionally doped, free electron concentrations ranging from 3.6×10^{17} to 8.4×10^{17} cm^{-3} were found in the samples by Hall effect measurements.

The PL spectra of these samples allowed us to observe the structure related to acceptors. The dependences of the PL spectra on the excitation power, temperature, and electron concentrations have been studied. A model describing the major experimental facts has been developed.

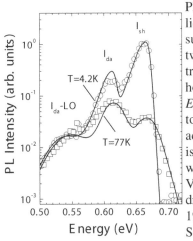

FIGURE 6.4

PL spectra of the thick sample Gs2060 at liquid-helium and liquid-nitrogen temperatures. The solid lines present the results of model fits. In calculations, maximum I_{sh} is due to two processes: (i) band-to-band recombination of free electrons and holes; and (ii) annihilation of free electrons and holes localized by shallow acceptors with binding energy $E_{sh} \approx 5$–10 meV. The maximum I_{da} is considered to be due to recombination of electrons with holes localized by deep acceptors ($E_{da} \approx 50-55$ meV), and the maximum $I_{da} - LO$ is a LO-phonon replica of the previous transition. Reprinted with permission from A. A. Klochikhin, V. Yu. Davydov, V. V. Emtsev, A. V. Sakharov, V. A. Kapitonov, B. A. Andreev, H. Lu, and W. J. Schaff, Physical Review B, 71, 195207 (2005). Copyright 2005 by the American Physical Society.

The PL spectra of sample Gs2060 at liquid-helium and liquid-nitrogen temperatures measured at steady state excitation are presented in Fig. 6.4. The most striking features of the spectra are the structure consisting of three peaks and the sensitivity to temperature variations.

The structure of the spectra of the high-quality samples can result from recombination of the holes trapped by impurity centers. The energy position of the middle peak agrees well with the data of Ref. [6.22] and can be attributed to annihilation of holes localized by deep acceptor states. The lowest-energy peak is shifted toward lower energies by the LO-phonon energy [6.25] and, therefore, can be regarded as a manifestation of the electron-phonon interaction. To interpret the high-energy peak, both shallow localized and band states of holes should be taken into account, as will be shown below.

In the paper of Wang *et al.* [6.26], a series of Mg-doped InN layers was characterized by low-temperature PL. Two peaks in the PL spectra of partially carrier-compensated Mg-doped InN were observed. The studies of the dependences of peak energies and intensities on the free carrier concentration and excitation power led the authors to the conclusion that the high-energy peak was due to band-to-band transitions and the low-energy peak resulted from free-to-acceptor transitions. These data showed that the PL band behaviors in lightly Mg-doped InN were similar to those reported for low carrier concentration *n*-type InN in which two kinds of PL peaks were observed [6.22, 6.24]. The Mg acceptor activation energy was estimated to be about 61 meV [6.26].

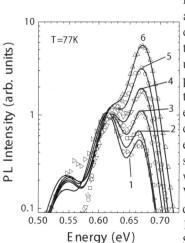

FIGURE 6.5

Excitation power-dependent PL spectra of sample Gs2060 at liquid-nitrogen temperature. Symbols are experimental data and the solid lines are results of model calculations (obtained for different positions of the hole chemical potential under the assumption of a quasi-equilibrium distribution of photoholes and increasing electron concentrations). Spectrum 1 was detected by an InSb photodiode at the minimal excitation power, spectra 2–6 were obtained with an InGaAs detector. Spectra 1–5 were obtained in the continuous wave excitation regime, and the pulsed excitation was used for spectrum 6. Spectra are normalized at 0.61 eV. Reprinted with permission from A. A. Klochikhin, V. Yu. Davydov, V. V. Emtsev, A. V. Sakharov, V. A. Kapitonov, B. A. Andreev, H. Lu, and W. J. Schaff, Physical Review B, 71, 195207 (2005). Copyright 2005 by the American Physical Society.

As can be seen from Fig. 6.4, the rise from liquid-helium to liquid-nitrogen temperature leads to intensity redistribution between two higher-energy peaks. This transformation of the spectrum with temperature may be explained in two ways.

The first one assumes an increase in the energy relaxation rate of non-equilibrium holes in shallow localized and band states followed by an increasing population of deep acceptor states.

The other version assumes that saturation of the localized states by photoholes is reached and an equilibrium or almost equilibrium distribution of holes is established. This version involves an increase in the non-radiative process rate with temperature followed by a shift of the photohole population boundary toward the band gap.

It is well known that the PL intensity decreases as temperature increases, which also points to the enhancement of non-radiative processes and makes the second explanation preferable. On the other hand, the second version means that the concentration of the localized states that can be populated by holes is low enough because the excitation powers at which the redistribution was observed were rather low.

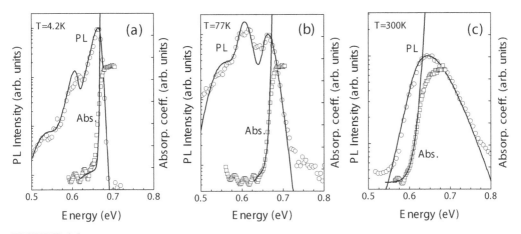

FIGURE 6.6

Temperature variations of the PL and absorption spectra of sample Gs2050. The symbols are the experimental data, and the solid lines are results of model calculations. (a) T = 4.2 K, (b) T = 77 K, and (c) T = 300 K. Reprinted with permission from A. A. Klochikhin, V. Yu. Davydov, V. V. Emtsev, A. V. Sakharov, V. A. Kapitonov, B. A. Andreev, H. Lu, and W. J. Schaff, Physical Review B, 71, 195207 (2005). Copyright 2005 by the American Physical Society.

An additional argument supporting the second version is given by the PL sensitivity to the excitation power. Figure 6.5 demonstrates the PL spectrum transformations for sample Gs2060 in a wide interval of excitation powers at liquid-nitrogen temperature. It can be seen from Fig. 6.5 that the spectra registered by InSb and InGaAs detectors coincide well in a wide range of energies above 0.6 eV. At low and middle excitation densities, an intensity redistribution between two high-energy peaks is observed. A further increase in the excitation power results in a high-energy shift of the spectrum and broadening of the high-energy peak due to the carrier concentration increase. A similar dependence was observed at the liquid-helium temperature. It is worth noting that the observed dependence can be understood only in the framework of the second version as a result of the population boundary shift.

Figures 6.6(a)-(c) show temperature dependences of the PL and absorption spectra in a wide temperature interval. The most dramatic transformation of the PL spectrum occurs as temperature increases from liquid-nitrogen to room temperature.

The calculations presented below show that these changes in PL spectra are due to the alteration of the microscopic PL formation mechanisms. At room temperature the dominant role is played by the band-to-band transitions resulting from the filling of the valence band states.

6.4   Model description of absorption and luminescence

6.4.1   Density of valence band states and photohole distribution

The InN samples of improved quality demonstrate more intense PL bands with exponentially decreasing tails formed by donor-acceptor annihilation of shallow localized electron and hole states [6.23]. The tails of this kind are known to appear in disordered systems, such as solid solutions and amorphous semiconductors, as a result of the random distribution of atoms over lattice sites or structural imperfections randomly scattered over the crystal. The formation of the tails of localized states leads to significant changes in the interband absorption and PL spectra. In the case of doped semiconductors, the spatial arrangement of doping centers plays the role of a random factor.

One can expect that, in the case of lower doping, the localized hole states of acceptor-type impurities and tail states of the conduction band will play a noticeable role in the formation of the interband luminescence. Then additional luminescence bands associated with recombination of electrons and deeply localized holes whose Bohr radius is smaller than the Thomas-Fermi screening radius $a_B \leq q_{TF}^{-1}$ can appear. Here, $q_{TF} = \sqrt{3}\omega_{pl}/v_F$ is the Thomas-Fermi wavevector, ω_{pl} is the frequency of the free carrier plasmon, and v_F is the velocity of electrons on the Fermi sphere of radius $k_F = p_F/\hbar = (3\pi^2 n_e)^{1/3}$ in reciprocal space. These bands must be red-shifted as compared with the band-to-band luminescence and can co-exist with Urbach tails of the band-to-band luminescence, like in GaAs and GaN.

We consider a model description of the density of states of the valence bands taking into account both the Urbach tail and acceptor states of different localization depths. The behavior of the density of states in the Urbach tail is commonly well described in a wide energy range by the exponential law [6.27, 6.28]. Taking the density of acceptor states in the Gaussian form, we write the model density of states

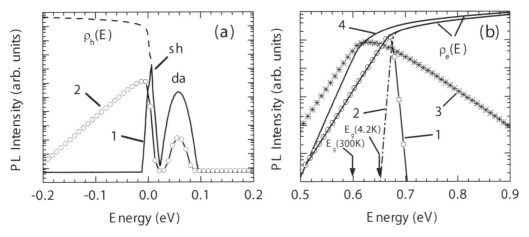

FIGURE 6.7

(a) Model density of the valence band states $\rho_h(E)$ (dashed curve) including the Urbach tail at $E_U^h = 1$ meV and acceptor states *sh* and *da*. Curves 1 and 2 are the hole populations at liquid-helium and room temperatures, respectively. (b) Model densities of the conduction band states $\rho_e(E)$ at two temperatures with the Urbach tail at $E_U^e = 15$ meV. $E_g(4.2K)$ and $E_g(300K)$ are positions of the band gap at liquid-helium and room temperatures, curves 1 and 3 are the densities of the populated states, and curves 2 and 4 are the densities of empty electron states at these temperatures, respectively. Reprinted with permission from A. A. Klochikhin, V. Yu. Davydov, V. V. Emtsev, A. V. Sakharov, V. A. Kapitonov, B. A. Andreev, H. Lu, and W. J. Schaff, Physical Review B, 71, 195207 (2005). Copyright 2005 by the American Physical Society.

for holes as

$$\rho^h(E) \sim \left[\frac{(2m_h)^{3/2} v_0}{4\pi^2 \hbar^3} \right] \sqrt{|E|} \tag{6.6}$$

below the valence band top, where E is negative and $|E| > E_U^h/2$. Then, at $E > -E_U^h/2$, the density of the hole states can be represented as

$$\rho^h(E) \sim \left[\frac{(2m_h)^{3/2} v_0}{4\pi^2 \hbar^3} \right] \sqrt{E_U^h/2} \exp\left(-(E - E_U^h/2)/E_U^h\right) +$$
$$+ \frac{N_{sh}}{(2\pi\gamma_{sh})^{1/2}} \exp\left\{ -\left[\frac{\hbar\omega - E_{sh}}{2\gamma_{sh}} \right]^2 \right\} + \frac{N_{da}}{(2\pi\gamma_{da})^{1/2}} \exp\left\{ -\left[\frac{\hbar\omega - E_{da}}{2\gamma_{da}} \right]^2 \right\}, \tag{6.7}$$

where E_U^h is the characteristic Urbach energy determining the typical localization energy and the radius of the bound state $(r_{loc} = \sqrt{[\hbar^2/(2m_h E_U^h)]})$; E_{sh}, γ_{sh}, N_{sh}, E_{da}, γ_{da}, and N_{da} are the localization energies, inhomogeneous broadenings, and dimensionless concentrations of the shallow and deep acceptors, respectively, and m_h is the hole effective mass. The Urbach parameter E_U^h in the calculations was taken to be 1 meV. This model density of states is given in Fig. 6.7.

It can be expected that the concentration of the localized valence band states in *n*-InN samples now available is still smaller than the electron concentration. If localized states are spread over a wide energy range then, first of all, the dependence on the filling of these states at relatively low excitation powers can influence the PL band. The steady-state luminescence depends on the distribution of the photoholes over the valence band states and, hence, their energy-relaxation rate and also on the excitation power.

In the limit of weak excitation of samples, characterized by a significant tail of localized states and fast relaxation rate, we can expect that only the holes localized by the deepest (spatially isolated from each other) states [6.23] will define the PL band shape. The stationary occupancy of these states is not equilibrium because they are isolated, and, at low temperatures, the number of holes at a given localization energy is simply proportional to the density of states. The PL mechanism of this kind was observed in spectra of disordered systems under weak excitation, when recombining excitons were captured by localized states of the tail [6.29]. The hole distribution in this limit is expected to be non-degenerate, and the temperature increase can lead to the Boltzmann distribution of populated states. The influence of the steady-state excitation power must be weak as long as the hole distribution remains non-degenerate.

The 'strong' excitation limit can be reached if the number of photoholes created by excitation is comparable to the number of localized states in the tail of the valence band. If in this case the energy relaxation time $\tau_E(T)$ is much less than the full temperature-dependent lifetime $\tau(T)$ of holes in radiative states, a degenerate distribution of the holes can be expected. We denote the generation rate of photoholes as G. Then the chemical potential of photoholes $\mu^h(G,T)$ that defines their equilibrium or almost equilibrium distribution can be found from

$$G\tau(T) = \int_{-\infty}^{\infty} \left[1 - n_E^h(G,T)\right] \rho_h(E)dE \qquad (6.8)$$

where the Fermi-distribution function for the valence band is

$$n_E^h(G,T) = \frac{1}{\exp[(E - \mu^h(G,T))/T] + 1}, \qquad (6.9)$$

and the temperature is expressed in energy units. Since $\tau(T)$ is temperature dependent, $\mu^h(G,T)$ must also depend on temperature. At not too strong excitations, $\mu^h(G,T)$ must be positive and lie above the top of the valence band. The hole chemical potential $\mu^h(G,T)$ at increasing excitation power at low temperatures becomes negative, resulting in the valence band state population. The really strong excitation limit will be achieved if the localized hole states are saturated and the electron concentration noticeably increases. It can be expected that the PL band profile in the strong excitation limit will be sensitive to the excitation power as well as to temperature.

In Fig. 6.7(a), the redistribution of the photohole population as the temperature is increased from 4.2 K to 300 K is presented. For simplicity, the position of the hole chemical potential just above the shallow acceptor state was not changed. However, a considerable depopulation of localized states takes place even under these conditions. As a result of similar redistribution, the localized states lose their role in the PL formation.

6.4.2   Conduction band states

In view of the strong nonparabolic behavior of the conduction band [6.9], we assume a linear increase of the effective mass with kinetic energy [6.30], so that $m_e(E) = m_\Gamma[1 + (E - E_g(n,T))/E_0]$. Then, the conduction band dispersion is described by

$$\varepsilon(k) = E_0 \left\{ \sqrt{\hbar^2 k^2 / 2m_\Gamma E_0 + 1/4} - 1/2 \right\}. \tag{6.10}$$

This expression was used in Ref. [6.31] to describe the conduction band dispersion and model the results in Ref. [6.32], using an electron effective mass at the $\Gamma-$point, $m_\Gamma = 0.07m_0$, and a nonparabolicity parameter, $E_0 = 0.4$ eV. The density of states of the conduction band $\rho^e(E)$ (see Fig. 6.7(b)) can be written as

$$\rho^e(E) = \left[\frac{(2m_\Gamma(1+\Delta(E)/E_0)^{3/2} v_0}{2\pi^2 \hbar^3} \right] \sqrt{\Delta(E)} \tag{6.11}$$

at $\Delta(E) = (E - E_g(n,T)) > E_U^e/2$ and

$$\rho^e(E) = \left[\frac{(2m_\Gamma(1+E_U^e/E_0)^{3/2} v_0}{2\pi^2 \hbar^3} \right] \sqrt{E_U^e/2} \exp\left(-(|\Delta(E))| - E_U^e/2)/E_U^e\right), \tag{6.12}$$

at $\Delta(E) = (E - E_g(n,T)) < E_U^e/2$, where E_U^e is the characteristic Urbach energy that determines the typical energy of electron localization and $E_g(n,T)$ is the band gap depending on the electron concentration and on temperature. The parameter E_U^e for the samples of high quality ranged from 7.5 to 15 meV.

The origin of the localized states that lie below the bottom of the conduction band may be associated with, for example, spatial fluctuations of the positions of shallow donors; if the Bohr radius of a Coulomb center is $a_B > q_{TF}^{-1}$, then the bound state on a single center will disappear, but a pair of such centers situated within a sphere of radius q_{TF}^{-1} may lead to the formation of a shallow localized state.

The dependence of the band gap $E_g(n,T)$ on the electron concentration, which is due to the Hartree-Fock exchange interaction, takes the form [6.7]

$$E_g(n,T) = E_g(0,T) - 20(n/10^{18})^{1/3}, \tag{6.13}$$

where $E_g(n,T)$ is the band gap expressed in meV at electron concentration n expressed in cm^{-3}, $E_g(0,T)$ is the band gap in the limit of zero electron concentration. The numerical factor was estimated in Ref. [6.7], as described in section 6.2.3. We assume that the temperature dependence of the band gap is universal and does not change with electron concentration.

The Fermi energy of the conduction band, $E_F = p_F^2/2m_e$ in meV, for the parabolic electron band is [6.7]

$$E_F = 3.58(m_0/m_e)(n/10^{18})^{2/3} \tag{6.14}$$

where m_e is the effective mass of an electron, m_0 is the free electron mass, and a numerical factor is found for a parabolic band by using the well known expression for E_F from Ref. [6.13].

FIGURE 6.8

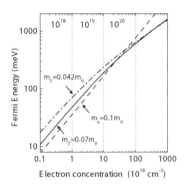

Model dependences of the Fermi energy on electron concentration for parabolic and nonparabolic dispersions of the conduction band. The dashed line corresponds to the parabolic band with $m_e = 0.1 m_0$. The dash-and-dot and solid lines correspond to the nonparabolic band with the linear dependence of effective mass on electron kinetic energy $m_e(E) = m_\Gamma (1 + E/E_0)$. The effective masses at the Γ-point are taken to be $0.042 m_0$ and $0.07 m_0$. Reprinted with permission from A. A. Klochikhin, V. Yu. Davydov, V. V. Emtsev, A. V. Sakharov, V. A. Kapitonov, B. A. Andreev, H. Lu, and W. J. Schaff, Physical Review B, 71, 195207 (2005). Copyright 2005 by the American Physical Society.

If the effective mass linearly increases with the electron kinetic energy, then the Fermi energy dependence on concentration can be described by

$$E_F = E_0 \left\{ \left[(3.58/E_0)(m_0/m_\Gamma)(n/10^{18})^{2/3} + 1/4 \right]^{1/2} - 1/2 \right\}. \quad (6.15)$$

The Fermi energy as a function of concentration for parabolic and nonparabolic bands is presented in Fig. 6.8.

6.4.3 Interband absorption and photoluminescence spectra

The interband optical absorption in doped crystals is formed by transitions from populated states of the valence bands to unpopulated states of the conduction band. This results in the Burstein-Moss shift of the optical absorption edge for the energy of the order of E_F [6.7].

The high-energy band of the interband photoluminescence arises due to annihilation of the photo-excited holes and electrons, which are still degenerate in the best InN samples available so far. Consequently, the PL band width at low temperatures is determined by E_F. The nonparabolicity of the electron band affects both the absorption coefficient profile and the luminescence band shape.

In the samples with the electron concentration $n \approx 1 \times 10^{18} \text{cm}^{-3}$, $E_F \approx 35$ meV for the parabolic band with $m_e = 0.1 m_0$ or 48 and 71 meV for the nonparabolic one with $m_\Gamma = 0.07 m_0$ and $0.042 m_0$, respectively. Under this condition, the shallow hole and electron states that form exponentially decaying Urbach tails of the density of states of the valence and conduction bands can play an important role in PL band shape formation. The luminescence bands formed by recombination of electrons and deeply localized holes are red-shifted, as in the case of GaAs and GaN. However, the tails of shallower states typically exist together with deep states.

6.4.3.1 Absorption

It should be noted that the dopants that are randomly distributed over the crystal and various defects of the crystal lattice affect the motion of electrons and holes. Scattering from a random potential created by impurities and defects breaks the momentum conservation law. Consequently, the sum of the electron $\mathbf{p_e}$ and hole $\mathbf{p_h}$ momenta are not equal to the momentum of a photon, which results in non-vertical interband transitions.

Then the energy of a pair created upon the absorption of a photon is

$$\hbar\omega = E_g(n,T) + \frac{p^2}{2\mu} + \frac{Q^2}{2M}, \tag{6.16}$$

where, $\mathbf{Q} = (\mathbf{p_e} - \mathbf{p_h})$ is the center of mass momentum and $\mathbf{p} = (\mathbf{p_e}\mu/m_e + \mathbf{p_h}\mu/m_h)$ is the momentum of relative motion, where μ and $M = (m_e + m_h)$ are the reduced and translation masses of the pair, respectively. The energy dependence of the band-to-band contribution to the dielectric susceptibility that determines the behavior of the absorption coefficient can be represented as

$$\chi''(\omega) \sim \frac{v_0}{(2\pi\hbar)^3} \int d^3p\, d^3Q\, n_{\mathbf{p}}^h \left(\frac{(\mathbf{p} - \mathbf{Q}\mu/m_e)^2}{2m_h} \right) \Delta(\mathbf{Q})\, \delta\left(\hbar\omega - E_g - \frac{p^2}{2\mu} - \frac{Q^2}{2M} \right) \times \left\{ 1 - n_{\mathbf{p}}^e \left(\frac{(\mathbf{p} + \mathbf{Q}\mu/m_h)^2}{2m_e} \right) \right\}, \tag{6.17}$$

where, $n_{\mathbf{p}}^e$ and $n_{\mathbf{p}}^h$ are the Fermi functions of electrons and holes. The spin quantum numbers are omitted for simplicity.

$\Delta(\mathbf{Q})$ is the squared, overlapping integral between the wave functions of the electron and hole that are created at the same point in space by a photon, and whose momentum of the center of mass is $\mathbf{Q}$

$$\Delta(\mathbf{Q}) = |\psi_{\mathbf{p}+\mathbf{Q}}^e(\mathbf{r})\psi_{\mathbf{p}-\mathbf{Q}}^h(\mathbf{r})|^2. \tag{6.18}$$

In the case of free motion of electrons and holes,

$$\Delta(\mathbf{Q}) \to \delta(\mathbf{Q}), \tag{6.19}$$

that is, the function reduces to a 3D delta-function that expresses the momentum conservation law. In this case only vertical transitions are allowed, and the kinetic energies of the created hole and electron depend only on the ratio of effective masses,

$$E_h = -(m_e/m_h)E_e. \tag{6.20}$$

The quantity $\mathbf{Q}$ may take a continuum set of values for different mechanisms of scattering of electrons and holes from impurities and defects, and it is necessary to perform a summation over these values. If $\Delta(\mathbf{Q})$ does not restrict the possible values of $\mathbf{Q}$, the integration with respect to d^3Q extends over the entire volume of the first Brillouin zone. The hole kinetic energy is restricted by the energy-conservation law alone,

$$0 \leq E_h \leq (\hbar\omega - E_g(n,T)). \qquad (6.21)$$

Since the population of the valence band in calculations of the absorption coefficient can be taken equal to unity, we present the energy dependence of the imaginary part of the interband susceptibility and of the absorption coefficient as

$$\chi''(\omega) \sim \rho^e(E) \left\{ 1 - \frac{1}{\left(\exp\left[\left(\frac{\mu}{m_e}(\hbar\omega - E_g(n,T)) - E_F \right) / T \right] + 1 \right)} \right\} \qquad (6.22)$$

in the case when the conservation law is obeyed.

In the intermediate case, in which the momentum conservation law is partially broken, the interpolation formula is

$$\chi''(\omega) \sim \int_{E_g(n,T)}^{\infty} \rho^e(E) dE$$

$$\times \int_{-\infty}^{0} dE_1 \rho^h(E_1) \Phi_{E,E_1} \delta(\hbar\omega - E_g(n,T) - E + E_1)[1 - n_E^e(T)], \qquad (6.23)$$

where the square of the overlapping integral takes the form,

$$\Phi_{E,E_1} = | \int d^3 r \psi_E^e(\mathbf{r}) \psi_{E_1}^h(\mathbf{r}) |^2 \qquad (6.24)$$

and the wave functions of electron $\psi_E^e(\mathbf{r})$ and hole $\psi_{E_1}^h(\mathbf{r})$ are no longer characterized by momenta. An accurate calculation of Eq. 6.24 requires knowledge of these wave functions. An approximate equation can be obtained by replacing Eq. 6.24 by the Gaussian function that has a maximum when the kinetic energies of particles satisfy Eq. 6.20, that is, at $|E_1| = (m_e/m_h)E$.

For a nonparabolic conduction band the susceptibility can also be calculated from Eq. 6.23. As demonstrated by Carrier and Wei [6.30], the nonparabolicity leads to an enhancement in the energy dependence of the absorption coefficient as compared with that for a parabolic conduction band.

6.4.3.2 Photoluminescence

Considering recombination, it is necessary to take into account, in addition to the factors that influence absorption processes, the additional factors of the distribution of photoholes and the electron-phonon interaction. The luminescence intensity produced by electron-hole recombination can be represented as

$$I_{PL}(\omega) \sim \int_{0}^{\infty} \rho^e(E) dE \int_{E_g(n,T)}^{-\infty} dE_1 \, \rho^h(E_1) \Phi_{E,E_1}$$

$$\times \delta(\hbar\omega - E_g(n,T) - E + E_1) \left[1 - n_{E_1}^h(G,T) \right] n_E^e(T), \qquad (6.25)$$

where,

$$\Phi_{E,E_1} = | \sum_{\lambda,\lambda_1} \int d^3 r \psi^e_{E_\lambda = E}(\mathbf{r}) \psi^h_{E_{\lambda_1} = E_1}(\mathbf{r}) |^2 \qquad (6.26)$$

is the overlapping integral between the electron and hole wave functions, including the localized acceptor states.

We can assume that the most effective electron-phonon interaction is realized in InN for the deeply localized states, as in other semiconductors. In the spectra of the InN samples studied, the recombination of the deep acceptor states is the most probable candidate for observing this effect. As shown in Ref. [6.19], the first stage of the energy relaxation of photoexcited carriers is very fast and can be attributed to the carrier interaction with longitudinal optical phonons. This feature is shared by many other crystals. Therefore, we can expect an LO-phonon replica of the recombination band of the deep acceptor. This effect can be simulated as an additional band in $\rho^h(E_1)$ proportional to the electron-phonon interaction constant.

6.5 Results of calculations and discussion

The calculations of the PL spectra were performed for a number of samples of different thicknesses having different electron concentrations estimated by conventional Hall measurements. The major goal of the calculations was to establish the origin of the PL and the fundamental parameters of InN that would be appropriate for the description of the experimental data on photoluminescence and absorption.

The variable parameters were the band gap $E_g(n,T)$ at $n = 0$ and $T = 0$, the electron effective mass at the Γ-point, the free carrier concentration, the band-gap shrinkage from liquid-helium to room temperatures, and the binding energies of shallow and deep acceptors.

In our analysis, the band gap $E_g(0,0)$ was varied from 0.665 to 0.675 eV. A simultaneous fitting of PL, absorption, and band-gap shrinkage leads to the conclusion that the band gap $E_g(0,0)$ should be taken to be equal or less than 0.670 eV, otherwise the temperature shrinkage should be taken larger than 60 meV, that is, too close to the value for the wide gap GaN crystal (see, for instance, Ref. [6.36]).

If the band-gap values are chosen to be within the interval 0.665–0.670 eV, then band-gap shrinkage between liquid-helium and room temperatures restricted within the range 55–60 meV and deep and shallow acceptors energies in the ranges 50–55 and 5–10 meV, respectively, are sufficient to describe the PL spectra. In this case the role of shallow acceptors can be attributed to the Urbach tail of the valence band with appropriate E_U^h. An analysis of the PL and absorption spectra shows that the band gap at room temperature is about 0.6 eV.

The electron effective mass at the Γ-point equal to $0.07m_0$ as suggested in Ref.

[6.30] can be considered as a best choice, though it leads to the electron concentrations differing from the Hall data.

The Hall concentrations give, under these conditions, Fermi energies that are too high and do not correspond to the observed PL band widths. Despite data [6.35–6.40] allowing one to assume a high enough inhomogeneity in the surface and interface layers, it looks reasonable to state that the minimum electron concentration in the high-quality samples is of the order of $1–2 \times 10^{17}$ cm^{-3}.

6.5.1 Structure of the photoluminescence spectra

The PL spectra of the InN samples exhibit a structure characterized by three stable features: the high-energy peak I_{sh}, the middle-energy peak I_{da}, and the weak low-energy peak $I_{da} - LO$ (Figs. 6.4–6.6). The high-energy peak is attributed to the overlapping bands produced by two mechanisms, namely, by transitions of degenerate electrons to shallow acceptor and/or Urbach tail states and by band-to-band transitions.

The second peak is likely to be attributable to the electron transitions to deep acceptor states, as suggested in Ref. [6.22]. The third feature is about 73 meV below the electron-deep-acceptor transitions. This energy coincides with the LO-phonon energy, according to Raman data [6.25]. To describe the third feature of the PL spectra, it is necessary to assume that the dimensionless electron-phonon constant is ≈ 0.02, which corresponds to a weak electron-phonon interaction. This value was found to be nearly independent of the carrier concentration.

The structure of the PL spectra becomes less pronounced with decreasing sample thickness (accompanied by the concentration increase) and disappears from the spectra of the samples with Hall concentrations above 10^{18} cm^{-3}.

6.5.2 Temperature dependence of photoluminescence and absorption spectra

The PL spectra exhibit strong variations in the range between liquid-helium and liquid-nitrogen temperatures at a constant excitation power (Figs. 6.4 and 6.7). The sensitivity of the PL spectrum to the temperature shows that the population of one of the two types of carriers forming the PL band is strongly influenced by temperature. On the other hand, the Hall data indicate that the electrons are still degenerate. Then, the relative decrease in the high-energy peak with temperature can be assigned to the redistribution of the population between the shallow and deep localized states of holes. Therefore, we can assume that the energy relaxation of photoholes is fast enough, so that it is possible to consider the hole distribution as equilibrium or almost equilibrium.

The spectral transformations cannot be ascribed to the simple thermal redistribution of holes between deep and shallow acceptors at a constant chemical potential of holes. However, they can be produced by a simultaneous shift of $\mu^h(G, T)$ toward the band gap. According to Eq. 6.8, this fact shows that the full relaxation time of holes $\tau(T)$ decreases with increasing temperature. For the high-quality samples, at liquid-helium temperature and at the highest excitation power applied, $\mu^h(G,T)$ is several meV above the shallow acceptor level.

The PL-band shapes are influenced by the inhomogeneous distribution of electrons. This results in a difference between the observed slope of the high-energy wing of the PL band in the low-temperature spectra and the slope that would follow from the Fermi function at a given temperature. Therefore, the observed slope at the liquid-helium temperature in spectra of the high-quality samples corresponds to an effective temperature equal to about 15 K. As a result, the effective temperature can be introduced instead of a more complex averaging procedure. The difference between the real and effective temperatures increases with increasing electron concentration. This can be attributed to the inhomogeneous distribution of the charge density.

The PL spectra experience transformations when samples are heated up to room temperature. A considerable shift and broadening of the PL band is accompanied by a disappearance of the structure. The model calculations show that increasing the temperature causes considerable broadening of both the electron and hole distributions (Fig. 6.7). In the high-quality samples, this results in a partial removal of the electron degeneration (see Fig. 6.7(b)). The thermal band-gap shrinkage shifts the PL-band to low energies, while an increase in the kinetic energies of carriers produces effects of the opposite sign. As a result of these two tendencies, the observed shift of the PL-band can take different values for samples of different qualities.

The microscopic mechanism regulating the temperature behavior of the PL-band shift is formed, on the one hand, by the difference in masses of the electron and heavy hole and, hence, by different values of their thermal momenta $p_T^h = \sqrt{2m_h T}$ and $p_T^e = \sqrt{2m_e T}$, respectively. On the other hand, under these conditions, an important question arises as to whether the momentum conservation law is obeyed in the annihilation processes or not. Depending on this, the interband transitions between states with different momenta can be allowed or restricted. For instance, the PL-band profiles of p-type samples of GaAs and GaSb are strongly influenced by the restriction following from the momentum conservation law, as demonstrated by Titkov *et al.* [6.33]. As a result, transitions involving hole states with large momenta are suppressed. An analogous restriction arises in samples of n-type InN if their quality is high enough. High thermal momenta of holes p_T^h corresponding to the maximum of their thermal distribution prevent their annihilation with electrons whose momenta are of the order of p_F or p_T^e. In this case, transitions of electron to the valence band states with $p^h \approx p^e$ occur, and the PL-band position is affected, as temperature increases, by band-gap shrinkage and also by the high-energy shift of the electron distribution. On the other hand, the breaking of the momentum conservation law involves the thermal holes in the recombination and additionally increases the PL-band energy by ≈ 30 meV. The interplay of different factors influencing the PL-band position leads to variations of the PL-band shift with temperature.

The other circumstance worth noting is the dependence of the PL-band maximum on the violation of the momentum conservation at a large difference between electron and the heavy hole masses even at constant temperature. As a result, it can be expected that the PL band of the sample containing a higher concentration of scatterers will be shifted toward higher energies at equal carrier concentrations and temperatures.

Simultaneous fits of PL and absorption spectra at liquid-helium, liquid-nitrogen, and room temperatures are presented in Figs. 6.6(a)–(c). The question arises as to whether it is possible to find the band gap or the band-gap shrinkage directly from the spectra. Note that the nonparabolicity of the conduction band, the influence of the band-filling effect, nonuniform carrier density, and breaking of the momentum conservation law in optical transitions make impossible the use of the simple $\alpha^2(E)$ extrapolation for finding the band gap and the band-gap shrinkage.

An attractive idea to use the sigmoidal equation [6.17, 6.34] to describe the spectral dependence of the absorption is also unsuitable for the accurate estimation of these characteristics. As a matter of fact, the absorption spectra of thick samples, similar to those presented in Figs. 6.6(a)–(c), demonstrate the point at which the Bragg interference disappears, which seems to be closer to the band gap than the other possible characteristics of spectra.

An interesting feature of the room-temperature luminescence is the disappearance of any structure associated with acceptors and phonon replica. This can be interpreted as the result of a large difference between densities of localized and band hole states.

6.5.3 Dependence of photoluminescence spectra on excitation power and excitation energy

Figure 6.5 demonstrates the dependence of the PL spectrum on the excitation power at liquid-nitrogen temperature which is typical for the samples of improved quality. The changes in the spectrum with decreasing excitation power are similar to those that occur with increasing temperature. The details of the dependences can be understood by taking into account the correlation between the number of electron-hole pairs excited by photons and, on the one hand, the number of localized hole states and, on the other hand, the number of free electrons.

If the excited-pair concentration is comparable with the concentration of localized hole states, but less than the electron concentration, then the behavior of the PL spectra can be understood in terms of a shift of the hole chemical potential (see Eq. 6.8) due to the generation-rate alterations. It is worth noting that in the samples of high quality at relatively low excitations only the magnitude redistribution between two high-energy maxima is observed, without any pronounced shifts of the high-energy PL-band maximum. This can be interpreted as a manifestation of the population redistribution between deep and shallow localized states of holes.

If the excitation power is enhanced up to a level when the created electron-hole pair concentration becomes comparable with the free-carrier concentration, then the broadening and the shift of the spectra can be observed, as demonstrated in Fig. 6.5. An excitation-induced increase of electron concentrations from 2.4×10^{17} at low excitation level to 4.2×10^{17} cm^{-3} at the maximum pulse excitation was established for sample Gs2060 (Fig. 6.5).

The measurements of PL spectra were performed for all of the samples using the lasers operating in the energy range from 2.41 eV to 0.81 eV as excitation sources in Ref. [6.24]. Figure 6.9 shows the luminescence spectra obtained at four laser energies for two InN samples with different carrier concentrations. It was found that the PL-spectral shapes are affected by excitation density, rather than by excitation energy. For all excitation energies, the changes in excitation density result in redistribution of PL intensity between the band-edge-related peak and the impurity-related peak for samples of different quality. These findings are typical of all direct band gap semiconductors.

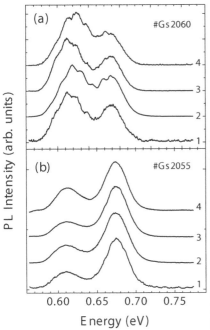

FIGURE 6.9

PL spectra of samples Gs2060 and Gs2055 at different excitation energies. Curves 1, 2, 3, and 4 were obtained with excitation energies 2.33, 1.55, 1.10, and 0.81 eV, respectively. The spectra were detected with an In-GaAs detector. Reprinted with permission from A. A. Klochikhin, V. Yu. Davydov, V. V. Emtsev, A. V. Sakharov, V. A. Kapitonov, B. A. Andreev, H. Lu, and W. J. Schaff, Physical Review B, 71, 195207 (2005). Copyright 2005 by the American Physical Society.

To summarize, the PL spectra of high-quality InN samples at liquid-helium and liquid-nitrogen temperatures are characterized by a well-resolved structure consisting of three peaks in the energy interval from 0.50 to 0.67 eV. Two lower-energy PL peaks can be attributed to the recombination of degenerate electrons with the holes trapped by deep acceptors and to the LO-phonon replica of this band. The higher-energy PL peak is considered to be a complex band formed by transitions of electrons to the states of shallow acceptors and/or the states of the Urbach tail populated by photoholes and also by the band-to-band recombination of free holes and electrons. Strong dependences of the intensities of the two higher-energy PL peaks on the excitation power and temperature have been found.

The interband PL and absorption spectra of n-InN samples with electron concentrations from 3.6×10^{17} to 6×10^{18} cm^{-3} have provided more accurate data on the band gap of InN than that from more heavily doped samples. The Urbach tails of the conduction and valence bands and the acceptor states were taken into account in the model calculations. The conduction band was assumed to be nonparabolic. The effective mass of an electron at the Γ-point was taken to be about 0.07 of the free electron mass. Calculations of PL and absorption spectra have shown that the band gap of InN in the limit of zero temperature and zero electron concentration is close to 0.665–0.670 eV and the band-gap reduction from zero to room temperature $(E_g(0K) - E_g(300K))$ is 55–60 meV.

6.6 Band bending of n-InN epilayers and exact solution of the classical Thomas-Fermi equation

Electron accumulation at an InN epilayer surface has been identified by different experimental techniques [6.37–6.43]. The statistical Thomas-Fermi theory is typically used as a background for description of an inhomogeneous electron gas (see, for instance, Ref. [6.44]). The theoretical approaches taking into account the 2D-confined states in accumulation layers [6.45–6.47] are based on model presentations of the accumulation layer potentials.

We consider here in detail the results obtained in Ref. [6.48] where it was demonstrated that the classical Thomas-Fermi equation for a planar accumulation layer of a degenerate semiconductor in the parabolic band approximation in the limit of zero temperature can be solved analytically. This solution gives simple and transparent expressions for the band bending potential and inhomogeneous electron density relating these functions to crystal parameters and external electrical charge. We show that the nonparabolic dispersion of the conduction band in InN can be taken into account as the next step. This solution can be used as a first approximation to the full quantum mechanical solution of the electron accumulation problem. This topic is covered in more detail in Chapter 12.

The next problem studied in Ref. [6.48] is the considerable discrepancies between conduction electron densities derived from Hall and optical measurements [6.24, 6.49, 6.50], especially in thin InN films. We present analysis of the PL spectrum of a thin InN epilayer and show that the surface and, probably, interface accumulation layers account for the difference between electron densities derived from the Hall and PL measurements. The samples studied in Ref. [6.48] were hexagonal InN epilayers grown by plasma-assisted molecular-beam epitaxy on Si(111) substrates using the double-buffer technique. Details of the growth process can be found elsewhere [6.51].

A typical model of the accumulation layer was considered. It was assumed that the accumulation layer arises due to a planar positive charge at the crystal surface. It was assumed that the semiconductor contains homogeneously distributed donors of density N_D, the charge of which is compensated for by free electrons of density n_e, so that $n_e = N_D$. Therefore, spatial fluctuations of the donor density were ignored. The planar positive charge of density Q_s is compensated for by additional free electrons $\delta n(z)$ inhomogeneously distributed within the crystal.

In accordance with the potential theory, the resulting inhomogeneous Coulomb potential can be written as

$$\phi(z) = \int d^2 r' dz' \frac{[-Q_s \delta(z')e + \delta n(z')e]}{\varepsilon |(z-z')^2 + (r')^2|^{1/2}}, \tag{6.27}$$

where $\delta(z')$ is the δ−function, the reference point for z is chosen to be at the surface, $\delta n(z')$ is the compensating electron density, and ε is the electron dielectric constant. Neither the homogeneous positive donor charge nor the homogeneous electron charge contributes to the inhomogeneous potential of the accumulation layer. We consider the potential normalized by the condition $\Phi(z) = \phi(z) - \phi(\infty)$. Then, integrating over $d^2 r'$ and assuming charge neutrality $Q_s = \int dz \delta n(z)$, we obtain for the potential

$$\Phi(z) = -\frac{2\pi e}{\varepsilon} \int z' dz' \delta n(z+z') \tag{6.28}$$

A similar expression for the potential can be found in Ref. [6.47]. The potential of Eq. 6.27 satisfies the Poisson equation. The Thomas-Fermi approach assumes that $\delta n(z)$ can be expressed through the Fermi distribution function. Since the surface plane is taken to be isotropic, the potential should have a cylindrical symmetry. Then the Thomas-Fermi equation can be presented as

$$\frac{d^2}{dz^2}[-e\Phi(z)] = \frac{2\pi}{\varepsilon} \frac{e^2}{3\pi^2} \left(\frac{2m^*}{\hbar^2}\right)^{3/2} [-e\Phi(z)]^{3/2}, \tag{6.29}$$

where m^* is the electron effective mass. The exact solution of the Thomas-Fermi equation in the case of spherical symmetry [6.44] is well known. In the case of a planar accumulation layer the Thomas-Fermi equation is one-dimensional, and the exact solution can be obtained in the entire range of z. The potential that is a solution of Eq. 6.29, the compensating electron density, and the external surface charge density are

$$-e\Phi(z) = R_H \alpha \left[\frac{a_B}{z+l}\right]^4, \quad \delta n(z) = \frac{10\alpha}{a_B^3}\left[\frac{a_B}{z+l}\right]^6, \quad Q_s = \frac{2\alpha}{a_B^2}\left[\frac{a_B}{l}\right]^5, \tag{6.30}$$

where, $R_H = e^2/2a_B$ is the hydrogen Rydberg, $a_B = \hbar^2/m_0 e^2$ is the hydrogen Bohr radius or atomic length unit, α is expressed via crystal characteristics

$$\alpha = \left[\frac{30\pi\varepsilon}{2}\right]^2 \left(\frac{m_0}{m^*}\right)^3, \tag{6.31}$$

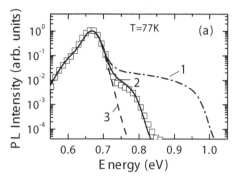 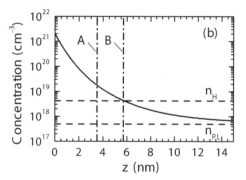

FIGURE 6.10

(a) PL band shapes for the 650-nm-thick InN sample #050118 with the Hall concentration $n_H = 4.2 \times 10^{18}$ cm^{-3}. The symbols represent experimental data. Lines 1 and 2 display the calculated PL band shapes when the accumulation layer is restricted by one of the two vertical lines (A and B, respectively) shown in Fig. 6.10(b). Line 3 shows the calculated PL band shape for the homogeneous electron distribution. In the electron density calculations the nonparabolic electron band with the effective mass at the Γ-point equal to $0.07m_0$ was assumed, the potential of Eq. 6.30 ($m^* = 0.20m_0$, $Q_s = 4 \times 10^{14}$ cm^{-2}, and $\varepsilon = 7.16$) was used. (b) Calculated inhomogeneous electron distribution near the surface (the solid line), the electron concentrations estimated from the Hall (n_H) and PL (n_{PL}) data. Vertical lines (A or B) show the boundaries between the sample regions participating or not in the PL. Reprinted with permission from A. A. Klochikhin, V. Yu. Davydov, I. Y. Strashkova, P. N. Brunkov, A. A. Gutkin, M. E. Rudinsky, H. Y. Chen, and S. Gwo, Physica Status Solidi Rapid Research Letters, 1, 159 (2007). Copyright 2007 by Wiley-VCH Publishers, Inc.

where m_0 is the free electron mass, and l is the characteristic length related to the surface charge by Eq. 6.30.

The classical solution obtained gives simple and transparent expressions for the band bending potential and inhomogeneous electron density in terms of crystal parameters and external electrical charge.

When the potential of Eq. 6.30 is found, the electron density for the nonparabolic conduction band can be calculated. Therefore, the nonparabolic dispersion of the electron band in InN can be approximately taken into account.

The PL-band shape for a 650-nm-thick InN sample is shown in Fig. 6.10. It can be seen that the shape is unusual, with a pronounced high-energy tail that can be attributed to radiation from the region of a high electron concentration. This shape was fitted using the approach given in [6.24] and taking into account the influence of the accumulation layer. We regard the epilayer of thickness $L = 650$ nm as a box filled with a free electron liquid and restricted by 6-eV-high walls. The potential inside the box is presented in the form symmetrical with respect to the sample surface and interface $\Phi(z) + \Phi(|z - L|) - 2\Phi(L/2)$, so that the potential and its derivative are equal to zero at $z = L/2$. The agreement of curve 2 with the experimental spectrum shown in Fig. 6.10 was obtained under the condition that the PL contribution from the surface accumulation layer (excluding the top 5.7-nm-thick part) is taken into account. The observation of the PL formed partially in the accumulation layer can be explained by localization of photoholes at deep traps because of which the electric field of the accumulation layer is not able to move them toward the center of the film. The electron concentration in the bulk region was found to be 0.5×10^{18} cm^{-3} by the procedure of fitting of the low-energy part of the PL-band, while the averaged concentration that takes into account both accumulation layers was estimated to be $n_{av} = 4 \times 10^{18}$ cm^{-3}, consistent with the Hall concentration $n_H = 4.2 \times 10^{18}$ cm^{-3}.

Thus, the exact solution of the Thomas-Fermi equation can provide such characteristics of the accumulation layer as its width, maximum electron concentration, and band bending with acceptable accuracy. The model calculations and analysis of PL and Hall data allow the free carrier content in an accumulation layer to be estimated.

6.7 Band bending effect on optical spectra of nano-size n-InN samples

Charge accumulation in surface layers and resulting band bending strongly affect optical characteristics of thin epilayers and nano-size samples of n-InN. We present model calculations and experimental data demonstrating that the PL-band shapes and temperature variations are sensitive to the presence of accumulation layers.

Vertically aligned wurtzite-InN nano-rods were grown on 3-inch-diameter Si(111) substrates by PAMBE at sample temperatures of 330°C (low temperature, LT) and 520°C (high temperature, HT) under nitrogen-rich growth conditions [6.52]. For the LT-InN nano-rods, the average rod diameter was quite uniform and could be varied from a few tens of nanometers to 100 nm (height/diameter ratio $\approx$ 10). Additionally, the InN epilayers on Si(111) substrates with thicknesses from 0.22 to 2.36 μm and Hall concentrations of 5×10^{18} cm^{-3} and 8×10^{18} cm^{-3} were studied.

The influence of an accumulation layer on the PL spectra can be revealed by studying the PL-band shape and temperature dependence. Figures 6.11(a) and 6.11(b) demonstrate the PL spectra of nano-rods of two types at different temperatures. The PL bands of both samples experience a considerable broadening with increasing temperature. The PL-band maximum of the HT sample does not shift with temperature, while the maximum of the PL band of the LT sample shifts toward higher energies with increasing temperature. These results differ from the behavior of the PL maximum typically observed for InN epilayers. For comparison, temperature variations of the PL spectra of two epilayers are presented in Figs. 6.11(c) and 6.11(d).

The temperature shift and broadening of the interband PL-band from epilayers is governed by three factors: a temperature shrinkage of the band gap; temperature broadening of the free electron distribution; and temperature variation in the photohole distribution. The hole distribution is of great importance because the

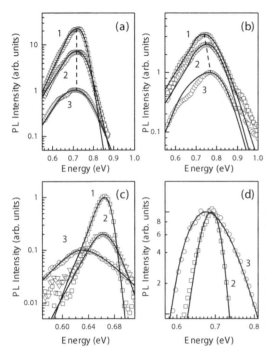

FIGURE 6.11

PL spectra of (a) HT nano-rods, (b) LT nano-rods, and (c) InN epilayers with thicknesses 0.65 μm and (d) 2.36 μm. Curves 1, 2, and 3 correspond to temperatures 4 K, 77 K, and 300 K, respectively. Symbols are the experimental data, and the solid lines are the results of calculations.

hole kinetic energy increases with temperature thereby shifting the PL maximum in the opposite direction as compared with the band-gap shrinkage. The difference in the temperature shifts (see Fig. 6.11) of the PL maxima of different samples can be explained by different roles of the momentum conservation law in interband transitions. If the momentum conservation law is broken and the indirect interband transitions dominate in the PL process, a considerable compensation of the band-gap shrinkage takes place (see Fig. 6.11(b) and 6.11(d)), which decreases the low-energy PL-band shift.

In case of low-dimensional samples, like nano-rods, an additional factor can play an important role, namely, an inhomogeneous distribution of the carrier concentra-

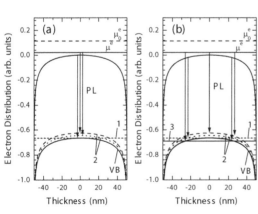

FIGURE 6.12

Schematic representation of the transformation of the PL process from (a) low temperature to (b) room temperature. Line 1 is the position of the valence band top, lines 2 denote the hole localized states, and line 3 in Fig. 6.11(b) shows the hole kinetic energy kT at room temperature. The arrows in (a) and (b) show the interband PL transitions at low and room temperatures. μ_0^e and μ^e are the chemical potentials without and with accumulation layers, respectively.

tion over the sample. At low temperature, photoholes populate the narrow central layer near the spatial top of the valence band, and PL is formed by transitions in this layer where the electron concentration is minimal (Fig. 6.12(a)). At room temperature, the holes are distributed within a broader layer where the hole potential energy is of the order of the thermal energy, kT. In this case, the PL transitions are possible in a broader range of the sample thickness where the electron concentration is larger (Fig. 6.12(b)) resulting in the high-energy shift and broadening of the PL band.

The electron concentration in doped n-InN samples can be estimated independently from PL data. These estimates are of great importance for nano-rods where Hall measurements are complicated. The free carrier concentration deduced from the PL data was found to increase from 3.75×10^{18} at 12 K to 7.75×10^{18} cm^{-3} at 300 K for the HT sample, and from 4.75×10^{18} at 12 K to 1.5×10^{19} cm^{-3} at 300 K for the LT sample. These data indicate that the sample region where the PL arises depends on temperature. The process of PL formation can be efficient in nano-rods because of the small sizes of these samples and because of a relatively large fraction of free carriers trapped by the surface accumulation layer at low temperatures.

Therefore, the influence of an accumulation layer on absorption of thin InN epi-layers and on PL spectra of InN nano-rods is strong, and important parameters of the accumulation layer can be deduced from optical spectra.

6.8 Vibrational spectroscopy of hexagonal InN and InGaN alloys

6.8.1 Introduction

The phonon spectrum is a fundamental characteristic of a crystal, which determines the thermodynamic properties of a material, kinetic properties of carriers, and phonon-assisted optical transitions. Such parameters of a phonon spectrum as the phonon dispersion curves and the phonon density-of-states (DOS) function reflect specific features of the crystal structure and interatomic interactions and provide valuable information on the crystal lattice dynamics. As a rule, information on the phonon dispersion curves and the phonon DOS function is obtained from neutron scattering experiments. However, modern technologies are not able to grow InN single crystals with the dimensions necessary for these experiments. Another valuable tool for the investigation of crystal lattice dynamics is Raman spectroscopy. Studies of first- and second-order Raman spectra yield information on phonon energies both at the center of the Brillouin zone and at its boundaries. An important advantage of Raman spectroscopy is the possibility to obtain information for objects as small as several tens of microns in size. Together with Raman spectroscopy, IR spectroscopy is also among the techniques that are most frequently used to study long-wavelength optical phonons.

TABLE 6.1

Selection rules and frequencies of optical phonons (cm^{-1}) in hexagonal InN at 300 K.

Symmetry of a vibration	E_2 (low)	A_1 (TO)	E_1 (TO)	E_2 (high)	A_1 (LO)	E_1 (LO)	B_1 (low)	B_1 (high)
Allowed scattering geometry	$z(yy)\bar{z}$ $z(xy)\bar{z}$ $y(xx)\bar{y}$	$y(zz)\bar{y}$ $y(xx)\bar{y}$	$y(xz)\bar{y}$	$z(yy)\bar{z}$ $z(xy)\bar{z}$ $y(xx)\bar{y}$	$z(yy)\bar{z}$	$x(yz)y$		
Experiment								
[6.54]				495	596			
[6.55]				491	590			
[6.56]	87	480	476	488	580	570	200	540
[6.25]	87	447	476	488	586	593	220	565
[6.57]		445	472	488	588			
[6.58]	88	440		490	590			
[6.59]		443	475	491	591			
[6.60]	88			490	590			
[6.61]		443	477	491	590			
[6.62]			477.9	491.1				
[6.63]				490.1	585.4			
[6.64]		448		490		598		
λ_{ex}=488nm*	87	449	475.8	490.6	583	592		
λ_{ex}=647nm*	87	449	475.8	490.6	592	598		
Calculation								
[6.57]	104	440	472	483			270	530
[6.58]	93	443	470	492	589	605	202	568
[6.65]	83	443	467	483	586	595	225	576
[6.66]	85	449	457	485	587	596	217	566

* Present work

6.8.2 Phonons in hexagonal InN. First-order Raman scattering

Hexagonal InN crystallizes in the wurtzite structure with four atoms in the unit cell and belongs to the C_{6v}^4 ($C6_3mc$) space group. According to the factor group analysis at the Γ-point, phonon modes in hexagonal InN belong to the following irreducible representations: $\Gamma_{ac} + \Gamma_{opt} = (A_1 + E_1) + (A_1 + 2B_1 + E_1 + 2E_2)$. Among the optical phonons, the modes of symmetry A_1 and E_1 are both Raman- and IR-active, the modes of symmetry E_2 are only Raman active, and the modes of symmetry B_1 are silent; that is, they are observed neither in Raman, nor in IR spectra [6.53]. Thus, six optical modes can be observed in the first-order Raman spectrum: A_1(TO), A_1(LO), E_1(TO), E_1(LO), E_2(high), and E_2(low).

The scattering geometries in which optical phonon modes of various types of symmetry can be observed are presented in Table 6.1. The Porto notation system is used in Table 6.1 and in what follows for describing the scattering geometry. The direction z is chosen parallel to the hexagonal axis, and x and y are mutually orthogonal and oriented in an arbitrary way in the plane perpendicular to the z direction.

The lattice dynamics of hexagonal InN have been studied experimentally by sev-

eral research groups using Raman scattering and infrared spectroscopic ellipsometry [6.25, 6.54–6.64]. Table 6.1 summarizes the data on the phonon mode frequencies at the Brillouin zone-center and the distribution of the modes over symmetry
types.

The majority of the Raman studies presented in Table 6.1 was performed on
InN films grown with *c*-axis normal to the substrate plane. Only three phonon
modes from the six Raman allowed modes can be observed in this case in the
backscattering configuration that is widely used for InN studies. They are E_2(low),
E_2(high), and A_1(LO) symmetry modes (see selection rules in Table 6.1).

InN samples with different orientations of the optical axis relative to the substrate
plane were prepared by PAMBE and studied in Ref. [6.25]. When the basal or *c*-
plane (0001) of sapphire was used, the optical axis of the InN layer was normal to
the substrate surface. However, when InN was deposited on the *r*-plane $(1\bar{1}02)$ of
sapphire, the epilayer hexagonal axis was parallel to the substrate plane and had
a fixed orientation. The availability of two sets of samples was important for the
observation of all allowed optical phonons and their symmetry assignment both in
Raman and IR measurements.

Room-temperature polarized first-order Raman spectra at an excitation energy of
2.54 eV for nominally undoped InN layers grown on sapphire substrates of two
orientations are shown in Figs. 6.13(a) and 6.13(b). As analysis of these spectra
showed, the measured first-order Raman spectra were consistent with the selection
rules for the wurtzite structure. The IR reflectance spectra were also measured for
InN samples grown on *c*- and *r*-sapphire substrates. The phonon frequencies of 448
cm^{-1} for A_1(TO) and 476 cm^{-1} for E_1(TO), which were obtained by means of a
Kramers-Kronig analysis of IR spectra, were in good agreement with the results of
Raman measurements (see inset in Fig. 6.13(b)).

Thus, all six Raman-active modes in InN were detected in Ref. [6.25] and the
symmetry of each mode was determined. Five phonon modes were measured on
the InN sample grown on an *r*-sapphire substrate. This allows one to obtain a
self-consistent pattern of the behavior of these phonons under various treatments
(for example, deformation, temperature, etc.). Only recently, the second paper has
appeared where the results of Raman measurements for an *a*-plane InN film grown
on an *r*-plane sapphire substrate by PAMBE were presented [6.64].

The difference in the frequencies for phonons of the same symmetry presented
in Table 6.1 can be attributed to different strain values in InN layers, since all the
films were grown on foreign substrates (sapphire, silicon, etc.). Strain-free values
of E_1(TO) and E_2(high) phonon modes equal to 477.9 cm^{-1} and 491.1 cm^{-1}, respectively, were determined by combining infrared spectroscopic ellipsometry and
Raman scattering measurements [6.62]. At the same time, Raman measurements
on a freestanding InN film grown by MBE yielded the strain-free Raman frequency
of the E_2(high) mode of 490.1 cm^{-1} [6.63]. Our Raman measurements on a 12-
micron-thick strain-free high-quality InN sample with an electron concentration of
3.6×10^{17} cm^{-3} (Gs2060) grown on *c*-plane sapphire at Cornell University (USA)

(a)
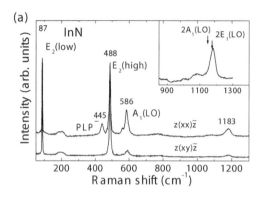

(b)
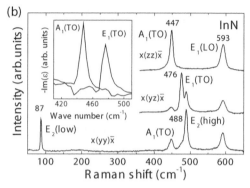

FIGURE 6.13

Polarized room-temperature first-order Raman spectra of nominally undoped InN samples grown on (a) a *c*-sapphire substrate and (b) an *r*-sapphire substrate. The spectra obtained at an excitation energy of $E_{exc} = 2.54$ eV. The inset of (a) shows the second-order spectrum. The inset of (b) shows the imaginary part of the contribution of $A_1(TO)$ and $E_1(TO)$ phonons to the dielectric lattice constant; the result was obtained by the Kramers-Kronig transformation of experimental data on reflectance at different orientations of the electric field vector E of the electromagnetic wave with respect to the c axis: $E \| \mathbf{c} - A_1(TO)$ and $E \perp \mathbf{c} - E_1(TO)$. Figure reproduced with permission from V. Yu. Davydov, V. V. Emtsev, I. N. Goncharuk, A. N. Smirnov, V. D. Petrikov, V. V. Mamutin, V. A. Vekshin, S. V. Ivanov, M. B. Smirnov, and T. Inushima, Applied Physics Letters, 75, 3297 (1999). Copyright 1999, American Institute of Physics.

[6.15] have given the value of the $E_2(\text{high})$ phonon mode equal to 490.6 cm^{-1}. We also studied the *a*-plane InN samples grown on *r*-sapphire by the Cornell University (USA) and National Tsing-Hua University (Taiwan) teams. The samples had electron concentrations of 1×10^{18} cm^{-3} and small strains (the measured $E_2(\text{high})$ phonon mode was at 490.5 cm^{-1}). So the frequencies of 449 cm^{-1} and 475.8 cm^{-1} measured on these samples can be considered to be strain-free values of $A_1(TO)$ and $E_1(TO)$ phonon modes, respectively.

Table 6.1 also lists data for zone-center optical phonons in hexagonal InN that were calculated by using different models [6.57, 6.58, 6.65, 6.66]. It can be seen that the calculated phonon frequencies are in satisfactory agreement with the experimental data.

6.8.3 Phonon-plasmon modes

According to the selection rules given in Table 6.1, the high-frequency lines in the Raman spectra of InN with wave numbers in the range between 580 and 596 cm^{-1}, were attributed by different authors to unscreened $A_1(LO)$ and $E_1(LO)$ phonon modes of hexagonal InN. However, the major part of Raman data presented in Table 6.1 were obtained for the InN samples having free-electron concentrations well over 10^{18} cm^{-3}. In this case the interpretation of a spectrum in the range of longitudinal optical vibrations and identification of the positions of unscreened LO phonons should include the consideration of the interaction of the LO phonons with collective excitations of free carriers.

In the presence of free carriers in heavily doped semiconductors, LO phonons couple strongly with plasmons. The unscreened LO phonon mode and the pure plasma oscillation mode are replaced by two LO-phonon-plasmon coupled modes (typically referred to as the L^- and L^+ mode) whose frequencies depend on the free carrier concentration. The carrier density dependence and the dispersion relations of the coupled modes are summarized in Ref. [6.67]. However, a pronounced mode is observed near the frequency of the unscreened A_1(LO)/E_1(LO) phonon in InN despite high free carrier concentrations in the films. Several authors have recently stated that this band is the large-wavevector L^- mode arising from wavevector nonconservation scattering processes [6.61, 6.68–6.70]. This mode approaches the frequency of an unscreened LO phonon from below if the wave vectors of both excitations are equal to or higher than the Thomas-Fermi screening wave vector [6.67].

As an example, we briefly describe the results obtained in Ref. [6.70]. A gauge invariant approach was developed to describe the Raman scattering from L^- and L^+ modes in heavily doped n-type semiconductors. This approach was realized for basic mechanisms responsible for the Raman processes, namely, allowed scattering mechanism (deformation potential and *interband* Fröhlich interactions), forbidden mechanism (*intraband* Fröhlich and Coulomb interactions), and the mechanism of charge density fluctuations. The sum rules controlling the charge conservation of electrons and ions were established for the Raman cross sections for these scattering mechanisms.

As a first step, typical features of the Raman cross sections for allowed, forbidden, and charge density fluctuation mechanisms were revealed under the assumption that the wavevector conservation law was only weakly broken due to absorption of exciting and scattered photons above the band gap. It was shown that the allowed mechanism leads to a maximum of the L^- mode near $\omega = \omega_{TO}$, whereas the forbidden mechanism gives a wide L^+_{PL} band above $\omega = \omega_{PL}$ as well as an additional sharp maximum L^+_{LO} at $\omega = \omega_{LO}$. The Raman scattering cross section due to charge density fluctuations will have a maximum for the L^- mode near $\omega = \omega_{TO}$ and a broad phonon-plasmon L^+_{PL}-band.

The role of the virtual electron-hole pair scattering from the impurities in the case of a strong breakdown of the momentum conservation law was analyzed for all three mechanisms of the Raman scattering in the range of transverse ω_{TO} and longitudinal ω_{LO} optical modes.

It has been shown that a common feature of the impurity-induced Raman scattering is the lack of a clearly seen band in the energy range of plasmon excitations because of the low density of states. At the same time, the band at the position of unscreened LO-phonon appears in the spectrum due to scattering of the virtual electron-hole pairs from the impurities characterized by a screened Coulomb or short-range potential. It has been demonstrated that the Raman LO-cross section is

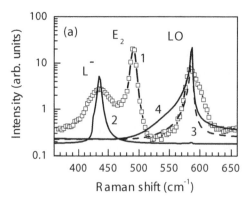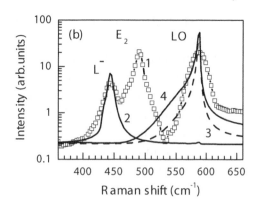

FIGURE 6.14

Raman spectra (curve 1) of InN samples with electron concentrations of (a) 6×10^{18} and (b) 1×10^{19} cm^{-3} obtained in the $z(xx)\bar{z}$ configuration at excitation $\lambda = 488$ nm. Calculations: curve 2 presents the allowed mechanism with momentum conservation broken by absorption of incident and scattered photons, curves 3 and 4 give scattering cross sections due to the forbidden mechanism with momentum conservation broken by screened charged centers or by the short-range potential with radius $a = \hbar/3p_F$, respectively. Estimated plasmon frequencies are $\Omega_{PL} = 1100$ cm^{-1} for (a) and ≥ 1250 cm^{-1} for (b). Maximum E_2 is due to the non-polar mode. Reprinted with permission from A. A. Klochikhin, V. Yu. Davydov, V. V. Emtsev, A. N. Smirnov, and R. van Baltz, physica status solidi (b), 242, R58 (2005). Copyright 2005 by Wiley-VCH Publishers, Inc.

proportional to the impurity concentration. In general, the cross section observed in experiments should include the terms with and without momentum conservation.

Transformation of the Raman spectrum of InN in the range of TO- and LO-phonons was analyzed in detail for samples of different quality (Fig. 6.14). It was found that the intensity of the LO-mode is highly sensitive to the type and concentration of impurities. A typical situation for as-grown InN samples is superposition of the allowed scattering with momentum conservation and the charged impurity-induced scattering without momentum conservation due to the Fröhlich electron-phonon interaction. The intensity of the latter scattering strongly depends on the doping level (see also section 6.8.5).

Thus, we can conclude that the spectral positions of the phonon modes in the range 580–596 cm^{-1} are close to those of unscreened phonons of symmetry $A_1(LO)$ and $E_1(LO)$. However, there is a very strong dependence of the positions of these bands on the excitation wave vector (see Table 6.1). This suggests that the problem of the identification of the true positions of the unscreened LO phonons in the InN Raman spectra demands additional experimental and theoretical studies.

The small-wavevector L^- mode was also observed in Raman spectra of InN. It has been found in Ref. [6.25] that, in addition to the E_2 and $A_1(LO)$ modes allowed by the selection rules for the $z(xx)\bar{z}$ configuration, one more band was observed at a frequency of 445 cm^{-1} in nominally undoped InN with electron concentration of $n = 2 \times 10^{20}$ cm^{-3}. A study of Mg-doped InN samples with different carrier concentrations has revealed that this band shifts toward lower frequencies when the free-electron concentration decreases (Fig. 6.15(a)). For this reason this band

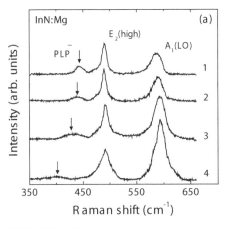
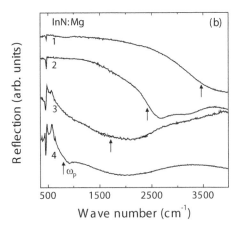

FIGURE 6.15

(a) Raman spectra and (b) IR reflectivity spectra for InN:Mg epilayers with different carrier concentrations in cm^{-3}: (1) 2×10^{20}, (2) 1×10^{20}, (3) 5×10^{19}, and (4) 1×10^{19}. Figure reproduced with permission from V. Yu. Davydov, V. V. Emtsev, I. N. Goncharuk, A. N. Smirnov, V. D. Petrikov, V. V. Mamutin, V. A. Vekshin, S. V. Ivanov, M. B. Smirnov, and T. Inushima, Applied Physics Letters, 75, 3297 (1999). Copyright 1999, American Institute of Physics.

was attributed to excitations that belong to the lower branch of mixed plasmon-LO-phonon modes (L^- mode). This interpretation was confirmed by the fact that the line associated with plasma oscillations in the IR reflection spectrum shifts from 3500 to 800 cm^{-1} as the carrier concentration decreases (Fig. 6.15(b)).

The transformations of the small-wavevector L^- mode and the large-wavevector L^- mode (A_1(LO) phonon line) in the proton irradiated InN were observed in Ref. [6.71]. The studies of the proton irradiated InN showed that the A_1(LO) phonon line intensity and the L^- coupled mode position in Raman spectra are dose-dependent. The growth in the A_1(LO) line intensity was attributed to the increase of the number of irradiation-induced defects responsible for the breaking of the momentum conservation law in Raman scattering. The high-frequency shift from 420 to 436 cm^{-1} of the L^- mode was explained by a substantial increase in the electron concentration in the irradiated samples. The last finding was in line with the Hall measurements, which indicated that the proton irradiation of InN led to a rapid increase in the electron concentration. At the same time, the x-ray and Raman data showed that irradiated layers preserve a wurtzite lattice structure and crystalline quality even at the end of the proton irradiation. A good correspondence of the changes in electron concentrations estimated from the PL spectra and Hall measurements in irradiated InN layers provides strong support to the attribution of infrared PL of this semiconductor to interband transitions.

6.8.4 Phonon dispersion in hexagonal InN

The InN lattice dynamics were simulated in Ref. [6.25] by using a phenomenological model based on short-range interatomic potentials and rigid-ion Coulomb

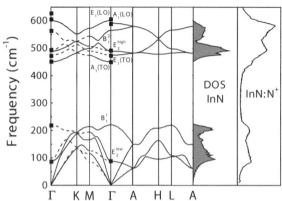

FIGURE 6.16

Calculated phonon dispersion curves and phonon DOS function for hexagonal InN. The disorder-induced Raman spectrum obtained at 7 K for N^+-implanted InN is also shown. Figure reproduced with permission from V. Yu. Davydov, V. V. Emtsev, I. N. Goncharuk, A. N. Smirnov, V. D. Petrikov, V. V. Mamutin, V. A. Vekshin, S. V. Ivanov, M. B. Smirnov, and T. Inushima, Applied Physics Letters, 75, 3297 (1999). Copyright 1999, American Institute of Physics.

interactions. The short-range potentials also accounted for the contributions from neighboring atoms of the second coordination sphere. The following parameters were used: the frequencies of optical phonons at the Γ-point, which were found from first-order Raman spectra; frequencies of silent modes, which were evaluated from spectra of structurally imperfect samples; the ion charges, whose values were determined on the basis of the experimentally observed LO-TO splitting; and the literature values of elastic constants.

The phonon dispersion curves calculated along the principal directions in the first Brillouin zone and the DOS function are shown in Fig. 6.16. It can be seen that the phonon spectrum of InN consists of two regions separated by a wide gap. The low-energy region (0–230 cm^{-1}) includes three acoustic and low-frequency silent B_1 and optical E_2(low) branches, whereas the upper region (450–600 cm^{-1}) is associated with high-frequency optical vibrations. On the whole, the phonon spectrum of InN, in contrast to AlN, is similar to that of GaN [6.72]. This is not surprising since the dynamics of both lattices is mainly governed by the motion of nitrogen atoms due to a large difference between the cation and anion masses.

It was shown in Ref. [6.72] that the phonon DOS function can be reconstructed for GaN and AlN on the basis of experimental Raman spectra of single crystals with a strongly distorted crystal lattice. The same approach was used for InN [6.25]. MBE-grown InN samples irradiated with nitrogen ions N^+ (energy 30 keV, dose 5×10^{14} ions/cm^2) were studied. As can be seen from Fig. 6.16, the main features of the calculated DOS function and the experimental DOS extracted from the Raman spectra of N^+-irradiated InN sample correlate well in the entire spectral region, thus proving the validity of model calculations.

In later papers [6.58, 6.65, 6.66], the calculations based on a modified valence-force model, a state-of-the-art density-functional perturbation theory, and an adiabatic bond charge model were employed to calculate the phonon dispersion relations and DOS function for wurtzite InN. The data obtained are in good agreement with the calculations in terms of the rigid-ion model [6.25].

Analysis of the dispersion relations for acoustic and optical phonons led to important conclusions concerning the possible anharmonic decay channels of optical phonons in InN. It was noted in Ref. [6.65] that the decay of the LO phonon into two acoustic phonons with equal energies and opposite wavevectors is forbidden in InN. This is because $\omega_{LO} > 2LA, TA$ over the entire spectral range. Recently, the lifetimes of LO phonons in InN and their dynamics have been studied by time-resolved Raman spectroscopy [6.73]. The authors demonstrated that both the $A_1(LO)$ and $E_1(LO)$ phonons decay primarily into a large wave vector TO phonon and a large wave vector TA/LA phonon, consistent with the phonon dispersion for wurtzite InN. Such a decay channel can affect the lifetime of LO phonons and thereby can govern the effects associated with hot-phonon-related processes, which play an important role in high-speed devices [6.65].

Using the calculated phonon DOS, the lattice specific heat of InN at a constant volume C_v and its temperature dependence were evaluated in Ref. [6.25]. The results of the calculations perfectly agreed with the experimental data reported in Ref. [6.74]. The Debye temperature, θ_D, as a function of temperature was calculated in a similar way to that in previous work on GaN [6.75]. It was found that the Debye temperature θ_D of InN at 0K is 370K. (Note that θ_D=800K for AlN and θ_D=570K for GaN [6.65]).

6.8.5 Raman studies of Mg-doped InN

Despite recent progress in the growth of epitaxial InN layers, p-type doping of InN is not yet well understood, which hinders the use of InN-based heterostructures in photonic and electronic applications. This problem can be solved only through complex investigations of doped InN using different techniques. Raman scattering was successfully used for a study of Mg-doped GaN, and valuable information on the correlation between Raman data and p-conduction of GaN was obtained [6.76]. However, there are few Raman experiments with Mg-doped InN [6.25, 6.77].

Recently, the results of experimental and theoretical studies of Mg-doped InN samples have been reported in Ref. [6.78]. Mg-doped 0.5–2.0-μm-thick InN layers were grown by plasma-assisted molecular-beam epitaxy on c-plane sapphire substrates and on Si(111) and r-plane sapphire substrates [6.15, 6.51]. All the samples had a wurtzite structure. The incorporation of Mg was verified by secondary ion mass spectrometry (SIMS). The Mg concentrations in the samples ranged from $N_{Mg} = 3.3 \times 10^{19}$ to 5.5×10^{21} cm^{-3}. According to the SIMS depth profiles, the Mg distribution was uniform throughout the films. Hall measurements of as-grown InN:Mg samples demonstrated n-type conduction, while thermal probe measurements indicated p-type conduction for some layers.

Spectra of the Mg-doped InN films are presented in Figs. 6.17(a) and 6.17(b) for the entire range of Mg concentrations. The spectra are normalized to the E_2(high) mode intensity and shifted along the vertical axis to simplify comparison. The most pronounced feature of the spectra is the enhancement of the LO-phonon band intensity with increasing Mg content. The integral intensity of the LO band was found to be similar for the samples grown on different substrates and having equal Mg content. In InN:Mg, this band was ascribed to the L^- mode resulting from

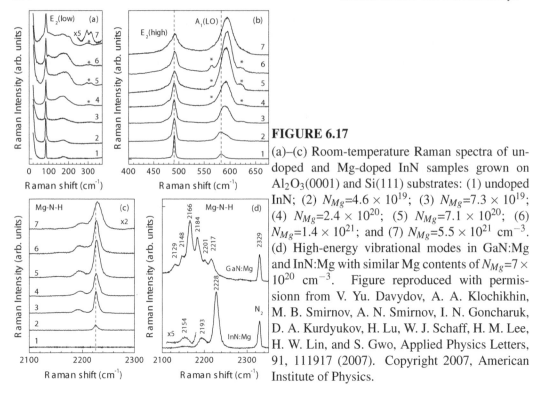

FIGURE 6.17
(a)–(c) Room-temperature Raman spectra of undoped and Mg-doped InN samples grown on
$Al_2O_3(0001)$ and $Si(111)$ substrates: (1) undoped
InN; (2) N_{Mg}=4.6 × 10^{19}; (3) N_{Mg}=7.3 × 10^{19};
(4) N_{Mg}=2.4 × 10^{20}; (5) N_{Mg}=7.1 × 10^{20}; (6)
N_{Mg}=1.4 × 10^{21}; and (7) N_{Mg}=5.5 × 10^{21} cm^{-3}.
(d) High-energy vibrational modes in GaN:Mg
and InN:Mg with similar Mg contents of N_{Mg}=7 ×
10^{20} cm^{-3}. Figure reproduced with permis-
sionn from V. Yu. Davydov, A. A. Klochikhin,
M. B. Smirnov, A. N. Smirnov, I. N. Goncharuk,
D. A. Kurdyukov, H. Lu, W. J. Schaff, H. M. Lee,
H. W. Lin, and S. Gwo, *Applied Physics Letters*,
91, 111917 (2007). Copyright 2007, American
Institute of Physics.

the Raman process with the breakdown of the wavevector conservation law due to
electron-hole pair scattering by Mg impurities.

The Raman cross section can be described in the linear approximation with respect to the impurity concentration, N_I, as long as the averaged spatial separation
of impurities, $\bar{r} = (3/4\pi N_I)^{1/3}$, is larger than the Thomas-Fermi screening length,
$1/q_{TF}$. If we regard the spatial size of an impurity potential well as a sphere of
radius $1/q_{TF}$, the nonlinear corrections to the cross section appear when some
spheres overlap and form clusters. Scattering of photo-excited electrons and holes
is most effective from single impurities when the transferred wave vector is of the
order of q_{TF}. An increase in the size of the potential wells of the clusters due to
overlapping leads to a decrease in the wave vector transfer in a scattering event,
and the effect of scattering from clusters can be neglected in the first approximation. Therefore, in order to find the dependence of Raman cross section on the
impurity concentration, we have to estimate the concentration of non-overlapping
potential wells in terms of the continuum percolation theory [6.79]. The number of
impurities with non-overlapping potential wells, N_I^1, both below and above the percolation threshold, can be written as $N_I^1 = \rho \, exp(-2\rho)$, where $\rho = [1/\bar{r}q_{TF}]^3 / 2$.

Figure 6.18 compares the experimental and model E_2(high) mode intensity as a
function of Mg content. It can be seen that below N_{Mg}=1 × 10^{21} cm^{-3} the variation
is close to linear, which means that Raman data for this concentration range can be
used to estimate quantitatively the Mg content in InN.

The intensities of two distinct peaks at 2193 and 2228 cm^{-1} in Raman spec-

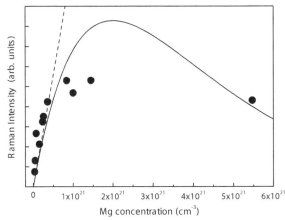

FIGURE 6.18

(a) Integral intensity of the LO-like mode of InN:Mg normalized to the E_2(high) mode intensity (symbols) and the number of impurities with non-overlapping potential wells (the solid line) as functions of Mg content. The dashed line is a linear approximation. Figure reproduced with permission from V. Yu. Davydov, A. A. Klochikhin, M. B. Smirnov, A. N. Smirnov, I. N. Goncharuk, D. A. Kurdyukov, H. Lu, W. J. Schaff, H. M. Lee, H. W. Lin, and S. Gwo, Applied Physics Letters, 91, 111917 (2007). Copyright 2007, American Institute of Physics.

tra of InN:Mg samples were found to correlate with the Mg content as well (see Figs. 6.17(c) and 6.17(d)). Similar high-energy vibrational modes in the Raman spectra of GaN:Mg were assigned to Mg-H bonds in the complexes with vacancies or interstitials [6.80]. However, the exact microscopic structure of these defect complexes is still unclear.

Detailed calculations of various Mg-N-Ga-H complexes in GaN:Mg have been made in Ref. [6.81]. It can be assumed that similar complexes also exist in InN:Mg. We choose the most symmetrical one among these hydrogen-metal complexes, that is, a linear H-Mg-H molecule, having the $D_{\infty h}$ symmetry of the point group. This choice is based on the fact that the Raman spectra of this complex will be sensitive to its orientation in the host lattice.

The group theory analysis gives information on the site symmetry of the defect complex in the lattice and also on its intrinsic symmetry. We assume that the H-Mg-H is an interstitial complex, having two main orientations, parallel or perpendicular to the hexagonal axis (Fig. 6.19). In the first case, the symmetry of the crystal with the defect reduces to the C_{3v} point group, whereas in the second case the symmetry lowers to C_s. Hence, the vibrational representations of the complexes at C_{3v} and C_s sites are given by $2A_1+E_1$ and $3A'+A''$, respectively. The IR and Raman selection rules for these modes are presented in Table 6.2. For C_s sites, we present both the selection rules for a complex lying in the plane being perpendicular to the *y*-

TABLE 6.2

IR and Raman selection rules for linear H-Mg-H defect complex.

Defect site symmetry	Mode	IR	Raman
C_{3v}	A_1	z	xx,yy,zz
	E_1	x,y	xx,yy,xy,xz,yz
C_s	A'	x,z(x,y,z)	xx,yy,zz,xz(xx,yy,zz,xy,xz,yz)
	A''	y(x,y)	xy,yz(xx,yy,xy,yz,xz)

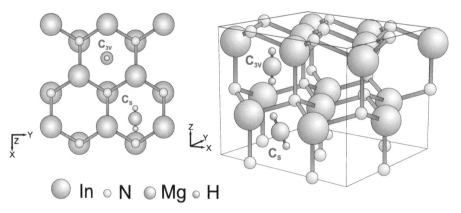

FIGURE 6.19

Crystal structures of hexagonal InN showing different symmetry sites of the linear H-Mg-H defect complex.

axis and averaged selection rules (in parentheses) because complexes are randomly distributed over 6 symmetry planes.

It can be seen from Fig. 6.20(a) and Fig. 6.20(b) that the line at 2228 cm^{-1} is Raman-active only in xx, yy, and zz scattering configurations. This behavior allows us to conclude that this line corresponds to the A_1 -mode of the defect complexes at C_{3v} sites rather than to the A' mode of the interstitial complexes at C_s sites because in the latter case the line would be observed in the yz geometry as well.

Comparison of the group-theory predictions with the experimental Raman spectra allows us to distinguish symmetric ($D_{\infty h}$) and non-symmetric ($C_{\infty V}$) linear H-Mg-H complexes. The comparison shows that these complexes are symmetric. Indeed, for a symmetric linear complex at the C_{3v} site, the $A_{1g} + A_{1u}$ modes of the free molecule complex transform into the $2A_1$ modes in the crystal. Since only an A_{1g} mode is Raman-active in a free molecule, we can assume that two $2A_1$ modes originating from the A_{1g} and A_{1u} modes will have different intensities in the Raman spectrum of the crystal. The ratio between the intensities of these $2A_1$ lines can be regarded as a measure of the interaction of the defect complex with the host lattice. It is this situation that is observed in our Raman spectra shown in Fig. 6.20(b). We can assign the intense line at 2228 cm^{-1} to the A_1 mode originating from the Raman-active A_{1g} mode of the free H-Mg-H molecule, whereas the weaker line at 2193 cm^{-1} can be assigned to the A_1 mode originating from the Raman-forbidden A_{1u} mode that becomes Raman-active in the crystal. The line at 2193 cm^{-1} is more intense in the zz configuration than in the xx and yy ones, which can be explained by the anisotropy of the Raman tensor of the molecule oriented along the z axis.

Weak lines at 293, 313, 565, and 622 cm^{-1}, which are shown in Figs. 6.17(a) and 6.17(b) by asterisks, can be assigned to local vibrational modes (LVM) of Mg that substitutes In. In order to verify this assignment the lattice dynamics of InN with substitutional impurities have been studied theoretically, and their Raman spectra were simulated. The model was described in Refs [6.25, 6.72]. In this

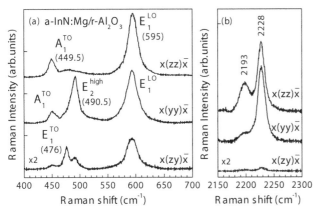

FIGURE 6.20

Polarized Raman spectra of Mg-doped *a*-plane InN grown on *r*-Al$_2$O$_3$ substrate, N_{Mg}=2.3 × 10^{20} cm^{-3}. The agreement with the selection rules for the wurtzite structure indicates a good crystalline quality of the *a*-plane InN layer. (λ_{ex}=488 nm). Reprinted with permission from V. Yu. Davydov *et al.*, physica status solidi (c), 5, 1648 (2008). Copyright 2008 by Wiley-VCH Publishers, Inc.

model, the Mg-N force constants were related to the In-N force constants as K(Mg-N) $= sK$(In-N). A scaling factor s was found by comparing the calculated and experimental spectra. The lattice spatial anisotropy was taken into consideration by using different force constants for the In-N bonds directed along and obliquely to the hexagonal axis. A similar procedure was employed earlier for Mg-doped GaN [6.82].

Several spectral lines whose frequencies exhibited a strong dependence on s have been found in the simulated Raman spectra. If s=1.1, two distinct lines in the frequency interval between the E_2(high) and LO modes appear. Frequencies of these lines are 557 and 624 cm^{-1}. Analysis of atomic displacements has shown that the mode at 624 cm^{-1} corresponds to the v_3 vibration of an isolated MgN_4 tetrahedron, whereas the mode at 557 cm^{-1} is a combination of v_3 with the v_4 "umbrella" vibration of the MgN_4 tetrahedron. It can be supposed that they correspond to the Raman lines observed at 565 and 622 cm^{-1}.

The simulated Raman spectrum has two s-independent lines at 303 and 315 cm^{-1} in the gap interval between the optical and acoustic InN branches. Eigenvector analysis allows us to assign these lines to LVMs coming from the v_4 vibration of the MgN_4 tetrahedron. The doublet structure of this vibration is due to a spatial anisotropy of the wurtzite-like lattice. The low-frequency and the high-frequency components correspond to the vibrations of the A- and E$_2$- symmetry, respectively. The lines observed in the experiment at 293 and 313 cm^{-1} can be tentatively assigned to these two modes.

The lattice dynamics of hexagonal InN with anion and cation vacancies were also investigated by using a similar approach. Several lines can be distinguished in the calculated Raman spectra in the regions 130–180 and 350–370 cm^{-1} for InN with nitrogen and indium vacancies, respectively. In addition, the spectra have several lines in the frequency interval between the E_2(high) and LO modes. Thus, the interpretation of the weak lines at 293, 313, 565, and 622 cm^{-1} given above can be regarded only as tentative. Further studies are needed to make a final conclusion.

6.8.6   Behavior of phonon modes in hexagonal InGaN alloys

The crystal lattice dynamics of hexagonal $In_xGa_{1-x}N$ alloys were theoretically analyzed in Ref. [6.83]. A one-mode type behavior was predicted by the modified random-element isodisplacement (MREI) model for all Raman-active modes. The one-mode behavior is characterized by a continuous and approximately linear variation of the frequencies of all modes between their values for GaN and InN with varying alloy composition.

Experimental investigations of the lattice dynamics of $In_xGa_{1-x}N$ alloys were carried out by several groups by using Raman spectroscopy [6.84–6.91]. The majority of groups concentrated on studies of two out of the six optical phonons allowed in Raman spectra of hexagonal $In_xGa_{1-x}N$ alloys (namely, E_2(high) and A_1(LO) symmetry phonons). The phonons of both types observed in Raman experiments were found to demonstrate a one-mode type behavior, which was consistent with the theoretical predictions [6.83]. At the same time, Alexson *et al.* [6.86] supposed a two-mode behavior of the E_2(high) mode based on its deviation from the linear dependence. The data on the behavior of E_1(TO) and A_1(TO) phonon modes in the compositional range $0.10 < x < 0.26$ were obtained by Kontos *et al.* [6.90]. Typically the investigations mentioned above employed $In_xGa_{1-x}N$ alloys having a limited range of compositions. The only exception was the work of Hernandez, who investigated E_2(high) and A_1(LO) optical phonons of $In_xGa_{1-x}N$ alloy over the entire composition range [6.91].

To obtain detailed information on the behaviors of as large a number of Raman allowed phonon modes as possible in a wide compositional range, we have studied a wide set of alloys with compositions ranging from GaN to InN.

The samples were hexagonal $In_xGa_{1-x}N$ epilayers grown on (0001) sapphire substrates. The alloy films in the compositional ranges $0.01 < x < 0.30$ and $0.35 < x < 0.98$ were grown by MOCVD at Ioffe Institute and PAMBE at Cornell University [6.15], respectively. The film thicknesses were around 0.5 μm in the MOCVD grown films and ranged from 0.2 to 0.5 μm in the MBE grown samples. All the alloys were nominally undoped films of n-type conductivity with a Hall carrier concentration $n = 1 - 5 \times 10^{18}$ cm^{-3}.

The alloy compositions were estimated from the data on Rutherford backscattering of deuterons. In addition, the energy spectra of protons and α-particles arising due to nuclear reactions of deuterons with oxygen and carbon atoms were used to estimate the oxygen and carbon concentrations in the alloys. The alloy compositions were also evaluated in high-resolution x-ray diffraction measurements.

Strain effects in $In_xGa_{1-x}N$ layers resulting from their incomplete relaxation and a difference in the thermal expansion coefficients of a substrate and alloy can lead to significant errors in the estimates of the compositions x. To obtain reliable values of x, x-ray studies involved estimation of two parameters of a unit cell of the wurtzite structure a and c by measuring symmetric Bragg (00.2), asymmetric Bragg (11.4), grazing incidence (11.0), or Laue reflections. The values of x and strain ε were calculated by using a set of equations

$$a = a(x)(1 + \varepsilon_\parallel), \tag{6.32}$$

$$c = c(x)(1 - \varepsilon_\parallel [p_{InN}x + p_{GaN}(1-x)]) \tag{6.33}$$

where $\varepsilon_\parallel = [a - a(x)]/a(x)$ is the in-plane isotropic strain in the lattice; a and c are the In$_x$Ga$_{1-x}$N layer parameters being measured; $a(x) = a_{InN}x + a_{GaN}(1-x)$ and $c(x) = c_{InN}x + c_{GaN}(1-x)$ are the lattice parameters determined by the composition (it is assumed that Vegard's law is valid); $p_{InN/GaN} = 2C_{13}/C_{33}$, where C_{13} and C_{33} are corresponding elastic constants for InN and GaN.

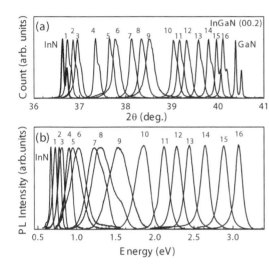

FIGURE 6.21

(a) X-ray diffraction data and (b) PL spectra of InN and In$_x$Ga$_{1-x}$N alloys with: (1) $x = 0.89$; (2) $x = 0.82$; (3) $x = 0.78$; (4) $x = 0.68$; (5) $x = 0.66$; (6) $x = 0.57$; (7) $x = 0.50$; (8) $x = 0.46$; (9) $x = 0.44$; (10) $x = 0.34$; (11) $x = 0.28$; (12) $x = 0.23$; (13) $x = 0.16$; (14) $x = 0.12$; (15) $x = 0.09$; and (16) $x = 0.06$.

For our studies, In$_x$Ga$_{1-x}$N samples with the highest structural and optical characteristics were selected. Figure 6.21 shows x-ray diffraction and PL spectra of the selected In$_x$Ga$_{1-x}$N samples. A distinguishing feature of these spectra is a considerable broadening of the x-ray and PL curves compared with those for GaN and InN. This indicates that the alloys are less perfect than binary crystals. However, no traces of phase separation or polymorphism have been revealed in the x-ray data or PL spectra of the samples.

Raman process in In$_x$Ga$_{1-x}$N can be described by different scattering mechanisms depending on the ratio between excitation energy and the In$_x$Ga$_{1-x}$N band gap that varies from 3.4 to 0.7 eV for samples with different compositions. The Raman measurements were performed at room temperature in a quasibackscattering geometry using lines of a He-Cd laser (3.81 and 2.81 eV), an Ar ion laser (2.41, 2.54, and 2.71 eV), a frequency doubled Nd:YAG laser (2.33 eV), a He-Ne laser (1.96 eV), and a Kr laser (1.92 and 1.83 eV) as an excitation source. The use of different excitation energies and scattering geometries where both the surface plane and edge of the samples were examined allowed us to get detailed information on four of six Raman allowed phonon modes.

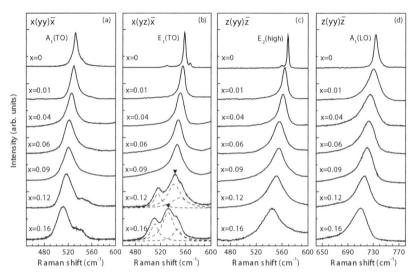

FIGURE 6.22

Room temperature polarized Raman spectra of A_1(TO), E_1(TO), E_2(high), and A_1(LO) symmetry modes of In$_x$Ga$_{1-x}$N in the compositional range $0.01 < x < 0.16$. The spectra corresponding to different In compositions are normalized to the maxima of the appropriate Raman lines.

As an example, Fig. 6.22 shows Raman spectra of A_1(TO), E_1(TO), E_2(high), and A_1(LO) symmetry modes in the compositional range $0.01 < x < 0.16$. The spectra corresponding to different In compositions are normalized to the maxima of the appropriate Raman lines. Good agreement between the polarized Raman spectra and the selection rules for the wurtzite structure points to a high crystalline quality of the In$_x$Ga$_{1-x}$N samples.

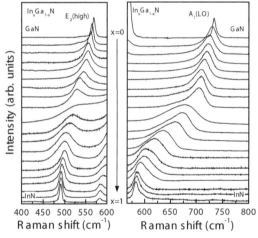

FIGURE 6.23

Room-temperature polarized Raman spectra for the E_2(high) and the A_1(LO) modes of the In$_x$Ga$_{1-x}$N epilayers with different In contents covering the entire alloy range. The spectra are normalized to the E_2(high) or A_1(LO) intensity.

Figure 6.23 shows Raman spectra recorded in the $z(xx)\bar{z}$ backscattering configuration for the entire set of the In$_x$Ga$_{1-x}$N epilayers. Here z is the direction of the hexagonal axis of the epilayer. In this configuration, only the A_1(LO) and the E_2(high) modes are allowed.

As can be seen from Fig. 6.23, the Raman spectra exhibit only one A_1(LO) peak over the entire composition range. This confirms the theoretical predictions of the one-mode-type behavior of A_1(LO) optical phonon in the In$_x$Ga$_{1-x}$N alloy [6.83]. It is evident that the A_1(LO) line is rather narrow for the samples with low concentrations of In or Ga; however, this mode exhibits considerable broadening for the samples with intermediate compositions. The line broadening points to a higher lattice disorder and the presence of composition fluctuations in the alloy (this issue is discussed in more detail below). It is also seen that for all the compositions x the A_1(LO) band shape is asymmetrical, with a long low-frequency tail and abrupt high-frequency edge. There is no additional high-frequency shoulder of the A_1(LO) mode which is typically related to spinodal decomposition. This suggests that there is no phase separation in the In$_x$Ga$_{1-x}$N alloys studied, which is consistent with x-ray and PL data.

The E_2(high) line in the Raman spectra in Fig. 6.23 is rather narrow for the samples with low In or Ga concentrations and experiences a drastic broadening for the samples with intermediate compositions. For this reason, no unambiguous conclusion on the type of behavior (one-mode or two-mode) of the E_2(high) line can be made.

Figure 6.24 compares compositional dependences of the frequencies of zone-center phonons in hexagonal In$_x$Ga$_{1-x}$N alloys calculated in Ref. [6.83] and the measured phonon frequencies which are corrected for the strains in In$_x$Ga$_{1-x}$N layers estimated from x-ray data.

In the case of biaxial strain, the Raman line shift is given by $\Delta\omega = 2a_\omega\varepsilon_\parallel + b_\omega\varepsilon_\perp$, where a_ω and b_ω are the deformation potentials of the corresponding phonon mode. The a_ω and b_ω parameters were calculated for In$_x$Ga$_{1-x}$N from the literature data on these parameters for GaN and InN as composition weighted averages: $a_\omega(x) = xa_\omega(InN) + (1-x)a_\omega(GaN)$ and $b_\omega(x) = xb_\omega(InN) + (1-x)b_\omega(GaN)$. The deformation potentials of phonon modes A_1(TO), E_1(TO), E_2(high) and A_1(LO) of GaN were taken from Refs [6.92, 6.93]. The deformation potentials of E_1(TO), E_2(high) and A_1(LO) modes of InN were taken from [6.62, 6.63]. Since no data on phonon deformation potentials of InN for the A_1(TO) mode are available in the literature, the deformation potentials of the A_1(TO) mode of GaN were used to estimate the strain-induced Raman shift for Ga-rich In$_x$Ga$_{1-x}$N alloys $(0 < x < 0.2)$. We did not correct for strain the frequencies of the A_1(TO) mode for In-rich In$_x$Ga$_{1-x}$N alloys $(0.6 < x < 1)$ because these frequencies were estimated from the L^- mode positions of the alloys with high electron concentrations $(n = 7\text{–}9 \times 10^{19}$ cm$^{-3})$ and can be regarded only as tentative.

It can be seen that the measured A_1(LO) phonon frequencies for Ga-rich and In-rich In$_x$Ga$_{1-x}$N alloys coincide well with the theoretically predicted linear compositional dependence for In$_x$Ga$_{1-x}$N alloys. Note that the linearity in the behavior of the A_1(LO) phonon frequencies is a favorable factor for development

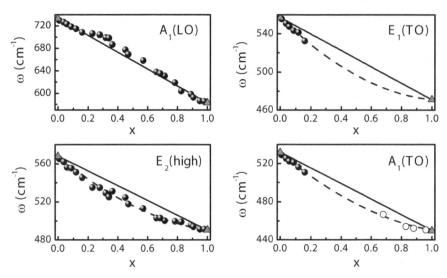

FIGURE 6.24

Frequencies of zone-center phonons in hexagonal $In_xGa_{1-x}N$ versus composition. The solid lines are calculated frequencies [6.83]. The full circles are experimental data obtained on $In_xGa_{1-x}N$ with low electron concentrations ($n = 1$–5×10^{18} cm^{-3}). The open circles are the positions of L^- mode obtained on $In_xGa_{1-x}N$ with high electron concentrations ($n = 7$–9×10^{19} cm^{-3}).

of the quantitative method for estimation of the $In_xGa_{1-x}N$ composition via Raman spectroscopy. The dependence of the $In_xGa_{1-x}N$ $A_1(LO)$ phonon mode on In content in the regions $0 < x < 0.2$ and $0.7 < x < 1$ can be approximated as $\omega(A_1(LO)) = 733 - 150x$. A similar expression for the compositional dependence of the $A_1(LO)$ phonon mode of relaxed $In_xGa_{1-x}N$ alloys for $0 < x < 0.3$ was obtained in Ref. [6.88].

However, the $A_1(LO)$ phonon frequencies for intermediate compositions ($0.3 < x < 0.6$) are above the linear extrapolation. A similar behavior of $A_1(LO)$ phonon frequencies was found also in Refs [6.87, 6.91]. One of the possible reasons for deviation of frequencies of $A_1(LO)$ phonons from a linear dependence can be a strong violation of the selection rules in $In_xGa_{1-x}N$ for this compositional range. As a consequence, the so-called quasi-LO modes can be observed in the Raman spectrum due to angular dispersion of polar phonons. The energies of these modes depend on the phonon propagation direction relative to the optical axis [6.94]. Since frequencies of the $E_1(LO)$ phonons for the entire range of $In_xGa_{1-x}N$ compositions are higher than the $A_1(LO)$ phonon frequencies, it can be expected that the maxima of such quasi-LO bands will be above the linear dependence calculated theoretically for the phonon of the $A_1(LO)$ symmetry.

The experimentally observed compositional dependence of the $E_2(high)$ phonon is obviously nonlinear (see Fig. 6.24). At the same time, there are no signs of the two-mode behavior of this phonon which is typically thought to be responsible for such deviations from linearity. A strong deviation from linearity is also observed for the $A_1(TO)$ and $E_1(TO)$ modes measured for $In_xGa_{1-x}N$ samples in the compo-

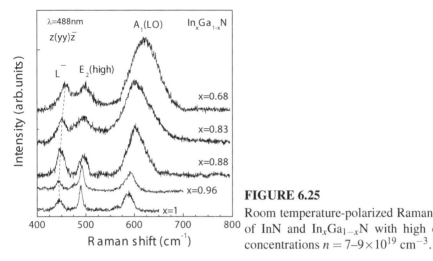

FIGURE 6.25

Room temperature-polarized Raman spectra of InN and In$_x$Ga$_{1-x}$N with high electron concentrations $n = 7$–9×10^{19} cm^{-3}.

sitional range $0.01 < x < 0.16$.

We also studied In$_x$Ga$_{1-x}$N alloys with a high In content ($0.4 < x < 1$) and a high electron concentration ($n = 7$–9×10^{19} cm^{-3}). The A_1(TO) phonon mode frequencies for In$_x$Ga$_{1-x}$N samples can be approximately estimated from the L^- mode position because it must be close in the spectra of these alloys to the position of the A_1(TO) phonon mode (see also section 6.8.3). These spectra are shown in Fig. 6.25, and the behavior of L^- mode is presented in Fig. 6.24. It is obvious that the latter also deviate from linearity.

The nonlinearity revealed in the behavior of the E_2(high), A_1(TO), and E_1(TO) modes in In$_x$Ga$_{1-x}$N indicates that the crystalline lattice dynamics of this alloy should be investigated more thoroughly both experimentally and theoretically.

6.8.7 Compositional dependence of Raman line broadening in InGaN

We have found that the phonon lines of the E_2(high) and A_1(LO) symmetry in Raman spectra of InGaN alloys experience an additional broadening depending on the alloy composition. This broadening can be attributed to the elastic scattering of phonons from compositional fluctuations in the alloy. In the case of low In or Ga concentrations the fluctuations can be described as isolated In and Ga atoms or clusters of relatively small sizes. The number of isolated atoms and clusters can be calculated exactly for many types of crystal lattices [6.95]. The concentration regions where the single atom effects will be dominating in the cation sublattice, which is close to the fcc sublattice, can be estimated as $x \leq 0.05$ or $(1-x) \leq 0.05$. At $x \approx 0.05$ and $(1-x) \approx 0.05$ the numbers of small isolated clusters and isolated atoms are comparable. Since the perturbation grows with the cluster size, it can be expected that scattering from clusters will play a leading role at higher In or Ga concentrations. The cluster approach is acceptable for concentrations up to $x < 0.1$–0.12 and $(1-x) < 0.1$–0.12 for the fcc sublattice.

The description of fluctuations for the intermediate concentration range $0.1 <$

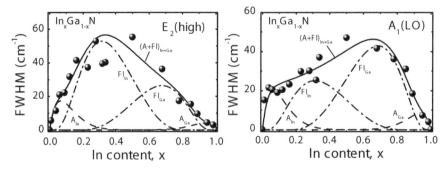

FIGURE 6.26

Experimental and calculated compositional dependences of phonon line broadening. Full circles are the experimental dependences of FWHMs on In content in $In_xGa_{1-x}N$ alloys for the E_2(high) and A_1(LO) modes. The dashed lines A_{In} and A_{Ga} represent isolated atom contributions; the dash-and-dot lines Fl_{In} and Fl_{Ga} are the contributions of fluctuations in the intermediate concentration region given by $F_{n,p}[x,(1-x)]G_{In}$ and $F_{n,p}[(1-x),x]G_{Ga}$. The solid line is the sum of all the contributions.

$x < 0.9$ must be different. As shown in Ref. [6.96], the fluctuations that are most efficient for phonon scattering can be found by considering a finite size of the In sublattice in which the concentration differs from the average concentration in the alloy.

The probability of such fluctuations strongly depends on their size n and the deviation of the atom numbers $n\,p$ within this volume from the average concentrations nx and $n(1-x)$ in the alloy. Therefore, a chosen size of the fluctuation and excessive concentration of In or Ga within this volume are the basic characteristics of such a fluctuation.

The In fluctuation probabilities characterized by chosen n and p are defined within the concentration interval $0 < x < (1-p)$. Similar fluctuations of Ga atoms can be found in the concentration interval $p < x < 1$. The appropriate choice of the size n and the deviation parameter p allows one to reconcile the region of intermediate concentrations with the regions of precisely calculated cluster fluctuations.

The study of the phonon line broadening as a function of alloy composition gives a possibility to estimate the fluctuation statistics, that is, the type, size, and number of fluctuations, and also the dependence of these parameters on the alloy composition. If both the Ga and In fluctuations are present, the composition dependence of phonon line broadening can be given by

$$\gamma_{j,\mathbf{k}-\mathbf{k'}}(\omega) \sim F_{n,p}[x,(1-x)]\,G_{In} + F_{n,p}[(1-x),x]\,G_{Ga} \qquad (6.34)$$

where G_{In} and G_{Ga} are the phonon DOS values which, in general, are functions of ω and of the alloy composition. The statistical factors, $F_{n,p}[x,(1-x)]$ and $F_{n,p}[(1-x),x]$ defined in Ref. [6.96] give the probability of fluctuations in the atom distribution.

The compositional dependences of E_2(high) and A_1(LO) line broadenings for

$In_xGa_{1-x}N$ alloys and the results of calculations using Eq. (6.34) are shown in Fig. 6.26. It can be seen that the dependences are characterized by an asymmetry with respect to the point $x = 0.5$. Such an asymmetry indicates that fluctuations of both types of atoms (In and Ga) are responsible for phonon scattering. It is also evident that there exists a wide range of concentrations where fluctuations of both types of atoms can coexist.

In calculations of the curves, single atom fluctuations and the fluctuations containing n sublattice sites were taken into account. In the intermediate concentration region it was assumed that n sublattice sites are filled with $n(x + p)$ In atoms and $n(1 - x - p)$ Ga atoms at $n = 15$–30 and at $p \approx 0.2$ which is close to the percolation threshold for the fcc sublattice. This corresponds to the linear size of the fluctuation approximately equal to 2–3 interatomic distances in the disordered sublattice. The obtained results allow one to estimate the number and size of fluctuations of a given type, and, therefore, the fraction of the crystal volume occupied by these fluctuations. For fluctuations of size $n = 15$–30 at $p \approx 0.2$, only about 1% of the whole volume is occupied by the fluctuations of the given type. This value is insufficient to form complexes due to overlapping of neighboring fluctuations of a given type, and most of these fluctuations remain spatially isolated.

At the same time, the experimental points have a considerable scatter as compared with smoothed theoretical curves. It can be supposed that these deviations reflect differences in the qualities of the samples used in the experiments. All the theoretical dependences were obtained for randomly distributed atoms In and Ga, and any deviation from randomness can cause a change in the broadening, resulting in the scatter of experimental points.

6.8.8 Resonant Raman scattering in InGaN alloys

Useful information on optical properties of alloys can be obtained from resonant Raman scattering [6.97]. Due to a strong electron-phonon Fröhlich interaction in polar crystals, a drastic enhancement of the light scattering from polar LO-phonons occurs under resonant conditions, that is, when the energy of the exciting photon is close to the band-gap energy of the semiconductor. As a result, the resonant Raman scattering can be used for the estimation of the band gap in alloys.

Recently, multi-phonon Raman scattering has been observed in the $In_xGa_{1-x}N$ alloy at $x = 0.27$ [6.91]. Multiple scattering from the longitudinal phonons is typical of the resonant scattering in polar semiconductors with the exciton structure. The resonant behavior of the multi-LO-phonon scattering in the $In_xGa_{1-x}N$ films which have high electron concentrations and no exciton structure involves the band-to-band excitations.

Resonant Raman scattering for the $In_xGa_{1-x}N$ alloys with In contents in the range from $x = 0.15$ to $x = 0.35$ was studied in Ref. [6.98]. The alloy samples were grown by MBE and MOVPE techniques at Cornell University [6.15] and Ioffe Institute, respectively. Nominally undoped films were degenerately n-type semiconductors at room temperature, with Hall concentrations in the range 1×10^{18} to 1×10^{19} cm^{-3}.

The multi-phonon Raman spectra which were obtained at different excitation energies above and below the band gap of the $In_{0.35}Ga_{0.65}N$ film are shown in Fig. 6.27(a). The behaviors of the spectra demonstrate that the Raman processes occur due to interband transitions from the top of the valence band to the conduction band followed by step-by-step emission of LO-phonons by the excited pairs. In this case a light electron can emit a number of LO-phonons and the cross sections with emission of n and $(n-1)$ LO-phonons can be comparable $\sigma_n/\sigma_{n-1} \approx w_{LO}\tau(E) \leq 1$ where w_{LO} is the probability of the electron-LO-phonon interaction and $\tau(E)$ is the full lifetime of the electron in the state with energy E [6.99].

The resonant single-phonon Raman cross section due to deformation potential interaction near the band gap can be described as $|\chi(\omega_i) - \chi(\omega_f)|^2$ [6.100], where $\chi(\omega)$ is the interband dielectric susceptibility tensor. The scattering amplitude of the forbidden 1LO Raman scattering due to the Fröhlich interaction is more resonant, and the cross section is proportional to $|\chi(\omega_i) - \chi(\omega_f)|^6$ [6.100]. If the momentum conservation law is broken due to scattering of electrons and holes from the short range potential, the 1LO forbidden scattering amplitude is similar to the LO+TO scattering amplitude calculated in Ref. [6.101], and in this case the cross section is proportional to $|\chi(\omega_i) - \chi(\omega_f)|^3$. Then the cross sections for different mechanisms can be written as

$$\sigma_{\beta\beta}(\omega) \sim \left|\chi_{\beta\beta}(\omega) - \chi_{\beta\beta}(\omega - \Omega_{LO})\right|^m \tag{6.35}$$

where β is the polarization of an incident or scattered photon, Ω_{LO} is the phonon frequency, and $\chi_{\beta\beta}(\omega)$ is the interband dielectric susceptibility tensor.

The calculated behavior of 1LO-cross sections given by Eq. 6.35 for m=6 and m=3 for different electron concentrations is presented in Fig. 6.27(b). The position of the Raman cross section maximum of Eq. 6.35 is defined by the band gap E_g and shifts toward higher energies with increasing Fermi energy. The Raman cross section maximum position is close to the optical absorption threshold and can be estimated as $E_{max} \approx E_g + E_F + \hbar\Omega_{LO}/2$. Therefore, E_g of $In_xGa_{1-x}N$ films can be estimated from the Raman data if the electron concentration and effective mass are known.

As an example, Fig. 6.27(c) and Fig. 6.27(d) present the experimental data on the Raman cross sectional profiles near the absorption threshold together with absorption and PL spectra of the samples $In_{0.35}Ga_{0.65}N$ and $In_{0.20}Ga_{0.80}N$, respectively. It can be seen that the positions of the Raman cross section maxima for both samples coincide well with the absorption thresholds, which is in line with theoretical predictions. Comparison of the calculated cross section profiles from Eq. 6.35 and experimental data shows that they coincide satisfactorily for m between $m = 2$ and $m = 3$.

Using the Hall data on electron concentrations and the effective electron masses obtained by linear interpolation between their values for InN and GaN ($m_e = 0.07m_0$ and $m_e = 0.22m_0$), E_g was found to be 1.90 eV and 2.40 eV for $In_{0.35}Ga_{0.65}N$ and

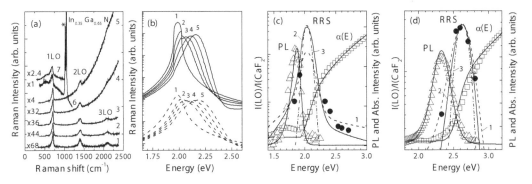

FIGURE 6.27

(a) Room-temperature Raman spectra of $In_{0.35}Ga_{0.65}N$ alloy at different excitation energies: (1) 2.71eV; (2) 2.6 eV; (3) 2.54 eV; (4) 2.41 eV; (5) 2.33 eV; (6) 1.92 eV; and (7) 1.83 eV. (b) Calculated Raman cross sections for $m = 6$ (solid lines) and $m = 3$ (dotted lines) for electron Fermi energies 0, 50, 100, 150, and 200 meV (curves 1, 2, 3, 4, and 5, respectively). (c)–(d) Experimental room-temperature absorption spectra (open squares), PL spectra (open triangles), and resonant Raman cross sectional profiles (full circles) of (c) $In_{0.35}Ga_{0.65}N$ and (d) $In_{0.20}Ga_{0.80}N$. The solid lines give results of model calculations. The Raman cross sectional profiles were calculated for $m = 2$, $m = 3$, and $m = 6$ (curves 1, 2 and 3, respectively). Reprinted with permission from V. Yu. Davydov, A. A. Klochikhin, I. N. Goncharuk, A. N. Smirnov, A. V. Sakharov, A. P. Skvortsov, M. A. Yagovkina, V. M. Lebedev, H. Lu, and W. J. Schaff, physica status solidi (b), 243, 1494 (2006). Copyright 2006 by Wiley-VCH Publishers, Inc.

$In_{0.20}Ga_{0.80}N$, respectively [6.98]. This shows that the resonant Raman experiments can be useful for studying the interband transitions of $In_xGa_{1-x}N$ alloys.

6.9 Concluding remarks

The method for optical data analysis described in this chapter is aimed at minimization of the uncertainty in basic parameters of InN. The conduction band of InN is assumed to be nonparabolic, and the effective electron mass at the Γ-point is taken to be about 0.07 of the free electron mass. Interband photoluminescence and absorption spectra of n-InN samples with electron concentrations of about 3.6×10^{17} cm^{-3} are considered in the framework of the model developed. The well- resolved structure and strong dependences of PL band intensities on the excitation power and temperature are explained. Calculations of PL and absorption spectra have shown that the band gap of InN in the limit of zero temperature and zero electron concentration is about 0.665–0.670 eV and the band-gap reduction as the temperature is increased from close to zero to room temperature $(E_g(0K) - E_g(300K))$ is 55–60 meV.

It is worth noting that InN samples are typically of the n type. However, many of them contain deep acceptors in different concentrations that affect the interband PL. In optical investigations, the photo-excited holes populate the acceptor states due to rapid relaxation and give rise to the conduction-band-acceptor PL which is shifted toward lower energies as compared with the band-to-band PL. Despite a deep localization, the acceptors lose holes as temperature increases to room temperature because of a large difference between the localized and band state densities. As a result, the photoholes populate the valence band states, and a high-energy PL shift occurs due to its transformation into the band-to-band process. This shift compensates partially the temperature shrinkage of the band gap and makes this effect dependent on the sample quality.

Attention should be paid also to the Urbach tail of the conduction band that can be revealed by analyzing in detail the PL band shape. The Urbach tail is a consequence of localization of electrons at the donor concentration fluctuations. In other words, there exist statistical donor "clusters", that is, the regions where donors are separated in space by the distances which are shorter than the Thomas-Fermi screening length. The potential well formed by such a "cluster" can localize an electron even at a high concentration of free carriers.

The chapter also treats the problem of the accumulation layer on the surface (interface) of InN epilayers the presence of which has been revealed by different experimental techniques. It is shown that the analytical solution of the classical Thomas-Fermi equation for a planar accumulation layer of a degenerate semiconductor gives simple expressions for the band bending potential and inhomogeneous electron density relating these functions to crystal parameters and external electrical charge. This solution can be used as a first approximation to the full quantum mechanical description of the electron accumulation layer. It is shown also that the surface electron accumulation layers and resulting band bending strongly affect optical characteristics of thin epilayers and nano-size n-InN samples. The model calculations and experimental data demonstrating that the PL-band shapes and temperature variations are presented.

Attention is also paid to results of experimental and theoretical studies of the lattice dynamics of hexagonal InN and InGaN alloys. Most of the experimental data have been obtained by means of Raman spectroscopy. It should be noted that in spite of considerable progress in studies of vibration properties of InN, several important questions have not yet been answered. For instance, a very strong dependence of the LO band positions in the Raman spectra of InN on the excitation wave vector has been found. This suggests that the problem of Γ-point LO-phonon energies, the knowledge of which is important for understanding of many phonon-assisted optical processes in InN, demands additional experimental and theoretical studies.

The problem of p-type doping of InN is not yet solved, which hinders the use of InN-based heterostructures in device applications. Raman scattering can be successfully used to obtain valuable information on the local structure of defect complexes formed at doping with acceptor impurities. However, experimental studies of the lattice dynamics of doped InN are at the initial stage. There are two Raman experiments with Mg-doped InN that indicate that Mg doping gives rise to interstitial defect complexes and Mg is not an acceptor. However, the theoretical calculations of the lattice dynamics of InN with substitutional impurities and anion/cation vacancies do not provide an unambiguous support of the interpretation of the experimental data. Joint experimental and theoretical research efforts in this direction can give valuable information on the microstructure and charge state of doping impurities, which would be helpful in solving the problem of p-type InN.

Few studies of the lattice dynamics of InGaN have been performed in spite of the fact that they are of not only fundamental but also applied importance. The experimentally observed nonlinear compositional dependence of the $A_1(TO)$, $E_1(TO)$ and $E_2(high)$ modes in InGaN indicates that the crystalline lattice dynamics of this alloy should be investigated more thoroughly both experimentally and theoretically. There is also a lack of detailed information on the phonon behavior of InAlN alloys. It can be expected that such investigations will result in development of the nondestructive quantitative method for determination of the InGaN and InAlN composition.

And finally, there are few experimental and theoretical studies of the phonon subsystem in the novel InN-based low-dimensional heterostructures and related quantum-size structures. Attention in future investigations should be focused on establishing the relation between the phonon spectra and structural characteristics of the low-dimensional systems, revealing the size effects in these structures, and studying the effects of strains and piezoelectric fields on lattice vibrations and the effects of compositional fluctuations of alloys on the phonon spectra of nanostructures. The light scattering effects caused by the presence of free carriers in nanostructures should be considered both theoretically and experimentally. Thus, new information on fundamental properties of the lattice dynamics of InN and related alloys is necessary for successful use of these compounds in electronic and photonic applications.

Acknowledgments

We wish to thank Dr. W. J. Schaff and Prof. S. Gwo for supplying InN and InGaN samples to Ioffe Institute, which allowed us to obtain the experimental data discussed in this chapter, and all the colleagues who made invaluable contributions to our studies of electronic states in InN and lattice dynamics of InN and InGaN. We also express our gratitude to A. Smirnov and N. Nazina for their help in preparation of the manuscript.

The work was supported by the Russian Foundation for Basic Research (RFBR Grant No. 09-02-01280), RFBR-NSC (Taiwan) under Project No. 08-02-92003-HHC, and the Programs "Quantum nanostructures" and "New materials and structures".

References

[6.1] T. L. Tansley and C. P. Foley, Optical band gap of indium nitride, *Journal of Applied Physics*, 59 (1986) 3241–3244.

[6.2] E. Burstein, Anomalous Optical Absorption Limit in InSb, *Physical Review*, 93 (1954) 632–633; T. S. Moss, The interpretation of the properties of indium antimonide, *Proc. Phys. Soc. B*, 67 (1954) 775–782.

[6.3] V. Yu. Davydov, A. A. Klochikhin, R. P. Seisyan, V. V. Emtsev, S. V. Ivanov, F. Bechstedt, J. Furthmüller, H. Harima, A. V. Mudryi, J. Aderhold, O. Semchinova, and J. Graul, Absorption and emission of hexagonal InN. Evidence of narrow fundamental band gap, *physica status solidi (b)*, 229 (2002) R1–R3.

[6.4] V. Yu. Davydov, A. A. Klochikhin, V. V. Emtsev, S. V. Ivanov, V. V. Vekshin, F. Bechstedt, J. Furthmüller, H. Harima, A. V. Mudryi, A. Hashimoto, A. Yamamoto, J. Aderhold, J. Graul, and E. E. Haller, Band gap of InN and In-Rich $In_xGa_{1-x}N$ alloys ($0.36 < x < 1$), *physica status solidi (b)*, 230 (2002) R4-R6.

[6.5] F. Bechstedt and J. Furthmüller, Do we know the fundamental energy gap of InN?, *Journal of Crystal Growth*, 246 (2002) 315–319.

[6.6] V. A. Tyagai, A. M. Evstigneev, A. N. Krasiko, A. F. Andreeva, and V. Y. Malakhov, Optical properties of indium nitride films, *Soviet Physics – Semiconductors*, 11 (1977) 1257–1259.

[6.7] V. Yu. Davydov, A. A. Klochikhin, V. V. Emtsev, D. A. Kudyukov, S. V. Ivanov, V. V. Vekshin, F. Bechstedt, J. Furthmüller, J. Aderhold, J. Graul, A. V. Mudryi, H. Harima, A. Hashimoto, A. Yamamoto, and E. E. Haller, Band gap of hexagonal InN and InGaN alloys, *physica status solidi (b)*, 234 (2002) 787–795.

[6.8] J. Wu, W. Walukiewicz, K. M. Yu, J. W. Ager III, E. E. Haller, H. Lu, W. J. Schaff, Y. Saito, and Y. Nanishi, Unusual properties of the fundamental band gap of InN, *Applied Physics Letters*, 80 (2002) 3967–3969.

[6.9] J. Wu, W. Walukiewicz, W. Shan, K. M. Yu, J. W. Ager III, E. E. Haller, H. Lu, W. J. Schaff, Effects of the narrow band gap on the properties of InN, *Physical Review B*, 66 (2002) 201403:1–4.

[6.10] T. Matsuoka, H. Okamoto, M. Nakao, H. Harima, and E. Kurimoto, Optical bandgap energy of wurtzite InN, *Applied Physics Letters*, 81 (2002) 1246–1248.

[6.11] Y. Nanishi, Y. Saito, and T. Yamaguchi, RF-molecular beam epitaxy growth and properties of InN and related alloys, *Japanese Journal Applied Physics, Part 1*, 42 (2003) 2549–2559.

[6.12] T. Inushima, M. Higashiwaki, and T. Matsui, Optical properties of Si-doped InN grown on sapphire (0001), *Physical Review B*, 68 (2003) 235204:1–7.

[6.13] D. Pines and P. Nozieres, *The Theory of Quantum Liquids*, Benjamin, New York, 1966.

[6.14] M. Gell-Mann and K. A. Brueckner, Correlation energy of an electron gas at high density, *Physical Review*, 106 (1957) 364–368.

[6.15] H. Lu, W. J. Schaff, J. Hwang, H. Wu, W. Yeo, A. Pharkya, and L. F. Eastman, Improvement on epitaxial grown of InN by migration enhanced epitaxy, *Applied Physics Letters*, 77 (2000) 2548–2550.

[6.16] H. Lu, W. J. Schaff, J. Hwang, H. Wu, G. Koley, and L. F. Eastman, Effect of an AlN buffer layer on the epitaxial growth of InN by molecular-beam epitaxy, *Applied Physics Letters*, 79 (2001) 1489–1491.

[6.17] J. Wu, W. Walukiewicz, W. Shan, K. M. Yu, J. W. Ager III, S. X. Li, E. E. Haller, H. Lu, and W. J. Schaff, Temperature dependence of the fundamental band gap of InN, *Journal of Applied Physics*, 94 (2003) 4457–4460.

[6.18] J. Wu, W. Walukiewicz, S. X. Li, R. Armitage, J. C. Ho, E. R. Weber, E. E. Haller, H. Lu, W. J. Schaff, A. Barcz, and R. Jakiela, Effects of electron concentration on the optical absorption edge of InN, *Applied Physics Letters*, 84 (2004) 2805–2807.

[6.19] F. Chen, A. N. Cartwright, H. Lu, and W. J. Schaff, Time-resolved spectroscopy of recombination and relaxation dynamics in InN, *Applied Physics Letters*, 83 (2003) 4984–4986.

[6.20] F. Chen, A. N. Cartwright, H. Lu, and W. J. Schaff, Temperature-dependent optical properties of wurtzite InN, *Physica E-Low-Dimensional Systems and Nanostructures*, 20 (2004) 308–312.

[6.21] F. Chen, A. N. Cartwright, H. Lu, and W. J. Schaff, Ultrafast carrier dynamics in InN epilayers, *Journal of Crystal Growth*, 269 (2004) 10–14.

[6.22] B. Arnaudov, T. Paskova, P. P. Paskov, B. Magnusson, E. Valcheva, B. Monemar, H. Lu, W. J. Schaff, H. Amano, and I. Akasaki, Energy position of near-band-edge emission spectra of InN epitaxial layers with different doping levels, *Physical Review B*, 69 (2004) 115216:1–5.

[6.23] V. Yu. Davydov and A. A. Klochikhin, The electron and vibrational states of InN and $In_xGa_{1-x}N$ solid solutions, *Semiconductors*, 38 (2004) 861–898, and references therein.

[6.24] A. A. Klochikhin, V. Yu. Davydov, V. V. Emtsev, A. V. Sakharov, V. A. Kapitonov, B. A. Andreev, H. Lu, and W. J. Schaff, Acceptor states in the photoluminescence spectra of n-InN, *Physical Review B*, 71 (2005) 195207:1–16.

[6.25] V. Yu. Davydov, V. V. Emtsev, I. N. Goncharuk, A. N. Smirnov, V. D. Petrikov, V. V. Mamutin, V. A. Vekshin, S. V. Ivanov, M. B. Smirnov, and T. Inushima, Experimental and theoretical studies of phonons in hexagonal InN, *Applied Physics Letters*, 75 (1999) 3297–3299.

[6.26] X. Wang, S. B. Che, Y. Ishitani, and A. Yoshikawa, Growth and properties of Mg-doped In-polar InN films, *Applied Physics Letters*, 90 (2007) 201913:1–3.

[6.27] A. A. Klochikhin and S. G. Ogloblin, Density of localized states in disordered solids, *Physical Review B*, 48 (1993) 3100–3115.

[6.28] A. A. Klochikhin, Fluctuation tail of valence bands in hydrogenated amorphous silicon, *Physical Review B*, 52 (1995) 10979–10992.

[6.29] A. Klochikhin, A. Reznitsky, S. Permogorov, T. Breitkopf, M. Grun, M. Hetterich, C. Klingshirn, V. Lyssenko, W. Langbein, and J. M. Hvam, Luminescence spectra and kinetics of disordered solid solutions, *Physical Review B*, 59 (1999) 12947–12972.

[6.30] P. Carrier and S. H. Wei, Theoretical study of the band-gap anomaly of InN, *Journal of Applied Physics*, 97 (2005) 033707:1–5.

[6.31] A. A. Klochikhin, V. Yu. Davydov, I. Y. Strashkova, and S. Gwo, Classical and quantum solutions of the planar accumulation layer problem within the parabolic effective-mass approximation, *Physical Review B*, 76 (2007) 235325:1–8.

[6.32] L. Colakerol, T. D. Veal, H. K. Jeong, L. Plucinski, A. DeMasi, T. Learmonth, P. A. Glans, S. Wang, Y. Zhang, L. F. J. Piper, P. H. Jefferson, A. Fedorov, T. C. Chen, T. D. Moustakas, C. F. McConville, and K. E. Smith, Quantized electron accumulation states in indium nitride studied by angle-resolved photoemission spectroscopy, *Physical Review Letters*, 97 (2006) 237601:1–4.

[6.33] A. N. Titkov, E. I. Chaikina, E. M. Komova, and N. G. Ermakova, Low-temperature luminescence of degenerate p-type crystals of direct-gap semiconductors, *Soviet Physics – Semiconductors*, 15 (1981) 198–202.

[6.34] F. B. Naranjo, M. A. Sànchez-Garcìa, F. Calle, E. Calleja, B. Jenichen, and K. H. Ploog, Strong localization in InGaN layers with high In content grown by molecular-beam epitaxy, *Applied Physics Letters*, 80 (2002) 231–233.

[6.35] W. J. Schaff, H. Lu, L. F. Eastman, W. Walukiewicz, K. M. Yu, S. Keller, S. Kurtz, B. Keyes, and L. Gevilas, in V. H. Ng and A. G. Baca, editors, *Fall 2004 ECS Meeting Proceedings, v. 2004-06 (ISBN 1-56677-419-5) State-of-the-Art Program on Compound Semiconductors XLI-and-Nitride and Wide Bandgap Semiconductors for Sensors, Photonics, and Electronics.*

[6.36] I. Vurgaftman, J. R. Meyer, and L. R. Ram-Mohan, Band parameters for III-V compound semiconductors and their alloys, *Journal of Applied Physics*, 89 (2001) 5815–5875.

[6.37] H. Lu, W. J. Schaff, L. F. Eastman, and C. E. Stutz, Surface charge accumulation of InN films grown by molecular-beam epitaxy, *Applied Physics Letters*, 82 (2003) 1736–1738.

[6.38] I. Mahboob, T. D. Veal, C. F. McConville, H. Lu, and W. J. Schaff, Intrinsic electron accumulation at clean InN surfaces, *Physical Review Letters*, 92 (2004) 036804:1–4.

[6.39] L. F. J. Piper, T. D. Veal, I. Mahboob, C. F. McConville, H. Lu, and W. J. Schaff, Temperature invariance of InN electron accumulation, *Physical Review B*, 70 (2004) 115333:1–6.

[6.40] C. H. Swartz, R. P. Tompkins, N. C. Giles, T. H. Myers, H. Lu, W. J. Schaff, and L. F. Eastman, Investigation of multiple carrier effects in InN epilayers using variable magnetic field Hall measurements, *Journal of Crystal Growth*, 269 (2004) 29–34.

[6.41] I. Mahboob, T. D. Veal, L. F. J. Piper, C. F. McConville, H. Lu, W. J. Schaff, J. Furthmüller, and F. Bechstedt, Origin of electron accumulation at wurtzite InN surfaces, *Physical Review B*, 69 (2004) 201307(R):1–4.

[6.42] S. X. Li, K. M. Yu, J. Wu, R. E. Jones, W. Walukiewicz, J. W. Ager III, W. Shan, E. E. Haller, H. Lu, and W. J. Schaff, Fermi-level stabilization energy in group III nitrides, *Physical Review B*, 71 (2005) 161201(R):1–4.

[6.43] W. Walukiewicz, J. W. Ager III, K. M. Yu, Z. Liliental-Weber, J. Wu, S. X. Li, R. E. Jones, and J. D. Denlinger, Structure and electronic properties of InN and In-rich group III-nitride alloys, *Journal of Physics D – Applied Physics*, 39 (2006) R83–R99.

[6.44] N. H. March, in S. Lundqvist and N. H. March, editors, *Theory of the inhomogeneous electron gas*, Plenum Press, NY and London, 1983.

[6.45] C. B. Duke, Optical absorption due to space-charge-induced localized states, *Physical Review*, 159 (1967) 632–644.

[6.46] J. A. Appelbaum and G. A. Baraff, Effect of magnetic field on the energy of surface bound states, *Physical Review B*, 4 (1971) 1235–1245.

[6.47] G. A. Baraff and J. A. Appelbaum, Effect of electric and magnetic fields on the self-consistent potential at the surface of a degenerate semiconductor, *Physical Review B*, 5 (1972) 475–497.

[6.48] A. A. Klochikhin, V. Yu. Davydov, I. Y. Strashkova, P. N. Brunkov, A. A. Gutkin, M. E. Rudinsky, H. Y. Chen, and S. Gwo, Band bending of n-InN epilayers and exact solution of the classical Thomas-Fermi equation, *physica status solidi, Rapid Research Letters*, 1 (2007) 159–161.

[6.49] B. Arnaudov, T. Paskova, P. P. Paskov, B. Magnusson, E. Valcheva, B. Monemar, H. Lu, W. J. Schaff, H. Amano, and I. Akasaki, Energy position of near-band-edge emission spectra of InN epitaxial layers with different doping levels, *Physical Review B*, 69 (2004) 115216:1–5.

[6.50] B. Arnaudov, T. Paskova, S. Evtimova, B. Monemar, H. Lu, and W. J. Schaff, Electron concentration and mobility profiles in InN layers grown by MBE, *physica status solidi (a)*, 203 (2006) 1681–1685.

[6.51] S. Gwo, C. L. Wu, C. H. Shen, W. H. Chang, T. M. Hsu, J. S. Wang, and J. T. Hsu, Heteroepitaxial growth of wurtzite InN films on Si(111) exhibiting strong near-infrared photoluminescence at room temperature, *Applied Physics Letters*, 84 (2004) 3765–3767.

[6.52] H. Y. Chen, C. H. Shen, H. W. Lin, C. H. Chen, C. Y. Wu, S. Gwo, V. Yu. Davydov, A. A. Klochikhin, Near-infrared photoluminescence of vertically aligned InN nanorods grown on Si(111) by plasma-assisted molecular-beam epitaxy, *Thin Solid Films*, 515 (2006) 961–966.

[6.53] C. A. Arguello, D. L. Rousseau, and S. P. S. Porto, First-Order Raman Effect in Wurtzite-Type Crystals, *Physical Review*, 181 (1969) 1351–1363.

[6.54] H. J. Kwon, Y. H. Lee, O. Miki, H. Yamano, and A. Yoshida, Raman spectra of indium nitride thin films grown by microwave-excited metalorganic vapor phase epitaxy on (0001) sapphire substrates. *Applied Physics Letters*, 69 (1996) 937–939.

[6.55] M. C. Lee, H. C. Lin, Y. C. Pan, C. K. Shu, J. Ou, W. H. Chen, and

W. K. Chen, Raman and x-ray studies of InN films grown by metalorganic vapor phase epitaxy, *Applied Physics Letters*, 73 (1998) 2606–2608.

[6.56] T. Inushima, T. Shiraishi, and V. Yu. Davydov, Phonon structure of InN grown by atomic layer epitaxy, *Solid State Communications*, 110 (1999) 491–495.

[6.57] J. S. Dyck, K. Kim, S. Limpijumnong, W. R. L. Lambrecht, K. Kash, and J. C. Angus, Identification of Raman-active phonon modes in oriented platelets of InN and polycrystalline InN, *Solid State Communications*, 114 (2000) 355–360.

[6.58] G. Kaczmarczyk, A. Kaschner, S. Reich, A. Hoffmann, C. Thomsen, D. J. As, A. P. Lima, D. Schikora, K. Lischka, R. Averbeck, and H. Riechert, Lattice dynamics of hexagonal and cubic InN: Raman-scattering experiments and calculations, *Applied Physics Letters*, 76 (2000) 2122–2124.

[6.59] F. Agulló-Rueda, E. E. Mendez, B. Bojarczuk, and S. Guha, Raman spectroscopy of wurtzite InN films grown on Si, *Solid State Communications*, 115 (2000) 19–21.

[6.60] E. Kurimoto, H. Harima, A. Hashimoto, and A. Yamamoto, MOCVD Growth of High-Quality InN Films and Raman Characterization of Residual Stress Effects, *physica status solidi (b)*, 228 (2001) 1–4.

[6.61] A. Kasic, M. Schubert, Y. Saito, Y. Nanishi, and G. Wagner, Effective electron mass and phonon modes in n-type hexagonal InN, *Physical Review B*, 65 (2002) 115206:1–7.

[6.62] V. Darakchieva, P. P. Paskov, E. Valcheva, T. Paskova, B. Monemar, M. Schubert, H. Lu, and W. J. Schaff, Deformation potentials of the E_1(TO) and E_2 modes of InN, *Applied Physics Letters*, 84 (2004) 3636–3638.

[6.63] X. Wang, S.-B. Che, Y. Ishitani, and A. Yoshikawa, Experimental determination of strain-free Raman frequencies and deformation potentials for the E_2(high) and A_1(LO) modes in hexagonal InN, *Applied Physics Letters*, 89 (2006) 171907:1–3.

[6.64] G. Shikata, S. Hirano, T. Inoue, M. Orihara, Y. Hijikata, H. Yaguchi, and S. Yoshida, RF-MBE growth of a-plane InN on r-plane sapphire with GaN underlayer, *Journal of Crystal Growth*, 301–302 (2007) 517–520.

[6.65] C. Bungaro, K. Rapcewicz, and J. Bernholc, *Ab initio* phonon dispersions of wurtzite AlN, GaN, and InN, *Physical Review B*, 61 (2000) 6720–6725.

[6.66] H. M. Tütüncü, G. P. Srivastava, and S. Duman, Lattice dynamics of the zinc-blende and wurtzite phases of nitrides, *Physica B*, 316 (2002) 190–194.

[6.67] G. Abstreiter, M. Cardona, and A. Pinczuk, Light scattering by free carrier excitations in semiconductors, in M. Cardona and G. Güntherodt (Eds), *Light Scattering in Solids IV*, Springer, Berlin, 1984, p. 12.

[6.68] J. S. Thakur, D. Haddad, V. M. Naik, R. Naik, G. W. Auner, H. Lu, and W. J. Schaff, $A_1(LO)$ phonon structure in degenerate InN semiconductor films, *Physical Review B*, 71 (2005) 115203:1–10.

[6.69] F. Demangeot, C. Pinquier, J. Frandon, N. Gaio, O. Briot, B. Maleyre, S. Ruffenach, and B. Gil, Raman scattering by the longitudinal optical phonon in InN: Wave-vector nonconserving mechanisms, *Physical Review B*, 71 (2005) 104305:1–6.

[6.70] A. A. Klochikhin, V. Yu. Davydov, V. V. Emtsev, A. N. Smirnov, and R. van Baltz, A gauge invariant approach to the Raman scattering in heavily doped crystals, *physica status solidi (b)*, 242 (2005) R58–R60.

[6.71] V. V. Emtsev, V. Yu. Davydov, A. A. Klochikhin, A. V. Sakharov, A. N. Smirnov, V. V. Kozlovskii, C. L. Wu, C. H. Shen, and S. Gwo, Effects of proton irradiation on electrical and optical properties of n-InN, *physica status solidi (c)*, 4 (2007) 2589–2592.

[6.72] V. Yu. Davydov, Y. E. Kitaev, I. N. Goncharuk, A. N. Smirnov, J. Graul, O. Semchinova, D. Uffmann, M. B. Smirnov, A. P. Mirgorodsky, and R. A. Evarestov, Phonon dispersion and Raman scattering in hexagonal GaN and AlN, *Physical Review B*, 58 (1998) 12899–12907. K. Karch and F. Bechstedt, *Ab initio* lattice dynamics of BN and AlN: Covalent versus ionic forces, *Physical Review B*, 56 (1997) 7404–7415.

[6.73] K. T. Tsen, J. G. Kiang, and D. K. Ferry, Direct measurements of the lifetimes of longitudinal optical phonon modes and their dynamics in InN, *Applied Physics Letters*, 90 (2007) 152107:1–3.

[6.74] S. Krukowski, A. Witek, J. Adamczyk, J. Jun, M. Bockowski, I. Grzegory, B. Lucznik, G. Nowak, M. Wroblewski, A. Presz, S. Gierlotka, S. Stelmach, B. Palosz, S. Porowski, and P. Zinn, Thermal properties of indium nitride, *Journal of Physics and Chemistry of Solids*, 59 (1998) 289-295.

[6.75] J. C. Nipko, C. K. Loong, C. M. Balkas, and R. F. Davis, Phonon density of states of bulk gallium nitride, *Applied Physics Letters*, 73 (1998) 34–36.

[6.76] H. Harima, Properties of GaN and related compounds studied by means of Raman scattering, *Journal of Physics: Condensed Matter*, 14 (2002) R967–R993, and references therein.

[6.77] V. V. Mamutin, V. A. Vekshin, V. Yu. Davydov, V. V. Ratnikov, Y. A. Kudriavtsev, B. Y. Ber, V. V. Emtsev, and S. V. Ivanov, Mg-doped hexagonal InN/Al$_2$O$_3$ films grown by MBE, *physica status solidi (a)*, 176 (1999) 373–378.

[6.78] V. Yu. Davydov, A. A. Klochikhin, M. B. Smirnov, A. N. Smirnov, I. N. Goncharuk, D. A. Kurdyukov, H. Lu, W. J. Schaff, H. M. Lee, H. W. Lin, and S. Gwo, Experimental and theoretical studies of lattice dynamics of Mg-doped InN, *Applied Physics Letters*, 91 (2007) 111917:1–3.

[6.79] S. W. Haan and R. Zwanzig, Series expansions in a continuum percolation problem, *Journal of Physics A – Mathematical and General*, 10 (1977) 1547–1555.

[6.80] A. Hoffmann, A. Kaschner, and C. Thomsen, Local vibrational modes and compensation effects in Mg-doped GaN, *physica status solidi (c)*, 0 (2003) 1783–1794, and references therein.

[6.81] F. A. Reboredo and S. T. Pantelides, Novel defect complexes and their role in the p-type doping of GaN, *Physical Review Letters*, 82 (1999) 1887–1890.

[6.82] G. Kaczmarczyk, A. Kaschner, A. Hoffmann, and C. Thomsen, Impurity-induced modes of Mg, As, Si, and C in hexagonal and cubic GaN, *Physical Review B*, 61 (2000) 5353–5357.

[6.83] H. Grille, C. Schnittler, and F. Bechstedt, Phonons in ternary group-III nitride alloys, *Physical Review B*, 61 (2000) 6091–6105.

[6.84] D. Behr, R. Niebuhr, H. Obloh, J. Wagner, K. H. Bachem, and U. Kaufmann, *Materials Research Society Symposium Proceedings*, 468 (1997) 213–218.

[6.85] H. Harima, E. Kurimoto, Y. Sone, S. Nakashima, S. Chu, A. Ishida, and H. Fujiyasu, Observation of phonon modes in bulk InGaN films by Raman scattering, *physica status solidi (b)*, 216 (1999) 785–788.

[6.86] D. Alexson, L. Bergman, R. J. Nemanich, M. Dutta, M. A. Stroscio, C. A. Parker, S. M. Bedair, N. A. El-Masry, and F. Adar, Ultraviolet Raman study of A_1(LO) and E_2 phonons in In$_x$Ga$_{1-x}$N alloys, *Journal of Applied Physics*, 89 (2001) 798–800.

[6.87] J. W. Ager III, W. Walukiewicz, W. Shan, K. M. Yu, S. X. Li, E. E. Haller, H. Lu, and W. J. Schaff, Multiphonon resonance Raman scattering in In$_x$Ga$_{1-x}$N, *Physical Review B*, 72 (2001) 155204:1–7.

[6.88] M. R. Correia, S. Pereira, J. Frandon, and E. Alves, Raman study of the A_1(LO) phonon in relaxed and pseudomorphic InGaN epilayers, *Applied Physics Letters*, 83 (2003) 4761–4763.

[6.89] V. Yu. Davydov, A. A. Klochikhin, V. V. Emtsev, A. N. Smirnov, I. N. Goncharuk, A. V. Sakharov, D. A. Kurdyukov, M. V. Baidakova, V. A. Vekshin, S. V. Ivanov, J. Aderhold, J. Graul, A. Hashimoto, and A. Yamamoto, Photoluminescence and Raman study of hexagonal InN and In-rich InGaN alloys, *physica status solidi (b)*, 240 (2003) 425–428.

[6.90] A. G. Kontos, Y. S. Raptis, N. T. Pelekanos, A. Georgakilas, E. Bellet-Amalric, and D. Jalabert, Micro-Raman characterization of $In_xGa_{1-x}N$/GaN/Al_2O_3 heterostructures, *Physical Review B*, 72 (2005) 1555336:1–10.

[6.91] S. Hernandez, R. Cosco, D. Pastor, L. Artus, K. P. O'Donnell, R. W. Martin, I. M. Watson, Y. Nanishi, and E. Calleja, Raman-scattering study of the InGaN alloy over the whole composition range, *Journal of Applied Physics*, 98 (2005) 013511:1–5.

[6.92] V. Yu. Davydov, N. S. Averkiev, I. N. Goncharuk, D. K. Nelson, I. P. Nikitina, A. S. Polkovnikov, A. N. Smirnov, M. A. Jacobson, and O. K. Semchinova, Raman and photoluminescence studies of biaxial strain in GaN epitaxial layers grown on 6H-SiC, *Journal of Applied Physics*, 82 (1997) 5097–5102.

[6.93] F. Demangeot, J. Frandon, P. Baules, F. Natali, F. Semond, and J. Massies, Phonon deformation potentials in hexagonal GaN, *Physical Review B*, 69 (2004) 155215:1–5.

[6.94] R. Loudon, The Raman Effect in Crystals, *Advances in Physics*, 13 (1964) 423–482.

[6.95] M. F. Sykes, D. S. Gaunt, and M. Glen, Percolation processes in 3 dimensions, *Journal of Physics A - Mathematical and General*, 9 (1976) 1705–1712.

[6.96] V. Yu. Davydov, I. N. Goncharuk, A. N. Smirnov, A. E. Nikolaev, W. V. Lundin, A. S. Usikov, A. A. Klochikhin, J. Aderhold, J. Graul, O. Semchinova, and H. Harima, Composition dependence of optical phonon energies and Raman line broadening in hexagonal $Al_xGa_{1-x}N$ alloys, *Physical Review B*, 65 (2002) 125203:1–13.

[6.97] M. Cardona, Resonance Phenomena, in G. Güntherodt and M. Cardona (Eds), *Light scattering in solids II, Topics in Applied Physics*, vol. 50, p. 19, Springer, Berlin, 1982.

[6.98] V. Yu. Davydov, A. A. Klochikhin, I. N. Goncharuk, A. N. Smirnov,

A. V. Sakharov, A. P. Skvortsov, M. A. Yagovkina, V. M. Lebedev, H. Lu, and W. J. Schaff, Resonant Raman scattering in InGaN alloys, *physica status solidi (b)*, 243 (2006) 1494–1498.

[6.99] A. A. Klochikhin, S. A. Permogorov, and A. N. Reznitsky, Multi-phonon processes in resonant scattering and exciton luminescence of crystals, *Soviet Physics – Zhurnal Eksperimentalnoi i Teoreticheskoi Fiziki*, 71 (1976) 2230–2251.

[6.100] J. Menendez and M. Cardona, Interference effects: A key to understanding forbidden Raman scattering by LO phonons in GaAs, *Physical Review B*, 31 (1985) 3696–3704.

[6.101] A. A. Abdumalikov and A. A. Klochikhin, The double resonance in two-phonon Raman scattering, *physica status solidi (b)*, 80 (1977) 43–50.

7

Optical properties of InN and related alloys

J. W. L. Yim and J. Wu

Department of Materials Science and Engineering, University of California; Materials Sciences Division, Lawrence Berkeley National Laboratory, Berkeley, CA 94720, United States of America

7.1 Introduction

Already at \$20 billion a year, the optoelectronics market is projected to expand rapidly in the next decades, driven largely by growth and advances in the solid-state lighting and laser technology sectors. The recent growth was fuelled by the introduction of high-brightness blue light-emitting diodes (LEDs) with $In_{1-x}Ga_xN$ as the active layer in the early 1990s [7.1, 7.2]. Research in this field has been heavily focused on Ga-rich $In_{1-x}Ga_xN$ and $Ga_{1-x}Al_xN$ alloys, whose band gaps cover the short wavelength visible and near ultraviolet parts of the electromagnetic spectrum. Since then, the rapid development of solid-state lighting technology has revolutionized the fields of optoelectronics and optics. Figure 7.1 shows the widely adopted International Commission on Illumination (CIE) chromaticity diagram which relates the color of light to the response of the human eye. As seen in this diagram, $In_{1-x}Ga_xN$ plays a dominant role in the fields covering from the blue to the green, corresponding to 400–530 nm in wavelength, or 3.1 to 2.3 eV in photon energy. The In molar fraction in these materials is limited to ~0.3.

On the other hand, much less effort has been devoted to InN and In-rich alloys. Earlier InN samples were synthesized using radio-frequency sputtering [7.3]. In most cases, this or similar methods produced polycrystalline samples with high electron concentrations [7.4] and significant oxygen contamination [7.5]. Such materials typically showed relatively low electron mobilities in the range of 10–100 cm^2/Vs. The optical absorption measured in these samples showed a strong absorption band in the infrared and an absorption edge at about 1.9 eV [7.6]. The value of 1.9 eV was thus widely quoted as the band gap of intrinsic InN [7.7]. One of the unexplained characteristics of this early synthesized InN was the lack of any light emission at or near the purported band edge. This was in stark contrast to GaN and Ga-rich InGaN, which show a strong luminescence despite the very large concentrations of point and extended defects typical for these materials.

Since then, the quality of nitride films has improved drastically with the advent of InGaN and InAlN films grown by metal-organic vapor phase epitaxy (MOVPE).

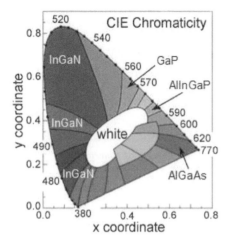

FIGURE 7.1

The CIE chromaticity diagram, where the wavelength is given in nm and the main LED materials are shown.

Correspondingly, a tremendous amount of effort has been put into the optical characterization of these films. The increased quality has translated to lower free carrier concentrations by unintentional doping, allowing the fundamental optical properties to be deconvoluted from band-filling dominated characteristics. Unexpectedly, it was discovered that the band gaps of InGaN decrease very rapidly with increasing In content, and fall well below 2 eV for compositions approaching 50% In [7.8, 7.9]. It was suggested that the band gap of InN could be much smaller than 1.9 eV [7.9]. In most cases, however, the rapid fall of the band gap versus the In molar fraction was attributed to an unusually large bowing parameter [7.8].

The major breakthrough in this field came about as a result of improved quality of InN films grown using molecular beam epitaxy (MBE) [7.10–7.12]. Growth of thick InN films with much reduced electron concentrations ($< 10^{18}$ cm^{-3}) and high electron mobilities (> 2000 cm^2/Vs) was essential to the progress in understanding the properties of this material. The room temperature fundamental band gap of this type of high-quality wurtzite InN was measured to be near 0.9 eV [7.11], 0.77 eV [7.13], and finally converged to 0.65 eV [7.14].

In this chapter, we review studies of optical properties of group III-nitride alloys in the context of this discovery. It is shown that the newly discovered low value of the energy gap of InN provides a basis for a consistent description of the electronic structure of InN and the InGaN and InAlN alloys over the entire composition range.

7.2 Narrow band gap of InN evidenced from optical measurements

High-quality wurtzite InN films with low electron concentration and high electron mobility have been produced using migration-enhanced MBE [7.15]. Figure 7.2 shows the optical characteristics of such a film, where the free electron concentration is 3.5×10^{17} cm^{-3} and the electron mobility is 2050 cm^2/Vs.

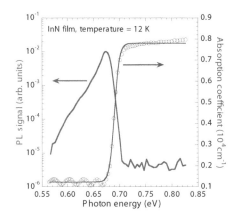

FIGURE 7.2

Absorption and photoluminescence of a high-quality, intrinsic InN film obtained at 12 K. The solid line through the absorption data points is a sigmoidal fit. Figure reproduced with permission from J. Wu, W. Walukiewicz, W. Shan, K. M. Yu, J. W. Ager III, S. X. Li, E. E. Haller, H. Lu, and W. J. Schaff, Journal of Applied Physics, 94, 4457 (2003). Copyright 2003, American Institute of Physics.

The optical absorption curve shows a strong onset slightly below 0.7 eV. There is no band gap feature in the 1.8–2.0 eV region, that is, in the energy range of previously reported band gaps [7.6]. The sample exhibits intense photoluminescence (PL) near the optical absorption edge, with a long tail toward the low energy side. The absorption coefficient rapidly increases to $\sim 10^4$ cm^{-1} above the onset, typical absorption intensity for direct band gap semiconductors. The PL signal weakens with increasing temperature, but is detectable even at room temperature. At low temperatures, a photomodulated reflectance (PR) spectrum was also observed [7.13] which shows a transition feature at the same energy with a profile that is characteristic of interband transitions in a direct band gap semiconductor. The simultaneous observation of the absorption edge, the PL, and the PR features at essentially the same energy indicates that this energy position corresponds to the transition across the fundamental band gap of InN.

Both the PL peak and absorption edge exhibit a redshift with increasing temperature (see Fig. 7.3), consistent with the expected behavior of a direct band gap semiconductor [7.14]. Earlier work on the anomalous temperature behavior of the band gap of InN can be explained by carrier thermalization from a degenerately doped material, which compensates for the intrinsic redshift. The absorption edge shifts to lower energies by ~ 47 meV as the temperature is increased from 12 K to room temperature. This change is significantly smaller than that of other group III-

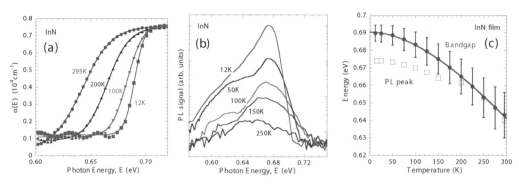

FIGURE 7.3

(a) Absorption and (b) PL (log scale) spectra of InN measured over a wide range of temperatures. (c) Temperature dependence of the PL peak and the band gap determined from the absorption curves. The solid curve shows a fit to the band gap with Varshni's equation. Figure reproduced with permission from J. Wu, W. Walukiewicz, W. Shan, K. M. Yu, J. W. Ager III, S. X. Li, E. E. Haller, H. Lu, and W. J. Schaff, Journal of Applied Physics, 94, 4457 (2003). Copyright 2003, American Institute of Physics.

nitrides. For example, the change is ∼72 and ∼92 meV for intrinsic GaN and AlN, respectively [7.16]. Also shown in Fig. 7.3 is the increasing difference between the absorption edge and the PL peak with decreasing temperature (∼16 meV at low temperature). This difference can be attributed to the fact that the low temperature PL is associated with transitions from low-density localized states, whereas the absorption edge is determined by the large-density band states. This difference is also sample specific, indicating that the shift of the PL peak energy cannot be used to accurately determine the temperature dependence of the fundamental band gap of InN.

The temperature dependence of the direct band gap, determined from the absorption edge, is well described by Varshni's equation,

$$E_g(T) = E_g(0) - \frac{\alpha T^2}{\beta + T} \tag{7.1}$$

The optical parameters of InN determined in this work are compared with those of GaN and AlN in Table 7.1. It can be seen that all these parameters show a monotonic chemical trend from AlN to GaN to InN. In Varshni's equation, β is physically associated with the Debye temperature of the crystal. The value of β = 454 K for InN is consistent with the calculated range of Debye temperatures for InN between 370 and 650 K [7.17]. α is much smaller than that of GaN and AlN, which suggests that the overall influence of thermal expansion and electron-phonon interaction on the fundamental band gap is much weaker in InN. This is expected considering the larger ionicity and weaker bonding in InN compared to AlN and GaN.

These optical characteristics were obtained from *c*-plane wurtzite InN films typically grown on (0001) sapphire substrates. In comparison, Fig. 7.4(a) shows the

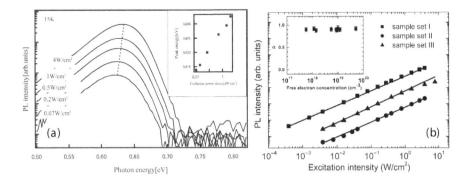

FIGURE 7.4

(a) PL spectra of an *a*-plane InN film at different laser excitation intensities at 15 K. The inset is the excitation power dependence of the peak energy position. Reprinted from Journal of Crystal Growth, 301–302, G. Shikata, S. Hirano, T. Inoue, M. Orihara, Y. Hijikata, H. Yaguchi, and S. Yoshida, 517–520, Copyright 2007, with permission from Elsevier. (b) Integrated PL intensity as a function of excitation intensity for a *c*-plane InN film. The solid lines are a power law fit. The inset shows the power law index being a constant regardless of free electron concentration. Reprinted from Solid State Communications, 137, S. P. Fu, Y. F. Chen, and K. Tan, 203–207, Copyright 2006, with permission from Elsevier.

PL spectra of an *a*-plane ($11\bar{2}0$) InN film grown on *r*-plane sapphire [7.18]. A strong emission is observed at similar photon energies as in the *c*-plane InN. The PL peak energy exhibits a blueshift with increasing excitation power, which can be attributed to the increasing population of the conduction and valence bands by photo-carriers at more intense excitation power. A similar blueshift of the PL peak has been observed in *c*-plane InN films as well, with the magnitude of shift depending on the excitation intensity and the free electron concentration [7.19]. In the same study, the authors also found a power law dependence of the integrated PL intensity on the excitation power (as shown in Fig. 7.4(b)). The power law is obeyed with a constant, near-unity index over a large range of excitation powers, free electron concentrations, and temperatures [7.19]. This behavior is consistent with an exciton-mediated electron-hole recombination mechanism.

TABLE 7.1

Comparison of physical parameters of wurtzite InN with those of AlN and GaN [7.16, 7.20, 7.21, 7.22]. Values not referenced were calculated from commonly accepted parameters.

Parameter	InN	GaN	AlN
$E_g(T=0)$ (eV)	0.690	3.507	6.230
α(meV/K)	0.414	0.909	1.799
β(K)	454	830	1462
m_e^*/m_0	0.07±0.02; 0.05	0.20	0.32
Exciton binding energy (meV)	4	33	50
Exciton Bohr radius (nm)	11.4	2.5	1.6
dE_g/dP (meV/kbar)	3.0±0.1; 2.7±0.1	3.9	4.9

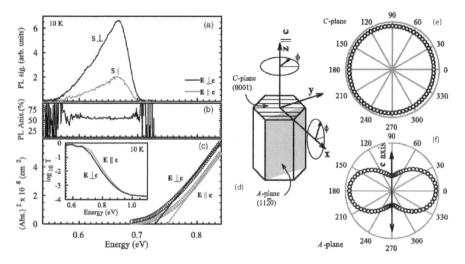

FIGURE 7.5

(a) PL spectra of an *a*-plane InN film with E ⊥ c and E ∥ c polarization, respectively. (b) Polarization anisotropy percentage, defined as $(S_\perp - S_\parallel)/(S_\perp + S_\parallel)$, of the spectra in (a). (c) Measured absorption coefficient squared of the *a*-plane film as a function of photon energy for the two polarizations. The inset shows the full transmission spectra. (d) Schematic of InN unit cell showing the relevant geometry. Variation of the integrated PL intensity with ϕ for (e) the *c*-plane and (f) the *a*-plane InN film, measured at 10 K. Figure reproduced with permission from J. Bhattachryya, S. Ghosh, M. R. Gokhale, B. M. Arora, H. Lu, and W. J. Schaff, Applied Physics Letters, 89, 151910 (2006). Copyright 2006, American Institute of Physics.

The PL signal is polarized when emitted from an *a*-plane film, as expected from a wurtzite material. Figure 7.5 shows the polarization dependence of the PL intensity recorded with a polarizer rotating with respect to the *c*-axis [7.23]. The PL polarization anisotropy is an intrinsic property of crystalline materials. In the case of InN, the PL anisotropy mimics the anisotropy of the wurtzite crystal structure which defines symmetries and angular momenta of the valence band wavefunctions, and consequently makes selection rules in interband optical transitions direction sensitive. The energy difference in the absorption edge for E ⊥ c and E ∥ c polarizations (Fig. 7.5(c)) is related to the crystal-field and spin-orbit splittings at the Γ-point in the valence bands [7.24]. A quantitative interpretation of these anisotropic effects requires the inclusion of possible anisotropic strain in the film induced by the lattice mismatch with the substrate [7.25]. However, this well-behaved optical anisotropy is an indication of the interband nature of the optical transitions near 0.7 eV.

As the narrow band gap of InN has been established through optical investigations, the carrier recombination mechanisms are naturally the next important subject to explore. Understanding of the physics and dynamics of recombination is crucial for the development of InN-based devices. The recombination physics can be studied by various time-resolved pump-probe techniques. Figure 7.6(a) shows the differential transmission transients of three InN films with distinctly different doping levels recorded at a range of temperatures [7.26]. The probe energy was

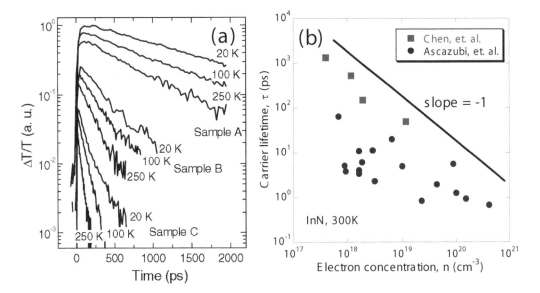

FIGURE 7.6

(a) Differential transmission of three samples with different electron concentrations under the pump fluence of 1 μJ/cm^2. The electron concentrations of the samples are: sample A, 1.3×10^{18} cm^{-3}; sample B, 2.7×10^{18}cm^{-3}; and sample C, 1.2×10^{19}cm^{-3}. Reprinted with permission from F. Chen, A. N. Cartwright, H. Lu, and W. J. Schaff, physica status solidi (a), 202, 768 (2005). Copyright 2005 by Wiley-VCH Publishers, Inc. (b) Recombination lifetime versus carrier concentration. The samples with high electron concentrations ($>2\times10^{19}$cm^{-3}) were Si-doped, while the rest were not intentionally doped.

tuned to the absorption edge ($\sim$0.65 eV). The pump pulse is shorter than 1 ps and its intensity was controlled such that the injected carrier density is much lower than the original equilibrium carrier concentration in the sample. It is clear that these transients can be described by a single exponential decay, with a non-equilibrium carrier lifetime (τ) depending on doping level and temperature. Similar dynamics were also probed using transient photo-reflectance [7.27]. The room-temperature carrier lifetime determined in these studies is plotted as a function of the electron concentration in Fig. 7.6(b). It can be seen, not surprisingly, that the carrier lifetime is approximately inversely proportional to the free electron concentration. This effect is analogous to the rapid decrease in electron mobility with increasing electron concentration in InN [7.28]; both effects depend on impurities and defects, which scatter free carriers, in the case of electrical transport, or mediate carrier recombination, in the case of optical transitions.

The photo-generated electrons and holes recombine mainly through three channels: non-radiative defect related, radiative interband, and non-radiative Auger recombination. The total recombination rate is written as,

$$\frac{1}{\tau} = \frac{1}{\tau_{\text{defect}}} + \frac{1}{\tau_{\text{rad}}} + \frac{1}{\tau_{\text{Auger}}} = \sigma \bar{v} N_{\text{defect}} + B_{\text{rad}} + B_{\text{Auger}} n^2, \qquad (7.2)$$

where σ, $\bar{v}$, and n are the capture cross section, mean speed, and concentration of free carriers, respectively. The fact that the overall slope of τ versus n is closer to -1 than to -2 in Fig. 7.6(b) suggests that the Auger effect is not the dominant recombination mechanism in InN, at least not for the pumping intensities used in these experiments.

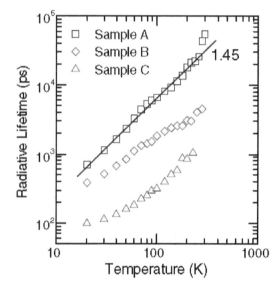

FIGURE 7.7

Radiative lifetime (τ_{rad}) of InN as a function of temperature. The solid line indicates a power law fit to sample A. Reprinted with permission from F. Chen, A. N. Cartwright, H. Lu, and W. J. Schaff, physica status solidi (a), 202, 768 (2005). Copyright 2005 by Wiley-VCH Publishers, Inc.

Using Eq. 7.2, Chen *et al.* estimated the radiative recombination coefficient to be $B_{rad}=(2.6\pm0.5)\times10^{-11}\mathrm{cm}^3\mathrm{s}^{-1}$ at 300 K for InN [7.26]. By relating the temperature dependence of PL intensity to the radiative recombination rate, they further extracted the temperature-dependent radiative recombination lifetime, τ_{rad}, as shown in Fig. 7.7, where the total lifetimes (τ) at 20 K and 300 K for sample A are also shown for comparison. It is clear from Fig. 7.7 that the defect-mediated non-radiative recombination is largely inactive at low temperatures, while it becomes dominant at higher temperatures. The derived τ_{rad} is strongly temperature dependent and scales with T^{γ}, where γ is close to 3/2. This is consistent with the Lasher-Stern model, which predicts that for bimolecular radiative recombination in direct-band gap semiconductors, $B_{rad} \sim T^{-3/2}$ when the momentum selection rule holds [7.26]. This behavior once again proves that the PL at $\sim$0.65 eV is indeed associated with the momentum-conserved interband transitions in InN.

Analyzing the differential transmission transient in Fig. 7.6(a) in energy domain (rather than time) reveals that although the signal at lower and peak energies exhibits a single-exponent decay which is recombination related as discussed above, for the high-energy shoulder, a fast decay ($\sim$10 ps) occurs prior to this carrier recombination (Fig. 7.8(a)) [7.29]. This fast decay is attributed to hot carrier relaxation. Assuming a Maxwell-Boltzmann distribution of hot electrons and holes, the

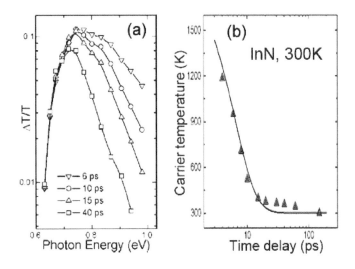

FIGURE 7.8

(a) Differential transmission spectrum at different time delays after the pump pulse is turned off. (b) Carrier temperature recorded at 300 K as a function of time delay calculated from (a). The solid curve is the expected behavior calculated using an LO phonon scattering model. Figure reproduced with permission from F. Chen, A. N. Cartwright, H. Lu, and W. J. Schaff, Applied Physics Letters, 83, 4984 (2003). Copyright 2003, American Institute of Physics.

carrier temperature can be derived from the slope of the high-energy shoulder in Fig. 7.8(a). The carrier temperature obtained by this means is plotted as a function of time in Fig. 7.8(b), showing a fast cooling of photo-excited carriers when the pumping source is turned off. This carrier cooling is explained by a thermalization process involving carrier-LO phonon scattering [7.29], as shown by the fitting curve in Fig. 7.8(b).

The narrow band gap of InN was not widely accepted immediately after the optical measurements, partly because the band gap of InN at 0.65 eV breaks the "common-cation rule" in semiconductors. The common-cation (anion) rule states that for isovalent, common-cation (anion) semiconductors, the direct band gap at the Γ-point increases as the anion (cation) atomic number decreases. The band gap of InN at 0.65 eV is not only much smaller than the previously reported band gap of InN at ~1.9 eV, but also smaller than the band gap of InP with E_g ~1.4 eV. However, the breakdown of the common-cation (anion) rule is not unusual in ionic semiconductors. As shown in Fig. 7.9, for Zn- and Cd-group VI compounds, the band gaps of the oxides are smaller than those of the sulphides.

Recent *ab initio* calculations of the band structure of InN have confirmed the unexpected narrow band gap of InN and illuminate its origin [7.30–7.32]. Two effects were found to be responsible [7.30]. First, the outermost atomic *s* orbital energy of N (−18.49 eV) is much lower than that of P (−14.09 eV) and other group V elements. Since the conduction band minimum at the Γ-point is mainly composed of *s* atomic orbitals, this effect lowers the conduction band minimum and

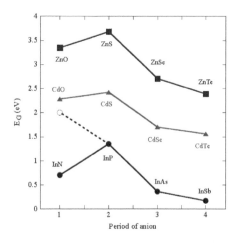

FIGURE 7.9

The breakdown of the common-cation rule in Zn-VI, Cd-VI, and In-V semiconductors.

electron affinity of nitrides. Secondly, the band gap deformation potential of InN was found to be much smaller in magnitude than that of InP (-3.7 eV versus -5.9 eV). This significantly weakens the atomic size effect which is the main mechanism behind the common-cation (anion) chemical rule. According to Wei *et al.*, the particularly small deformation potential in InN is due to the combined effects of (i) a large difference between the cation In $5s$ and anion N $2s$ orbital energies; (ii) a strong repulsion between the N $2p$ and the high-lying In $4d$ orbitals; and (iii) a long In-N bond length. A similar situation also exists in some II-VI semiconductors, which explains the breakdown of the common-cation rule in Zn- and Cd-group VI semiconductors.

7.3 Conduction band structure of InN and its effects on optical properties

A direct consequence of the narrow gap of InN is the strong nonparabolicity of the lowest conduction band. It is well known that for narrow direct gap semiconductors, such as InSb [7.33], the conduction band dispersion takes a nonparabolic form as a result of the $\mathbf{k} \cdot \mathbf{p}$ interaction across the narrow gap between the conduction and valence bands. A simple analytical form of the nonparabolic dispersion for the conduction band of InN can be obtained from Kane's $\mathbf{k} \cdot \mathbf{p}$ Hamiltonian [7.21],

$$E_c(k) = E_g + \frac{\hbar k^2}{2m_0} + \frac{1}{2}\left(\sqrt{E_g^2 + 4E_P \cdot \frac{\hbar k^2}{2m_0}} - E_g \right), \qquad (7.3)$$

where $E_g = 0.65$ eV is the intrinsic band gap energy, m_0 is the electron rest mass, and $E_P = 2|\langle S|P_z|Z\rangle|^2/m_0$ is an energy parameter related to the $\mathbf{k} \cdot \mathbf{p}$ matrix element, and is typically $\sim$10–15 eV. Equation 7.3 neglects the spin-orbit and crystal-field splitting energies in the valence bands since they are typically extremely small in nitrides [7.16]. Perturbation due to the conduction band minimum of other remote bands is also ignored, as they are at least 4 eV away.

As shown in Eq. 7.3, the nonparabolicity of the conduction band is more pronounced for small E_g (that is, narrow-gap semiconductors) and/or large E_P due to the fact that the conduction band feels stronger perturbation from the valence bands when E_g is smaller. At small k values (that is, close to the Γ-point), Eq. 7.3 is simplified into a parabolic band,

$$E_c(k) = E_g + \frac{\hbar k^2}{2m_e^*(0)}, \tag{7.4}$$

where the effective electron mass at the conduction band minimum $m_e^*(0)$ is

$$\frac{m_e^*(0)}{m_0} = \left(1 + \frac{E_P}{E_g}\right)^{-1}. \tag{7.5}$$

Figure 7.10 shows the calculated conduction band dispersion using Eq. 7.3 (nonparabolic) and Eq. 7.4 (parabolic), respectively. The parabolic dispersion deviates severely from the nonparabolic one when k $> \sim 0.05$ Å^{-1}, or, as shown below, when the electron concentration $n >\sim 10^{19}$ cm^{-3} so that the Fermi level, E_F, is displaced deep into the conduction band.

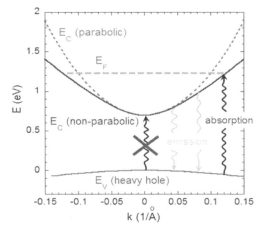

FIGURE 7.10

Calculated conduction band dispersion of InN using the Kane model.

In the case of degenerate doping, optical absorption is forbidden for transitions below the Fermi level. Therefore, the onset of the optical absorption overestimates the intrinsic band gap, leading to a phenomenon known as the Burstein-Moss effect [7.34]. Optical emission below the Fermi level, such as PL, is still possible but significantly broadened compared to the intrinsic band-edge emission, as illustrated schematically in Fig. 7.10. This implies that for heavily doped semiconductors, such as earlier, sputter-grown InN films, the absorption and luminescence spectra should be interpreted with caution.

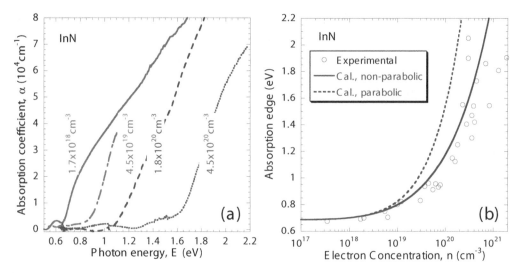

FIGURE 7.11

(a) Absorption curves for InN films with a wide range of free electron concentrations. (b) Absorption edge (optical gap) plotted as a function of electron concentration [7.34]. The $\mathbf{k} \cdot \mathbf{p}$ calculated results are also shown. Figure reproduced with permission from J. Wu, W. Walukiewicz, S. X. Li, R. Armitage, J. C. Ho, E. R. Weber, E. E. Haller, H. Lu, W. J. Schaff, A. Barcz, and R. Jakiela, Applied Physics Letters, 84, 2805 (2004). Copyright 2004, American Institute of Physics.

The Burstein-Moss shift illustrated in Fig. 7.11(a) well explains the discrepancy between the earlier 1.9 eV and the newly established 0.65 eV band gap. High-quality InN films have been grown by MBE where the free electron concentration is varied over several orders of magnitude by controlled Si doping [7.34]. The absorption edge was determined from optical absorption experiments and plotted as a function of free electron concentration in Fig. 7.11(b). Clearly, the absorption edge, or what is sometimes referred to as the "optical gap", varies continuously from 0.65 eV, the intrinsic band gap of InN, to ~2 eV for samples with $n > 5 \times 10^{20}$ cm^{-3} free electrons. Band gaps of earlier, sputter-grown InN films fall accurately onto this dependence, and are therefore well explained by the Burstein-Moss effect. The high density of free electrons can be donated from unintentional dopants, such as oxygen and native defects. Electron concentrations as high as 2×10^{21} cm^{-3} have been reported in InN. The extreme n-type propensity of InN is a direct consequence of its low-lying conduction band minimum [7.35]. For a detailed discussion on this topic, readers are referred to Chapter 4 and also to Chapters 10 and 11 on experimental and theoretical studies of defects in InN.

The increase in absorption edge in Fig. 7.11(b) with increasing electron concentration was calculated by the dispersion relation in Eq. 7.3 (nonparabolic) and Eq. 7.4 (parabolic) evaluated at the Fermi wavevector $k_F = (3\pi^2 n)^{1/3}$, neglecting the thermal broadening of the Fermi distribution. In the calculation, conduction band renormalization effects due to the electron-electron interaction and the electron-ionized impurity interaction have been taken into account [7.34] (see also

Chapters 6 and 9). These effects become significant at high electron concentrations, resulting in a conduction band downshift of greater than 0.15 eV per decade of increase in n when $n > \sim 10^{19}$ cm^{-3}. The calculated dependences are compared to the experimental data and to each other in Fig. 7.11(b). The calculated optical band gap, assuming a parabolic conduction band, shows a Burstein-Moss shift that is too large at a given carrier concentration to describe the experimental data.

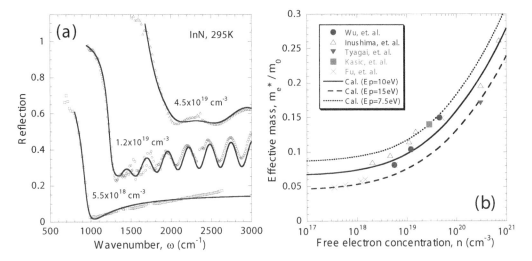

FIGURE 7.12

(a) Infrared reflection curves of three InN samples with different free electron concentrations. The second and third curves are vertically offset for clarity. The solid lines are a theoretical fit using a standard complex dielectric function model [7.21]. Reprinted with permission from J. Wu, W. Walukiewicz, W. Shan, K. M. Yu, J. W. Ager III, E. E. Haller, H. Lu, and W. J. Schaff, Physical Review B, 66, 201403(R) (2004). Copyright 2004 by the American Physical Society. (b) Effective electron mass as a function of electron concentration. The curves are calculated dependences based on the nonparabolic dispersion using different E_P values [7.21, 7.22].

Another consequence of the conduction band nonparabolicity is the k-dependent density-of-states electron effective mass given by

$$m_e^*(k) \equiv \frac{\hbar^2 k}{dE_c(k)/dk}.$$ (7.6)

As k approaches 0, Eq. 7.6 gives the conduction band minimum effective mass defined in Eq. 7.6. The electron effective mass of InN has been measured by several groups using plasma reflection spectroscopy [7.21, 7.22] and infrared spectroscopic ellipsometry. One example is shown in Fig. 7.12. The infrared reflection curves are fitted with a standard dielectric function model,

$$R(\omega) = \left| \frac{\sqrt{\varepsilon(\omega)} - 1}{\sqrt{\varepsilon(\omega)} + 1} \right|^2,$$ (7.7)

where the complex dielectric function is given by the classical dielectric model,

$$\varepsilon(\omega)/\varepsilon_\infty = 1 - \frac{\omega_p^2}{\omega^2 + i\omega\gamma}, \tag{7.8}$$

and the plasma frequency is

$$\omega_p = \sqrt{\frac{ne^2}{\varepsilon_0\varepsilon_\infty m_e^*}}. \tag{7.9}$$

The plasma edge clearly shifts to higher energy as n is increased. The effective mass values obtained from the infrared reflection spectroscopy and Hall effect measurements using Eq. 7.9 are shown in Fig. 7.12(b) and compared with calculated results from Eq. 7.6. It can be seen that although the data were reported by different groups for InN films grown by different methods, the calculations based on the nonparabolic conduction band using $E_g = 0.65$ eV and $E_P \sim 10$ eV show good agreement with the measured effective masses. Note that the same value of $E_P \sim 10$ eV was also used in calculating the Burstein-Moss shift in Fig. 7.11. Therefore a good consistent picture is established in describing the conduction band of InN based on the $\mathbf{k \cdot p}$ model. The extrapolation of the curve in Fig. 7.12(b) leads to a small effective mass of $m_e^*(0)/m_0 = 0.07 \pm 0.02$ at the bottom of the conduction band. Recently, Fu *et al.* reported a lower value of $m_e^*(0)/m_0 \approx 0.05$ using infrared reflection measurements [7.22]. A larger E_P value of 15 eV would be needed to fit with this conduction band minimum effective mass.

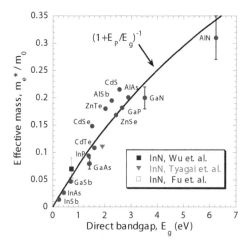

FIGURE 7.13

Effective electron mass at the conduction band minimum as a function of the direct band gap at the Γ-point in various semiconductors. The solid line is a fit to Eq. 7.5, which leads to a universal $E_P = 11.9$ eV.

It has been proposed that the interaction element E_P is material insensitive and is nearly a constant, between 10 and 15 eV, much larger than the band gap of most semiconductors and close to the value expected in the empty lattice model [7.36]. Therefore, according to Eq. 7.5, $m_e^*(0)/m_0$ is approximately proportional to E_g for direct-gap group III-V and II-VI semiconductors. In Fig. 7.13 we show this relationship where a universal $E_P = 11.9$ eV was used. This dependence is approximately linear for small E_g materials. Interestingly, the previously reported $m_e^* = 0.11m_0$ measured from sputter-grown InN films with an "optical gap" at 1.9 eV also falls near this dependence [7.4]. This is because for those degenerately doped films, both m_e^* and the "optical gap" were measured at a Fermi level deep into the conductance band. In this case, the "apparent" E_g is raised by the Burstein-Moss effect, but m_e^* is increased by the band nonparabolicity as well, and, as a result, they fall at a position near the universal curve, as shown in Fig. 7.13.

7.4   Optical properties of InGaN and InAlN alloys

The re-evaluation of the band gap of InN brings new insights on studies of group III-nitride alloys that were already intensively investigated for their application in optoelectronics. Figure 7.14 shows the room-temperature absorption curves for In-rich InGaN and InAlN alloys over a wide range of compositions [7.28, 7.37]. The absorption edge shows a rapid blueshift from the band gap of InN with increasing Ga or Al content.

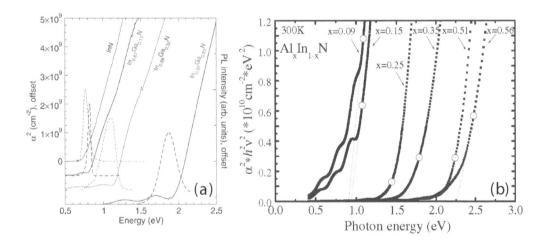

FIGURE 7.14

(a) PL spectrum (dashed) and absorption squared (solid) for InN and In$_{1-x}$Ga$_x$N with different compositions. Some of these curves are vertically offset for clarity. Reproduced with permission of IoP Publishing Limited from W. Walukiewicz, J. W. Ager III, K. M. Yu, Z. Liliental-Weber, J. Wu, S. X. Li, R. E. Jones, and J. D. Denlinger, Journal of Physics D: Applied Physics, 39, R83 (2006). Copyright 2006, Institute of Physics. (b) Absorption curves for In$_{1-x}$Al$_x$N alloys with a wide range of compositions, from which the band gap is determined as a function of x. Reproduced from W. Terashima, S. B. Che, Y. Ishitani, and A. Yoshikawa, Japanese Journal of Applied Physics, 45, L539 (2006) by copyright permission of Institute of Pure and Applied Physics.

The band gaps of $In_{1-x}Ga_xN$ and $In_{1-x}Al_xN$ are plotted as a function of x in Fig. 7.15(a). Representative data on the Ga- or Al-rich side from the literature are also shown for completion of the entire composition range [7.38, 7.39]. It can be seen that the data on the In-rich side make a smooth transition to the data points on the Ga- or Al-rich side. One of the significant aspects of Fig. 7.15(a) is that it demonstrates that the fundamental band gap of the group III-nitride ternary alloy system covers a wide spectral region ranging from the near infrared at $\sim 1.9\ \mu$m (0.65 eV for InN) to the ultraviolet at $\sim 0.36\ \mu$m (3.4 eV for GaN) or 0.2 μm (6.2 eV for AlN). As shown by the solid lines in Fig. 7.15(a), the band gap of $In_{1-x}Ga_xN$ and $In_{1-x}Al_xN$ over the entire composition range can be fitted well by the following standard bowing equation,

$$E_g(x) = E_g(0) \cdot (1-x) + E_g(1) \cdot x - b \cdot x \cdot (1-x). \qquad (7.10)$$

The bowing parameter is found to be $b = 1.4 \pm 0.1$ eV [7.38] for $In_{1-x}Ga_xN$ and 5.0 ± 0.5 eV for $In_{1-x}Al_xN$ [7.37]. The bowing for $In_{1-x}Ga_xN$ is relatively small. For example, a bowing parameter as large as 2.63 eV is needed to describe the composition dependence of the band gap on the Ga-rich side if an InN band gap of 1.9 eV is assumed. On the other hand, the large b for $In_{1-x}Al_xN$ is related to its much wider band gap range, and the large uncertainty of b comes from the more scattered data measured in this wide range of E_g values.

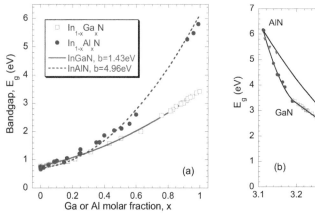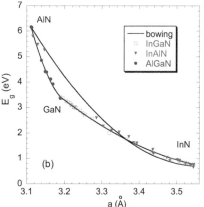

FIGURE 7.15

Band gap of InGaN [7.38] and InAlN [7.37] as a function of Ga or Al molar fraction. The solid and dashed lines are bowing curves (Eq. 7.10) with best-fit bowing parameters. (b) The band gap of InGaN, InAlN, and AlGaN plotted as a function of in-plane lattice constant a. The solid lines show the bowing dependence using the best-fit bowing parameters determined from (a).

It is informative to plot the band gaps as a function of lattice constant instead of composition, so that the three alloys $In_{1-x}Ga_xN$, $In_{1-x}Al_xN$, and $Al_{1-x}Ga_xN$ can be plotted on the same coordinates. To do so, we assume a linear dependence of the lattice constant on composition following Vegard's law. The band gap bowing in Fig. 7.15(a) is consequently mapped to a bowing as a function of in-plane lattice constant, as shown in Fig. 7.15(b). Interestingly, for the available range of experimental data, the band gap of $In_{1-x}Al_xN$ falls onto the same curve of $In_{1-x}Ga_xN$. Therefore, InGaN and InAlN will have the same band gap if their compositions are tuned separately so that they have the same lattice constant. This indicates that Al, Ga, and In affect the band gap of these alloys predominantly through their atomic size and bonding length, at least in the composition range for which experimental data is shown in Fig. 7.15.

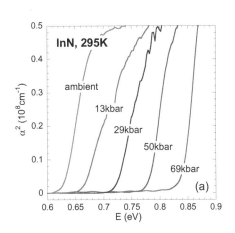 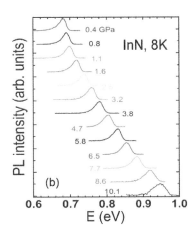

FIGURE 7.16

(a) Absorption squared of a high-quality InN film at different hydrostatic pressures. Figure reproduced with permission from S. X. Li, J. Wu, E. E. Haller, W. Walukiewicz, W. Shan, H. Lu, W. J. Schaff, Applied Physics Letters, 83, 4963 (2003). Copyright 2003, American Institute of Physics. (b) PL signal at different hydrostatic pressures reported by a different group. Reprinted with permission from A. Kamińska, G. Franssen, T. Suski, I. Gorczyca, N. E. Christensen, A. Svane, A. Suchocki, H. Lu, W. J. Schaff, E. Dimakis, and A. Georgakilas, Physical Review B, 76, 075203 (2007). Copyright 2007 by the American Physical Society.

It is intriguing to note that the band gaps of both $In_{1-x}Ga_xN$ and $In_{1-x}Al_xN$ are solely functions of their lattice constants, which suggests the alloying atom size is very important in determining the band gaps. An effective experimental means to decouple the chemical influence from the atomic size effect on the band gap is by the application of hydrostatic pressure.

There have been only a few experimental studies of the pressure behavior of GaN [7.40, 7.41], Ga-rich $In_{1-x}Ga_xN$ alloys [7.40, 7.42], and AlN [7.43]. Although there is a relatively good consensus on the band gap pressure coefficients of GaN and AlN, much less has been known about the pressure dependence of the energy gaps in In containing group III-nitride alloys. A wide range of band gap pressure coefficients has been found even in the most extensively studied Ga-rich $In_{1-x}Ga_xN$ alloys. Figure 7.16 shows the absorption curves [7.20] and PL signal [7.44] of high-

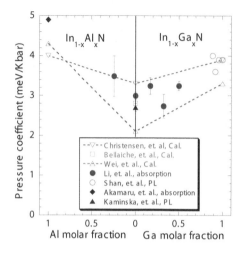

FIGURE 7.17

Pressure coefficients of the band gaps of InN, GaN, AlN, and their alloys.

quality InN films measured at room temperature and low temperature, respectively. Both the absorption edge and the PL signal show a strong blueshift under the application of hydrostatic pressure. These curves shift in a parallel fashion without significant changes in the lineshape, indicating that the high crystal quality is preserved within the range of applied pressure.

The pressure coefficients of the band gaps (dE_g/dP) of InN and In-rich InGaN and InAlN alloys in comparison with those of AlN and GaN are shown in Fig. 7.17. dE_g/dP of InN has been measured to be between 2.7 [7.44] and 3.0 [7.20] meV/kbar. For $In_{1-x}Ga_xN$ alloys the pressure coefficient is close to that of pure InN and in between the two theoretical predictions by Wei and Zunger [7.45] and Christensen and Gorczyca [7.46]. Considering the pressure coefficient of GaN to be 4 meV/kbar, dE_g/dP of $In_{1-x}Ga_xN$ increases with increasing x at a low rate of about 0.01 meV/kbar per % Ga molar fraction. dE_g/dP of $In_{0.75}Al_{0.25}N$ was measured to be $\sim$3.5 meV/kbar, and extrapolates to a value of $\sim$5 meV/kbar at $x = 1$; this agrees well with the pressure coefficient of AlN determined from absorption experiments [7.43]. Note that the pressure coefficients of the group III-nitrides are much smaller than those of other IIIV compounds. For example, dE_g/dP of 11 meV/kbar for GaAs is almost three times larger than that for GaN. It has been argued that this trend can be attributed to the larger ionicity of the group III-nitrides due to the high electronegativity of N. In group III-V semiconductors, higher ionicity typically leads to smaller pressure coefficients [7.47]. The Phillips ionicity is 0.31 for GaAs and is significantly smaller than the value of 0.50 for GaN. The trend applies to group-III nitrides as well, among which larger cations give higher ionicity (AlN $f_i = 0.449$, GaN $f_i = 0.500$, and InN $f_i = 0.578$ [7.48]) and thus smaller pressure coefficients.

7.5   Optical properties of related nanostructures

In recent years, semiconductor nanostructures have come under extensive investigation for applications in high-performance electronic [7.49, 7.50] and optical [7.51] devices. Such semiconductor structures offer a distinct way to study electrical, photonic, and thermal transport phenomena as a function of dimensionality and size reduction. In particular, the wide range of demonstrated and potential applications has made semiconductor nanowires (NWs) and quantum dots (QDs) a rapidly growing focus of research. With the interesting properties of In-rich group III-nitrides described in previous sections, it is natural that there has been much exploration into the growth and characterization of InN, $In_{1-x}Ga_xN$, and $In_{1-x}Al_xN$ nanostructures.

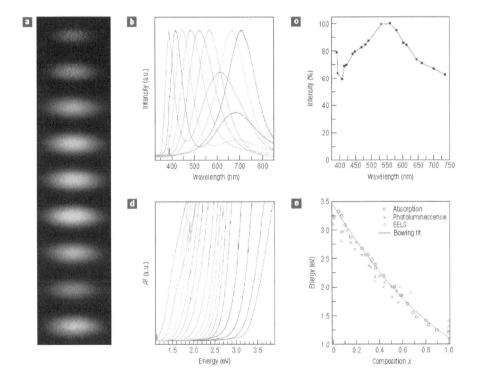

FIGURE 7.18

(a) Charge-coupled device images (color in the original version) of the emission from $In_xGa_{1-x}N$ nanowires; (b) visible PL emission ($x = 0$–0.6); (c) corrected PL peak intensities; (d) optical absorption spectra ($x = 0$–1) of the $In_xGa_{1-x}N$ nanowire arrays with varying composition x; and (e) band gap plotted as a function of In fraction x for PL, absorption and EELS, and bowing equation fit to absorption data. Reprinted by permission from Macmillan Publishers Ltd.: T. Kuykendall, P. Ulrich, S. Aloni, P. Yang, Nature Materials, 6, 951 (2007), copyright 2007.

Figure 7.18 shows the optical characterization of recently synthesized $In_xGa_{1-x}N$ NWs by Kuykendall *et al.* by using a combinatorial CVD approach in a four-zone furnace [7.52]. X-ray and electron diffraction prove that these nanowires were grown as single crystals over the entire composition range of $0 < x < 1$ across a single substrate with no phase separation. It was argued that these nanowires were grown via a self-catalyzed process enabled by the low temperature (550°C) and high growth rate which promotes and stabilizes the formation of the thermodynamically unstable product.

Despite the strong "yellow luminescence" for Ga-rich compositions which is typical for Ga-rich InGaN thin films as well, these nanowires exhibit band gap values spanning from the infrared (1.2 eV) to the ultraviolet (3.4 eV) spectral range. This range is consistent with the narrow band gap of InN, and shows great potential for fabricating nano-scale full-color light emitting and light harvesting devices. As with thin films, the greater than 0.65 eV optical gap for the InN NWs is attributed to the well-known Burstein-Moss shift caused by the high unintentional doping [7.34]. One of the remarkable properties of these nanowires is the high quantum efficiency over a wide range of compositions (Fig. 7.18(c)), which overcomes the well-known "valley of death" drop-off in PL efficiency for InGaN thin films when the composition moves away from GaN [7.53]. The authors suggest that this enhancement is due to the unique growth mechanism and geometry of nanowires; these help relax strain and eliminate threading dislocations, which usually act as non-radiative recombination centers in InGaN thin films.

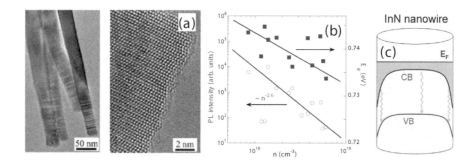

FIGURE 7.19

(a) Low and high magnification TEM images of InN nanowires grown by MBE. (b) Correlation of the computed electron concentration with the PL integral intensity and band gap E_g, respectively. (c) A schematic band diagram illustrating the electron-hole recombination between the conduction band (CB) and valence band (VB) of an InN nanowire, where an electron accumulation layer exists on the surface. Reproduced with permission from T. Stoica, R. J. Meijers, R. Calarco, T. Richter, E. Sutter, and H. Luth, Nano Letters, 6, 1541 (2006). Copyright 2006, American Chemical Society.

InN NWs have also been grown using low-temperature MBE. Figure 7.19 shows the TEM images and optical properties of these NWs [7.54]. The electron concentration (n) and fundamental band gap in these nanowires were derived from fitting the PL lineshape with a model which included the convolution of the electron and hole distributions in the nanowires. The PL intensity was found to scale as $\sim n^{-2.6}$, a stronger dependence than for a dominant Auger recombination, for which a dependence of n^{-1} is expected. The authors thus conclude that the decrease in PL efficiency arises not only as a result of the increase of the electron concentration, but also because other nonradiative recombination processes increase, such as recombination at the wire surface. It is now well established that there exists an intrinsic electron accumulation layer on the surface of InN as a result of the surface Fermi level pinning deep into the conduction band (Fig. 7.19(c)) [7.55–7.57]. The fundamental band gap (E_g) is reduced in this layer, an effect known as band gap renormalization [7.21], and the average free electron concentration of the system is increased. In small-diameter nanowires, these effects become more prominent due to the large surface-to-volume ratio. They are responsible for the decrease in E_g with increasing n in Fig. 7.19(b). More thorough discussion of the issue of InN surfaces can be found in Chapters 12 and 13.

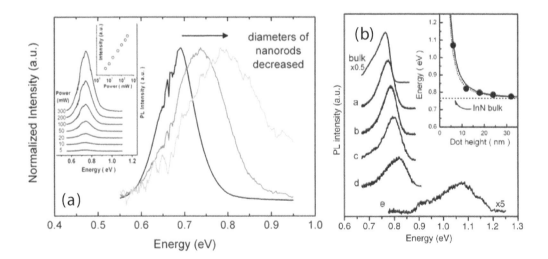

FIGURE 7.20

(a) PL spectra of InN nanorods. As the size of the nanorods decreases, the peak shows a blueshift. The inset shows that the PL intensity grows linearly with the excitation power. Reproduced with permission of IoP Publishing Limited from C. K. Chao, H. S. Chang, T. M. Hsu, C. N. Hsiao, C. C. Kei, S. Y. Kuo, and J. I. Chyi, Nanotechnology, 17, 3930 (2006). Copyright 2006, Institute of Physics. (b) PL spectra measured at 17 K for InN bulk and InN quantum dots with different heights. The insert shows the peak energy as a function of dot height. The solid line (dotted line) is calculated within the effective mass approximation using $0.042m_0$ ($0.07m_0$) as the electron effective mass. Figure reproduced with permission from W. C. Ke, C. P. Fu, C. Y. Chen, L. Lee, C. S. Ku, W. C. Chou, W. H. Chang, M. C. Lee, W. K. Chen, W. J. Lin, and Y. C. Cheng, Applied Physics Letters, 88, 191913 (2006). Copyright 2006, American Institute of Physics.

As in other semiconductor nanostructures, the 'optical gap' of InN increases in nanowires (nanorods) and quantum dots in the presence of substantial quantum confinement. The characteristic length scale for this to happen can be set at the exciton Bohr radius of InN which is of the order of 10 nm. Figure 7.20(a) shows the PL spectra at different excitation powers and with different nanorod diameters [7.58]. The linear dependence of the PL intensity on the excitation power over two orders of magnitude (5–300 mW), as shown in the inset of Fig. 7.20(a), indicates the interband nature of the luminescence. The observed blueshift of the PL peak is attributed mostly to the quantum size effect.

Similar blueshift in PL has been observed in InN quantum dots as well. Figure 7.20(b) shows the PL spectra of self-assembled InN quantum dots embedded in GaN grown by MOCVD [7.59]. The QD height was measured by AFM, and defines the degree of confinement in the QDs. The peak energies shift systematically from 0.78 to 1.07 eV as the average dot height was reduced from 32.4 to 6.5 nm. The inset shows that the PL peak shift can be well explained by a standard quantum confinement model within the effective mass approximation [7.59].

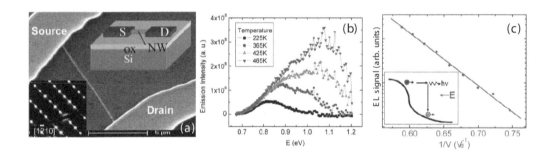

FIGURE 7.21

(a) SEM image of a representative InN nanowire FET. The left inset shows an electron diffraction pattern. (b) The electroluminescence (EL) spectra from an InN nanowire under variable temperatures. (c) EL emission intensity (log scale) versus $1/V_d$ from an InN transistor. The inset shows a schematic of the proposed hot carrier emission mechanism. Reproduced with permission from J. Chen, G. Cheng, E. Stern, M. A. Reed, and P. Avouris, Nano Letters, 7, 2276 (2007). Copyright 2007, American Chemical Society.

Electroluminescence (EL) is another powerful tool to probe the electron-photon interactions in semiconductors. It is especially suitable for studying the emission properties of single nanowires in the field-effect transistor (FET) geometry where the nanowire is electrically biased between a source electrode and a drain electrode. Spectrally resolved light emission is recorded as a function of either the biasing voltage and/or back gate voltage. Figure 7.21 shows such an EL experiment on single InN nanowires [7.60]. The EL spectrum shows a peak near E_g of InN and blue shifts at higher temperature. The EL intensity increases with increasing biasing voltage V_d as $\exp(-V_0/V_d)$, where V_0 depends on the optical phonon scattering length. These observations agree well with an impact excitation mechanism which is depicted in the inset of Fig. 7.21(b). Under the high biasing electric field, an electron is accelerated to a sufficiently high energy; if it does not lose that energy to optical phonons, it can impact-excite electron-hole pairs that then decay radiatively, giving rise to the EL signal at E_g. The electron that lost energy to the generated exciton regains kinetic energy from the biasing field and continues this process. The surface electron accumulation layer was found to enhance the radiative electron-hole recombination in these InN nanowires through a plasmon-exciton coupling mechanism.

Devices harnessing the wide-spectrum optoelectronic possibilities of group III-nitrides have only recently been fabricated at the laboratory scale. Multi-color LEDs and high-mobility FETs have been demonstrated with Ga-rich InGaN and AlGaN nanowires and nanowire heterostructures [7.49, 7.61]. In these nanowire devices, the greatest In content used was 0.35, corresponding to ~600 nm orange-color emission. With the rapid progress in nanowire synthesis and integration, extension of the spectrum to infrared by increasing the In composition is expected. This will open a wide avenue to a new class of nitride-based, high-efficiency, wide-spectrum, and long-lifetime optoelectronic devices, such as detectors, lasers, and solar cells.

7.6 Conclusion and outlook

We have presented an up-to-date review of the optical properties of InN and related group III-nitride alloys. On the basis of the discovery of the narrow band gap of InN, the fundamental optical parameters of InN and In-rich InGaN and InAlN have been re-evaluated. The narrow band gap of InN results in a strongly nonparabolic conduction band, with a small electron effective mass at the conduction band minimum. The band gaps of group III-nitride alloys span a wide spectral region, ranging from the near infrared to the ultraviolet.

These findings are expected to have significant impact on the device applications of group III-nitrides. The widely tunable direct energy gaps of the nitride alloys offer potential for new applications. Most interestingly, the band gaps of $In_{1-x}Ga_xN$ alloys ranging from 0.65 to 3.42 eV provide an almost perfect match to the solar spectrum. This opens up an interesting opportunity for using these alloys in high-efficiency, multijunction solar cells. Current multijunction cells are based on three different semiconductors with fixed band gaps, Ge (0.7 eV), GaAs (1.4 eV), and GaInP (1.9 eV). The main advantage of the InGaN alloys would be the flexibility in the choice of the gaps, which would allow optimization of the performance of such cells. Superior radiation resistance has also been reported for these alloys, making them especially suitable for device applications in harsh environments, such as in outer space and nuclear reactors.

InN-related group III-nitride alloys may also find applications in other device applications, such as infrared LEDs, components for 1.55 μm fiber optics, terahertz radiation emitters, high-power and high-speed transistors, and thermoelectric devices. Of course, realization of most of these devices relies on the availability of p-type InN and related In-rich alloys which has not been practically achieved. However, the recently reported evidence of p-type activity in Mg-doped InN [7.57, 7.62] makes these novel applications highly promising in the near future. In these doping investigations, optical techniques are powerful and complementary to electrical characterization, as described in detail in Chapter 10. For example, recent photoluminescence measurements on Mg-doped InN demonstrated for the first time an acceptor ionization energy of $\sim$60 meV [7.63, 7.64].

Acknowledgments

We wish to gratefully acknowledge collaborations with Dr. W. Walukiewicz, Dr. Hai Lu, Dr. W. J. Schaff, Prof. E. E. Haller, Prof. Y. Nanishi, Dr. K. M. Yu, Dr. J. W. Ager III, and Dr. S. X. Li. J. W. acknowledges support from the National Science Foundation under Grant No. EEC-0425914, and an LDRD grant from the Lawrence Berkeley National Laboratory, and J. W. L. Y. acknowledges the NSF Graduate Research Fellowship.

References

[7.1] S. Nakamura, InGaN/AlGaN blue-light-emitting diodes, *Journal of Vacuum Science & Technology A*, 13 (1995) 705–710.

[7.2] S. Nakamura, The roles of structural imperfections in InGaN-based blue light-emitting diodes and laser diodes, *Science*, 281 (1998) 956–961.

[7.3] H. J. Hove and J. J. Cuomo, Electrical and optical properties of rf-sputtered GaN and InN, *Applied Physics Letters*, 20 (1972) 71–73.

[7.4] V. A. Tyagi, A. M. Eustigneev, A. N. Krasilo, A. F. Andreeva, and V. Y. Malatidiou, Optical properties of indium nitride films, *Soviet Physics – Semiconductors*, 11 (1977) 1257–1259.

[7.5] K. L. Westra, R. P. W. Lawson, and M. J. Brett, The effects of oxygen contamination on the properties of reactively sputtered indium nitride films, *Journal of Vacuum Science & Technology A*, 6 (1988) 1730–1732.

[7.6] T. L. Tansley and C. P. Foley, Optical band gap of indium nitride, *Journal of Applied Physics*, 59 (1986) 3241–3244.

[7.7] H. Morkoç, *Nitride Semiconductors and Devices*, Springer, Heidelberg (1999).

[7.8] S. Yamaguchi, M. Kariya, S. Nitta, T. Takeuchi, C. Wetzel, H. Amano, and I. Akasaki, Anomalous features in the optical properties of $Al_{1x}In_xN$ on GaN grown by metal organic vapor phase epitaxy, *Applied Physics Letters*, 76 (2000) 876–878.

[7.9] T. Matsuoka, H. Tanaka, T. Sasaki, and A. Katsui, Wide-gap semiconductor (In,Ga)N, Proceedings of the 16th International Symposium on GaAs and Related Compounds, Karuizawa, Japan, 1989, *Institute of Physics Conference Series* vol. 106, Institute of Physics, Bristol, 1990, p. 141.

[7.10] H. Lu, W. J. Schaff, J. Hwang, H. Wu, G. Koley, and L. F. Eastman, Effect of an AlN buffer layer on the epitaxial growth of InN by molecular-beam epitaxy, *Applied Physics Letters*, 79 (2001) 1489–1491.

[7.11] V. Y. Davydov, A. A. Klochikhin, R. P. Seisyan, V. V. Emtsev, S. V. Ivanov, F. Bechstedt, J. Furthmüller, H. Harima, A. V. Mudryi, J. Aderhold, O. Semchinova, and J. Graul, Absorption and emission of hexagonal InN. Evidence of narrow fundamental band gap, *physica status solidi (b)*, 229 (2002) R1–R3.

[7.12] Y. Nanishi, Y. Saito, and T. Yamaguchi, RF-molecular beam epitaxy growth and properties of InN and related alloys, *Japanese Journal of Applied Physics, Part 1*, 42 (2003) 2549–2559.

[7.13] J. Wu, W. Walukiewicz, K. M. Yu, J. W. Ager III, E. E. Haller, H. Lu, W. J. Schaff, Y. Saito, and Y. Nanishi, Unusual properties of the fundamental band gap of InN, *Applied Physics Letters*, 80 (2002) 3967–3969.

[7.14] J. Wu, W. Walukiewicz, W. Shan, K. M. Yu, J. W. Ager III, S. X. Li, E. E. Haller, H. Lu, and W. J. Schaff, Temperature dependence of the fundamental band gap of InN, *Journal of Applied Physics*, 94 (2003) 4457–4460.

[7.15] H. Lu, W. J. Schaff, J. Hwang, H. Wu, W. Yeo, A. Pharkya, and L. F. Eastman, Improvement on epitaxial grown of InN by migration enhanced epitaxy, *Applied Physics Letters*, 77 (2000) 2548–2550.

[7.16] I. Vurgaftman, J. R. Meyer, and L. R. Ram-Mohan, Band parameters for IIIV compound semiconductors and their alloys, *Journal of Applied Physics*, 89 (2001) 5815–5875.

[7.17] V. Y. Davydov, V. V. Emtsev, I. N. Goncharuk, A. N. Smirnov, V. D. Petrikov, V. V. Mamutin, V. A. Vekshin, S. V. Ivanov, M. B. Smirnov, and T. Inushima, Experimental and theoretical studies of phonons in hexagonal InN, *Applied Physics Letters*, 75 (1999) 3297–3299.

[7.18] G. Shikata, S. Hirano, T. Inoue, M. Orihara, Y. Hijikata, H. Yaguchi, and S. Yoshida, RF-MBE growth of a-plane InN on r-plane sapphire with a GaN underlayer, *Journal of Crystal Growth*, 301–302 (2007) 517–520.

[7.19] S. P. Fu, Y. F. Chen, and K. W. Tan, Recombination mechanism of photoluminescence in InN epilayers, *Solid State Communications*, 137 (2006) 203–207.

[7.20] S. X. Li, J. Wu, E. E. Haller, W. Walukiewicz, W. Shan, H. Lu, and W. J. Schaff, Hydrostatic pressure dependence of the fundamental bandgap of InN and In-rich group III nitride alloys, *Applied Physics Letters*, 83 (2003) 4963–4965.

[7.21] J. Wu, W. Walukiewicz, W. Shan, K. M. Yu, J. W. Ager III, E. E. Haller, H. Lu, and W. J. Schaff, Effects of the narrow band gap on the properties of InN, *Physical Review B*, 66 (2002) 201403(R):1–4.

[7.22] S. P. Fu and Y. F. Chen, Effective mass of InN epilayers, *Applied Physics Letters*, 85 (2004) 1523–1525.

[7.23] J. Bhattachryya, S. Ghosh, M. R. Gokhale, B. M. Arora, H. Lu, and W. J. Schaff, Polarized photoluminescence and absorption in a-plane InN films, *Applied Physics Letters*, 89 (2006) 151910:1–3.

[7.24] R. Goldhahn, P. Schley, A. T. Winzer, M. Rakel, C. Cobet, N. Esser, H. Lu, and W. J. Schaff, Critical points of the band structure and valence band ordering at the Γ point of wurtzite InN, *Journal of Crystal Growth*, 288 (2006) 273–277.

[7.25] Sandip Ghosh, Pranob Misra, H. T. Grahn, Bilge Imer, Shuji Nakamura, S. P. DenBaars, and J. S. Speck, Polarized photoreflectance spectroscopy of strained A-plane GaN films on R-plane sapphire, *Journal of Applied Physics*, 98 (2005) 026105:1–3.

[7.26] F. Chen, A. N. Cartwright, H. Lu, and W. J. Schaff, Temperature dependence of carrier lifetimes in InN, *physica status solidi (a)*, 202 (2005) 768–772.

[7.27] R. Ascazubi, I. Wilke, S. Cho, H. Lu, and W. J. Schaff, Ultrafast recombination in Si-doped InN, *Applied Physics Letters*, 88 (2006)

112111:1–3.

[7.28] W. Walukiewicz, J. W. Ager III, K. M. Yu, Z. Liliental-Weber, J. Wu, S. X. Li, R. E. Jones, and J. D. Denlinger, Structure and electronic properties of InN and In-rich group III-nitride alloys, *Journal of Physics D — Applied Physics*, 39 (2006) R83–R99.

[7.29] F. Chen, A. N. Cartwright, H. Lu, and W. J. Schaff, Time-resolved spectroscopy of recombination and relaxation dynamics in InN, *Applied Physics Letters*, 83 (2003) 4984–4986.

[7.30] S. H. Wei, X. Nie, I. G. Batyrev, and S. B. Zhang, Breakdown of the band-gap-common-cation rule: The origin of the small band gap of InN, *Physical Review B*, 67 (2003) 165209:1–4.

[7.31] Y. H. Li, X. G. Gong, and S. H. Wei, *Ab initio* all-electron calculation of absolute volume deformation potentials of IV-IV, III-V and II-VI semiconductors: The chemical trends, *Physical Review B*, 73 (2006) 245206:1–5.

[7.32] D. Segev, A. Janotti, C. G. Van de Walle, Self-consistent band-gap corrections in density functional theory using modified pesudopotentials, *Physical Review B*, 75 (2007) 035201:1–9.

[7.33] E. O. Kane, Band structure of indium antimonide, *Journal of Physics and Chemistry of Solids*, 1 (1957) 249-261.

[7.34] J. Wu, W. Walukiewicz, S. X. Li, R. Armitage, J. C. Ho, E. R. Weber, E. E. Haller, H. Lu, W. J. Schaff, A. Barcz, and R. Jakiela, Effects of electron concentration on the optical absorption edge of InN, *Applied Physics Letters*, 84 (2004) 2805–2807.

[7.35] J. Wu, W. Walukiewicz, K. M. Yu, W. Shan, J. W. Ager III, E. E. Haller, H. Lu, W. J. Schaff, W. K. Metzger, and S. Kurtz, Superior radiation resistance of $In_{1x}Ga_xN$ alloys: Full-solar-spectrum photovoltaic material system, *Journal of Applied Physics*, 94 (2003) 6477–6482.

[7.36] P. Y. Yu and M. Cardona, Fundamentals of semiconductors: physics and materials properties (Springer, Berlin, 1999).

[7.37] W. Terashima, S. B. Che, Y. Ishitani, and A. Yoshikawa, Growth and characterization of AlInN tenary alloys in whole composition range and fabrication of InN/AlInN multiple quantum wells by RF molecular beam epitaxy, *Japanese Journal of Applied Physics*, 45 (2006) L539–L542.

[7.38] J. Wu, W. Walukiewicz, K. M. Yu, J. W. Ager III, E. E. Haller, H. Lu, and W. J. Schaff, Small band gap bowing in $In_{1-x}Ga_xN$ alloys, *Applied Physics Letters*, 80 (2002) 4741–4743.

[7.39] J. Wu, W. Walukiewicz, K. M. Yu, J. W. Ager, S. X. Li, E. E. Haller, H. Lu, and W. J. Schaff, Universal bandgap bowing in group-III nitride alloys, *Solid State Communications*, 127 (2003) 411–414.

[7.40] W. Shan, W. Walukiewicz, E. E. Haller, B. D. Little, J. J. Song. M. D. McCluskey, N. M. Johnson, Z. C. Feng, M. Schurman, and R. A. Stall, Optical properties of $In_xGa_{1x}N$ alloys grown by metalorganic chemical vapor deposition, *Journal of Applied Physics*, 84 (1998) 4452–4458.

[7.41] P. Perlin, L. Mattos, N. A. Shapiro, J. Kruger, W. S. Wong, T. Sands, N. W. Cheung, and E. R. Weber, Reduction of the energy gap pressure coefficient of GaN due to the constraining presence of the sapphire substrate, *Journal of Applied Physics*, 85 (1999) 2385–2389.

[7.42] T. Suski, H. Teisseyre, S. P. Lepkowski, P. Perlin, H. Mariette, T. Kitamura, Y. Ishida, H. Okumura, and S. F. Chichibu, Light emission versus energy gap in group-III nitrides: hydrostatic pressure studies, *physica status solidi (b)*, 235 (2003) 225–231.

[7.43] H. Akamaru, A. Onodera, T. Endo, and O. Mishima, Pressure dependence of the optical-absorption edge of AlN and graphite-type BN, *Journal of Physics and Chemistry of Solids*, 63 (2002) 887–894.

[7.44] A. Kamińska, G. Franssen, T. Suski, I. Gorczyca, N. E. Christensen, A. Svane, A. Suchocki, H. Lu, W. J. Schaff, E. Dimakis, and A. Georgakilas, Role of conduction-band filling in the dependence of InN photoluminescence on hydrostatic pressure, *Physical Review B*, 76 (2007) 075203:1–5.

[7.45] S. H. Wei and A. Zunger, Predicted band-gap pressure coefficients of all diamond and zinc-blende semiconductors: Chemical trends, *Physical Review B*, 60 (1999) 5404–5411.

[7.46] N. E. Christensen and I. Gorczyca, Optical and structural properties of III-V nitrides under pressure, *Physical Review B*, 50 (1994) 4397–4415.

[7.47] S. Adachi, GaAs, AlAs, and $Al_xGa_{1-x}As$ material parameters for use in research and device applications, *Journal of Applied Physics*, 58 (1985) R1–R29.

[7.48] J. C. Phillips, Bonds and bands in semiconductors (Academic Press, New York, 1973).

[7.49] Y. Li, F. Qian, J. Xiang, and C. M. Lieber, Nanowire electronic and optoelectronic devices, *Materials Today*, 9(10) (2006) 18–27.

[7.50] C. Thelander, P. Agarwal, S. Brongersma, J. Eymery, L. F. Feiner, A. Forchel, M. Scheffler, W. Riess, B. J. Ohlsson, U. Gösele, and

L. Samuelson, Nanowire-based one-dimensional electronics, *Materials Today*, 9(10) (2006) 28–35.

[7.51] R. Agarwal and C. M. Lieber, Semiconductor nanowires: optics and optoelectronics, *Applied Physics A*, 85 (2006) 209–215.

[7.52] T. Kuykendall, P. Ulrich, S. Aloni, and P. Yang, Complete composition tunability of InGaN nanowires using a combinatorial approach, *Nature Materials*, 6 (2007) 951–956.

[7.53] D. Fuhrmann, Optimization scheme for the quantum efficiency of GaInN-based green light-emitting diodes, *Applied Physics Letters*, 88 (2006) 071105:1–3.

[7.54] T. Stoica, R. J. Meijers, R. Calarco, T. Richter, E. Sutter, and H. Luth, Photoluminescence and intrinsic properties of MBE-grown InN nanowires, *Nano Letters*, 6 (2006) 1541–1547.

[7.55] H. Lu, W. J. Schaff, L. F. Eastman, and C. E. Stutz, Surface charge accumulation of InN films grown by molecular-beam epitaxy, *Applied Physics Letters*, 82 (2003) 1736–1738.

[7.56] I. Mahboob, T. D. Veal, C. F. McConville, H. Lu, and W. J. Schaff, Intrinsic electron accumulation at clean InN surfaces, *Physical Review Letters*, 92 (2004) 036804:1–4.

[7.57] J. W. L. Yim, R. E. Jones, K. M. Yu, J. W. Ager III, W. Walukiewicz, W. J. Schaff, and J. Wu, Effects of surface states on electrical characteristics of InN and $In_{1-x}Ga_xN$, *Physical Review B*, 76 (2007) 041303(R):1–4.

[7.58] C. K. Chao, H. S. Chang, T. M. Hsu, C. N. Hsiao, C. C. Kei, S. Y. Kuo, and J. I. Chyi, Optical properties of indium nitride nanorods prepared by chemical-beam epitaxy, *Nanotechnology*, 17 (2006) 3930–3932.

[7.59] W. C. Ke, C. P. Fu, C. Y. Chen, L. Lee, C. S. Ku, W. C. Chou, W. H. Chang, M. C. Lee, W. K. Chen, W. J. Lin, and Y. C. Cheng, Photoluminescence properties of self-assembled InN dots embedded in GaN grown by metal organic vapor phase epitaxy, *Applied Physics Letters*, 88 (2006) 191913:1–3.

[7.60] J. Chen, G. Cheng, E. Stern, M. A. Reed, and P. Avouris, Electrically excited infrared emission from InN nanowire transistors, *Nano Letters*, 7 (2007) 2276–2280.

[7.61] F. Qian, S. Gradecak, Y. Li, and C. M. Lieber, Core/multishell nanowire heterostructures as multicolor, high-efficiency light-emitting diodes, *Nano Letters*, 5 (2005) 2287–2291.

[7.62] R. E. Jones, K. M. Yu, S. X. Li, W. Walukiewicz, J. W. Ager,

E. E. Haller, H. Lu, and W. J. Schaff, Evidence for *p*-type doping of InN, *Physical Review Letters*, 96 (2006) 125505:1–4.

[7.63] X. Wang, S. B. Che, Y. Ishitani, and A. Yoshikawa, Growth and properties of Mg-doped In-polar InN films, *Applied Physics Letters*, 90 (2007) 201913:1–3.

[7.64] N. Khan, N. Nepal, A. Sedhain, J. Y. Lin, and H. X. Jiang, Growth and properties of Mg-doped In-polar InN films, *Applied Physics Letters*, 91 (2007) 012101:1–3.

8

Theory of InN bulk band structure

J. Furthmüller, F. Fuchs, and F. Bechstedt

Institut für Festkörpertheorie und -optik, Friedrich-Schiller-Universität and European Theoretical Spectroscopy Facility (ETSF), Max-Wien-Platz 1, D-07743 Jena, Germany

8.1 Introduction

8.1.1 A short history

Indium nitride (InN) is a key material for GaN-based light-emitting devices, since InN incorporation is necessary for adjusting the near ultraviolet pure-GaN band gap to the blue spectral region [8.1]. Despite the technological importance of InN, its fundamental physical parameters were controversial until about 2005 [8.2, 8.3]. In the 1970s and 1980s, the direct band gap of InN was estimated as being approximately 2 eV [8.4, 8.5]. Among the group-III nitrides InN remained the most mysterious compound due mainly to the difficulty in growing high-quality crystals because of the extremely high equilibrium vapor pressure of nitrogen. Early absorption measurments were therefore restricted to polycrystalline films grown by DC discharge [8.4], reactive cathodic sputtering [8.6], RF sputtering [8.5, 8.7], or metalorganic vapor phase epitaxy (MOVPE) [8.8].

There were also limitations for the electronic-structure calculations. The *ab initio* electronic-structure theory as a combination of density functional theory (DFT) [8.9] in the local density approximation (LDA) [8.10] or in a semilocal approximation including generalized gradient corrections (GGA) [8.11, 8.12] and the quasiparticle (QP) theory taking into account the exchange-correlation (XC) self-energy in Hedin's GW approximation [8.13] could not be used to predict a plausible gap value. Usually, including the In $4d$ electrons as valence electrons, the DFT-LDA or DFT-GGA gives an overlap of conduction and valence band states, that is, a negative gap [8.14–8.20]. However, after correcting the overestimation of the p-d repulsion and hence the tendency for closing the gap due to the shallow In $4d$ electrons [8.21], the perturbative treatment of the GW corrections could be applied. Thus, theory [8.20, 8.22] predicted a fundamental gap of less than 1 eV for InN.

Simultaneously, the growth techniques for crystalline InN layers have been significantly improved, especially by application of plasma-assisted molecular beam epitaxy (MBE). Resulting InN layers have shown photoluminescence (PL) at photon energies below 0.9 eV [8.23]. These results were immediately confirmed using high-quality crystalline layers of wurtzite InN grown by MBE [8.24, 8.25] or MOVPE [8.26] with a tendency for gap values closer to 0.7 eV.

8.1.2   Deviation of InN from the common cation rule

The observed very small gap value for InN is surprising. With respect to the chemical trends observed for III-V compounds, InN exhibits an unexpected anomalous behavior [8.27]. Within the compounds InX (X=Sb,As,P,N), InN violates, for example, the usual common cation rule for the band gap. From X=Sb down to X=N one would expect a monotonic increase of the fundamental band gap from 0.24 eV (InSb), 0.42 eV (InAs), and 1.46 eV (InP) to InN [8.28]. An InN gap value of about 1.9 eV would be plausible. However, rather than InN, it is InP that possesses the largest gap of all In-V compounds. Only within the XN group (X=Al,Ga,In) InN shows the expected behavior, having the smallest gap among all III-nitrides [8.29]. The apparently anomalous behavior of InN has to be traced back to a combination of several effects reflecting the complexity of the electronic structure of In compounds in general [8.28, 8.29, 8.30]. Among them are the high electronegativity of nitrogen [8.31] and, consequently, the small band-gap deformation potential [8.28], the low-lying In $5s$ level [8.29], and the *p-d* repulsion [8.20, 8.22, 8.30] which should be strongest within the In-V compounds (see also Fig. 8.1).

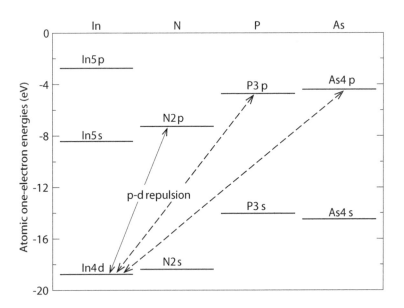

FIGURE 8.1

Energy levels of valence electrons for atoms forming InX compounds (X=As,P,N). The energies are taken from DFT-LDA calculations of free atoms using a scalar-relativistic approach [8.32, 8.33].

8.1.3 Gap reduction due to *p-d* repulsion effects

Besides general chemical trends there is indeed an important effect that contributes to the closing of the band gap, not only in III-V compounds, but in general in many compounds involving heavy elements, the so-called *p-d* repulsion between *p* valence states and shallow *d* core states. This effect has been described and studied in detail in a famous work of Wei and Zunger [8.21] for many standard semiconductors and insulators. The *p-d* repulsion results in a pushing up of *p*-type bands toward higher energies. Since usually the valence band maxima of semiconductors with Γ_6/Γ_1 (wurtzite) or Γ_{15} (zinc-blende) symmetry possess dominantly *p*-character, this means that the valence band maximum is shifted to higher energies. If the conduction band minimum possesses dominantly *s*-character, as is usually the case for direct-gap semiconductors as for most III-V or II-VI compounds, this means effectively an extra closing of the gap because the *s* valence states remain uninfluenced by the shallow *d* core states. For the majority of semiconductors one finds a rather moderate *p-d* repulsion effect (a typical order of magnitude is about 0.1–0.3 eV). However, for a few semiconductors, in particular strongly ionic ones like some II-VI compounds, one may also find very strong *p-d* repulsion effects of the order of about 1 eV [8.21]. No values have been given yet for InN in the literature. Taking the Wei and Zunger model, one basically needs to know the amount of *d* character in the *p*-like valence band maximum (or vice versa the amount of *p* character in the shallow *d* core state bands) and the matrix element of the momentum operator between the valence band maximum and the *d* bands. Both quantities can be easily calculated with standard *first-principles* methods. The trend follows the energetic distance between the In 4*d* level and the anion *np* level with $n = 2$ (N), $n = 3$ (P), and $n = 4$ (As) as shown in Fig. 8.1. Taking calculated values for the required model parameters and inserting them into the model of Wei and Zunger one obtains an estimate for the *p-d* repulsion effect which amounts to about 0.8 eV [8.20, 8.30]. This is an exceptionally large value among all III-V compounds, also much larger than the value for GaN which seems to be about 0.2 eV according to Fiorentini *et al.* [8.34]. We trace this large value back to the very strong bond ionicity $g = 0.85$ of InN which is comparable to ionicities of group-IIB oxides like ZnO exhibiting similarly large *p-d* repulsion effects. In contrast, the ionicity g of the other III-V compounds is much weaker. For instance, GaAs has a value of $g = 0.32$ [8.35].

8.2 Methods of parameter-free band structure calculations

8.2.1 Density functional theory

In general, the density-functional theory [8.9] in the local-density approximation [8.10] or in a generalized gradient approximation [8.11, 8.12] is a useful tool for the calculation of structural and thermodynamic properties of solids. Standard packages like the Vienna Ab-initio Simulation Package (VASP) [8.32, 8.33] have implemented these tools in a very efficient way, on the basis of ultrasoft pseudopotentials or the projector-augmented wave (PAW) method [8.36]. The fundamental theorem of DFT is the Hohenberg-Kohn theorem [8.9] according to which the ground state of a system can be described uniquely using the ground-state density of the electrons $n(\mathbf{r})$ only, that is, there exists a unique energy functional $E[n(\mathbf{r})]$ which takes a minimum at the ground-state density which itself is uniquely determined by a given external potential v_{ext}.

Because of the difficulty to obtain any reasonable description of the kinetic energy in terms of the particle density, Kohn and Sham [8.10] introduced a representation employing single-particle wave functions belonging to a virtual system of non-interacting particles possessing the same ground-state density and ground-state energy as the interacting system. Such a virtual system of non-interacting particles can be achieved by introducing an effective potential acting on the particles which does not only include the external potential but is an effective density-dependent potential containing also the interactions among the particles. This gives rise to the well-known Kohn-Sham (KS) equation

$$\left\{ -\frac{\hbar^2}{2m}\Delta_{\mathbf{r}} + v_{\text{ext}}(\mathbf{r}) + v_{\text{H}}(\mathbf{r}) + v_{\text{XC}}(\mathbf{r}) \right\} \varphi_i(\mathbf{r}) = \varepsilon_i \varphi_i(\mathbf{r}) \tag{8.1}$$

with the electron density

$$n(\mathbf{r}) = \sum_{i \ (\text{occ.})} |\varphi_i(\mathbf{r})|^2 . \tag{8.2}$$

The Hartree potential $v_{\text{H}}(\mathbf{r}) = \int d^3\mathbf{r}' \, v(\mathbf{r}-\mathbf{r}') \, n(\mathbf{r}')$ is directly generated by the electron density. The exchange-correlation potential v_{XC}, the functional derivative of the exchange-correlation energy E_{XC} of the system, $v_{\text{XC}} = \delta E_{\text{XC}}[n(\mathbf{r})]/\delta n(\mathbf{r})$, is thereby the only unknown quantity which requires approximations as the local-density aproximation or (generalized) gradient approximations account also for inhomogeneities of the electron gas.

The KS equation has to be solved self-consistently with respect to the density via the KS orbitals $\varphi_i(\mathbf{r})$. The non-interacting particles are characterized by a set of quantum numbers i. In the case of translationally invariant crystals it holds $i \equiv v\mathbf{k}$ with v as the band index and $\mathbf{k}$ as a wave vector from the Brillouin zone (BZ). The Kohn-Sham eigenvalues are from a mathematical point of view Lagrange parameters which have to be introduced to obey the orthonormality constraints of the Kohn-Sham wave functions. However, from a physical point of view the $\varepsilon_i \equiv \varepsilon_v(\mathbf{k})$ also describe the band structure of the virtual system of non-interacting particles. This band structure is not necessarily the same as the true single-particle band structure of the interacting system, which may be measurable by exciting the electron system and creating quasiparticles [8.13]. There should be a loose connection to the true quasiparticle (QP) band structure in a sense that the qualitative picture should be correct because the effective potential entering the Kohn-Sham equation reflects the interactions between the particles. For that reason people often like to interpret this Kohn-Sham band structure as a good approximation to the real band structure. In most cases, apart from wrong energetical positions and deviations of the band widths with respect to experimental findings, band dispersions are displayed correctly. Effective band masses and deformation parameters can be calculated fairly well within DFT-LDA or DFT-GGA. Even concerning optical properties (dielectric constant) one obtains often a qualitatively reasonable description (apart from wrong peak positions and sometimes also wrong peak heights) and reasonable static dielectric constants. This often justifies the physical interpretation of the Kohn-Sham band structure.

Since DFT is, however, a pure ground-state theory, there must be failures in the quantitative description of excitation processes, that is, the proper description of single-(quasi)particle band structures and optical spectra. The major weakness is that the Kohn-Sham band structure describes eigenvalue (and hence excitation energy) spectra which correspond to a *frozen* charge density. However, excitations also always introduce a *perturbation* of the system giving rise to *polarization* and *relaxation* effects, that is, changes in the electron density but also screening effects on (quasi)particle interactions. None of this is accounted for on the ground-state DFT level. Therefore, in order to remove the remaining deficiencies one has to account for quasiparticle effects within the framework of many-body perturbation theory, for example, within the so-called Hedin's GW approximation [8.37]. Going beyond a single-particle description, solving the Bethe-Salpeter equation (BSE) [8.38, 8.39], it is even possible to obtain a correct description of optical spectra with full inclusion of electron-hole Coulomb interactions, that is, of excitonic effects.

8.2.2 Quasiparticle band structure theory within the GW approach

A widely used tool is the many-body perturbation theory based on a Green's function approach [8.39, 8.40]. Since we have to describe polarization and relaxation effects due to the perturbation introduced by the excitation, it is natural to describe the response of the system by a Green's function G. The perturbation is basically a Coulomb interaction; however, since we perturb a polarizable system it is a screened interaction. The screening itself can be described by a dielectric function which itself can also be derived within the Green's function approach. The final outcome is a coupled set of fundamental equations first written down by Hedin [8.37, 8.40]. The most simple approximation within this system of equations is the so-called GW approximation. Here one introduces an exchange-correlation self-energy operator $\Sigma = iGW$, where G denotes the Green's function of the system and W the screened Coulomb interaction $\varepsilon^{-1}v$ derived from the bare Coulomb interaction v screened by the dielectric function ε which is wavevector- and frequency dependent. Formally, one has to solve the quasiparticle equation [8.13]

$$\left\{ -\frac{\hbar^2}{2m}\Delta_{\mathbf{r}} + v_{\text{ext}}(\mathbf{r}) + v_{\text{H}}(\mathbf{r}) \right\} \Phi_i^{\text{QP}}(\mathbf{r}) + \int d^3\mathbf{r}' \Sigma(\mathbf{r},\mathbf{r}';\varepsilon_i^{\text{QP}}) \Phi_i^{\text{QP}}(\mathbf{r}') = \varepsilon_i^{\text{QP}} \Phi_i^{\text{QP}}(\mathbf{r})$$

$$(8.3)$$

which has a similar structure to the Kohn-Sham equation of DFT (Eq. 8.1). The basic difference is that the exchange-correlation potential of DFT is now replaced by the non-Hermitian, non-local, energy-dependent self-energy operator Σ, resulting in quasiparticle eigenvalues $\varepsilon_i^{\text{QP}}$ and quasiparticle wave functions $\Phi_i^{\text{QP}}(\mathbf{r})$. Since the Hamiltonian becomes non-Hermitian, the QP eigenvalues are usually complex with the real parts describing resonance energies (excitation energies) and the imaginary parts describing lifetimes of excited states. In principle, also the quasiparticle equation has to be solved self-consistently since even in the most simple approximation, first of all, the Green's function depends on the quasiparticle wave functions and eigenvalues, and, secondly, also the dielectric function (and hence the screened Coulomb interaction) depends on the quasiparticle wave functions and eigenvalues if the dielectric response is given in an independent-(quasi)particle approximation [8.41].

The most common approach, however, avoids self-consistency. The main motivation for this step is the extreme numerical effort necessary to calculate the self-energy operator (and its matrix elements). Since empirically one often finds that the quasiparticle wave functions are similar to the Kohn-Sham wave functions, it is natural to take them as a starting point for the GW calculation and hence to calculate quasiparticle corrections in the sense of a perturbation (described by the difference between Σ and the DFT XC potential) treated non-self-consistently within first-order perturbation theory. One uses the approximation

$$\varepsilon_i^{\text{QP}} = \varepsilon_i + \langle \varphi_i | \Sigma(\varepsilon_i^{\text{QP}}) - v_{\text{XC}} | \varphi_i \rangle \tag{8.4}$$

inserting the DFT dielectric function ε into the random-phase approximation to obtain an approximate screened Coulomb interaction W_0, starting with the DFT Green's function G_0. This non-self-consistent approach usually works very well and gives very good results for the GW eigenvalues (band gaps, band widths, band dispersions) [8.13].

8.2.3   Improved starting point for GW

There are a number of systems where, within LDA or GGA, DFT results in a fundamentally wrong picture for the band structure (and hence also fundamentally wrong excitation properties). Unfortunately, InN is among these systems. Therefore, InN poses several challenges to theory which cannot be handled by standard tools employing a DFT starting point and (non-self-consistent) GW treatment only. This is clearly demonstrated in Fig. 8.2 showing QP gaps based on a DFT starting point with a semilocal GGA XC potential [8.42, 8.43] for a large variety of semiconductors and insulators. Despite the good overall success, for some systems, such as

InN or ZnO, we find disastrously bad results. This holds also for the energetical positions of shallow *d* core states, which are much too high.

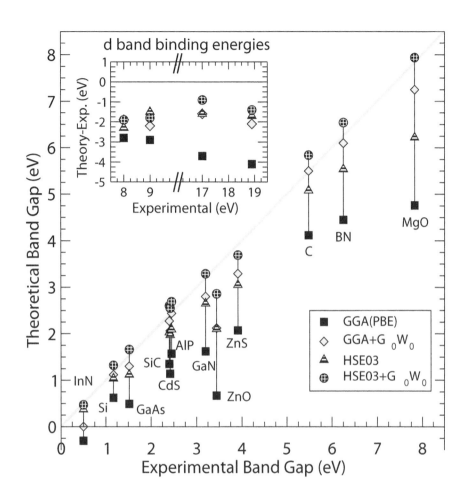

FIGURE 8.2

Computed band gaps vs. measured ones for several semiconductors and insulators [8.42]. In addition, the differences of experimental and theoretical binding energies of shallow *d* core electrons are given for ZnO, ZnS, InN, and GaAs. Besides the results derived from KS eigenvalues (Eq. 8.1) using a semilocal GGA XC potential or a non-local HSE03 XC potential, gaps based on QP eigenvalues (Eq. 8.4) are also presented.

The fundamental problem is the vanishing (or as many people like to say negative) gap of InN in DFT using LDA or GGA. The usual behavior for typical III-V compounds with a direct gap (for example, GaN but also most other III-V compounds) is to find a valence band maximum which is threefold degenerate (for zinc-blende) or split into twofold and non-degenerate levels (for wurtzite) at Γ, possessing predominantly *p* character, and a conduction band minimum which is non-degenerate at Γ, possessing predominantly *s* character. This *s*-like lowest con-

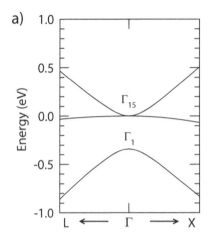 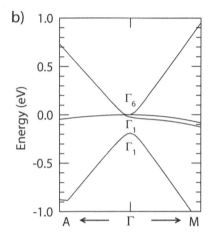

FIGURE 8.3

The band structure of *zb*-InN (a) and *w*-InN (b) around the Fermi level (used as energy zero) close to the Γ-point. They are computed within DFT-LDA [8.30] and treating the In 4*d* electrons as valence electrons.

duction band shows in most cases a very pronounced deep "valley" at Γ (that is, the minimum gap is much smaller than the "average gap"). Experimentally, InN of course shows the expected behavior. Qualitatively, also for InN one should obtain the common picture found for all direct-gap III-V compounds. In contrast to experiment, theory finds a zero-gap semiconductor with an "inverted" band structure that takes, not only an unrealistic, but rather weird form close to Γ. Without spin-orbit interaction, the level sequence is Γ_{6v}, Γ_{1v}, and Γ_{1c} for wurtzite (*w*) or Γ_{15v} and Γ_{1c} for zinc-blende (*zb*). The negative eigenvalue difference ("gap") $\Gamma_{1c} - \Gamma_{6v}$ ($\Gamma_{1c} - \Gamma_{15v}$) amounts to −0.21 eV (−0.36 eV) for the 2H (3C) polymorph [8.30]. Figure 8.3 illustrates this for the zinc-blende and wurtzite structure of InN. Since for zinc-blende InN one effectively finds a threefold degenerate state again (which may be interpreted as Γ_{15v} state), now with a non-degenerate state about 0.3 eV below (which may be interpreted as Γ_{1c} state), people often like to call it a negative gap. Looking at Fig. 8.3, however, we think "zero gap" semiconductor might be a better description of the situation (also for the wurtzite case).

The zero gap found in DFT-LDA or DFT-GGA, together with the inverted band structure, has dramatic consequences: Neither useful effective masses nor useful volume deformation potentials can be given around Γ. The zero gap gives rise to a diverging dielectric function. The completely wrong ordering of the bands around Γ also prohibits the simple use of the commonly applied non-self-consistent GW approach. Already the diverging dielectric function, resulting in an extreme overscreening of the screened Coulomb interaction *W*, makes such a GW approach useless. Therefore, one has to think about alternative approaches or even tricks to overcome this serious problem.

Another problem indirectly related to the band gap problem for InN is the very bad description of the In $4d$ bands. A common problem of DFT is that in particular shallow core levels (like In $4d$) are usually bound too weakly, that is, having a binding energy too low even for free atoms. As a result, the corresponding d-bands also occur at positions energetically much too high in the bulk band structure. For InN, for example, the In $4d$ related bands occur close to or even above the N $2s$ related bands although experiment clearly finds a position below the N $2s$ bands. As a further consequence, the d-bands also occur much closer to the p-like valence band maximum. Hence, the p-d repulsion effect is strongly overestimated, resulting in an even stronger gap-closing effect than in nature. So in fact, one has not only to solve the band gap problem, but also an In $4d$ band position problem because the exceptionally large p-d repulsion demands also a reasonable In $4d$ band position in order to obtain a reasonably good description of the real p-d repulsion.

The strong p-d repulsion pushing the p-like valence-band maximum (VBM) states toward the s-like conduction-band minimum (CBM) basically closes the gap. In a proper DFT ground-state calculation one has to treat the In $4d$ states as valence electrons. Freezing them into the core yields quite bad structural and thermodynamic properties because the strong polarization effects on the In $4d$ shell are not taken into account [8.17, 8.18]. However, from the viewpoint of a pure band structure calculation, freezing the In $4d$ electrons into the core may be a useful trick to open a finite gap because removing the In $4d$ states from the valence electrons implicitly also has the effect of "switching off" the p-d repulsion [8.20, 8.22]. Indeed, with the In $4d$ electrons frozen into the core (performing all calculations at the lattice constants obtained from a proper DFT calculation treating the In $4d$ as valence electrons) one finds positive gaps of 0.58 eV (0.43 eV) for the wurtzite (zinc-blende) polytype. Astonishingly, the resulting DFT band structures do not only offer a good starting point for a GW calculation but also exhibit qualitatively a good overall description of many band structure parameters [8.30]. One does not only obtain reasonable effective masses and deformation parameters, but also optical properties (dielectric functions) are described reasonably.

A further possible approach is the application of the so-called LDA+U (or GGA +U) method [8.44]. Using this approach also results in a downshift of the energetic position of the d-bands. For large enough on-site U parameters, even opening of the gap is possible [8.45]. However, it is completely unclear how to choose a proper U parameter. Since all interactions are screened in a solid, it is clear that the bare atomic U parameter should not be used. On the other hand, just scaling down the U parameter with the bulk (long-wavelength) dielectric constant may also not be a good approach because the rather strongly localized character of the core d states would require something as an effective dielectric constant for intermediate to larger wave vectors which is expected to be smaller than the bulk (long-wavelength) dielectric constant. A proper estimate of the necessary screening constant is, however, very complicated because it requires the calculation of the full wavevector-dependent dielectric function and afterwards a weighted averaging corresponding to the Fourier transforms of the d-states. Hence, the appropriate

U parameter is effectively unknown and should be something in between the bare atomic and the fully (bulk) screened parameter. Janotti *et al.* [8.45] estimated an effective on-site parameter $U=1.9$ eV by means of an atomic Coulomb interaction energy of 14.4 eV and an electronic dielectric constant $\varepsilon_\infty =7.6$. In their approach, this gives a gap opening from -0.18 eV to 0.03 eV for 2H-InN. A stronger gap opening requires yet larger values for U (for example, some of our unpublished test results show that values of about $U=6.7$ eV have a similar effect as freezing the In $4d$ into the core).

Of course, taking the band structure for InN with frozen In $4d$ as a starting point for GW (or even the BSE) still ends up in a disaster because we get reasonable quasiparticle shifts but start now from a much too large DFT-LDA or DFT-GGA gap. Depending on the U parameter this may also hold for the LDA+U or GGA+U starting point. In principle, one can perform a GW (and even BSE) calculation but one has afterwards to re-correct the result for the now completely neglected p-d repulsion effect. This can be done by adding the difference of a Kohn-Sham band structure obtained with frozen and unfrozen In $4d$ states (with an alignment procedure aligning the unshifted pure s states). Since we know that DFT-LDA or DFT-GGA with its bad In $4d$ band position overestimates the p-d repulsion one can even put an empirical correction [8.20, 8.22] to the resaturation of the p-d repulsion effect, taking only 85% to 90% of the calculated effect into account instead of the full effect (the ratio to be used is given by the ratio of the calculated distance of the In$4d$ bands from the VBM and the measured distance taken from experiment). Such a three-step approach (freezing the In $4d$ into the core, applying GW and BSE, and re-correcting for 85–90% of the p-d repulsion effect) has been applied with great success in a recent paper, basically reproducing all experimental findings for the band gaps and optical spectra [8.30].

Although the aforementioned three-step approach was very successful it is still an unsatisfactory approach since one has to leave the firm grounds of "*ab initio*" theory. But even LDA+U or GGA+U is also somehow empirical because we lack knowledge about an appropriate U value which in fact becomes then an empirical parameter as well. Therefore, alternative true *first-principles* approaches are necessary to fix all problems. Performing fully self-consistent GW calculations (maybe also including vertex corrections) is a real nightmare task! They yield a fundamental gap below 1 eV [8.46], but at extreme computational costs. A much more affordable approach is to replace the DFT-LDA or DFT-GGA starting point to solve the QP equation (Eqs. 8.3 and 8.4) by an electronic structure much closer to the desired QP result. There are two alternative approaches in the literature. One combines GW with exact-exchange DFT calculations in the optimized-effective potential framework [8.47]. The starting gaps are 1.0 eV (2H) and 0.81 eV (3C). With QP corrections gap values of 0.69 eV (2H) and 0.53 eV (3C) result [8.48]. However, the crystal-field (CF) splitting between Γ_{6v} and Γ_{1v} is, at 66 meV, much larger than experimental values of 19–24 meV [8.49]. Another GW treatment is based on so-called generalized Kohn-Sham schemes [8.50].

Such generalized Kohn-Sham schemes have in common that, motivated and guided by quantum chemistry (see, for example, Ref. [8.51]), non-local exchange correlation potentials involving partial or screened exact exchange are inserted into the Kohn-Sham equation (combined with local DFT potentials). These "hybrid" exchange-correlation energy functionals (and resulting potentials) have been mainly developed by researchers from the quantum chemistry community (see, for example, Ref. [8.42] and references therein). They have proven to be a great success in providing a highly improved description of thermodynamic data, in particular also in situations where bond breaking and bond formation plays a role, for example, the calculation of reaction barriers. Some of them are also successful to provide a much improved description of single-particle band structures. From the viewpoint of standard total-energy and force calculations they represent a useful tool somewhere in between DFT and Hartree-Fock (or even Hartree-Fock with Configuration Interactions). However, from the viewpoint of band structure calculations these "generalized Kohn-Sham schemes" can also be re-interpreted in a completely different way. Changing from a quantum chemist's to a physicist's view, one can also consider these schemes as very oversimplified GW schemes. Thereby the XC self-energy operator Σ in the QP equation (Eq. 8.3) is replaced by a spatially nonlocal XC potential.

GW in its most common formulation does not divide exchange-correlation simply into pure (exact) exchange plus correlation, but rather in a more clever way, into a screened exchange (SEX) part complemented by a Coulomb hole (COH) part and dynamic corrections [8.40]. This is the famous (dynamical) COHSEX approximation to the GW self-energy as, for example, introduced in the long write-up of Hybertsen and Louie [8.52]. The COH part can thereby be expressed by a purely local potential. The basic point is that partial or screened exchange is also one of the ingredients of the hybrid exchange-correlation functionals used in the generalized Kohn-Sham schemes. This partial or screened exchange is complemented by a DFT-like functional giving rise to a local DFT-like potential. Altogether it reflects the basic structure of a *static* COHSEX approach with a very crude model dielectric function (and missing dynamical screening effects). Of course, a priori the very approximate model-GW character [8.53] does not necessarily guarantee full success but it makes it very likely that such a scheme yields quite reasonable results which might be already very close to GW results. Since the eigenvalues and eigenfunctions resulting from the self-consistent solution of the generalized Kohn-Sham equation (Eq. 8.1, where v_{XC} is replaced by an integral operator) are close to the expected QP quantities in Eq. 8.3, the eigenvalues obtained by the simple perturbation-theory treatment, Eq. 8.4, should even tend toward results of self-consistent GW calculations [8.46].

In this spirit we have recently examined various such generalized Kohn-Sham (gKS) schemes as alternative starting points for a GW calculation (which is expected to yield only small additional corrections, implying that again a simple non-self-consistent "one-shot" GW approach is sufficient) – or for quick reference even as an approximate full replacement for GW calculations. Indeed, for several hybrid XC functionals, very promising results have been obtained, not only for InN, but also many other semiconductors and insulators [8.42]. Among the variety of non-local exchange functionals described in the literature we found the HSE03 functional [8.54] very useful and most accurate. The authors, Heyd, Scuseria, and Ernzerhof, combine parts of bare and screened exchange with an explicit local density functional. Figure 8.2 demonstrates the outstanding quality of such an approach. Taking HSE03 results as a starting point for GW yields excellent HSE03+G_0W_0 results also for very "difficult" materials such as InN or ZnO. This holds, by the way, not only for band gaps, but also for the energetical positions of shallow d-core states, which are highly improved as well. Therefore, in section 8.3 we will present (and compare to experiment) results obtained on the level of HSE03 plus GW only. These results should represent the best possible theoretical values that can be given.

8.2.4 Bethe-Salpeter equation and excitonic spectra

A computationally even more demanding and challenging task is the treatment of excitons and resulting optical spectra. In principle, the full excitonic problem requires the setting up and diagonalization of an electron-hole Hamiltonian $\hat{H}$. Within Hedin's GW approach and restricting to static screening this Hamiltonian reads, in matrix form, as [8.39, 8.55]

$$
\begin{aligned}
\hat{H}(vc\mathbf{k}, v'c'\mathbf{k}') =& \left[\varepsilon_c^{QP}(\mathbf{k}) - \varepsilon_v^{QP}(\mathbf{k}) \right] \delta_{vv'} \delta_{cc'} \delta_{\mathbf{k},\mathbf{k}'} \\
& - \int d^3\mathbf{r} \int d^3\mathbf{r}' \, \varphi_{c\mathbf{k}}^*(\mathbf{r}) \, \varphi_{c'\mathbf{k}'}(\mathbf{r}) \, W(\mathbf{r},\mathbf{r}') \, \varphi_{v\mathbf{k}}(\mathbf{r}') \, \varphi_{v'\mathbf{k}'}^*(\mathbf{r}') \\
& + 2 \int d^3\mathbf{r} \int d^3\mathbf{r}' \, \varphi_{c\mathbf{k}}^*(\mathbf{r}) \, \varphi_{v\mathbf{k}}(\mathbf{r}) \, \bar{v}(\mathbf{r},\mathbf{r}') \, \varphi_{c'\mathbf{k}'}(\mathbf{r}') \, \varphi_{v'\mathbf{k}'}^*(\mathbf{r}'),
\end{aligned}
\tag{8.5}
$$

where matrix elements between KS or gKS wave functions of conduction (c) band states and valence (v) band states occur. Contributions to Eq. 8.5 which correspond to antiresonant coupling or destroy particle number conservation have been omitted. The first term in Eq. 8.5 alone describes non-interacting electron-hole pairs in terms of single-quasiparticle eigenvalues that have to be taken from a GW calculation. It is corrected for the screened electron-hole Coulomb attraction described by the second term. The third contribution governed by the non-singular part of the bare Coulomb interaction $\bar{v}(\mathbf{r} - \mathbf{r}')$ (where the bar shall indicate that the Coulomb singularity, that is, the $\mathbf{G}=0$ component of its Fourier transform, has been removed), represents the electron-hole exchange in the case of singlet pairs [8.38] or the crystal local-field effects [8.56, 8.57]. After diagonalization of the matrix, using the eigenvalues E_Λ and eigenfunctions $A_\Lambda(vc\mathbf{k})$ of pair states Λ, one can obtain the frequency-dependent dielectric function as

$$\varepsilon_{\alpha\alpha}(\omega) = \delta_{\alpha\alpha} + \frac{16\pi e^2 \hbar^2}{V} \sum_{\Lambda} \left| \sum_{cv\mathbf{k}} \frac{\langle c\mathbf{k} | \mathrm{v}_\alpha | v\mathbf{k} \rangle}{\varepsilon_c(\mathbf{k}) - \varepsilon_v(\mathbf{k})} A_\Lambda(vc\mathbf{k}) \right|^2 \times \qquad (8.6)$$

$$\times \left[\frac{1}{E_\Lambda - \hbar(\omega + i\gamma)} + \frac{1}{E_\Lambda + \hbar(\omega + i\gamma)} \right],$$

where v_α is the corresponding Cartesian component of the single-particle velocity operator and γ the damping constant. V denotes the crystal volume. The details of the standard approach using the direct diagonalization of Eq. 8.5 have been discussed elsewhere [8.58, 8.59].

Since the rank of the Hamiltonian matrix in Eq. 8.5 given by the number of valence bands times the number of conduction bands times the number of $\mathbf{k}$ points square (with a rather bad convergence with respect to the number of $\mathbf{k}$ points) is extremely large, a straightforward diagonalization of this matrix [8.58, 8.59] is often not possible due to excessive CPU and memory requirements. Therefore, sometimes the Haydock recursion method is applied [8.60]. Alternatively, Hahn *et al.* have formulated a new approach [8.55, 8.61] based on the calculation of the time evolution of the excitonic states followed by a Fourier transform on time domain. This approach is only able to deliver optical spectra directly, but not exciton eigenvalues or wave functions. However, it has the big advantage that it never requires the explicit diagonalization of the Hamiltonian matrix because it needs only to evaluate the exciton Hamiltonian matrix times the current wave function for every time step. This results in massive CPU-time savings and a greatly improved scaling of the calculational effort with system size. Therefore, we mainly present results based on this time-evolution approach. Unfortunately, the computation of the wave functions and eigenvalues within the HSE03 gKS scheme at a minimum of 4204 $\mathbf{k}$ points in the BZ of the wurtzite structure required for reasonable convergence of the results is impossible for CPU-time reasons. Therefore, in section 8.4 we use the three-step procedure described above [8.30]. It is based on Kohn-Sham results obtained within the DFT-LDA framework and freezing the In $4d$ electrons into the core. Resulting DFT-LDA and GW-QP eigenvalues have been re-corrected for p-d repulsion effects, adding the differences between eigenvalues computed with and without taking the In $4d$ electrons into account multiplied with a certain scaling factor.

8.3 Quasiparticle bands and density of states

8.3.1 Band structures

In Fig. 8.4, we present the quasiparticle band structures for the wurtzite and zincblende polytypes of InN calculated within the HSE03+G_0W_0 approach without spin-orbit coupling. Figure 8.5 shows some more details in a narrow energy range around the fundamental gap. In order to complement the band structure plots we have also listed the quasiparticle eigenvalues at different high-symmetry **k** points in the Brillouin zone for both polytypes in Tables 8.1 and 8.2.

The band structures of the valence electrons can be divided into three main regions. From the global band structure picture (Fig. 8.4), we see that the In $4d$-derived bands still penetrate the bottom of the N $2s$-derived valence bands. This is in contrast to the separation believed from early experimental data [8.62], but in agreement with recent results from x-ray photoemission spectroscopy [8.63]. Nevertheless, the accurate description of the localized In $4d$ and N $2s$ orbitals is a challenge using a pseudopotential approach. The variety of bands can be best explained in the zinc-blende case (Fig. 8.4(b)). At **k** points away from Γ, the In $4d$-orbital and N $2s$-orbital symmetries both support the irreducible representation a_1, creating an a_1-a_1 splitting by an s-d repulsion [8.64]. Due to the proximity of the corresponding orbital energies (cf. Fig. 8.1), the s-d coupling is strong and the s-like valence band splits into an upper (a_1^u) and a lower (a_1^l) band with the In $4d$ bands in between [8.64]. In the case of the wurtzite structure (Fig. 8.4 (a)) the low-lying energy bands are additionally influenced by a strong crystal-field splitting of more than 3 eV.

The lower bands are separated from the uppermost three (zb) or six (w) valence bands in the range from 0 to -6 eV by the so-called ionicity gap [8.65]. InN has a large ionicity gap of about 7 eV due to the large electronegativity difference of about 1.26 for In and N [8.31]. Most interesting are the bands near Γ which are strongly influenced by a nonparabolicity and, in the wurtzite case (Fig. 8.5(b)), by a crystal-field splitting of roughly $\Delta_{cf} = 20$ meV between the uppermost twofold-degenerate Γ_{6v} and the lower Γ_{1v} state. In contrast to earlier studies [8.30], we find a positive splitting. This is traced back to the three-step procedure used in Ref. [8.30] and the fact that also p-d repulsion influences the sign of the crystal-field splitting. The computed HSE03+G_0W_0 value $\Delta_{cf} = 22.9$ meV approaches the experimental one which varies in the range $\Delta_{cf} = 19$–23 meV [8.49]. The influence of the spin-orbit coupling on the VBM is small due to compensating effects between p and d electrons [8.66].

The lowest conduction bands exhibit an extremely large dispersion near the Brillouin zone (BZ) center Γ (Figs. 8.4 and 8.5). As a consequence, extremely small direct InN gaps $E_g(\Gamma_{1c} - \Gamma_{6v}) = 0.71$ eV (wurtzite) or $E_g(\Gamma_{1c} - \Gamma_{15v}) = 0.47$ eV (zinc-blende) result. Similar to other compounds with first-row elements, like GaN or ZnO, the cubic polytype possesses a gap which is about 0.2–0.3 eV smaller than that of the hexagonal modification. Interestingly, the quasiparticle opening of the band gaps is relatively small. The InN gap values of the gKS theory using the HSE03 approach [8.42] amount to $E_g(\Gamma_{1c} - \Gamma_{6v}) = 0.59$ eV (wurtzite) and $E_g(\Gamma_{1c} - \Gamma_{15v}) = 0.37$ eV (zinc-blende). The HSE03+G_0W_0 gaps approach recent measured values of 0.70 ± 0.05 eV for wurtzite [8.67] and 0.6 eV for zinc-blende [8.68, 8.69] InN. The CBM at Γ is much lower in energy than the conduction-band

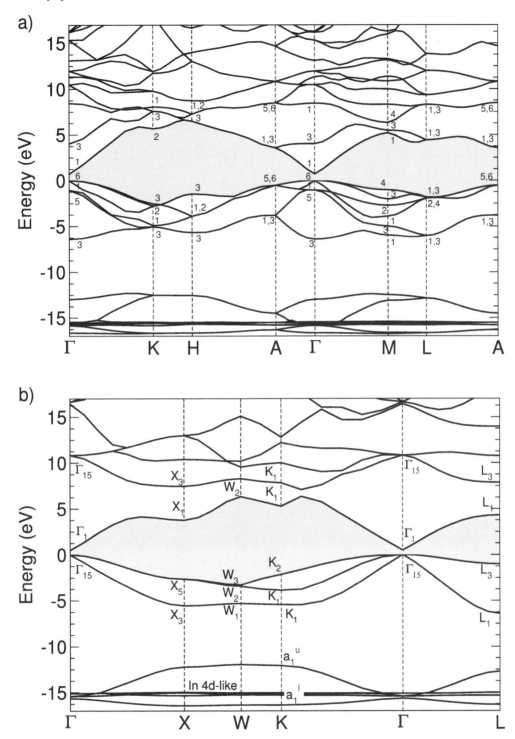

FIGURE 8.4

Quasiparticle band structure of *w*-InN (a) and *zb*-InN (b) calculated within the HSE03+G_0W_0 approach. The valence band maximum Γ_{6v} (*w*) or Γ_{15v} (*zb*) is taken as energy zero. The shaded area indicates the fundamental gap region.

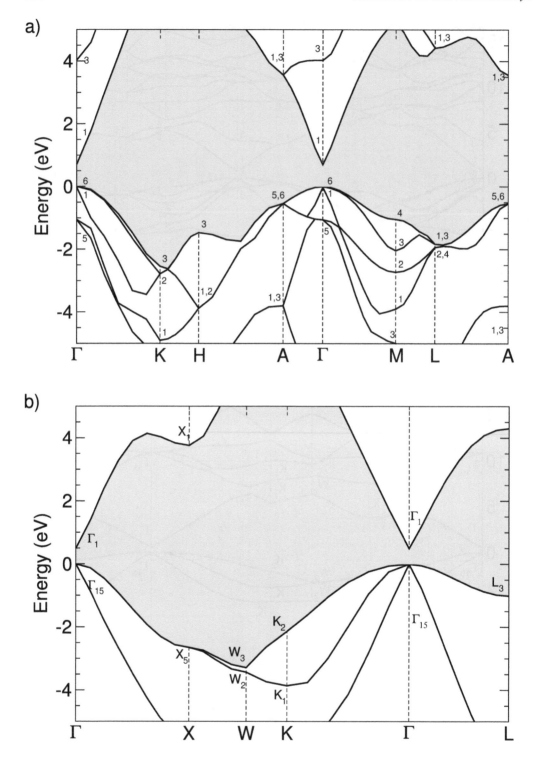

FIGURE 8.5

Quasiparticle band structure of *w*-InN (a) and *zb*-InN (b) from Fig. 8.4 in a narrow energy region around the fundamental gap (shaded area).

TABLE 8.1

Quasiparticle eigenvalues of the lowest eight conduction ($c1$–$c8$) and highest eight valence ($v1$–$v8$) bands of w-InN at high-symmetry points in the hexagonal BZ. All values are in eV.

Band	$v8$	$v7$	$v6$	$v5$	$v4$	$v3$	$v2$	$v1$
Γ	−15.46	−12.97	−6.41	−1.05	−1.05	−0.02	0.00	0.00
K	−12.52	−12.52	−5.05	−5.05	−4.90	−2.78	−2.78	−2.54
M	−13.10	−12.48	−5.97	−4.99	−3.90	−2.73	−2.04	−1.04
A	−14.52	−14.52	−3.80	−3.80	−0.54	−0.54	−0.54	−0.54
H	−12.54	−12.54	−5.67	−5.67	−3.88	−3.88	−1.47	−1.40
L	−12.85	−12.85	−6.04	−6.04	−1.94	−1.94	−1.83	−1.83

Band	$c1$	$c2$	$c3$	$c4$	$c5$	$c6$	$c7$	$c8$
Γ	0.71	4.03	8.43	10.47	10.47	11.34	11.96	11.96
K	5.68	7.58	7.58	8.04	9.78	9.78	11.73	11.93
M	5.20	5.46	6.32	8.30	10.00	10.86	11.37	13.17
A	3.56	3.56	8.48	8.48	10.82	10.82	10.82	10.82
H	6.49	6.49	7.26	7.26	8.72	8.73	13.03	13.03
L	4.42	4.42	8.29	8.29	9.35	9.35	12.00	12.00

TABLE 8.2

Quasiparticle eigenvalues of the lowest eight conduction ($c1$–$c8$) and highest eight valence ($v1$–$v8$) bands of zb-InN at high-symmetry points in the cubic BZ. All values are in eV.

Band	$v8$	$v7$	$v6$	$v5$	$v4$	$v3$	$v2$	$v1$
Γ	−15.42	−15.42	−15.42	−15.11	−15.11	0.00	0.00	0.00
X	−15.35	−15.04	−15.04	−14.99	−12.28	−5.53	−2.64	−2.64
L	−15.34	−15.34	−15.02	−15.02	−12.75	−6.40	−1.01	−1.01
K	−15.30	−15.12	−15.14	−15.04	−12.12	−5.41	−3.86	−2.13
W	−15.27	−15.24	−15.07	−15.04	−12.00	−5.28	−3.43	−3.29

Band	$c1$	$c2$	$c3$	$c4$	$c5$	$c6$	$c7$	$c8$
Γ	0.47	10.83	10.83	10.83	16.43	16.62	16.62	16.62
X	3.72	7.48	10.41	13.03	13.03	19.02	20.78	23.21
L	4.30	7.88	10.75	10.75	13.98	18.16	18.16	21.37
K	5.27	7.86	10.05	12.25	12.82	19.72	20.72	21.59
W	6.33	8.36	9.54	10.21	15.13	17.41	20.70	21.62

edge at other points in **k**-space. The next band minima are more than 3 eV higher in energy. The strong conduction-band dispersion has a consequence for the branch point energy or the charge neutrality level E_B [8.70, 8.71]. It lies within the conduction band and not in the fundamental gap. The branch point energy is defined as the average midgap energy across the entire BZ [8.71]. It can be determined by calculating the half-way point between the mean value of the lowest conduction band and the mean value of the highest valence band. Using the band structures in Fig. 8.4, such a calculation yields a branch point energy E_B that lies about 1 eV above the CBM in agreement with other estimates [8.72]. Measurements probing the conduction band electron plasma confirm the existence of electron accumulation at InN surfaces, with a surface Fermi level location near 1.58 ± 0.10 eV above the VBM [8.73].

8.3.2 Effective masses

Having a closer look at the band structure data for the highest valence bands of zinc-blende InN (Fig. 8.5 (b)) and the lowest conduction band, one finds a very strong wavevector dispersion of the lowest conduction band around Γ, along with a pronounced nonparabolicity which can also be found in one of the three valence bands. These findings can be traced back to a strong coupling of valence and conduction bands. A simple description can be obtained using Kane's **k·p** four-band model [8.74], neglecting for simplicity the crystal-field splitting in the wurtzite case and also the spin-orbit splitting which are both of the order of a few meV only. We present in the following the theory for the zinc-blende polytype, but it is approximately also valid for the wurtzite case, since three of the six uppermost valence bands do not couple to the lowest conduction band for symmetry reasons. Introducing the effective coupling constant $P = \frac{\hbar}{im} \langle \Gamma_{1c} | \hat{p}_i | \Gamma_{15v,i} \rangle$ between the s-like Γ_{1c} conduction band state at Γ and the three p_x-, p_y-, and p_z-like $\Gamma_{15v,i}$ states at Γ and the quantity $E_k = \frac{\hbar^2 k^2}{2m}$, one can write a four-band **k·p** Hamiltonian of the form [8.74]

$$\hat{H}_{4\times4} = \begin{pmatrix} E_g + E_k & iPk_x & iPk_y & iPk_z \\ -iPk_x & E_k & 0 & 0 \\ -iPk_y & 0 & E_k & 0 \\ -iPk_z & 0 & 0 & E_k \end{pmatrix}. \tag{8.7}$$

With $E_p = \frac{2m}{\hbar^2} P^2$, this results in Bloch bands

$$\varepsilon_{c,v_2}(\mathbf{k}) = \frac{1}{2}\left(E_g \pm \sqrt{E_g^2 + 4E_p E_k}\right) + E_k, \tag{8.8}$$

$$\varepsilon_{v_1,v_3}(\mathbf{k}) = -E_k.$$

These bands are isotropic but two of them, ε_c and ε_{v_2}, exhibit a strong nonparabolicity. This nonparabolicity makes it difficult to measure a well-defined effective mass because it is expected to depend on the position of the Fermi level, that is, on the concentration of free electrons. Anyway, for the bottom of the conduction band at Γ, an effective electron-mass

$$\frac{m^*}{m_0} = \frac{1}{1 + E_p/E_g} \tag{8.9}$$

can be defined. With a gap value of $E_g = 0.7$ eV and typical coupling strengths $E_p = 10...15$ eV one derives electron masses of $m^* = 0.045...0.065 m_0$. For larger $\mathbf{k}$ values the $\mathbf{k}$-dependent terms in the conduction band (Eq. 8.8) become more important and the effective mass increases. For wurtzite InN with $E_k = 1$ eV the electron mass increases to values close to the free electron mass. For that reason one may state good agreement with experimental values of $0.05 m_0$ [8.75], $0.07 m_0$ [8.76], or $0.085 m_0$ [8.77] determined for w-InN samples containing free electrons.

8.3.3 Density of states

In Fig. 8.6 we show the total and s-, p-, and d-projected density of states (DOS) of the wurtzite and zinc-blende polymorphs. In all figures the valence band maximum has been taken as the energy zero. In the energy region from -17 to -12 eV of the N $2s$ and In $4d$ bands the influence of the polytype is negligible. This is due to the localization of the corresponding atomic orbitals and the similar tetrahedrally bonded configurations which only differ for distances larger than the third nearest neighbors. The central peak is related to the In $4d$ electrons. Also a low-energy peak is visible related to the N $2s$ states hybridized with In $4d$ ones, that is, the a_1^l band. The s-d hybridization leads to a rather extended a_1^u band in Fig. 8.4 and, hence, to a broad peak in the DOS corresponding to this band.

The DOS in the energy region from -6.5 to 0 eV of the uppermost valence bands in Fig. 8.6 is dominated by p states. Substantial s contributions are only visible for the lower DOS peak. The d contributions due to the p-d repulsion are weak. The shape of the uppermost part depends very much on the interaction of the third nearest or more distant neighbors. For zb-InN (Fig. 8.6(b)), due to the relatively flat nature of the Γ_{15v} bands, the DOS rises rapidly below the VBM, peaking around the critical point L_3. However, between this point and X_5, the bands vary almost linearly in $\mathbf{k}$ (see Fig. 8.4 (b)), leading to a rather constant DOS, that is, a plateau. The following small peak arises from the turning points in the band structure around K_1. In the wurtzite case (Fig. 8.6(a)), the onset of the DOS corresponds to Γ_{6v} bands followed by the Γ_{1v} one. Instead of rising to a plateau, the DOS continues to rise rapidly due to a number of turning points in the band structure (see Fig. 8.4 (a)), such as Γ_5, A_6, or M_4. The lowest predominant peak in the DOS occurs largely due to the critical points at H_3 and in the Σ (Γ-M) direction between Γ_3 and M_1 (for wurtzite) or the almost degenerate critical points X_3, W_1, and K_1 (for zinc-blende). The comparison of the broadened DOS with results of x-ray photoemission measurements in Fig. 8.7 [8.63] confirms the characteristic features discussed above. This concerns the peak structure and the differences between the spectra for w- and zb-InN. The only discrepancy between the calculated DOS and the experimental

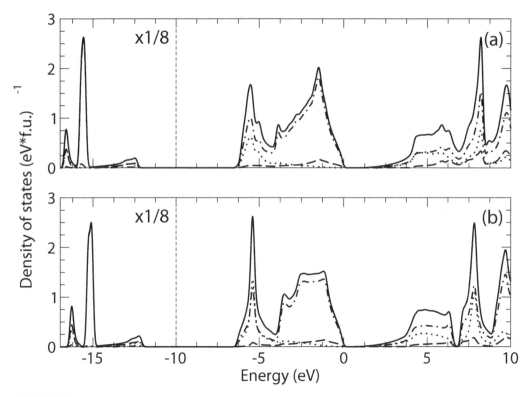

FIGURE 8.6

Quasiparticle density of states per formula unit (f.u.) for *w*-InN (a) and *zb*-InN (b). The VBM is used as the energy zero. Besides the total DOS (solid line), the partial densities of states resolved with respect to *s* (dotted lines), *p* (dot-dashed lines), or *d* (dashed lines) symmetry are shown.

spectra is that the low-energy peak occurs at slightly lower binding energies in the experimental spectra.

The DOS in Fig. 8.6 in the energy region of the conduction bands up to 8 eV is equally dominated by *s* and *p* states. However, there are also contributions from *d* states, at least for the higher energies. Most important for several properties is the lower part of the usually unoccupied DOS which is rather similar for *w*- and *zb*-InN. A common characteristic is the smooth rather non-square-root-like onset of the density of states for the conduction bands resulting from the strongly dispersive and highly nonparabolic lowest conduction band near Γ. It can be understood in terms of the nonparabolic band structure. According to the **k·p** eigenvalues (Eq. 8.8) one computes a DOS $D(\varepsilon)$ close to the onset of the conduction bands at energies ε near the gap value E_g to be ($\varepsilon > E_g$)

$$D(\varepsilon) \propto \frac{f(\varepsilon) - E_p}{f(\varepsilon) + E_p} \sqrt{2\left(\varepsilon - E_g\right) + \left(E_g + E_p\right) - f(\varepsilon)}, \qquad (8.10)$$

$$f(\varepsilon) = \sqrt{\left(E_g + E_p\right)^2 + 4E_p\left(\varepsilon - E_g\right)}.$$

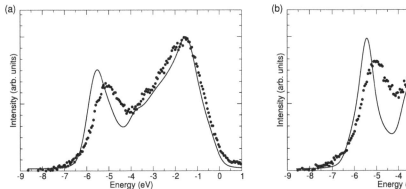

FIGURE 8.7

Calculated DOS with an additional Lorentzian broadening of 0.2 eV (solid line) to account for lifetime and 0.45 eV Gaussian broadening to account for the instrumental resolution and x-ray photoemission spectra (points) for w-InN (a) and zb-InN (b) [8.63]. The photoemission spectra have been shifted to align the VBM at 0 eV, as for the calculations.

For $\varepsilon \to E_g$ (very close to the CBM) it still shows the well-known square-root behavior $\propto (\varepsilon - E_g)^{1/2}$. However, for larger energies $E_g < \varepsilon - E_g \ll E_p$ the DOS increases more strongly $\propto (\varepsilon - E_g)^{3/2}$. Only at energies around 4 eV, a stronger contribution of the lowest conduction band from other **k**-space regions of the Brillouin zone with weaker dispersions emerges, resulting in a broad and initially almost plateau-like structure. It is followed by stronger peaks at higher energies (8 eV and beyond) resulting from critial points or weakly dispersive parts of the higher conduction bands.

The density of states of the conduction bands is hardly directly accessible to experimental spectroscopies. Figure 8.8 shows x-ray absorption spectra (XAS) of w-InN recorded in the total electron yield (TEY) mode for two different incident angles, 20° and 70°, with respect to the c-axis of the hexagonal crystal [8.78]. For light polarization **e** perpendicular to the c-axis (with the c-axis parallel to the z-axis) the optical transitions from the initial N $1s$ states into the N $2p_z$ states are dipole forbidden. Likewise, for light polarization **e** parallel to the c-axis (z-axis), transitions from N $1s$ initial states into the N $2p_{x,y}$ states are forbidden. Assuming that excitonic effects do not change the spectra, the experimental spectra are compared with the partial densities of states projected onto the N $2p_z$ and N $2p_{x,y}$ states. In order to account for the selection rules and the chosen incidence angles (different from 0° and 90°), we compare explicitly to weighted linear combinations of projected densities of states corresponding to $\frac{2}{9}$N $2p_{x,y} + \frac{7}{9}$N $2p_z$ (Fig. 8.8(a)) and $\frac{7}{9}$N $2p_{x,y} + \frac{2}{9}$N $2p_z$ (Fig. 8.8(b)). The XAS spectrum recorded at an incidence angle of 20° is dominated by two peaks at approximately 8 and 10 eV above the VBM. If the angle of incidence is changed to 70° the intensity of the 8 eV feature increases dramatically while that of the 10 eV peak is reduced. The characteristic peak behavior is also represented in the calculated linear combinations of N $2p_z$ and N $2p_{x,y}$ con-

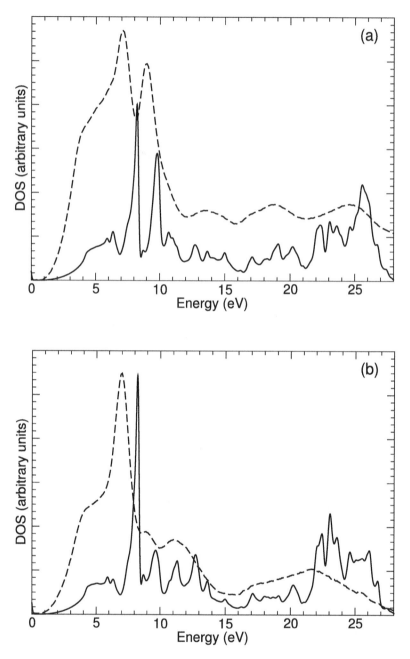

FIGURE 8.8

X-ray absorption spectra (dashed line) taken in the total electron yield mode for the N K-edge and incidence angles of (a) 20° and (b) 70° with respect to the c-axis of w-InN [8.78]. They are compared with calculated unoccupied partial densities of states (solid line) projected onto N $2p_z$ states or N $2p_{x,y}$ states. Accounting best for the selection rules of N $1s \rightarrow$ N $2p$ transitions for different light polarization **e** and for the used incidence angles of 20° and 70°, we display calculated spectra corresponding to weighted projections (a) $\frac{2}{9}N2p_{x,y} + \frac{7}{9}N2p_z$ and (b) $\frac{2}{9}N2p_z + \frac{7}{9}N2p_{x,y}$. The theoretical spectra have been reasonably broadened to account for lifetime and instrumental broadening. A common energy zero at the VBM has been used.

tributions to the unoccupied partial densities of states shown in Fig. 8.8. However, there is an almost rigid shift of about 1.3 eV of the calculated peaks toward higher energies. This value may be interpreted as the binding energy of the N $1s$-N $2p$ core-conduction-band excitons not considered in the calculations.

8.4 Optical properties and excitons

8.4.1 Overall spectra: Excitonic effects and critical points

Figures 8.9 and 8.10 show the imaginary parts of the macroscopic optical dielectric function (Eq. 8.6) including excitonic and quasiparticle effects (Eq. 8.5). In the wurtzite case we have to discuss two components, the ordinary and extraordinary dielectric functions for light polarization parallel and perpendicular to the hexagonal c-axis, and hence a resulting anisotropy of the spectra. In the figures we also show experimental data of Goldhahn *et al.* [8.66, 8.67, 8.77–8.79] derived from spectroscopic ellipsometry. The overall agreement between theoretical and experimental data is excellent. Some deviations, in particular in the plateau-like region between 1 and 4 eV, have to be traced back to numerical problems in the description of such a plateau. Finite **k**-point samplings and broadenings lead to artificial oscillatory features. Test calculations on a lower level of approximations for the quasiparticle effects and the electron-hole interaction, but with an extremely fine **k**-point sampling and using the tetrahedron integration method versus spectra with fewer **k** points and using a smearing method, instead showed a much better representation of the plateau for the numerically improved computation. Unfortunately, the extreme effort necessary to treat excitons does not allow for arbitrary dense **k**-point samplings and in addition the time-evolution scheme also imposes a finite smearing on the spectra due to the finite simulation time (even if a tetrahedron integration would be used). Anyway, such small numerical deviations do not alter the overall picture and do not affect the other energy regions. The deviations in the absolute values of the imaginary parts may be traced back to several reasons. For both the computational treatment and the measurement, the smallness of the dielectric function may induce an accuracy problem. For example, in the case of silicon, the imaginary part of the dielectric function is larger by a factor of about 10 (theory) or 5 (experiment) [8.55].

In both polytypes one can find a rather sharp absorption onset at the gap energy followed by a plateau with practically constant imaginary part of the dielectric function up to about 4 eV. In this energy region, only transitions from the highest three valence bands into the lowest conduction band contribute to the dielectric function. For photon energies above 4.5 eV, also transitions into the higher conduction bands occur. Similar to the density of states, one also finds some pronounced peaks around 5, 6, and 9 eV. They can be partly attributed to interband transitions

TABLE 8.3

Positions of the main high-energy spectral features in the imaginary parts of the dielectric function obtained from spectroscopic ellipsometry [8.68, 8.81] and spectra calculations [8.30] for (a) w-InN (ordinary), (b) w-InN (extraordinary), and (c) zb-InN from Figures 8.9 and 8.10. In addition, also peak positions Theor. (QP) in the independent-quasiparticle spectra without excitonic effects shown in Figs. 8.11 and 8.12 are presented. For the purpose of comparison differences of interband quasiparticle transition energies [8.30] have also been given. All values are in eV.

System	Spectral feature	Exp.	Theor.	Theor. (QP)	QP difference
(a)	shoulder	4.88	4.7	5.0	4.10 ($A_{5/6} \rightarrow A_{1/3}$)
w-InN					5.05 ($\Gamma_5 \rightarrow \Gamma_3$)
(ord.)	peak	5.35	5.5	6.0	6.21 ($M_4 \rightarrow M_1$)
					6.39 ($L_{2/4} \rightarrow L_{1/3}$)
	shoulder	6.05	6.1	6.8	6.63 ($M_4 \rightarrow M_3$)
					6.78 ($\Gamma_3 \rightarrow \Gamma_1$)
	peak	-	7.2	7.3	7.14 ($A_3 \rightarrow A_1$)
	peak	7.87	7.8	8.1	8.09 ($M_2 \rightarrow M_3$)
	peak	8.60	8.7	9.2	8.53 ($K_{1/3} \rightarrow K_2$)
					8.91 ($H_{1/2} \rightarrow H_3$)
					9.16 ($A_{1/3} \rightarrow A_{1/3}$)
(b)	shoulder	-	4.8	4.9	(high-symmetry lines?)
w-InN	peak	5.38	5.6	6.2	6.36 ($L_{2/4} \rightarrow L_{1/3}$)
(ext.)	shoulder	-	6.7	6.8	7.06 ($M_3 \rightarrow M_1$)
					7.14 ($A_{1/3} \rightarrow A_{1/3}$)
	peak	7.63	7.6	7.6	8.01 ($H_3 \rightarrow H_{1/2}$)
			7.9	8.2	8.28 $K_3 \rightarrow K_2$)
	peak	9.44	9.4	9.7	9.86 ($L_{1/3} \rightarrow L_{2/4}$)
					9.89 ($K_{1/3} \rightarrow K_{1/3}$)
(c)	peak	5.09	5.1	5.6	-
zb-InN	shoulder	-	5.5	6.1	-
	peak	6.14	6.25	6.6	6.40 (E_2: $X_5 \rightarrow X_1$)
					6.42 (E_1: $L_3 \rightarrow L_1$)
	shoulder	-	7.1	7.4	-
	shoulder	7.38	7.75	7.9	7.50 ($K_2 \rightarrow K_1$)
	shoulder	-	8.7	8.9	9.01 ($L_3 \rightarrow L_1$)
					9.05 ($K_1 \rightarrow K_1$)
					9.22 ($X_3 \rightarrow X_1$)
	peak	9.28	9.25	9.7	9.43 ($W_3 \rightarrow W_2$)
					9.68 ($K_2 \rightarrow K_1$)
					9.72 ($W_2 \rightarrow W_2$)
					9.82 ($X_5 \rightarrow X_3$)

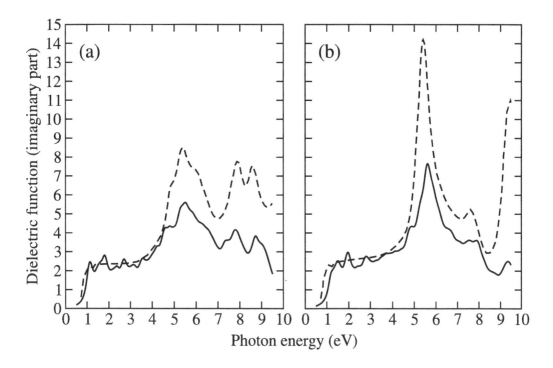

FIGURE 8.9

Imaginary parts of the dielectric function of *w*-InN calculated for Coulomb-correlated electron-hole pairs (solid lines) [8.30] and measured by means of spectroscopic ellipsometry at room temperature (dashed lines) [8.79, 8.80]. Results for the (a) ordinary and (b) extraordinary light polarization are presented. A broadening of 0.2 eV is used in the theoretical spectra.

at critical points. Therefore, we give in Table 8.3 a list of quasiparticle transition energies at high-symmetry points of the Brillouin zone for both polytypes, which have been derived from the three-step procedure described in section 8.2, along with the denotation to which band pair each transition belongs. One can see some approximate correspondence between these values and the peak structures of the dielectric function. However, there are also discrepancies due to the weights by the dipole matrix elements as well as a general redshift due to electron-hole attraction. For example, one expects that the double-peak structure in the absorption spectra of the cubic polytype (Fig. 8.10) may be essentially related to the E_1 ($L_3 \rightarrow L_1$) and E_2 ($X_5 \rightarrow X_1$) transitions (for denotation see Ref. [8.82]) which always give rise to pronounced absorption peaks for tetrahedrally coordinated semiconductors crystallizing in zinc-blende structure [8.82]. The HSE03+G_0W_0 eigenvalues in Table 8.2 with differences $\hbar\omega = 5.31$ eV (E_1) and $\hbar\omega = 6.36$ eV (E_2) confirm this picture. Unfortunately, according to the identification of the optical transition energies in Table 8.3(c), this is not the case using the quasiparticle band structure obtained within the three-step procedure [8.30]. The inclusion of an almost rigid *p-d* repulsion overestimates the energy distance of (especially) the L_1 conduction band with respect to the VBM. Using this treatment of the *p-d* repulsion, the E_1 and E_2 van

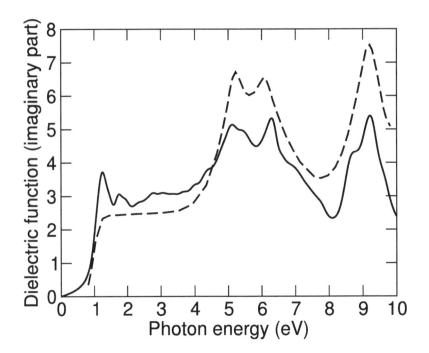

FIGURE 8.10

Imaginary parts of the dielectric function of zb-InN calculated for Coulomb-correlated electron-hole pairs (solid line) and measured by means of spectroscopic ellipsometry (dashed line) [8.68, 8.81]. A broadening of 0.2 eV is used in the theoretical spectrum.

Hove singularities fall together and essentially form the high-energy peak above 6 eV. Nevertheless, not only high-symmetry points, but in general whole **k**-space regions where bands with weaker dispersion occur contribute to the peak structures. This holds, for example, for the first rather strong and broad peak. The peak somewhat above $\hbar\omega = 5$ eV cannot be related to a transition at a high-symmetry point within the band structure resulting within the three-step procedure. However, the band structure plotted in Figure 8.5(b) indicates possible strong contributions to the joint density of states from **k**-space regions around 0.6 ΓX and 0.7 ΓL. Therefore, a classical assignment of peak structures in the dielectric function to critical points alone is not really allowed or possible for InN. Table 8.3 can only provide a guide for which transitions should result from which **k**-point region. More important is the excellent agreement found for measured (Exp.) and calculated (Theor.) peak positions in Table 8.3. There are only deviations of the peak positions in the range 0.0 to 0.2 eV. Also the relative intensities and hence the identification of the spectral features as peaks and shoulders in Figs. 8.9 and 8.10 indicate the excellence of the theoretical description of the optical properties of both bulk InN polytypes, taking into account the necessary many-body effects, such as quasiparticle shifts and electron-hole attraction.

The mentioned discrepancies between peak positions in both theoretical and ex-

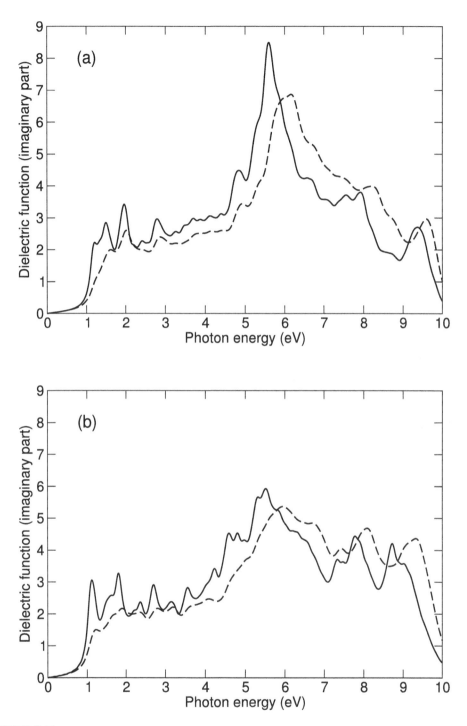

FIGURE 8.11

Imaginary parts of the dielectric function of *w*-InN computed with Coulomb attraction and local fields (solid lines, as in Fig. 8.9) as well as without excitonic effects (dashed lines), that is, for uncorrelated quasielectrons and quasiholes. A broadening of 0.2 eV is used in the spectra for ordinary (a) and extraordinary (b) light polarization.

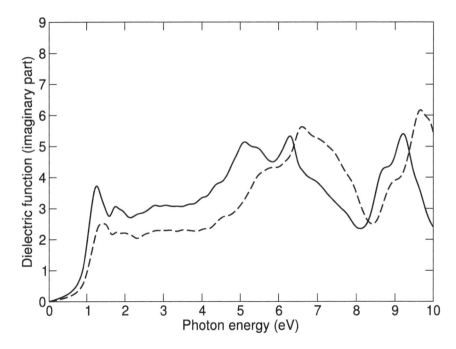

FIGURE 8.12

Imaginary part of the dielectric function of *zb*-InN with (solid line, as in Fig. 8.10) and without (dashed line) excitonic effects. An artificial broadening of 0.2 eV is applied.

perimental absorption spectra in Figures 8.9 and 8.10 and excitation energies at critical points derived from the quasiparticle band structure due to excitonic effects are more clearly demonstrated in Figures 8.11 and 8.12. Indeed, the dashed lines representing uncorrelated electron-hole pairs exhibit maxima and other spectral features at energies which correspond to critical points in the joint density of states (see Table 8.3). At a first glance one expects that these structures can be directly identified with features in the experimental spectra in Figs. 8.9 and 8.10. However, this is obviously not the case for photon energies $\hbar\omega \gtrsim 5$ eV. The comparison with spectra that include Coulomb attraction and local fields (electron-hole exchange) in Figs. 8.11 and 8.12 show drastic excitonic effects which violate the simplified interpretation of optical spectra in terms of critical points [8.82]. In the energy range of the high-energy critical points, the excitonic effects lead to a significant red shift of about 0.5 eV. In addition, a drastic redistribution of spectral strength occurs from higher toward lower energies. This is clearly visible from the increase of the central peak around $\hbar\omega = 5.5$–5.6 eV in the absorption spectra of not only *w*-InN but also of *zb*-InN. Additionally, several shoulders become much more pronounced. Consequently, we state the presence of significant excitonic effects also for higher electron-hole pair energies despite the huge energy distance between electrons and holes and although these effects do not lead to bound excitons or at least bound excitons belonging to high-energy critical points which are resonant with electron-hole continua from other transitions. The main reason for the observed effect is the mixing of different electron-hole pairs $vc\mathbf{k}$ in a defined pair state Λ by the electron-hole pair amplitude $A_\Lambda(vc\mathbf{k})$ in the macroscopic dielectric function $\varepsilon(\omega)$ (Eq. 8.6).

8.4.2 Plateau-like region

The most interesting behavior in the lineshape is surely the onset and the plateau of the imaginary part of the dielectric function. Therefore, we have to discuss how a detailed analysis of the dielectric function can yield a lot of valuable information and how one can understand the low-energy part of the absorption spectra. As already discussed for the band structure, a simple four-band $\mathbf{k} \cdot \mathbf{p}$ Hamiltonian (Eq. 8.7) can explain the band structure near the gap and around Γ. We could derive analytical expressions for the band dispersions (Eq. 8.8) based on two fundamental parameters, E_g and E_p. From the band dispersions it is of course immediately possible to derive the transition energies. But from $\mathbf{k} \cdot \mathbf{p}$ theory one can also (starting from matrix element P at Γ) derive an expression for the $\mathbf{k}$-dependence of the optical transition matrix elements. Surprisingly, not only the band structure, but also the dielectric function in the energy region below 4–5 eV (where only transitions from the three topmost valence bands into the lowest conduction band occur) can be explained. This holds not only qualitatively but also quantitatively. Taking the simple four-band model explained in detail above, one obtains transition energies from Eq. 8.8

$$E_{cv_{1,3}}(\mathbf{k}) = \frac{1}{2}\left(E_g + \sqrt{E_g^2 + 4E_p E_k}\right), \tag{8.11}$$

$$E_{cv_2}(\mathbf{k}) = \sqrt{E_g^2 + 4E_p E_k}.$$

For the optical transition matrix elements in Eq. 8.6 one obtains

$$\langle c\mathbf{k}|\frac{\mathbf{p}}{m}|v_{1,3}\mathbf{k}\rangle = \frac{im}{\hbar}P\left\{\frac{1}{2}\left(1 + \frac{E_g}{\sqrt{E_g^2 + 4E_p E_k}}\right)\right\}^{\frac{1}{2}}\mathbf{n}_{1,3}, \tag{8.12}$$

$$\langle c\mathbf{k}|\frac{\mathbf{p}}{m}|v_2\mathbf{k}\rangle = -\frac{m}{\hbar}P\frac{\mathbf{k}}{k}$$

with orthonormal unit vectors $\mathbf{n}_{1,3}$ perpendicular to $\mathbf{k}$. Entering these matrix elements and transition energies into the Ehrenreich-Cohen formula [8.41] (corresponding to Eq. 8.6, but neglecting excitonic effects) for the dielectric function, one finds for the imaginary part (with Bohr radius a_B) [8.30]

$$\Im\,\varepsilon(\omega) = \frac{1}{3}\sqrt{\frac{e^2}{2a_B E_p}}\sqrt{1-x}\left(\sqrt{1+x}+8\right)\Theta(1-x), \tag{8.13}$$

where x is given as $x = E_g/\hbar\omega$.

Close to the fundamental gap this gives rise to a square-root-like onset $\Im\varepsilon(\omega) \propto (\hbar\omega - E_g)^{1/2}$, as expected for parabolic bands. Far away from the absorption edge approximately a constant is obtained, $\Im\varepsilon(\omega) \approx 3\sqrt{e^2/2a_B E_p}$. This explains the plateau-like character of the imaginary part of the dielectric function observed in the energy region below the onset of transitions into higher conduction bands. Even the absolute value obtained by *first-principles* calculations and found in measurments, $\Im\varepsilon(\omega) \approx 2.86...3.50$, is reproduced correctly for $E_p = 10...15$ eV.

8.4.3   Absorption edge: Energy splittings

Extending the simple four-band model described above for crystal-field splittings in the wurtzite case, one can even describe the anisotropy of the dielectric function. But even more, taking also spin-orbit coupling into account, resulting in an eight-band Hamiltonian, one can explain more details of the optical spectra near the absorption onset. Chuang and Chang [8.83] have formulated the necessary full eight-band $\mathbf{k}\cdot\mathbf{p}$ Hamiltonian to model band structures of wurtzite semiconductors. Introducing the crystal-field splitting parameter Δ_{cf} splitting off two (Γ_{6v} and Γ_{1v}) bands from the originally threefold degenerate valence band edge Γ_{15v} as well as the spin-orbit splitting parameter Δ_{so}, the Hamiltonian becomes more complicated but shall not be displayed here (for details see Ref. [8.83]). The anisotropy of the spin-orbit interaction will be neglected here.

At the Γ-point, diagonalization of the $\mathbf{k}\cdot\mathbf{p}$ Hamiltonian gives rise to the eigenvalue E_c for the conduction band Γ_{7c} (with spin-orbit interaction) and the three valence band eigenvalues Γ_{9v}, Γ_{7+v}, and Γ_{7-v}

$$\varepsilon_{v_1} = E_v + \Delta_{cf} + \frac{1}{3}\Delta_{so} , \tag{8.14}$$

$$\varepsilon_{v_2} = E_v + \frac{\Delta_{cf}}{2} - \frac{\Delta_{so}}{6} + \sqrt{\left(\frac{\Delta_{cf}}{2} - \frac{\Delta_{so}}{6}\right)^2 + \frac{2}{9}\Delta_{so}^2} ,$$

$$\varepsilon_{v_3} = E_v + \frac{\Delta_{cf}}{2} - \frac{\Delta_{so}}{6} - \sqrt{\left(\frac{\Delta_{cf}}{2} - \frac{\Delta_{so}}{6}\right)^2 + \frac{2}{9}\Delta_{so}^2} .$$

The limit $\Delta_{cf}=0$ recovers the results for the zinc-blende case. The energy splittings $\varepsilon_{v_1} - \varepsilon_{v_2}$ and $\varepsilon_{v_1} - \varepsilon_{v_3}$ given by

$$\varepsilon_{v_1} - \varepsilon_{v_2} = \frac{1}{2}\left[\Delta_{cf} + \Delta_{so} - \sqrt{(\Delta_{cf} + \Delta_{so})^2 - \frac{8}{3}\Delta_{cf}\Delta_{so}}\right] , \tag{8.15}$$

$$\varepsilon_{v_1} - \varepsilon_{v_3} = \frac{1}{2}\left[\Delta_{cf} + \Delta_{so} + \sqrt{(\Delta_{cf} + \Delta_{so})^2 - \frac{8}{3}\Delta_{cf}\Delta_{so}}\right] ,$$

where in the cubic case ε_{v_1} and ε_{v_2} become degenerate (Γ_{8+v}) and ε_{v_3} (Γ_{7-v}) is split off by Δ_{so}, can be seen in optical measurements, for example, studying the fine structure of the spectra close to the absorption edge.

A quantitative analysis is difficult for InN because the smallness of Δ_{cf} and Δ_{so} requires a high energy resolution of the order of meV in the measured spectra. Despite this fact, Goldhahn *et al.* [8.49] made an attempt to analyze their spectroscopic ellipsometry data in terms of the detailed **k·p** results (Eqs. 8.14 and 8.15), taking also into account the different selection rules for the accompanying optical transitions. For *a*-plane oriented *w*-InN, they found an energetical shift of about 25 meV between the absorption curves for the ordinary and for the extraordinary polarization. The splitting can be explained by the characters of the eigenfunctions. The highest valence band state (Γ_{9v}) possesses pure $p_{x,y}$ character (with the *z*-axis assumed parallel to the *c*-axis of the hexagonal crystal). For positive crystal-field splittings Δ_{cf} which are larger than or equal to the spin-orbit splitting Δ_{so}, the second-highest valence band state (Γ_{7+v}) also possesses predominantly $p_{x,y}$ character with a small p_z component, whereas the third-highest valence band state (Γ_{7-v}) possesses predominantly p_z character with a small $p_{x,y}$ component. According to the characters of the eigenfunctions, the transition with energy $E_c - \varepsilon_{v_1}$ ($\Gamma_{9v} \rightarrow \Gamma_{7c}$) is dipole forbidden for the extraordinary light polarization and the transition with energy $E_c - \varepsilon_{v_2}$ ($\Gamma_{7+v} \rightarrow \Gamma_{7c}$) is allowed but very weak for the extraordinary light polarization (due to the smallness of the p_z component in the Γ_{7+v} eigenfunction), whereas for the ordinary light polariztion both transitions are allowed and strong. Conversely, the transition with energy $E_c - \varepsilon_{v_3}$ ($\Gamma_{7-v} \rightarrow \Gamma_{7c}$) is allowed but very weak for the ordinary light polarization (due to the smallness of the $p_{x,y}$ component in the Γ_{7-v} eigenfunction), but allowed and strong for the extraordinary light polarization. Therefore, the onset of the absorption curve for the ordinary polarization is dominated by the two energetically lowest transitions (with energies $E_c - \varepsilon_{v_1}$ and $E_c - \varepsilon_{v_2}$), whereas for the extraordinary light polarization the onset of the absorption curve is dominated by the third-lowest transition with energy $E_c - \varepsilon_{v_3}$. Assuming that the very weak second-lowest transition with energy $E_c - \varepsilon_{v_2}$ is not visible in the measured spectra for the extraordinary light polarization, one should find an energy shift between the absorption curves for the ordinary and the extraordinary light polarizations which amounts to the energy difference $\varepsilon_{v_1} - \varepsilon_{v_3}$ [8.49]. The smaller energy difference $\varepsilon_{v_1} - \varepsilon_{v_2}$ that should only be detectable in the form of a shoulder-like feature close to the onset of the absorption for the ordinary light polarization is seemingly not visible in the spectra measured by Goldhahn *et al.* due to the limited spectral resolution and due to the fact that the measured curves end at energies above the fundamental gap because of the Burstein-Moss effect (and have been extrapolated down to $\Im\varepsilon(\omega) = 0$). Therefore, it still remains slightly unclear whether the observed energy shift is exactly given by $\varepsilon_{v_1} - \varepsilon_{v_3}$, or maybe (as a lower bound) by $\varepsilon_{v_2} - \varepsilon_{v_3}$ only, or by some average value in between.

Goldhahn *et al.* [8.49] have compared their results to theoretical values $\Delta_{cf} = 19$ meV of Carrier and Wei [8.28] and $\Delta_{so} = 13$ meV of Cardona and Christensen [8.66] and estimated a very small energy difference $\varepsilon_{v_1} - \varepsilon_{v_2}$ of about 3 meV which is difficult to resolve in the absorption curve for the ordinary light polarization. For the energy difference $\varepsilon_{v_1} - \varepsilon_{v_3}$ (or as lowest bound $\varepsilon_{v_2} - \varepsilon_{v_3}$) which should determine the energetical shift between the absorption curves for the ordinary and the extraodinary light polarizations, they estimated values which are in excellent agreement with the measured shift of 25 meV. This remains true if the calculated HSE03 value of $\Delta_{so} = 21.3$ meV for *zb*-InN (assuming that the value for *w*-InN is almost the same) and the HSE03+G_0W_0 value of $\Delta_{cf} = 22.9$ meV for *w*-InN are inserted. One still gets a small energy difference $\varepsilon_{v_1} - \varepsilon_{v_2}$ of 9.3 meV and an energy difference $\varepsilon_{v_1} - \varepsilon_{v_3}$ of 35 meV (or about 26 meV for the energy difference $\varepsilon_{v_2} - \varepsilon_{v_3}$) which is also in agreement with the measured shift of Goldhahn *et al.* of 25 meV. Due to the fact that it is seemingly not yet possible to extract the energy difference $\varepsilon_{v_1} - \varepsilon_{v_2}$ from measured spectra, at the moment the analysis is restricted to a simple consistency check only with respect to available theoretical data for Δ_{so} and Δ_{cf}. Also no data for the zinc-blende polytype exists yet. Therefore, a more extensive discussion is not possible and a fully quantitative analysis of experimental data requires more detailed theoretical and experimental studies with the highest possible spectral resolution and also taking into account the anisotropy of the spin-orbit interaction in the wurtzite case.

8.4.4   Absorption edge: Excitonic effects

An interesting detail is the influence of many-body effects on the adsorption edge. Besides general overall tendencies related to the Coulomb interaction between electrons and holes, such as a redshift of spectra with respect to pure single-quasiparticle spectra neglecting excitonic effects and redistribution of oscillator strengths, it is also expected that bound electron-hole states exist below the fundamental gap as in other semiconductors. The first peak above $\hbar\omega = 1$ eV in the imaginary part of the dielectric function $\varepsilon(\omega)$ calculated for *zb*-InN (Fig. 8.10) seems to indicate such excitonic effects. However, the peak is a consequence of the high joint density of states in this energy region (cf. Fig. 8.5 (b)) and occurs in the spectra without excitonic effects (Fig. 8.12). According to the simple theory of Wannier and Mott [8.84, 8.85] for extended electron-hole pair bound states one should find a hydrogen-atom-like series of bound states below the fundamental gap with a characteristic effective Rydberg constant which is downscaled from the true hydrogen Rydberg constant by introducing the background screening of the crystal by reducing the Coulomb attraction with the static (electronic) dielectric constant ε_∞ or the static dielectric constant of the crystal ε_0 (taking into account lattice polarization) and by replacing the electron mass by a reduced mass resulting from the electron and hole effective masses. These bound states should be visible by a corresponding hydrogen-like optical transition series appearing close to the adsorption edge [8.82]. This is indeed predicted by the analytic theory for parabolic bands [8.86].

In experiments, limited sample quality and more or less strong line broadenings still inhibit the observation of such bound exciton spectra for InN. Even more important is the influence of the free electrons with typical concentrations 10^{17}–10^{19} cm^{-3} in MBE-grown samples [8.66, 8.67, 8.77–8.79]. Usually the electron concentrations in these samples are above the Mott transition density. Consequently, the excitons should be dissociated and no bound states are visible in the absorption spectra. On the other hand, the theory should be able to describe these effects for InN which is not influenced by free carriers. From this viewpoint it is astonishing that no such features are visible in the theoretical spectra as well. The solution of this mystery is simple: Numerically, both the time-evolution scheme used for the calculation of the excitonic spectra as well as finite **k**-point mesh densities impose an artificial broadening onto the spectra which completely smears out such features. Only at numerical resolutions much better than the exciton binding energy, these bound states become visible. Such a resolution can be obtained by diagonalizing the exciton Hamiltonian (Eq. 8.5) explicitly instead of using the more efficient but numerically less accurate time-evolution scheme and by using **k**-point samplings that are as dense as possible.

In a recent strong effort we could indeed identify bound exciton states applying *ab initio* electronic-structure calculations. This has recently been demonstrated for biaxially strained zinc oxide [8.87] with resulting exciton binding energies of about 60 meV, close to the experimental values. Still, for numerical reasons, one cannot find the full set of bound states, but only the lowest few bound states. In addition, numerical effects also affect the position of the lowest bound state (defining the exciton binding energy). However, from calculations with different **k**-point samplings, it is possible to achieve a rather reliable extrapolation of the results toward an infinitely dense **k**-point mesh, that is, toward a numerical error of zero. This method has been also applied to InN [8.88]. We found a tendency for very small exciton binding energies of about 4 meV. Such small values may be also understandable in the Wannier-Mott framework. Neglecting the nonparabolicity of the bands with a reduced mass of about $0.05m_0$ and a static dielectric constant (including lattice polarization) of about $\varepsilon_0 = 13$ [8.29], a binding energy of 4 meV is estimated using the exciton Rydberg formula.

8.5 Outlook

Although existing theoretical and experimental work already provides a consistent and rather detailed picture of electronic and optical properties of InN, some of the fine details are still lacking. In particular, experimental or theoretical data about the influence of spin-orbit coupling are still rare and the inclusion of the spin-orbit interaction is not yet possible on the GW level. One has to rely on approximate estimates based on the DFT level. Also experimentally it still remains a delicate task. It is in principle possible to determine the spin-orbit splitting at the VBM as well as the crystal-field splitting in the wurtzite case from the onset behavior of optical spectra. Future work should focus particularly on a precise determination of the crystal-field splitting and spin-orbit splitting in order to be able to give precise **k·p** parameters for a detailed 8-band Kane model. Efforts into this direction have already started from the theoretical side and experiment has already demonstrated that this is possible, at least in principle [8.49].

Besides the study of binary InN (or other binary group-III-nitrides like GaN or AlN) alloys of group-III-nitrides, in particular InGaN but also InGaAlN, are of high technological importance. Alloying allows, on the one hand, a simple band-gap engineering via the variation of the fundamental gap with alloy composition. On the other hand, the (average) lattice constant of alloys also varies with alloy composition, and hence changing the alloy composition allows the reduction of the lattice constant mismatches between group-III-nitride alloy films and growth substrates, in order to reduce elastic strain. In particular, quaternary alloys like InGaAlN offer an excellent possibility to tune the gap and lattice constant independently. Although several studies have already been published on the thermodynamic properties of such alloys (phase diagram, stability), a proper theoretical treatment as well as extensive experimental studies on electronic and optical properties, for example, densities of states or dielectric functions, of such alloys are still widely missing. Therefore, future activities should also focus on this direction.

Another very important but widely unexplored problem is the influence of strain on the electronic and optical properties of InN (and other nitrides). Usually no thick bulk crystals are available, but rather thin nitride films are grown on various substrates. Due to unavoidable lattice constant mismatches, the film-substrate boundary has many defects, for example, lattice-mismatch dislocations, but maybe even cracks that decrease the sample quality, or the films are rather defect-free and perfect, but biaxially strained with a planar lattice constant of the nitride parallel to the substrate surface being identical to the substrate lattice constant. Such biaxial strain may on the one hand introduce artificial strain-induced anisotropies, even for cubic polytypes, but, due to the finite volume deformation potentials, also changes in the band structure and hence in the dielectric function have to be expected. This holds in particular for such subtle details like the spin-orbit splitting or even worse the crystal-field splitting in the wurtzite case. For example, recently it was demonstrated for zinc oxide that biaxial strain can completely invert the band ordering of the topmost valence bands at Γ, that is, change the sign of the crystal-field splitting [8.87]. Therefore a special focus should also be put on this topic, in particular since all experiments suffer in a certain way from strain effects and knowing the influence of strain is vital for understanding apparently inconsistent, maybe even contradicting, or at least widely scattered experimental results.

Acknowledgments

We would like to thank G. Kresse, R. Goldhahn, L. F. J. Piper, T. D. Veal, P. D. C. King, and V. Yu. Davydov for valuable discussions and communications. The authors acknowledge financial support from the Deutsche Forschungsgemeinschaft

(Project No. Be 1346/18-1,2) and the European Community in the framework of the network of excellence NANOQUANTA (Contract No. NMP4-CT-2004-500198).

References

[8.1] H. Morkoç, *Nitride Semiconductors and Devices*, Springer, New York, 1999.

[8.2] T. V. Shubina, S. V. Ivanov, V. N. Jmerik, D. D. Solnyshkov, V. A. Vekshin, P. S. Kop'ev, A. Vasson, J. Leymarie, A. Kavokin, H. Amano, K. Shimono, A. Kasic, and B. Monemar, Mie resonances, infrared emission, and the band gap of InN, *Physical Review Letters*, 92 (2004) 117407:1–4.

[8.3] F. Bechstedt, J. Furthmüller, O. Ambacher, and R. Goldhahn, Comment on "Mie resonances, infrared emission, and the band gap of InN", *Physical Review Letters*, 93 (2004) 269701:1.

[8.4] K. Osamura, S. Naka, and Y. Murakami, Preparation and optical properties of $Ga_{1-x}In_xN$ thin films, *Journal of Applied Physics*, 46 (1975) 3432–3437.

[8.5] T. L. Tansley and C. P. Foley, Optical band gap of indium nitride, *Journal of Applied Physics*, 59 (1986) 3241–3244.

[8.6] N. Puychevrier and M. Menoret, Synthesis of III-V semiconductor nitrides by reactive cathodic sputtering, *Thin Solid Films*, 36 (1976) 141–145.

[8.7] K. L. Westra, R. P. W. Lawson, and M. J. Brett, The effects of oxygen contamination on the properties of reactively sputtered indium nitride films, *Journal of Vacuum Science and Technology A*, 6 (1988) 1730–1732.

[8.8] Q. Guo and A. Yoshida, Temperature dependence of band gap change in InN and AlN, *Japanese Journal of Applied Physics*, 33 (1994) 2453–2456.

[8.9] P. Hohenberg and W. Kohn, Inhomogeneous Electron Gas, *Physical Review*, 136 (1964) B864–B871.

[8.10] W. Kohn and L. J. Sham, Self-consistent equations including exchange and correlation effects, *Physical Review*, 140 (1965) A1133–A1138.

[8.11] J. P. Perdew and Y. Wang, Accurate and simple density functional for the electronic exchange energy: Generalized gradient approximation, *Physical Review B*, 33 (1986) 8800–8802.

[8.12] J. P. Perdew, Unified theory of exchange and correlation beyond the local density approximation, in P. Ziesche and H. Eschrig (Eds) *Electronic Structure of Solids '91*, Akademie Verlag, Berlin, 1991, pp. 11–20.

[8.13] W. G. Aulbur, L. Jönsson, and J. W. Wilkins, Quasiparticle calculations in solids, in F. Seitz, D. Turnbull, and H. Ehrenreich (Eds) *Solid State Physics*,

Vol. 54, Academic Press, New York, 2000, pp. 1–218.

[8.14] N. E. Christensen and I. Gorczyca, Optical and structural properties of III-V nitrides under pressure, *Physical Review B*, 50 (1994) 4397–4415.

[8.15] L. Bellaiche, K. Kunc, and J. M. Besson, Isostructural phase transition in InN wurtzite, *Physical Review B*, 54 (1996) 8945–8949.

[8.16] M. Ueno, M. Yoshida, A. Onodera, O. Shimomura, and K. Takemura, Stability of the wurtzite-type structure under high pressure: GaN and InN, *Physical Review B*, 49 (1994) 14–21.

[8.17] A. F. Wright and J. S. Nelson, Consistent structural properties for AlN, GaN, and InN, *Physical Review B*, 51 (1995) 7866–7869.

[8.18] U. Grossner, J. Furthmüller, and F. Bechstedt, Bond-rotation versus bond-contraction relaxation of (110) surfaces of group-III nitrides, *Physical Review B*, 58 (1998) R1722–R1725.

[8.19] C. Stampfl and C. G. Van de Walle, Density-functional calculations for III-V nitrides using the local-density approximation and the generalized gradient approximation, *Physical Review B*, 59 (1999) 5521–5535.

[8.20] F. Bechstedt and J. Furthmüller, Do we know the fundamental energy gap of InN? *Journal of Crystal Growth*, 246 (2002) 315–319.

[8.21] S. H. Wei and A. Zunger, Role of metal d states in II-VI semiconductors, *Physical Review B*, 37 (1988) 8958–8981.

[8.22] F. Bechstedt, J. Furthmüller, M. Ferhat, L. K. Teles, L. M. R. Scolfaro, J. R. Leite, V. Yu. Davydov, O. Ambacher, and R. Goldhahn, Energy gap and optical properties of $In_xGa_{1-x}N$, *physica status solidi (a)*, 195 (2003) 628–633.

[8.23] V. Yu. Davydov, A. A. Klochikhin, R. P. Seisyan, V. V. Emtsev, S. V. Ivanov, F. Bechstedt, J. Furthmüller, H. Harima, A. V. Mudryi, J. Aderhold, O. Semchinova, and J. Graul, Absorption and emission of hexagonal InN. Evidence of narrow fundamental band gap, *physica status solidi (b)*, 229 (2002) R1–R3.

[8.24] J. Wu, W. Walukiewicz, K. M. Yu, J. W. Ager III, E. E. Haller, H. Lu, W. J. Schaff, Y. Saito, and Y. Nanishi, Unusual properties of the fundamental band gap of InN, *Applied Physics Letters*, 80 (2002) 3967–3969.

[8.25] V. Yu. Davydov, A. A. Klochikhin, V. V. Emtsev, S. V. Ivanov, V. V. Vekshin, F. Bechstedt, J. Furthmüller, H. Harima, A. V. Mudryi, A. Hashimoto, A. Yamamoto, J. Aderhold, J. Graul, and E. E. Haller, Band Gap of InN and In-Rich $In_xGa_{1-x}N$ alloys ($0.36 < x < 1$), *physica status solidi (b)*, 230 (2002) R4–R6.

[8.26] T. Matsuoka, H. Okamoto, M. Nakao, H. Harima, and E. Kurimoto, Optical bandgap energy of wurtzite InN, *Applied Physics Letters*, 81 (2002) 1246–1248.

[8.27] B. R. Nag, Comment on "Band Gap of InN and In-Rich $In_xGa_{1-x}N$ Alloys $(0.36 < x < 1)$", *physica status solidi (b)*, 233 (2002) R8–R9.

[8.28] P. Carrier and S. H. Wei, Theoretical study of the band-gap anomaly of InN, *Journal of Applied Physics*, 97 (2005) 033707:1–5.

[8.29] F. Bechstedt, Nitrides as seen by a theorist, in B. Gil (Ed.) *Low-Dimensional Nitride Semiconductors*, Oxford University Press, Oxford, 2002, pp. 11–56.

[8.30] J. Furthmüller, P. H. Hahn, F. Fuchs, and F. Bechstedt, Band structures and optical spectra of InN polymorphs: Influence of quasiparticle and excitonic effects, *Physical Review B*, 72 (2005) 205106:1–14.

[8.31] *Table of Periodic Properties of the Elements*, Sargent-Welch Scientific Company, Skokie, IL, 1980.

[8.32] G. Kresse and J. Furthmüller, Efficient iterative schemes for ab initio total-energy calculations using a plane-wave basis set, *Physical Review B*, 54 (1996) 11169–11186.

[8.33] G. Kresse and J. Furthmüller, Efficiency of ab-initio total energy calculations for metals and semiconductors using a plane-wave basis set, *Computational Materials Science*, 6 (1996) 15–50.

[8.34] V. Fiorentini, M. Methfessel, and M. Scheffler, Electronic and structural properties of GaN by the full-potential linear muffin-tin orbitals method: The role of the *d* electrons, *Physical Review B*, 47 (1993) 13353–13362.

[8.35] A. Garcia and M. L. Cohen, First-principles ionicity scales. I. Charge asymmetry in the solid state, *Physical Review B*, 47 (1993) 4215–4220.

[8.36] G. Kresse and D. Joubert, From ultrasoft pseudopotentials to the projector augmented-wave method, *Physical Review B*, 59 (1999) 1758–1775.

[8.37] L. Hedin, New method for calculating the one-particle Green's function with application to the electron-gas problem, *Physical Review*, 139 (1965) A796–A823.

[8.38] L. J. Sham and T. M. Rice, Many-particle derivation of the effective-mass equation for the Wannier exciton, *Physics Review*, 144 (1966) 708–714.

[8.39] G. Strinati, Application of the Green's functions method to the study of the optical properties of semiconductors, *Rivista del Nuovo Cimento*, 11 (1988) 1–86.

[8.40] L. Hedin and S. Lundqvist, Effects of electron-electron and electron-phonon interaction on the one-electron states of solids, in H. Ehrenreich, F. Seitz, and D. Turnbull (Eds) *Solid State Physics*, Vol. 23, Academic Press, New York,

1969, pp. 1–181.

[8.41] B. Adolph, V. I. Gavrilenko, K. Tenelsen, F. Bechstedt, and R. Del Sole, Nonlocality and many-body effects in the optical properties of semiconductors, *Physical Review B*, 53 (1996) 9797–9808.

[8.42] F. Fuchs, J. Furthmüller, F. Bechstedt, M. Shishkin, and G. Kresse, Quasiparticle band structure based on a generalized Kohn-Sham scheme, *Physical Review B*, 76 (2007) 115109:1–8.

[8.43] M. Shishkin and G. Kresse, Self-consistent GW calculations for semiconductors and insulators, *Physical Review B*, 75 (2007) 235102.

[8.44] V. I. Anisimov, J. Zaanen, and O. K. Andersen, Band theory and Mott insulators: Hubbard U instead of Stoner I, *Physical Review B*, 44 (1991) 943–954.

[8.45] A. Janotti, D. Segev, and C. G. Van de Walle, Effects of cation d states on the structural and electronic properties of III-nitride and II-oxide wide-band-gap semiconductors, *Physical Review B*, 74 (2006) 045202:1–9.

[8.46] M. Van Schilfgaarde, T. Kotani, and S. Faleev, Quasiparticle self-consistent GW theory, *Physical Review Letters*, 96 (2006) 226402:1–4.

[8.47] P. Rinke, A. Qteish, J. Neugebauer, C. Freysoldt, and M. Scheffler, Combining GW calculations with exact-exchange density-functional theory: an analysis of valence-band photoemission for compound semiconductors, *New Journal of Physics*, 7 (2005) 126:1–35.

[8.48] P. Rinke, M. Winkelnkemper, A. Qteish, D. Bimberg, J. Neugebauer, and M. Scheffler, Consistent set of band parameters for the group-III nitrides AlN, GaN, and InN, *Physical Review B*, 77 (2008) 075202:1–15.

[8.49] R. Goldhahn, P. Schley, A. T. Winzer, M. Rakel, C. Cobet, N. Esser, H. Lu, and W. J. Schaff, Critical points of the band structure and valence band ordering at the Γ point of wurtzite InN, *Journal of Crystal Growth*, 288 (2006) 273–277.

[8.50] A. Seidl, A. Görling, P. Vogl, J. A. Majewski, and M. Levy, Generalized Kohn-Sham schemes and the band-gap problem, *Physical Review B*, 53 (1996) 3764–3774.

[8.51] J. Muscat, A. Wander, and N. M. Harrison, On the prediction of band gaps from hybrid functional theory, *Chemical Physics Letters*, 342 (2001) 397-401.

[8.52] M. S. Hybertsen and S. G. Louie, Electron correlation in semiconductors and insulators: Band gaps and quasiparticle energies, *Physical Review B*, 34 (1986) 5390–5413.

[8.53] F. Bechstedt, Quasiparticle corrections for energy gaps in semiconductors, in U. Rössler (Ed.) *Advances in Solid State Physics*, Vol. 32, Springer,

Berlin/Heidelberg, 1992, pp. 161-177.

[8.54] J. Heyd, G. E. Scuseria, and M. Ernzerhof, Hybrid functionals based on a screened Coulomb potential, *Journal of Chemical Physics*, 118 (2003) 8207–8215.

[8.55] W. G. Schmidt, S. Glutsch, P. H. Hahn, and F. Bechstedt, Efficient $O(N^2)$ method to solve the Bethe-Salpeter equation, *Physical Review B*, 67 (2003) 085307:1–7.

[8.56] W. Hanke and L. J. Sham, Local-field and excitonic effects in the optical spectrum of a covalent crystal, *Physical Review B*, 12 (1975) 4501–4511.

[8.57] W. Hanke and L. J. Sham, Many-particle effects in the optical spectrum of a semiconductor, *Physical Review B*, 21 (1980) 4656–4673.

[8.58] S. Albrecht, L. Reining, R. Del Sole, and G. Onida, *Ab Initio* Calculation of excitonic effects in the optical spectra of semiconductors, *Physical Review Letters*, 80 (1998) 4510–4513.

[8.59] M. Rohlfing and S. G. Louie, Excitons and optical spectrum of the Si(111)-(2×1) surface, *Physical Review Letters*, 83 (1999) 856–859.

[8.60] L. X. Benedict, E. L. Shirley, and R. B. Bohn, Optical absorption of insulators and the electron-hole interaction: An *ab initio* calculation, *Physical Review Letters*, 80 (1998) 4514–4517.

[8.61] P. H. Hahn, W. G. Schmidt, and F. Bechstedt, Bulk excitonic effects in surface optical spectra, *Physical Review Letters*, 88 (2002) 016402.

[8.62] Q. X. Guo, M. Nishio, H. Ogawa, A. Wakahara, and A. Yoshida, Electronic structure of indium nitride studied by photoelectron spectroscopy, *Physical Review B*, 58 (1998) 15304–15306.

[8.63] P. D. C. King, T. D. Veal, C. F. McConville, F. Fuchs, J. Furthmüller, F. Bechstedt, J. Schörmann, D. J. As, K. Lischka, H. Lu, and W. J. Schaff, Valence band density of states of zinc-blende and wurtzite InN from x-ray photoemission spectroscopy and first-principles calculations, *Physical Review B*, 77 (2008) 115213:1–7.

[8.64] C. Persson and A. Zunger, *s-d* coupling in zinc-blende semiconductors, *Physical Review B*, 68 (2003) 073205.

[8.65] R. A. Pollak, L. Ley, S. Kowalczyk, D. A. Shirley, J. D. Joannopoulos, D. J. Chadi, and M. L. Cohen, X-ray photoemission valence-band spectra and theoretical valence-band densities of states for Ge, GaAs, and ZnSe, *Physical Review Letters*, 29 (1972) 1103–1105.

[8.66] M. Cardona and N. E. Christensen, Spin-orbit splittings in AlN, GaN and InN, *Solid State Communications*, 116 (2000) 421–425.

[8.67] K. M. Yu, Z. Liliental-Weber, W. Walukiewicz, W. Shan, J. W. Ager III, S. X. Li, R. E. Jones, E. E. Haller, H. Lu, and W. J. Schaff, On the crystalline structure, stoichiometry and band gap of InN thin films, *Applied Physics Letters*, 86 (2005) 071910:1–3.

[8.68] J. Schörmann, D. J. As, K. Lischka, P. Schley, R. Goldhahn, S. F. Li, W. Löffler, M. Hetterich, and H. Kalt, Molecular beam epitaxy of phase pure cubic InN, *Applied Physics Letters*, 89 (2006) 261903:1–3.

[8.69] P. Schley, R. Goldhahn, C. Napierla, G. Gobsch, J. Schörmann, D. J. As, K. Lischka, M. Feneberg, and K. Thonke, Dielectric function of cubic InN from the mid-infrared to the visible spectral range, *Semiconductor Science and Technology* 23 (2008) 055001:1–6.

[8.70] W. Mönch, *Semiconductor Surfaces and Interfaces*, Springer, Berlin, 2001.

[8.71] J. Tersoff, Schottky barriers and semiconductor band structures, *Physical Review B*, 32 (1985) 6968–6971.

[8.72] P. D. C. King, T. D. Veal, P. H. Jefferson, S. A. Hatfield, L. F. J. Piper, C. F. McConville, F. Fuchs, J. Furthmüller, F. Bechstedt, H. Lu, and W. J. Schaff, Determination of the branch-point energy of InN: Chemical trends in common-cation and common-anion semiconductors, *Physical Review B*, 77 (2008) 045316:1–6 and references therein.

[8.73] I. Mahboob, T. D. Veal, L. F. J. Piper, C. F. McConville, H. Lu, W. J. Schaff, J. Furthmüller, and F. Bechstedt, Origin of electron accumulation at wurtzite InN surfaces, *Physical Review B*, 69 (2004) 201307:1–4.

[8.74] C. Hamaguchi, *Basic Semiconductor Physics*, Springer, Berlin, 2001.

[8.75] S. P. Fu and Y. F. Chen, Effective mass of InN epilayers, *Applied Physics Letters*, 85 (2004) 1523–1525.

[8.76] J. Wu, W. Walukiewicz, W. Shan, K. M. Yu, J. W. Ager III, E. E. Haller, H. Lu, and W. J. Schaff, Effects of the narrow band gap on the properties of InN, *Physical Review B*, 66 (2002) 201403(R):1–4.

[8.77] T. Inushima, M. Higashiwaki, and T. Matsui, Optical properties of Si-doped InN grown on sapphire (0001), *Physical Review B*, 68 (2003) 235204:1–7.

[8.78] L. F. J. Piper, L. Colakerol, T. Learmonth, P. A. Glans, K. E. Smith, F. Fuchs, J. Furthmüller, F. Bechstedt, T. C. Chen, T. D. Moustakas, and J. H. Guo, Electronic structure of InN studied using soft x-ray emission, soft x-ray absorption, and quasiparticle band structure calculations, *Physical Review B*, 76 (2007) 245204:1–5.

[8.79] R. Goldhahn, S. Shokhovets, V. Cimalla, L. Spiess, G. Ecke, O. Ambacher, J. Furthmüller, F. Bechstedt, H. Lu, and W. J. Schaff, Dielectric function of "narrow" band gap InN, in C. Wetzel, E. T. Yu, J. S. Speck, A. Rizzi,

Y. Arakawa (Eds) *GaN and Related Alloys*. MRS Symposia Proceedings, Vol. 743, L5.9, Materials Research Society, Warrendale, PA, 2003.

[8.80] R. Goldhahn, A. T. Winzer, V. Cimalla, O. Ambacher, C. Cobet, W. Richter, N. Esser, J. Furthmüller, F. Bechstedt, H. Lu, and W. J. Schaff, Anisotropy of the dielectric function for wurtzite InN, *Superlattices and Microstructures*, 36 (2004) 591-597.

[8.81] R. Goldhahn, P. Schley, J. Schörmann, D. J. As, K. Lischka, F. Fuchs, F. Bechstedt, C. Cobet, N. Esser, Dielectric function and band structure of cubic InN, in K. Godehusen (Ed.) *BESSY — Annual Report 2006*, Berliner Elektronenspeicherring-Gesellschaft für Synchrotronstrahlung m.b.H. (BESSY), 2007, pp. 529–531.

[8.82] P. Y. Yu and M. Cardona, *Fundamentals of Semiconductors*, Springer, Berlin, 1996.

[8.83] S. L. Chuang and C. S. Chang, $\mathbf{k} \cdot \mathbf{p}$ method for strained wurtzite semiconductors, *Physical Review B*, 54 (1996) 2491–2504.

[8.84] G. H. Wannier, The structure of electronic excitation levels in insulating crystals, *Physical Review*, 52 (1937) 191–197.

[8.85] N. F. Mott, Conduction in polar crystals. II. The conduction band and ultraviolet absorption of alkali-halide crystals, *Transactions of the Faraday Society*, 34 (1938) 500–506.

[8.86] R. J. Elliott, Intensity of optical absorption by excitons, *Physical Review*, 108 (1957) 1384–1389.

[8.87] A. Schleife, C. Rödl, F. Fuchs, J. Furthmüller, and F. Bechstedt, Strain influence on valence-band ordering and excitons in ZnO: An *ab initio* study, *Applied Physics Letters*, 91 (2007) 241915:1–3.

[8.88] F. Fuchs, C. Rödl, A. Schleife, and F. Bechstedt, Efficient $O(N^2)$ approach to solve the Bethe-Salpeter equation for excitonic bound states, *Physical Review B*, 78 (2008) 085103:1–13.

9

Ellipsometry of InN and related alloys

R. Goldhahn and P. Schley

Institute of Physics and Institute of Micro- and Nanotechnologies, Technical University of Ilmenau, PF 100565, 98684 Ilmenau, Germany

M. Röppischer

Institute for Analytical Sciences, Department Berlin, Albert-Einstein-Str. 9, 12489 Berlin, Germany

9.1 Introduction

9.1.1 A brief review on recent ellipsometry results

Indium nitride represents the least studied compound among the group-III nitride materials. For a long time it was accepted that the band gap of the hexagonal polymorph with wurtzite crystal structure (w-InN) amounts to 1.89 eV at room temperature (RT) [9.1]. The value refers to transmission studies of polycrystalline needle-like crystallites which were prepared by reactive sputtering. The repeated observation [9.2–9.4] of an absorption edge considerably below 1 eV for high-quality single-crystalline films, grown by plasma-induced molecular beam epitaxy (PI-MBE) or metalorganic vapor phase epitaxy (MOVPE), has initiated ongoing intensive research work.

Many optical studies of InN focus on the determination of the absorption coefficient $\alpha(\hbar\omega)$ as a function of photon energy ($\hbar\omega$) from transmission studies. Typical absorption coefficients above $1 \times 10^4 \, \text{cm}^{-1}$ constrict the maximum film thickness of those experiments and, in particular, only the small energy range around the absorption edge is accessible. These restrictions can be overcome by applying spectroscopic ellipsometry (SE). As a reflection based method, SE allows the determination of the real (ε_1) and imaginary part (ε_2) of the complex dielectric function (DF, $\bar{\varepsilon} = \varepsilon_1 + i \cdot \varepsilon_2$) independent of the layer thickness for every photon energy. The DF represents one of the fundamental quantities because its shape sensitively depends on the material quality and its analysis and interpretation are directly linked to the theory. The spectral dependence of $\bar{\varepsilon}$ shows some peculiarities which are unambiguously related to the band structure, that is, the dispersion of the conduction $E_c(\mathbf{k})$ and valence bands $E_v(\mathbf{k})$ as a function of the wave vector $\mathbf{k}$. For a qualitative interpretation and a comparison to theoretical results, it is in the primary step not essential whether the DF was calculated in the one-electron picture [9.5, 9.6] or if electron-hole interaction (exciton effects) [9.7] was consequently taken into account (see Chapter 8 for a comprehensive discussion).

The sharp onset of ε_2 in the near infrared (NIR) spectral region is attributed to transitions between valence bands (VB) and the conduction band (CB) in the vicinity of the Γ-point of the Brillouin zone. It defines the absorption edge of the investigated sample, but it does not necessarily represent the fundamental gap of the material due to the carrier-induced Burstein-Moss shift (BMS) in the highly degenerate InN films [9.8–9.10]. It is obvious that the shape of the absorption coefficient obtained from transmission studies is likewise influenced by the BMS [9.11] due to the proportionality $\alpha \sim \varepsilon_2$. Beside the absorption edge, the spectral dependence of ε_2 in the visible (VIS), ultraviolet (UV), and vacuum-ultraviolet (VUV) regions shows pronounced resonances known as *critical points* (CP) or *Van Hove* singularities [9.12]. The CP features are only observed if high-quality crystalline material is studied. The SE studies of Goldhahn *et al.* [9.13] in the photon energy range above 4 eV demonstrated for the first time that MBE-grown material with a fundamental absorption edge considerably below 1 eV shows the characteristic peak structure of hexagonal InN. In contrast, the photon energy dependence of ε_2 for sputtered films with an absorption edge of about 1.9 eV [9.13] deviates considerably from both the data for crystalline layers and the theoretical results.

The preliminary results on the DF for crystalline material allowed a demonstration [9.14] that Mie resonances due to scattering or absorption of light in InN layers containing metallic indium clusters are not responsible for the low absorption edge which was proposed by Shubina *et al.* [9.15]. The onset of the host (InN) absorption had to be chosen at 0.9 eV in the calculations [9.14] in order to get excellent agreement with the experimental data of Ref. [9.15] for the In-cluster containing films. It should be noted that the rather high value of 0.9 eV is fully consistent with an intrinsic band gap for w-InN of ~ 0.7 eV. The shift of the absorption edge by 200 meV corresponds exactly to the expected BMS for an electron concentration of $N_e \approx 2 \times 10^{19}$ cm^{-3} (a detailed discussion will be given in the following sections). SE studies over an extended photon energy range confirmed the impact of In-clusters on the optical response of a film [9.16], but the results obtained for layers without metallic clusters emphasized the low band gap of InN.

A distinct influence of oxygen on the optical properties of single-crystalline InN can be excluded from the *in situ* growth monitoring by SE. Pseudo-dielectric functions taken during MOVPE [9.17–9.19] or MBE [9.20, 9.21] deposition exhibit all the characteristic features found in the *ex situ* studies. Moreover, the *in situ* ellipsometry provides a very sensitive tool to demonstrate how thermal treatment [9.18] or exposure to hydrogen [9.22] changes the optical properties of InN.

All the above mentioned studies were carried out on (0001)- or (000$\overline{1}$)-oriented c-plane w-InN layers. SE measurements on those films allow, however, only the determination of a DF which is very close to the ordinary dielectric tensor component ($\overline{\varepsilon}_o$) corresponding to electric field (E) polarization perpendicular to the c-axis ($E \perp c$) of the optically anisotropic material. Non-polar surface orientations are required for determining the extraordinary dielectric tensor component ($\bar{\varepsilon}_e$) corresponding to E-polarization parallel to the c-axis ($E \parallel c$). The ellipsometry measurements [9.23] on ($11\bar{2}0$)-oriented a-plane InN, covering the whole spectral range from the NIR to the VUV, yielded for the first time both dielectric tensor components. The optical anisotropy found in the range of the high-energy CPs excellently agrees with the recently published theoretical results [9.7]. Moreover, a difference in the absorption edge energies for $E \perp c$ and $E \parallel c$ was reported which can only be interpreted in terms of the VB splitting around the Γ-point of the Brillouin zone; it represents a distinguishing feature of all wurtzite nitrides. The energy region in which this polarization behavior is found marks without doubt the band gap range of the material. The similarity of $\bar{\varepsilon}_o$ as obtained from the studies of c- and a-plane films was subsequently demonstrated [9.24]. Finally, SE also revealed optical anisotropy of the DF in the energy range from 16 to 28 eV where the properties of the In $4d$ core levels have to be taken into account [9.25].

The verification of the spectral dependence of the DF from the NIR into the UV by many groups provided the basis for a detailed analysis of the DF. In particular, the transition energies for the UV/VUV CPs as obtained from the ordinary DF were reported [9.10, 9.16, 9.26]. They undergo a continuous shift to higher energies with decreasing In-content x in the alloy systems which was demonstrated by SE of In-rich wurtzite In$_x$Al$_{1-x}$N [9.9] and In$_x$Ga$_{1-x}$N [9.10] epitaxial films.

Almost phase-pure zinc-blende (zb-) InN only recently became available [9.27]. The theoretical calculations [9.6, 9.7, 9.28] predict for the cubic compound a lower band gap than for w-InN. The recently published ellipsometry results [9.29] revealed indeed a gap of zb-InN below 0.6 eV. It is considerably lower than the previously assumed value of 1.8 eV which was used for the estimation of the band gap bowing parameter in the zb-InGaN alloy system [9.30]. The new results demand a reevaluation of this quantity.

Although only a few studies have been reported so far, ellipsometry in the mid-infrared (MIR) [9.26, 9.29, 9.31, 9.32] will gain much attention in the future. Those studies yield the transversal-optical (TO) phonon frequencies and the longitudinal-optical (LO) phonon-plasmon coupled modes. The free-carrier plasma frequency ω_p is directly derived from the latter. Because ω_p depends on the ratio of the electron concentration and the averaged optical electron effective mass, one gets direct access to these quantities by a careful data analysis [9.29]. In comparison to Hall measurements, MIR-SE allows a much more accurate determination of N_e in the bulk-like part of the InN films [9.29, 9.32].

9.1.2 Open questions

It is now widely accepted that the fundamental gap E_0 for w-InN is below 0.7 eV at RT. However, in contrast to GaN, for example, the exact value of E_0 is still under debate for several reasons:

(i) Most studied films suffer from the high unintentional electron concentrations N_e. It causes a N_e-dependent band-gap renormalization (BGR), but the two approaches

[9.8, 9.33] currently used to account for the phenomena yield different magnitudes of the effect. An additional uncertainty arises from the determination of N_e via Hall measurements due to the formation of electron accumulation layers [9.34, 9.35] at the surface/interface. The extracted sheet carrier concentration divided by the layer thickness yields too high values for the bulk-like part of the sample [9.34, 9.36–9.39].

(ii) Conduction band filling causes a concentration-dependent blue shift of the absorption edge [9.8] which is known as the Burstein-Moss shift. Under those condition, neither the renormalized gap E_{ren} nor the transition energy at the Fermi wave vector $E_F(k_F)$ can be deduced from the experimental data using any analytical formula. It requires numerical calculation of the optical response [9.9, 9.10].

(iii) Even if low carrier density samples in the $10^{16}\,\mathrm{cm}^{-3}$ range became available, the commonly used extrapolation methods cannot be applied for determining E_0 from the photon energy dependence of the absorption coefficient. The low band gap causes a strong nonparabolicity of the conduction band [9.8] and the density of states does not show a square root dependence. Furthermore, the refractive index is not constant around the gap [9.40]. As for any other semiconductor [9.12, 9.41], electron-hole interaction modifies the shape of $\alpha(\hbar\omega)$ in comparison to the one-electron picture, giving rise to the observation of free excitonic lines (bound exciton states) below and a pronounced enhancement of the absorption due to continuum exciton states above the gap. The latter effect is known as the Sommerfeld enhancement factor [9.12]. Excitonic contributions combined with CB nonparabolicity demand a more elaborate analysis of the experimental data [9.42, 9.43].

(iv) Finally, the influence of biaxial in-plane strain [9.44–9.46], caused by the mismatch of the lattice constants and the thermal expansion coefficients of the epitaxial layers and the substrate, has been mostly disregarded. It is well known from the GaN system, for example, that strain shifts the bands and therefore the transition energies in a characteristic manner [9.47]. This aspect was only recently addressed for w-InN [9.38, 9.48, 9.49].

Independent of the aforementioned necessary details of the data analysis, a band gap of w-InN at RT between 0.63 eV [9.38] and 0.68 eV [9.10, 9.11] seems very likely.

9.1.3 Outline

The determination of the intrinsic dielectric function for wurtzite and zinc-blende InN would require bulk crystals with a carrier concentration in the low $10^{16}\,\mathrm{cm}^{-3}$ range and an excellent surface quality. Those samples are currently not available and all data reported so far are from the investigation of epitaxial films. In this case the shape of the DFs extracted from SE measurements depends critically on the models employed for fitting the multi-layer systems [9.50]. The peculiarities of the surface, such as roughness, oxide layers, or organic contamination, were not taken into account in some recent publications. These data represent therefore more or less a specific sample property.

The extraction of fundamental band structure parameters is the second step of the DF analysis. It turns out that some important film properties were not reported, such as strain, or are questionable (for example, bulk electron concentration). It makes a direct comparison of the results rather difficult, in particular in the photon energy range around the gap. Finally, different approaches were applied in order to estimate the zero-density band gap.

For all these reasons, the properties of only a few samples, for which the structural data are available, will be discussed in the current chapter. The SE measurements were carried out under identical conditions followed by a multi-layer analysis for determining the DFs. All steps of the parameter extraction will be given in detail. It provides the reader with a better insight into the dependencies.

The outline of the remaining part of the chapter is as follows. An overview on the theoretical background of the optical properties and methods is given in the next section. Having defined the optical quantities, the influence of strain on the band structure and oscillator strength is demonstrated. Then the approach used for describing band-gap renormalization is described. The experimental results for wurtzite InN and their interpretation follow in Section 9.3. We show how MIR-SE can be used to get more reliable values for the electron concentrations. The dependence of the transition energies on the composition for hexagonal InGaN and InAlN is summarized in the fourth section. The recent findings on the properties of cubic InN are reported in the final part.

9.2 Theoretical background

9.2.1 Fundamental relations

Strain-free InN with wurtzite structure belongs to the $P6_3mc(C_{6v}^4)$ space group and is an optically uniaxial material. So far, almost all optical studies of epitaxial InN were carried out on so-called c-plane films with (0001) (In-face) or $(000\bar{1})$ (N-face) orientations known. In this case, the optic axis (c-axis) is oriented normal to the surface (here x-y plane) along the z-direction. The dielectric tensor takes the form

$$\overleftrightarrow{\varepsilon} = \begin{pmatrix} \varepsilon_x & 0 & 0 \\ 0 & \varepsilon_y & 0 \\ 0 & 0 & \varepsilon_z \end{pmatrix} = \begin{pmatrix} \varepsilon_o & 0 & 0 \\ 0 & \varepsilon_o & 0 \\ 0 & 0 & \varepsilon_e \end{pmatrix}. \tag{9.1}$$

The principal tensor components ε_o (ordinary) and ε_e (extraordinary) describe the optical response of the semiconductor to linearly polarized light. They correspond to electric field polarization either perpendicular ($E \perp c$) or parallel ($E \parallel c$) to the optic axis, respectively. Only the ordinary component is accessible in the polar c-plane orientation under normal incidence of light. In-plane isotropic strain changes the spectral dependence of ε_o and ε_e, but the relation $\varepsilon_x = \varepsilon_y = \varepsilon_o$ holds true.

Both, ε_o and ε_e, become accessible if instead $(11\bar{2}0)$- *a*-plane or $(1\bar{1}00)$-oriented *m*-plane bulk crystals are investigated. Epitaxial films with these orientations, deposited on foreign substrates, show in many cases anisotropy of the in-plane strain with the result that all three components ε_x, ε_y, and ε_z differ from each other [9.48, 9.49]. This particular case will not be discussed in the current chapter due to the lack of sufficient experimental data.

The dependence of the tensor components on the photon energy, known as complex dielectric function, is given in the form $\bar{\varepsilon}_j(\omega) = \varepsilon_{1,j}(\omega) + i \cdot \varepsilon_{2,j}$ (j = o,e). The imaginary part of the DF can be readily obtained from any form of an one-electron band structure, for example, from the quasi-particle band structure [9.7], via Fermi's Golden Rule. Of course, the calculated shape of $\varepsilon_{2,j}$ differs from work to work depending on the approach used to get the band structure. The integral over the whole Brillouin zone (BZ)

$$\varepsilon_{2,j}(\omega) = \frac{\pi \hbar e^2}{\varepsilon_0 \omega m_0} \frac{1}{8\pi^3} \sum_{c,v} \int_{BZ} f_{cv,j}(\mathbf{k}) \delta(E_c(\mathbf{k}) - E_v(\mathbf{k}) - \hbar\omega) d^3 k, \qquad (9.2)$$

has to be calculated (m_0 - free electron mass, ε_0 - permittivity of free space, e - elementary charge). The valence and conduction band energies as a function of wave vector $\mathbf{k}$ are represented by $E_v(\mathbf{k})$ and $E_c(\mathbf{k})$, respectively. The optical anisotropy originates from the orientation-dependent dimensionless oscillator strength $f_{cv,j}(\mathbf{k}) = 2|P_{cv,j}|^2/(m_0 \hbar\omega)$ which is proportional to the momentum matrix element $|P_{cv,j}|^2$ (for a detailed discussion see, for example, Ref. [9.51]).

Critical points (CP) also known as *Van Hove* singularities are found at certain $\mathbf{k}$-points of the band structure. They are defined by the condition $\nabla_{\mathbf{k}}(E_c(\mathbf{k}) - E_v(\mathbf{k})) = 0$. The spectral dependence of $\varepsilon_{2,j}(\omega)$ shows peculiarities in the photon energy range around $\hbar\omega = E_c(\mathbf{k}) - E_v(\mathbf{k})$ like resonances or sharp changes of the absorption depending on the dimension and number of negative reduced effective masses [9.12]. As demonstrated in chapter 8, the necessary inclusion of electron-hole interaction via solving the Bethe-Salpeter equation for the DF calculation leads to shift of the $\varepsilon_{2,j}(\omega)$ resonances to lower energies and a redistribution of oscillator strength with respect to the one-electron DF. Although the term *CP* refers strictly only to the band structure it will be used hereafter in order to denote the main (excitonic) transition energy.

Since the real and imaginary parts of the DF for each polarization direction obey the Kramers-Kronig (KK) relation

$$\varepsilon_{1,j}(\omega) = 1 + \frac{2}{\pi} \wp \int_0^{+\infty} \frac{\omega' \cdot \varepsilon_{2,j}(\omega')}{\omega'^2 - \omega^2} d\omega', \qquad (9.3)$$

($\wp$ - the principal value of the integral), it is obvious that the singularities strongly influence the shape of $\varepsilon_{1,j}(\omega)$ far below the CP transition energy. Therefore, a discussion of nitride optical properties should not only focus on the band gap region but also include the range of the high energetic CPs. A detailed review can be found elsewhere [9.40].

The optical properties of semiconductors are often expressed in terms of the complex refractive index $\bar{N}_{\mathrm{j}}(\omega)$ (again along the principal axis of the crystal). It is related to $\bar{\varepsilon}_{\mathrm{j}}(\omega)$ via

$$\bar{N}_{\mathrm{j}}(\omega) = n_{\mathrm{j}}(\omega) + i\kappa_{\mathrm{j}}(\omega) = \sqrt{\bar{\varepsilon}_{\mathrm{j}}(\omega)}. \tag{9.4}$$

Here, $n_{\mathrm{j}} = \{0.5[\varepsilon_{1,\mathrm{j}} + (\varepsilon_{1,\mathrm{j}}^2 + \varepsilon_{2,\mathrm{j}}^2)^{1/2}]\}$ and $\kappa_{\mathrm{j}} = \{0.5[\varepsilon_{1,\mathrm{j}} - (\varepsilon_{1,\mathrm{j}}^2 + \varepsilon_{2,\mathrm{j}}^2)^{1/2}]\}$ denote the index of refraction and the extinction coefficient, respectively. Finally, for the estimation of the light penetration depth, for example, the knowledge of the absorption coefficient $\alpha_{\mathrm{j}}(\omega)$ is important which is related to $\varepsilon_{2,\mathrm{j}}$ by

$$\alpha_{\mathrm{j}}(\omega) = \frac{\omega}{n_{\mathrm{j}}(\omega)c_0}\varepsilon_{2,\mathrm{j}}(\omega), \tag{9.5}$$

where c_0 is the velocity of light in vacuum. It should be noticed that the spectral dependence of $\alpha_{\mathrm{j}}(\omega)$ does not only depend on $\varepsilon_{2,\mathrm{j}}(\omega)$ but also on $n_{\mathrm{j}}(\omega)$ which is generally not constant and in particular not around the band gap! The effect influences the determination of band gaps from transmission studies, but it is disregarded in most works.

The metastable phase of the group-III nitrides is the zinc-blende crystal structure belonging to the $F4_3m(T_d^2)$ space group. Unstrained layers show isotropic optical behavior, that is, all principal components of the dielectric tensor are identical ($\varepsilon_x(\omega) = \varepsilon_y(\omega) = \varepsilon_z(\omega) = \varepsilon(\omega)$), and only the scalar quantity $\bar{\varepsilon}(\omega)$ has to be determined.

9.2.2 Influence of strain on the optical properties

The optical properties of all hexagonal nitrides around the band gap are strongly correlated to the valence band structure around the Γ-point of the Brillouin zone and the symmetry of the corresponding wave functions. Crystal field (Δ_{cf}) and spin-orbit interactions (Δ_{so}) split the threefold (neglecting spin up and down) degenerated VB maximum with Γ_{15}^{v} symmetry into one Γ_9^{v} and two Γ_7^{v} bands. The quasi-cubic model (also called the Hopfield model) [9.52] yields for the energetic position of Γ_9^{v} relative to the two other bands

$$\Gamma_9^{\mathrm{v}} - \Gamma_{7\pm}^{\mathrm{v}} = \frac{\Delta_{\mathrm{cf}} + \Delta_{\mathrm{so}}}{2} \mp \frac{1}{2}\sqrt{(\Delta_{\mathrm{cf}} + \Delta_{\mathrm{so}})^2 - \frac{8}{3}\Delta_{\mathrm{cf}}\Delta_{\mathrm{so}}} . \tag{9.6}$$

A spin-orbit energy of 13 meV was calculated for InN [9.53]. Using this value, Goldhahn *et al.* estimated a Δ_{cf} energy of 19 meV from their experimental studies [9.9]. For comparison, Carrier and Wei [9.54] reported theoretical values of $\Delta_{\mathrm{so}} = 5$ meV and $\Delta_{\mathrm{cf}} = 19$ meV. The proximity of the valence bands and the narrow band gap cause strong deviations from a parabolic character for all bands. In order to describe the effects correctly at least an 8×8 $\mathbf{k} \cdot \mathbf{p}$ model should be applied.

However, fully consistent band structure parameters have not been reported so far; the recently published ones of Rinke *et al.* [9.55] appear to be unsuitable because the spin-orbit interaction was neglected.

A more convenient approach, being sufficient for demonstrating the general behavior, is to treat the nonparabolicity of the CB separately and to describe the VBs within the 6×6 framework. It can also be used to study the influence of in-plane strain on the optical properties. Previous work on GaN revealed that not only the transition energies are shifted in a characteristic manner [9.47], but the oscillator strengths are also modified [9.56]. A detailed calculation of the w-InN optical properties for any possible strain combination has been published elsewhere [9.49]; in the current chapter only results for isotropic in-plane strain (being typical for the growth of c-plane films on c-plane sapphire substrate) will be presented.

We adopt the six band $\mathbf{k} \cdot \mathbf{p}$ model derived by Chuang and Chang [9.57] for treating the influence of strain on the band structure of wurtzite semiconductors. The full 6×6 VB Hamiltonian is given by

$$H^v(\mathbf{k}) = \begin{bmatrix} F & -K^* & -H^* & 0 & 0 & 0 \\ -K & G & H & 0 & 0 & \Delta \\ -H & -H* & \lambda & 0 & \Delta & 0 \\ 0 & 0 & 0 & F & -K & H \\ 0 & 0 & \Delta & -K^* & G & -H^* \\ 0 & \Delta & 0 & H^* & -H & \lambda \end{bmatrix} = \begin{bmatrix} H^U_{3\times3}(\mathbf{k}) & 0 \\ 0 & H^L_{3\times3}(\mathbf{k}) \end{bmatrix}, \quad (9.7)$$

which can be block-diagonalized into two 3×3 matrices in the form

$$H^U_{3\times3}(\mathbf{k}) = \begin{bmatrix} F & K & -iH \\ K & G & \Delta - iH \\ iH & \Delta + iH & \lambda \end{bmatrix} \quad \text{and} \quad H^L_{3\times3}(\mathbf{k}) = \begin{bmatrix} F & K & iH \\ K & G & \Delta + iH \\ -iH & \Delta - iH & \lambda \end{bmatrix}.$$

The lower 3×3 Hamiltonian is the complex conjugate of the upper one and both have exactly the same eigenvalues since the energies are real. The wave functions of the upper Hamiltonian are the complex conjugate of the corresponding wave functions of the lower Hamiltonian. The elements of the Hamiltonian take the form

$$F = \Delta_1 + \Delta_2 + \lambda + \theta \tag{9.8}$$

$$G = \Delta_1 - \Delta_2 + \lambda + \theta$$

$$K = \frac{\hbar^2}{2m_0} A_5 (k_x + ik_y)^2 + D_5 (\varepsilon_{xx} + 2i\varepsilon_{xy} - \varepsilon_{yy})$$

$$H = \frac{\hbar^2}{2m_0} A_6 k_z (k_x + ik_y) + D_6 (\varepsilon_{zx} + i\varepsilon_{yz})$$

$$\lambda = \frac{\hbar^2}{2m_0} \left[A_1 k_z^2 + A_2 k_t^2 \right] + D_1 \varepsilon_{zz} + D_2 (\varepsilon_{xx} + \varepsilon_{yy})$$

$$\theta = \frac{\hbar^2}{2m_0} \left[A_3 k_z^2 + A_4 k_t^2 \right] + D_3 \varepsilon_{zz} + D_4 (\varepsilon_{xx} + \varepsilon_{yy})$$

$$\Delta = \sqrt{2}\Delta_3.$$

The in-plane wave vector is given via $k_t^2 = k_x^2 + k_y^2$. The parameters Δ_1, Δ_2, and Δ_3 are related to the crystal field and spin-orbit energy by $\Delta_1 = \Delta_{cf}$ and $\Delta_2 = \Delta_3 = \Delta_{so}/3$ (cubic approximation). The terms containing A_i represent the contributions of remote bands which are calculated by Löwdin's perturbation theory [9.58]. These parameters are similar to the Luttinger parameters in zinc-blende crystals; a summary of widely used A_i parameters is given in Table 9.1. Even in Refs. [9.59] and [9.60], two sets of parameters were reported. For the subsequent calculations, we used the data from the right column of Table 9.1.

TABLE 9.1
A_i parameters for wurtzite InN.

	Ref. [9.59]	Ref. [9.59]	Ref. [9.60]	Ref. [9.60]
A_1	−9.470	−10.841	−9.620	−9.280
A_2	−0.641	−0.651	−0.720	−0.600
A_3	8.771	10.100	8.970	8.680
A_4	−4.332	−4.864	−4.220	−4.340
A_5	−4.264	−4.825	−4.350	−4.320
A_6	−5.546	−6.556	-	−6.080

The components of the strain tensor to be used in Eq. (9.8) are denoted by ε_{ii}, and D_i are the deformation potentials for the wurtzite crystals. The growth direction for c-plane InN films is along the z-direction, which is parallel to the c-axis. So the film is free to expand or contract along this direction. The strain in the growth plane is usually isotropic ($\varepsilon_{xx} = \varepsilon_{yy}$) when the films are grown on substrates having hexagonal symmetry such as c-plane SiC or c-plane sapphire. So the strain tensor contains only the following non-vanishing diagonal elements:

$$\varepsilon_{xx} = \varepsilon_{yy} = \frac{a - a_0}{a} \tag{9.9}$$

$$\varepsilon_{zz} = \frac{c - c_0}{c_0} = -\frac{2C_{13}}{C_{33}}\varepsilon_{xx} , \qquad (9.10)$$

where C_{13} and C_{33} denote the stiffness constants, and a_0 and a are the lattice constants for strain-free InN and the strained film, respectively.

Let us initially consider the VB structure of unstrained w-InN around the Γ-point of the Brillouin zone. The eigenvalues of Eq. (9.8) represent the three closely spaced top valence-band states. These are often labeled too as heavy hole (HH), light hole (LH), and crystal-field split-off hole (CH). Their dispersion along and perpendicular to the *c*-axis (along k_z and k_t directions, respectively) is shown in Fig. 9.1. The bands take the symmetry Γ_9^v, Γ_{7+}^v, and Γ_{7-}^v at $k = 0$. The energy of the uppermost VB (Γ_9^v) amounts to $\Delta_1 + \Delta_2 = \Delta_{cf} + \Delta_{so}/3 = 23.3\,\text{meV}$. The splitting of the two Γ_7^v VBs with respect to Γ_9^v is simply given by Eq. (9.6) yielding 6.4 meV and 25.6 meV for $\Gamma_9^v - \Gamma_{7+}^v$ and $\Gamma_9^v - \Gamma_{7-}^v$, respectively. The HH valence band is characterized by an almost smooth dispersion; the LH and CH bands show the typical repulsion behavior for small k_t values.

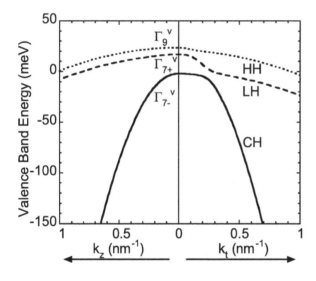

FIGURE 9.1

Valence band dispersion for unstrained wurtzite InN crystals in the vicinity of the Γ-point of the Brillouin zone.

The energetic position and the dispersion of the CB is needed in addition for the calculation of the optical properties. It is characterized by Γ_7^c symmetry at the center of the BZ. Disregarding the small orientation anisotropy of the conduction band, which is responsible for the slightly different electron effective masses parallel and perpendicular to the *c*-axis [9.32], the dependence on the **k** is given by

$$E_c(\mathbf{k}) = \frac{E_0}{2} + \frac{\hbar^2 k^2}{2m_0} + \frac{1}{2}\sqrt{E_0^2 + 4E_P\frac{\hbar^2 k^2}{2m_0}} + \Delta_1 + \Delta_2 + \alpha_{||}\varepsilon_{zz} + \alpha_\perp(\varepsilon_{xx} + \varepsilon_{yy}) .$$

$$(9.11)$$

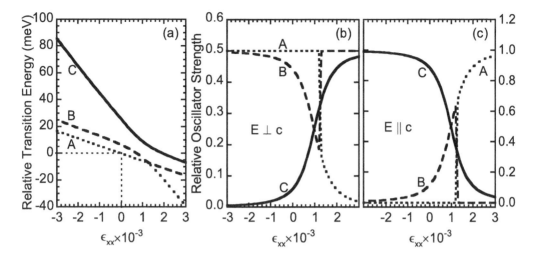

FIGURE 9.2

(a) Splitting of the transition energies at the Γ- point of the Brillouin zone and their shift as a function of biaxial in-plane strain. All data are plotted as the difference to the value for E_A in the strain-free case (corresponding to E_0) as explained in the text. The relative oscillator strength of the three transitions for light polarization $\mathbf{E} \perp \mathbf{c}$ and $\mathbf{E} \parallel \mathbf{c}$ is shown in (b) and (c), respectively.

The first three terms account for the nonparabolicity of the CB in the framework of Kane's two-band $\mathbf{k} \cdot \mathbf{p}$ model [9.61]; the fundamental strain-free zero-density gap is denoted by E_0. E_P is an energy parameter being proportional to the momentum matrix element; we adopt for w-InN the value of $E_P = 10\,\text{eV}$ reported in Ref. [9.8]. The last two terms in Eq. (9.11) describe the strain-induced shift of the conduction band; the coefficients $\alpha_{\parallel}$ and $\alpha_{\perp}$ denote the conduction band deformation potential parallel and perpendicular to the c-axis, respectively.

Finally, the deformation potentials are needed for the calculations. We follow the approach of Bhattacharyya *et al.* [9.48, 9.49] and apply the quasi-cubic approximation yielding $D_1 = 3.7\,\text{eV}, D_2 = 4.5\,\text{eV}, D_3 = -(D_1 - D_2) = 8.2\,\text{eV}, D_4 = -D_3/2 = -4.1\,\text{eV}$ and $D_5 = -4.0\,\text{eV}$ for the VBs and $\alpha_{\parallel,\perp} = -7.2\,\text{eV}$ for the CB.

In the following part, optical transitions from the HH (Γ_9^v at $k = 0$), LH (Γ_{7+}^v), and CH (Γ_{7-}^v) bands into the CB (Γ_7^c) are labeled A, B, and C, while the transition energies at $k = 0$ are labelled E_A, E_B, and E_C, respectively. For demonstrating how strain shifts the transition energies and changes the VB splitting, the energy difference $\Gamma_7^c - \Gamma_9^v = E_A = E_0$ for strain-free material is used as a reference point, that is, the behavior of $E_{A,B,C} - E_0$ is discussed. The calculated results as a function of isotropic in-plane strain are plotted in Fig. 9.2(a). For compressive in-plane strain all transition energies undergo a continuous shift to higher energies. The splitting between E_A and E_B remains almost constant and amounts to $\sim 9\,\text{meV}$. The third transition C exhibits a much stronger shift. Toward the tensile strain direction, the behavior shows a peculiarity. At first, the $E_A \leftrightarrow E_B$ difference decreases followed by an another increase. The slope of curves changes in addition at this point which is attributed to the strain-induced valence band structure modification. Figure 9.1 shows that the lowest energy difference between the HH and LH bands is found at small k_t wave vectors. Tensile strain lowers the splitting furthermore and, at a value of $\varepsilon_{xx} = 1.3 \times 10^{-3}$, a distinct anticrossing between the HH and LH bands due to strong coupling occurs. It is accompanied by an exchange of the nature of the wave functions. This effect becomes very important for the discussion of the oscillator strength.

With the knowledge of the band energies and wave functions, the contribution of each transition to the imaginary part of the DF can be calculated. The method is described in detail in Refs. [9.56] and [9.57]. It was demonstrated in the former reference that the general behavior does not change if electron-hole interaction is additionally taken into account. The sum of the oscillator for any polarization direction is always unity; that is, for the two polarization directions discussed in the current chapter, one gets

$$f_{A,j} + f_{B,j} + f_{C,j} = 1, \qquad j = \text{o}, \text{e} . \tag{9.12}$$

The relative contributions change with strain as depicted in Fig. 9.2(b) and (c) for $E \perp c$ and $E \parallel c$, respectively. Let us consider at first the results for the ordinary polarization configuration. According to Fig. 9.2(a), the fundamental absorption edge is always defined by the energy of E_A. Its oscillator strength under compressive strain is constant and amounts to $f_{A,\text{o}} = 0.5$. Transition B also appears strong, but the low splitting between E_B and E_A of only 6 meV makes it unlikely to observe B separately. The contribution of C is only small for compressive strain. Pronounced interchanges of the oscillator strength should be noticed for the tensile strain range. The anticrossing of HH and LH leads to a sharp drop of $f_{A,\text{o}}$ accompanied by an enhancement of $f_{B,\text{o}}$ to 0.5. Further increase of the tensile strain causes an additional increase of the splitting of E_A and E_B and some peculiarities around the absorption edge should be observed. A weak onset of absorption is measured at E_A which is followed by a second but now strong increase of $\varepsilon_{2,\text{o}}$ (or α_o) at E_B. The SE measurements on $(11\bar{2}0)$-oriented w-InN films yield the spectral dependence of $\bar{\varepsilon}_e$ corresponding to $E \parallel c$ for which the oscillator strength results of Fig. 9.2(c) have to be applied. The layers proved to be almost relaxed. Then, $f_{A,e}$ is zero, $f_{B,e}$ is very small, and the main absorption edge is found at E_C. A detailed analysis of the experimental results will be given in section 9.3.5.

It should be noted that the above discussed properties are restricted to a limited region around the Γ-point. The oscillator strengths become k-dependent for all wurtzite semiconductors if a wider range is considered [9.51]. However, the reliability of the experimental data for w-InN is not yet sufficient to demonstrate this behavior.

9.2.3   Valence band ordering and optical selection rules for wurtzite alloys

It was shown in the previous section that the wurtzite nitrides exhibit a pronounced absorption anisotropy in the vicinity of the fundamental band gap caused by the valence band ordering around the Γ-point of the Brillouin zone and the symmetry of the corresponding wave functions. Obviously, similar behavior should be expected for the $In_xGa_{1-x}N$ and $In_xAl_{1-x}N$ alloys. The spin-orbit energies of all nitrides are positive; values of 13 meV [9.53], 18 meV [9.56], and 19 meV [9.54] are very likely for InN, GaN, and AlN, respectively. With these values, Δ_{cf} energies of 19 meV [9.9], 10 meV [9.56], and -230 meV [9.62] are estimated from experimental studies. In the case that both quantities, Δ_{so} and Δ_{cf}, are positive (InN, GaN, and their alloys), the uppermost valence band is always formed by Γ_9^v states.

The behavior for the $In_xAl_{1-x}N$ alloys is more complicated. The Γ_{7+}^v band becomes the uppermost one with the sign change of Δ_{cf}. This is pointed out in Fig. 9.3, where the relative energy position of the three valence bands is plotted as a function of Δ_{cf} for $\Delta_{so} = 13$ meV. Assuming a linear dependence of Δ_{cf} on the alloy composition (no data on the accurate dependence have been reported so far), the band crossing of Γ_9^v and Γ_{7+}^v should occur at an Al content of $x = 0.076$.

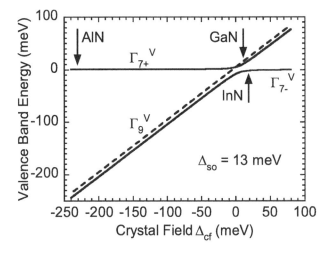

FIGURE 9.3
Relative valence band energies at the Γ-point of the Brillouin zone as a function of crystal field for a spin-orbit energy of 13 meV.

Optical transitions from Γ_9^v into the Γ_7^c conduction band (again denoted by A) are only allowed in the configuration $E \perp c$, while $\Gamma_{7+}^v \longrightarrow \Gamma_7^c$ (B) and $\Gamma_{7-}^v \longrightarrow \Gamma_7^c$ (C) transitions contribute to both, $\bar{\varepsilon}_o$ and $\bar{\varepsilon}_e$, but with a polarization dependent transition probability as displayed in Fig. 9.4. Together with the band energies of Fig. 9.3 the following fundamental properties should be noted: (i) a positive Δ_{cf} value leads to a larger extraordinary absorption edge energy (transition B) with respect to the ordinary one (transition A); (ii) the relative contribution of B to $\bar{\varepsilon}_e$ becomes weaker with increasing Δ_{cf}, which means that only a pronounced feature due to C is observed (for example, in the case of InN); and (iii) independent of polarization, the lowest observable absorption edge for negative Δ_{cf} energies is always found at the energy E_B! The appearance of B is strong for $E \parallel c$ but only weak for $E \perp c$ with a dominating transition A. With decreasing spin-orbit energies (decrease from AlN to InN), this effect becomes more distinct.

The previous considerations strictly apply only for strain-free nitrides. The pseu-

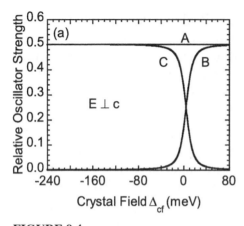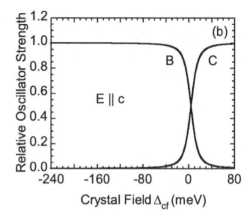

FIGURE 9.4

Relative oscillator strength of transitions involving the three valence bands for light polarization $E \perp c$ (a) and $E \parallel c$ (b) calculated for a spin-orbit energy of 13 meV.

domorphic film growth on rather thick bulk-like layers or the mismatch in the thermal expansion coefficients introduces, however, in-plane biaxial strain. It shifts the transition energies and alters the transition probabilities in a characteristic manner (for example, see Ref. [9.56]). The inclusion of electron-hole interaction (excitonic effects), which is inevitable due to the large effective Rydberg energies of the nitrides, does not change the general dependence of the relative oscillator strength compared to the single-electron picture [9.56].

9.2.4 Carrier-induced band gap renormalization and Burstein-Moss shift

High electron concentration (N_e) causes a change of the band structure in the vicinity of the Γ-point of the Brillouin zone and a change of the absorption properties which is illustrated in Fig. 9.5. Only the uppermost HH valence band is shown beside the CB for the sake of simplicity. Two effects have to be considered in order to attain the zero-density band gaps E_0 for InN and the In-rich InGaN and InAlN alloys. First, carrier-induced many-body interaction leads to gap shrinkage Δ_{BGR}, also known as band-gap renormalization. The lower or *renormalized* gap is in the following denoted by E_{ren}. Secondly, all CB states are occupied up to the Fermi energy (E_{F}) if zero temperature is assumed. Excitation of holes from the VBs into these states is forbidden. Based on **k**-conservation for the main absorption processes, the onset of ε_2 is shifted to higher photon energies (Burstein-Moss shift) and found at $E_{\mathrm{F}}(k_{\mathrm{F}})$. A quantitative description of both effects is given below.

Two approaches have been proposed in order to describe BGR for w-InN. The first one is based on a semi-empirical formula in the form

$$\Delta_{\mathrm{BGR}}^{(1)} = -20\,\mathrm{meV}(N_e/10^{18}\,\mathrm{cm}^{-3})^{1/3} \,. \tag{9.13}$$

The prefactor was adjusted to match experimental absorption and photoluminescence data. It was recently corrected from the earlier proposed value of -11.1

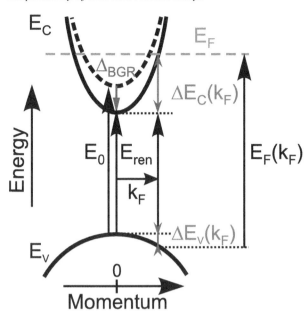

FIGURE 9.5
Schematic illustration of the band alignment in the vicinity of the Γ-point of the BZ induced by band-gap renormalization Δ_{BGR} and band filling effects $\Delta E_{\mathrm{c}}(k_{\mathrm{F}})$ and $\Delta E_{\mathrm{v}}(k_{\mathrm{F}})$ as explained in the text. Note that $E_{\mathrm{F}}(k_{\mathrm{F}})$ represents the valence-conduction-band spacing at the Fermi wave vector k_{F}.

[9.2] to -20 [9.63]. The second method calculates separately the shifts due to electron-electron ($\Delta E_{\mathrm{e-e}}$) and electron-ionized impurity ($\Delta E_{\mathrm{e-i}}$) interaction [9.8]. The total effect amounts to

$$\Delta_{\mathrm{BGR}}^{(2)} = \Delta E_{\mathrm{e-e}} + \Delta E_{\mathrm{e-i}} \ . \tag{9.14}$$

The downward shift of the CB resulting from the electron-electron interaction is given by

$$\Delta E_{\mathrm{e-e}} = -\frac{2e^2 k_{\mathrm{F}}}{4\pi^2 \varepsilon_0 \varepsilon_{\mathrm{r}}} - \frac{e^2 k_{\mathrm{TF}}}{8\pi \varepsilon_0 \varepsilon_{\mathrm{r}}} \left[1 - \frac{4}{\pi} \arctan\left(\frac{k_{\mathrm{F}}}{k_{\mathrm{TF}}}\right) \right], \tag{9.15}$$

where $k_{\mathrm{F}} = (3\pi^2 N_e)^{1/3}$ is the Fermi wave vector, $k_{\mathrm{TF}} = (2/\sqrt{\pi})(k_{\mathrm{F}}/a_{\mathrm{B}})^{1/2}$ denotes the Thomas-Fermi screening wave vector, $a_{\mathrm{B}} = 0.53$ Å$\times 10^{-10} \varepsilon_{\mathrm{r}} m_0/m_{\mathrm{e}}^*(N_e)$ represents the effective Bohr radius, and ε_{r} is the static dielectric constant. The average electron effective mass $m^*(N_e)$ accounts for the nonparabolic dispersion of the CB for a low band gap material and has to be calculated via [9.64]:

$$\frac{1}{m^*(N_e)} = \frac{1}{12\pi^3 \hbar^2 N_e} \int d\mathbf{k} \frac{\partial^2 E_{\mathrm{c}}(k)}{\partial k^2} f(E_{\mathrm{c}}) \ , \tag{9.16}$$

where $f(E_{\mathrm{c}})$ is the Fermi distribution function for the CB electrons. The integral runs over all occupied states in the CB. The position of the Fermi energy E_{F} follows from the fit of the $\varepsilon_2(\omega)$ curves as described in section 9.3.5. The dispersion of the CB $E_{\mathrm{c}}(\mathbf{k})$ should be modified with respect to Eq. (9.11). BGR lowers the gap and enhances the interaction between the conduction and valence bands. Consequently, we applied the formula given by Kane's two-band $\mathbf{k} \cdot \mathbf{p}$ model [9.61] and inserted the renormalized band gap ($E_{\mathrm{ren}} = E_0 + \Delta_{\mathrm{BGR}}$) instead of the fundamental band gap E_0 in order to account for the enhanced nonparabolicity of the CB [9.10, 9.29]:

$$E_c(\mathbf{k}) = \frac{E_{\mathrm{ren}}}{2} + \frac{\hbar^2 k^2}{2m_0} + \frac{1}{2}\sqrt{E_{\mathrm{ren}}^2 + 4E_P \frac{\hbar^2 k^2}{2m_0}}. \tag{9.17}$$

Our approach was recently emphasized by DFT band-structure calculations of degenerately doped semiconductors [9.65]. Consequently and in accordance with the recent experimental results of Kamińska *et al.* [9.66], E_{ren} also contains the influence of strain. Furthermore, a linear dependence of E_P on the alloy composition is assumed between the values for InN and GaN of 10 eV [9.8] and 13.2 eV [9.67], respectively, or for InN and AlN (18 eV [9.55]).

The contribution of the electron-ion interaction to the gap shrinkage can be written as

$$\Delta E_{\mathrm{e-i}} = -\frac{e^2 N_e}{\varepsilon_0 \varepsilon_r a_B k_{\mathrm{TF}}^3}. \tag{9.18}$$

For the subsequent calculations, an ε_r value of 9.5 was employed [9.5] being appropriate for hexagonal InN with a band gap of 0.67 eV. It is also used as a first approximation for alloys with $x > 0.67$. For comparison, $\varepsilon_r = 10.4$ was reported [9.68] for GaN with a much higher gap of $E_0 = 3.447$ eV.

Figure 9.6 provides a comparison of the two dependencies $\Delta_{\mathrm{BGR}}^{(1)}$ and $\Delta_{\mathrm{BGR}}^{(2)}$ on the bulk electron concentration for the hexagonal InN films. Both approaches yield a markedly different size of the BGR, especially for high N_e values. It turns out that the BGR correction has a strong influence on the determination of the zero-density gap.

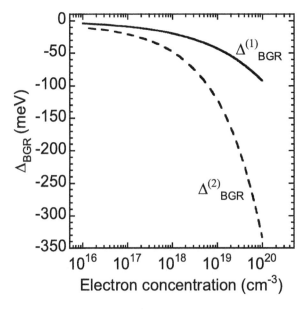

FIGURE 9.6

Comparison of the BGR contributions as a function of the electron concentration for two different approaches. The solid and the dashed lines describe the CB shift due to band-gap renormalization according to Ref. [9.63] and Ref. [9.8], respectively.

The second effect which has to be considered in the interpretation of the optical data is the Burstein-Moss shift. In degenerate semiconductors, the Fermi level E_F is located above the CB minimum $[\Delta E_c(k_F)]$ as illustrated in Fig. 9.5. As a consequence, the onset of absorption does not occur at the Γ-point (at $k = 0$) of the BZ but around (at finite temperatures) the Fermi wave vector k_F. Here, the energy of the valence band at k_F is lower in comparison to the Γ-point [this quantity is denoted by $\Delta E_v(k_F)$]. This means that a small amount of the Burstein-Moss shift $[\Delta E_{cv}(k_F) = \Delta E_c(k_F) - \Delta E_v(k_F) = E_F(k_F) - E_{ren}]$ is caused by the curvature of the VB. The smooth heavy hole VB dispersion ($E_v(\mathbf{k})$) can be described by a parabolic dependence with a proposed effective mass for the holes of $m_h = 0.5m_0$ [9.69]. Note that BGR lowers the blueshift of $E_{cv}(k_F)$ with N_e in comparison to a rigid band structure, even a redshift becomes possible for low electron concentrations.

9.2.5 Analysis of ellipsometric data

The ellipsometric measurements presented in the current chapter were carried out in three spectral ranges. A Fourier-transform-based spectroscopic ellipsometer was used to investigate the optical properties in the MIR spectral range (350–2000 cm^{-1}). The spectral resolution was set to 1 cm^{-1}. The room temperature spectra were taken at angles of incidence ϕ of 60°, 65°, and 70°. A commercial lab ellipsomter was used in the NIR to UV range from 0.54 up to 5.5 eV. Multiple angles ϕ were measured. For the UV to VUV range we employed the ellipsometer attached to the Berlin Electron Storage Ring for Synchrotron Radiation (BESSY II). In this case, the angle of incidence was fixed to $\sim$68°. The latter set-up allowed us to carry out in addition temperature-dependent studies.

It is commonly assumed that ellipsometry determines directly the real and imaginary parts of the complex DF. This statement is not fully correct and leads sometimes to misinterpretations of the experimental data. In the SE experiments, one starts with linearly-polarized light shining on the sample under an angle of incidence ϕ. The two electric field components perpendicular $E_{s,i}$ and parallel $E_{p,i}$ to the plane of incidence undergo a different reflection which is expressed by the complex Fresnel coefficients $\bar{r}_s$ and $\bar{r}_p$. The field components of the reflected light are given by

$$E_{s,r} = \bar{r}_s \cdot E_{s,i} = r_s e^{i\delta_s} \cdot E_{s,i}, \qquad E_{p,r} = \bar{r}_p \cdot E_{p,i} = r_p e^{i\delta_p} \cdot E_{p,i} . \qquad (9.19)$$

The ratio of $E_{p,r}$ to $E_{s,r}$ defines the ellipsometric parameters Ψ and Δ which are given by

$$\bar{\rho} = \frac{\bar{r}_p}{\bar{r}_s} = \frac{r_p}{r_s} e^{i(\delta_p - \delta_s)} = \frac{r_p}{r_s} e^{i(\Delta)} = \tan\Psi e^{i\Delta} . \qquad (9.20)$$

Only for a semi-infinite, smooth sample, the complex reflection coefficient $\bar{\rho}$ can be directly converted into the desired quantity

$$\bar{\varepsilon} = \sin^2\phi \left(1 + \tan^2\phi \left(\frac{1-\rho}{1+\rho} \right)^2 \right) , \qquad (9.21)$$

the complex DF. The application of Eq. (9.21) in any other case yields only a pseudo-dielectric function which is commonly denoted by $\langle \bar{\varepsilon} \rangle$. Organic contamination alone is enough to alter the optical response, as demonstrated in Fig. 9.7. A hexagonal and a cubic InN film were measured before and after heating under UHV conditions, which is known to remove the organic molecules from the surface. A strong increase of $\langle \varepsilon_2 \rangle$ is found after treatment. This influence has to be taken into account when samples are measured in air.

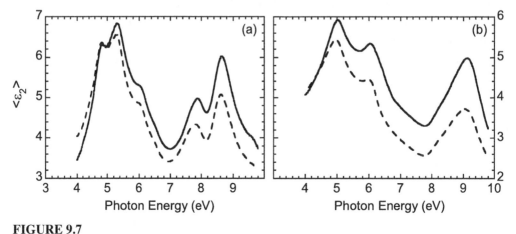

FIGURE 9.7

Comparison of the imaginary parts of the pseudo-DF before (dashed lines) and after (solid lines) heating under UHV conditions for (a) w-InN (sample In100) and (b) zb-InN (sample A). The data were recorded at room temperature.

Oxides [9.16] or surface roughness [9.16, 9.19, 9.50] cause similar behavior, an increase of $\langle \varepsilon_2 \rangle$ and a decrease of $\langle \varepsilon_1 \rangle$ (not presented here). Therefore, in order to get the DF of the material, modeling of the optical response must follow the SE measurements. The real and the imaginary parts of the complex DF in the range of 0.54 – 9.5 eV were obtained by fitting the experimental data Ψ and Δ using a multi-layer model similar to the procedure reported in Ref. [9.50, 9.70]. For the interpretation of ellipsometry data, the roughness of the surface (3–5 nm) was treated in the Bruggeman effective-medium approximation (EMA) assuming 50% of voids in a matrix of the underlying material. The reliability and accuracy of this approach for obtaining bulk DFs have been demonstrated elsewhere [9.71]. In this work, ε_1 and ε_2 were independently fitted at every photon energy, that is, without making any assumption concerning the spectral dependence of $\bar{\varepsilon}$. The Kramers-Kronig consistency of the optical data was proven by numerical integration of ε_2. For films with (0001) orientation, that is, with c-axis normal to the surface, the extracted data are very close to the ordinary DF. A comparison of the DF shape for a-plane and c-plane w-InN films [9.23, 9.24] revealed only small deviations in magnitude for photon energies above 8 eV. The effect has to be attributed to the decreasing index of refraction and therefore increasing angle of refraction in this range. It means that the electric field component parallel to the *c*-axis, probing $\bar{\varepsilon}_e$, increases. As a result, the determined DF is increasingly influenced by the extraordinary DF which shows here a much larger oscillator strength.

9.3 Experimental results for wurtzite InN

9.3.1 Samples

The w-InN films studied in this work were grown [9.3, 9.72–9.74] by PI-MBE either on *c*-plane or $(1\bar{1}02)$-oriented (*r*-plane) sapphire substrates in order to obtain *c*- or *a*-plane orientation, respectively. Growth of the (0001)-oriented layer with In-face polarity (sample In100) was initiated by the deposition of 10 nm *c*-plane AlN, continued with a thick (~200 nm) *c*-plane GaN buffer, and finished by the deposition of the InN film at a temperature of 470°C [9.3]. The procedure to obtain the high-quality $(000\bar{1})$-oriented InN films with N-face polarity can be summarized as follows. Prior to growth of sample N100a, the *c*-plane sapphire substrate was thermally cleaned at 940°C followed by a nitridation of the surface at 400°C for 45 minutes. Then, a high temperature N-polar GaN epilayer was grown with a thickness of 500 nm. This N-polar GaN template resulted in the growth of a N-polar InN film at 580°C [9.72]. The polarity was also confirmed by coaxial impact collision ion scattering spectroscopy measurements [9.75]. The N-polar InN film of sample N100b was grown at 530°C on top of a low-temperature-grown 60-nm-thick InN buffer layer which was directly deposited on the sapphire substrate [9.73]. Finally, sample *a*100 consists of an *a*-plane InN film grown on an *r*-plane sapphire substrate by using a 12 nm AlN nucleation layer followed by a 300- nm-thick GaN buffer layer. In this case, *a*-plane AlN and GaN were formed on *r*-plane sapphire first, leading to the formation of *a*-plane InN [9.74]. The GaN buffer layer was confirmed to be $(11\bar{2}0)$-oriented and the InN film, grown on top at $T = 470°C$, follows the orientation of the GaN buffer. The thickness of investigated InN layers as well as the electron concentrations and mobilities resulting from preliminary Hall measurements are given in Table 9.2.

The lattice constants (*a* and *c*) of the w-InN layers were determined by employing

TABLE 9.2

Layer thicknesses, electrical properties from Hall measurements as well as lattice constants and in-plane strain values of the investigated w-InN films.

Sample	d_{InN} (nm)	N_e(Hall) ($\times 10^{18}$cm^{-3})	μ_e(Hall) (cm^{2}V^{-1}s^{-1})	a (Å)	c (Å)	ε_{xx} ($\times 10^{-4}$)
In100	1000	1.5	1200	3.5400	5.6990	6.50
N100a	2300	3.4	1450	3.5295	5.7081	−23.18
N100b	550	11	780	3.5385	5.7044	2.26
*a*100	670	6.0	270	3.5378	5.7036	0.10

x-ray diffraction (XRD) measurements. In the case of the a-plane sample (a100), the lattice parameters along the [11$\bar{2}$0] and [0001] directions of the crystal are given. As demonstrated in the following sections, not only band-filling effects, but also in-plane strain plays an important role for the evaluation of the experimental data around the absorption edge. The strain-free values of the w-InN lattice constants were determined by Paszkowicz *et al.* [9.76] on InN powder and amount to $a_0 = 3.5377$ Å and $c_0 = 5.7037$ Å. Therefore, using the strain-free value a_0, we gain the in-plane strain ε_{xx} of the w-InN films. Those values are summarized in Table 9.2 together with the determined lattice constants of the w-InN films.

9.3.2 Infrared spectroscopic ellipsometry

Hall measurements are commonly employed to determine electrically the electron concentration of thin w-InN films. An uncertainty arises from the presence of a strong surface electron accumulation layer [9.35, 9.77] in these films. An averaged, but too high volume concentration is calculated from the obtained sheet electron density (including accumulation layer) divided by the layer thickness.

Infrared spectroscopic ellipsometry (IR-SE) represents one of the complementary measurement methods to obtain the electron density of the layer by optical investigations. Particularly, it offers an opportunity to determine N_e in the bulk-like part of the sample and represents an alternative, if Hall measurements are not possible because of conducting buffer layers or substrates.

The electron concentration is obtained from the analysis of the coupled phonon-plasmon modes in the MIR determining the shape of the complex DF of hexagonal InN. The anisotropic DF $\bar{\varepsilon}_j(\omega)$ (for the electric field polarization along $j =$ " $\perp$", " $\parallel$" with respect to the optical axis) in the MIR range can be described by a factorized model based on the anharmonic coupling effects between free-carrier plasmons and longitudinal-optical phonons [9.31, 9.78]:

$$\bar{\varepsilon}_j(\omega) = \varepsilon_{\infty,j} \frac{\prod_{i=1}^{2} \left(\omega^2 + i\gamma_{\text{LPP},ij}\omega - \omega^2_{\text{LPP},ij} \right)}{\left(\omega^2 + i\gamma_{\text{p},j}\omega \right) \left(\omega^2 + i\gamma_{\text{TO},j}\omega - \omega^2_{\text{TO},j} \right)}, \tag{9.22}$$

where $\omega_{\text{LPP},ij}$ and $\gamma_{\text{LPP},ij}$ are the eigenfrequency and the broadening value of the ith longitudinal-phonon-plasmon (LPP) modes, respectively. The $\gamma_{\text{p},j}$ parameters are considered as the broadening values of the plasma excitation. $\omega_{\text{TO},j}$ and $\gamma_{\text{TO},j}$ are the frequency and the broadening value of the transversal-optical lattice mode, respectively. The high-frequency dielectric constants $\varepsilon_{\infty,j}$ account for the remaining contribution of interband transitions. They depend for a degenerate semiconductor on the carrier density which will be demonstrated in section 9.3.4. The values to be used for w-InN films presented here are summarized in Table 9.3. The $\omega_{\text{LPP},ij}$ eigenfrequencies are related to the TO phonon frequency $\omega_{\text{TO},j}$, the LO phonon frequency $\omega_{LO,j}$, and the plasma frequency $\omega_{p,j}$ via:

$$\omega_{LPP,ij}^2 = \frac{1}{2}\left[\omega_{LO,j}^2 + \omega_{p,j}^2 + (-1)^i\sqrt{(\omega_{LO,j}^2 + \omega_{p,j}^2)^2 - 4\omega_{p,j}^2\omega_{TO,j}^2}\right].\qquad(9.23)$$

Due to the crystal-axis orientation of the InN films, not all of the model parameters can be obtained here. In particular, the MIR-SE data from the c-plane-oriented films have slight sensitivity to resonance frequencies with polarization vector parallel to the optical axis. The $\omega_{TO,\parallel}$ [A_1(TO)] and $\omega_{LO,\parallel}$ [A_1(LO)] frequencies were taken from Raman measurements [9.79] and kept constant at 447 cm^{-1} and 581 cm^{-1}, respectively. The broadening ($\gamma_{TO,\parallel}$) of the A_1(TO) phonon mode was assumed to be equal to the E_1(TO) broadening value $\gamma_{TO,\perp}$. The E_1(LO) resonance frequency $\omega_{LO,\perp}$ was adopted as 593 cm^{-1} [9.79]. In addition, the $\varepsilon_{\infty,j}$, $\omega_{p,j}$, and $\gamma_{p,j}$ were treated isotropically in the MIR data analysis. Hence, particularly the frequency and broadening parameters of the $\omega_{TO,\perp}$ [E_1(TO)] phonon mode and the plasma frequency are the adjustable parameters in the MIR data analysis.

The determination of the electron concentrations and mobilities from the fitted ω_p and γ_p values represents the final step of the data analysis. The quantities are related to each other by the following equations [9.64]:

$$\omega_{p,j}^2 = \frac{N_e e^2}{\varepsilon_0 \varepsilon_{\infty,j} m^*(N_e)} \qquad \text{and} \qquad (9.24)$$

$$\mu_{e,j} = \frac{e}{m^*(N_e)\,\gamma_{p,j}}.\qquad(9.25)$$

The averaged effective electron mass $m^*(N_e)$ is calculated via Eq. (9.16). Note only the quantity $N_e/m^*(N_e)$ can be directly determined from the plasma frequency [9.63, 9.80]. The whole problem has therefore to be solved self-consistently because all required quantities E_{ren}, E_F, and $m^*(N_e)$ depend on N_e via the BGR or BMS as described in section 9.2.4. In particular, E_F has to be determined from the fit of the experimental $\varepsilon_2(\omega)$ curves around the gap as described in section 9.3.5.

Figure 9.8 demonstrates the excellent agreement between the fit (solid lines) and the ellipsometric Ψ [(a), (b)] and Δ [(c), (d)] spectra (dashed lines) for samples N100a (left) and In100 (right) in the mid-infrared region (based on the multi-layer model w-InN/w-GaN/w-AlN/sapphire). The buffer layers exhibit no significant free-carrier excitations. For the sapphire substrate the transverse and longitudinal optical phonon frequencies as reported by Schubert *et al.* [9.81] were used. Beside the distinct InN spectral features, those due to the AlN nucleation layer, the GaN buffer layer, and the sapphire substrate are observed.

The fitted $\omega_{TO,j}$, $\omega_{p,j}$, and $\gamma_{p,j}$ are summarized in Table 9.3. For the former we obtain values between 476.2 cm^{-1} and 477.3 cm^{-1} which match well with IR-SE results from Kasic *et al.* [9.26, 9.31] and Raman measurements from Davydov *et al.* [9.82]. MIR reflectance measurements as shown in Figs. 9.8(e) and (f) support the spectral dependence of $\bar{\varepsilon}(\omega)$ according to Eq. (9.22). The excellent agreement of measured and calculated data emphasizes the validity of the model used

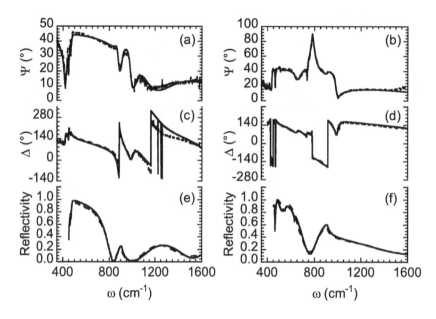

FIGURE 9.8

Measured (dashed lines) and modeled (solid lines) Ψ [(a) and (b)] and Δ [(c) and (d)] spectra at 65° angle of incidence as well as the IR reflectance spectrum [(e) and (f)] for samples N100a (left) and In100 (right).

TABLE 9.3

Summary of the IR-SE determined phonon and plasma frequencies as well as broadening parameters, electron concentrations and mobilities, and sample-dependent high-frequency dielectric constants of investigated w-InN films. The average electron effective masses $m^*(N_e)$ are given in addition.

Sample	ω_{TO} (cm^{-1})	ω_p (cm^{-1})	γ_p (cm^{-1})	N_e(IR-SE) ($\times 10^{18}$cm^{-3})	μ_e(IR-SE) (cm^{2}V^{-1}s^{-1})	$\varepsilon_\infty(N_e)$	$m^*(N_e)$ /m_0
In100	476.2	450	143	1.2	950	7.84	0.0686
N100a	477.3	692	110	3.0	1200	7.77	0.0706
N100b	477.1	1083	198	7.5	627	7.61	0.0752

and proves the quality of our extracted parameters. The comparison of the obtained electron concentrations N_e(IR-SE) summarized in Table 9.3 with the Hall data N_e(Hall) from Table 9.2 indicates that the optical studies yield in all cases lower values in the bulk-like part of the samples. The observation is consistent with results of magneto-optical ellipsometry investigations [9.32] as well as with the analysis of photoluminescence spectra [9.36].

9.3.3   Interband dielectric function of wurtzite InN

The aim of this section is to provide an overview on the dielectric function of wurtzite InN in the spectral range from 0.55 to 9.5 eV for which interband excitations determine the shape. Below, the results of an *a*- and *c*-plane sample will be compared, demonstrating the fundamental properties of the material. A detailed analysis of the different energy ranges follows in the subsequent sections.

The ($11\bar{2}0$)-oriented w-InN film was measured by SE in two c-axis orientations, either parallel or perpendicular to the plane of incidence. Figure 9.9 shows a comparison of (a) the extracted imaginary and (b) real parts of the DF for an a-plane layer. The solid and dashed curves represent the ordinary and extraordinary dielectric tensor components of w-InN. First of all, a characteristic optical anisotropy is observed over the whole investigated range which was confirmed by reflectance anisotropy measurements published elsewhere [9.23]. The imaginary part of both polarizations increases strongly between 0.7 and 1 eV followed by a plateau-like behavior up to 4 eV.

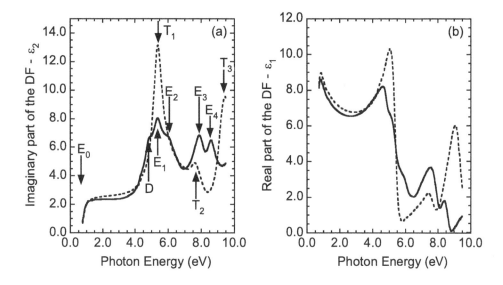

FIGURE 9.9

(a) Imaginary and (b) real parts of the DF for the a-plane w-InN sample a100 obtained from fitting ellipsometric data Ψ and Δ. The ordinary and extraordinary components are drawn by solid and dashed lines, respectively. The transition energies for both polarization directions are marked by arrows and labeled with E and T, respectively, as explained in the text.

The subsequent energy range up to 9.5 eV reveals pronounced features in the spectra which are related to CPs of the band structure. These peaks and shoulders appear in the ordinary (labeled as D and $E_1 - E_4$) as well as in the extraordinary (labeled as $T_1 - T_3$) dielectric tensor component. While at least five CPs can be observed in $\bar{\varepsilon}_o$, only three features dominate the lineshape of the $\bar{\varepsilon}_e$ spectra in the UV–VUV range. The CP transition energies can be obtained from a refined analysis fitting the third derivatives of the DFs as will be presented below in section 9.3.7.

Characteristic for all wurtzite nitrides is the strong enhancement of ε_2 in the vicinity of E_1 for $E \parallel c$ compared to $E \perp c$. The ratio amounts to $\sim 13.1/8$ for InN. The enhancement effect is the major reason for the larger $\varepsilon_{1,e}$ values in the low-energy part of the spectra (apart from around the gap) compared to $\varepsilon_{1,o}$ due to KK relation between ε_1 and ε_2 (see Eq. (9.3)). The experimental results on the anisotropy are consistent with *ab initio* calculations [9.7] in the case that electron-hole interaction was successively taken into account. The excitonic effects cause an overall redshift of the peaks and a redistribution of oscillator strength with respect to the single-particle band structure and DF (a more detailed discussion on InN is provided in Chapter 8). The theoretical and experimental results for w-GaN and w-AlN [9.83] emphasize the universality of this observation.

SE measurements on *c*-plane w-InN layers with an optical axis normal to the surface allow only the determination of a DF which should be very close to the ordinary dielectric tensor component $\bar{\varepsilon}_o$. Figure 9.10(a) shows the real and imaginary parts of this DF obtained from the measurements of a thick *c*-plane InN film with N-face polarity (sample N100a). The shape does not differ from the results found for (0001)-oriented In-face material which was discussed in detail in Ref. [9.10]. It provides further proof that N-polar material can be grown with high optical quality.

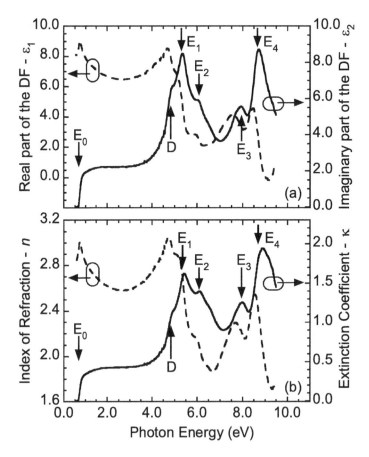

FIGURE 9.10

Ordinary DF (a) and complex refractive index (b) of hexagonal InN as obtained from the ellipsometry studies of a $(000\bar{1})$-oriented layer. The real and imaginary parts are plotted by dashed and solid lines, respectively.

Analogous to the *a*-plane film, the imaginary part increases sharply between 0.7 and 1 eV followed by a plateau-like behavior up to 4 eV and pronounced features such as peaks or shoulders as indicated by arrows (D and E_1-E_4) in the high-energy part of the spectrum.

As already mentioned, when SE is employed to study *c*-plane films only an effective ordinary DF can be extracted by the evaluation of the experimental data because a small fraction of the *p*-polarized light is always aligned parallel to the *c*-axis. This influence is weak as long as the refractive index is large. The comparison of the ordinary DF of the *a*-plane film in Fig. 9.9 with the effective ordinary DF of the *c*-plane one in Fig. 9.10(a) illustrates the difference. First, the E_1/E_2 ratio differs slightly, and second, only the magnitude of E_4 is enhanced in the effective ordinary DF. Therefore, the effective DFs from *c*-plane film investigations are treated here as equivalent to $\bar{\varepsilon}_o$ in the low-energy range.

The DF can be converted into the real (n) and imaginary (κ) part of the complex refractive index $\bar{N}$ (see Eq. (9.4)). The result for the N-face sample is displayed in Fig. 9.10(b). It demonstrates that $n_j(\omega)$ is generally not constant and in particular not around the band gap. The absorption coefficient $\alpha_j(\omega)$ is related to $\varepsilon_{2,j}(\omega)$ as defined in Eq. (9.5) and also depends on $n_j(\omega)$. Therefore, it is more accurate to analyze directly ε_2 around the absorption edge in order to attain the fundamental band gap E_0 as will be demonstrated in section 9.3.5 than to fit only the absorption coefficient and to disregard the spectral dependence of the refractive index.

9.3.4 Experimental data around the band gap

Figure 9.11(a) shows a magnification of the imaginary parts of the DF around the band gap for three w-InN films as well as the corresponding emission spectra (PL) recorded at room temperature. First, the spectra indicate a Stokes shift which is defined as the difference between the emission and the absorption spectra. Second, the shape of ε_2 around the absorption edge is very similar for both, N- and In-face, polarities. It reveals a sharp increase up to ~1 eV followed by a plateau-like behavior. The onset of absorption is shifted to higher energies from sample In100 over N100a to N100b indicating (i) an increasing Burstein-Moss shift due to band filling, and/or (ii) a different size of the in-plane strain. Similar to our results, Kasic *et al.* observed a distinct blueshift of the absorption edge with increasing electron concentration for w-InN films [9.26]. Furthermore, N_e influences also the shape of ε_1. As shown in Fig. 9.11(b), the peak position shifts due to Kramers-Kronig consistency of ε_1 and ε_2, and the high-frequency dielectric constant ε_∞ (extrapolation of ε_1 to zero photon energy) becomes sample dependent. In order to obtain this quantity the spectral dependence of ε_1 below the peak maximum has to be fitted. For this range we applied the analytic formula [9.40, 9.84]

$$\varepsilon_1(\hbar\omega) = 1 + \frac{2}{\pi}\left(\frac{A_0}{2}\ln\frac{E_1^2-(\hbar\omega)^2}{E_G^2-(\hbar\omega)^2} + \frac{A_1 E_1}{E_1^2-(\hbar\omega)^2}\right), \tag{9.26}$$

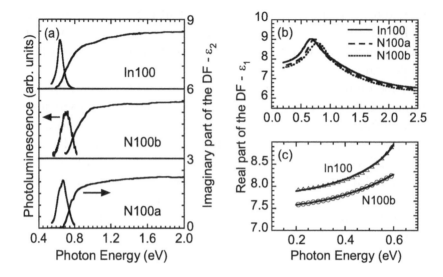

FIGURE 9.11

(a) Room temperature PL spectra and imaginary parts of the DF for three w-InN samples with varying electron densities. The ε_2 data are vertically shifted, and PL curves are normalized to equal amplitude for the sake of clarity. The real parts of the Kramers-Kronig-transformed ε_2 data are shown in (b) down to 0.2 eV. (c) Fit results (solid lines) of ε_1 below the gap according to Eq. (9.26) as described in the text.

which was derived by the Kramers-Kronig transformation of ε_2 which has a constant value A_0 between the effective band gap E_G (roughly corresponding to the absorption edge here) and the delta function at E_1 with magnitude A_1. The latter summarizes the contributions of all high-energy critical-point transitions [9.84]. The fits are given in Fig. 9.11(c) for the samples with the lowest (In100) and highest (N100b) expected electron concentration. The calculation yielded, for all samples, values of $E_1 = (4.04 \pm 0.09)$ eV, $A_1 = (24.28 \pm 3.04)$ eV, and $A_0 = (2.64 \pm 0.41)$ while E_G increases from 0.69 to 0.79 eV from sample In100 to N100b, respectively. With these values ε_∞ can be calculated via Eq. (9.26) in the case of $\hbar\omega \to 0$. The determined sample dependent ε_∞ values are listed in Table 9.3; they range between 7.84 and 7.61. The lower value is found for sample N100b which exhibits the highest carrier concentration causing a shift of the absorption edge to higher energies. It should be noticed that Kasic *et al.* [9.31] proposed a much lower value of only $\varepsilon_\infty = 6.7 \pm 0.1$ which is adopted in many studies in order to analyze plasma frequencies.

9.3.5 Analysis of the DF around the band gap

In the following, the method for analyzing ε_2 will be demonstrated by means of two examples representing different electron concentrations: an In-face (In100) and a N-face (N100a) sample with $N_e = 1.2 \times 10^{18}$ cm^{-3} and $N_e = 3.0 \times 10^{18}$ cm^{-3}, respectively. The analysis of the experimental data starts from a calculation of the imaginary part of the DF that is proportional to the joint density of states via

$$\varepsilon_2(\hbar\omega) \sim \frac{1}{(\hbar\omega)^2} \frac{2}{(2\pi)^3} \int_{BZ} |P_{cv}|^2 [1 - f(E_c)] \times \delta(E_c(k) - E_v(k) - \hbar\omega) d^3k. \quad (9.27)$$

The term $[1-f(E_c)]$ takes into account that the absorption requires empty states in the CB. The CB nonparabolicity is included by inserting Eq. (9.17) for $E_c(k)$. The integration is carried out in reciprocal (k) space over the whole BZ. The energetic position of the Fermi level above the conduction band minimum $[\Delta E_c(k_F)]$ at the Fermi wave vector k_F follows from the calculation of the electron density via

$$N_e = \int_{E_{ren}}^{+\infty} f(E_c) g(E) dE. \quad (9.28)$$

The density of states $g(E)$ is obtained from Eq. (9.17). $\Delta E_c(k_F)$ and E_{ren} are varied until the integral matches self-consistently the experimental electron concentrations of Table 9.4, and the spectral dependence of ε_2 according to Eq. (9.27) agrees with the experimental data. The results of this procedure for the In- and N-face InN films are plotted in Fig. 9.12(a) and (b), respectively. The calculated (experimental) dependencies of $\varepsilon_2(\omega)$ are plotted by the dashed (solid) lines. Excellent overall agreement is achieved for both samples. The Fermi distribution function is given for comparison. For a value of $f(E_c) = 0.5$ (vertical line), ε_2 values of 0.47 (In100) and 0.67 (N100a) correspond to the optical transitions at k_F. The CB-VB spacing amounts to $E_F(k_F) = 0.677$ eV and $E_F(k_F) = 0.732$ eV, respectively, that is, a distinct blueshift is found with increasing N_e. We determine the renormalized band gaps E_{ren} with 0.619 eV and 0.613 eV for the In100 and N100a samples, respectively, from which a total Burstein-Moss shift $[\Delta E_{cv}(k_F) = \Delta E_c(k_F) - \Delta E_v(k_F) = E_F(k_F) - E_{ren}]$ of 58 and 119 meV is deduced. Thereby, the Fermi energy is determined to be 52 meV (In100) and 105 meV (N100a) above the CB minimum. The energy position of the VB at k_F $[\Delta E_v(k_F)]$ is found to be shifted -6 meV (In100) and -14 meV (N100a) in comparison to those at the Γ-point.

TABLE 9.4
Experimentally extracted Fermi energy from the ordinary tensor component of the DF as well as contributions to Burstein-Moss shift used in order to determine the renormalized band gaps E_{ren}.

Sample	N_e (10^{18} cm^{-3})	$E_F(k_F)$ (eV)	$\Delta E_c(k_F)$ (meV)	$\Delta E_v(k_F)$ (meV)	$\Delta E_{cv}(k_F)$ (meV)	E_{ren} (eV)
In100	1.2	0.677	52	-6	58	0.619
N100a	3.0	0.732	105	-14	119	0.613
N100b	7.5	0.785	192	-27	219	0.566
a100 ($E \perp c$)	6.0	0.780	168	-23	191	0.589

The E_{ren} data of all studied samples are summarized in Table 9.4. The values decrease by approximately 50 meV between the lowest and highest electron con-

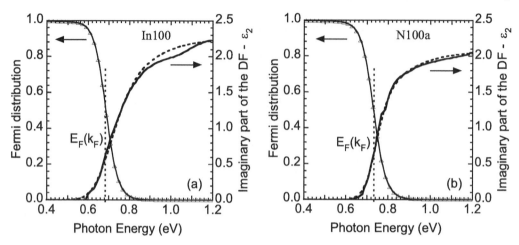

FIGURE 9.12

Calculated shape of the imaginary part of the DF (dashed lines) in comparison to the experimental data (solid lines) for the In-face (a) and the N-face InN film (b) with $N_e = 1.2 \times 10^{18}\,\mathrm{cm}^{-3}$ and $N_e = 3.0 \times 10^{18}\,\mathrm{cm}^{-3}$, respectively. Band filling and band-gap renormalization are taken into account for the calculation as explained in the text. The Fermi distribution function (open triangles) is given for comparison.

centration InN films. This finding is mainly attributed to BGR; the influence of residual strain is less pronounced. Both effects will be discussed in the following section.

Another important conclusion can be drawn from Fig. 9.12. A plot of $(\alpha\hbar\omega)^2$ or $(\varepsilon_2\hbar^2\omega^2)^2$ versus photon energy $\hbar\omega$ and the extrapolation to zero yields under no circumstances $E_F(k_F)$; the extracted value comes close to E_{ren} for low electron concentrations.

Figure 9.13 shows for the $(11\bar{2}0)$-oriented InN film (sample $a100$) a magnification of the ordinary (solid line) and extraordinary (dashed line) ε_2 data from Fig. 9.9. The comparison of the $\varepsilon_{2,\mathrm{e}}$ and $\varepsilon_{2,\mathrm{o}}$ curves unambiguously indicates that the onset of absorption for $\mathbf{E} \parallel \mathbf{c}$ is detected at higher photon energies. The effect is the direct verification of the expected polarization anisotropy of the oscillator strengths displayed in Fig. 9.2. It is deduced from the XRD data that this sample exhibits almost no in-plane strain which would modify the relative oscillator strengths with respect to strain-free material. Therefore, the strong increase of absorption in the ordinary and extraordinary data is caused by transition A and C, respectively. Applying the same approach, as described above, to the data of both polarization directions, the shape in the band gap region can be theoretically reproduced. As depicted in Fig. 9.13, the Fermi energy at k_F was found to be shifted by 25 meV to higher energies for $\varepsilon_{2,\mathrm{e}}$. Hence, the Fermi energy of the extraordinary part amounts to 0.805 eV. Due to a lack of data for the $E_p^{\parallel}$ value and since only a small anisotropy in the effective mass has been observed [9.85], we apply the same set of parameters for the $\varepsilon_{2,\mathrm{e}}$ evaluation as for the ordinary one. That leads to a renormalized gap of 0.614 eV for the experimental configuration $\boldsymbol{E} \parallel \boldsymbol{c}$ in com-

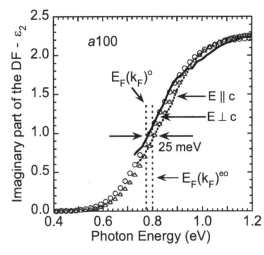

FIGURE 9.13

Calculated shape of the ordinary (circles) and extraordinary (triangles) ε_2 components of the DF in comparison to the experimental data around the band gap. The solid and dashed lines refer to ordinary and extraordinary data, respectively.

parison to 0.589 eV for $\mathbf{E} \perp \mathbf{c}$. Bhattacharyya *et al.* studied the optical anisotropy of a $(11\bar{2}0)$-oriented film by polarized transmission and photoluminescence spectroscopy [9.48]. The splitting of the ordinary and extraordinary absorption edge was determined to be 20 meV. Due to anisotropic in-plane strain, they identified the main contributing transitions as B and C in this case, being consistent with an A-C transition energy splitting of 25 meV. In addition, the PL studies revealed a polarization anisotropy with clear twofold symmetry; the emission for $\mathbf{E} \perp \mathbf{c}$ was found to be more than three times stronger than that for $\mathbf{E} \parallel \mathbf{c}$. The authors proposed a strain-free band gap of $E_\mathrm{A} = 0.7 \pm 0.02$ eV at $T = 10$ K.

9.3.6 Zero-density strain-free gap of wurtzite InN

In the final step, BGR and the influence of residual strain inside the InN layers have to be considered in order to attain the zero-density strain-free gap E_0 of wurtzite InN. The in-plane biaxial strain ε_{xx} was deduced from XRD and Raman measurements. Compressive (N100a) and tensile (In100, N100b) in-plane strain were found for three samples, whereas $a100$ should be almost relaxed (see Table 9.5). The energy dispersion of the CB as a function of in-plane strain follows from Eq. (9.11). As explained in section 9.2.2, for compressive in-plane strain the band transition energies at the Γ-point of the BZ increase; for tensile strained layers the energies analogously shift to smaller values. With knowledge of ε_{xx}, one obtains the shift of the CB energy ΔE_strain that amounts to -4 meV and 12 meV for sample In100 and N100a, respectively.

Now, with knowledge of the carrier-dependent BGR contributions we can estimate the strain-free zero-density band gap of the w-InN films. As mentioned in section 9.2.4, two approaches ($\Delta_\mathrm{BGR}^{(1)}$ and $\Delta_\mathrm{BGR}^{(2)}$) have been suggested for describing BGR formalism. So the strain-free zero-density band gaps as a function of the employed BGR approach result from $E_0^{(1)} = E_\mathrm{ren} - \Delta_\mathrm{BGR}^{(1)} - \Delta E_\mathrm{strain}$ and $E_0^{(2)} = E_\mathrm{ren} - \Delta_\mathrm{BGR}^{(2)} - \Delta E_\mathrm{strain}$ in case of methods (1) and (2), respectively. The results are listed in Table 9.5. Due to the stronger $E_0^{(1)}$ spread of 38 meV in comparison to only 13 meV for $E_0^{(2)}$ values, $\Delta_{\text{BGR}}^{(2)}$ is preferred for the calculation of the strain-free zero-density band gap $E_0^{(2)} := E_0$. For the In-face and N-face sample In100 and N100a, respectively, the E_0 value is determined to be equal and amounts to 0.674 eV. The yielded values range within 0.671 eV and 0.684 eV for all studied w-InN samples (see Table 9.5) and the averaged strain-free zero-density gap of 0.675 eV agrees well with the theoretically calculated one of 0.67 eV from Persson *et al.* [9.86]. In addition, using the renormalized gap of 0.614 eV resulting from the analysis of the extraordinary absorption edge (see Fig. 9.13), we obtain a value of 0.709 eV for the transition E_{C}.

TABLE 9.5
In-plane strain values (ε_{xx}) and the corresponding caused CB shift ΔE_{Strain} as well as two different contributions to band-gap renormalization $\Delta_{\text{BGR}}^{(1)}$ and $\Delta_{\text{BGR}}^{(2)}$ utilized for determining the strain-free zero-density band gaps $E_0^{(1)}$ and $E_0^{(2)}$, respectively.

Sample	ε_{xx} ($\times 10^{-4}$)	ΔE_{strain} (meV)	$\Delta_{\text{BGR}}^{(1)}$ (meV)	$E_0^{(1)}$ (eV)	$\Delta_{\text{BGR}}^{(2)}$ (meV)	$E_0^{(2)}$ (eV)
In100	6.50	-4	-21	**0.644**	-51	**0.674**
N100a	-23.18	12	-29	**0.630**	-73	**0.674**
N100b	2.26	-1	-39	**0.606**	-104	**0.671**
a100 ($E \perp c$)	-0.50	~ 0	-36	**0.625**	-95	**0.684**

Figure 9.14(a) shows again the BGR contributions of the two approaches, but now as a function of the third root of the ratio $N_e/1 \times 10^{18} \, \text{cm}^{-3}$. Using the achieved strain-free zero-density band gap of $E_0 = 0.675$ eV, the BGR values as a function of each sample electron density can be calculated by $\Delta_{\text{BGR}} = E_{\text{ren}} - E_0 - \Delta E_{\text{Strain}}$. These results are depicted as diamonds in Fig. 9.14(a) and furthermore confirm the excellent accordance with the second approach.

At least, we consider transition energies as a function of in-plane strain. Figure 9.2(a) displays only the relative transition energies E_{B} and E_{C} as the difference to the value for E_{A} in the strain-free case. Setting the above obtained value of $E_0 = 0.675$ eV as the strain-free one, Fig. 9.14(b) shows the calculated energies for the transitions A, B, and C from the three VBs into the CB as function of in-plane strain. In this context, the three transition energies can be determined via $E_{\text{ren}} - \Delta_{\text{BGR}}^{(2)} = E_{\text{A,B,C}}$. This was done for the above listed a-plane and three c-plane films. The a-plane one allows the determination of the transition energies of the ordinary ($\hateq E_{\text{A}}$) and extraordinary ($\hateq E_{\text{C}}$) component, while only E_{A} can be obtained from c-plane films. The attained energies E_{A} (circles) and E_{C} (rectangle) are displayed in Fig. 9.14(b) as a function of the in-plane strain. The deviation between the experimental data and the theoretical dependence is always below 10 meV.

In summary, the above presented self-consistent approach of optically deter-

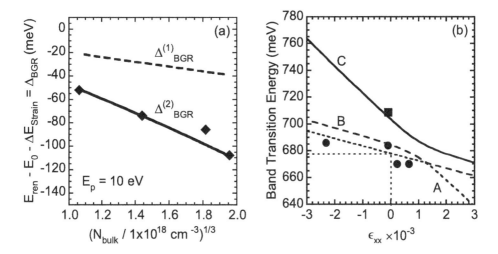

FIGURE 9.14

(a) BGR contributions as a function of the electron concentration for the two different approaches $\Delta^{(1)}_{BGR}$ (dashed line) and $\Delta^{(2)}_{BGR}$ (solid line). The diamonds represent the calculated BGR values as explained in the text. (b) Calculated transition energies (solid lines) from the three valence bands into the conduction band as a function of biaxial in-plane strain. The symbols mark the experimental data as obtained from the analysis of the ordinary and extraordinary DF.

mining the electron densities, the position of the Fermi energies, and the renormalized gaps from the spectral dependence of the dielectric function leads to a zero-density strain-free band gap of InN at room temperature of $E_0 = 0.675\,\text{eV}$. How does it compare with the results of other groups? It turns out that BGR and BMS compensate each other for an electron concentration of approximately $N_e = 1.1 \times 10^{18}\,\text{cm}^{-3}$. Analyzing the absorption data for a $1\,\mu\text{m}$ thick film with $N_e = 1 \times 10^{18}\,\text{cm}^{-3}$, Walukiewicz *et al.* determined a gap of 0.68 eV at room temperature [9.11]. Transmission measurements on films with $N_e \sim 3 \times 10^{17}\,\text{cm}^{-3}$, for which the BMS is smaller than the BGR, yielded absorption edges of 0.64 eV [9.87] and 0.65 eV [9.38]. It might be possible that the BMS corrected gap would be slightly larger.

Finally, the Kramers-Kronig transformation of the calculated ordinary $\varepsilon_2(\omega)$ curve for the strain-free zero-density gap of $E_A = E_0 = 0.675\,\text{eV}$ leads to an $\varepsilon^{\perp}_{\infty}$ of 7.83, whereas the KK transformation of the calculated extraordinary $\varepsilon_2(\omega)$ curve ($E_C = 0.700\,\text{eV}$) gives an $\varepsilon^{\parallel}_{\infty}$ of 8.03.

9.3.7 High-energy critical points of the band structure

Besides the gap structure, pronounced features appear in the high-energy part (above 4.5 eV) of the DF. They are related to the peculiarities of the joint density of states in the vicinity of critical points, also known as Van Hove singularities.

In the past, transition energies of the CPs for w-InN were estimated either from

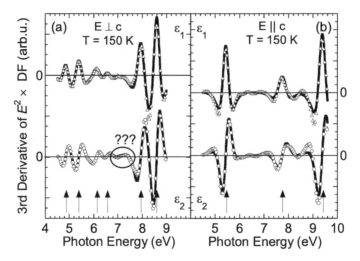

FIGURE 9.15

Fit of the third derivative of the ordinary (a) and extraordinary (b) DF for the *a*-plane InN film *a*100 at $T = 150$ K. The triangles and circles represent the experimental data ε_1 and ε_2, respectively. The solid lines are the best fits. The energetic position of the transitions at the Van Hove singularities are labeled by arrows.

digitizing peaks and shoulders of the ε_2-spectra [9.16, 9.19, 9.23, 9.24] or by applying a model DF lineshape [9.26]. Recently, we presented a more accurate approach to evaluate the Van Hove singularities by calculating the third derivative of the point-by-point DF multiplied by the square of the photon energy E [9.9, 9.10]. The resulting spectra can be fitted [9.88] via

$$\frac{d^3}{dE^3}\left(E^2\bar{\varepsilon}\right) = \sum_j e^{i\phi_j} \frac{C_j}{\left(E + i\Gamma_j - E_{\mathrm{CP},j}\right)^{n/2}}, \qquad (9.29)$$

where ϕ_j, C_j, Γ_j, and $E_{\mathrm{CP},j}$ denote the phase angle, the magnitude, the broadening energy, and the transition energy of the j-th CP, respectively. Both the real and imaginary parts are fitted in order to increase the reliability of the results. As described in section 9.3.3, the determination of the ordinary and extraordinary dielectric tensor components from SE experiments becomes possible if films are studied for which the optical axis (*c*-axis) is oriented off-normal (for example, the *a*-plane configuration). Besides the anisotropy at the Γ-point of the BZ, such films reveal as well an anisotropic behavior in the high-energy part of the spectra that is indicated by a different lineshape of $\bar{\varepsilon}_o$ and $\bar{\varepsilon}_e$ corresponding to an electric field perpendicular and parallel to the *c*-axis, respectively.

Applying Eq. (9.29) to the data of the *c*-plane samples and to the ordinary and extraordinary dielectric tensor components of the *a*-plane one, transition energies can be extracted from the fits. For the latter one, as an example, the obtained curves at $T = 150$ K for both polarization directions are shown in Fig. 9.15(a) and (b). All transitions of both components are nicely reproduced with $n = 8$ that accounts for excitonic-type CPs. Up to seven transitions can be resolved in the third derivative spectra. Note the additional D_2 and D_3 features at $\sim6.6\,\mathrm{eV}$ and $\sim7.3\,\mathrm{eV}$, respectively, become visible only in the derivative spectra. A comparison of the $\bar{\varepsilon}_o$ CP transition energies for w-InN at room temperature with results from Refs. [9.26] and [9.16] is given in Table 9.6. The critical-point transition energies at $T = 150\,\mathrm{K}$ as well as the theoretically predicted values are summarized in Table 9.7.

TABLE 9.6

Comparison of transition energies at Van Hove singularities for InN at room temperature.

Sample	D (eV)	E_1 (eV)	E_2 (eV)	D_2 (eV)	D_3 (eV)	E_3 (eV)	E_4 (eV)
$a100$	4.80	5.32	6.09	6.51	7.1–7.4[c]	7.89	8.54
In100	4.81	5.38	6.12	6.58	7.1–7.4[c]	7.95	8.57
N100a	4.84	5.37	6.12	6.56	7.29	7.96	8.62
N100b	4.82	5.39	6.12	6.58	7.1–7.4[c]	7.95	8.60
InN[a]	4.84	5.41	6.10				
InN[b]	4.85	5.38	6.18				

[a]Reference [9.26].
[b]Reference [9.16].
[c]Only a weak structure in the third derivative spectra.

TABLE 9.7

Comparison of determined critical-point transition energies for InN at $T = 150\,\mathrm{K}$ to results of *ab initio* calculations.

Sample	D (eV)	E_1 (eV)	E_2 (eV)	D_2 (eV)	D_3 (eV)	E_3 (eV)	E_4 (eV)
$a100$	4.85	5.39	6.12	6.55	7.1–7.4[a]	7.95	8.59
In100	4.86	5.43	6.15	6.60	7.32	7.99	8.61
N100a	4.88	5.40	6.15	6.58	7.34	7.99	8.64
N100b	4.86	5.41	6.14	6.59	7.34	7.99	8.61
Theory[b] $(T = 0\,\mathrm{K})$	4.7	5.5	6.1	~6.5[c]	7.2	7.8	8.7

[a]Only a weak structure in the third derivate spectra, cannot be precisely resolved.
[b]Ref. [9.7].
[c]Estimated from Fig. 8 of Ref. [9.7].

First, the third derivative analysis with its higher accuracy yields a slightly lower value especially for transition D. This is not surprising because it appears only as a low-energy shoulder of E_1 as can be seen in Fig. 9.10. Digitizing the shoulder leads to an overestimation of the transition energy. Furthermore, InN films with c-plane orientation but different polarities show excellent agreement of their CP transition energies within 30 meV, that is, the results do not depend on the polarity and therefore growth conditions of the films. Comparing the c- and a-plane samples, the D and E_2 energies match within a few tens of meV, whereas E_1, D_2, E_3, and E_4

exhibit a slightly larger spread. This can be related to the influence of the extraordinary component on the effective ordinary DF of the c-plane films. Secondly, the low-temperature spectra of the DF are characterized by an average blueshift of the transition energies of about 30 meV in comparison to room temperature. The improved crystal quality of the c-plane films and the decreased thermal broadening at $T = 150$ K lead to a more precise resolution of CP D_2, whereas D_3 can be unambiguously resolved in the low-temperature third derivative spectra. Its transition energy is determined to be 7.32–7.34 eV.

TABLE 9.8

Transition energies at Van Hove singularities of the extraordinary DF tensor component ($\bar{\varepsilon}_e$) for an a-plane InN film at room temperature and 150 K in comparison to *ab initio* calculations.

Sample $a100$	T_1 (eV)	T_2 (eV)	T_3 (eV)
295 K	5.39	7.72	9.36
150 K	5.44	7.75	9.38
Theory[a]	5.6	7.6/7.9	9.4

[a]Reference [9.7].

Table 9.8 presents the results of Van Hove singularities for the extraordinary component of the DF. As for the ordinary spectra, transition energies shift by $\sim$30 meV from RT to 150 K. The measured and calculated peak positions deviate by less than 0.2 eV. At present, the theoretically predicted double peak at 7.6/7.9 eV cannot be resolved in the experimental spectra, where only one peak at 7.72 eV is observed.

9.4 Properties of In-rich wurtzite alloys

Although the research work on the application of hexagonal InN for devices is still in its earliest stages, its alloys with GaN and AlN are already widely used for light-emitting and electronic devices. $In_xGa_{1-x}N$ and $In_xAl_{1-x}N$ films are employed as the active region, as optical wave-guides or as barrier layers for which the fundamental properties should be very precisely known in order to optimize the performance of the devices. It is worth mentioning that the band-gap revision for hexagonal InN has a strong impact on how to represent the compositional dependence of band-structure parameters providing the bases for modeling. The bowing parameter of the main absorption edge b_0, which describes the deviation of $E_0(x)$

from a linear shift with x, is probably the most intensively studied one among these quantities. An elaborate analysis, however, should distinguish whether $E_0(x)$ has to be attributed to $E_A(x)$ or $E_B(x)$ or to both according to the VB ordering and optical selection rules described in section 9.2.3. And as for any other semiconductor alloy system, the transition energies of all Van Hove singularities should also exhibit a characteristic shift. This dependence was discussed in detail in Refs. [9.10] and [9.9] for the $In_xGa_{1-x}N$ and $In_xAl_{1-x}N$ alloys, respectively. With knowledge of the dielectric function for the binary nitrides and the corresponding bowing parameters of the high-energy critical points, a model can be developed which allows the DF to be calculated for any composition [9.40].

The current chapter cannot cover all these aspects. Therefore we initially discuss the properties of the $In_xAl_{1-x}N$ system demonstrating the common dependencies. Then a short summary of essential $In_xGa_{1-x}N$ properties follows.

9.4.1 Interband dielectric function of InAlN alloys

The $In_xAl_{1-x}N$ alloys represent the most challenging system for data analysis because their gaps cover the whole spectral range accessible with nitride compounds. The $\Gamma_7^c - \Gamma_9^v = E_A$ splitting at room temperature increases from 0.675 eV (InN) to 6.213 eV (AlN), while the $\Gamma_7^c - \Gamma_{7+}^v = E_B$ dependence from 0.681 eV to 5.989 eV [9.62, 9.89] is slightly weaker. Note these values refer to the spacing of the bands in the single-particle BS and can be directly compared to the calculations. The absorption edge of AlN is observed at considerably lower energies due to the reported high exciton binding energies between 48 and 71 meV [9.62, 9.89]; it has to be taken into account if data for the Al-rich alloys are analyzed. Those films are of particular interest because they can be grown nearly lattice-matched to GaN for an InN molar fraction close to $16 - 18\%$ [9.90] depending on the properties of the buffer. No deviation of the lattice constants from Vegard's law was detected which was theoretically predicted [9.91]. High-quality Al-rich alloys only recently became available, and the optical investigations are still in their infancy [9.92], that is, ellipsometry studies cannot yet be presented.

In order to demonstrate the systematic change of the DF with increasing Al-content, a sample set of three $\sim$400 nm thick $In_xAl_{1-x}N$ films ($x = 0.91$, $x = 0.83$, $x = 0.71$) was grown on c-sapphire substrates by PI-MBE using a $\sim$200 nm AlN buffer layer. The growth temperature was about 470 °C. The Al atomic fraction was determined by XRD measurements. The analysis shows that almost strain-free InAlN epitaxial layers were formed with *c*-axis orientation normal to the surface. The composition was obtained from the lattice parameters by applying Vegard's law. Hall measurements at room temperature revealed electron concentrations of 2.4×10^{18} cm^{-3} ($x = 0.91$), 3.0×10^{18} cm^{-3} ($x = 0.83$), and 3.85×10^{18} cm^{-3} ($x = 0.71$).

Figure 9.16 provides an overview of (a) the real and (b) the imaginary parts of the DFs for In-rich InAlN alloys over the whole investigated range up to 9.5 eV. Due to the *c*-axis orientation normal to the substrate, the DFs are very close to the

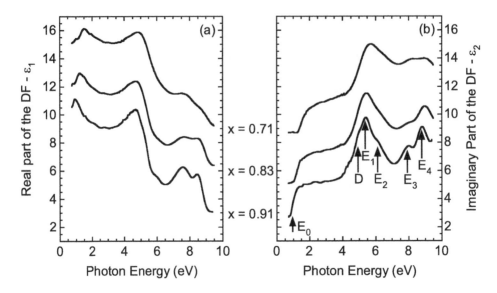

FIGURE 9.16

(a) Real and (b) imaginary parts of the almost ordinary DF of w-$In_xAl_{1-x}N$ alloys obtained from analysis of ellipsometry measurements. The alloy data are vertically shifted proportional to the aluminum content for sake of clarity, that is, by 2.7, 5.1, and 8.7.

ordinary dielectric tensor component which was discussed in more detail for InN in section 9.3.7. Beside the gap structure, the shape of the DF is analogous to the binary compounds, InN and AlN, that is, it is strongly influenced by critical points of the band structure in the energy range above 4.5 eV as marked in Fig. 9.16(b). Despite a larger broadening in comparison to InN, the Van Hove singularities (D and $E_1 - E_4$) are well resolved, confirming the good crystalline quality of the InAlN films. As expected for an alloy system, they undergo a continuous shift to higher energies with increasing Al-content. While for composition of $x = 0.91$ the shape of the DF is similar to the InN one, for an Al mole fraction of 29% the spectra already become more AlN-like [9.40] which manifests in a decreasing oscillator strength of the transition D.

9.4.2   Analysis of the InAlN DF around the band gap

The obtained ε_2 data around the absorption edge of the films in comparison to the emission spectra recorded at RT are shown in Fig. 9.17(a). Each InAlN layer shows pronounced PL. Both the onset of ε_2 and the PL peak undergo a clear shift to higher energies with decreasing In-content, but the absorption edge broadening increases as well. For the studied compositions, the $E_B(x)$ transition defines the onset of absorption, but its oscillator strength $f_{B,o}$ is much lower than $f_{A,o}$ in the studied configuration with the optic axis normal to the surface. Therefore, absorption is attributed to A-transitions while emission is mainly due to $E_B(x)$.

Despite the larger broadening of the data around the absorption edge with in-

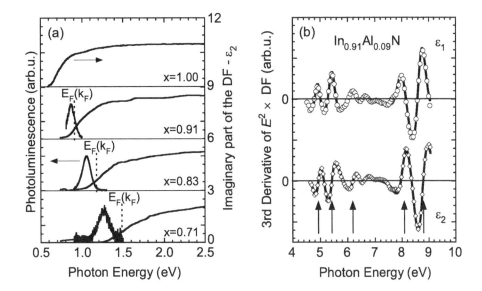

FIGURE 9.17

(a) PL spectra and imaginary parts of the DF at room temperature for three w-InAlN alloys with varying In mole fraction. An ε_2 spectrum of a w-InN film is given for comparison. The ε_2 data are vertically shifted, and PL curves are normalized to equal amplitude for the sake of clarity. (b) Fit of the third derivative of the $In_{0.91}Al_{0.09}N$ film at $T = 295\,K$. The circles and diamonds represent the experimental ε_1 and ε_2 data, respectively. The solid lines are the best fits. The energetic position of the transitions at the Van Hove singularities are labeled by arrows.

creasing Al concentration, the position of the Fermi energy can be estimated for all samples in the same way as discussed in sections 9.3.5 and 9.3.6 for w-InN. An ε_r value of 9.5 was proposed [9.5] for wurtzite InN with a band gap of 0.67 eV. It is also used as an approximation for alloys with $x > 0.71$. For comparison, $\varepsilon_r = 10.12$ was reported [9.93] for AlN with a much higher gap. For the E_P parameter we assume a linear dependence on the alloy composition between the values for InN and AlN of 10 eV and 18 eV [9.55], respectively. The analysis provides E_F values which are drawn in Fig. 9.17(a) as vertical dashed lines for each film. By calculating contributions to the BMS and BGR the fundamental band gaps of the alloys amount to 0.88, 1.14, and 1.46 eV. A summary of all obtained values is given in Ta-

TABLE 9.9

Experimentally extracted Fermi energy from the effective ordinary DFs as well as contributions to Burstein-Moss shift and band-gap renormalization in order to determine the zero-density band gaps of the alloys.

x	N_e (10^{18} cm^{-3})	$E_F(k_F)$ (eV)	$\Delta E_c(k_F)$ (meV)	$\Delta E_v(k_F)$ (meV)	$\Delta E_{cv}(k_F)$ (meV)	E_{ren} (eV)	Δ_{BGR} (meV)	E_0 (eV)
0.91	2.4	0.905	77	−13	90	0.815	−65	0.88
0.83	3.0	1.160	80	−15	95	1.065	−75	1.14
0.71	3.85	1.480	92	−18	110	1.370	−90	1.46

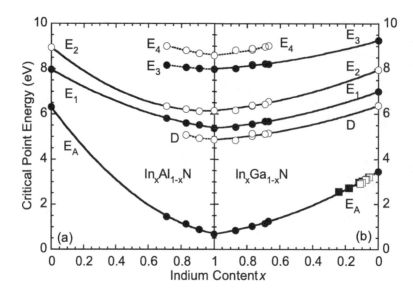

FIGURE 9.18

Room temperature critical point energies of (a) $In_xAl_{1-x}N$ and (b) $In_xGa_{1-x}N$ alloys as a function of the indium content x. The open and full circles refer to our work. Open and full squares display measured band-gap values adapted from Refs. [9.100] and [9.102], respectively. The solid lines represent the fit results for determining the bowing parameters.

ble 9.9. Taking into account the values of $E_A = 0.675\,eV$ (InN) and $E_A = 6.213\,eV$ (AlN), the composition dependence of the $E_A(x)$ gap can be described by

$$E_A^{InAlN}(x) = xE_A^{InN} + (1-x)E_A^{AlN} - b_Ax(1-x). \tag{9.30}$$

The fit leads to a bowing parameter of $b_A = (4.0 \pm 0.3)\,eV$, which is shown in Fig. 9.18(a) by the lowest solid line. This value lies in between other published values of $b_0 = 3.0\,eV$ [9.94], $4.96\,eV$ [9.95], and $5.3\,eV$ [9.96] for which the absorption edge energy extracted from transmission studies was analyzed.

9.4.3 High-energy critical points of InAlN alloys

Due to the promising results in the case of the studied indium nitride films, the high-resolution determination of CP transition energies is now applied to the InAlN alloys. In the same manner, we calculate the third derivative of each alloy point-by-point DF multiplied by the square of the photon energy E. Then, the resulting spectra are fitted by Eq. (9.29). We obtain excellent agreement to the experimental data which is exemplarily demonstrated for the $In_{0.91}Al_{0.09}N$ film in Fig. 9.17(b). A smooth CP energy shift with the Al concentration to higher values is observed as depicted in Fig. 9.18(a). Note the weak dependence of E_3 on the In content which is expected by the observed merging of E_2 and E_3 for AlN [9.40]. Furthermore, the InAlN alloys do not exhibit additional D_2 and D_3 features in between the energy range of E_2 and E_3 compared to InN. An overview of the fitted critical point ener-

TABLE 9.10

Comparison of CP transition energies for InAlN alloys at room temperature.

x	E_A (eV)	D (eV)	E_1 (eV)	E_2 (eV)	E_3 (eV)	E_4 (eV)
1.00	0.675	4.82	5.37	6.09	7.95	8.60
0.91	0.88	4.92	5.52	6.12	8.02	8.68
0.83	1.14	5.07	5.60	6.20	8.06	8.81
0.71	1.46		5.81	6.33	8.16	9.01
0.00	6.213		7.97	8.95		

gies of all studied samples is given in Table 9.10. The averaged values of Table 9.6 for InN and the data of Ref. [9.97] for AlN of $E_1 = 7.97$ eV, $E_2 = 8.95$ eV are included. The bowing parameters for the high-energy CPs of the InAlN material system are fitted in analogy to b_A via Eq. (9.30). We attain bowing parameters as indicated in Fig. 9.18(a) by solid lines of $b_1 = 1.8$ eV and $b_2 = 2.7$ eV for E_1 and E_2, respectively. No estimations can be made for the other critical points due to the lack of corresponding data for AlN.

9.4.4 Properties of InGaN alloys

Until 2002, most studies focused on the analysis of the absorption edge for Ga-rich InGaN alloys. A band gap of 1.89 eV for InN was assumed in order to determine the bowing parameter. Wetzel *et al.* proposed a composition independent b_0 of 3.2 eV from photoreflectance spectroscopy [9.98], while a composition dependent value of $4.8 - -3.8$ eV for $0 \leq x \leq 0.1$ was proposed in Ref. [9.99]. Applying later on an E_0 value of 0.8 eV instead of 1.89 eV and taking strain into account, McCluskey *et al.* corrected [9.100] the previously published result to a composition independent b_0 of 2.6 eV. Subsequent absorption investigations [9.101] for the composition range of $0.5 \leq x \leq 1.0$ yielded a much smaller value of $b_0 = 1.4$ eV.

Based on a comprehensive ellipsometry study, we compared the ordinary DFs of metal- and nitrogen-face In-rich ($x \geq 0.67$) $In_xGa_{1-x}N$ alloys [9.10]. The general behavior is similar to the dependencies summarized above for InAlN. As expected, no dependence of the optical properties on the film polarity was found. For a comparison to InAlN, all fitted InGaN CP energies are plotted in Fig. 9.18(b). Special attention was paid to the determination of the band gap as a function of composition. A bowing parameter $b_A = 1.72$ was obtained for the $\Gamma_7^c - \Gamma_9^v = E_A$ transition, based on $E_A = 0.68$ eV for InN and the data from Ref. [9.102]. It decreases slightly to $b_A = 1.7$ with the strain-corrected band gap of $E_A = 0.675$ eV. The bowing energies for all CPs are given in Table 9.11. Both InAlN and InGaN reveal considerably lower bowing parameters of high-energy CPs in comparison to those for the band gap. These data provide an accurate basis for a parameterized model, allowing the calculation of the ordinary DF for all compositions [9.40].

TABLE 9.11

Bowing parameters for the CP energies
of the ternary nitrides InAlN and
InGaN. All energies are in eV.

	b_A	b_D	b_1	b_2	b_3
$In_xAl_{1-x}N$	4.0		1.8	2.7	
$In_xGa_{1-x}N$	1.7	0.8	1.1	1.0	0.7

9.5   Experimental results for zinc-blende InN

In comparison to the hexagonal counterpart, the properties of cubic InN are much less intensively investigated so far. It is attributed to the difficulties in growing the meta-stable compound with high phase purity. Due to the lack of reliable experimental data, a band gap value of 1.8 eV was used in order to determine a bowing parameter [9.30] from the early ellipsometry studies of zb-$In_xGa_{1-x}N$ alloys [9.103]. With respect to the above summarized recent findings for w-InN, this E_0 is, of course, much too high. There is no doubt that the gap for the zinc-blende material always lies below the wurtzite one. The difference for GaN, for example, amounts to ~200 meV and calculations suggest 80 to 230 meV for InN [9.6, 9.7, 9.54, 9.55, 9.86]. Based on the progress in growing single-crystalline cubic InN [9.27], another similarity of the two polymorphs was demonstrated recently, the formation of a surface electron accumulation layer [9.104]. Consequently, the analysis of the optical data for zb-InN should be carried out by the same approach as for the wurtzite material.

9.5.1   Samples

The three zb-InN samples were grown on 3C-SiC substrates ($N_e \approx 5 \times 10^{17} \text{cm}^{-3}$) by rf plasma-assisted molecular beam epitaxy at growth temperatures of 434°C (sample A), 431°C (sample B), and 419°C (sample C). All films were nominally undoped. Prior to the growth of the 127 nm (A), 122 nm (B), and 75 nm (C) thick zb-InN layers, ~600 nm thick zb-GaN buffer layers ($N_e \approx 2 \times 10^{17} \text{cm}^{-3}$) were deposited. High-resolution X-ray diffraction measurements were carried out and reciprocal space maps were recorded to determine the phase purity of the zb-InN layers [9.27]. With decreasing growth temperature a decrease of hexagonal inclusions was observed in the $\omega - 2\theta$-scans. Those amount to 11% (A), 10% (B), and 5% (C), respectively, and are some of the lowest reported for zb-InN. Bragg peaks were found at 35.8°, 39.9°, and 41.3° corresponding to zb-InN (002), zb-GaN (002), and 3C-SiC (002), respectively. The lattice constants obtained from the $\omega - 2\theta$-scan are (5.01 ±0.01) Å, which are in good agreement to other published values of 4.98 Å [9.105] and 4.986 Å [9.106], and also with theoretical calculations [9.7, 9.28]. Further details related to growth and structural characterization have been published elsewhere [9.27, 9.29]. The conducting substrate and GaN buffer layer did not allow reliable sheet electron concentrations to be determined.

9.5.2   Infrared spectroscopic ellipsometry

Infrared ellipsometry is employed for determining N_e in the bulk-like part of the zb-InN layers. The isotropic DF $\bar{\varepsilon}(\omega)$ (contrary to the hexagonal counterpart, no preferred optical axis exists) is obtained by fitting the experimental Ψ and Δ data in the MIR range using a multi-layer model zb-InN/zb-GaN/3C-SiC with surface roughness in the range of 4–6 nm on top. In analogy to w-InN, Eqs. (9.22) and (9.23) are used to describe the coupled plasmon-phonon modes. The high-frequency dielectric constant of each zb-InN film depends again on N_e which will be analyzed in section 9.5.4. The employed ε_∞ values are summarized in Table 9.12. As in previous ellipsometric MIR studies [9.26], ω_{LO} was kept constant. Here, a value of $588\,\mathrm{cm}^{-1}$ was adopted which was obtained from Raman measurements of zb-InN films [9.107]. It means only the ω_p and ω_{TO} frequencies and the broadening parameters have to be adjusted. Figure 9.19 shows the excellent agreement between the fit and the ellipsometric Ψ (a) – (c) and Δ (d) – (f) spectra of all cubic samples. MIR reflectance measurements (dashed lines) as depicted in Figs. 9.19(g) – (i) emphasize the spectral dependence of $\bar{\varepsilon}(\omega)$ for the cubic films. The good accordance to the calculated data (solid lines) confirms the validity of the model used and proves the quality of our extracted parameters.

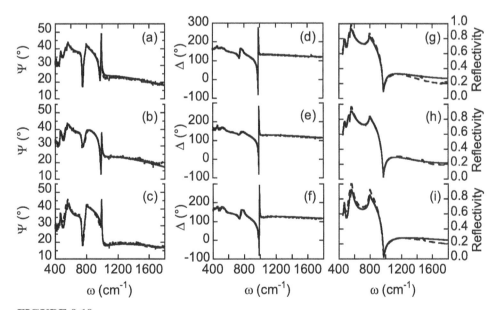

FIGURE 9.19

Measured (dashed lines) and modeled (solid lines) Ψ [(a) – (c)] and Δ [(d) – (f)] spectra at 65° angle of incidence for samples A–C (from top to bottom). The calculated (solid lines) IR reflectance spectra from the obtained MIR-DF are in excellent agreement with the measured (dashed lines) reflectivities [(g) – (i)].

The fitted ω_{TO}, ω_p, and γ_p values are provided in Table 9.12. For the former we get values between 468.9 cm^{-1} and 470.0 cm^{-1} which match well to the Raman result of $\omega_{TO} = 470$ cm^{-1} [9.107]. The average effective electron masses were calculated via Eq. (9.16). Hence, the electron concentrations and mobilities of the cubic films can also be calculated via Eqs. (9.24) and (9.25). The data correspond to N_e values between 2.1 and 3.8×10^{19} cm^{-3} while μ_e is found between 835 and 580 cm^2 V^{-1} s^{-1} (see Table 9.12).

TABLE 9.12
Summary of the determined phonon and plasma frequencies as well as broadening parameters, electron concentrations, mobilities, and high-frequency dielectric constants of zb-InN. The average electron effective masses $m^*(N_e)$ are also given.

Sample	ω_{TO} [cm^{-1}]	ω_p [cm^{-1}]	γ_p [cm^{-1}]	N_e [10^{19}cm^{-3}]	μ_e [cm^2/Vs]	$\varepsilon_\infty(N_e)$	$m^*(N_e)$ /m_0
A	468.9	2025	167	2.1	835	6.97	0.067
B	469.3	2061	190	2.2	720	6.95	0.068
C	470.0	2599	212	3.8	580	6.64	0.076

9.5.3 Interband dielectric function of zinc-blende InN

Figure 9.20(a) shows the real and imaginary parts of the DF for zb-InN over the whole investigated photon energy range; the data correspond to sample A exhibiting an electron concentration of 2.1×10^{19} cm^{-3}. First, the spectral dependence of ε_2 up to 4 eV is very similar to the behavior found for w-InN (see Fig. 9.10). We observe a sharp increase of ε_2 in the energy range from 0.8 up to 1.3 eV followed by a plateau. The nearly constant value of the imaginary part up to 3 eV is again a clear indication for a nonparabolic conduction band. Second, the high-energy part of the cubic spectrum also reveals distinctive features (correlated to Van Hove singularities of the band structure), but the shape of ε_2 is markedly different from both the ordinary and extraordinary DF of hexagonal InN. A pronounced double-peak structure is found between 5 eV and 6.5 eV followed by a shoulder at ~8 eV and a third peak at ~9.5 eV. The experimentally determined spectral shape is in good agreement with the *ab initio* results displayed in Fig. 9.20(b). The theoretical curve was calculated for Coulomb-correlated electron-hole pairs as described in chapter 8. The correct peak positions are only reproduced by including the exciton effects. The determination of the (excitonic) CP transition energies as well as their assignment to the high-symmetry points of the BS will be discussed in section 9.5.5. Almost identical results were obtained for the other films; only the onset of absorption depends on the carrier concentration of the films which is analyzed in detail below.

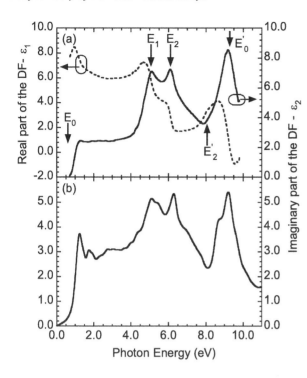

FIGURE 9.20
(a) Dielectric function of zinc-blende InN with an electron concentration of $2.1 \times 10^{19}\,\text{cm}^{-3}$. The real (dashed line) and imaginary part (solid line) are obtained from fitting the room-temperature ellipsometric data of sample A. (b) Imaginary part of the DF of zb-InN calculated for Coulomb-correlated electron-hole pairs. The result is taken from Chapter 8.

9.5.4 Analysis of the DF around the band gap

Figure 9.21(a) shows the imaginary parts of the DF around the band gap for three zb-InN films at room temperature as well as the corresponding emission spectra (PL) recorded at 10 K. First, the shape of ε_2 around the absorption edge is very similar to w-InN. It reveals a sharp increase in the energy range from 0.8 up to ~1.3 eV followed by a plateau-like behavior. The onset of absorption for sample C is shifted by 160 meV to higher energies in comparison to samples A and B. It indicates that the BMS is strongest for sample C, while a lower but similar electron concentration is expected for the two other samples. Secondly, N_e also influences the shape of ε_1 as already demonstrated for w-InN in section 9.3.4. Here, we observe the same behavior. The peak position shifts due to Kramers-Kronig consistency of ε_1 and ε_2, and the high-frequency dielectric constant ε_∞ becomes sample dependent, as can be seen in Fig. 9.21(b). For obtaining the ε_∞ values, the ε_1 spectra below the peak maximum has to be fitted via Eq. (9.26).

The fit results for the samples with the lowest (A) and highest (C) expected electron concentration are presented in Fig. 9.21(c). The calculation yielded for all samples values of $E_1 = (4.35 \pm 0.22)$ eV, $A_1 = (20.26 \pm 1.13)$ eV, and $A_0 = (3.1 \pm 0.1)$ while E_G increases from 0.97 to 1.14 eV from sample A to C, respectively. The determined sample dependent ε_∞ values are listed in Table 9.12; a clear lowering is found for sample C.

The fit of the $\varepsilon_2(\omega)$ curves was carried out in analogy to the self-consistent approach for w-InN presented in sections 9.2.4 and 9.3.5. An E_P parameter of 14 eV [9.29] is inserted into Eq. (9.17) in order to describe the dispersion relation of the

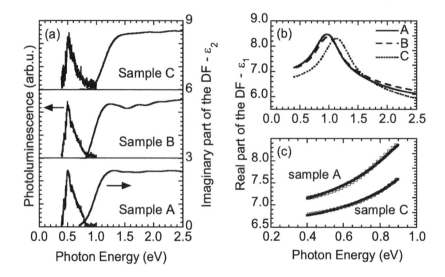

FIGURE 9.21

(a) PL spectra at $T = 10\,$K and imaginary parts of the DF at room temperature for zb-InN samples with different electron densities. The ε_2 data are vertically shifted, and PL curves are normalized to equal amplitude for the sake of clarity. (b) Real parts of the Kramers-Kronig-transformed ε_2 data. (c) Fit results (solid lines) of ε_1 below the gap via Eq. (9.26) for the samples showing the lowest (rectangles) and highest (circles) electron concentration.

nonparabolic CB; the VB is again assumed to be parabolic with an effective hole mass of $m_h = 0.5m_0$ [9.29]. A comparison of the good correspondence between calculated and experimental $\varepsilon_2(\omega)$ curves for samples with the lowest (sample A) and highest (sample C) electron concentration is shown in Fig. 9.22(a) and (b), respectively. The onset of absorption for sample C is shifted by about 160 meV to higher energies in comparison to sample A due to carrier-induced BMS.

A summary of all determined characteristic values is given in Table 9.13. It can be seen that the renormalized band gaps (E_{ren}) for the highly degenerate zb-InN films are nearly identical. One assumes, similarly to w-InN, a temperature-dependent band-gap shift of 35 – 50 meV from RT to 10 K [9.87]; the onset of the low temperature PL spectra (see Fig. 9.21) is indeed found at approximately these energies. All three samples show nearly the same PL signal, further confirming the equality of our extracted renormalized band gaps given in Table 9.13.

Now, the zero-density band gap E_0 can be estimated from E_{ren} and the calculated Δ_{BGR}. The calculation of the BGR is carried out with a value of 12.3 [9.29] for the static dielectric constant ε_r of zb-InN via Eqs. (9.15) and (9.18). The detailed analysis yields E_0 values of 0.595, 0.591, and 0.603 eV for sample A, B, and C, respectively. For comparison, an E_P of only 10 eV as for w-InN [9.10, 9.8] increases E_{ren} up to 100 meV and the difference between the three E_0 values becomes much larger; therefore a zero-density band gap of 0.595 eV with a corresponding effective electron mass of $0.041\,m_0$ at the CB minimum is very likely. Note that our

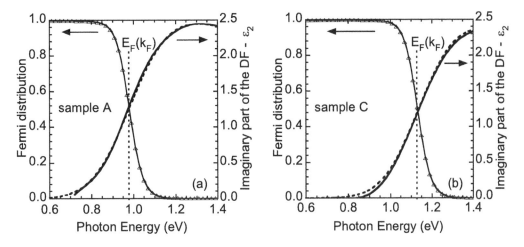

FIGURE 9.22

Calculated shape of the imaginary part of the DF (dashed lines) in comparison to the experimental data (solid lines) for samples A and C, that is, typical examples for samples with lower and higher carrier density, respectively. Band filling and band-gap renormalization are taken into account in the calculations, as explained in the text. The Fermi distribution function (open triangles) is also given for comparison.

results agree well with the theoretically predicted zb-InN fundamental band gap of 0.59 eV by Persson *et al.* [9.86]. It is thus 80 meV lower than for the hexagonal counterpart. Such a difference was also predicted by the calculations presented in Refs. [9.54] and [9.86].

Finally, the Kramers-Kronig transformation of the calculated $\varepsilon_2(\omega)$ curve for $E_0 = 0.595$ eV leads to an ε_∞ of 7.84.

TABLE 9.13

Fermi energies as well as contributions to Burstein-Moss shift and band-gap renormalization used for determining the zero-density band gaps E_0 of cubic InN.

Sample	$E_F(k_F)$ [eV]	$\Delta E_c(k_F)$ [meV]	$\Delta E_v(k_F)$ [meV]	$\Delta E_{cv}(k_F)$ [meV]	E_{ren} [eV]	ΔE_{e-i} [meV]	ΔE_{e-e} [meV]	Δ_{BGR} [meV]	E_0 [eV]
A	0.975	464	−56	520	0.455	−86	−54	−140	**0.595**
B	0.982	475	−57	532	0.450	−87	−54	−141	**0.591**
C	1.130	618	−82	700	0.430	−108	−65	−173	**0.603**

9.5.5 High-energy critical points of the band structure

The evaluation of CP transition energies for zb-InN is carried out with the same approach successfully applied to w-InN. The third derivative of the point-by-point DF multiplied by the square of the photon energy was numerically calculated. The obtained data for the real and imaginary part are simultaneously fitted via Eq. (9.29) in order to increase the reliability of the results. Figure 9.23 shows a comparison of

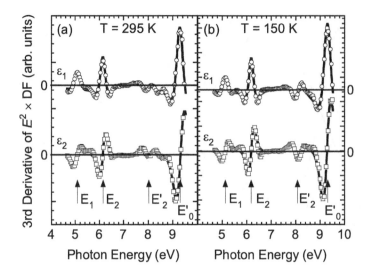

FIGURE 9.23

Fit of the third derivative of the DF for a zb-InN film at (a) room temperature and (b) $T = 150\,\mathrm{K}$. The circles and rectangles represent the experimental ε_1 and ε_2 data, respectively. The solid lines are the best fits. The transition energies of the Van Hove singularities are labeled by arrows.

the experimental data (symbols) to the fit (solid lines) presented for one cubic InN film for $T = 295\,\mathrm{K}$ (a) and $T = 150\,\mathrm{K}$ (b). All transitions are nicely reproduced with $n = 8$ that accounts for excitonic-type CPs. The cubic sample set shows good agreement of its CP energies within 20 meV. A comparison of the (excitonic) CP transition energies of zb-InN at room temperature, $T = 150\,\mathrm{K}$, and the theoretically calculated ones is given in Table 9.14. The experimentally determined energy shift between both temperatures amounts to approximately 40 meV, which is very close to the results for w-InN. Furthermore, the excellent agreement of the $T = 150\,\mathrm{K}$ data and the theoretical values should be noticed.

As already mentioned, the Coulomb correlation of electrons and holes causes an overall redshift of the ε_2 peaks and a redistribution of oscillator strength with respect to the independent quasi-particle DF. Knowing this shift, the excitonic transi-

TABLE 9.14

A comparison of transition energies at Van Hove singularities for cubic InN at room temperature and $T = 150\,\mathrm{K}$ to theoretical calculations.

	E_1 (eV)	E_2 (eV)	E_2' (eV)	E_0' (eV)
295 K	5.08	6.14	8.20	9.28
150 K	5.12	6.20	8.24	9.31
Theory[a] ($T = 0\,\mathrm{K}$)	5.12	6.25	8.30	9.25

[a]See Chapter 8.

tion energies can be traced back to the Van Hove singularities of the single-particle band structure. The transition E_1 is found to be likely located at the L-point, whereas E_2 and E_2' are probably correlated to optical transitions from the upper-most VB into a lower and an upper CB at the X-point of the BZ, respectively. The highest CP energy deduced from the third derivative spectra is proposed to be assigned to the E_0' transition at the Γ-point of the BZ for the following reasons: (i) the optical properties of any nitride semiconductor are strongly influenced by the p-like bonding (valence band) and anti-bonding (conduction band) states. Transitions from the former into the s-like CB at the Γ- point of the BZ determine the fundamental band gap E_0 of the material. Due to mixing effects, transitions between the p-like bonding and anti-bonding states become partly allowed giving rise to a Van Hove singularity (usually denoted by E_0') which is detected in the imaginary part of the DF at much higher photon energies than the E_0 feature [9.108]; (ii) a tight-binding estimation with the parameters from Ref. [9.109] was used and makes clear why the splitting between the bonding and anti-bonding states is much higher for InN compared to other In-compound semiconductors. The splitting between the p-like bonding VB (Γ_v^{15}) and anti-bonding CB (Γ_c^{15}) can be calculated from

$$\Delta E = E_0' = \Gamma_c^{15} - \Gamma_v^{15} = \sqrt{\left(E_p^c - E_p^a\right)^2 + \left(2E_{xx}\right)^2}, \qquad (9.31)$$

where E_p^c and E_p^a denote the p-atomic orbital energies of the cation and the anion, respectively. The exchange matrix element E_{xx} results from

$$E_{xx} = \frac{4}{3}\left[V_{pp\sigma} + 2V_{pp\pi}\right] \approx 9.76\,\mathrm{eV} \times \left(\frac{\text{\AA}}{d_{ac}}\right), \qquad (9.32)$$

where $V_{pp\sigma}$ and $V_{pp\pi}$ are the matrix elements of the interaction Hamiltonian between the p orbitals along (σ) and perpendicular (π) to the p-orbital direction, respectively, and are usually referred to as the overlap parameters. The distance of the cation and anion is given by d_{ac}; and (iii) the following peculiarities of zb-InN should be noted. First, the difference between the p-atomic orbitals ($E_p^c - E_p^a$) for indium and nitrogen amounts to 8.47 eV, and corresponding values of the other nitrides are almost identical. Second, the exchange matrix element increases due to the reduced interatomic spacing for InN (d_{ac} = 2.15 Å). So in summary, the tight-binding approach leads to a splitting between Γ_v^{15} and Γ_c^{15} states of 9.46 eV. We determined the critical point transition energy to be 9.28 eV at room temperature (9.31 eV at T = 150 K) which is in excellent agreement with the result of the tight-binding approach. Therefore, we conjecture that this CP is related to the E_0' transition.

9.5.6 Bowing parameter of zinc-blende InGaN

Only a few studies of zb-In$_x$Ga$_{1-x}$N alloys have been reported so far (for example, Refs. [9.30, 9.103, 9.110]). In a previous paper [9.30], the DFs of highly-degenerate

Ga-rich $In_xGa_{1-x}N$ alloys ($x \leq 0.18$; N_e in the range of $10^{19}\,cm^{-3}$) were determined by SE. From an analysis of the DFs around the absorption edge, transition energies were obtained which are plotted by full triangles in Fig. 9.24 as a function of the In mole fraction x. Using 1.80 and 3.23 eV as E_0 values for the cubic binary compounds InN and GaN, respectively, a band gap bowing of $b = 1.4$ eV was determined.

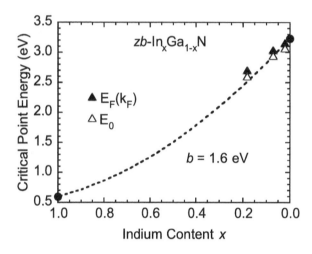

FIGURE 9.24

Band gap energies of zinc-blende In-GaN alloys as a function of indium content. Full circles represent the fundamental band gaps of zinc-blende InN and GaN. Triangles are assigned to previous results of zb-InGaN alloys as explained in the text.

The considerably improved data analysis of ε_2 in the presence of high carrier concentrations indicates, however, that the previously published InGaN data correspond rather to $E_F(k_F)$ and not to the intrinsic band gaps of the alloys. Applying the models introduced in sections 9.2.4 and 9.3.5 as well as carrier effective masses and E_P parameter for GaN and Ga-rich alloys from Rinke *et al.* [9.55], we are now able to calculate the influence of N_e on the CB filling and the band-gap renormalization. Based on these data, we get intrinsic band gaps as indicated by the open triangles in Fig. 9.24. Of course, the redshift is comparably small (due to the high effective masses for the carriers), but it has an influence on the reevaluation of the bowing parameter. For the two binary compounds we use the revised E_0 values of 0.595 eV (InN) and 3.23 eV (GaN) [9.30]. Then, the fit yields the dashed line in Fig. 9.24 with a new bowing parameter for the cubic $In_xGa_{1-x}N$ alloy system of $b_0 = 1.6$ eV. For comparison, Gorczyca *et al.* [9.111] calculated a composition-dependent bowing ranging from 1.95 eV for $x = 0.031$ to 1.65 eV for $x = 0.25$ and 1.63 eV for $x = 0.5$. They used for the gaps of InN and GaN values of 0.62 eV and 3.28 eV.

9.6   Summary and outlook

Spectroscopic ellipsometry from the mid-infrared via the near-infrared to the vacuum-ultraviolet spectral region has been proven to be a very powerful tool for studying the absorption related properties of InN and its alloys with GaN and AlN. Taking for the data analysis surface roughness and interface layers into account, a reliable dielectric function is obtained in the primary step which reflects the essential features of the materials. Except for the MIR range, no assumption was made concerning the spectral dependence of the DF, that is, the values of the real and imaginary parts were fitted for each photon energy. The interpretation of the spectral dependence represents the second step of the studies from which sample-dependent parameters such as the frequencies of the coupled phonon-plasmon modes, the spacing between the conduction and valence bands at the Fermi wave vector, or the transition energies at the Van Hove singularities are determined. The intrinsic properties of the compounds were attained in the final step by correcting these data for carrier-induced band-gap renormalization and the Burstein-Moss shift as well as for strain.

The MIR data can be successfully represented by a coupled plasmon-phonon dielectric function model; the accuracy of this DF was verified by comparison with the results of reflectance studies. The frequencies of the lower and upper polariton branches in conjunction with the longitudinal- and transversal phonon energies yield the plasma frequency. Taking into account the averaged effective optical mass of the collective electron excitation and the density-dependent high-frequency dielectric constant, this quantity is used for an accurate determination of the electron concentration in the bulk-like part of the layers.

No analytic formula exists allowing the fit of ε_2 around the fundamental interband absorption edge in the NIR range in the presence of high electron concentrations and band nonparabolicity. The NIR data have to be self-consistently analyzed (coupled to the MIR results) by calculating the imaginary part of the DF via Fermi's Golden rule and treating the Fermi energy and the renormalized band gap as adjustable parameters. Two approaches to account for the electron-induced gap shrinkage were compared. The presently preferred method yields 0.675 eV for the zero-density strain-free gap of hexagonal InN at room temperature; the observed pronounced optical anisotropy in this photon range caused by valence band ordering emphasizes the low band gap of InN. The value for the cubic counterpart is 80 meV lower and was determined to be 0.595 eV. Analyzing the DF of a large number of InGaN and InAlN alloys in a similar way, the bowing parameters of the E_A (hexagonal) and E_0 (cubic) gaps were obtained.

A main result of this work is the determination of the DFs in the UV-VUV spectral region and their comparison with the results of the calculations summarized in Chapter 8. It turns out that the theoretical results on the imaginary part of the ordinary and extraordinary DF for hexagonal InN and on the isotropic DF for cubic InN are only in excellent agreement with the experimental data if electron-hole interaction (exciton effects) is taken into account by solving the Bethe-Salpeter equation. The Coulomb correlation leads to redshift of the absorption peaks and a redistribution of oscillator strength in comparison to the independent quasi-particle DF. Only with this approach is the experimental polarization anisotropy of the hexagonal polymorph fully reproduced. The results of other groups for hexagonal GaN and AlN emphasize the universality of the finding. A fit of the third derivatives of the DFs yields the excitonic transition energies which can be traced back to the Van Hove singularities in the single-particle band structure. For the hexagonal InGaN and InAlN alloys, the bowing parameters of the high-energy excitonic transitions were reported.

Despite the convincing results presented in this chapter, the discussion of many topics has to be continued in the future. First of all, the determination of the zero-density strain-free gap should be mentioned. Repeated reports of an absorption edge around 0.64 eV (RT) for hexagonal InN with an electron concentration of $\sim 3 \times 10^{17} \, \text{cm}^{-3}$ should not be disregarded. A possible explanation for our higher value of 0.675 eV might be that the band-gap renormalization approach preferred here overestimates the effect. In order to answer this question, input from the theory is demanded because it is at present not known how to take the nonparabolicity of the conduction band correctly into account. Considerable progress in this respect will, however, only be achieved by a clear definition of which energy was and is attained from the experimental data.

The influence of strain on the shape of the absorption represents another important aspect. It changes the position of the band edges and the relative oscillator strengths. The analysis of the effect is still in the very earliest stages because the lattice parameters were often not reported and the fundamental quantities such as crystal-field, spin-orbit coupling, and the deformation potentials are not exactly known yet. Ellipsometric studies of alloys with non-polar surface orientation, although difficult to grow, can add valuable information. Both the ordinary and extraordinary DF will be obtained, exhibiting a clear splitting of the main absorption edges for the two polarization directions. In particular for the InAlN system, the size of the splitting depends, via the crystal-field parameter, strongly on the composition. It is at present not clear whether the crystal-field changes linearly with the Al-content.

Finally, the presented results on cubic InN are very promising. There is no doubt, that this material exhibits the lowest band gap among all nitride compounds. In order to make use of the advantages, however, the residual electron concentration has to be reduced by a further optimization of the growth process. In zinc-blende devices, for example, the huge polarization-induced built-in electric fields of polar hexagonal heterostructures limiting the radiative efficiency can be avoided. More data on the properties of the cubic alloys are required in order to allow an optimized design of such devices.

Acknowledgments

Part of this work was supported by the Thüringer Ministerium für Wirtschaft, Technologie und Arbeit (B509-04011). R. G. acknowledges financial support by Deutsche Forschungsgemeinschaft (DFG). The ellipsometric studies at BESSY were supported by the Bundesministerium für Bildung und Forschung (BMBF) (grants 05KS4KTB/3 and 05ES3XBA/5). We would like to thank our co-workers A. T. Winzer, C. Buchheim, C. Napierala, S. Shokhovets, J. Pezoldt, and G. Gobsch (Ilmenau) for their input. The authors are very grateful to H. Lu and W. J. Schaff (Cornell University) for providing the In- and M-face InN, InAlN, and InGaN samples, respectively, and to M. Kurouchi, H. Naoi, Y. Nanishi (Ritsumeikan University), X. Wang, and A. Yoshikawa (Chiba University) for supplying the N-face InN and InGaN samples. We would also like to thank J. Schörmann, D. J. As, and K. Lischka (University of Paderborn) for the support with zinc-blende InN samples. Furthermore, we gratefully acknowledge M. Rakel, C. Cobet, C. Werner, and N. Esser of the ISAS Berlin for their assistance with BESSY measurements. The excellent collaboration with the theory group of the FSU Jena (F. Bechstedt, F. Fuchs, and J. Furthmüller) helped very much for the interpretation of the experimental data. We are particularly indebted to V. Cimalla and O. Ambacher (IAF Freiburg), M. Feneberg, and K. Thonke (University of Ulm) in this respect. Finally, we would like to thank T. D. Veal and C. F. McConville (University of Warwick) for many stimulating discussions on the properties of InN.

References

[9.1] T. L. Tansley and C. P. Foley, Optical band gap of indium nitride, *Journal of Applied Physics*, 59 (1986) 3241–3244.

[9.2] V. Yu. Davydov, A. A. Klochikhin, R. P. Seisyan, V. V. Emtsev, S. V. Ivanov, F. Bechstedt, J. Furthmüller, H. Harima, A. V. Mudryi, J. Aderhold, O. Semchinova, and J. Graul, Absorption and emission of hexagonal InN. Evidence of narrow fundamental band gap, *physica status solidi (b)*, 229 (2002) R1–R3.

[9.3] J. Wu, W. Walukiewicz, K. M. Yu, J. W. Ager III, E. E. Haller, H. Lu, W. J. Schaff, Y. Saito, and Y. Nanishi, Unusual properties of the fundamental band gap of InN, *Applied Physics Letters*, 80 (2002) 3967–3969.

[9.4] T. Matsuoka, H. Okamoto, M. Nakao, H. Harima, and E. Kurimoto, Optical bandgap energy of wurtzite InN, *Applied Physics Letters*, 81 (2002) 1246–1248.

[9.5] C. Persson, R. Ahuja, A. Ferreira da Silva, and B. Johansson, First-principle calculations of the dielectric function of zinc-blende and wurtzite InN, *Journal of Physics: Condensed Matter*, 13 (2001) 8945–8950.

[9.6] F. Bechstedt, J. Furthmüller, M. Ferhat, L. K. Teles, L. M. R. Scolfaro, J. R. Leite, V. Yu. Davydov, O. Ambacher, and R. Goldhahn, Energy gap and optical properties of $In_xGa_{1-x}N$, *physica status solidi (a)*, 195 (2003) 628–633.

[9.7] J. Furthmüller, P. H. Hahn, F. Fuchs, and F. Bechstedt, Band structures and optical spectra of InN polymorphs: Influence of quasiparticle and excitonic effects, *Physical Review B*, 72 (2005) 205106:1–14.

[9.8] J. Wu, W. Walukiewicz, W. Shan, K. M. Yu, J. W. Ager III, E. E. Haller, H. Lu, and W. J. Schaff, Effects of the narrow band gap on the properties of InN, *Physical Review B*, 60 (2002) 201403:1–4.

[9.9] R. Goldhahn, P. Schley, A. T. Winzer, G. Gobsch, V. Cimalla, O. Ambacher, M. Rakel, C. Cobet, N. Esser, H. Lu, and W. J. Schaff, Detailed analysis of the dielectric function for wurtzite InN and In-rich InAlN alloys, *physica status solidi (a)* 203 (2006) 42–49.

[9.10] P. Schley, R. Goldhahn, A. T. Winzer, G. Gobsch, V. Cimalla, O. Ambacher, H. Lu, W. J. Schaff, M. Kurouchi, Y. Nanishi, M. Rakel, C. Cobet, and N. Esser, Dielectric function and Van Hove singularities for In-rich $In_xGa_{1-x}N$ alloys: comparison of N- and metal-face materials, *Physical Review B*, 75 (2007) 205204:1–8.

[9.11] W. Walukiewicz, J. W. Ager III, K. M. Yu, Z. Lilienthal-Weber, J. Wu, S. X. Li, R. E. Jones, and J. D. Denlinger, Structure and electronic properties of InN and In-rich group III-nitride alloys, *Journal of Physics D: Applied Physics*, 39 (2006) R83-R99.

[9.12] P. Y. Yu and M. Cardona, *Fundamentals of Semiconductors*, Springer, Berlin, 2001, p. 245ff.

[9.13] R. Goldhahn, S. Shokhovets, V. Cimalla, L. Spieß, G. Ecke, O. Ambacher, J. Furthmüller, F. Bechstedt, H. Lu, and W. J. Schaff, Dielectric function of "narrow" bandgap InN, *Materials Research Society Symposium Proceedings*, 743 (2003) 361–366.

[9.14] F. Bechstedt, J. Furthmüller, O. Ambacher, and R. Goldhahn, Comment on "Mie resonances, infrared emission, and the band gap of InN", *Physical Review Letters*, 93 (2004) 269701-1.

[9.15] T. V. Shubina, S. V. Ivanov, V. N. Jmerik, D. D. Solnyshkov, V. A. Vekshin, P. S. Kopev, A. Vasson, J. Leymarie, A. Kavokin, H. Amano, K. Shimono, A. Kasic, and B. Monemar, Mie resonances, infrared emission, and the band gap of InN, *Physical Review Letters*, 92 (2004) 117407:1–4.

[9.16] M. Losurdo, G. Bruno, T.-H. Kim, S. Choi, and A. Brown, Study of the dielectric function of hexagonal InN: Impact of indium clusters and of native oxide, *Applied Physics Letters*, 88 (2006) 121928:1–3.

[9.17] T. Schmidtling, M. Drago, U. W. Pohl, and W. Richter, Spectroscopic ellipsometry during metalorganic vapor phase epitaxy of InN, *Journal of Crystal Growth*, 248 (2003) 523–527.

[9.18] M. Drago, T. Schmidtling, C. Werner, M. Pristosek, U. W. Pohl, and W. Richter, InN growth and annealing investigations using in-situ spectroscopic ellipsometry, *Journal of Crystal Growth*, 272 (2004) 87–93.

[9.19] M. Drago, P. Vogt, and W. Richter, MOVPE growth of InN with ammonia on sapphire, *physica status solidi (a)*, 203 (2006) 116–126.

[9.20] M. Yoshitani, K. Akasaka, X. Wang, S.-B. Che, Y. Ishitani, and A. Yoshikawa, In situ spectroscopic ellipsometry in plasma-assisted molecular beam epitaxy of InN under different surface stoichiometries, *Journal of Applied Physics*, 99 (2006) 044913:1–6.

[9.21] M. Losurdo, T.-H. Kim, S. Choi, P. Wu, M. M. Giangregorio, G. Bruno, and A. Brown, Real time optical monitoring of molecular beam epitaxy of InN on SiC substrates, *Journal of Vacuum Science and Technology B*, 25 (2007) 1014–1018.

[9.22] M. Losurdo, M. M. Giangregorio, G. Bruno, T.-H. Kim, S. Choi, A. Brown, G. Pattinari, M. Capizzi, and A. Polimeni, Behavior of hydrogen in InN investigated in real time exploiting spectroscopic ellipsometry, *Applied Physics Letters*, 91 (2007) 081917:1–3.

[9.23] R. Goldhahn, A. T. Winzer, V. Cimalla, O. Ambacher, C. Cobet, W. Richter, N. Esser, J. Furthmüller, F. Bechstedt, H. Lu, and W. J. Schaff, Anisotropy of the dielectric function for wurtzite InN, *Superlattices and Microstructures*, 36 (2004) 591–597.

[9.24] R. Goldhahn, P. Schley, A. T. Winzer, M. Rakel, C. Cobet, N. Esser, H. Lu, and W. J. Schaff, Critical points of the band structure and valence band ordering at the Γ point of wurtzite InN, *Journal of Crystal Growth*, 288 (2006) 273–277.

[9.25] M. Rakel, C. Cobet, N. Esser, F. Fuchs, F. Bechstedt, R. Goldhahn, W. G. Schmidt, and W. Schaff, GaN and InN conduction-band states studied by ellipsometry, *Physical Review B*, 77 (2008) 115120:1–8.

[9.26] A. Kasic, E. Valcheva, B. Monemar, H. Lu, and W. J. Schaff, InN dielectric function from the midinfrared to the ultraviolet range, *Physical Review B*, 70 (2004) 115217:1–8.

[9.27] J. Schörmann, D. J. As, K. Lischka, P. Schley, R. Goldhahn, S. F. Li, W. Löffler, M. Hetterich, and H. Kalt, Molecular beam epitaxy of phase pure cubic InN, *Applied Physics Letters*, 89 (2006) 261903:1–3.

[9.28] D. Bagayoko, L. Franklin, and G. L. Zhao, Predictions of electronic, structural, and elastic properties of cubic InN, *Journal of Applied Physics*, 96 (2004) 4297–4301.

[9.29] P. Schley, R. Goldhahn, C. Napierala, G. Gobsch, J. Schörmann, D. J. As, K. Lischka, M. Feneberg, and K. Thonke, Dielectric function of cubic InN from the mid-infrared to the visible spectral range, *Semiconductor Science and Technology*, 23 (2008) 055001:1–6.

[9.30] R. Goldhahn, J. Scheiner, S. Shokhovets, T. Frey, U. Köhler, D. J. As, and K. Lischka, Refractive index and gap energy of cubic $In_xGa_{1-x}N$, *Applied Physics Letters*, 76 (2000) 291–293.

[9.31] A. Kasic, M. Schubert, Y. Saito, Y. Nanishi, and G. Wagner, Effective electron mass and phonon modes in n-type hexagonal InN, *Physical Review B*, 65 (2002) 115206:1–7.

[9.32] T. Hofmann, V. Darakchieva, B. Monemar, H. Lu, W. J. Schaff, and M. Schubert, Optical Hall effect in hexagonal InN, *Journal of Eelectronic Materials*, 37 (2008) 0611–0615.

[9.33] V. Yu. Davydov, A. A. Klochikhin, V. V. Emtsev, D. A. Kurdyukov, S. V. Ivanov, V. A. Vekshin, F. Bechstedt, J. Furthmüller, F. Aderhold, J. Graul, A. V. Mudryi, H. Harima, A. Hashimoto, A. Yamamoto, and E. E. Haller, Band gap of hexagonal InN and InGaN alloys, *physica status solidi (b)*, 234 (2002) 787–795.

[9.34] H. Lu, W. J. Schaff, L. F. Eastman, and C. E. Stutz, Surface charge accumulation of InN films grown by molecular-beam epitaxy, *Applied Physics Letters*, 82 (2003) 1736-1738.

[9.35] I. Mahboob, T. D. Veal, C. F. McConville, H. Lu, and W. J. Schaff, Intrinsic electron accumulation at clean InN surfaces, *Physical Review Letters*, 92 (2004) 036804:1–4.

[9.36] B. Arnaudov, T. Paskova, P. P. Paskov, B. Magnusson, E. Valcheva, B. Monemar, H. Lu, W. J. Schaff, H. Amano, and I. Akasaki, Energy position of near-band-edge emission spectra of InN epitaxial layers with different doping levels, *Physical Review B*, 69 (2004) 115216:1–5.

[9.37] V. Cimalla, M. Niebelschütz, G. Ecke, V. Lebedev, O. Ambacher, M. Himmerlich, S. Krischok, J. A. Schaefer, H. Lu, and W. J. Schaff, Surface band bending at nominally undoped and Mg-doped InN by Auger Electron Spectroscopy, *physica status solidi (a)*, 203 (2006) 59–65.

[9.38] C. S. Gallinat, G. Koblmüller, J. S. Brown, S. Bernardis, J. S. Speck, G. D. Chern, E. D. Readinger, H. Shen, and M. Wraback, In-polar InN grown by plasma-assisted molecular beam epitaxy, *Applied Physics Letters*,

89 (2006) 032109:1–3.

[9.39] G. Koblmüller, C. S. Gallinat, S. Bernardis, J. S. Speck, G. D. Chern, E. D. Readinger, H. Shen, and M. Wraback, Optimization of the surface and structural quality of N-face InN grown by molecular beam epitaxy, *Applied Physics Letters*, 89 (2006) 071902:1–3.

[9.40] R. Goldhahn, C. Buchheim, P. Schley, A. T. Winzer, and H. Wenzel, Optical constants of bulk nitrides, in *Nitride semiconductor devices: Principles and simulation*, ed. J. Piprek, Wiley, Weinheim, 2007, pp. 95–115.

[9.41] E. Y. Lin, T. S. Lay, and T. Y. Chang, Accurate model including Coulomb-enhanced and Urbach-broadened absorption spectrum of direct-gap semiconductors, *Journal of Applied Physics*, 102 (2007) 123511:1–10.

[9.42] S. Shokhovets, G. Gobsch, and O. Ambacher, Excitonic contribution to the optical absorption in zinc-blende III-V semiconductors, *Physial Review B*, 74 (2006) 155209:1–11.

[9.43] S. Shokhovets, O. Ambacher, and G. Gobsch, Conduction-band dispersion relation and electron effective mass in III-V and II-VI zinc-blende semiconductors, *Physical Review B*, 76 (2007) 125203:1–18.

[9.44] S. Yamaguchi, M. Kariya, S. Nutto, T. Takeuchi, C. Wetzel, H. Amano, and I. Akasaki, Structural properties of InN on GaN grown by metalorganic vapor-phase epitaxy, *Journal of Applied Physics*, 85 (1999) 7682–7688.

[9.45] E. Dimakis, E. Iliopoulos, K. Tsagaraki, A. Adikimenakis, and A. Georgakilas, Biaxial strain and lattice constants of InN(0001) films grown by plasma-assisted molecular beam epitaxy, *Applied Physics Letters*, 88 (2006) 191918:1–3.

[9.46] X. Wang, S.-B. Che, Y. Yoshitani, and A. Yoshikawa, Experimental determination of strain-free Raman frequencies and deformation potentials for the E_2 high and $A-1(LO)$ modes in hexagonal InN, *Applied Physics Letters*, 89 (2006) 171907:1–3.

[9.47] H. Y. Peng, M. D. McCluskey, Y. M. Gupta, M. Kneissl, and N. M. Johnson, Shock-induced band-gap shift in GaN: Anisotropy of the deformation potentials, *Physical Review B*, 71 (2005) 115207:1–5.

[9.48] J. Bhattacharyya, S. Ghosh, M. R. Gokhale, B. M. Arora, H. Lu, and W. J. Schaff, Polarized photoluminescence and absorption in A-plane InN films, *Applied Physics Letters*, 89 (2006) 151910:1–3.

[9.49] J. Bhattacharyya and S. Ghosh, Electronic band structure of wurtzite InN around the fundamental gap in the presence of biaxial strain, *physica status solidi (a)*, 204 (2007) 439-446.

[9.50] R. Goldhahn and S. Shokhovets, Optical constants of III-Nitrides—

Experiments, in *III-Nitride semiconductors: Optical properties II*, ed. M. O. Manasreh and H. X. Jiang, Taylor & Francis, New York, 2002, pp. 73–113.

[9.51] S. Shokhovets, O. Ambacher, B. K. Meyer, and G. Gobsch, Anisotropy of the momentum matrix element, dichroism, and the conduction-band dispersion relation of wurtzite semiconductors, *Physical Review B*, 78 (2008) 035207:1–18.

[9.52] J. J. Hopfield, Fine structure in the optical absorption edge of anisotropic crystals, *Journal of Physics and Chemistry of Solids*, 15 (1960) 97–107.

[9.53] M. Cardona and N. E. Christensen, Spinorbit splittings in AlN, GaN and InN, *Solid State Commununications*, 116 (2000) 421–425.

[9.54] P. Carrier and S.-H. Wei, Theoretical study of the band-gap anomaly of InN, *Journal of Applied Physics*, 97 (2005) 033707:1–5.

[9.55] P. Rinke, M. Winkelnkemper, A. Qteish, D. Bimberg, J. Neugebauer, and M. Scheffler, Consistent set of band parameters for the group-III nitrides AlN, GaN, and InN, *Physical Review B*, 77 (2008) 075202:1–15.

[9.56] B. Gil, Stress effects on optical properties, in *Gallium Nitride II*, Vol. 57, Semiconductors and Semimetals, eds J. I. Pankove and T. D. Moustakas, Academic Press, San Diego, 1999, p. 209.

[9.57] S. L. Chuang and C. S. Chang, k·p method for strained wurtzite semiconductor, *Physical Review B*, 54 (1996) 2491–2504.

[9.58] P. O. Löwdin, A note on the quantum-mechanical perturbation theory, *Journal of Chemical Physics*, 19 (1951) 1396–1401.

[9.59] D. J. Dugdale, S. Brand, and R. A. Abram, Direct calculation of k · p parameters for wurtzite AlN, GaN, and InN, *Physical Review B*, 61 (2000) 12933–12938.

[9.60] Y. C. Yeo, T. C. Chong, and M. F. Li, Electronic band structures and effective-mass parameters of wurtzite GaN and InN, *Journal of Applied Physics*, 83 (1998) 1429–1435.

[9.61] E. O. Kane, Band structure of indium antimonide, *Journal of Physics and Chemistry of Solids*, 1 (1957) 249–261.

[9.62] L. Chen, B. J. Skromme, R. F. Dalmau, R. Schlesser, Z. Sitar, C. Chen, W. Sun, J. Yang, and M. A. Khan, Band-edge exciton states in AlN single crystals and epitaxial layers, *Applied Physics Letters*, 85 (2004) 4334–4336.

[9.63] A. A. Klochikhin, V. Yu. Davydov, V. V. Emtsev, A. V. Sakharov, V. A. Kapitonov, B. A. Andreev, H. Lu, and W. J. Schaff, Acceptor states in the photoluminescence spectra of *n*-InN, *Physical Review B*, 71 (2005) 195207:1–16.

[9.64] M. P. Hasselbeck and P. M. Enders, Electron-electron interactions in the nonparabolic conduction band of narrow-gap semiconductors, *Physical Review B*, 57 (1998) 9674–9681.

[9.65] A. Walsh, J. L. F. Da Silva, and S.-H. Wei, Origins of band-gap renormalization in degenerately doped semiconductors, *Physical Review B*, 78 (2008) 075211:1–5.

[9.66] A. Kamińska, G. Franssen, T. Suski, I. Gorczyca, N. E. Christensen, A. Svane, A. Suchoki, H. Lu, W. J. Schaff, E. Dimakis, and A. Georgakilas, Role of conduction-band filling in the dependence of InN photoluminescence on hydrostatic pressure, *Physical Review B*, 76 (2007) 075203:1–5.

[9.67] S. Shokhovets, G. Gobsch, and O. Ambacher, Momentum matrix element and conduction band nonparabolicity in wurtzite GaN, *Applied Physics Letters*, 86 (2005) 161908:1–3.

[9.68] A. S. Barker and M. Ilegems, Infrared lattice vibrations and free-electron dispersion in GaN, *Physical Review B*, 7 (1973) 743–750.

[9.69] I. Vurgaftman and J. R. Meyer, Band parameters for nitrogen-containing semiconductors, *Journal of Applied Physics*, 94 (2003) 3675–3696.

[9.70] R. Goldhahn, Dielectric function of nitride semiconductors: Recent experimental results, *Acta Physica Polonica A*, 104 (2003) 123–147.

[9.71] R. Goldhahn, S. Shokhovets, J. Scheiner, G. Gobsch, T. S. Cheng, C. T. Foxon, U. Kaiser, G. D. Kipshidze, and W. Richter, Determination of group III nitride film properties by reflectance and ellipsometric spectroscopy studies, *physica status solidi (a)*, 177 (2000) 107–115.

[9.72] X. Wang, S.-B. Che, Y. Yoshitani, and A. Yoshikawa, Effect of epitaxial temperature on N-polar InN films grown by molecular beam epitaxy, *Journal of Applied Physics*, 99 (2006) 073512:1–5.

[9.73] T. Araki, Y. Saito, T. Yamaguchi, M. Kurouchi, Y. Nanishi, and H. Naoi, Radio frequency-molecular beam epitaxial growth of InN epitaxial films on (0001) sapphire and their properties, *Journal of Vacuum Science and Technology B*, 22 (2004) 2139–2143.

[9.74] H. Lu, W. J. Schaff, L. F. Eastman, J. Wu, W. Walukiewicz, V. Cimalla, and O. Ambacher, Growth of a-plane InN on r-plane sapphire with a GaN buffer by molecular-beam epitaxy, *Applied Physics Letters*, 83 (2003) 1136–1138.

[9.75] X. Wang and A. Yoshikawa, Molecular beam epitaxy growth of GaN, AlN and InN, *Progress in Crystal Growth and Characterization of Materials*, 48/49 (2004) 42–103.

[9.76] W. Paszkowicz, R. Černý, and S. Krukowski, Rietveld refinement for indium nitride in the 105–295 K range, *Powder Diffraction*, 18 (2003) 114–121.

[9.77] L. F. J. Piper, T. D. Veal, I. Mahboob, C. F. McConville, H. Lu, and W. J. Schaff, Temperature invariance of InN electron accumulation, *Physical Review B*, 70 (2004) 115333:1–6.

[9.78] A. A. Kukharskii, Plasmon-phonon coupling in GaAs, *Solid State Communications* 13 (1973) 1761–1765.

[9.79] V. Yu. Davydov, V. V. Emtsev, I. N. Goncharuk, A. N. Smirnov, V. D. Petrikov, V. V. Mamutin, V. A. Vekshin, S. V. Ivanov, M. B. Smirnov, and T. Inushima, Experimental and theoretical studies of phonons in hexagonal InN, *Applied Physics Letters*, 75 (1999) 3297–3299.

[9.80] Y. Ishitani, T. Ohira, X. Wang, S.-B. Che, and A. Yoshikawa, Broadening factors of $E_1(LO)$ phonon-plasmon coupled modes of hexagonal InN investigated by infrared reflectance measurements, *Physical Review B*, 76 (2007) 045206:1–7.

[9.81] M. Schubert, T. E. Tiwald, and C. M. Herzinger, Infrared dielectric anisotropy and phonon modes of sapphire, *Physical Review B*, 61 (2000) 8187–8201.

[9.82] V. Yu. Davydov, A. A. Klochikhin, M. B. Smirnov, V. V. Emtsev, V. D. Petrikov, I. A. Abroyan, A. I. Titov, I. N. Goncharuk, A. N. Smirnov, V. V. Mamutin, S. V. Ivanov, and T. Inushima, Phonons in hexagonal InN. Experiment and theory, *physica status solidi (b)*, 216 (1999) 779–783.

[9.83] L. X. Benedict, T. Wethkamp, K. Wilmers, C. Cobet, N. Esser, E. L. Shirley, W. Richter, and M. Cardona, Dielectric function of wurtzite GaN and AlN thin films, *Solid State Commununications*, 112 (1999) 129–133.

[9.84] S. Shokhovets, R. Goldhahn, G. Gobsch, S. Piekh, R. Lantier, A. Rizzi, V. Lebedev, and W. Richter, Determination of the anisotropic dielectric function for wurtzite AlN and GaN by spectroscopic ellipsometry, *Journal of Applied Physics*, 94 (2003) 307–312.

[9.85] T. Hofmann, T. Chavdarov, V. Darakchieva, H. Lu, W. J. Schaff, and M. Schubert, Anisotropy of the Γ-point effective mass and mobility in hexagonal InN, *physica status solidi (c)*, 3 (2006) 1854–1857.

[9.86] C. Persson and A. Ferreira da Silva, Linear optical response of zinc-blende and wurtzite III-N (III = B, Al, Ga, and In), *Journal of Crystal Growth*, 305 (2007) 408–413.

[9.87] J. Wu, W. Walukiewicz, W. Shan, K. M. Yu, J. W. Ager III, S. X. Li, E. E. Haller, H. Lu, and W. J. Schaff, Temperature dependence of the fundamental band gap of InN, *Journal of Applied Physics*, 94 (2003) 4457–4460.

[9.88] D. E. Aspnes, Direct verification of the third-derivative nature of electroreflectance spectra, *Physical Review Letters*, 28 (1972) 168–171.

[9.89] E. Silveira, J. A. Freitas, Jr., O. J. Glembocki, G. A. Slack, and J. A. Schowalter, Excitonic structure of bulk AlN from optical reflectivity and cathodoluminescence measurements, *Physical Review B*, 71 (2005) 041201:1–4.

[9.90] V. Darakchieva, M. Beckers, M.-Y. Xie, L. Hultman, B. Monemar, J.-F. Carlin, E. Feltin, M. Gonschorek, and N. Granjean, Effects of strain and composition on the lattice parameters and applicability of Vegard's rule in Al-rich $Al_{1-x}In_xN$ films grown on sapphire, *Journal of Applied Physics*, 103 (2008) 103513:1–7.

[9.91] Z. Dridi, B. Bouhafs, and P. Ruterana, First-principles investigation of lattice constants and bowing parameters in wurtzite $Al_xGa_{1-x}N$, $In_xGa_{1-x}N$ and $In_xAl_{1-x}N$ alloys, *Semiconductor Science and Technology*, 18 (2003) 850–856.

[9.92] R. Butté, J.-F. Carlin, E. Feltin, M. Gonschorek, S. Nicolay, G. Christmann, D. Simeonov, A. Castiglia, J. Dorsaz, H. J. Buehlmann, S. Christopoulos, G. Baldassarri Höger von Högersthal, A. J. D. Grundy, M. Mosca, C. Pinquier, M. A. Py, F. Demangeot, J. Frandon, P. G. Lagoudakis, J. J. Baumberg, and N. Grandjean, Current status of AlInN layers lattice-matched to GaN for photonics and electronics, *Journal of Physics D: Applied Physics*, 40 (2007) 6328–6344.

[9.93] V. Yu. Davydov and S. K. Tikhonov, Pressure dependence of the dielectric and optical properties of wide-gap semiconductors, *Semiconductors*, 32 (1998) 947–949.

[9.94] J. Wu, W. Walukiewicz, K. M. Yu, J. W. Ager III, S. X. Li, E. E. Haller, H. Lu, and W. J. Schaff, Universal bandgap bowing in group-III nitride alloys, *Solid State Communications*, 127 (2003) 411–414.

[9.95] W. Terashima, S.-B. Che, Y. Ishitani, and A. Yoshikawa, Growth and characterization of AlInN ternary alloys in whole composition range and fabrication of InN/AlInN multiple quantum wells by RF molecular beam epitaxy, *Japanese Journal of Applied Physics*, 45 (2006) L539–L542.

[9.96] T. S. Oh, J. O. Kim, H. Jeong, Y. S. Lee, S. Nagarajan, K. Y. Lim, C.-H. Hong, and E.-K. Suh, Growth and properties of Al-rich $In_xAl_{1-x}N$ ternary alloy grown on GaN template by metalorganic chemical vapour deposition, *Journal of Physics D: Applied Physics*, 41 (2008) 095402:1–5.

[9.97] C. Buchheim, R. Goldhahn, M. Rakel, C. Cobet, N. Esser, U. Rossow, D. Fuhrmann, and A. Hangleiter, Dielectric function and critical points of the band structure for AlGaN alloys, *physica status solidi (b)*, 242 (2005) 2610–2616.

[9.98] C. Wetzel, T. Takeuchi, S. Yamaguchi, H. Katoh, H. Amano, and I. Akasaki,

Optical band gap in $Ga_{1-x}In_xN$ $(0 < x < 0.2)$ on GaN by photoreflection spectroscopy, *Applied Physics Letters*, 73 (1998) 1994–1996.

[9.99] M. D. McCluskey, C. G. Van de Walle, C. P. Master, L. T. Romano, and N. M. Johnson, Large band gap bowing in $In_xGa_{1-x}N$ alloys, *Applied Physics Letters*, 72 (1998) 2725–2727.

[9.100] M. D. McCluskey, C. G. Van de Walle, L. T. Romano, B. S. Krusor, and N. M. Johnson, Effect of composition on the band gap of strained $In_xGa_{1-x}N$ alloys, *Journal of Applied Physics*, 93 (2003) 4340–4342.

[9.101] J. Wu, W. Walukiewicz, K. M. Yu, J. W. Ager III, E. E. Haller, H. Lu, and W. J. Schaff, Small band-gap bowing in $In_{1-x}Ga_xN$ alloys, *Applied Physics Letters*, 80 (2002) 4741–4743.

[9.102] S. Pereira, M. R. Correia, T. Monteiro, E. Pereira, E. Alves, A. D. Sequeira, and N. Franco, Compositional dependence of the strain-free optical band gap in $In_xGa_{1-x}N$ layers, *Applied Physics Letters*, 78 (2001) 2137–2139.

[9.103] R. Goldhahn, J. Scheiner, S. Shokhovets, T. Frey, U. Köhler, D. J. As, and K. Lischka, Determination of optical constants for cubic $In_xGa_{1-x}N$ layers, *physica status solidi (b)*, 216 (1999) 265–268.

[9.104] P. D. C. King, T. D. Veal, C. F. McConville, F. Fuchs, J. Furthmüller, F. Bechstedt, P. Schley, R. Goldhahn, J. Schörmann, D. J. As, K. Lischka, D. Muto, H. Naoi, Y. Nanishi, H. Lu, and W. J. Schaff, Universality of electron accumulation at wurtzite *c*- and *a*-plane and zinc-blende InN surfaces, *Applied Physics Letters*, 91 (2007) 092101:1–3.

[9.105] I. Vurgaftman, J. R. Meyer, and L. R. Ram-Mohan, Band parameters for III–V compound semiconductors and their alloys, *Journal of Applied Physics*, 89 (2001) 5815–5875.

[9.106] V. Cimalla, J. Pezoldt, G. Ecke, R. Kosiba, O. Ambacher, L. Spieß, G. Teichert, H. Lu, and W. J. Schaff, Growth of cubic InN on *r*-plane sapphire, *Applied Physics Letters*, 83 (2003) 3468–3470.

[9.107] G. Kaczmarczyk, A. Kaschner, S. Reich, A. Hoffmann, C. Thomsen, D. J. As, A. P. Lima, D. Schikora, K. Lischka, R. Averbeck, and H. Riechert, Lattice dynamics of hexagonal and cubic InN: Raman-scattering experiments and calculations, *Applied Physics Letters*, 76 (2000) 2122–2124.

[9.108] P. Lautenschlager, M. Garriga, and M. Cardona, Temperature dependence of the interband critical-point parameters of InP, *Physical Review B*, 36 (1987) 4813–4820.

[9.109] W. A. Harrison, *Elementary Electronic Structure*, World Scientific, Singapore, 1999.

[9.110] D. G. Pacheco-Salazar, J. R. Leite, F. Cerdeira, E. A. Meneses, S. F. Li,

D. J. As, and K. Lischka, Photoluminescence measurements on cubic InGaN layers deposited on a SiC substrate, *Semiconductor Science and Technology*, 21 (2006) 846–851.

[9.111] I. Gorczyca, N. E. Christensen, A. Svane, K. Laaksonen, and R. M. Nieminen, Electronic band structure of $In_xGa_{1-x}N$ under pressure, *Acta Physica Polonica A*, 112 (2007) 203–208.

10

Electronic properties of InN and InGaN: Defects and doping

W. Walukiewicz[1], **K. M. Yu**[1], **J. W. Ager III**[1], **R. E. Jones**[1,2], **and N. Miller**[1,2]

[1] *Materials Sciences Division, Lawrence Berkeley National Laboratory, Berkeley, California 94720, United States of America*

[2] *Department of Materials Science and Engineering, University of California, Berkeley, California 94720, United States of America*

10.1   Introduction

The last decade has witnessed an unprecedented growth in fundamental studies and practical applications of the group III-nitrides and their alloys. GaN and Ga-rich InGaN and AlGaN thin films are currently used in a variety of commercial optoelectronic devices, including green, blue, and ultraviolet light emitting diodes (LEDs) and lasers (see, for example, Ref. [10.1]). Group III-nitrides have also found applications in other electronic devices. For example, advanced GaN/AlGaN high power microwave transistors are now commercially available. This spectacular progress was made possible by rapid advances in bulk materials synthesis and epitaxial growth techniques.

However, the success in applications of group III-nitrides has been limited to Ga-rich alloys, that is, devices incorporating $In_xGa_{1-x}N$ and $Al_xGa_{1-x}N$ at relatively small values of x. It was discovered very early on that alloying GaN with InN has a detrimental effect on electrical and optical properties of the resulting alloys. The difficulties in fabricating high-quality InGaN epitaxial films with large In-content remains a major problem in achieving light emitters in the visible spectrum. Although InN was synthesized more than 35 years ago [10.2], only a limited amount of often contradictory information was available on the properties and electronic structure of this material. In general, these initial attempts to synthesize InN resulted in highly n-type conducting material with a diffuse optical absorption edge and no detectable luminescence [10.3, 10.4]. Many of these attempts resulted in material with very high oxygen contamination levels. The previously generally accepted energy gap of 1.9 eV was measured in an RF sputtered sample [10.5]. However, there were also reports indicating a large spread of the value of the band gap. Also, a study of the composition dependent band gap in InGaN alloys indicated that the energy gap of InN should be much smaller than 1.9 eV [10.6].

Several years ago, research groups in the US, Russia and Japan were able to grow much better quality InN using molecular beam epitaxy [10.7–10.12]. The new films had much lower residual electron concentrations and all optical characteristics indicated an energy gap of only about 0.7 eV [10.13]. This discovery generated a controversy, but also a great expansion of research on InN and In-rich group III-nitride alloys. In view of these new developments, it was questioned why it took so long to determine such a basic parameter as the band gap and what makes InN such an unusual and difficult material to characterize. Subsequent research efforts by several groups have demonstrated that most of the remarkable characteristics of InN can be attributed to the extraordinarily high electron affinity and low energy gap of this material. Various aspects of the structural and electronic properties of InN and In-rich group III-nitrides are extensively discussed in different chapters of this book. This chapter reviews the status of research on doping and defects in InN and the In-rich group III-nitrides. It is not intended to be a comprehensive review of this field. It is rather an attempt to provide a single, unifying approach to defect and doping problems based on the Amphoteric Defect Model (ADM). Section 10.2 introduces the ADM, discusses intentional and unintentional *n*-type doping and presents results of recent theoretical calculations of electron mobility. Surface electron concentration is discussed in Section 10.3. The effect of native point defects introduced by high energy particle irradiation on the electrical and optical properties of group III-nitrides is presented in Section 10.4. Finally, Section 10.5 is devoted to a presentation of the recent progress in achieving *p*-type doping.

10.2   *n*-type doping and electron transport

10.2.1   Amphoteric defect model

The behavior of native defects and the degree to which semiconductors can be doped *n*- and *p*-type can be understood within the ADM [10.14–10.16]. This model provides guidance on the nature (donor or acceptor) and the concentrations of charged native point defects. It is based on the observation that the energies of the charge transition states of localized defects such as dangling bonds and vacancies are insensitive to the locations of the valence and conduction band edges in different semiconductors. The energies line up on an absolute scale, relative to the vacuum level. For defects capable of supporting multiple charges, the formation energy of each charge state strongly depends on the position of the Fermi level, E_F. For example, it has been demonstrated that in some instances a specific defect can undergo Fermi-level-induced transformation from donor (acceptor) to acceptor (donor). The transformation occurs when E_F crosses a universal energy reference, the Fermi level stabilization energy (E_{FS}), located at about 4.9 eV below the vacuum level. The location of E_F relative to E_{FS} determines the nature of the dominant native point defects. Thus for $E_F < E_{FS}$ ($E_F > E_{FS}$), the formation energy of native donor (acceptor) defects is lower than that of acceptors (donors). The formation energies of both donors and acceptors are equal when E_F reaches E_{FS}. Consequently, for high enough defect concentrations the Fermi energy will be stabilized (pinned) at E_{FS}. This result has very important implications for understanding the properties of semiconductors in the presence of a large concentration of native point defects. Intentional introduction of defects through high energy particle irradiation will result in a high resistivity material when E_{FS} is located in the band gap (for example, GaAs) or conducting material for E_{FS} located in a band (for example, InAs). In addition, since all practically available semiconductor surfaces have a large concentration of defects in the form of dangling bonds, E_{FS} determines surface Fermi level pinning. Consequently, surface charge depletion is observed in materials with E_{FS} in the band gap (GaAs) and surface charge accumulation or inversion is found in materials with E_{FS} in a band (InAs).

It is important to note that the formation energy for donors (acceptors) increases as E_F moves upward (downward) toward E_{FS}, and thus there is a maximum in charged defect formation energy for $E_F = E_{FS}$. From this it is easy to see why the location of E_{FS} with respect to the conduction and valence band edges is a key factor determining the efficiency of n- and p-type doping. For example, a material with E_{FS} located close to (far away from) the conduction band edge will be easy (difficult) to dope n-type, because the formation energy of compensating acceptor defects will be higher (lower) in n-type material. Analogous arguments apply to the effect of the location of E_{FS} relative to the valence band edge on efficiency of p-type doping [10.16].

The amphoteric defect model has been successful in explaining doping behavior of a wide variety of group III-V semiconductors [10.14, 10.16, 10.17] and a concept analogous to the ADM has been used to explain doping behavior of group II-VI semiconductors [10.18]. In order to illustrate how the ADM can be used to evaluate the doping behavior we consider the case of n-type doping of GaN_xAs_{1-x}. GaAs has its conduction band at $E_{FS} + 0.9$ eV and its valence band at $E_{FS} - 0.5$ eV and thus is predicted to exhibit limitations on the maximum free electron concentration but less significant limitation on the maximum hole concentration. Experimentally, a maximum free hole concentration as high as 10^{21} cm^{-3} has been reported in GaAs [10.19], while the maximum free electron concentration (n_{max}) achievable under equilibrium growth conditions is limited to the mid-10^{18} cm^{-3} range, corresponding to E_F located approximately 0.1 eV above the conduction band edge or 1 eV above E_{FS} [10.20]. We note that non-equilibrium techniques such as Se+Ga co-implantation [10.21] and pulse electron beam irradiation techniques [10.22] have been used to achieve n_{max} values in GaAs up to 2×10^{19} cm^{-3}, corresponding to a maximum Fermi level at $E_{FS} + 1.3$ eV.

According to the ADM, a shifting of the conduction or valence band toward E_{FS} should improve n- or p-type doping, respectively. Since alloying of GaAs with

GaN to form dilute GaN_xAs_{1-x} results in a large downward shift of the conduction band edge [10.23, 10.24], it was predicted that a higher free electron concentration could be achieved in GaN_xAs_{1-x} compared with GaAs. This prediction was fully confirmed by experiments [10.25]. As is shown in Fig. 10.1, n_{max} increases strongly with the N content (x) in $Ga_{1-3x}In_{3x}N_xAs_{1-x}$ alloys with a maximum observed value of 7×10^{19} cm^{-3} for $x = 0.033$. This value is ~ 20 times greater than that observed for a GaAs film (3.5×10^{18} cm^{-3}) grown under the same conditions.

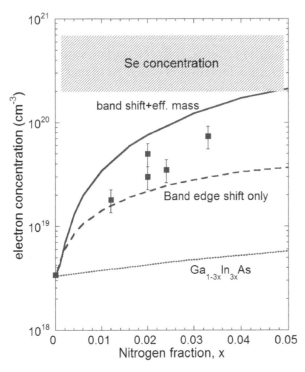

FIGURE 10.1

A comparison of the measured maximum electron concentration with the calculated values as a function of N fraction in $Ga_{1-3x}In_{3x}N_xAs_{1-x}$. Calculated n_{max} including effects of downward shift of the conduction band only (dashed curve) and considering both the band shift and the enhancement of the density of states effective mass (solid curve) are shown. The calculated n_{max} for samples with no N are also shown in the figure (dotted curve). The shaded area indicates the range of Se concentration in these samples. Reprinted with permission from K. M. Yu *et al.*, *Physical Review B*, 61 (2000) R13337. Copyright 2000 by the American Physical Society.

10.2.2   Unintentional and intentional *n*-type doping

In order to evaluate the doping behavior of group III-nitrides and their alloys, one needs to know the locations of the conduction and valence band edges relative to E_{FS}. This requires a detailed knowledge of the electron affinities, band gaps, and the bowing parameters for the conduction and valence band edges. There has been a large range in the valence band offset values between the end-point compounds (InN, GaN, and AlN) reported in the literature. For example, for InN (top)/GaN (bottom) heterointerfaces with In/Ga-polarity, the range of experimentally reported valence band offsets spans from 0.5 to 1.05 eV [10.26–10.30], while calculated values range from 0.3 to 1.8 eV [10.31–10.33]. The III-N band offsets are discussed in Chapter 4. Our previous work has used the values found by Martin *et al.* of 1.05 eV for the InN/GaN and 1.81 eV for the InN/AlN valence band offset, to calculate the conduction and valence band edge positions of the alloys (Fig. 10.2). In this calculation the band gap bowing was assumed to split between the conduction and valence bands in proportion to their respective offsets, and an electron affinity of 4.1 eV for GaN was used [10.34]. It should be noted, however, that there is still uncertainty regarding the band alignment. For a value at the bottom end of the previously reported InN/GaN valence band offsets (0.5 eV), the electron affinity of InN would be 6.35 eV, which is considerably higher than the value of 5.8 eV that we use here.

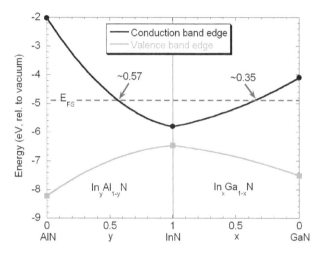

FIGURE 10.2
Conduction and valence band edges for $In_yAl_{1-y}N$ and $In_xGa_{1-x}N$. The position of the Fermi stabilization energy (E_{FS}) at -4.9 eV is indicated with a dashed line. E_{FS} is above the conduction band edge for $0 \leq y < 0.57$ and $0 \leq x < 0.35$.

It is seen from Fig. 10.2 that the energy difference between E_{FS} and band edges varies greatly with composition. In InN and the In-rich alloys, E_{FS} is located deeply in the conduction band or close to the conduction band edge. This is consistent with the extreme propensity of these materials for *n*-type conductivity.

The location of E_{FS} in the conduction band or close to the conduction band edge suggests that donors are the dominant defects in InN, InGaN, GaN and In-rich In-AlN. This is consistent with experimental observations showing that undoped material is always *n*-type. Although impurity atoms, particularly H and O, do contribute to the innate *n*-type conductivity in InN [10.35], secondary ion mass spectrometry (SIMS) results have shown that their concentrations cannot always account for the high electron concentrations [10.36]. A striking feature of the band alignment shown in Fig. 10.2 is that E_{FS} is located well above the valence band edge of all group III-nitrides, confirming the well recognized difficulties with *p*-type doping of these materials. In the extensively studied case of GaN, the valence band edge is located at 2.7 eV below E_{FS} and the free hole concentration is limited to about 10^{18} cm^{-3} [10.37]. The issues related to *p*-type doping of InN and InGaN alloys are discussed in Section 10.5 of this chapter.

The extremely low position of the conduction band edge of InN with respect to E_{FS} results in a large reduction of the formation energy of native donor, compared to acceptor, defects. This helps to explain why undoped InN films are always *n*-type with reported electron concentrations in single-crystal wurtzite films ranging from mid-10^{17} cm^{-3} to 10^{21} cm^{-3}. The lowest reported electron concentration to date for single crystal InN is 3.15×10^{17} cm^{-3}, in a film with an electron mobility of 2160 cm^2/Vs [10.38]. An even lower electron concentration of 5.3×10^{16} cm^{-3} and a mobility of 2700 cm^2/Vs was reported for a polycrystalline film produced by RF sputtering [10.39], but these values have not since been reproduced in either a polycrystalline or single crystal film.

Intentional *n*-type doping of GaN and InN has been achieved by silicon and oxygen. However, it was recognized that these dopants introduce a significant amount of strain in the GaN layers which at high doping levels ($> 10^{19}$ cm^{-3}) can cause cracking [10.40–10.42]. Germanium is also known to be a shallow donor in GaN with an ionization energy of 19 meV [10.43]. It is believed that Ge doping can be advantageous due to the similarity in size between Ga (In) and Ge, and therefore film cracking due to strain can be mitigated. Moreover, the lower melting point of Ge (1210 K) compared to Si (1683 K) enables the more efficient doping of InGaN without affecting the growth temperature of the film. Although Ge is a suitable candidate for *n*-type doping of the group III nitrides, only a few reports are published on Ge doping of GaN by MOCVD [10.44, 10.45] and MBE [10.46].

Hageman *et al.* [10.46] investigated the *n*-type doping of GaN using Ge in plasma-assisted MBE growth. They found that Ge is an excellent element for an *n*-type dopant in GaN with a linear correlation between vapor pressure and carrier concentration. The solubility limit of Ge in GaN was found to be $\sim 4 \times 10^{20}$ cm^{-3}. Above the solubility limit, degradation in surface morphology and crystal quality along with secondary phase (Ge and Ge$_3$N$_4$) formation was observed. Ge-doped In$_x$Ga$_{1-x}$N films with *x* in the range of 0.3–0.4 were also investigated [10.47], and the results showed that a high doping efficiency can be achieved up to $n \sim 8 \times 10^{20}$ cm^{-3} with good crystalline quality.

10.2.3 Electron mobility

There has been a large variation in mobility values reported for InN in the literature. Figure 10.3 shows measured electron mobilities plotted against electron concentration in the as-grown undoped InN samples grown by MBE that we have studied to date. The scatter in the data shown in Fig. 10.3 reflects differences in quality among InN films and may be attributed to surface effects, scattering by dislocations [10.48] and three-dimensional defects, and differences in the degree of crystallinity of the films. Also, as will be discussed later, some of the electrons may originate from multiply charged native defects that act as very efficient scattering centers, while others may originate from singly charged impurities.

Theoretical analysis of the electron mobility has been carried out by considering all of the main electron scattering mechanisms, including optical phonon, acoustic piezoelectric, acoustic deformation potential, and ionized center scattering. The calculations employed a two-band Kane model in which the conduction band struc-

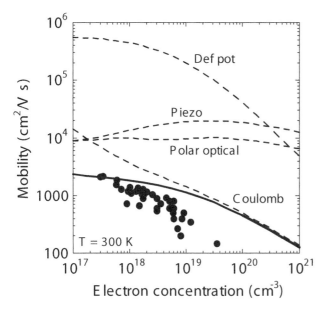

FIGURE 10.3
Experimentally measured electron mobilities plotted as a function of electron concentration (circles) in as-grown, undoped InN films grown by MBE. Theoretical electron mobilities limited by Coulomb scattering by triply charged defects, polar optical phonon scattering, acoustic piezoelectric, and acoustic deformation potential are plotted as dashed lines. The solid line is the theoretical mobility accounting for all of these mechanisms. Reprinted with permission from L. Hsu, R. E. Jones, S. X. Li, K. M. Yu, and W. Walukiewicz, Journal of Applied Physics, 102 (2007) 073705. Copyright 2007, American Institute of Physics.

ture is determined by the $\mathbf{k} \cdot \mathbf{p}$ interaction between the conduction band and the valence band [10.49]. This interaction couples the *p*-like valence band to *s*-like conduction band states, resulting in mixed-symmetry conduction band states. It produces a nonparabolic conduction band, described by the dispersion relation

$$E(k) = -\frac{E_g}{2} + \left[\left(\frac{E_g}{2}\right)^2 + \frac{E_g \hbar^2 k^2}{2m_0^*} \right]^{1/2} + \frac{\hbar^2 k^2}{2m_0}, \tag{10.1}$$

where the energy is referenced to the bottom of the conduction band, m_0^* is the effective mass at the Γ-point, E_g is the band gap, and m_0 is the electron rest mass. A band-edge electron effective mass of $0.07m_0$ was used in the following calculations [10.50]. One of the most important features of the nonparabolic dispersion relation (Eq. 10.1) is the resulting energy-dependent effective mass that affects the energy dependence of the density of states and that reduces the mobility compared to the parabolic case.

To calculate the electron mobility limited by ionized center scattering, a theoretical scheme that was developed for InSb and other narrow gap semiconductors by Zawadzki and Szymańska [10.51] was followed. The equation for the energy-dependent electron mobility limited by ionized defect scattering was modified to include the potential for a defect charge state greater than one. The adapted equation is given by

$$\mu_i(k) = \frac{\varepsilon^2}{2\pi e^3 \hbar Z^2 N_i F_i} \left(\frac{\mathrm{d}E}{\mathrm{d}k}\right)^2 k, \tag{10.2}$$

where ε is the static dielectric constant (a value of 9.3 was used [10.52]), e is the electron charge, Z is the charge of the ionized defect centers, N_i is the effective concentration of ionized defects, and F_i is a k-dependent function that takes into account free electron screening effects as well as the reduction of the scattering rates resulting from the mixed nature of the conduction band wavefunctions [10.51]. The effective concentration of scattering centers is $N_i = n/Z$, where n is the electron concentration. The macroscopic mobility is obtained by averaging the microscopic mobility (Eq. 10.2) with the Fermi-Dirac distribution function. Figure 10.3 shows that, at low electron concentrations, the electron mobility is well described by a combination of polar optical phonon and ionized impurity scattering [10.53]. At higher concentrations, the phonon scattering becomes less significant. The experimental mobilities decrease very rapidly with increasing electron concentration and cannot be explained by Coulomb scattering from singly charged centers. However, since native defects are most likely responsible for the mobility reduction at high electron concentrations in undoped samples, one has to consider the possibility that defects can be multiply charged and act as much more efficient scattering centers. In group III-V compounds vacancies can support up to three charges. This would greatly enhance the scattering rate since, as is evident from Eq. 10.2, the mobility is inversely proportional to the square of the defect charge Z.

The results of the calculations of mobility limited only by scattering from singly-charged and triply-charged donor defects with no compensation are represented by the upper dot and lower dash curves, respectively, in Fig. 10.4. Mobility data from InN films grown by more conventional MBE and a newly developed energetic neutral atomic beam lithography/epitaxy (ENABLE) technique are shown. ENABLE utilizes a collimated beam of ~ 2 eV N atoms as the active species that are reacted with thermally evaporated Ga and In metals [10.54, 10.55]. The technique provides a larger N atom flux compared to MBE and reduces the need for high substrate temperatures, making isothermal growth over the entire InGaN alloy composition range possible.

Figure 10.4 shows clearly that the experimental values for the MBE-grown InN are closer to the triply charged defect limit while the mobility in the ENABLE-grown materials is much closer to the singly charged donor limit. This is consistent with the assumption that due to the high N flux in the ENABLE technique there is better control of the crystal stoichiometry and a reduction in the concentration of vacancy-like defects. Therefore in ENABLE-grown InN, the electrons are believed to originate from singly charged O donors. This was confirmed by SIMS measurements on an ENABLE-grown InN film in which the residual O and electron concentrations are very similar. In contrast, the donors in the MBE-grown films are primarily due to native defects, which have been shown to have a 3^+ charge state [10.56]. A study of charge transport properties in InGaN and InAlN films grown by MBE shows that electron mobility is determined primarily by ionized defect and alloy disorder scattering [10.53].

Monte Carlo simulations have also been used to estimate electron mobility in InN as a function of electron concentration [10.57]. Ionized center scattering was assumed to be from singly charged scattering centers, and phonon scattering mech-

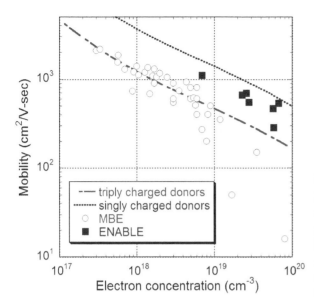

FIGURE 10.4

Measured electron mobilities plotted as a function of electron concentration in as-grown undoped InN films grown by MBE and ENABLE methods. The theoretical electron mobilities limited by singly and triply charged donor defect scattering are also shown.

anisms were also included. The authors used a static dielectric constant of 15.3 and an effective mass of $0.04m_0$. Both the Brooks-Herring and Conwell-Weisskopf approaches were used, and they agreed well for electron concentrations below 10^{16} cm^{-3}. Above this concentration, the approaches begin to diverge, creating a large range of possible electron mobilities. This is discussed in more detail in Chapter 5.

10.3 Surface electron accumulation

According to the ADM analysis above, it is expected that the surface Fermi energy will be pinned close to E_{FS} by dangling bond defects [10.16]. Referring to Fig. 10.2, a surface Fermi level pinning ~ 1 eV above the conduction band edge (CBE) is predicted in InN, which would create an electron accumulation layer in n-type (undoped) films and an inversion layer in p-type material. The surface band bending and electron and hole concentrations can be predicted for this pinning configuration by solving Poisson's equation,

$$\frac{\mathrm{d}^2 V(z)}{\mathrm{d}z^2} = \frac{1}{\varepsilon \varepsilon_0}\rho(z) \tag{10.3}$$

where

$$\rho(z) = e\left[N_D - n(z) - N_A + p(z)\right] \tag{10.4}$$

where V is the potential, ε is the static dielectric constant of InN, ρ is the net space-charge, n is the electron concentration, p is the hole concentration, N_D is the donor concentration, and N_A is the acceptor concentration. The boundary conditions for Eq. 10.3 are $E_F(z = 0)$, which is the pinning level, and $E_F(z = \infty)$, which is set by $N_D - N_A$ in the bulk of the film. Self-consistent calculations [10.58] for the *n*-type and *p*-type cases are shown in Fig. 10.5; a band-edge effective mass of $0.07m_0$ and nonparabolic conduction band parameters from Ref. [10.50] were used. Similar calculations have been reported by Veal *et al.* [10.59], who additionally included the effect of the carrier wavefunctions tending to zero amplitude at the surface, and Yim *et al.* [10.60], who additionally considered effects due to narrowing of the band gap near the surface.

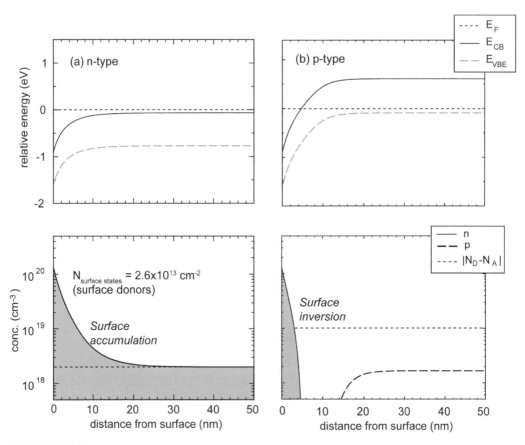

FIGURE 10.5

Calculated band bending (top) and free carrier concentrations (bottom) for (a) *n*-type ($N_D = 2 \times 10^{18}$ cm^{-3}) and (b) *p*-type ($N_A = 1 \times 10^{19}$ cm^{-3}) InN obtained by numerically solving Eqs. 10.3 and 10.4. The surface Fermi level pinning was assumed to be 0.9 eV above the conduction band edge. Surface electron accumulation is predicted for *n*-type InN and a surface inversion layer is predicted for *p*-type InN, with a maximum surface electron concentration of over 10^{20} cm^{-3} in both cases.

Qualitative evidence for this accumulation layer comes from electrolyte-based capacitance voltage (CV) measurements [10.61, 10.62] which showed that the electron concentration has a maximum near the surface that rapidly decreases farther into the film, appearing to saturate at a value close to the bulk electron concentration measured by the Hall effect. High-Resolution Electron-Energy-Loss Spectroscopy (HREELS) provides direct quantitative measure of the surface electron accumulation layer in InN [10.63, 10.64] (see also Chapter 12). Other evidence for the surface electron accumulation has been reported from angle-resolved photoelectron spectroscopy [10.65], scanning tunneling spectroscopy [10.66], and valence-band x-ray photoemission spectroscopy [10.67]. Based on the position of E_{FS} with respect to the conduction band edge of the InAlN and InGaN alloys (Fig. 10.2), it is expected that the surface Fermi level will be pinned above the conduction band edge for $x > 0.35$ in $In_xGa_{1-x}N$ and $y > 0.57$ in $In_yAl_{1-y}N$, resulting in surface accumulation of electrons for these compositions, in approximate agreement with recent experimental measurements [10.68–10.70].

10.4 *n*-type doping by high energy particle irradiation

The relationship between electronic properties and native defects has been extensively investigated by intentionally introducing native point defects into InN and $In_xGa_{1-x}N$ samples in a controlled fashion by energetic particle irradiation with high energy electrons, protons, and $^4He^+$ particles. Because E_{FS} is well above the CBE in InN and In-rich InGaN, donor-like point defects initially form as a result of the irradiation. As the irradiation dose increases, the electron concentration is expected to increase until E_F approaches E_{FS}. At this point both donor- and acceptor-type defects are formed at similar rates, and compensate each other, leading to a stabilization of E_F and a saturation of the electron concentration. Hence, a large increase and then saturation in the Burstein-Moss shift of the optical absorption edge is predicted.

The initial free electron concentrations of the $In_xGa_{1-x}N$ samples used in the irradiation study ranged from the low 10^{18} cm^{-3} to low 10^{17} cm^{-3} and the mobility ranged from 7 cm^2/Vs ($x = 0.24$) to above 1500 cm^2/Vs ($x = 1$). An MOCVD-grown GaN sample (3 μm thick) with an electron concentration of 7.74×10^{17} cm^{-3} and mobility of 189 cm^2/Vs and *n*-type GaAs samples (10–13 μm thick, $n \sim 8 \times 10^{16}$ cm^{-3}) were also included in this study.

The samples were irradiated with 1 MeV electrons, 2 MeV protons, and 2 MeV $^4He^+$ particles. The fluences of electrons ranged from 5×10^{15} to 1×10^{17} cm^{-2} and those of protons and $^4He^+$ particles were between 1.1×10^{14} and 2.7×10^{16} cm^{-2}. In all cases, the particle penetration depth greatly exceeded the film thickness, assuring a homogeneous damage distribution.

The displacement damage dose methodology developed by the Naval Research Laboratory for modeling solar cell degradation in space environments was used to scale the irradiation damage. It creates a single scale for measuring the irradiation damage from different particles, making it useful for this study [10.71, 10.72]. The displacement damage dose (D_d, in units of MeV/g) is defined as the product of the non-ionizing energy loss (NIEL) and the particle fluence. For the films irradiated here, the NIEL was either obtained from the tables in Ref. [10.73] or from the SRIM (Stopping and Range of Ions in Matter) program [10.74].

To avoid effects due to sample inhomogeneity and variations in the properties of the metal contact in the Hall effect measurements, the evaluation of the proton and $^4He^+$ particle-irradiation damage was done sequentially at progressively higher radiation doses on the same samples. Near-surface carrier concentration profiles of InGaN were measured with the Electrochemical Capacitance-Voltage (ECV) technique with 0.2 M NaOH:EDTA as the electrolyte.

10.4.1   Point defect generation

Point defects including interstitials and vacancies are generated during $^4He^+$ irradiation due to energy transfer from the $^4He^+$ ions to the lattice atoms. SRIM calculations [10.74] show that the number of In vacancies created by the irradiation is about twice as high as the number of N vacancies. Figure 10.6 shows the calculated In and N vacancy distribution in InN. Dynamic annealing and vacancy annihilation processes are not taken into account. Since the InN films used in these studies are typically less than 2 μm thick, the vacancy distributions can be considered uniform.

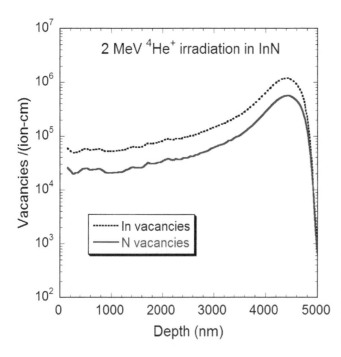

FIGURE 10.6

Calculated In and N vacancy distributions in InN created by 2 MeV $^4He^+$ irradiation by SRIM, assuming a displacement energy of 15 keV for In and N.

Figure 10.7 shows the results of ion channeling and x-ray diffraction investigations on InN films irradiated by 2 MeV ^{4}He$^+$ ions. The normalized yield (χ) is the ratio of the Rutherford back scattering (RBS) yields in the [0001] aligned direction and the random orientation. The normalized yields as a function of depth in Fig. 10.7(b) show that at the surface, χ increases from 0.04 in an as-grown InN sample to 0.10 after 1.8×10^{16} cm^{-2} of ^{4}He$^+$ irradiation, indicating that the InN film remains single crystalline in spite of the high concentration of radiation-induced defects. The rapid increase of the χ value in the irradiated sample suggests that a high density of defects is present in the film.

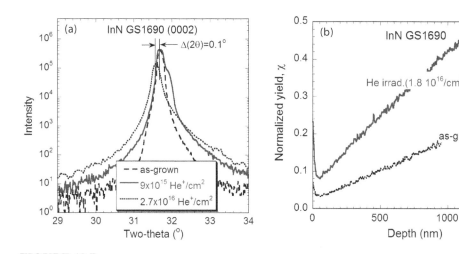

FIGURE 10.7

(a) XRD measurements and (b) normalized yield of ion channeling in InN samples irradiated by 2 MeV ^{4}He$^+$ to a total ^{4}He$^+$ dose of 9×10^{15}–2.7×10^{16} cm^{-2}.

The (0002) diffraction peaks shown in Fig. 10.7(a) reveal that after irradiation with a ^{4}He$^+$ fluence of 9×10^{15} cm^{-2} ($D_d = 6.1 \times 10^{15}$ MeV/g), no observable change in the diffraction peak position is observed. When the film was irradiated with a much higher ^{4}He$^+$ fluence of 2.7×10^{16} cm^{-2} ($D_d = 1.8 \times 10^{16}$ MeV/g) the lattice parameter of the film increased by $\sim 0.3\%$. Since extended crystalline defects, such as dislocations and twins, do not alter the lattice parameter of a crystal, the increase in lattice parameter can be attributed to a high density of point defects in the irradiated sample.

Cross-sectional transmission electron microscopy of ^{4}He$^+$ irradiated InN films shows no significant increase in extended defects other than dislocation loops in the InN sample irradiated with a ^{4}He$^+$ fluence of 2.7×10^{16} cm^{-2} (see Chapter 14). Therefore we believe that point defects, which are not observable in TEM except agglomerated as dislocation loops, are responsible for the structural changes observed in channeling and XRD. Moreover, the radiation-induced point defects in InN also have a profound effect on the electrical and optical properties of the material.

The nature of the point defects in irradiated GaN and InN was also investigated by slow positron annihilation spectroscopy [10.75]. In ^{4}He$^+$-irradiated GaN, it was found that the Ga vacancies act as important compensating centers and these Ga vacancies were introduced at a rate of 3600 cm^{-1}. In InN, however, negative In vacancies were introduced at a significantly lower rate of 100 cm^{-1}, making them negligible in the compensation of the radiation-induced n-type conductivity. On the other hand, negative non-open volume defects were introduced at a rate higher than 2000 cm^{-1}. These defects were believed to be related to N interstitials and it was suggested that they play a role in limiting the free-electron concentration at the highest irradiation fluences.

10.4.2   Effect on electron concentration

Figure 10.8 shows the free electron concentration in three In$_x$Ga$_{1-x}$N samples as a function of irradiation dose by 1 MeV electrons, 2 MeV protons and 2 MeV ^{4}He$^+$. The dose of the various irradiating species was converted to displacement damage dose D_d [10.71, 10.72] so that the electron concentration as a function of damage irrespective of irradiation conditions can be established. The electron concentrations in InN and In$_{0.4}$Ga$_{0.6}$N increase with irradiation dose and then saturate at a

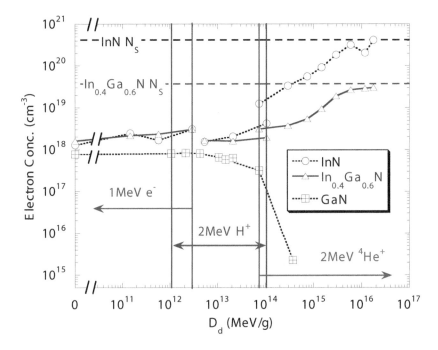

FIGURE 10.8

Free electron concentration in In$_x$Ga$_{1-x}$N as a function of displacement damage dose. The ranges of damage by different particles (electron, proton, and ^{4}He$^+$) are labeled and the calculated saturation concentrations for InN and In$_{0.4}$Ga$_{0.6}$N are also marked.

certain maximum value at a sufficiently high D_d ($> 10^{16}$ MeV/g). In contrast to In-rich InGaN, the electron concentration in GaN decreases with irradiation dose, and the sample eventually becomes semi-insulating.

The opposite responses to irradiation damage between In-rich $In_xGa_{1-x}N$ and GaN can be attributed to the relative position of their band edges with respect to E_{FS}. From the band edge positions of $In_xGa_{1-x}N$ shown in Fig. 10.2, it is expected that the radiation-induced defects in $In_xGa_{1-x}N$ with $x > 0.35$ are mostly donor-like (unless the $In_xGa_{1-x}N$ sample is heavily n-type to start with) and thus raise the electron concentration. Once the electron concentration is high enough that E_F reaches E_{FS}, it is pinned as the formation rate of donor-like and acceptor-like defects becomes equal. On the other hand, in GaN (or $In_xGa_{1-x}N$ with $x < 0.35$), E_{FS} is inside the band gap. In an n-type material, $E_F > E_{FS}$, and irradiation-induced defects are primarily acceptor-like and thus lower the electron concentration. The saturation concentration (N_S) of $In_xGa_{1-x}N$ in heavily irradiated samples can be calculated by setting E_F at E_{FS} in the equation for the concentration [10.51]:

$$N_s = \frac{1}{3\pi^2} \int_{E_{CE}}^{\infty} \frac{\exp\left(\frac{E-E_{FS}}{k_BT}\right)}{\left[1+\exp\left(\frac{E-E_{FS}}{k_BT}\right)\right]} k_c(E)^{3/2}\, dE \tag{10.5}$$

where E_{CE} is the energy of the conduction band edge and $k_c(E)$ is found by solving Eq. 10.1 for k. In Fig. 10.9, the calculated electron saturation concentration (solid line) is plotted together with experimental measurements (squares) of $In_xGa_{1-x}N$ [10.62]. The excellent agreement between calculation and the experimental results suggests that particle irradiation can be an effective and simple method to control the doping (electron concentration) in In-rich $In_xGa_{1-x}N$ via native point defects.

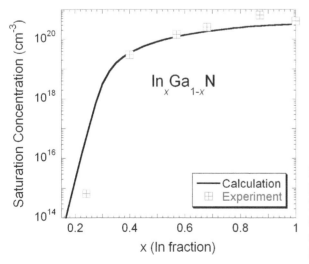

FIGURE 10.9

Electron saturation concentration in $In_xGa_{1-x}N$ as a function of In fraction. Square data points are measurements of the irradiation experiments and the solid line is calculation.

10.4.3 Triply-charged native defects

The effect of particle irradiation on the electron mobility provides information about the charge state of the native defects. As expected, the increase in donor defects and associated increase in electron concentration with irradiation create a corresponding decrease in the electron mobility. As discussed in Section 10.2.3, ionized impurity scattering is the dominant scattering mechanism in InN films with high electron concentrations, as well as in In-rich InGaN and InAlN [10.53]. Figure 10.10 shows the electron mobility as a function of electron concentration in irradiated InN films, along with the theoretical mobility limited by triply- and singly-charged donor scattering. The excellent agreement between the experimental data and the theoretical mobility for triply-charged donor defects provides strong evidence that the dominant defects produced by high energy particle irradiation are triply-charged donors.

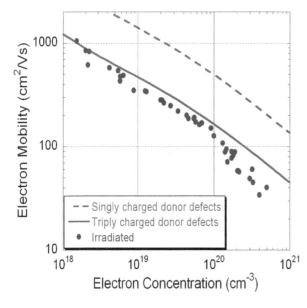

FIGURE 10.10

Electron mobility in InN films subjected to 2 MeV ^{4}He$^+$ irradiation (filled circles). Higher electron concentrations correspond to higher irradiation doses. Theoretical mobilities limited by scattering from triply- and singly-charged defects are also shown.

Monte Carlo simulation (SRIM) [10.74] shows that 2 MeV He$^+$ irradiation creates large concentrations of N and In vacancies (Fig. 10.6). Our calculations find evidence only for triply charged donors, with no compensation, suggesting that both types of vacancies have this nature. It is not surprising that the nitrogen vacancy would be a donor, based on electron counting considerations. Early calculations of defect energies predicted the N vacancy to be stable in the triple charge state in *p*-type material, but singly charged in *n*-type material [10.76]. It is less obvious how In vacancies could be donors. In fact, Stampfl *et al.* [10.76] predict that indium vacancies would be triply charged acceptors. The behavior of Ga vacancies in GaAs provides insight into how In vacancies could also be triple donors.

In GaAs, when E_F is below E_{FS}, both vacancy sites act as donors. The Ga vacancy relaxes into an As antisite and an As vacancy, which form a triply charged donor complex. While E_F lies below E_{FS} only in p-type GaAs, it is below E_{FS} in all of the InN films considered here. Thus, it could be expected that both the V_{In} and the V_N be triple donors, with the V_{In} relaxing via the reaction: $V_{In} \rightarrow (N_{In} + V_N)^{3+}$.

Thermal annealing of irradiated films has been shown to create high electron concentrations with high mobilities. The mobilities in such films approach those of the as-grown films, while films that were annealed without prior irradiation show little change (Fig. 10.11). The sharp increase in the electron mobility cannot be explained by a simple removal of the radiation-generated defects by annealing as it is not commensurate with the relatively modest reduction of the electron concentration. Instead, it has been found that a spatial ordering of the defects could be energetically favorable, due to their high concentrations and triple charge state which repels them from each other [10.56]. Annealing provides the thermal energy for defect diffusion, and they move toward a partially ordered state. Full ordering would not be possible due to the high concentrations of dislocations and other defects in the materials. The partial ordering reduces the scattering efficiency, increasing the mobility.

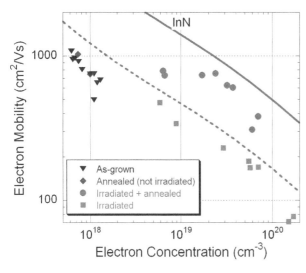

FIGURE 10.11

Electron mobilities and concentrations of InN films, as-grown and after irradiation and/or thermal annealing treatments. Theoretical mobility limited by singly- and triply-charged donor defect scattering is also shown.

10.4.4 Effect on optical properties

As discussed above, high energy particle irradiation can be used to produce InN with electron concentrations in a wide range from mid-10^{17} cm^{-3} to mid-10^{20} cm^{-3}. This can be done on the same sample, eliminating all the other factors affecting electron concentration through intentional doping. Figure 10.12 shows the evolution of the optical absorption spectra of InN and In$_{0.4}$Ga$_{0.6}$N with increasing 2 MeV

$^4\text{He}^+$ irradiation dose. In both samples, the absorption coefficient is of the order of 5×10^4 cm^{-1} at 0.5 eV above the absorption onset, which is typical for direct band gap semiconductors. With increasing irradiation the absorption edges show a blue shift. More specifically, the absorption edge shifts to higher energy, while the baseline and the slope above the absorption edge remain unchanged. The shift is composition dependent, with smaller shifts found in samples with higher Ga content. From as-grown to the highest dose of $^4\text{He}^+$ irradiation of 2.7×10^{16} cm^{-2} ($D_d = 1.84 \times 10^{16}$ MeV/g), the absorption edge is shifted by 1.05 eV in InN, by 0.71 eV in In$_{0.7}$Ga$_{0.3}$N (data not shown), but by only 0.15 eV in In$_{0.4}$Ga$_{0.6}$N. The shift is also observed to saturate with radiation dose. For example, as seen in Fig. 10.12(a) for InN, the blue shift slows down as the irradiation dose increases and eventually becomes insensitive to further irradiation at a sufficiently high $^4\text{He}^+$ dose (typically $> 10^{16}$ cm^{-2}). The behavior of the more Ga-rich ($x < 0.30$) material is different. As illustrated in Fig. 10.13, irradiation of GaN does not affect the fundamental absorption edge energy at ~ 3.4 eV but rather produces a new sub-band gap absorption feature at ~ 2.7 eV. Both the strength and the linewidth of the absorption peak increase with increasing irradiation dose. Clearly unfilled or partially-filled defect states are formed inside the band gap of GaN as a result of the irradiation. Notably, this absorption feature is near the position of E_{FS}.

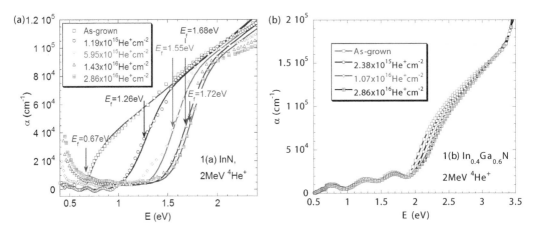

FIGURE 10.12
The evolution of InN and In$_{0.4}$Ga$_{0.6}$N absorption spectra with increasing doses of 2 MeV $^4\text{He}^+$ irradiation. In (a), the experimental results of InN are shown as data points and the numerical fits of the optical absorption spectra (see text) are shown as solid lines. The Fermi levels obtained from the fitting are marked by the arrows. Reprinted with permission from S. X. Li *et al.*, Applied Physics Letters, 87 (2005) 161905. Copyright 2005, American Institute of Physics.

To demonstrate the agreement between the optical and electrical properties of irradiated InN, the absorption spectra are numerically analyzed to obtain E_F. The absorption was modeled after Kane [10.49] to account for the nonparabolicity of the conduction band. The Burstein-Moss shift of the absorption edge due to filling

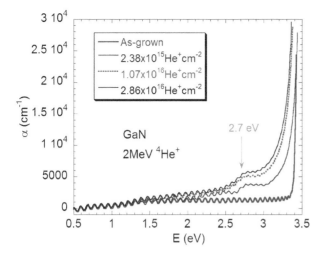

FIGURE 10.13

The absorption spectrum of GaN with increasing doses of 2 MeV ^{4}He$^+$ irradiation. Reprinted with permission from S. X. Li *et al.*, Applied Physics Letters, 87 (2005) 161905. Copyright 2005, American Institute of Physics.

of lower conduction band states by electrons was incorporated by multiplying the ideal energy dependent absorption coefficient for InN (α_0) by the Fermi distribution function. To account for inhomogeneous broadening effects, a Gaussian function was convoluted with α_0,

$$\alpha(E) = \frac{1}{\Delta\sqrt{\pi}} \int_{-\infty}^{\infty} \alpha_0(E') \left[1 + \exp\left(\frac{E_F - E'}{k_B T}\right)\right]^{-1} \exp\left[-\left(\frac{E' - E}{\Delta}\right)^2\right] dE', (10.6)$$

where Δ is the Gaussian broadening parameter. The best fits, which are plotted in Fig. 10.12(a) as solid lines, are obtained by adjusting Δ and the Fermi level (E_F). The corresponding E_F values are labeled by arrows; the Δ values do not vary significantly and lie consistently between 0.21 and 0.23 eV. The E_F values of InN and the In$_x$Ga$_{1-x}$N alloys derived from absorption spectra show excellent agreement with the electron concentrations from room-temperature Hall effect measurements (Fig. 10.8), for a different set of irradiated samples. The electrical and optical results are consistent with the prediction of the amphoteric defect model: native point defects introduced by irradiation are incorporated as donors in In-rich In$_x$Ga$_{1-x}$N ($x > 0.35$), but as acceptors in Ga-rich In$_x$Ga$_{1-x}$N ($x < 0.35$). The native donors increase the electron concentration and cause a blue shift of the absorption edge in In-rich In$_x$Ga$_{1-x}$N, while the native acceptors lower the electron concentration and form defect states inside the band gap of Ga-rich In$_x$Ga$_{1-x}$N, as observed by optical absorption. The observed large blue shift of the absorption edge is a clear manifestation of the Burstein-Moss shift. Furthermore, the consistency and repeatability of the measurements suggest that the irradiation is a dependable method with which to control the doping and optical properties of In$_x$Ga$_{1-x}$N alloys [10.77, 10.78].

The photoluminescence (PL) properties of InGaN alloys are also affected by irradiation. Figure 10.14 summarizes the irradiation dose dependence of the relative integrated PL intensities, normalized to the corresponding as-grown samples, for In$_x$Ga$_{1-x}$N alloys over the entire composition range. For comparison, the relative

PL intensities of GaAs and In$_{0.5}$Ga$_{0.5}$P, which are materials in current tandem solar cells, are also shown. The results in Fig. 10.14 indicate a superior radiation resistance of In$_x$Ga$_{1-x}$N compared to standard tandem solar cell materials. For example, the PL intensities of In$_{0.5}$Ga$_{0.5}$P and GaAs decreased rapidly to 0.4% and 1.1% of the original intensity, respectively, at $D_d \sim 10^{12}$ MeV/g. The PL intensity of GaN, on the other hand, decreased to 14% of the original intensity at $D_d \sim 2 \times 10^{13}$ MeV/g. In striking contrast, the PL intensity of InN and In$_x$Ga$_{1-x}$N alloys increased slightly with irradiation dose, and did not decrease significantly until $D_d > 10^{15}$ MeV/g.

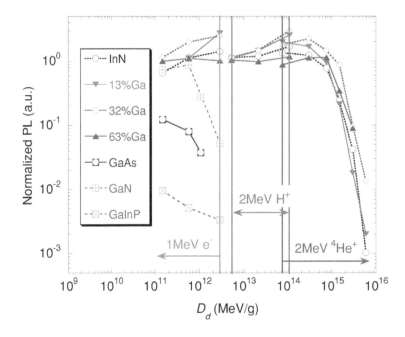

FIGURE 10.14

Summary of the effect of energetic particle irradiation on the normalized integrated PL intensity of In$_x$Ga$_{1-x}$N, GaAs, and Ga$_{0.51}$In$_{0.49}$P. For references the D_d ranges of 1 MeV electrons, 2 MeV protons, and 2 MeV ^{4}He$^+$ are labeled. Reprinted with permission from S. X. Li *et al.*, Applied Physics Letters, 88 (2006) 151101. Copyright 2006, American Institute of Physics.

It is widely recognized that defect levels near mid-gap are the most effective non-radiative recombination centers, as they have a high probability of capturing both electrons and holes [10.15, 10.79, 10.80]. Therefore, the rapid radiation-induced quenching of the PL intensity in GaAs, In$_{0.5}$Ga$_{0.5}$P, and Ga-rich In$_x$Ga$_{1-x}$N can be attributed to the nonradiative recombination centers inside the band gap formed by the radiation damage. On the other hand, in In-rich In$_x$Ga$_{1-x}$N, the radiation-induced defect levels are located above the CBE and there is a barrier for trapping of electrons by the defect centers, preventing the defects from acting as non-radiative recombination centers. At small radiation doses, the non-radiative recombination rate does not change significantly due to such a barrier. The PL intensity increases due to the increasing concentration of the majority carriers (electrons). However, as the electron concentration continues to rise, E_F approaches E_{FS} and the barrier for electron trapping reduces. The radiation-induced native defects become more effective non-radiative recombination centers, quenching the PL intensity. In Fig. 10.15, the relative PL intensity of an InN sample is plotted as a function of E_F, derived from the carrier concentration obtained from Hall effect measurements. As expected, the PL intensity increases slightly at low values of E_F, and then drastically diminishes when $E_F > 0.9$ eV above the valence band edge.

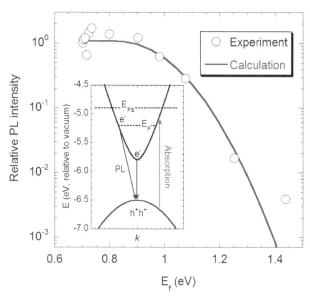

FIGURE 10.15

The relative PL intensity of InN from experiments (circles) and from calculations (line). A diagram depicting the broadening and blue shift of the PL signal as a result of the breakdown of k-conservation is shown inset. The Fermi level on the x-axis is referenced to the valence band edge. Reprinted with permission from S. X. Li *et al.*, Applied Physics Letters, 88 (2006) 151101. Copyright 2006, American Institute of Physics.

The calculated PL intensity in InN is plotted in Fig. 10.15. The calculation shows good agreement with experimental results. As is schematically shown in the inset of Fig. 10.15, when the momentum conservation rule is relaxed, all conduction band electrons can radiatively recombine with photoexcited minority holes that relax to the top of the valence band. This leads to the electron concentration-dependent broadening and shift of the emission line. The best fit obtained for the broadening (Δ) was 0.20 eV, which is consistent with the broadening parameter used to describe the shape of the fundamental absorption edge in irradiated InN [10.81].

A numerical analysis of the effect of n-type doping on the shape and energy position of the PL in InN can be found in Ref. [10.82]. The shift of the PL emission peak has sometimes been confused with the blue shift of the fundamental absorption edge (the Burstein-Moss shift) [10.83, 10.84]. In fact, the shift of the absorption edge associated with radiative transitions to the empty states above E_F in the conduction band [10.77] is much larger than the shift of the PL emission associated with transitions from the occupied states below E_F.

10.5 *p*-type doping

Although the first thin films of single crystal GaN were grown in 1969 [10.85], it took 20 years for effective *p*-type doping with Mg to be achieved [10.86]. Once this crucial breakthrough occurred, the first effective optoelectronic devices were announced 6 years later [10.87], and this has led to commercial applications with LEDs and lasers [10.1]. With the 2002 "discovery" of the 0.67 eV InN band gap (Chapters 6 and 7), the direct band gap tuning range of $In_xGa_{1-x}N$ was shown to extend from the near-IR to the ultraviolet. However, the current practical device applications with $In_xGa_{1-x}N$ use a maximum x of about 20% (quantum well structures can have somewhat larger x values). Clearly, the range of optoelectronic applications in this ternary alloy system has not yet been realized. As with the endpoint case of GaN discussed above, a critical issue is *p*-type doping.

As described in Section 10.2, due to the extremely high electron affinity of InN, strongly *n*-type behavior in as-grown material is expected. This creates challenges in obtaining *p*-type material due to the low formation energy for compensating native donors. However, it has been possible to drive E_F down in *p*-GaN to a level that, if achievable in InN or in InGaN, would produce *p*-type material. This section describes progress toward achieving *p*-type conductivity in InN and $In_xGa_{1-x}N$ alloys with large values of x (that is, for those compositions with the CBE lying below E_{FS}, see Fig. 10.2). While a number of dopants have been tried (for example, Be), we will restrict the discussion to Mg-doping, which has been the most effective *p*-type dopant for GaN.

10.5.1   Challenges of standard characterization

The electron-rich surface layer in InN prevents direct measurement of the bulk electrical properties of InN doped with Mg, as depicted schematically in Fig. 10.16. Single field Hall effect measurements performed on a series of MBE-grown, Mg-doped ($1-5 \times 10^{20}$ cm^{-3}) InN films are shown in Fig. 10.17. The sign of the Hall coefficient suggests the films are *n*-type. However, in these measurements only the surface layer is measured and the electrically isolated Mg-doped bulk is not evaluated. We note that the sheet concentrations of free electrons in the Mg-doped samples determined in these Hall effect measurements ranged from 5×10^{13} to

n-type inversion

InN:Mg

FIGURE 10.16

Schematic drawing of InN:Mg, showing the *n*-type inversion layer and the depletion region (light color) below it.

7×10^{14} cm^{-2}, which is relatively consistent with the surface state densities in the low 10^{13} cm^{-2} range predicted by the surface E_F pinning (Fig. 10.5) and measured by HREELS [10.63, 10.64], particularly if conduction along the sides of the sample and at the interface with the GaN buffer layer is considered. The measured mobilities were in the range of 15 to 90 cm^2/Vs. If the films were entirely n-type, the average electron concentrations would range from 1×10^{18} to 1×10^{19} cm^{-3}. As shown in Fig. 10.17(a), the corresponding electron mobility values in these samples would be roughly one order of magnitude lower than those of undoped InN at similar electron concentrations. However, if we assume that the measurement probes the accumulation layer only, we derive higher concentrations in the mid-10^{20} cm^{-3} range and the agreement with the expected mobility trend in n-type InN is excellent (Fig. 10.17(b)). Thus, the surface inversion layers prevent the direct detection of p-type activity in InN by single field Hall effect characterization. We note that evidence of parallel hole conduction in Mg-doped InN films has been reported from Hall effect measurements using variable magnetic fields up to 12 T and modeling with a quantitative mobility spectrum analysis (QMSA) algorithm [10.88].

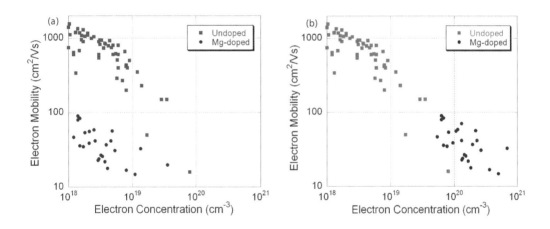

FIGURE 10.17
Comparison of the electron mobility measured by the Hall effect for undoped and Mg-doped InN. The electron concentrations for the Mg-doped samples were calculated by (a) assuming the films were entirely n-type and (b) assuming only the thin surface layer contributes to the Hall measurement of the Mg-doped films.

10.5.2 Evidence of p-type InN:Mg

10.5.2.1 Capacitance-voltage measurements

The surface accumulation/inversion layer in InN has been resistant to many attempts to use physical and chemical methods to remove or passivate it [10.89]. With only one unrepeated exception [10.90], metals form Ohmic contacts to the surface layer in undoped and Mg-doped InN. However, the double layer which forms when the InN surface comes into contact with a liquid electrolyte is insulating, and can be used to make a blocking contact to InN films. The resulting device is similar to a metal-insulator-semiconductor (MIS) diode and potentials up to ± 1 V can be applied to the surface without current flow, as illustrated in Fig. 10.18 [10.91]. Capacitance due to near-surface space-charge can be measured and negative potentials applied to the surface can be used to deplete the accumulation/inversion layer and measure space-charge due to ionized donors and/or acceptors in the bulk [10.91, 10.92]. Capacitance-voltage (CV) data in the form of Mott-Schottky plots (C^{-2} vs. V) for undoped and Mg-doped InN films are shown in Fig. 10.19. While a full analysis requires numerical modeling (see below), a qualitative picture can be developed within the depletion approximation [10.93]. In this, dC^{-2}/dV is inversely proportional to the net space-charge at the edge of the depletion region (for full depletion, this is $N_A - N_D$). For the undoped films, the low-slope region at lower bias reflects the depletion of the heavily doped surface accumulation layer while the linear, higher-slope region at higher bias reflects the lower net space-charge in the bulk. We observe that in some films, a sufficiently high bias can be applied without breaking down the insulating double layer and the surface can be inverted, similar to effects observed in MIS structures [10.94], leading to a slope change in the Mott-Schottky plot [10.60]. In contrast, for the Mg-doped films, a change in the sign of the slope in the C^{-2} vs. V data is seen as the surface voltage is increased to deplete the surface inversion layer. While the low-slope region at low voltages was similar to n-type films and reflects the space-charge associated with the electron-rich surface layer, the constant slope region at higher voltage has the opposite sign, indicating a net concentration of acceptors in the bulk.

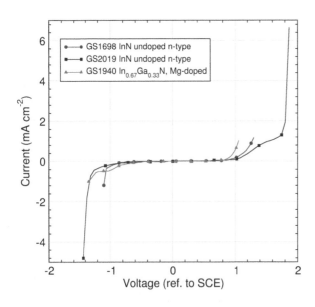

FIGURE 10.18

Current voltage measurements for undoped InN films and for a Mg-doped InGaN film using an electrolyte contact. The current is blocking for ± 1 V with respect to saturated calomel electrode (SCE); voltages in this range can be used to deplete the surface accumulation/inversion layers. Reprinted with permission from J. W. Ager III, *et al.*, physica status solidi (b), 244, 1820 (2007). Copyright 2007 by Wiley-VCH Publishers, Inc.

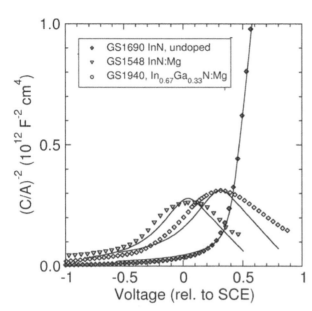

FIGURE 10.19

Mott-Schottky plot of capacitance-voltage data obtained from InGaN films using an electrolyte contact. The solid lines are from the Poisson equation analysis discussed in the text. The Fermi level pinning was set to 0.9 eV above the conduction band edge. For the undoped film, the CV analysis yields $N_D - N_A = 2 \times 10^{18}$ cm^{-3} which is comparable to the value obtained from the Hall effect (1.2×10^{18} cm^{-3}). For the Mg-doped InN and InGaN films, $N_A - N_D$ values of 2.2×10^{19} cm^{-3} and 1.9×10^{19} cm^{-3} are obtained. Reprinted with permission from J. W. Ager III, R. E. Jones, D. M. Yamaguchi, K. M. Yu, W. Walukiewicz, S. X. Li, E. E. Haller, H. Lu, and W. J. Schaff, physica status solidi (b), 244, 1820 (2007). Copyright (2007) by Wiley-VCH Publishers, Inc.

Quantitative modeling of the CV data was performed by solving the Poisson equation (Eq. 10.3): as shown by the solid lines in Fig. 10.19, the Poisson equation solution is in near-quantitative agreement with the experimental data. Effects due to near-surface band-gap narrowing can be considered in the modeling and these provide a small correction to the analysis [10.60]. For *n*-type films, there is good agreement between $N_D - N_A$ obtained from the CV analysis and the electron concentration determined from the Hall effect. For Mg-doped films, the $N_A - N_D$ values in the bulk obtained from the analysis are in the low-10^{19} cm^{-3} range. These results provide definitive proof of a net concentration of ionized acceptors below the *n*-type surface layer [10.92, 10.95].

10.5.2.2 Thermopower

Measurement of the thermopower (Seebeck coefficient) is an established method for determining the majority carrier type in semiconductors [10.96]; indeed, it is the physical basis for the "hot-probe"-type determination methods. Qualitative evidence of hole conduction in Mg-doped In$_x$Ga$_{1-x}$N for $0 < x < 0.88$ has been reported from thermoelectric hot probe measurements [10.97]. The Seebeck coefficient in a semiconductor increases with effective mass and with decreasing free carrier concentration. In InN, both of these factors favor the detection of hole conductivity in the presence of a surface inversion layer (the surface electrons have a lower mass and a higher concentration than the holes in the bulk). Thermopower measurements have been performed for undoped InN and Mg-doped InN [10.98], and the results are shown in Fig. 10.20. Undoped InN has a negative Seebeck coefficient, as expected for *n*-type material. A positive Seebeck coefficient is observed for InN:Mg, showing that holes are mobile in this material. No evidence of carrier "freeze-out" was observed for temperatures down to 200 K; this is consistent with degenerate conduction in the highly doped films.

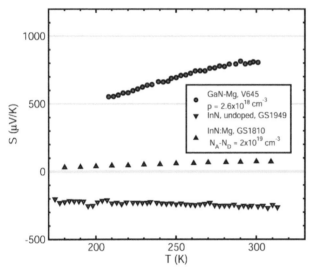

FIGURE 10.20

Thermopower data for undoped InN and Mg-doped InN. A positive thermopower, indicating hole conduction, is observed in Mg-doped InN. Data from Mg-doped GaN is included for comparison. Adapted from Ref. [10.98].

10.5.2.3 Compensation of *p*-InN by irradiation

As we have discussed in the last section, the native defects created by ion irradiation are donors in InN [10.62]. Since E_F lies far below E_{FS} in *p*-type InN, high energy irradiation can be used to compensate acceptors in Mg-doped *p*-type InN. In undoped InN, the increase in electron concentration is proportional to the irradiation dose, and the electron mobility decreases with dose [10.99]. The Mg-doped samples show a strikingly different behavior, as depicted in Fig. 10.21. At 2 MeV $^4He^+$ doses below mid-10^{14} cm^{-2}, the rate of increase in the electron concentration is much less than in undoped InN, and electron mobility remains approximately constant. At higher doses, the electron concentration increases continuously up to the saturation level of $\sim 4 \times 10^{20}$ cm^{-3}, whereas mobility shows a nonmonotonic dependence on the irradiation dose. In some samples, a maximum in the measured mobility can be observed, as shown in Fig. 10.21(b). It is believed that this occurs when the radiation-generated donors just over compensate the electrically-active Mg acceptors, and *n*-type transport first occurs throughout the film rather than only in the inversion layers at the surface and interface. At this point, the Hall effect measurements begin to reveal the electrical properties of the entire film. At the *p*-to *n*-type conversion threshold, the defect concentration in the bulk is lower than that at the surface, and electron mobility in the bulk is higher. Thus, the electron mobility determined from the Hall effect increases at this point because the bulk electrons, which see fewer scattering centers than the surface electrons, now contribute to (and at these thicknesses dominate) the measurement. With further damage, more donor defects are introduced throughout the film and electron mobility decreases correspondingly, as in the undoped sample. When the concentration of ionized defects in the entire film becomes comparable to that in the surface layer, the electron mobility becomes comparable to the initial mobility.

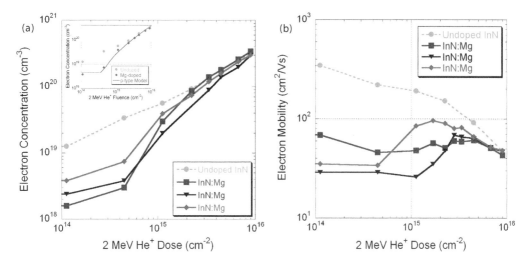

FIGURE 10.21

(a) Electron concentration, measured by Hall effect, as a function of the 2 MeV ^{4}He$^+$ dose in the three Mg-doped films, as well as an undoped InN film. The inset shows the experimental data for one Mg-doped film (GS1810) plotted with the *p*-type model. (b) Electron mobility, measured by Hall effect, as a function of the 2 MeV ^{4}He$^+$ dose in the three Mg-doped InN films, as well as an undoped InN film.

The dependence of the electron concentration and mobility on irradiation dose has been modeled using a parallel conduction approach [10.100]. In *p*-type material only the surface inversion layer contributes to the charge transport. However, after high enough irradiation dose to compensate all acceptors, the *p*-type bulk converts to *n*-type and two parallel *n*-type layers on the surface and in the bulk contribute to the charge transport. The relative contributions of each layer to the conductivity and the Hall effect are given by

$$N_{Hall}\mu_{Hall} = N_{surf}\mu_{surf} + N_{bulk}\mu_{bulk} \qquad (10.7)$$

and

$$N_{Hall}^2\mu_{Hall}^2 = N_{surf}^2\mu_{surf}^2 + N_{bulk}^2\mu_{bulk}^2 \qquad (10.8)$$

where N represents the sheet concentration. Because of the Fermi level pinning close to E_{FS} at the surface, we assumed a constant contribution from the surface and interface layers of 6×10^{13} electrons cm^{-2}. We assumed a surface electron mobility of 42 cm^2/Vs, which was the initial experimentally measured electron mobility. The bulk electron mobility and free carrier concentration depend on the irradiation dose and were calculated using the data for 2 MeV ^{4}He$^+$ irradiation of undoped InN, while including the effect of compensation from the Mg acceptors. As seen in Fig. 10.22, the modeling is in good qualitative agreement with the experimental data and explains a counter-intuitive increase of the effective Hall mobility with increasing sample damage. The increase is just a reflection of the fact that, at a ^{4}He$^+$ dose of about 4×10^{14} cm^{-2}, the bulk of the crystal becomes *n*-type and starts to contribute to the conductivity.

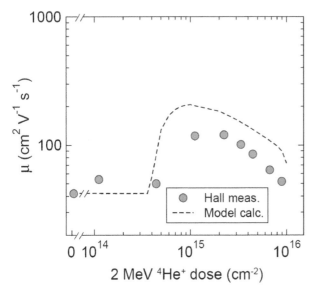

FIGURE 10.22
Modeling of mobility using Eqs. 10.7 and 10.8 for compensation of Mg-doped InN by irradiation induced donors. Reprinted with permission from J. W. Ager III, N. Miller, R. E. Jones, K. M. Yu, J. Wu, W. J. Schaff, and W. Walukiewicz, *physica status solidi* (b), 245, 873 (2008). Copyright 2008 by Wiley-VCH Publishers, Inc.

10.5.2.4 Photoluminescence in Mg-doped InN

It has been reported that Mg-doped InN does not exhibit photoluminescence (PL) when the Mg concentration is higher than the residual donor concentration. For InN samples with lower Mg concentrations, PL features are observed which have been used to estimate a Mg acceptor activation energy in the range of 60–100 meV [10.88, 10.101, 10.102]. The effect of producing additional donors with irradiation on un-doped and Mg-doped films was investigated. Nearly all undoped InN films with electron concentrations less than 2×10^{20} cm^{-3} exhibit a PL signal. The PL of these films is insensitive to the ^{4}He$^+$ irradiation and is completely quenched only for 2 MeV ^{4}He$^+$ fluences of at least 4.5×10^{15} cm^{-2}. In contrast to undoped InN, as-grown, Mg-doped films with *p*-type bulk conductivity do not exhibit PL. As illustrated in Fig. 10.23, a production of compensating donors causes the PL signal to be "recovered" following irradiation with a 2 MeV He$^+$ dose of 1.1×10^{15} cm^{-2}. This corresponds to a threshold dose that produces $\sim 4 \times 10^{19}$ cm^{-3} electrons, and is sufficient to convert the bulk of the sample from *p*- to *n*-type. The irradiation produces an InN material with features analogous to those of undoped *n*-type InN with a moderate electron concentration. Irradiation with higher doses at first increases the PL signal; however, at a ^{4}He$^+$ fluence of 6.7×10^{15} cm^{-2}, the PL is quenched in a manner similar to that observed in *n*-type samples.

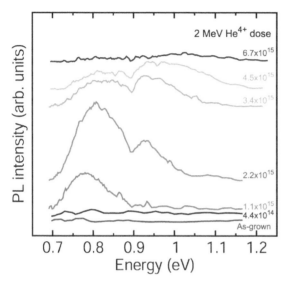

FIGURE 10.23

Photoluminescence (PL) spectra of one Mg-doped InN film as a function of the 2 MeV ^{4}He$^+$ dose, showing the onset of PL after a dose of 1.1×10^{15} cm^{-2}, followed by its quenching after a dose of 6.7×10^{15} cm^{-2}. Reprinted with permission from J. W. Ager III, N. Miller, R. E. Jones, K. M. Yu, J. Wu, W. J. Schaff, and W. Walukiewicz, *physica status solidi (b)*, 245, 873 (2008). Copyright 2008 by Wiley-VCH Publishers, Inc.

In order to overcome the challenge of surface inversion layers in InN that prevent the direct detection of *p*-type activity in InN by single field Hall effect characterization, a wide range of experimental approaches have been carried out including electrolyte-based CV measurements, thermopower, transport measurements of as-grown and ^{4}He$^+$-irradiated films, and photoluminescence spectroscopy. The experimental data provided direct evidence for the *p*-type activity in Mg-doped InN films.

10.5.3   *p*-type InGaN alloys

Compared to GaN and InN, *p*-type doping in In$_x$Ga$_{1-x}$N, particularly in the range for which surface inversion is expected ($0.34 < x < 1.0$, see Fig. 10.2), is less well investigated by direct electrical measurements. X-ray photoemission spectroscopy (XPS) measurements of the surface band bending in In$_x$Ga$_{1-x}$N have been reported recently which support the existence of *p*-type activity across the entire composition range [10.103]. Capacitance-voltage measurements also support this picture. Capacitance-voltage data (Mott-Schottky plot, C^{-2} vs. V) for three Mg-doped InGaN films are shown in Fig. 10.24. For $x = 0.95$ and 0.67, the data are qualitatively similar to InN:Mg. There is a region of shallow positive slope at low bias corresponding to space-charge due to electrons from the surface donors. At increasing bias, the inversion layer is depleted and the slope of the Mott-Schottky plot changes to become negative. At higher applied biases, capacitance corresponding to the depletion edge of the *p*-type bulk is observed. While the depletion approximation is not valid for the low-bias regions, our Poisson equation modeling has shown that an estimate of $N_A - N_D$ (net acceptor concentration) can be obtained from a linear fit to the data in the negative slope region. Using this method, net acceptor concentrations in the 10^{19} cm^{-3} range are obtained. We note that similar net acceptor concentrations are obtained (see above) in InN:Mg films which had Mg concentrations in the 10^{20}–10^{21} cm^{-3} range as measured by SIMS. It is possible that the measured acceptor concentrations in InN:Mg and InGaN:Mg indicate a substitutional limit for Mg in the 10^{19} cm^{-3} range. The third measured sample was of composition In$_{0.19}$Ga$_{0.81}$N. Referring to Fig. 10.2, the conduction band edge of In$_{0.19}$Ga$_{0.81}$N lies above E_{FS}. Therefore, a surface inversion layer is not expected for this composition. This is consistent with the monotonic C^{-2} vs. V data, which shows a net acceptor concentration throughout the measurement.

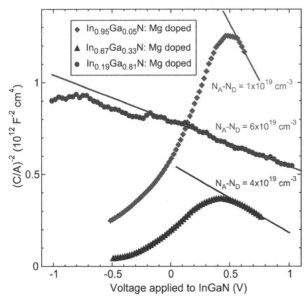

FIGURE 10.24
Mott-Schottky plot of electrolyte-contacted capacitance data for three InGaN compositions. The net acceptor concentration was estimated using the depletion approximation in regions where the surface inversion layer (if present) had been depleted. Reprinted with permission from J. W. Ager III, N. Miller, R. E. Jones, K. M. Yu, J. Wu, W. J. Schaff, and W. Walukiewicz, *physica status solidi* (b), 245, 873 (2008). Copyright 2008 by Wiley-VCH Publishers, Inc.

10.6   Conclusions and outlook

Results presented in this chapter clearly demonstrate that native defects play a crucial role in determining properties of InN and In-rich group III-nitrides. The extreme propensity of InN to *n*-type conductivity is unmatched by any other semiconductors. Although there has been significant progress in reducing the unintentional doping levels, the minimum electron concentration achieved in InN is still higher than 10^{17} cm^{-3}. Despite several years of extensive studies, the origin of this residual doping is still unclear. Native defects, extrinsic impurities [10.104], and extended defects [10.105, 10.106] were proposed as sources of electrons. Similarly the origin of electron accumulation on the surfaces of InN and In-rich group III-nitride alloys is still a hotly debated issue. Although native defects provide a straightforward explanation of this effect, it has also been suggested that it is caused by surface specific defects. However, the latter explanation appears to be inconsistent with the exceptional resistance of the surface electron accumulation to any surface treatments. Also it is very unlikely that the same stabilization energy on the surface and in heavily damaged bulk could have such different origins.

The exceptionally large span of direct energy gap values in the group III-nitrides has generated great interest in using these materials for a variety of device applications. The realization that the energy gaps of InGaN alloys provide a perfect match to the solar spectrum and an intrinsic resilience of the group III-nitrides to high energy particle damage have generated a huge interest in using these alloys in full-spectrum, high-efficiency multijunction solar cells for space applications [10.107]. Full realization of such devices would require fabrication of *p/n* junctions in the whole composition range. This is a difficult task especially for In-rich alloys where the conduction in the heavily *n*-type surface electron accumulation layer short circuits *p/n* junctions. Currently most of the efforts with group III-nitride solar cells concentrate on fabrication of Ga-rich InGaN based solar cells with an energy gap of about 2.4 eV, to avoid the surface electron accumulation. However, most of the reports indicate relatively poor device performance that is most likely associated with a difficulty in eliminating current leakage. A better understanding of the crystal structure and control of the stoichiometry of these alloys will be needed for success in this area.

Another unique feature of group III-nitride alloys is the very large range of the conduction band offsets, allowing for formation of extremely deep quantum wells: ~ 2 eV for InN/GaN and as much as ~ 4 eV for the InN/AlN system. This could potentially be used for a range of electronic devices including high power heterojunction bipolar transistors. However, the grand challenge facing this technology is the broad range of temperatures required for epitaxial growth of the alloys in the whole composition range. This makes preparation of compositional superlattices or quantum wells rather difficult. In addition, it is still not clear if there is an accumulation of charge at interfaces with In-rich alloys in the same way as happens on free surfaces. Such accumulation would be detrimental to the device performance.

The discovery of the narrow gap of InN several years ago led to a considerable increase in the fundamental research of nitride semiconductors. A great progress in the understanding of the basic properties of InN was a result of the work of many research groups worldwide. A long term sustained research effort will be required to fully understand and develop practical applications for this complex material system.

Acknowledgments

This work was supported the Director, Office of Science, Office of Basic Energy Sciences, Division of Materials Sciences and Engineering, of the U.S. Department of Energy under Contract No. DE-AC02-05CH11231. We thank W. J. Schaff for providing MBE films and M. A. Hoffbauer for providing films grown by the ENABLE method. We also acknowledge S. X. Li, J. Wu, Z. Liliental-Weber, and J. D. Denlinger for contributing some of the results presented in this chapter.

References

[10.1] S. Nakamura, S. J. Pearton, and G. Fasol, *The blue laser diode: the complete story* (Springer, New York, 2000).

[10.2] H. J. Hovel and J. J. Cuomo, Electrical and optical properties of rf-sputtered GaN and InN, *Applied Physics Letters*, 20 (1972) 71–73.

[10.3] V. A. Tyagai, A. M. Evstigneev, A. N. Krasiko, A. F. Andreeva, V. Y. Malakhov, Optical properties of indium nitride films, *Soviet Physics – Semiconductors*, 11 (1977) 1257–1259.

[10.4] K. L. Westra, R. P. W. Lawson, and M. J. Brett, The effects of oxygen contamination on the properties of reactively sputtered indium nitride films, *Journal of Vacuum Science and Technology A*, 6 (1988) 1730–1732.

[10.5] T. L. Tansley and C. P. Foley, Optical bandgap of indium nitride, *Journal of Applied Physics*, 59 (1986) 3241–3244.

[10.6] T. Matsuoka, H. Tanaka, T. Sasaki, and A. Katsui, Wide-gap semiconductor (In,Ga)N, Proceedings of the 16th International symposium on GaAs and related compounds, Karuizawa, Japan, 1989, *Institute of Physics Conference Series* vol. 106, Institute of Physics, Bristol, 1990, p. 141.

[10.7] H. Lu, W. J. Schaff, J. Hwang, H. Wu, W. Yeo, A. Pharkya, and L. F. Eastman, Improvement on epitaxial grown of InN by migration enhanced epitaxy, *Applied Physics Letters*, 77 (2000) 2548–2550.

[10.8] H. Lu, W. J. Schaff, J. Hwang, H. Wu, G. Koley, and L. F. Eastman, Effect of an AlN buffer layer on the epitaxial growth of InN by molecular-beam epitaxy, *Applied Physics Letters*, 79 (2001) 1489–1491.

[10.9] V. Y. Davydov, A. A. Klochikhin, R. P. Seisyan, V. V. Emtsev, S. V. Ivanov, F. Bechstedt, J. Furthmüller, H. Harima, A. V. Mudryi, J. Aderhold, O. Semchinova, and J. Graul, Absorption and emission of hexagonal InN. Evidence of narrow fundamental band gap, *physica status solidi (b)*, 229 (2002) R1–R3.

[10.10] K. Sugita, H. Takatsuka, A. Hashimoto, and A. Yamamoto, Photoluminescence and optical absorption edge for MOVPE-grown InN, *physica status solidi (b)*, 240 (2003) 421–424.

[10.11] Y. Saito, T. Yamaguchi, H. Kanazawa, K. Kano, T. Araki, Y. Nanishi, N. Teraguchi, and A. Suzuki, Growth of high-quality InN using low-temperature intermediate layers by RF-MBE, *Journal of Crystal Growth*, 237 (2002) 1017–1021.

[10.12] Y. Saito, H. Harima, E. Kurimoto, T. Yamaguchi, N. Teraguchi, A. Suzuki, T. Araki, and Y. Nanishi, Growth temperature dependence of indium nitride crystalline quality grown by RF-MBE, *physica status solidi (b)*, 234 (2002) 796–800.

[10.13] J. Wu, W. Walukiewicz, K. M. Yu, J. W. Ager III, E. E. Haller, H. Lu, W. J. Schaff, Y. Saito, and Y. Nanishi, Unusual properties of the fundamental band gap of InN, *Applied Physics Letters*, 80 (2002) 3967–3969.

[10.14] W. Walukiewicz, Amphoteric native defects in semiconductors, *Applied Physics Letters*, 54 (1989) 2094–2096.

[10.15] D. D. Nolte, Surface recombination, free-carrier saturation, and dangling bonds in InP and GaAs, *Solid State Electronics*, 33 (1990) 295–298.

[10.16] W. Walukiewicz, Intrinsic limitations to the doping of wide-gap semiconductors, *Physica B*, 302 (2001) 123–134.

[10.17] W. Walukiewicz, *Materials Research Society Symposium Proceedings*, 300 (1993) 421.

[10.18] S. B. Zhang, S. H. Wei, and A. Zunger, A phenomenological model for systematization and prediction of doping limits in II-VI and I-III-VI$_2$ compounds, *Journal of Applied Physics*, 83 (1998) 3192–3196.

[10.19] K. M. Yu, W. Walukiewicz, T. Wojtowicz, I. Kuryliszyn, X. Liu, Y. Sasaki, and J. K. Furdyna, Effect of the location of Mn sites in ferromagnetic Ga$_{1-x}$Mn$_x$As on its Curie temperature, *Physical Review B*, 65 (2002) 201303(R):1–4.

[10.20] W. Walukiewicz, Diffusion, interface mixing and Schottky barrier formation, *Materials Science Forum*, 143–147 (1993) 519–530.

[10.21] T. Inada, S. Kato, T. Hara, and N. Toyoda, Ohmic contacts on ion-implanted *n*-type GaAs layers, *Journal of Applied Physics*, 50 (1979) 4466–4468.

[10.22] T. Inada, K. Tokunaga, and S. Taka, Pulsed electron-beam annealing of selenium-implanted gallium arsenide *Applied Physics Letters*, 35 (1979) 546–548.

[10.23] M. Weyers, M. Sato, and H. Ando, Red shift of photoluminescence and absorption in dilute GaAsN alloy layers, *Japanese Journal of Applied Physics – Part 2*, 31 (1992) L853–L855.

[10.24] W. Shan, W. Walukiewicz, J. W. Ager III, E. E. Haller, J. F. Geisz, D. J. Friedman, J. M. Olson, and S. R. Kurtz, Band anticrossing in GaInNAs alloys, *Physical Review Letters*, 82 (1999) 1221–1224.

[10.25] K. M. Yu, W. Walukiewicz, W. Shan, J. W. Ager III, J. Wu, E. E. Haller, J. F. Geisz, D. J. Friedman, and J. M. Olson, Nitrogen-induced increase of the maximum electron concentration in group III-N-V alloys, *Physical Review B*, 61 (2000) R13337–R13340.

[10.26] G. Martin, A. Botchkarev, A. Rockett, and H. Morkoç, Valence-band discontinuities of wurtzite GaN, AlN, and InN heterojunctions measured by x-ray photoemission spectroscopy, *Applied Physics Letters*, 68 (1996) 2541–2543.

[10.27] C. F. Shih, N. C. Chen, P. H. Chang, and K. S. Liu, Band offsets of InN/GaN interface, *Japanese Journal of Applied Physics – Part 1*, 44 (2005) 7892–7895.

[10.28] C.-L. Wu, H.-M. Lee, C.-T. Kuo, S. Gwo, and C.-H. Hsu, Polarization-induced valence-band alignments at cation- and anion-polar InN/GaN heterojunctions, *Applied Physics Letters*, 91 (2007) 042112:1–3; C.-L. Wu, H.-M. Lee, C.-T. Kuo, C.-H. Chen, and S. Gwo, Cross-sectional scanning photoelectron microscopy and spectroscopy of wurtzite InN/GaN heterojunction: Measurement of "intrinsic" band lineup, *Applied Physics Letters*, 92 (2008) 162106:1–3.

[10.29] Z. H. Mahmood, A. P. Shah, A. Kadir, M. R. Gokhale, S. Ghosh, A. Bhattacharya, and B. M. Arora, Determination of InN–GaN heterostructure band offsets from internal photoemission measurements, *Applied Physics Letters*, 91 (2007) 152108:1–3.

[10.30] P. D. C. King, T. D. Veal, C. E. Kendrick, L. R. Bailey, S. M. Durbin, and C. F. McConville, InN/GaN valence band offset: High-resolution x-ray photoemission spectroscopy measurements, *Physical Review B*, 78 (2008) 033308:1–4.

[10.31] C. G. Van de Walle and J. Neugebauer, Small valence-band offsets at GaN/InGaN heterojunctions, *Applied Physics Letters*, 70 (1997) 2577–2579.

[10.32] S.-H. Wei and A. Zunger, Valence band splittings and band offsets of AlN, GaN, and InN, *Applied Physics Letters*, 69 (1996) 2719–2721.

[10.33] Ö Akıncı, H. Hakan Gürel, and H. Ünlü, Modelling of bandgap and band offset properties in III-N related heterostructures, *Superlattices and Microstructures*, 36 (2004) 685–692.

[10.34] V. Bougrov, M. Levinshtein, S. Rumyantsev, and A. Zubrilov, in M. E. Levinshtein, S. L. Rumyantsev, and M. S. Shur, editors, *Properties of advanced semiconductor materials: GaN, AlN, InN, BN, SiC, SiGe* (Wiley-Blackwell, 2001), p. 1–30.

[10.35] C. S. Gallinat, G. Koblmüller, J. S. Brown, S. Bernardis, J. S. Speck, G. D. Chern, E. D. Readinger, H. Shen, and M. Wraback, In-polar InN grown by plasma-assisted molecular beam epitaxy, *Applied Physics Letters*, 89 (2006) 032109:1–3.

[10.36] J. Wu, W. Walukiewicz, S. X. Li, R. Armitage, J. C. Ho, E. R. Weber, E. E. Haller, H. Lu, W. J. Schaff, A. Barcz, and R. Jakiela, Effects of electron concentration on the optical absorption edge of InN, *Applied Physics Letters*, 84 (2004) 2805–2807.

[10.37] U. Kaufmann, P. Schlotter, H. Obloh, K. Köhler, and M. Maier, Hole conductivity and compensation in epitaxial GaN:Mg layers, *Physical Review B*, 62 (2000) 10867–10872.

[10.38] H. Lu, W. J. Schaff, L. F. Eastmann, J. Wu, W. Walukiewicz, D. C. Look, and R. J. Molnar, Growth of thick InN by molecular beam epitaxy, *Materials Research Society Symposium Proceedings*, 743 (2003) L4.10.1–L4.10.6.

[10.39] T. L. Tansley and C. P. Foley, Electron mobility in indium nitride, *Electronics Letters*, 20 (1984) 1066–1068.

[10.40] A. Krost, A. Dadgar, G. Strassburger, and R. Clos, GaN-based epitaxy on silicon: stress measurements, *physica status solidi (a)*, 200 (2003) 26–35.

[10.41] J. Sánchez-Páramo, J. M. Calleja, M. A. Sáanchez-García, and E. Calleja, Optical investigation of strain in Si-doped GaN films, *Applied Physics Letters*, 78 (2001) 4124–4126.

[10.42] L. T. Romano, C. G. Van de Walle, J. W. Ager III, W. Götz, and R. S. Kern, Effect of Si doping on strain, cracking, and microstructure in GaN thin films grown by metalorganic chemical vapor deposition, *Journal of Applied Physics*, 87 (2000) 7745–7752.

[10.43] W. Götz, R. S. Kern, C. H. Chen, H. Liu, D. A. Steigerwald, and R. M. Fletcher, Hall-effect characterization of III-V nitride semiconductors for high efficiency light emitting diodes, *Materials Science and Engineering B*, 59 (1999) 211–217.

[10.44] S. Nakamura, T. Mukai, and M. Senoh, Si- and Ge-Doped GaN Films Grown with GaN Buffer Layers, *Japanese Journal of Applied Physics – Part 1*, 31 (1992) 2883–2888.

[10.45] J. K. Sheu and G. C. Chi, The doping process and dopant characteristics of GaN, *Journal of Physics: Condensed Matter*, 14 (2002) R657–R702.

[10.46] P. R. Hageman, W. J. Schaff, J. Janinski, and Z. Liliental-Weber, n-type doping of wurtzite GaN with germanium grown with plasma-assisted molecular beam epitaxy, *Journal of Crystal Growth*, 267 (2004) 123–128.

[10.47] K. M. Yu, L. A. Reichertz, I. Gherasoiu, J. W. Ager III, and W. Walukiewicz, *unpublished*.

[10.48] D. C. Look, H. Lu, W. J. Schaff, J. Jasinski, and Z. Liliental-Weber, Donor and acceptor concentrations in degenerate InN, *Applied Physics Letters*, 80 (2002) 258–260.

[10.49] E. O. Kane, Band structure of indium antimonide, *Journal of Physics and Chemistry of Solids*, 1 (1957) 249–261.

[10.50] J. Wu, W. Walukiewicz, W. Shan, K. M. Yu, J. W. Ager III, E. E. Haller, H. Lu, and W. J. Schaff, Effects of the narrow band gap on the properties of InN, *Physical Review B*, 66 (2002) 201403(R):1–4.

[10.51] W. Zawadzki and W. Szymańska, Elastic electron scattering in InSb-type semiconductors, *physica status solidi (b)*, 45 (1971) 415–432.

[10.52] V. Y. Davydov and A. A. Klochikhin, Electronic and vibrational states in InN and $In_xGa_{1-x}N$ solid solutions, *Semiconductors*, 38 (2004) 861–898.

[10.53] L. Hsu, R. E. Jones, S. X. Li, K. M. Yu, and W. Walukiewicz, Electron mobility in InN and III-N alloys, *Journal of Applied Physics*, 102 (2007) 073705:1–6.

[10.54] A. H. Mueller, E. A. Akhadov, and M. A. Hoffbauer, Low-temperature growth of crystalline GaN films using energetic neutral atomic-beam lithography/epitaxy, *Applied Physics Letters*, 88 (2006) 041907:1–3.

[10.55] E. A. Akhadov, D. E. Read, A. H. Mueller, J. Murray, and M. A. Hoffbauer, Innovative approach to nanoscale device fabrication and low-temperature nitride film growth, *Journal of Vacuum Science and Tech-*

nology B, 23 (2005) 3116–3119.

[10.56] R. E. Jones, S. X. Li, E. E. Haller, H. C. M. van Genuchten, K. M. Yu, J. W. Ager III, Z. Liliental-Weber, W. Walukiewicz, H. Lu, and W. J. Schaff, High electron mobility InN, *Applied Physics Letters*, 90 (2007) 162103:1–3.

[10.57] V. M. Polyakov and F. Schwierz, Low-field electron mobility in wurtzite InN, *Applied Physics Letters*, 88 (2006) 032101:1–3.

[10.58] nextnano3, available from http://www.wsi.tum.de/nextnano3.

[10.59] T. D. Veal, L. F. J. Piper, W. J. Schaff, and C. F. McConville, Inversion and accumulation layers at InN surfaces, *Journal of Crystal Growth*, 288 (2006) 268–272.

[10.60] J. W. L. Yim, R. E. Jones, K. M. Yu, J. W. Ager III, W. Walukiewicz, W. J. Schaff, and J. Wu, Effects of surface states on electrical characteristics of InN and In$_{1-x}$Ga$_x$N, *Physical Review B*, 76 (2007) 041303(R):1–4.

[10.61] H. Lu, W. J. Schaff, L. F. Eastman, and C. E. Stutz, Surface charge accumulation of InN films grown by molecular-beam epitaxy, *Applied Physics Letters*, 82 (2003) 1736-1738.

[10.62] S. X. Li, K. M. Yu, J. Wu, R. E. Jones, W. Walukiewicz, J. W. Ager III, W. Shan, E. E. Haller, H. Lu, and W. J. Schaff, Fermi-level stabilization energy in group III nitrides, *Physical Review B*, 71 (2005) 161201(R):1–4.

[10.63] I. Mahboob, T. D. Veal, L. F. J. Piper, C. F. McConville, H. Lu, W. J. Schaff, J. Furthmüller, and F. Bechstedt, Origin of electron accumulation at wurtzite InN surfaces, *Physical Review B*, 69 (2004) 201307(R):1–4.

[10.64] I. Mahboob, T. D. Veal, C. F. McConville, H. Lu, and W. J. Schaff, Intrinsic electron accumulation at clean InN surfaces, *Physical Review Letters*, 92 (2004) 036804:1–4.

[10.65] L. Colakerol, T. D. Veal, H. K. Jeong, L. Plucinski, A. DeMasi, T. Learmonth, P. A. Glans, S. Wang, Y. Zhang, L. F. J. Piper, P. H. Jefferson, A. Fedorov, T. C. Chen, T. D. Moustakas, C. F. McConville, and K. E. Smith, Quantized electron accumulation states in indium nitride studied by angle-resolved photoemission spectroscopy, *Physical Review Letters*, 97 (2006) 237601:1–4.

[10.66] T. D. Veal, L. F. J. Piper, M. R. Phillips, M. H. Zareie, H. Lu, W. J. Schaff, and C. F. McConville, Doping-dependence of subband energies in quantized electron accumulation at InN surfaces, *physica status solidi (a)*, 204 (2007) 536–542.

[10.67] P. D. C. King, T. D. Veal, C. F. McConville, F. Fuchs, J. Furthmüller, F. Bechstedt, P. Schley, R. Goldhahn, J. Schörmann, D. J. As, K. Lischka, D. Muto, H. Naoi, Y. Nanishi, H. Lu, and W. J. Schaff, Universality of electron accumulation at wurtzite *c*- and *a*-plane and zincblende InN surfaces, *Applied Physics Letters*, 91 (2007) 092101:1–3.

[10.68] T. D. Veal, P. H. Jefferson, L. F. J. Piper, C. F. McConville, T. B. Joyce, P. R. Chalker, L. Considine, H. Lu, and W. J. Schaff, Transition from electron accumulation to depletion at InGaN surfaces, *Applied Physics Letters*, 89 (2006) 202110:1–3.

[10.69] P. D. C. King, T. D. Veal, H. Lu, P. H. Jefferson, S. A. Hatfield, W. J. Schaff, and C. F. McConville, Surface electronic properties of *n*- and *p*-type InGaN alloys, *physica status solidi (b)*, 245 (2008) 881–883.

[10.70] P. D. C. King, T. D. Veal, A. Adikimenakis, H. Lu, L. R. Bailey, E. Iliopoulos, A. Georgakilas, W. J. Schaff, and C. F. McConville, Surface electronic properties of undoped InAlN alloys, *Applied Physics Letters*, 92 (2008) 172105:1–3.

[10.71] S. R. Messenger, G. P. Summers, E. A. Burke, R. J. Waters, and M. A. Xapsos, Modeling solar cell degradation in space: A comparison of the NRL displacement damage dose and the JPL equivalent fluence approaches, *Progress in Photovoltaics*, 9 (2001) 103–121.

[10.72] S. R. Messenger, E. A. Burke, G. P. Summers, M. A. Xapsos, R. J. Walters, E. M. Jackson, and B. D. Weaver, Nonionizing energy loss (NIEL) for heavy ions, *IEEE Transactions on Nuclear Science*, 46 (1999) 1595–1602.

[10.73] J. F. Ziegler, J. P. Biersack, and U. Littmark, *The Stopping and Range of Ions in Solids* (Pergamon Press, New York, 1985).

[10.74] SRIM-2008, available from http://www.srim.org/.

[10.75] F. Tuomisto, A. Pelli, K. M. Yu, W. Walukiewicz, and W. J. Schaff, Compensating point defects in ^{4}He$^+$-irradiated InN, *Physical Review B*, 75 (2007) 193201:1–4.

[10.76] C. Stampfl, C. G. Van de Walle, D. Vogel, P. Krüger, and J. Pollmann, Native defects and impurities in InN: First-principles studies using the local-density approximation and self-interaction and relaxation-corrected pseudopotentials, *Physical Review B*, 61 (2000) R7846–R7849.

[10.77] S. X. Li, E. E. Haller, K. M. Yu, W. Walukiewicz, J. W. Ager III, J. Wu, W. Shan, H. Lu, and W. J. Schaff, Effect of native defects on optical properties of In$_x$Ga$_{1-x}$N alloys, *Applied Physics Letters*, 87

(2005) 161905:1–3.

[10.78] S. X. Li, K. M. Yu, J. Wu, R. E. Jones, W. Walukiewicz, J. W. Ager III, W. Shan, E. E. Haller, H. Lu, and W. J. Schaff, Native defects in $In_xGa_{1-x}N$ alloys, *Physica B*, 376 (2006) 432–435.

[10.79] J. M. Langer and W. Walukiewicz, Surface recombination in semiconductors, *Materials Science Forum*, 196–201 (1995) 1389–1394.

[10.80] D. E. Aspnes, Recombination at semiconductor surfaces and interfaces, *Surface Science*, 132 (1983) 406–421.

[10.81] S. X. Li, R. E. Jones, E. E. Haller, K. M. Yu, W. Walukiewicz, J. W. Ager III, Z. Liliental-Weber, H. Lu, and W. J. Schaff, Photoluminescence of energetic particle-irradiated $In_xGa_{1x}N$ alloys, *Applied Physics Letters*, 88 (2006) 151101:1–3.

[10.82] B. Arnaudov, T. Paskova, P. P. Paskov, B. Magnusson, E. Valcheva, B. Monemar, H. Lu, W. J. Schaff, H. Amano, and I. Akasaki, Free-to-bound radiative recombination in highly conducting InN epitaxial layers, *Superlattices and Microstructures*, 36 (2004) 563–571.

[10.83] K. S. A. Butcher and T. L. Tansley, InN, latest development and a review of the band-gap controversy, *Superlattices and Microstructures*, 38 (2005) 1–37.

[10.84] J. C. Ho, P. Specht, Q. Yang, X. Xu, D. Hao, and E. R. Weber, Effects of stoichiometry on electrical, optical, and structural properties of indium nitride, *Journal of Applied Physics*, 98 (2005) 093712:1–5.

[10.85] H. P. Maruska and J. J. Tietjen, The preparation and properties of vapor-deposited single-crystal-line GaN, *Applied Physics Letters*, 15 (1969) 327–329.

[10.86] H. Amano, M. Kito, K. Hiramatsu, and I. Akasaki, P-type conduction in Mg-doped GaN treated with low-energy electron beam irradiation (LEEBI), *Japanese Journal of Applied Physics*, 28 (1989) L2112–L2114.

[10.87] S. Nakamura, M. Senoh, N. Iwasa, and S. Nagahama, High-brightness InGaN blue, green and yellow light-emitting diodes with quantum well structures, *Japanese Journal of Applied Physics – Part 2*, 34 (1995) 797-799.

[10.88] P. A. Anderson, C. H. Swartz, D. Carder, R. J. Reeves, S. M. Durbin, S. Chandril, and T. H. Myers, Buried *p*-type layers in Mg-doped InN, *Applied Physics Letters*, 89 (2006) 184104:1–3.

[10.89] V. Cimalla, G. Ecke, M. Niebelschtz, O. Ambacher, R. Goldhahn, H. Lu, and W. J. Schaff, Surface conductivity of epitaxial InN, *physica status solidi (c)*, 2 (2005) 2254–2257.

[10.90] V.-T. Rangel-Kuoppa, S. Suihkonen, M. Sopanen, and H. Lipsanen, Metal contacts on InN: Proposal for Schottky contact, *Japanese Journal of Applied Physics*, 45 (2006) 36–39.

[10.91] J. W. Ager III, R. E. Jones, D. M. Yamaguchi, K. M. Yu, W. Walukiewicz, S. X. Li, E. E. Haller, H. Lu, and W. J. Schaff, p-type InN and In-rich InGaN, *physica status solidi (b)*, 244 (2007) 1820–1824.

[10.92] R. E. Jones, K. M. Yu, S. X. Li, W. Walukiewicz, J. W. Ager III, E. E. Haller, H. Lu, and W. J. Schaff, Evidence for *p*-Type Doping of InN, *Physical Review Letters*, 96 (2006) 125505:1–4.

[10.93] P. Blood and J. W. Orton, *The electrical characterization of semiconductors: majority carriers and electron states* (Academic Press, London, 1992), p. 268–278.

[10.94] S. M. Sze, *Physics of semiconductor devices* (Wiley, New York, 1981), p. 362–430.

[10.95] X. Wang, S. B. Che, Y. Ishitani, and A. Yoshikawa, Systematic study on *p*-type doping control of InN with different Mg concentrations in both In and N polarities, *Applied Physics Letters*, 91 (2007) 242111:1–3.

[10.96] G. S. Nolas, J. Sharp, and H. J. Goldsmid, *Thermoelectrics: Basic principles and new materials developments* (Springer, 2001).

[10.97] K. D. Matthews, X. Chen, D. Hao, W. J. Schaff, and L. F. Eastman, MBE growth and characterization of Mg-doped InGaN and InAlN, *physica status solidi (c)*, 5 (2008) 1863–1865.

[10.98] J. W. Ager III, N. Miller, R. E. Jones, K. M. Yu, J. Wu, W. J. Schaff, and W. Walukiewicz, Mg-doped InN and InGaN – Photoluminescence, capacitance-voltage and thermopower measurements, *physica status solidi (b)*, 245 (2008) 873–877.

[10.99] R. E. Jones, S. X. Li, L. Hsu, K. M. Yu, W. Walukiewicz, Z. Liliental-Weber, J. W. Ager III, E. E. Haller, H. Lu, and W. J. Schaff, Native-defect-controlled n-type conductivity in InN, *Physica B*, 376 (2006) 436–439.

[10.100] R. L. Petritz, Theory of an experiment for measuring the mobility and density of carriers in the space-charge region of a semiconductor surface, *Physical Review*, 110 (1958) 1254–1262.

[10.101] X. Wang, S. B. Che, Y. Ishitani, and A. Yoshikawa, Growth and properties of Mg-doped In-polar InN films, *Applied Physics Letters*, 90

(2007) 201913:1–3.

[10.102] N. Khan, N. Nepal, A. Sedhain, J. Y. Lin, and H. X. Jiang, Mg acceptor level in InN epilayers probed by photoluminescence, *Applied Physics Letters*, 91 (2007) 012101:1–3.

[10.103] P. D. C. King, T. D. Veal, P. H. Jefferson, C. F. McConville, H. Lu, and W. J. Schaff, Variation of band bending at the surface of Mg-doped InGaN: Evidence of p-type conductivity across the composition range, *Physical Review B*, 75 (2007) 115312:1–7.

[10.104] A. Janotti and C. G. Van de Walle, Sources of unintentional conductivity in InN, *Applied Physics Letters*, 92 (2008) 032104:1–3.

[10.105] V. Cimalla, V. Lebedev, F. M. Morales, R. Goldhahn, and O. Ambacher, Model for the thickness dependence of electron concentration in InN films, *Applied Physics Letters*, 89 (2006) 172109:1–3.

[10.106] L. F. J. Piper, T. D. Veal, C. F. McConville, H. Lu, and W. J. Schaff, Origin of the *n*-type conductivity of InN: The role of positively charged dislocations, *Applied Physics Letters*, 88 (2006) 252109:1–3.

[10.107] J. Wu, W. Walukiewicz, K. M. Yu, W. Shan, J. W. Ager III, E. E. Haller, H. Lu, W. J. Schaff, W. K. Metzger, and S. Kurtz, Superior radiation resistance of $In_{1x}Ga_xN$ alloys: Full-solar-spectrum photovoltaic material system, *Journal of Applied Physics*, 94 (2003) 6477–6482.

11

Theory of native point defects and impurities in InN

A. Janotti and C. G. Van de Walle

Materials Department, University of California, Santa Barbara, CA 93106-5050, United States of America

11.1 Introduction

The electrical and optical properties of a semiconductor material, such as InN, are remarkably sensitive to the presence of even small concentrations of defects and impurities, down to one defect or impurity in 10^9 host atoms ($\sim 10^{14}$ cm^{-3}). Defects can greatly affect carrier mobility, lifetime and recombination, as well as luminescence efficiency. Defects also assist the diffusion mechanisms involved in growth, processing, and device degradation [11.1–11.3]. The deliberate incorporation of impurities, doping, forms the basis of much of semiconductor technology and can also be drastically affected by the presence of native point defects such as vacancies, self-interstitials, and antisites. Such defects may cause self-compensation: for instance, in an attempt to dope the material *p*-type, certain native defects which act as donors may spontaneously form and compensate the deliberately introduced acceptors [11.4]. Therefore, controlling defects and impurities is a decisive factor in enabling any semiconductor, InN included, to be used in electronic or optoelectronic devices.

InN is a semiconductor with a narrow direct band gap of ~ 0.7 eV [11.5–11.7] and an exceptionally high electron mobility. Owing to very favorable velocity/field characteristics, InN is a promising material for high-electron-mobility transistors, with potentially enhanced performance over GaN-based devices [11.8]. Another application is in terahertz devices [11.9]. InN can also be mixed with GaN (direct band gap of 3.4 eV) in the form of InGaN alloys with direct band gaps spanning a wide range of the spectrum, from infrared (0.7 eV) to ultraviolet (3.4 eV) [11.10, 11.11]. InGaN alloys are, therefore, promising materials for applications in optoelectronic devices such as high-efficiency solar cells, light-emitting diodes (LEDs), and laser diodes.

Although GaN and Ga-rich InGaN alloys are already in use in commercially available blue LEDs and lasers [11.12], many questions still exist about the basic electronic and optical properties of InN and In-rich InGaN alloys. High-quality epitaxial films of InN have recently been synthesized that allowed for the determination of band gap and electron effective mass [11.5–11.7, 11.10, 11.13]. However, these epitaxial films of InN invariably exhibit high *n*-type conductivity, with carrier concentrations ranging from 10^{17} to 10^{20} cm^{-3}. The cause of this unintentional *n*-type conductivity is still widely debated [11.7, 11.14]. Nitrogen vacancies and unintentional incorporation of impurities, such as oxygen or interstitial hydrogen [11.7, 11.10, 11.14–11.17], are among the proposed causes.

The identification and characterization of defects and impurities in semiconductors is a formidable task, and usually requires the integration of many different experimental techniques, such as electrical, optical, magnetic, and mass spectroscopy techniques [1.1–1.3]. Computational studies have been playing an increasing role in the investigation of point defects in semiconductors. In particular, first-principles calculations based on the density functional theory [11.18–11.20] have proved to be a powerful tool for studying the structural and electronic properties of native defects and impurities in semiconductors [11.21]. Stable and metastable positions of native defects and impurities in the crystal lattice and the surrounding local structural relaxations can be determined by minimizing total energies with respect to the atomic coordinates [11.19, 11.20]. A formalism has been developed that allows for calculations of formation energies as a function of Fermi-level position and chemical potentials [11.4, 11.21]. Accordingly, defect and impurity concentrations can be predicted in *n*-type and *p*-type material; and, in the case of compound materials such as InN, concentrations can be predicted in N-rich, In-rich, or any condition in between. Electronic transition levels associated with impurities and defects can also be determined from formation energies, and the electrical activity of impurities and defects can be analyzed as a function of Fermi-level position [11.21]. In addition, frequencies of local vibrational modes can be calculated by diagonalizing calculated force-constant matrices [11.21], and can be directly compared with infrared or Raman spectroscopy measurements.

In this chapter we review the theory of native point defects and impurities in InN, focusing on results of first-principles calculations based on density functional theory and pseudopotentials. We discuss the structural and electronic properties of the most relevant native defects and selected impurities, and their impact on the electrical properties of InN. In particular, we discuss the possible causes of unintentional *n*-type conductivity in as-grown InN, and the prospects for *p*-type doping. In Section 11.2 we describe the theoretical methods, and in Section 11.3 we address the basic electronic and structural properties of InN. In Section 11.4 we present the results for point defects and in Section 11.5 the results for Si, O, Mg, and H impurities in InN. Section 11.6 summarizes the chapter.

11.2   Theoretical approach

First-principles calculations based on density functional theory (DFT) and the local density approximation (LDA) [11.18] are now standard for investigating the ground-state properties of solids, molecules, and atoms. The DFT-LDA method allows for calculations of total energies, lattice parameters, atomic positions, and forces (derivatives of total energy with respect to atomic positions); structural parameters are typically predicted to within a few percent of the experimental values [11.19, 11.20]. By using supercells with periodic boundary conditions, it is possible to investigate the electronic structure of defects and impurities in solids, search for equilibrium atomic positions and local structural relaxations, and calculate migration paths and energy barriers for diffusion [11.19, 11.20]. From total energies, one can compute formation energies of defects and impurities, and, accordingly, equilibrium concentrations and electronic transition levels [11.21].

In most cases, the electronic problem is solved by separating the chemically inert core electrons from the chemically active valence electrons. The latter are assumed to move in an effective potential given by the nuclei and the respective core electrons in a frozen core approximation. These ions (nucleus plus core electrons) are then represented by pseudopotentials [11.22] or, in more recent implementations, by projector-augmented wave (PAW) potentials [11.23]. An additional complication arises in the case of InN (as well as GaN): the indium $4d$ states occur at about 15 eV below the valence-band maximum in InN, and therefore significantly interact with the valence states. Although they are quite localized, they cannot legitimately be regarded as core states, and therefore are referred to as "semicore states". These d states can significantly affect the structural and electronic properties, and therefore need to be included in the computational treatment, either explicitly by being treated as valence electrons, or approximately with the nonlinear core-correction formalism [11.24].

Native defects or impurities are simulated by adding and/or removing atoms to/from a sufficiently large supercell (a piece of the crystal that is periodically repeated, in order to restore translational periodicity). The wavefunctions are expanded in a plane-wave basis set, and integrations over the Brillouin zone are replaced by sums over a finite and regular mesh of special k-points. Typical calculations for InN discussed in this chapter employ a 96-atom supercell, a $2 \times 2 \times 2$ mesh of special k-points, and an energy cutoff of 400 eV for the plane-wave basis expansion. The nuclei and core electrons are represented by PAW potentials as implemented in the VASP code [11.23, 11.25]. Tests as a function of plane-wave cutoff and k-point sampling show that the results for neutral defects and impurities are numerically converged to within 0.1 eV; somewhat larger errors may occur for charged defects due to defect-defect interactions in neighboring supercells.

11.2.1 Formation energies

The formation energy of point defects and impurities and, therefore, their concentration, depend on the growth or annealing conditions [11.21]. For example, in InN the formation energy of a nitrogen vacancy is determined by the relative abundance of In and N atoms, as expressed by the chemical potentials μ_{In} and μ_{N}, respectively. If the vacancy is electrically charged, the formation energy further depends on the position of the Fermi level (E_F), which is the energy of the electron reservoir in the crystal, that is, the electron chemical potential. Forming a nitrogen vacancy requires the removal of one nitrogen atom and, accordingly, the formation energy is given by

$$E^f(V_N^q) = E_{tot}(V_N^q) - E_{tot}(\text{InN}) + \mu_N + q(E_F + E_v), \qquad (11.1)$$

where $E_{tot}(V_N^q)$ is the total energy of the supercell containing the nitrogen vacancy in the charge state q, and $E_{tot}(\text{InN})$ is the total energy of the InN perfect crystal in the same supercell. The nitrogen chemical potential μ_N depends on the experimental growth conditions, which can be either In-rich, N-rich, or anything in between. It is therefore regarded as a variable, with *bounds* imposed by thermodynamic equilibrium conditions. μ_N is subject to an upper bound given by the energy of N in a N_2 molecule, $E_{tot}(N_2)$, corresponding to extreme N-rich conditions. Similarly, the indium chemical potential μ_{In} is subject to an upper bound given by the energy of In in bulk indium, $E_{tot}(\text{In})$, corresponding to extreme In-rich conditions. It should be kept in mind that μ_N and μ_{In}, which are free energies, are temperature and pressure dependent.

The upper bounds defined above also lead to lower bounds given by the thermodynamic stability condition for InN,

$$\mu_{In} + \mu_N = \Delta H_f(\text{InN}), \qquad (11.2)$$

where $\Delta H_f(\text{InN})$ is the enthalpy of formation of InN crystal (negative for a stable compound). The upper limit on the indium chemical potential [$\mu_{In}^{max} = E_{tot}(\text{In})$] then results in a lower limit on the nitrogen chemical potential: $\mu_N^{min} = E_{tot}(N_2) + \Delta H_f(\text{InN})$. Conversely, the upper limit on the nitrogen chemical potential [$\mu_N^{max} = E_{tot}(N_2)$] results in a lower limit on the indium chemical potential: $\mu_{In}^{min} = E_{tot}(\text{In}) + \Delta H_f(\text{InN})$. The calculated enthalpy of formation of InN in the ground-state wurtzite phase is $\Delta H_f(\text{InN}) = -0.4$ eV, compared to the experimental value of -0.3 ± 0.1 eV at 298 K [11.26]. Therefore, the range over which the chemical potentials μ_N and μ_{In} can vary is quite narrow.

The Fermi level E_F in Eq. (11.1) is not an independent parameter, but is determined by the condition of charge neutrality. In principle, equations such as (11.1) can be formulated for every native defect and impurity in the material; the complete problem, including free-carrier concentrations in valence and conduction bands, can then be solved self-consistently, imposing charge neutrality. However, it is instructive to examine formation energies as a function of E_F in order to analyze the behavior of defects when the doping level changes. We reference E_F with respect to the valence-band maximum E_v, and allow E_F to vary from 0 to E_g, where E_g is the fundamental band gap. Note that the valence-band maximum E_v is taken from a calculation of a perfect-crystal supercell, corrected by the alignment of the electrostatic potential in the perfect-crystal supercell and in a region far from the defect in the supercell containing the defect, as described by Van de Walle and Neugebauer [11.21]. No additional corrections to address interactions between charged defects are included here. It has become clear that the frequently employed Makov-Payne correction [11.27] often significantly overestimates the correction, to the point of producing results that are less accurate than the uncorrected numbers [11.21, 11.28]. In the absence of a more rigorous approach we feel it is better to refrain from applying poorly understood correction schemes.

Expressions similar to Eq. (11.1) apply to all native defects and impurities. In the case of impurities in InN, the formation energy also depends on the impurity chemical potential μ_X. The latter is referred to the elemental phase of the impurity X, but can be further restricted by the formation of the possible In_mX_n or X_mN_n phases [$m\mu_{In} + n\mu_X < \Delta H_f(In_mX_n)$ or $m\mu_X + n\mu_N < \Delta H_f(X_mN_n)$, respectively]. For example, the hydrogen chemical potential μ_H is restricted by the formation of NH_3 molecules in the N-rich limit, and the magnesium chemical potential μ_H is restricted by the formation of bulk Mg_3N_2. In the case of an oxygen impurity in InN, μ_O is restricted by the formation of In_2O_3 in the In-rich limit.

11.2.2 Defect transition levels

Defects often introduce levels in the band gap of semiconductors [11.1, 11.2]; these levels involve transitions between different charge states of the same defect and can be derived from formation energies. The transition levels are not to be confused with the Kohn-Sham states that result from band-structure calculations. The transition level $\varepsilon(q/q')$ is defined as the Fermi-level position for which the formation energies of charge states q and q' are equal. $\varepsilon(q/q')$ can be obtained from

$$\varepsilon(q/q') = \frac{E^f(D^q) - E^f(D^{q'})}{q' - q}, \tag{11.3}$$

where $E^f(D^q)$ is the formation energy of the defect D in the charge state q when the Fermi level is at the valence-band maximum ($E_F=0$). The experimental significance of this level is that for Fermi-level positions below $\varepsilon(q/q')$, charge state q is stable, while for Fermi-level positions above $\varepsilon(q/q')$, charge state q' is stable. Transition levels can be observed in experiments where the final charge state can fully relax to its equilibrium configuration after the transition, such as in deep-level transient spectroscopy (DLTS) [11.1–11.3]. Transition levels correspond to thermal ionization energies. Conventionally, if a defect transition level is positioned near the band edge such that the defect is likely to be thermally ionized at room temperature (or at device operating temperatures), this transition level is called a shallow level; if it is unlikely to be ionized at room temperature, it is called a deep level. Note that shallow centers may occur in two cases: first, if the transition level in the band gap is close to one of the band edges (valence-band maximum (VBM) for an acceptor, conduction-band minimum (CBM) for a donor); second, if the transition level is actually a *resonance* in either the conduction or valence band. In that case, the defect necessarily becomes ionized, because an electron (or hole) can find a lower-energy state by transferring to the CBM (VBM). This carrier can still be coulombically attracted to the ionized defect center, being bound to it in a "hydrogenic effective mass state". This second case coincides with what is normally considered to be a "shallow center" (and is the more common scenario). Note that in this case the hydrogenic effective mass levels that are experimentally determined are *not* directly related to the calculated transition level, which is a resonance above (below) the CBM (VBM).

11.2.3 Defect and impurity concentrations

The concentration of a point defect or impurity depends on its formation energy. In thermodynamic equilibrium and in the dilute regime (that is, neglecting defect-defect interactions), the concentration of a point defect is given by

$$c = N_{\text{sites}} \exp \left(-\frac{E^f}{k_B T} \right), \tag{11.4}$$

where E^f is the defect or impurity formation energy, N_{sites} the number of sites the defect or impurity can be incorporated on, k_B the Boltzmann constant, and T the temperature. Equation (11.4) shows that defects with *high* formation energies will occur in *low* concentrations. Here we neglect the contributions from the formation volume and the formation entropy. The former is related to the change in the volume when the defect is introduced into the system; it is negligible in the dilute regime and tends to become important only under high pressure. The formation entropy is related mainly to the change in the vibrational entropy. Formation entropies of point defects and impurities are typically of the order of a few k_B, and therefore become important only at very high temperatures. Moreover, vibrational entropy contributions largely cancel out when comparing different defects or different configurations for impurities, or when assessing solubilities [11.21].

11.2.4 LDA deficiency and possible ways to overcome it

DFT-LDA calculations have become standard tools for computing structural and electronic properties of solids. However, an important limitation is that DFT-LDA seriously underestimates band gaps. Since defects and impurities often induce fully occupied or semi-occupied states across the band gap, we expect that their formation energy and associated transition levels will also be underestimated by LDA. The larger the conduction-band character, the larger is the error. In order to fully explore the predictive power of DFT calculations, the band-gap problem in LDA must be carefully considered and possibly overcome. We note that use of the generalized gradient approximation (GGA) does not substantially improve the situation. Any observed effects on the gap are usually attributable to differences in the calculated lattice constants: LDA typically underestimates lattice constants, while GGA tends to overestimate them. Larger lattice constants result in smaller band gaps for direct-gap semiconductors. When calculated at the same lattice constant, LDA and GGA band gaps are usually comparable in magnitude.

In the case of InN, the LDA band gap error is particularly severe because LDA underestimates the binding energy of the In $4d$ electrons. The In $4d$ states form a narrow band that overlaps with the valence band in InN. The In d states couple with the nitrogen p states that form the top of the valence band. This coupling pushes the valence-band maximum upwards, reducing the band gap [11.29]. In the LDA the underbinding of the d states causes the p-d coupling to be unphysically strong, resulting in too large a reduction of the band gap [11.30].

A few approaches that address the LDA band-gap problem have been applied to the study of point defects in InN. These approaches are: (i) self-interaction and relaxation corrected pseudopotentials (SIRC) [11.31], (ii) the LDA+U [11.32], and (iii) modified pseudopotentials (MPP) [11.33]. A common feature of the SIRC and LDA+U is the improved description of the semicore In $4d$ states that affects not only the N-$2p$ derived top of the valence band (VBM) through the p-d coupling, but also the In-$5s$ derived bottom of the conduction band (CBM) (through better screening by the more localized In $4d$ states [11.32, 11.33]).

In the SIRC approach, pseudopotentials that incorporate atomic self-interaction corrections and electronic relaxations are constructed and utilized in the calculation of solids [11.34]. For InN, the SIRC method leads to larger binding energy of the semicore In $4d$ states and a wider band gap [11.31]. Unfortunately, the SIRC approach does not allow evaluation of total energies (and forces), and the LDA corrections of formation energies have to be estimated based on the changes in the band structure.

The LDA+U method was conceived to be applied to partially filled bands associated with d states of transition metals [11.35–11.37]. It consists of including an on-site Coulomb correlation interaction that opens a gap between occupied and unoccupied bands associated with transition-metal d states. It was recognized that LDA+U also improves the description of fully occupied semicore d states in semiconductors [11.30, 11.32, 11.38]. In this approach an orbital-dependent correction that accounts for electronic correlations in the narrow bands derived from semicore d states is added to the LDA potential. In the case of InN, this on-site Coulomb correlation corrects the position of the narrow bands derived from the In d states, and affects both the valence-band maximum and conduction-band minimum. It opens the band gap by shifting the valence band downwards and the conduction band upwards [11.30, 11.32]. The LDA+U thus provides a partial correction of the band gap, the remaining correction arising from the intrinsic LDA error in predicting the position of the conduction band.

As an advantage over the SIRC approach, the LDA+U allows for calculation of total energies and forces. Therefore, by combining self-consistent calculations based on LDA and LDA+U it is possible to calculate the dependence of transition levels and formation energies on the theoretical band gap, and extrapolate the results to the experimental band gap. The transition levels and formation energies are thus corrected according to their valence- and conduction-band character in a consistent and quantitative fashion [11.39–11.41]. Note that the value of the U

parameter can be derived entirely from first principles [11.30], by calculating U for the atom and screening this value by using the high-frequency dielectric constant $\varepsilon(\infty)$ of the solid. For InN, this procedure gives $U=1.9$ eV for the semicore In d states. Further discussion and justification of this approach is contained in Refs. [11.30] and [11.32].

In the MPP approach [11.33], an atom-centered repulsive potential of Gaussian shape is added at the all-electron stage of the pseudopotential generation, within LDA and the norm-conserving scheme. This potential acts primarily on the lowest-energy $1s$ state, and affects higher-lying states through orthogonality of the wave functions. As a consequence, the semicore In $4d$ states are less screened and become more localized. Judicious choice of the Gaussian potentials can result in band structures in close agreement with experiment, while preserving the ability of the pseudopotentials to produce structures and energies that are as accurate as those obtained with regular pseudopotentials. When applied to defects, the approach allows for performing fully self-consistent calculations of total energies and atomic relaxations, while simultaneously obtaining band structures in which defect levels are calculated within a band gap that is very close to the experimental value.

Other methods that overcome the LDA deficiency, such as GW [11.42, 11.43], Quantum Monte Carlo [11.44], and hybrid functionals [11.45], have also recently been applied to the calculation of defects in semiconductors. These methods are much more computationally intensive that the SIRC, LDA+U, and MPP described above. It is important to note that GW has to be combined with the exact exchange method in order to properly describe the band structure of InN [11.46–11.48]. The combination of exact exchange with GW is extremely computationally expensive and, therefore, impractical for systematic studies of point defects in InN. Moreover, current implementations of GW do not allow for calculation of total energies. Quantum Monte Carlo calculations are also prohibitively expensive from a computational point of view for calculations that involve systems with semicore d electrons. Hybrid functionals are a promising approach, allowing for calculation of total energies and providing both accurate band gaps and structural parameters. They are less computationally expensive than GW and QMC, and will play an important role in defect calculations in the near future. However, to our knowledge, calculations of point defects in InN using hybrid functionals have not been performed to date.

Finally, although the LDA describes InN as a semimetal (the conduction band and valence band overlap at the Γ point), it should be noted that the closure of the gap occurs only near the Γ point and affects only a very small portion of the Brillouin zone. At the special k points used for integrations over the Brillouin zone, the material still behaves like a semiconductor, and it is still possible, even in LDA, to perform calculations for point defects in which defect states are correctly occupied in the respective charge states.

11.3 Results and discussion

11.3.1 Basic structural and electronic properties of InN

InN crystallizes in the wurtzite structure. Within LDA the calculated lattice parameters are $a = 3.51$ Å, $c = 5.67$, compared with the experimental values of $a = 3.54$ Å, $c = 5.70$ [11.49]. The calculated band structures according to LDA and LDA+U are shown in Fig. 11.1. The band gap separates the valence-band maximum, which is composed mostly of N $2p$ states, from the conduction-band minimum, which is composed mostly of In $5s$ states. The LDA band gap is -0.18 eV (that is, the CBM lies below the VBM), while the LDA+U band gap is 0.03 eV [11.30]. Recent exact exchange-GW calculations give a band gap of 0.7 eV [11.47, 11.48], in good agreement with the reported experimental values ranging from 0.65 to 0.75 eV [11.5–11.7, 11.13].

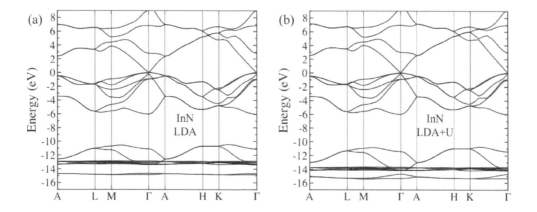

FIGURE 11.1

Calculated electronic band structure of wurtzite InN using (a) LDA and (b) LDA+U. The zero of energy was placed at the valence-band maximum for both cases. LDA gives a negative band gap of -0.18 eV, while LDA+U gives a small positive band gap of 0.03 eV. Reprinted with permission from A. Janotti, D. Segev, and C. G. Van de Walle, Physical Review B, 74 (2006) 045202. Copyright 2006 by the American Physical Society.

11.3.2 Native point defects in InN

In compound semiconductors such as InN there are six possible native point defects: vacancies, which consist of missing atoms in the regular sublattices of N or In (V_N and V_{In}); self-interstitials, which consist of extra nitrogen or indium atoms occupying interstitial sites in the InN lattice (N_i and In_i); and antisites, where nitrogen or indium atoms occupy sites in the "wrong" sublattice (N_{In} and In_N), that is, an extra nitrogen atom replacing an indium atom or vice versa.

11.3.2.1 Tight-binding calculations

Defects in InN were previously studied by Jenkins and Dow [11.50], who used parametrized tight-binding calculations to investigate native defects and doping. They assumed a band gap of 1.9 eV for InN. Jenkins and Dow suggested that the nitrogen vacancy would be responsible for the observed *n*-type conductivity in as-grown InN, and that the In antisite would explain an optical absorption peak attributed to a defect level located in the gap. Their results for nitrogen vacancies are inconsistent with results from first-principles methods, described in the next section. For instance, nitrogen vacancies are known to induce large local lattice relaxations that greatly affect the position of transition levels, as described below. This feature was not captured in the early calculations [11.50].

Based on the Jenkins and Dow calculations, Tansley and Egan [11.51] discussed their results of optical measurements on radio-frequency reactively sputtered polycrystalline films of InN. Tansley and Egan obtained a band gap of 1.94 eV and attributed defect levels in the upper half of the band gap to nitrogen vacancies and nitrogen antisites. Since the currently accepted value of the InN fundamental band gap is 0.7 eV, we believe that the results of Ref. [11.51] are artifacts related to the poor crystalline quality and/or high impurity content of their samples.

In an early review on nitrides, Strite and Morkoç [11.52] carefully pointed out that it was believed, but not proven, that the observed *n*-type conductivity in InN was caused by nitrogen vacancies.

11.3.2.2 DFT-LDA calculations

More recently, Stampfl *et al.* [11.31] performed first-principles DFT-LDA calculations of native point defects in zinc-blende InN. Although zinc blende is not the ground-state phase of InN, the local tetrahedral structure and the electronic properties of zinc-blende InN are very similar to those of the ground-state wurtzite InN. Therefore, it is expected that the defect physics in wurtzite InN can be quantitatively described by calculations for the more symmetric zinc-blende InN. Indeed, calculations have confirmed that formation energies of defects and impurities in the wurtzite phase are similar to those in the zinc-blende phase, except for interstitial configurations, for which the atomic environments are quite different: the geometry of the wurtzite structure allows for octahedral and tetrahedral interstitial sites, whereas the zinc-blende structure allows for tetrahedral and hexagonal interstitial sites.

The results obtained by Stampfl *et al.* are shown in Fig. 11.2, where the formation energy of each point defect is plotted as a function of the Fermi level, for In-rich conditions. Figure 11.2 follows the standard convention in which only Fermi-level values within the band gap are shown, but of course the formation energies [see Eq. (11.1)] also apply to Fermi-level positions above the CBM. The results show that the defects with lowest formation energies for any Fermi-level position in the band gap are nitrogen vacancies and indium interstitials. Note that V_N and In$_i$ are most favorable to form in the In-rich limit. However, due to the small formation enthalpy of InN (-0.4 eV), V_N and In$_i$ are also the defects with the lowest formation energies under N-rich conditions.

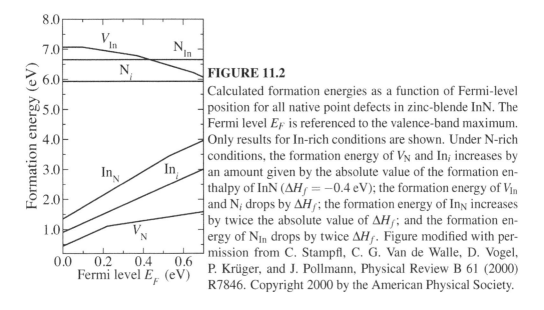

FIGURE 11.2

Calculated formation energies as a function of Fermi-level position for all native point defects in zinc-blende InN. The Fermi level E_F is referenced to the valence-band maximum. Only results for In-rich conditions are shown. Under N-rich conditions, the formation energy of V_N and In$_i$ increases by an amount given by the absolute value of the formation enthalpy of InN ($\Delta H_f = -0.4$ eV); the formation energy of V_{In} and N$_i$ drops by ΔH_f; the formation energy of In$_N$ increases by twice the absolute value of ΔH_f; and the formation energy of N$_{In}$ drops by twice ΔH_f. Figure modified with permission from C. Stampfl, C. G. Van de Walle, D. Vogel, P. Krüger, and J. Pollmann, Physical Review B 61 (2000) R7846. Copyright 2000 by the American Physical Society.

Nitrogen interstitials (N$_i$), indium vacancies (V_{In}), and both nitrogen and indium antisites (N$_{In}$ and In$_N$) were found to have high formation energies. Therefore, these defects will be present in insignificant concentrations in as-grown InN. Due to the exceedingly high formation energies of N$_i$ and N$_{In}$ (Fig. 11.2) and the fact that these defects are unlikely to influence the unintentional *n*-type conductivity in InN, Stampfl *et al.* did not investigate all possible charge states, limiting themselves to the neutral charge state.

The slopes in the formation energy versus Fermi level curves in Fig. 11.2 indicate the charge state of the respective defects. For each defect, only the lowest-energy charge state is shown for a particular position of the Fermi level. The kinks indicate the transition levels, that is, the Fermi-level values for which two charge states of the same defect have equal formation energies [Eq. (11.3)]. For example, the nitrogen vacancy V_N has a transition level at 0.2 eV above the VBM. It is stable in the 3+ charge state for Fermi levels below 0.20 eV, and in the + charge state for Fermi levels above 0.20 eV; hence, V_N is a shallow donor. Note that the 2+ charge state is not stable for any Fermi-level position in the band gap; the reason for this so-called "negative-*U*" behavior will be discussed below. The indium interstitial In$_i$ is stable exclusively in the 3+ charge state, and is therefore also a shallow donor. The indium antisite In$_N$ is a shallow donor, too. It is stable in the 4+ charge state for Fermi-level values below 0.55 eV, and in the 3+ charge state for Fermi levels above 0.55 eV.

The indium vacancy V_{In} is an acceptor defect. It is stable in the neutral charge state for Fermi-level values below 0.10 eV, in the $-$ charge state for Fermi levels between 0.10 eV and 0.35 eV, in the 2$-$ charge state for Fermi levels between 0.35 eV and 0.65 eV, and in the 3$-$ charge state for Fermi levels above 0.65 eV. The formation energy of the indium vacancy is very high (Fig. 11.2); we therefore expect the concentration of indium vacancies to be very low in good-quality material. Indeed, positron annihilation spectroscopy [11.53] has confirmed an absence of indium vacancies in InN grown by molecular beam epitaxy. The high formation energy may also explain why irradiation with ^{4}He$^+$ introduces indium vacancies at a much lower rate [11.54] than the rate for gallium vacancies in GaN, which have a significantly lower formation energy (in *n*-type material) [11.21].

Although Stampfl *et al.* did not investigate charge states other than neutral for the nitrogen antisite N_{In} and nitrogen interstitial N_i, we expect N_{In} to be a deep donor from the electronic states that are induced in the band gap; similarly, we expect N_i to be a deep acceptor when occupying the octahedral interstitial site or electrically neutral when in the split interstitial form where two N atoms share the same N lattice site.

11.3.2.3 Beyond-LDA results

Using SIRC pseudopotentials, Stampfl *et al.* [11.31] also calculated the change in the defect-induced states in InN as the band gap is corrected. V_N induces a fully symmetric a_1 state near the VBM, and a threefold degenerate t_2 state above the CBM. In the neutral vacancy V_N^0 the a_1 state is doubly occupied and the t_2 would be singly occupied. Since the t_2 state is above the CBM, the system can lower its energy by transferring the electron from the t_2 to the CBM. Therefore, the neutral charge state is never stable. In the SIRC pseudopotential calculations, it is found that the t_2 state is shifted with the CBM as the gap is corrected. The fact that defect states above the CBM shift upward with the conduction band is a very important finding. It confirms that V_N acts as a shallow donor, even when a band-gap correction is applied. The singlet state near the VBM is also shifted to higher energy, but by an amount that is much smaller than the band-gap correction. Furthermore, an inspection of the wave functions of the defect-induced states indicated that they are very similar in SIRC and LDA.

Although the SIRC approach does not allow evaluation of total energies, it is possible to estimate how the calculated changes in the band structure affect the total energy. For V_N^0, the total energy should reflect the shift in the triplet state located above the CBM, that is, an increase in the formation energy of V_N^0. This keeps the transition state above the CBM and confirms that the nitrogen vacancy is still a shallow donor even when band-gap corrections are applied. This contribution to the energy is absent for V_N^+, for which the t_2 states are unoccupied. The formation energy of V_N^+ is mostly affected by the shift of the occupied a_1 state near the VBM. From the SIRC results, this shift is much smaller than that of the t_2 states, and is expected to increase the formation energy of V_N^+ by ~ 0.4 eV.

The indium vacancy creates a doubly occupied fully symmetric a_1 state that is resonant in the valence band, and a threefold t_2 state close to the VBM. In the neutral charge state, V_{In}^0, the t_2 is occupied by three electrons (in this case the system is expected to be subjected to a Jahn-Teller distortion). The t_2 state can accept up to three more electrons, making the indium vacancy a triple acceptor (V_{In}^{3-}). The position of this state with respect to the VBM shifts by less than 0.01 eV in SIRC compared to LDA, and therefore, only a small increase in the formation energy of the indium vacancy is expected as the band gap is corrected.

The nitrogen antisite N_{In} creates a doubly occupied singlet state in the band gap, as well as a higher-lying triplet state that can accept up to six electrons. In the SIRC calculations, the singlet state is higher in the band gap by about 0.3 eV. This implies a higher formation energy of this defect than obtained with the LDA calculations. Since the LDA energy was already so high as to imply very small concentrations, this correction would not change the conclusion that N_{In} is unlikely to affect the electrical properties of InN. The empty triplet state follows the CBM as the band gap is corrected and, because it is empty, it does not affect the formation energy of the neutral N_{In}.

The indium antisite induces a doubly occupied singlet state resonant in the valence band, and a triplet state, occupied by four electrons, above the CBM. The SIRC calculations show that this state shifts upwards with respect to the VBM, but by an amount smaller than the shift in the CBM. With the band-gap correction using the SIRC pseudopotentials, the triplet state still remains well above the CBM, and the formation energy of neutral In_N^0 would be higher than that given by the LDA calculations. Because the triplet state is above the CBM, its four electrons are transferred to the conduction band, making the indium antisite stable exclusively in the 4+ charge state (In_N^{4+}). The doubly occupied singlet state is resonant in the valence band, and we expect it to shift by a small amount with the band-gap correction. Hence its contribution to the formation energy of In_N^{4+} is expected to be small.

The indium interstitial (In_i) is stable as a shallow donor that exists exclusively in the 3+ charge state (In_i^{3+}) for all Fermi-level positions in the band gap. Since In_i only introduces states above the CBM, we expect the formation energy of In_i^{3+} to be only slightly affected by the band-gap correction. In n-type InN (E_F at the CBM), the formation energy of In_i^{3+} is ~ 3 eV under the most favorable In-rich conditions. This indicates that the concentration of In_i^{3+} in n-type as-grown InN is very low, well below detection limits. Therefore, we expect that indium interstitials do not contribute to the observed unintentional n-type conductivity in InN.

Nitrogen interstitials (N_i) have high formation energies, as shown in Fig. 11.2. They introduce states in the band gap that have mostly N $2p$ character, and therefore, should follow the VBM as the band gap is corrected. In this case their formation energy is well described by the LDA, and we expect the equilibrium concentration of N_i to be insignificant in as-grown InN.

We conclude that the nitrogen vacancy V_N is by far the lowest energy defect in InN under any conditions. The properties of V_N were therefore investigated in more detail in wurtzite InN, using 96-atom supercells and based on both LDA and LDA+U [11.55]. Since LDA+U provides a partial correction to the band gap, transition levels and formation are consistently corrected according to the valence- versus conduction-band character of the states in the band gap. This method was previously successfully applied to the calculation of oxygen vacancies in ZnO, where transition levels and optical excitations were found in good agreement with experiment [11.41, 11.56].

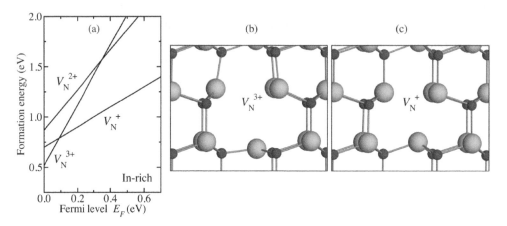

FIGURE 11.3

(a) Calculated formation energies as a function of Fermi level for the nitrogen vacancy in wurtzite InN, under In-rich conditions. Results for all three charge states, V_N^+, V_N^{2+}, and V_N^{3+}, are shown. E_F is referenced to the valence-band maximum. The formation energies were corrected using LDA and LDA+U calculations as described in the text. Panels (b) and (c) show local structural relaxations around V_N^+ and V_N^{3+}. Indium atoms are represented by large balls and N atoms by small balls.

Formation energies as a function of Fermi level for all charge states of V_N in wurtzite InN are shown in Fig. 11.3(a), and the local structural relaxations for the 3+ and + charge states are shown in Figs. 11.3(b) and (c). The results for V_N are consistent with Stampfl *et al.*'s calculations [11.31] shown in Fig. 11.2. V_N is stable in the 3+ and + charge states, with the $\varepsilon(3+/+)$ transition level at 0.1 eV above the VBM.

The nitrogen vacancy V_N in InN is a negative-U system, in which the 2+ charge state is unstable for any position of the Fermi level in the band gap. This can be explained by the large difference in local structural relaxations around V_N for different charge states. In the positive charge state (+), the a_1 state near the VBM is doubly occupied. Hence, V_N^+ is stabilized by an inward relaxation of the four surrounding In atoms, maximizing their bonding, as shown in Fig. 11.3(b). As the electrons are removed from the a_1 state, the four In nearest neighbors strongly relax outward, strengthening their bonding with their remaining N neighbors, as shown in Fig. 11.3(b). These relaxations lower the formation energies of V_N^+ and V_N^{3+} with respect to that of V_N^{2+}, so that the latter is only metastable. Accordingly, the only relevant transition level is that between the 3+ and + charge states, $\varepsilon(3+/+)$.

Finally, we discuss the possible role of nitrogen vacancies in the unintentional n-type conductivity in as-grown InN. The formation energy of V_N in n-type material, for example, with the Fermi level E_F at the CBM, is 1.4 eV. This corresponds to an equilibrium concentration of V_N^+ of 2×10^{13} cm^{-3} at a typical growth temperature of $\sim 500°C$ [11.17]. The observed electron concentrations in as-grown InN are several orders of magnitude higher, and therefore cannot be explained by the presence of nitrogen vacancies. We note, however, that the formation energy of V_N for E_F at the valence-band maximum is relatively low, and we expect V_N to be a possible source of compensation in p-type InN.

11.3.3 Impurities in InN

Having established that native point defects are unlikely to cause the observed unintentional n-type conductivity in as-grown InN, we now discuss the behavior of selected impurities in InN. We focus on oxygen, silicon, and hydrogen as possible donors. In addition, we also discuss the effects of Mg incorporation and its role as a p-type dopant in InN.

11.3.3.1 Donors (O and Si)

Oxygen is a common unintentional impurity in nitrides, and a shallow donor in GaN. Silicon is an n-type dopant in GaN and Al$_x$Ga$_{1-x}$N. In InN, oxygen is expected to occupy the N site (O$_N$), while Si is expected to occupy the In site (Si$_{In}$). Stampfl *et al.* [11.31] calculated the formation energies of Si$_{In}$ and O$_N$ in In-rich InN. Their results are shown in Fig. 11.4. The O and Si chemical potentials were assumed to be determined by thermal equilibrium with Al$_2$O$_3$ and Si$_3$N$_4$, respectively; this constitutes an upper limit on the solubility. Both Si$_{In}$ and O$_N$ are shallow donors and have much lower formation energies than the nitrogen vacancy V_N. Based on the calculated formation energies of native defects shown in Fig. 11.2, we expect that there will be no compensation of these impurities by acceptor-type native defects. Therefore, n-type doping of InN is easily attainable through extrinsic doping.

11.3.3.2 Acceptors (Mg and C)

Stampfl *et al.* [11.31] also investigated the effects of incorporation of Mg acceptors in InN (Mg$_{In}$). These results are also included in Fig. 11.4. The Mg chemical potential was assumed to be limited by the formation of Mg$_3$N$_2$. The position of the acceptor level, that is, the transition level between neutral and negative charge states, is at about 0.2 eV above the VBM. This value is very similar to the Mg acceptor level in GaN [11.57]. The formation energy of Mg in InN is lower than in GaN, indicating higher solubility for Mg in InN than in GaN. Just like in GaN,

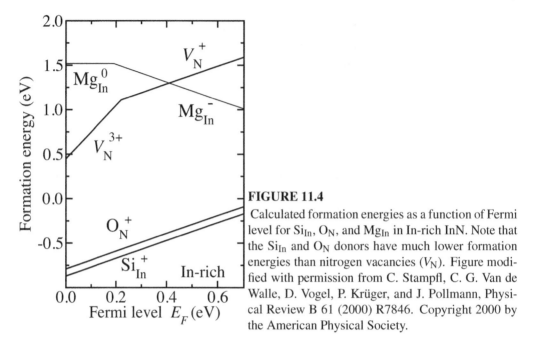

FIGURE 11.4

Calculated formation energies as a function of Fermi level for Si_{In}, O_N, and Mg_{In} in In-rich InN. Note that the Si_{In} and O_N donors have much lower formation energies than nitrogen vacancies (V_N). Figure modified with permission from C. Stampfl, C. G. Van de Walle, D. Vogel, P. Krüger, and J. Pollmann, Physical Review B 61 (2000) R7846. Copyright 2000 by the American Physical Society.

nitrogen vacancies are a potential source of compensation in Mg-doped InN. Jones *et al.* [11.58] have recently succeeded in probing a *p*-type Mg-doped layer underneath a surface electron accumulation layer in InN. Difficulties in overcoming the latter precluded a direct quantitative evaluation of the *p*-type conductivity in InN.

Using LDA calculations, Ramos *et al.* [11.59] investigated acceptor carbon impurities in InN (C_N). Their results indicate that C_N is a shallow acceptor with low formation energy and suggest that C incorporation in the N sublattice in InN is likely to occur even in the presence of a *p*-type background doping. Despite the prediction, to our knowledge there are no experimental reports on the acceptor behavior of C impurities in InN.

11.3.3.3 Hydrogen in InN

We now turn to the effect of hydrogen on the properties of InN. Hydrogen is a common impurity in semiconductors, found in almost all growth and processing environments [11.60]. The presence of hydrogen is known to have a strong effect on the electronic properties of many materials. Hydrogen has been primarily thought of as an interstitial impurity that passivates intrinsic defects and other impurities, thereby significantly improving the electronic properties. In most semiconductors, interstitial hydrogen is amphoteric: it is stable as a donor in *p*-type and as an acceptor in *n*-type material, always counteracting the prevailing conductivity [11.60]. This behavior has been well established in GaN [11.61]. In contrast, interstitial hydrogen (H_i) in InN was predicted to be stable exclusively as a donor [11.15, 11.60, 11.62]. H_i strongly bonds to nitrogen, causing a breaking (or at least weakening) of a N-In chemical bond. Interstitial hydrogen can be found in the bond-center or in the antibonding configurations, as shown in Fig. 11.5(a) and (b). Despite the strength of the N-H bond, calculations have shown that interstitial hydrogen is a fast diffuser in InN, with a migration barrier of 1.1 eV, causing it to be mobile at relatively modest temperatures [11.55]. Interstitial hydrogen that is incorporated during growth or processing will therefore still be mobile while the samples are cooled down.

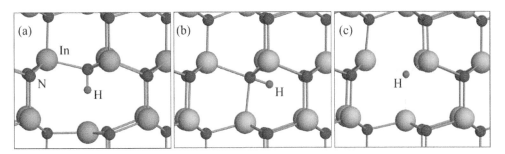

FIGURE 11.5
Local structure of interstitial hydrogen in (a) the bond-center configuration and (b) the antibonding configuration. (c) Substitutional hydrogen on a nitrogen site. Figure reproduced with permission from A. Janotti and C. G. Van de Walle, Applied Physics Letters 92 (2008) 032104. Copyright 2008, American Institute of Physics.

In addition to being incorporated as an interstitial, hydrogen can also occupy a substitutional site in InN. This behavior is similar to what has been found in ZnO and MgO [11.63]. Substitutional hydrogen in InN was investigated using LDA and LDA+U calculations [11.55]. It was found that atomic hydrogen occupies a nitrogen site and forms a multicenter bond with its four In nearest neighbors, as discussed below. The calculated formation energies of interstitial and substitutional hydrogen are shown in Fig. 11.6. The formation energy of the nitrogen vacancy is also shown for comparison. Figure 11.6 shows that the formation energy of H_N is lower than that of V_N^+ for all Fermi-level positions within the band gap. Substitutional hydrogen is stable exclusively in the 2+ charge state, H_N^{2+}. For E_F at the CBM, the formation energy of H_N^{2+} is 1.1 eV, that is, 0.3 eV lower than that of V_N. This formation energy corresponds to an equilibrium concentration of 2×10^{15} cm^{-3} at $T = 500°$C. Note that this estimate is based on equilibrium with H_2; in reality, equilibrium is more likely with an adsorbed species on the surface, which lowers the formation energy and raises the solubility. Substitutional hydrogen is therefore a plausible cause of unintentional n-type conductivity in InN. We note that even in InN grown by molecular beam epitaxy, hydrogen concentrations exceeding 10^{18} cm^{-3} have been found [11.17].

Frequencies of the local vibrational modes related to substitutional hydrogen in InN were calculated with the goal of facilitating future experimental detection and identification. H_N^{2+} gives rise to three almost degenerate local vibration modes, with calculated frequencies close to 540 cm^{-1} [11.55]. Comparing with values of 570 cm^{-1} and 590 cm^{-1} for the highest longitudinal and transverse optical phonons in

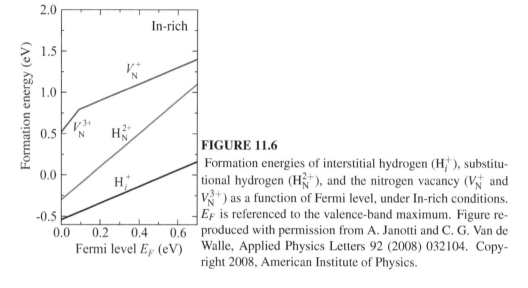

FIGURE 11.6

Formation energies of interstitial hydrogen (H_i^+), substitutional hydrogen (H_N^{2+}), and the nitrogen vacancy (V_N^+ and V_N^{3+}) as a function of Fermi level, under In-rich conditions. E_F is referenced to the valence-band maximum. Figure reproduced with permission from A. Janotti and C. G. Van de Walle, Applied Physics Letters 92 (2008) 032104. Copyright 2008, American Institute of Physics.

InN [11.49], we expect strong coupling with bulk modes that will make experimental observation of the H_N^{2+} local vibrational modes very challenging.

The fact that hydrogen can act as a double donor seems counterintuitive, since it has only one electron. Its electronic structure can be understood in a simple tight-binding or molecular orbital picture, as shown in Fig. 11.7. We start from the nitrogen vacancy. Removing a nitrogen atom from the InN lattice leaves four In dangling bonds (DBs), occupied with three electrons. In the near-tetrahedral environment of the wurtzite structure, these DBs combine into a doubly occupied fully symmetric a_1 state located in the band gap, plus three almost degenerate states located at ~ 2 eV above the CBM. The electron that would occupy the lowest of these three states is transferred to the CBM, resulting in the positive charge state of the nitrogen vacancy (V_N^+) as shown in Fig. 11.7. This is indeed consistent with the results of first-principles calculations (Figs. 11.2 and 11.3). When hydrogen is placed on the nitrogen site, the hydrogen $1s$ state combines with the a_1 state from V_N^+, resulting in a bonding state at 6.5 eV below the VBM, and an antibonding state at 6 eV above the CBM. The electron that would occupy this antibonding state is transferred to the CBM, turning substitutional hydrogen into a double donor (H_N^{2+}). Note that the three almost degenerate V_N states remain unaltered and constitute the non-bonding states in Fig. 11.7.

To complete the survey of potential donors, we return to interstitial hydrogen, which is stable exclusively in the positive charge state (Fig. 11.6). The calculated frequencies of the local vibrational modes are 3050 cm^{-1} for the stretching mode (including anharmonicity) and 626 cm^{-1} for the wagging modes [11.55]. The formation energy of H_i^+ is lower than that of either H_N^{2+} or V_N^+. In order to address the overall stability of interstitial and substitutional hydrogen, one also has to consider their migration barriers. The calculated migration barrier of interstitial hydrogen is 1.1 eV and corresponds to H breaking a bond with N and forming a bond with a nearest neighbor N. Using the calculated vibrational frequency as an attempt fre-

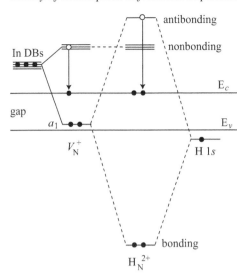

FIGURE 11.7

Schematic illustration of the electronic structure of substitutional hydrogen H_N^{2+} in InN. The a_1 state of V_N^+ and the hydrogen $1s$ state combine into bonding/antibonding states. The bonding state is identified as a multicenter bond involving H and the four In neighbors. The antibonding state located in the conduction band will be unoccupied. Figure reproduced with permission from A. Janotti and C. G. Van de Walle, Applied Physics Letters 92 (2008) 032104. Copyright 2008, American Institute of Physics.

quency, we estimate that H_i^+ will become mobile at temperatures around 100°C. Although H_i is mobile at relatively modest temperatures, this does not necessarily imply that it easily diffuses out of the InN samples. Present InN films exhibit a high electron accumulation on the surface that is associated with a pinning of the Fermi level at $\sim$0.8 eV above the CBM due to intrinsic surface states [11.64–11.66]. The associated band bending creates a potential barrier that impedes outdiffusion of positively charged impurities. We therefore expect that a significant concentration of interstitial hydrogen can be trapped inside the material and contribute to the *n*-type conductivity.

It is also relevant to discuss the stability of substitutional hydrogen H_N^{2+} in InN. While interstitial hydrogen migrates by breaking and forming H-N bonds, H_N^{2+} can migrate via two distinct processes: (1) migration assisted by a nitrogen vacancy, where H jumps to a nearby V_N, leaving a vacancy behind; or (2) migration by a concerted exchange with a nearest-neighbor N atom. The activation energy for the vacancy-assisted process (1) is the sum of $E^f(V_N^+)$ (1.4 eV for E_F at the CBM) and the barrier for the exchange process $H_N^{2+} \leftrightarrow V_N^+$ (estimated to be higher than 1 eV), resulting in a barrier of at least 2.4 eV [11.55]. A direct calculation of the energy barrier for the concerted exchange mechanism (2) gives 2.2 eV [11.55]. These barriers indicate that H_N^{2+} would be mobile at ~ 500°C. However, it is also important to consider the dissociation of substitutional hydrogen into a nitrogen vacancy and a hydrogen interstitial: $H_N^{2+} \rightarrow H_i^+ + V_N^+$. The activation energy for this dissociation process is 1.6 eV, based on the sum of the binding energy $E_b = E^f(H_i^+) + E^f(V_N^+) - E^f(H_N^{2+}) = 0.5$ eV and the migration barrier of H_i^+ (1.1 eV). Using the calculated vibrational frequency of 540 cm^{-1} for H_N^{2+} as an attempt frequency and an activation energy of 1.6 eV, substitutional hydrogen is expected to be stable up to at least 300°C [11.55]. Note that H_N^{2+} is subject to the same barriers to outdiffusion as discussed for H_i above. Therefore, both substitutional and interstitial H are potential causes of unintentional *n*-type conductivity in InN.

11.4   Summary

The current theory of native point defects and impurities in InN has been reviewed with an emphasis on first-principles calculations based on density functional theory. The results of these calculations indicate that native point defects cannot explain the universally observed *n*-type conductivity in as-grown InN. The nitrogen vacancy, which has been commonly invoked to explain the *n*-type conductivity in as-grown InN, is a shallow donor but has a high formation energy in *n*-type InN. The predicted equilibrium concentration is several orders of magnitude lower than the observed electron concentrations. The other native point defects have even higher formation energies and therefore will not occur in observable concentrations in *n*-type InN. Unintentional incorporation of impurities, such as oxygen and hydrogen, may well be responsible for the observed levels of *n*-type conductivity in InN. Hydrogen is a stable donor when present in an interstitial configuration, but it can also be stable as a substitutional impurity. It replaces nitrogen, forms a multicenter bond with its four In nearest neighbors, and, counterintuitively, acts as a double donor. Both interstitial and substitutional hydrogen have low formation energies, making them plausible candidates for unintentional *n*-type dopants.

Acknowledgments

This work was supported by the NSF MRSEC Program under award No. DMR05-20415 and by the UCSB Solid State Lighting and Energy Center. It also made use of the CNSI Computing Facility under NSF grant No. CHE-0321368. We acknowledge helpful discussions with C. Gallinat, G. Koblmüller, and J. Speck.

References

[11.1] M. Lannoo and J. Bourgoin, in *Point defects in semiconductors I: Theoretical aspects* (Springer, Berlin, 1981); M. Lannoo and J. Bourgoin, in *Point defects in semiconductors II: Experimental aspects* (Springer, Berlin, 1983).

[11.2] S. T. Pantelides (Ed.), *Deep centers in semiconductors: A state-of-the-art approach*, 2nd ed. (Gordon and Breach Science Publishers, Yverdon, 1992).

[11.3] M. Stavola (Ed.), *Identification of defects in semiconductors*, *Semiconductors and semimetals* vol. 51B (Academic Press, San Diego, 1999).

[11.4] D. B. Laks, C. G. Van de Walle, G. F. Neumark, and S. T. Pantelides, Role of native point defects in wide-band-gap semiconductors, *Physical Review Letters*, 66 (1991) 648–651.

[11.5] V. Y. Davydov, A. A. Klochikhin, R. P. Seisyan, V. V. Emtsev, S. V. Ivanov, F. Bechstedt, J. Furthmüller, H. Harima, A. V. Mudryi, J. Aderhold, O. Semchinova, and J. Graul, Absorption and emission of hexagonal InN. Evidence of narrow fundamental band gap, *physica status solidi (b)*, 229 (2002) R1–R3.

[11.6] V. Y. Davydov, A. A. Klochikhin, V. V. Emtsev, S. V. Ivanov, V. V. Vekshin, F. Bechstedt, J. Furthmüller, H. Harima, A. V. Mudryi, A. Hashimoto, A. Yamamoto, J. Aderhold, O. Semchinova, J. Graul, and E. E. Haller, Bandgap of InN and In-rich In$_x$Ga$_{1-x}$N alloys $(0.36 < x < 1)$, *physica status solidi (b)*, 230 (2002) R4–R6.

[11.7] J. Wu, W. Walukiewicz, K. M. Yu, J. W. Ager, E. E. Haller, H. Lu, W. J. Schaff, Y. Saito, Y. Nanishi, Unusual properties of the fundamental band gap of InN, *Applied Physics Letters*, 80 (2002) 3967–3969.

[11.8] E. Bellotti, B. K. Doshi, K. F. Brennan, J. D. Albrecht, and P. P. Ruden, Ensemble Monte Carlo study of electron transport in wurtzite InN. *Journal of Applied Physics*, 85 (1999) 916–923.

[11.9] G. D. Chern, E. D. Readinger, H. Shen, M. Wraback, C. S. Gallinat, G. Koblmüller, and J. S. Speck, Excitation wavelength dependence of terahertz emission from InN and InAs, *Applied Physics Letters*, 89 (2006) 141115:1–3.

[11.10] A. G. Bhuiyan, A. Hashimoto, and A. Yamamoto, Indium nitride (InN): A review on growth, characterization, and properties, *Journal of Applied Physics*, 94 (2003) 2779–2808.

[11.11] W. Walukiewicz, J. W. Ager III, K. M. Yu, Z. Liliental-Weber, J. Wu, S. X. Li, R. E. Jones, and J. D. Denlinger, Structure and electronic properties of InN and In-rich group III-nitride alloys, *Journal of Physics D: Applied Physics*, 39 (2006) R83–R99.

[11.12] *Introduction to nitride semiconductor blue lasers and light emitting diodes*, edited by S. Nakamura and S. F. Chichibu (Taylor and Francis, London, 2000).

[11.13] G. Koblmüller, C. S. Gallinat, S. Bernardis, J. S. Speck, G. D. Chern, E. D. Readinger, H. Shen, and M. Wraback, Optimization of the surface and structural quality of N-face InN grown by molecular beam epitaxy, *Applied Physics Letters*, 89 (2006) 071902:1–3.

[11.14] L. F. J. Piper, T. D. Veal, C. F. McConville, H. Lu, and W. J. Schaff, Origin of the n-type conductivity of InN: The role of positively charged dislocations, *Applied Physics Letters*, 88 (2006) 252109:1–3.

[11.15] S. Limpijumnong and C. G. Van de Walle, Passivation and doping due to hydrogen in III-nitrides, *physica status solidi (b)*, 228 (2001) 303–307.

[11.16] D. C. Look, H. Lu, W. J. Schaff, J. Jasinski, and Z. Liliental-Weber, Donor and acceptor concentrations in degenerate InN, *Applied Physics Letters*, 80 (2002) 258–260.

[11.17] C. S. Gallinat, G. Koblmüller, J. S. Brown, S. Bernardis, J. S. Speck, G. D. Chern, E. D. Readinger, H. Shen, and M. Wraback, In-polar InN grown by plasma-assisted molecular beam epitaxy, *Applied Physics Letters*, 89 (2006) 032109:1–3.

[11.18] P. Hohenberg and W. Kohn, Inhomogeneous Electron Gas, *Physical Review*, 136 (1964) B864; W. Kohn and L. J. Sham, Self-consistent equations including exchange and correlation effects, *Physical Review*, 140 (1965) A1133–A1138.

[11.19] M. C. Payne, M. P. Teter, D. C. Allan, T. A. Arias, and J. D. Joannopoulos, Iterative minimization techniques for ab initio total-energy calculations: molecular dynamics and conjugate gradients, *Reviews of Modern Physics*, 64 (1992) 1045–1097.

[11.20] R. M. Martin, *Electronic Structure: Basic Theory and Methods* (Cambridge University Press, Cambridge, 2004).

[11.21] C. G. Van de Walle and J. Neugebauer, First-principles calculations for defects and impurities: Applications to III-nitrides, *Journal of Applied Physics*, 95 (2004) 3851–3879.

[11.22] M. Bockstedte, A. Kley, J. Neugebauer, and M. Scheffler, Density-functional theory calculations for poly-atomic systems: electronic structure, static and elastic properties and ab initio molecular dynamics, *Computer Physics Communications*, 107 (1997) 187–222.

[11.23] P. E. Blöchl, Projector augmented-wave method, *Physical Review B*, 50 (1994) 17953; G. Kresse and J. Joubert, From ultrasoft pseudopotentials to the projector augmented-wave method, *Physical Review B*, 59 (1999) 1758–1775.

[11.24] S. G. Louie, S. Froyen, and M. L. Cohen, Nonlinear ionic pseudopotentials in spin-density-functional calculations, *Physics Review B*, 26 (1982) 1738–1742.

[11.25] G. Kresse and J. Furthmüller, Efficient iterative schemes for ab initio total-energy calculations using a plane-wave basis set, *Physical Review B*, 54 (1996) 11169–11186; G. Kresse and J. Furthmüller, Efficiency of ab-initio total energy calculations for metals and semiconductors using a plane-wave basis set, *Computational Material Science*, 6 (1996) 15–50.

[11.26] M. R. Ranade, F. Tessier, A. Navrotsky, and R. Marchand, Calorimetric determination of the enthalpy of formation of InN and comparison with AlN and GaN, *Journal of Materials Research*, 16 (2001) 2824–2831.

[11.27] G. Makov and M. C. Payne, Periodic boundary conditions in ab initio calculations, *Physical Review B*, 51 (1995) 4014–4022.

[11.28] J. Shim, E. K. Lee, Y. J. Lee, and R. M. Nieminen, Density-functional calculations of defect formation energies using supercell methods: Defects in diamond, *Physical Review B*, 71 (2005) 035206:1–12.

[11.29] S.-H. Wei and A. Zunger, Role of metal d states in II-VI semiconductors, *Physical Review B*, 37 (1988) 8958–8981.

[11.30] A. Janotti, D. Segev, and C. G. Van de Walle, Effects of cation d states on the structural and electronic properties of III-nitride and II-oxide wide-band-gap semiconductor, *Physical Review B*, 74 (2006) 045202:1–9.

[11.31] C. Stampfl, C. G. Van de Walle, D. Vogel, P. Krüger, and J. Pollmann, Native defects and impurities in InN: First-principles studies using the local-density approximation and self-interaction and relaxation-corrected pseudopotentials, *Physical Review B*, 61 (2000) R7846–R7849.

[11.32] A. Janotti and C. G. Van de Walle, Absolute deformation potentials and band alignment of wurtzite ZnO, MgO, and CdO, *Physical Review B*, 75 (2007) 121201(R):1–4.

[11.33] D. Segev, A. Janotti, and C. G. Van de Walle, Self-consistent band-gap corrections in density functional theory using modified pseudopotentials, *Physical Review B*, 75 (2007) 035201:1–9.

[11.34] D. Vogel, P. Krüger, and J. Pollmann, Structural and electronic properties of group-III nitrides, *Physical Review B*, 55 (1997) 12836–12839, and references therein.

[11.35] V. I. Anisimov, I. V. Solovyev, M. A. Korotin, M. T. Czyzyk, and G. A. Sawatzky, Density-functional theory and NiO photoemission spectra, *Physical Review B*, 48 (1993) 16929–16934; A. I. Liechtenstein, V. I. Anisimov, and J. Zaanen, Density-functional theory and strong interactions: Orbital ordering in Mott-Hubbard insulators, *Physical Review B*, 52 (1995) R5467–R5470.

[11.36] V. I. Anisimov, F. Aryasetiawan, and A. I. Lichtenstein, First-principles calculations of the electronic structure and spectra of strongly correlated systems: the LDA+U method, *Journal of Physics: Condensed Matter*, 9 (1997) 767–808.

[11.37] O. Bengone, M. Alouani, P. Blöchl, and J. Hugel, Implementation of the projector augmented-wave LDA+U method: Application to the electronic structure of NiO, *Physical Review B*, 62 (2000) 16392–16401.

[11.38] C. Persson and A. Zunger, s-d coupling in zinc-blende semiconductors,

Physical Review B, 68 (2003) 073205:1–4.

[11.39] A. Janotti and C. G. Van de Walle, Oxygen vacancies in ZnO, *Applied Physics Letters*, 87 (2005) 122102:1–3.

[11.40] A. Janotti and C. G. Van de Walle, New insights into the role of native point defects in ZnO, *Journal of Crystal Growth*, 287 (2006) 58–65.

[11.41] A. Janotti and C. G. Van de Walle, Native point defects in ZnO, *Physical Review B*, 76 (2007) 165202:1–22.

[11.42] L. Hedin, New method for calculating the one-particle Green's function with application to the electron-gas problem, *Physical Review*, 139 (1965) A796–A823.

[11.43] M. Hedström, A. Schindlmayr, G. Schwarz, and M. Scheffler, Quasiparticle corrections to the electronic properties of anion vacancies at GaAs(110) and InP(110), *Physical Review Letters*, 97 (2006) 226401:1–4.

[11.44] W. K. Leung, R. J. Needs, G. Rajagopal, S. Itoh, and S. Ihara, Calculations of silicon self-interstitial defects, *Physical Review Letters*, 83 (1999) 2351–2534.

[11.45] E. R. Batista, J. Heyd, R. G. Hennig, B. P. Uberuaga, R. L. Martin, G. E. Scuseria, J. Umrigar, and J. W. Wilkins, Comparison of screened hybrid density functional theory to diffusion Monte Carlo in calculations of total energies of silicon phases and defects, *Physical Review B*, 74 (2006) 121102 (R):1–4.

[11.46] P. Rinke, A. Qteish, J. Neugebauer, C. Freysoldt, and M. Scheffler, Combining GW calculations with exact-exchange density-functional theory: an analysis of valence-band photoemission for compound semiconductors, *New Journal of Physics*, 7 (2005) 126:1–35.

[11.47] P. Rinke, M. Scheffler, A. Qteish, M. Winkelnkemper, D. Bimberg, and J. Neugebauer, Band gap and band parameters of InN and GaN from quasiparticle energy calculations based on exact-exchange density-functional theory, *Applied Physics Letters*, 89 (2006) 161919:1–3.

[11.48] P. Rinke, M. Winkelnkemper, A. Qteish, D. Bimberg, J. Neugebauer, and M. Scheffler, Consistent set of band parameters for the group-III nitrides AlN, GaN, and InN, *Physical Review B*, 77 (2008) 075202:1–15.

[11.49] *Semiconductors – Data Handbook*, 3rd ed., edited by O. Madelung (Springer, Berlin, 2004).

[11.50] D. W. Jenkins and J. D. Dow, Electronic structures and doping of InN, $In_xGa_{1-x}N$, and $In_xAl_{1-x}N$, *Physical Review B*, 39 (1989) 3317–3329.

[11.51] T. L. Tansley and R. J. Egan, Point-defects in the nitrides of aluminum, gallium, and indium, *Physical Review B*, 45 (1992) 10942–10950.

[11.52] S. Strite and H. Morkoç, GaN, AlN, and InN: a review, *Journal of Vacuum Science and Technology B*, 10 (1992) 1237–1266.

[11.53] J. Oila, A. Kemppinen, A. Laakso, K. Saarinen, W. Egger, L. Liszkay, P. Sperr, H. Lu, and W. J. Schaff, Influence of layer thickness on the formation of In vacancies in InN grown by molecular beam epitaxy, *Applied Physics Letters*, 84 (2004) 1486–1488.

[11.54] F. Tuomisto, A. Pelli, K. M. Yu, W. Walukiewicz, and W. J. Schaff, Compensating point defects in ^{4}He$^+$-irradiated InN, *Physics Review B*, 75 (2007) 193201:1–4.

[11.55] A. Janotti and C. G. Van de Walle, Sources of unintentional conductivity in InN, *Applied Physics Letters*, 92 (2008) 032104:1–3.

[11.56] A. Janotti and C. G. Van de Walle, Oxygen vacancies in ZnO, *Applied Physics Letters*, 87 (2005) 122102:1–3.

[11.57] J. Neugebauer and C. G. Van de Walle, Role of hydrogen in doping of GaN, *Applied Physics Letters*, 68 (1996) 1829–1831.

[11.58] R. E. Jones, K. M. Yu, S. X. Li, W. Walukiewicz, J. W. Ager, E. E. Haller, H. Lu, and W. J. Schaff, Evidence for p-type doping of InN, *Physical Review Letters*, 96 (2006) 125505:1–4.

[11.59] L. E. Ramos, J. Furthmüller, L. M. R. Scolfaro, J. R. Leite, and F. Bechstedt, Substitutional carbon in group-III nitrides: Ab initio description of shallow and deep levels, *Physical Review B*, 66 (2002) 075209:1–9.

[11.60] C. G. Van de Walle and J. Neugebauer, Universal alignment of hydrogen levels in semiconductors, insulators, and solutions, *Nature*, 423 (2003) 626–628.

[11.61] J. Neugebauer and C. G. Van de Walle, Theory of hydrogen in GaN, In Hydrogen in semiconductors II, edited by N. H. Nickel, Semiconductors and semimetals, Vol. 61, edited by R. K. Willardson and E. R. Weber (Academic Press, Boston, 1999), p. 479.

[11.62] E. A. Davis, S. F. J. Cox, R. L. Lichti, and C. G. Van de Walle, Shallow donor state of hydrogen in indium nitride, *Applied Physics Letters*, 82 (2003) 592–594.

[11.63] A. Janotti and C. G. Van de Walle, Hydrogen multicentre bonds, *Nature Materials*, 6 (2007) 44–47.

[11.64] I. Mahboob, T. D. Veal, C. F. McConville, H. Lu, and W. J. Schaff, Intrinsic electron accumulation at clean InN surfaces, *Physical Review Letters*, 92 (2004) 036804:1–4.

[11.65] T. D. Veal, L. F. J. Piper, I. Mahboob, H. Lu, W. J. Schaff, and C. F. McConville, Electron accumulation at InN/AlN and InN/GaN interfaces, *physica status solidi (c)*, 2 (2005) 2246–2249.

[11.66] D. Segev and C. G. Van de Walle, Origins of Fermi-level pinning on GaN and InN polar and nonpolar surfaces, *Europhysics Letters*, 76 (2006) 305–311.

12

Surface electronic properties of InN and related alloys

T. D. Veal, P. D. C. King, and C. F. McConville

Department of Physics, University of Warwick, Coventry, CV4 7AL, United Kingdom

12.1   Introduction

One of the most remarkable properties of InN to be discovered so far is the presence of a native electron accumulation layer at its surface [12.1]. While all other common n-type III-V semiconductors, with the exception of InAs, and the vast majority of all semiconductors, have depletion layers at their surfaces, the surface of InN has a charge accumulation layer with a sheet density in excess of 10^{13} e/cm^2. This property influences the most basic electrical characterization of InN films; using the single-field Hall effect technique does not give the properties of the 'bulk' of an InN film. Instead, the average electrical properties of the entire InN film are obtained, including the effects of parallel conduction through the surface, bulk and interface regions, as discussed in Chapters 4 and 5. The presence of a related inversion layer at the surface of p-type InN makes the p-type region beneath the surface electron-rich region difficult to characterize. Surface electron accumulation also results in almost all metal/InN contacts exhibiting Ohmic behavior and has implications for doping behavior and potential device applications, such as solar cells, transistors, generation of terahertz radiation and gas and chemical sensors.

In this chapter the results of investigations of the electronic and structural properties of InN surfaces are presented. Evidence of surface electron accumulation from high-resolution electron-energy-loss spectroscopy and valence band x-ray photoemission spectroscopy (XPS) is reviewed. The surface space-charge properties are described by solutions of the Poisson equation within the modified Thomas-Fermi approximation, incorporating conduction band nonparabolicity. The surface Fermi level position associated with the electron accumulation is independent of surface orientation for a-plane and In- and N-polarity c-plane wurtzite InN. Zinc-blende InN also exhibits surface electron accumulation. The quantized subband electron states associated with the extreme downward band bending at the InN surface have been investigated using electron tunneling spectroscopy and angle resolved photoemission spectroscopy. The one-electron potential, carrier concentration profile, quantized subband state energies, and parallel dispersion relations are calculated for the surface electron accumulation layer by numerically solving the Schrödinger equation for the potential well obtained from solving the Poisson equation. The inversion space-charge layer at the surface of *p*-type InN is also described. The overriding origin of the surface electron accumulation is identified as the particularly low Γ-point conduction band minimum in the bulk band structure that enables the existence of positively charged donor-type surface states. The various candidates for the microscopic origin of the donor surface states are discussed. The structure of InN surfaces, including the presence of indium adlayers, is investigated using core-level XPS, electron diffraction and ion scattering spectroscopy. Finally, the results are presented of valence band XPS studies of the surface electronic properties of undoped and Mg-doped InGaN and InAlN thin films as a function of alloy composition across the entire composition range.

12.2   Electron accumulation at InN surfaces

12.2.1   Initial evidence

The first indications of electron accumulation at InN surfaces and interfaces were only observed once epitaxial growth methods had been optimized sufficiently for free electron densities significantly below 10^{20} cm^{-3} to be realized in InN thin films. Evidence of a two dimensional electron gas at an AlN/InN interface was obtained from capacitance-voltage (C-V) profiling and no Schottky barrier was found for Hg contacts on InN [12.2].

Shortly afterwards, a series of InN films with thicknesses of between 10 and 80 nm were grown on either an AlN or GaN buffer. It was found that for InN films on both buffers, their sheet electron density increased linearly with film thickness, which means that there is a roughly constant carrier density in the 'bulk' of all of the InN thin films, with carrier density found from the slope. However, by extrapolating the fitted line to zero film thickness, the intercepts were found to be non-zero. As shown in Chapter 5, Fig. 5.9, for InN films on an AlN buffer, the residual sheet charge was found to be 4.3×10^{13} cm^{-2}, while for InN films on a GaN buffer, the residual sheet charge is about 2.5×10^{13} cm^{-2} [12.3, 12.4]. This result means that carriers are not uniformly distributed in the film; there must be surface charge accumulation or interface-related electron density [12.3, 12.4]. C-V profiling, performed using a KOH-based electrolyte, indicated the presence of electron accumulation at the surface of InN [12.4]. Integrating the carrier concentration versus depth curve, a surface sheet density of 1.6×10^{13} cm^{-2} was obtained, a value of the same order, but nevertheless lower than the excess sheet density obtained from the Hall effect results.

The presence of electron accumulation was confirmed by synchrotron radiation photoemission spectroscopy of an InN surface with Ti deposited on it [12.5]. Additionally, Ohmic I-V characteristics were measured for Ti, Al and Ni contacts and a Hg probe on InN [12.4]. Several more recent studies using I-V characterization of contacts [12.6, 12.7] and electrochemical C-V profiling [12.8–12.11], have all provided additional evidence of InN's surface electron accumulation. Additional film-thickness-dependent Hall effect measurements [12.12, 12.13] and multiple-field Hall effect measurements [12.14–12.17] are discussed further in Chapter 4 in the context of both surface electron accumulation and interface-related electron density.

12.2.2 Intrinsic electron accumulation at clean surfaces

Once the existence of electron accumulation at metal/InN interfaces had been established, the question naturally followed of whether electron accumulation is an intrinsic property of InN that would therefore be exhibited by its clean surface. High-resolution electron-energy-loss spectroscopy and high-resolution x-ray photoemission spectroscopy were both used to answer this question.

12.2.2.1 High-resolution electron-energy-loss spectroscopy

The space-charge layer on the surface of clean InN was first investigated with high-resolution electron-energy-loss spectroscopy (HREELS) [12.1, 12.18]. In this technique that is represented schematically in Fig. 12.1, low-energy (5–100 eV) monochromated probing electrons that follow a specular scattering trajectory exchange energy via a Coulombic interaction with polarization fields arising from the collective excitations of both the lattice (phonons) and the electrons in the conduction band (plasmons). By changing the energy of the probing electrons, the entire space-charge region can be surveyed. Electron spectroscopies are the preferred methods to investigate the properties of clean InN surfaces, as, unlike other techniques, such as capacitance-voltage profiling, no contacts to the surface are required. Furthermore, rather than probing the entire film, HREELS only probes the near-surface space-charge region, allowing this to be distinguished from any charge accumulated at the InN/buffer layer interface.

The InN(0001) thin film used for the HREELS study was grown by molecular-beam epitaxy (MBE) at Cornell University to a thickness of 200 nm on top of a 200 nm GaN buffer layer. A further 200 nm AlN layer was grown between the buffer layer and the *c*-plane sapphire substrate. Such films were confirmed to be In polarity by co-axial impact-collision ion scattering spectroscopy, as discussed further in section 12.4. Single-field Hall effect measurements indicated an average conduction electron density, n, of 6×10^{18} cm^{-3} and an average mobility of 600 cm^{2}V^{-1}s^{-1}. A clean InN surface was achieved *in situ* by atomic hydrogen cleaning (AHC) in order to remove the atmospheric contaminants [12.19]. The InN(0001) sample exhibited a (1×1) low energy electron diffraction (LEED) pattern after cleaning, indicating a well-ordered surface, as shown in the inset of Fig. 12.2. The removal of atmospheric contaminants was confirmed by the absence in HREELS of vibrational modes associated with adsorbed hydrocarbons and native oxides.

A series of normalized HREEL spectra recorded from the clean InN(0001) sur-

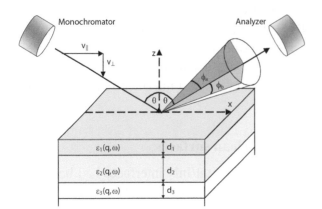

FIGURE 12.1
Schematic of a specular electron-energy-loss experiment. Electrons from a cathode pass through a monochromator, are reflected at the sample and are then energy analyzed. Along their trajectory, the probing electrons exchange energy with polarization fields arising from the plasmons and phonons of the sample. To simulate the measured electron-energy-loss spectra, the sample is modeled as a series of layers, each with its own thickness, d, and dielectric function, $\varepsilon(q, \omega)$.

face with a range of probing energies is shown in Fig. 12.2, along with semi-classical dielectric theory simulations. Three distinct features are observed in the HREEL spectra. The intense feature at zero loss energy is due to elastically scattered electrons. The first loss feature at $\sim$66 meV is assigned to Fuchs-Kliewer surface phonon excitations [12.20]. The second loss feature at $\sim$250 meV is due to conduction band electron plasmon excitations. The plasmon peak undergoes a $\sim$30 meV downward dispersion as the energy of the probing electrons is increased from 10 to 30 eV. This can be understood in terms of a surface layer of higher plasma frequency than that of the bulk and provides direct evidence for the existence of an electron accumulation layer at the InN surface.

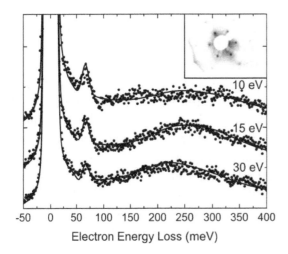

FIGURE 12.2
Specular HREEL spectra recorded at 300 K from an atomic hydrogen cleaned InN(0001)-(1×1) surface with incident electron energies of 10, 15 and 30 eV (points) and the corresponding semi-classical dielectric theory simulations (solid lines). The (1×1) LEED pattern, recorded with a beam energy of 164 eV, is shown in the inset. Reprinted with permission from T. D. Veal, I. Mahboob, L. F. J. Piper, C. F. McConville, H. Lu, and W. J. Schaff, Journal of Vacuum Science and Technology B, 22 (2004) 2175. Copyright 2004, American Vacuum Society.

The existence of an electron accumulation layer at the InN surface is quantitatively confirmed by simulating the HREEL spectra. The energy exchange between the probing electrons and the polarization field arising from the collective excitations of the free carriers in the conduction band results from the long-ranged Coulomb interaction. The dipole fields associated with plasmon excitations of wavevector q decay exponentially with a characteristic length q^{-1}. Therefore, by altering the kinetic energy of the incident electrons, both the wavevector transfer parallel to the surface and the resulting probing depth are varied. The HREEL spectra can be simulated using semi-classical dielectric theory within the methodology developed by Lambin *et al.* [12.21–12.23]. In the first part of the 'semi-classical' dielectric theory, the classical loss probability spectrum is computed by approximating the paths of the probing electrons as classical trajectories with initial velocities parallel and perpendicular to the surface, $v_{\parallel}$ and $v_{\perp}$, and incident angle of θ with respect to the surface normal. The quantum mechanical part consists of accounting for the Bose-Einstein distribution of multiple-scattering losses and gains at non-zero temperature. The final part of the simulation process consists of a convolution of the calculated electron-energy-loss spectrum with a model of the spectrometer response function to account for instrumental broadening and the finite acceptance angles (ϕ_a and ϕ_b) of the electron analyzer. Simulations of the spectra are required to obtain the true plasma frequency from the spectra because the observed plasmon peak position is influenced by the band bending, spatial dispersion, and plasmon damping. A four-layer model (depicted in Fig. 12.3(a)) was required to simulate the HREEL spectra, where each layer has its own frequency- and wavevector-dependent hydrodynamic dielectric function [12.21]. A plasma dead layer, that is, a carrier-free layer of 3 Å was required, both to simulate the variation in the phonon peak intensity and approximate the quantum mechanical effect of the surface potential barrier [12.24]. Two further layers of enhanced plasma frequency were also needed to reproduce the plasmon tail at high loss energy. Finally, a bulk layer with a plasma frequency of 211 meV was used to obtain the correct plasmon peak position. This layer profile was necessary to reproduce the dispersion of the plasmon peak. The results of the HREELS simulations are shown in Fig. 12.2, where all the spectra were simulated using the same plasma frequency profile.

Knowledge of the conduction band dispersion relation is required in order to translate the plasma frequencies extracted from HREELS into carrier concentrations. Kane's **k·p** model was utilized to calculate the nonparabolic dispersion of the conduction band [12.25, 12.26], as described in section 12.2.3. The interdependence of the plasma frequency, ω_p, the carrier concentration, n, and the average effective mass, m_{av}^*, can consequently be calculated from knowledge of the conduction band dispersion. These calculations were used to translate the plasma frequencies extracted from the HREELS simulations into carrier concentrations. The resulting conduction electron-depth profile, determined from the HREELS simulations for InN, is presented in Fig. 12.3. A maximum electron density of $\sim 2.8 \times 10^{20}$ cm^{-3} occurs in the near surface, declining to the bulk carrier concentration of

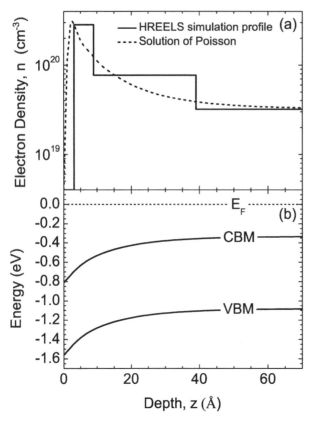

FIGURE 12.3

(a) The layered charge profile used in the HREELS simulations (solid line) and the corresponding smooth charge profile, $n(z)$, calculated by solving the Poisson equation within the MTFA (dotted line). (b) The band bending profile, showing the CBM and VBM (solid lines) with respect to the Fermi level (dashed line).

3.2×10^{19} cm^{-3}. This analysis clearly confirms the presence of an intrinsic electron accumulation layer on the clean InN surface.

Realistic smooth charge profiles were calculated to determine the surface state density, the band bending and the position of the Fermi level at the surface. The smooth charge profile that most closely resembles the HREELS simulation profile is shown in Fig. 12.2, along with the corresponding band bending profile. The charge profiles were calculated by solving the Poisson equation within the modified Thomas-Fermi approximation (MTFA) (described in detail in section 12.2.3). The carrier concentration as a function of depth, $n(z)$, depends on the local Fermi level which is determined by the bulk Fermi level and the value of the band bending potential, $V(z)$, which, in turn, is given by the solution of the Poisson equation. This charge profile calculation yields a surface state density, N_{ss}, of $\sim 2.5 \pm 0.2 \times 10^{13}$ cm^{-2} and band bending, V_{bb}, of ~ 0.56 eV. As a result of this band bending, the surface Fermi level, E_{Fs}, is located $\sim 1.64 \pm 0.10$ eV above the valence band maximum (VBM).

12.2.2.2 High-resolution x-ray photoemission spectroscopy

In addition to using the variation with depth of the conduction band electron plasma frequency in the near-surface region, the space-charge properties of the surface of a semiconductor can also be determined by comparing the surface and bulk Fermi levels. The surface Fermi level can be directly measured by XPS. The bulk Fermi can be derived from the Hall effect free electron density or infrared reflectivity-determined bulk plasma frequency using semiconductor statistics or, for the case of degenerate doping, more directly estimated from optical absorption spectra.

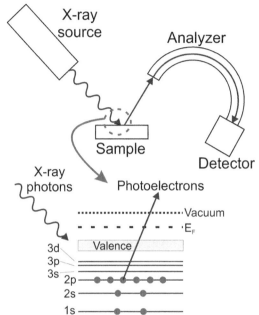

FIGURE 12.4
Schematic of an x-ray photoemission experiment. Monochromated x-rays impinge upon the sample, photoelectrons are emitted from core-levels or the valence band and those that are not prevented from leaving the sample by inelastic scattering are energy analyzed.

XPS analysis was performed using a Scienta ESCA300 spectrometer at the National Centre for Electron Spectroscopy and Surface analysis, Daresbury Laboratory, UK. This consists of a rotating anode Al-K$_\alpha$ x-ray source (hν = 1486.6 eV), x-ray monochromator and 300 mm mean radius spherical-sector electron energy analyzer with parallel electron detection system. The analyzer was operated with 0.8 mm slits and at a pass energy of 150 eV. Gaussian convolution of the analyzer broadening with an effective line width of 0.27 eV for the x-ray source gives an effective instrument resolution of 0.45 eV. The Fermi level position (zero of the binding energy scale) was calibrated using the Fermi edge of an ion bombarded silver reference sample. The XPS technique is illustrated schematically in Fig. 12.4. The valence band maximum (VBM) is determined by linearly extrapolating the leading edge of the valence band to the baseline in order to account for the instrumental resolution-induced tail [12.27]. While the x-rays penetrate deep into the sample, the surface sensitivity of XPS is due to the inelastic scattering of the photoelectrons. Since the mean free path, λ, of electrons with kinetic energies of ~1485 eV is ~25 Å in InN, XPS is higly surface sensitive. For XPS using the normal emission geometry, the probability, P, of a photoelectron originating from a depth, d, escaping from the sample to be energy analyzed diminishes exponentially according to P = exp($-d/\lambda$).

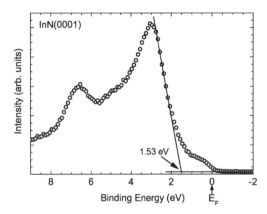

FIGURE 12.5

The valence band XPS spectrum of InN in the region of the valence band maximum. The binding energy scale is with respect to the Fermi level, E_F. The valence band maximum occurs at the intersection of a linear fit to the linear portion of the leading edge and the background. The valence band maximum is estimated to lie 1.53±0.10 eV below the surface Fermi level.

A photoemission spectrum in the region of the valence band maximum from a 350-nm-thick atomic hydrogen cleaned InN(0001) film is shown in Fig. 12.5(a). The linear extrapolation of the leading edge of the valence band emission and the baseline are also shown. The intersection of these two lines indicates that the VBM is at 1.53 eV below the surface Fermi level. Since InN has a band gap of ∼0.64 eV at room temperature, this indicates that the surface Fermi level is ∼0.89 eV above the conduction band minimum (CBM). The bulk Fermi level can be estimated from the transport properties of the film. From single-field Hall effect measurements, the average electron concentration and mobility of the film are 1.9×10^{18} cm^{-3} and 1030 cm^{2}V^{-1}s^{-1}, respectively. The results of **k·p** carrier statistics calculations (as outlined in section 12.2.3) show that this electron density corresponds to a bulk Fermi level of ∼0.10 eV above the conduction band minimum and 0.74 eV above the valence band maximum. Therefore, the XPS measurement indicates that the surface Fermi level is 1.53 − 0.74 = 0.79 eV above the bulk Fermi level. The resulting band bending profile and electron concentration versus depth curve, calculated by solving the Poisson equation within the MTFA (see section 12.2.3), are shown in Fig. 12.6. The density of positively charged surface states determined from the calculation is 1.6×10^{13} cm^{-2}, the same as the value determined from C-V profiling [12.4] and similar to the value of 2.5×10^{13} cm^{-2} determined from the HREELS measurements. To maintain charge neutrality, this density of surface states is compensated by an equal density of conduction electrons in the accumulation layer.

12.2.3   Surface space-charge calculations: solution of the Poisson equation

Semiconductor surface space-charge calculations provide a link between the band bending profile and the resulting carrier density profile. As illustrated in section 12.2.2.1, such calculations can be used to obtain the surface state density, the band bending and the position of the Fermi level at the surface from HREELS measurements. Similarly, as shown in section 12.2.2.2, knowledge of the surface Fermi level from XPS and an estimated bulk electron density from Hall effect or optical measurements enables the carrier density as a function of depth to be

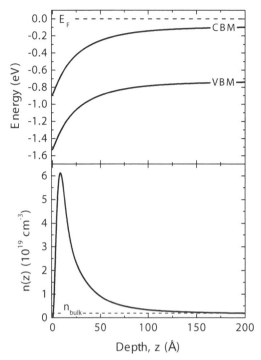

FIGURE 12.6

The surface space-charge layer properties, calculated by solving the Poisson equation within the modified Thomas-Fermi approximation. The XPS-determined surface Fermi level value of 1.53 eV above the VBM was used, along with the Hall effect-measured bulk electron density of 1.9×10^{18} cm^{-3}. The band bending profile, showing the CBM and VBM with respect to the Fermi level, and the free electron concentration are plotted as a function of depth below the surface.

calcualted. The method employed here, solving the Poisson equation within the MTFA [12.24, 12.28], takes account of the surface potential barrier reducing the carrier density to zero at the surface. Importantly, this approach also allows the conduction band nonparabolicity to be incorporated straightforwardly without requiring full self-consistent solutions of the Schrödinger and Poisson equations. In this section, the equations to be solved in order to calculate surface space-charge profiles are described. This Poisson-MTFA formalism is subsequently used in section 12.2.5 as the first stage of nonparabolic coupled Poisson-Schrödinger calculations of the subband energies and dispersions in quantized electron accumulation layers.

The starting point for calculating surface space charge layer properties is the Poisson equation. The one-electron potential, $V(z)$, that describes the band bending as a function of depth, z, below the semiconductor surface satisfies the Poisson equation

$$\frac{d^2 V(z)}{dz^2} = -\frac{e}{\varepsilon(0)\varepsilon_0} [N_D^+ - N_A^- - n(z) + p(z)], \qquad (12.1)$$

where $\varepsilon(0)$ is the static dielectric constant, N_D^+ (N_A^-) is the bulk donor (acceptor) concentration and $n(z)$ ($p(z)$) is the electron (hole) concentration at depth, z, below the surface. The potential must also satisfy the following boundary conditions

$$V(z) \to 0 \text{ as } z \to \infty, \qquad (12.2)$$

and

$$\left.\frac{dV}{dz}\right|_{z=0} = \frac{e}{\varepsilon(0)\varepsilon_0}N_{ss} \tag{12.3}$$

where N_{ss} is the surface state density.

The electron and hole concentrations in the Poisson equation are given by

$$n(z) = \int_0^\infty g_c(E) f_{FD} f(z) dE, \tag{12.4}$$

and

$$p(z) = \sum_i \int_{E_{v_i}}^{-\infty} g_{v_i}(E)[1 - f_{FD}(E)] f(z) dE, \tag{12.5}$$

where the sum over i denotes the sum over a number of valence bands, $g_c(E)$ [$g_{v_i}(E)$] is the density of states for the nonparabolic conduction band [i^{th} valence band], f_{FD} is the Fermi-Dirac function, including the dependence on the potential

$$f_{FD}(E) = \frac{1}{1 + \exp\{[E - E_F + V(z)]/k_B T\}}, \tag{12.6}$$

where E_F is the Fermi level, k_B is Boltzmann's constant and T is temperature, and $f(z)$ is the MTFA factor to account for the interference between incident and reflected electron waves at the surface potential barrier

$$f(z) = 1 - \text{sinc}\left[\frac{2z}{L}\left(\frac{E}{k_B T}\right)^{1/2}\left(1 + \frac{E}{E_g}\right)^{1/2}\right], \tag{12.7}$$

where, if the semiconductor is nondegenerately doped in the bulk, L is the thermal length $L = \hbar/(2m_0^* k_B T)^{1/2}$ and, if degenerately doped, $L = 1/k_F$ is the Fermi length, where k_F is the Fermi wavevector.

While the density of states for each valence band in Eq. 12.5 takes the well known form for a parabolic band here, the expression for the density of states for the conduction band in Eq. 12.4 requires further description due to the effects of nonparabolicity. The nonparabolic conduction band dispersion can be approximated using Kane's **k·p** model [12.25], described by the Hamiltonian, $\mathcal{H}$, given by

$$\mathcal{H} = \begin{bmatrix} \tilde{\mathcal{H}} & 0 \\ 0 & \tilde{\mathcal{H}} \end{bmatrix}, \tag{12.8}$$

where

$$\tilde{\mathcal{H}} = \begin{bmatrix} E_s & 0 & kP & 0 \\ 0 & E_p - \Delta_{so}/3 & \sqrt{2}\Delta_{so}/3 & 0 \\ kP & \sqrt{2}\Delta_{so}/3 & E_p & 0 \\ 0 & 0 & 0 & E_p + \Delta_{so}/3 \end{bmatrix}, \tag{12.9}$$

where E_s (E_p) is the conduction (valence) band edge energy, Δ_{so} is the spin-orbit splitting, k is the wavevector, and P is Kane's matrix element. Here, for InN, the simplifications of zero anisotropy and zero crystal field splitting have been made because the crystal field splitting [12.29] and anisotropy of the electron effective mass [12.30] have both been found to be small. The conduction band dispersion is therefore given, to a reasonable approximation, by

$$E_c(k) = E' + E_k, \tag{12.10}$$

where

$$E_k = \frac{\hbar^2 k^2}{2m_0}, \tag{12.11}$$

where m_0 is the free electron mass and E' is the largest eigenvalue of the Hamiltonian [Eqs. 12.8 and 12.9] given by the largest solution of

$$E'(E' + E_g)(E' + E_g + \Delta_{so}) - k^2 P^2(E' + E_g + 2\Delta_{so}/3) = 0, \tag{12.12}$$

where $E_s = 0$ and $E_p = -E_g - \Delta_{so}/3$ have been used in order to define the zero of energy at the conduction band minimum. Kane's matrix element is given by

$$P^2 = \frac{3\hbar^2(1/m_0^* - 1/m_0)}{2[2/E_g + 1/(E_g + \Delta_{so})]}, \tag{12.13}$$

where m_0^* is the conduction band minimum effective mass. Since $\Delta_{so} \ll E_g$, as a simplifying approximation, the spin-orbit splitting can be set to zero, so that the solution of Eq. 12.13 takes the two band $\mathbf{k} \cdot \mathbf{p}$ analytical form for the conduction band dispersion

$$E_c(k) = \frac{1}{2} \left[-E_g + \sqrt{E_g^2 + 4k^2 P^2} \right] + E_k, \tag{12.14}$$

with Kane's matrix element simplifying to

$$P^2 = \frac{\hbar^2}{2m_0} \left(\frac{m_0}{m_0^*} - 1 \right) E_g, \tag{12.15}$$

and the density of states is given by

$$g_c(k) = \frac{k^2}{\pi^2} \left[\frac{dE_c(k)}{dk} \right]^{-1} = \frac{k/\pi^2}{4P^2[E_g^2 + 4k^2 P^2]^{-1/2} + (\hbar^2/m_0)}. \tag{12.16}$$

Having established all of the expressions required to determine the various quantities contained in the Poisson equation, the procedure for solving it can be described. Solution of Eqs. 12.1, 12.4 and 12.5, for a given bulk Fermi level (or equivalently the bulk carrier density) and surface state density, proceeds by choosing a trial amount of band bending and then numerically integrating the Poisson equation to determine whether the boundary conditions (Eqs. 12.2 and 12.3) are satisfied. Interval bisection is employed to converge to the value of the band bending that does satisfy the boundary conditions. This allows the band bending potential to be obtained throughout the space charge region, enabling the surface space-charge profiles to be determined (as shown in Figs. 12.3 and 12.6). Additionally, rather than starting with a given surface state density and finding the corresponding amount of band bending at the surface, it is equivalent to start with a given amount of band bending (the separation between the bulk and surface Fermi levels) and determine the associated surface state density by solving the Poisson equation.

12.2.4   Universal electron accumulation

Having established that electron accumulation is an inherent property of clean InN(0001) surfaces, the question naturally arises as to whether the phenomenon is present, and to what extent, for other wurtzite InN surface orientations and for the zinc-blende polymorph. Here, wurtzite non-polar a-plane, both In- and N-polarity c-plane and zinc-blende (001) surfaces are investigated and the universal nature of the electron accumulation at InN surfaces is demonstrated [12.31]. All samples were grown by plasma-assisted molecular beam epitaxy. The c-plane wurtzite samples were grown on c-plane sapphire substrates, with the In- and N-polarity [(0001) and (000$\bar{1}$), respectively] samples incorporating a GaN/AlN and low-temperature InN buffer layer, respectively. The a-plane (11$\bar{2}$0) sample was grown on an r-plane (1$\bar{1}$02) sapphire substrate incorporating a GaN/AlN buffer layer. The (001) zinc-blende sample was grown on a (001) 3C-SiC substrate incorporating a zinc-blende GaN buffer layer, resulting in an estimated 95% zinc-blende phase InN film [12.32].

12.2.4.1   Wurtzite InN

The surface Fermi level of each atomic hydrogen cleaned sample was determined from XPS analysis and the bulk Fermi level was estimated from single-field Hall effect measurements. The position of the VBM at the surface is often calculated by extrapolating a linear fit to the leading edge of the valence band photoemission, as descibed above in section 12.2.2.2. Here, however, an alternative method is demonstrated, whereby the surface Fermi level to VBM separation is determined from the shift in energy required to align the valence band photoemission spectra with quasiparticle-corrected density functional theory (QPC-DFT) calculations (using the HSE03 hybrid exchange-correlation potential) of the valence band density of states (VB-DOS) which, by convention, have the VBM at 0 eV. Details of the calculations are reported in Chapter 8. For comparison with the experimental results, the QPC-DFT is broadened by a 0.2 eV full width at half maximum (FWHM) Lorentzian and a 0.45 eV FWHM Gaussian to account for lifetime and instrumental broadening, respectively.

The valence band photoemission spectra for the wurtzite samples are shown, with the QPC-DFT VB-DOS calculations, in Fig. 12.7. All the photoemission spectra are coincident in energy, indicating the same VBM to surface Fermi level separation. This is determined, from the shift in peak position compared to the QPC-DFT calculations, to be 1.53±0.10 eV for all three samples. This value is the same as that determined for InN(0001) using the linear extrapolation method in section 12.2.2.2. The differences in intensity of the ~3 eV peak between the samples have been shown to be a signature of the film orientation [12.33].

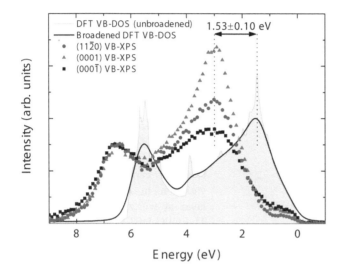

FIGURE 12.7

Valence band photoemission spectra (points) for $(11\bar{2}0)$, (0001), and $(000\bar{1})$ InN surfaces, relative to the Fermi level. QPC-DFT VB-DOS is shown without (shaded) and with lifetime and instrumental broadening. The indicated shift in peak position gives the VBM to surface Fermi level separation. Reprinted with permission from P. D. C. King, *et al.*, Applied Physics Letters, 91 (2007) 092101. Copyright 2007, American Institute of Physics.

The position of the surface Fermi level, well above the CBM, indicates a downward band bending relative to the Fermi level at the surface, leading to electron accumulation for all wurtzite samples measured. To quantify this further, the band bending profile and corresponding accumulation of electrons in the surface space-charge region was determined by solving Poisson's equation numerically within the MTFA, following the method described in section 12.2.3. The profiles are shown in Fig. 12.8, and the relevant parameters listed in Table 12.1.

TABLE 12.1

Bulk electron concetration, n, determined from single-field Hall effect measurements, and corresponding bulk Fermi level above the CBM, E_{F_b}, calculated using Fermi-Dirac carrier statistics, and the **k·p** Kane model to account for conduction band nonparabolicity. The band bending, V_{bb}, is calculated from the difference between the surface and bulk Fermi levels. Poisson-MTFA calculations give the surface state density, N_{ss}.

Orientation	n (cm^{-3})	E_{F_b} (eV)	V_{bb} (eV)	N_{ss} cm^{-2}
$(11\bar{2}0)$	4.8×10^{18}	0.164	−0.725	1.66×10^{13}
(0001)	3.0×10^{18}	0.124	−0.765	1.64×10^{13}
$(000\bar{1})$	6.7×10^{18}	0.200	−0.689	1.65×10^{13}

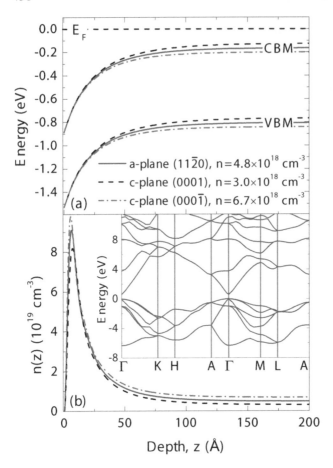

FIGURE 12.8
(a) Band bending relative to the Fermi level and (b) the resulting carrier concentration variation in the accumulation layer at wurtzite InN surfaces. The QPC-DFT bulk band structure is shown in the inset. Reprinted with permission from P. D. C. King, *et al.*, Applied Physics Letters, 91 (2007) 092101. Copyright 2007, American Institute of Physics.

Despite differences in bulk Fermi level positions, the pinning of the surface Fermi level at the same energy for *a*-plane and both polarities of *c*-plane InN means that the band bending close to the surface is very similar [Fig. 12.8(a)], resulting in similar near-surface charge profiles [Fig. 12.8(b)]. Indeed, the calculated surface state density is the same for all samples [Table 12.1], indicating the universality of the electron accumulation at wurtzite InN surfaces.

In addition to this universality of the electron accumulation at the clean surfaces of different orientation InN films, evidence of very similar electron accumulation has been observed at the oxidized surface of a wide variety of *c*-plane InN samples grown by different methods in several different laboratories. The valence band XPS spectra from both In-polarity and N-polarity InN samples grown by MBE at Canterbury and Ritsumeikan Universities are shown in Fig. 12.9, along with that from a further In-polarity sample grown by MBE at Cornell University and a N-polarity sample grown by MOVPE at Penn State University. The surface Fermi level is found to be 1.40 ± 0.10 eV above the VBM for all six samples. Since this value is significantly higher than the bulk Fermi level in these samples, these results confirm that InN surfaces oxidized by exposure to the atmosphere exhibit electron accumulation. The fact that the surface Fermi level at oxidized surfaces is slightly lower than the value of 1.53 eV measured at clean InN surfaces suggests that the presence of electronegative oxygen at the surface may, via charge transfer, act to reduce the amount of electron accumulation at the InN surfaces. This appears to contradict the interpretation presented in Chapter 5, where it was suggested that surface oxygen atoms may be the source of the electrons in the accumulation layer for air-exposed samples. However, oxygen atoms may exhibit different electrical behavior in different lattice sites.

These valence band XPS results on samples from different laboratories grown using both MBE and MOVPE indicate that the existence and extent of the surface electron accumulation are not sensitive to the details of the growth method. This is further confirmed by our studies of InN samples grown at University of California at Santa Barbara (UCSB) which demonstrate that the surface electron accumulation is invariant for In- and N-polarity InN samples grown by MBE under different In:N ratios, ranging from In-rich to N-rich [12.34].

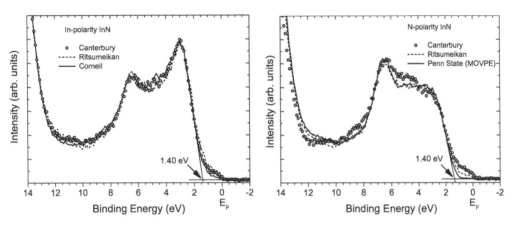

FIGURE 12.9

Valence band XPS spectra from In-polarity InN samples grown by MBE at Canterbury, Ritsumeikan and Cornell Universities and from N-polarity InN samples grown by MBE at Canterbury and Ritsumeikan Universities and by MOVPE at Penn State University. All six samples exhibit a surface Fermi level 1.40±0.10 eV above the VBM, indicating that they have similar surface electron accumulation.

12.2.4.2   Zinc-blende InN

The valence band photoemission is shown for zinc-blende InN(001) in Fig. 12.10(a). The density of states is somewhat different for zinc-blende to wurtzite polymorphs, and the detailed agreement of the XPS with the QPC electronic structure calculations is described in Chapter 8 and in Ref. [12.35]. Alignment of the experimental and calculated VB-DOS is obtained when the VBM to surface Fermi level separation is set to a value of 1.38±0.10 eV. Poisson-MTFA calculations for InN(001), using a room temperature band gap of 0.56 eV [12.32] and a band-edge electron effective mass of $0.039m_0$ (based on the empirical relation $m_0^* \sim 0.07E_g$ [12.36]), are shown in Fig. 12.10. Due to growth on a conducting substrate, single-field Hall effect measurements could not be used to determine the carrier concentration in the InN layer. The bulk Fermi level was therefore estimated from the imaginary part of the dielectric function (determined by spectroscopic ellipsometry as described in Chapter 9), shown in the inset of Fig. 12.10, to be 1.05 eV above the VBM, corresponding to a bulk electron concentration (from carrier statistics) of 3.2×10^{19} cm^{-3}. Despite the higher bulk carrier density than in the wurtzite samples, a distinct electron accumulation is still observed, although the Fermi level appears to pin slightly lower above the VBM and the surface state density from Poisson-MTFA calculations (N_{ss}=9.11 $\times$ 10^{12} cm^{-2}) is somewhat lower for the zinc-blende than the wurtzite case.

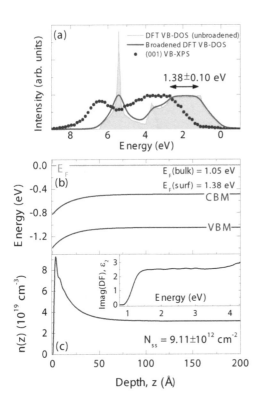

FIGURE 12.10

(a) Valence band photoemission spectrum from InN(001) (points) and zinc-blende QPC-DFT VB-DOS shown without (shaded) and with lifetime and instrumental broadening (solid black line). Poisson-MTFA calculations yield (b) band bending and (c) carrier concentration profiles. The bulk Fermi level was determined from the imaginary part of the dielectric function, measured by spectroscopic ellipsometry (shown in the inset). Reprinted with permission from P. D. C. King, *et al.*, Applied Physics Letters, 91 (2007) 092101. Copyright 2007, American Institute of Physics.

12.2.5   Quantized electron subbands in the accumulation layer

The presence of such extreme downward band bending at InN surfaces over a depth of a few nanometers immediately leads to the expectation that the electrons in the accumulation layer will be quantized in the deep and narrow surface potential well. Indeed, initial evidence of the quantized conduction electron subbands in the InN surface accumulation layer has been obtained by both electron tunneling spectroscopy [12.37, 12.38] and angle-resolved photoemission spectroscopy [12.39] of InN. Here, a methodology for solving the Schrödinger equation for the Poisson-MTFA-determined one-electron potential and a nonparabolic conduction band is presented. The energies of the subband minima and the subband dispersions are determined and compared with the electron tunneling and angle-resolved photoemission data.

12.2.5.1 Solutions of the Schrödinger equation for quantized accumulation layers

The numerical solution of the Schrödinger equation for the one-electron potential determined by the solution of the Poisson-MTFA calculations (see section 12.2.3) proceeds via a Fourier-series representation [12.26, 12.40, 12.41]. The one-electron (band bending) potential breaks the translational symmetry of the crystal. It is therefore appropriate to express the Schrödinger equation in terms of envelope functions made up of Wannier functions $\Psi(\mathbf{r}_\parallel, z)$, where $\mathbf{r}_\parallel(z)$ is the parallel (normal) component of the position vector. In this representation, the Schrödinger equation is given by [12.41, 12.42]

$$[E_c(-i\nabla) + V(z)]\Psi(\mathbf{r}_\parallel, z) = E\Psi(\mathbf{r}_\parallel, z), \tag{12.17}$$

where the eigenfunction for a subband j and a given parallel wavevector $\mathbf{k}_\parallel$ is

$$\Psi_{\mathbf{k}_\parallel, j}(\mathbf{r}_\parallel, z) = \frac{1}{\sqrt{A}} \exp(i\mathbf{k}_\parallel \cdot \mathbf{r}_\parallel) \psi_{\mathbf{k}_\parallel, j}(z), \tag{12.18}$$

where A is a normalization factor and $\psi_{\mathbf{k}_\parallel, j}(z)$ is the component of the eigenfunction normal to the surface for a given subband and parallel wavevector.

Imposing the boundary condition that the wave functions vanish at the surface and assuming a system of length ℓ, such that the wave functions decay to zero by $z = \ell$, $\psi_{\mathbf{k}_\parallel, j}(z)$ can be expanded as a Fourier sine series

$$\psi_{\mathbf{k}_\parallel, j}(z) = \sum_{\nu=1}^{\infty} \sqrt{\frac{2}{\ell}} a_\nu^{\mathbf{k}_\parallel, j} \sin\left(\frac{\nu\pi}{\ell} z\right). \tag{12.19}$$

Substituting this into the Schrödinger equation (Eq. 12.17) gives the matrix representation of the problem for a given $\mathbf{k}_\parallel$,

$$\mathbf{M}^{\mathbf{k}_\parallel} \mathbf{a}^{\mathbf{k}_\parallel} = E^{\mathbf{k}_\parallel} \mathbf{a}^{\mathbf{k}_\parallel}, \tag{12.20}$$

where the matrix elements are given by

$$[\mathbf{M}]_{\nu\nu'} = E_c(k_\nu)\delta_{\nu\nu'} + \frac{2}{\ell} \int_0^\ell V(z)\sin\left(\frac{\nu\pi z}{\ell}\right)\sin\left(\frac{\nu'\pi z}{\ell}\right) dz, \tag{12.21}$$

where $k_\nu = \sqrt{\mathbf{k}_\parallel^2 + (\nu\pi/\ell)^2}$ and $\delta_{\nu\nu'}$ is the Kronecker delta function. The eigenvalues and eigenfunctions of $\mathbf{M}$ can therefore be used to determine the subband energies and wave functions confined normal to the surface for a particular one-electron potential given by the Poisson-MTFA solution. In practice, the infinite sum in Eq. 12.19 is truncated at $v = v_{max}$, giving a ($v_{max} \times v_{max}$) matrix in Eqs. 12.20 and 12.21. Setting v_{max} to a value of 500 was found to be sufficient to ensure convergence for all cases considered here.

This method has been used to determine the one-electron potential, carrier concentration profile, quantized subband state energies, and parallel dispersion relations for an InN accumulation layer [12.26]. Solutions of the coupled Poisson and Schrödinger equations are shown in Fig. 12.11(a) for a 'typical' bulk electron density of $n_b = 2.5 \times 10^{18}$ cm^{-3} and the 'universal' surface state density described in section 12.2.4 of 1.6×10^{13} cm^{-2}. Two quantized conduction electron subband states are obtained for the current example. The normal components of the eigenfunctions are shown superimposed on the subbands in Fig. 12.11. The wave functions are equal to zero at the surface (z=0) due to the imposed boundary condition. Also, the wave functions decay into the potential barrier, such that a long way from the surface, they have zero amplitude. It is the peak of these wave functions close to the surface that is responsible for the peak in the electron density in the accumulation layer [Fig. 12.11(b)].

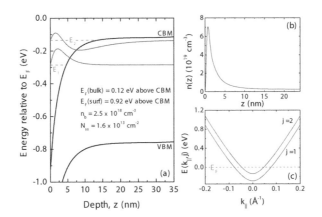

FIGURE 12.11

(a) Downward band bending at an InN surface with a surface sheet density of $N_{ss} = 1.6 \times 10^{13}$ cm^{-2} and bulk carrier density of $n_b = 2.5 \times 10^{18}$ cm^{-3} and (b) corresponding near-surface carrier density. Two quantized states form (E_1 and E_2) and their corresponding (a) wave functions and (c) parallel dispersion relations are shown. Reprinted with permission from P. D. C. King, T. D. Veal, and C. F. McConville, Physical Review B, 77 (2008) 125305. Copyright 2008 by the American Physical Society.

The parallel dispersion of the subbands is also obtained from the model, as shown in Fig. 12.11(c). These dispersions become rather linear with increasing $k_{\parallel}$, indicating a distinct nonparabolicity. However, this nonparabolicity is not simply described by a Kane-like dispersion. Instead, the full numerical solution of the Schrödinger equation for different values of $k_\parallel$ is required.

Although the Fermi level is strongly pinned at clean InN surfaces, it may be possible to induce changes in the pinning position in a number of ways. As described in section 12.2.6, a large increase in the bulk doping level has been observed to increase the Fermi level pinning position, in order to maintain charge neutrality [12.43]. The changes in the measured conductivity of InN when exposed to a number of solvents and gases have been attributed to changes in the surface electronic properties [12.44–12.46]. Additionally, the deposition on the surface of metals of varying electronegativity can cause a variation in the Fermi level pinning position [12.47]. The variation in surface state density and corresponding downward band bending associated with these modifications to the clean surface will cause a pronounced variation in the number and confinement energies of the subbands.

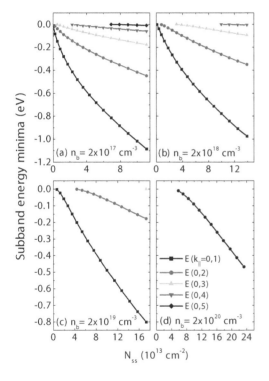

FIGURE 12.12

Variation in energy of the subband minima relative to the bulk CBM (top of the potential well) with surface state density for bulk carrier concentrations of (a) 2×10^{17} cm^{-3}, (b) 2×10^{18} cm^{-3}, (c) 2×10^{19} cm^{-3}, and (d) 2×10^{20} cm^{-3}. The points show the calculated subband energy minima for a downward band bending (right to left) of 3.3, 3.1, 2.9, 2.7, ..., eV. Reprinted with permission from P. D. C. King, T. D. Veal, and C. F. McConville, Physical Review B, 77 (2008) 125305. Copyright 2008 by the American Physical Society.

This is illustrated by the variation in subband energy minima for a variety of surface state densities and bulk carrier densities shown in Fig. 12.12. With increasing surface state density, the amount of downward band bending increases, and so the potential well becomes deeper at the surface. This causes the number of subbands bound within the well to increase and the subbands to become confined deeper within the well. Conversely, with increasing bulk doping, fewer subbands are confined for a given amount of band bending. As the bulk doping level increases, the screening length becomes shorter, and so the potential well formed at the surface becomes narrower. This acts to increase the energy of the subband minima above the conduction band edge at the surface, leading to fewer bound subbands for a given amount of downward band bending. By control of both the surface Fermi level pinning position and the bulk doping, it is therefore possible to control the number and binding energies of the subbands within the potential well.

12.2.5.2 Electron tunneling spectroscopy

Electron tunneling spectroscopy using a Pt-Ir tip in contact with a native oxide at the InN surface has been employed to investigate the binding energies of the subbands in the electron accumulation layers at InN and In-rich InGaN surfaces [12.37, 12.38]. Dips in the normalized conductance $(dI/dV)/(I/V)$, where I is the tunneling current and V the applied bias voltage, were attributed to tunneling into subband states. The normalized conductance for a sample with a bulk carrier concentration of $n = 2.5 \times 10^{18}$ cm^{-3} is shown in Fig. 12.13. The energy of the subbands can be compared with those calculated for a 'typical' InN accumulation layer with the same bulk carrier density in Fig. 12.11.

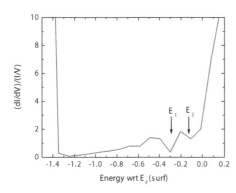

FIGURE 12.13

Normalized conductance from electron tunneling spectroscopy of an InN surface with a bulk carrier concentration of $n = 2.5 \times 10^{18}$ cm^{-3}. The energy is referenced to the surface Fermi level. The arrows show the calculated subband energy minima for a typical InN accumulation layer with this bulk carrier concentration (shown in Fig. 12.11). Reprinted with permission from P. D. C. King, T. D. Veal, and C. F. McConville, Physical Review B, 77 (2008) 125305. Copyright 2008 by the American Physical Society.

The energetic positions of the calculated subband minima for this accumulation layer profile are shown as vertical arrows in Fig. 12.13. These agree well with the subband features in the normalized conductance within the (rather low) resolution of the experimental data. In order to make a more quantitative comparison, however, higher resolution experimental data would be needed. Also, perturbing effects, such as tip-induced band bending [12.48], when a bias is applied and the modification of the surface Fermi level pinning at an oxidized surface may need to be addressed. These very preliminary electron tunneling spectroscopy results demonstrate that in vacuo scanning tunneling spectroscopy may be a valuable approach to elucidate the properties of the quantized electron accumulation at InN surfaces in the future.

12.2.5.3 Angle-resolved photoemission spectroscopy

The subband states at InN surfaces have also been investigated using high resolution angle-resolved photoemission spectroscopy (ARPES) [12.39]. Quantized states were observed at the surface, and the in-plane dispersion of these states was determined from photocurrent intensity maps as a function of parallel wavevector and binding energy. The photocurrent intensity map along the Σ ($\Gamma \rightarrow M$) direction for binding energies up to 1.2 eV below the Fermi level for an InN sample prepared by two 10 min cycles of 500 eV Ar^+ ion bombardment and annealing at 300°C is shown in Fig. 12.14.

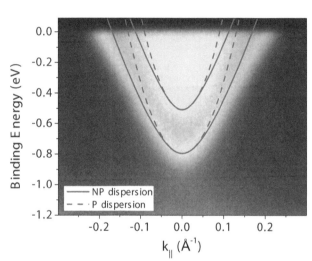

FIGURE 12.14

ARPES photocurrent intensity map of the parallel dispersion (along Σ ($\Gamma \rightarrow M$) in the surface plane) of two quantized subbands in an InN electron accumulation layer. The data was recorded at 60 K using a photon energy of 70 eV and the Fermi level is at 0 eV. Simulated nonparabolic (NP) and parabolic (P) subband dispersions for InN with a surface state density of $N_{ss} = 8.1 \times 10^{13}$ cm^{-2} and bulk density of $n_b = 3.7 \times 10^{19}$ cm^{-3} corresponding to a downward band bending of 1.8 eV are shown (solid line). Reprinted with permission from P. D. C. King, T. D. Veal, and C. F. McConville, Physical Review B, 77 (2008) 125305. Copyright 2008 by the American Physical Society.

The two dispersions observed in the photocurrent map have been attributed to the parallel dispersion of two subbands. Simulations for InN with a surface state density of $N_{ss} = 8.1 \times 10^{13}$ cm^{-2} and bulk density of $n_b = 3.7 \times 10^{19}$ cm^{-3}, corresponding to a downward band bending of 1.8 eV, result in two confined subbands with their minima located 0.80 and 0.51 eV below the Fermi level at the surface, in agreement with the experimental results.

The dispersions resulting from these calculations are shown relative to the experimental dispersions in Fig. 12.14. At low parallel wavevector, there is good agreement between the calculated and experimental dispersions, although these diverge somewhat at higher wavevector. This may be due to small errors in the bulk conduction band edge effective mass used, failure of the Kane **k·p** bulk dispersion relations at higher wave vector, neglecting the small anisotropy of the conduction band dispersion in wurtzite InN, or the fact that many body effects are not included.

Despite the slight divergence of the calculated and experimental dispersions at high parallel wave vectors, a distinct nonparabolicity is evident in both the experimental and model dispersions. The subband minimum momentum effective masses, determined from the calculated subband dispersions, are $0.080m_0$ and $0.055m_0$ for the first (E_1) and second (E_2) subbands, respectively. The equivalent parabolic dispersion for these effective masses is also shown in Fig. 12.14, indicating the importance of including nonparabolicity in the calculations.

It should be noted that, while the surface state density required to reproduce the experimental data is within the range of previously measured experimental values, it is somewhat higher than those typically derived from XPS data from clean InN surfaces. The reasons for this may be due to the surface preparation treatment, which involved Ar^+ ion bombardment and annealing at 300°C. This may lead to surface In enrichment [12.49] or the introduction of near-surface donor defects [12.50]. Additionally, many body effects, not accounted for in the modeling, may also be responsible for the apparent discrepancy between the sheet densities derived from XPS and ARPES.

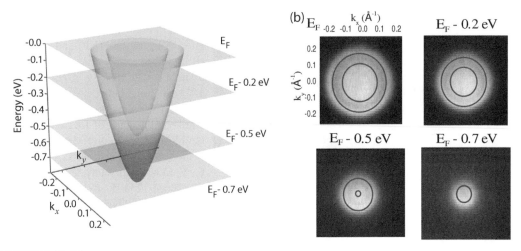

FIGURE 12.15

(a) Calculated two-dimensional subbands in the quantized accumulation layer, with slices through at the Fermi level, E_F, and at $E_F - 0.2$ eV, $E_F - 0.5$ eV and $E_F - 0.7$ eV, corresponding to (b) the experimental (photoemission maps, recorded at 60 K using a photon energy of 70 eV) and calculated (solid lines) constant energy contours.

As the confining potential in the accumulation layer is only in the direction normal to the surface, the subbands that form are two-dimensional. The calculated subband dispersion is shown, as a function of both k_x and k_y, in Fig. 12.15. The two-dimensional nature of the subbands was further investigated with ARPES by performing constant energy cuts through the subbands at the Fermi level and at 0.2, 0.5 and 0.7 eV below the Fermi level. Here, the binding energy was kept constant relative to the Fermi level, while k_x and k_y were simultaneously varied. These constant energy contours are shown in Fig. 12.15. Approximately circular contours were obtained, which decrease in radius with increasing binding energy below the Fermi level, consistent with the calculated contours. The approximately circular nature of the contours reveals no significant anisotropy in the conduction band dispersion. At the Fermi level, the constant energy contour is analogous to the Fermi surface of a metal. Using the simple relation

$$N_{2D} = \frac{k_F^2}{2\pi},$$ (12.22)

the areal density of the subband states can be estimated from the experimental contours. The radii (Fermi wavevector, k_F) of the constant energy contours at the Fermi level are estimated to be ~ 0.1 Å^{-1} and ~ 0.2 Å^{-1} for the first and second band, respectively, consistent with the Fermi wavevector of the subband dispersions shown in Fig. 12.14, corresponding to an areal density of 1.6×10^{13} cm^{-2} and 6.4×10^{13} cm^{-2}, respectively. Thus, the total areal density of the electrons in the subband states is approximately 8.0×10^{13} cm^{-2}, in agreement with the concentration required to neutralize the density of positively charged surface states determined from the coupled Poisson-Schrödinger calculations.

12.2.6 Origin of electron accumulation – the low Γ-point CBM

The observed electron accumulation at the surface of n-type InN is due to the presence of positively charged donor-type surface states. The existence of such surface states requires that the following conditions are satisfied. First, in order to have predominantly donor-type character, the surface states must lie below the branch-point energy or charge neutrality level, E_{CNL}. This is the cross-over point from states higher in the gap that are mainly of conduction band character (acceptor-type) to states lower in energy that are mainly of valence band character (donor-type) [12.51, 12.52]. This charge neutrality level falls close to the center of the band gap in the complex band structure [12.53]. Second, the surface states must be at least partly above the Fermi level, since valence band (donor-type) states are positively charged when unoccupied and neutral when occupied. The surface Fermi level can be pinned above the Γ-point CBM by unoccupied, positively charged donor-type surface states. These donors acquire a positive surface charge by donating electrons into the conduction band. This results in downward band bending and electron accumulation. This combined requirement that the surface states are ionized donors and lie above the Fermi level can only be achieved in n-type semiconductors when the Γ-point CBM lies significantly below E_{CNL}. For the Γ-point CBM to lie below the average mid-gap energy of the entire Brillouin zone, it must be particularly low in energy relative to the rest of the conduction band. The wurtzite InN band structure calculated using density functional theory, as described in Chapter 8, indicates that the Γ-point CBM is indeed much lower than the overall conduction band. Moreover, the results of these calculations also provide information from which the charge neutrality level of InN can be estimated. This allows the existence of ionized donor-type surface states to be explained, to a first approximation, in terms of the bulk band structure.

Since the Γ-point conduction band energy bears little relation to the conduction band edge as a whole, Tersoff's approximate, semi-empirical method for obtaining the charge neutrality level of a semiconductor uses the *indirect* CBM and an effective VBM that takes account of the effect of spin-orbit splitting [12.53]. Using the indirect CBM at the A-point of the InN band structure (as calculated using the methods described in Chapter 8) locates the charge neutrality level, E_{CNL}, at 1.78 eV above the VBM and ~ 1.1 eV above the Γ-point CBM. To achieve overall charge neutrality, the Fermi level at the surface is expected to lie in the vicinity of this branch-point energy. The surface Fermi level of InN was found from the photoemission measurements described in sections 12.2.2.2 and 12.2.4 to be ~ 1.53 eV above the VBM and therefore close to, but below, E_{CNL}. For the donor-type surface states to be charged, and thus explain the observed electron accumulation, they must be at least partially unoccupied and therefore lie above the Fermi level. Details of the structure and chemistry of the surface are also important in determining the exact position of the surface Fermi level, but their importance is secondary to that of the bulk band structure. The microscopic origins of the surface states are discussed in section 12.4.1. The phenomenon of a low Γ-point CBM also explains the electron accumulation that occurs at InAs and In_2O_3 surfaces [12.54, 12.55].

12.2.6.1 Determination of the charge neutrality level position in InN

The location of the charge neutrality level with respect to the Γ-point band extrema in InN has been investigated via studies of undoped and Si-doped InN samples using a combination of high-resolution x-ray photoemission spectroscopy, Hall effect measurements and optical absorption spectroscopy [12.43]. The InN samples were grown at Cornell University on *c*-plane sapphire substrates by plasma-assisted molecular-beam epitaxy. The InN layer thicknesses ranged from 250 to 2000 nm and the growth temperature was $\sim 480°C$. The carrier concentrations and mobilities (from single-field Hall effect measurements) vary from 2.0×10^{18} to 6.6×10^{20} cm^{-3} and 1100 to 38 $cm^2 V^{-1} s^{-1}$, respectively. All except the lowest carrier concentration sample were doped with Si. The optical absorption measurements were performed using a Perkin-Elmer Lambda 25 UV-visible spectrophotometer and a Perkin-Elmer Spectrum GX Fourier transform infrared spectrometer for energies above and below 1.2 eV, respectively. The XPS spectrometer used in this study is described in section 12.2.2.2.

The carrier concentration, from single-field Hall effect measurements, increases with increasing Si-cell temperature used during growth, indicating that Si is being incorporated into the InN host and is electrically active, acting as a donor. This is confirmed by XPS measurements of the Si $2p$ core-level peak, shown in Fig. 12.16, which increases in intensity with increasing carrier concentration. Additionally, the binding energy of the peak (~ 102 eV) is indicative of Si-N bonding (the peak is chemically shifted from its elemental position in Si of ~ 99 eV), confirming that the

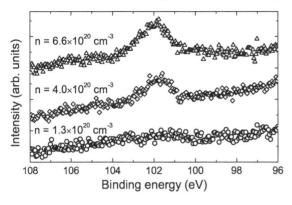

FIGURE 12.16

XPS spectra of the Si $2p$ region from Si-doped InN as a function of electron concentration.

Si preferentially occupies the In site, therefore acting as a donor.

XPS was employed to determine the pinning position of the surface Fermi level as a function of bulk carrier concentration; the leading edges of the valence-band photoemission spectra are shown in Fig. 12.17. Increasing the doping shifts the leading edge of the valence-band photoemission to higher binding energies, indicating an increase in the VBM to Fermi level separation at the surface. The values of the surface Fermi level position determined by extrapolating the leading edge of the valence-band photoemission to the baseline are shown in the left panel of Fig. 12.18, revealing a stabilization of the surface Fermi level at 1.83±0.10 eV above the VBM with increasing doping.

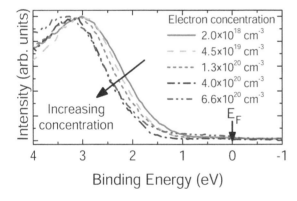

FIGURE 12.17

Valence band photoemission spectra from InN as a function of electron concentration. Reprinted with permission from P. D. C. King, *et al.*, Physical Review B, 77 (2008) 045316. Copyright 2008 by the American Physical Society.

Additionally, optical absorption spectra (not shown) indicate the effects of doping on the bulk Fermi level [12.43]. A significant increase in the absorption edge energy is observed with increasing Si-doping; this is attributed to the Burstein-Moss (or band-filling) effect [12.56, 12.57], whereby the (degenerate) Fermi level shifts to higher energies with increasing doping. An exponential Urbach tail is seen below this absorption edge, and the extent of this tail increases with doping concentration due to the increase in band-tailing effects. The bulk Fermi level position can be determined from the electron concentration in the samples and carrier statistics calculations, and is shown in Fig. 12.18. It should be noted that, except for

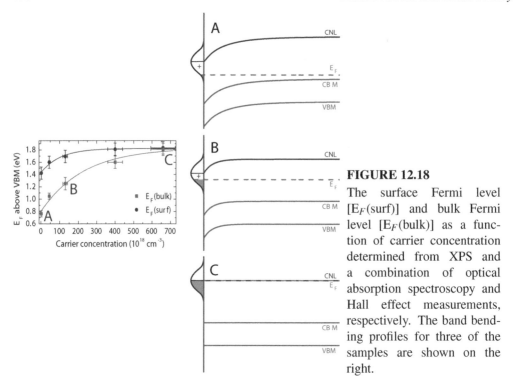

FIGURE 12.18

The surface Fermi level [E_F(surf)] and bulk Fermi level [E_F(bulk)] as a function of carrier concentration determined from XPS and a combination of optical absorption spectroscopy and Hall effect measurements, respectively. The band bending profiles for three of the samples are shown on the right.

the most heavily doped sample, the electron concentrations determined from the single-field Hall effect measurements contain a contribution from the surface electron accumulation and the interface region in addition to the bulk. This introduces an error in the calculated bulk Fermi level based on the single-field Hall effect concentrations, as represented by the error bars in Fig. 12.18. Initially, an increase in carrier concentration leads to a rapid increase in bulk Fermi level, although the rate of this reduces with increasing carrier concentration due to the conduction band nonparabolicity. In Fig. 12.18, the surface and bulk Fermi levels are seen to converge as the carrier concentration is increased. Any difference between the bulk and surface Fermi level positions must be incorporated via a bending of the bands relative to the Fermi level, which therefore tends smoothly to zero with increasing bulk carrier concentration. The evolution of the band bending is shown in parts A, B, and C of Fig. 12.18 in order of increasing bulk electron density.

For a given bulk Fermi level, the surface Fermi level position is determined by the considerations of charge neutrality. If the surface Fermi level is located below the charge neutrality level, some donor surface states will be unoccupied and hence positively charged. This surface charge must be balanced by a space charge due to downward band bending, leading to an increase in the near-surface electron density (an accumulation layer). For nominally undoped (low carrier concentration) InN (Fig. 12.18 A), the downward band bending is extreme, resulting in the observed large accumulation of electrons at the surface. There is a relatively large energy separation between the surface Fermi level and charge neutrality level and corresponding high density of unoccupied, positively-charged surface states. For initial increases in bulk Fermi level, the change in the space charge can be accommodated by very small shifts in the surface Fermi level position; the Fermi level is strongly pinned at the surface. However, as the bulk Fermi level increases further, the reduction in band bending means that the space charge is no longer sufficient to balance the surface state charge, causing the surface Fermi level to move closer to the charge neutrality level in order that fewer donor surface states are unoccupied, reducing the surface state charge (Fig. 12.18 B). As the bulk Fermi level approaches the charge neutrality level, the surface Fermi level must also therefore approach the charge neutrality level, causing both the band bending to tend to zero and the donor surface states to become occupied and neutral (Fig. 12.18 C). By creating this zero band bending 'flat band' condition by bulk doping in this way, it is possible to use XPS to directly measure the energy position of the charge neutrality level with respect to the valence band maximum. The XPS indicates a stabilization of the surface Fermi level upon heavy *n*-type doping at 1.83 ± 0.10 eV. Therefore, the charge neutrality level lies 1.83 eV above the VBM.

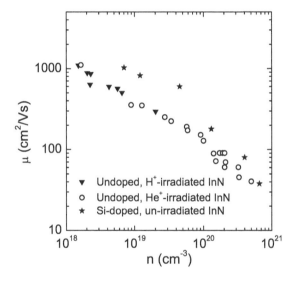

FIGURE 12.19

Electron mobility in Si-doped InN films (stars) compared with that in InN films subjected to H^+ (triangles) or He^+ irradiation (open circles). Higher electron concentrations correspond to higher irradiation doses.

The derived charge neutrality level value is also consistent to within experimental error with the stabilized bulk Fermi level position in heavily He^+ ion irradiated InN [12.58]. A bulk Fermi level of 1.72 eV was observed by optical absorption spectroscopy in the most heavily irradiated sample, as shown in Fig. 10.12 of Chapter 10. In addition to stabilizing the Fermi level at the charge neutrality level, Si-doping and irradiation also lead to the electron mobility converging in the limit of high free electron density, as shown in Fig. 12.19. This indicates that the mobility of the most heavily Si-doped samples is limited by the same scattering resulting from the presence of native defects as in the irradiated samples.

The charge neutrality level value measured here also agrees very well with previous theoretical calculations. Van de Walle and Neugebauer [12.59] located the charge neutrality level at 1.88 eV above the VBM in InN using ab initio calculations. Mönch used empirical tight-binding calculations and obtained a value of 1.5 eV above the VBM. [12.62]. Additionally, Green's functions calculations by Robertson and Falabretti [12.60] give a branch point 1.87 eV above the VBM, and they also determined a value of 1.88 eV using the theoretical calculations of Wei and Zunger [12.61]. And finally, the measured value also compares well with the aforementioned value of 1.78 eV determined from applying Tersoff's method to the DFT band structure.

The location of the charge neutrality level 1.83 eV above the VBM is plotted on the calculated band structure in Fig. 4.3 in Chapter 4 and can be seen to lie approximately at the middle of the band gap across the entire Brillouin zone. Additionally, the conduction band minimum at the Γ-point can be seen to lie very low in energy compared with the rest of the conduction band edge and below the charge neutrality level. As discussed above, it is this property of the bulk band structure of InN that gives rise to the electron accumulation.

12.3   Inversion layers at the surfaces of *p*-type InN

The band structure of InN, with the Γ-point CBM lying far below the charge neutrality level, also has implications for the properties of *p*-type InN surfaces. Not only will the pinning of the surface Fermi level high in the conduction band result in electron accumulation in *n*-type InN, but it will also produce a strong inversion layer at the surface of *p*-type InN. An example of an inversion layer charge profile and the associated downward band bending is shown in Fig. 12.20. The charge profile is calculated by solving Poisson's equation within the MTFA, using a nonparabolic conduction band and parabolic heavy hole, light hole and crystal hole valence bands [12.63]. The procedure is the same as that described in section 12.2.3.

The inversion layer profile shown in Fig. 12.20 is representative of typical properties determined so far for InN samples that have been doped with the correct amount of Mg to apparently exhibit *p*-type conductivity. For example, electrochemical capacitance-voltage (ECV) profiling of Mg-doped InN samples has been used to probe the acceptor density [12.64–12.66], as described in Chapters 3 and 10. Additionally, quantitative mobility spectrum analysis of multiple-field Hall effect measurements of Mg-doped InN has provided evidence of both 'light' and 'heavy' free holes [12.65].

In Fig. 12.21, the valence band photoemission from a Mg-doped InN film is compared with that from undoped InN. The Mg-doped InN sample is the same one for which a net acceptor density of 2×10^{19} cm^{-3} was estimated from ECV profiling [12.64]. Before surface cleaning, surface Fermi levels of 1.35$\pm$0.10 eV and 1.14$\pm$0.10 eV above the VBM were determined for the undoped and Mg-doped

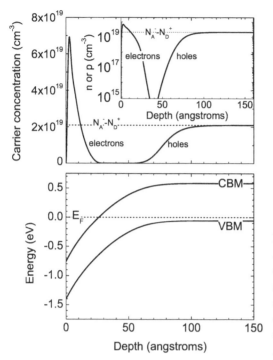

FIGURE 12.20

Surface inversion space-charge profile plotted on a linear scale and on a logarithmic scale (inset), calculated to correspond to an InN film doped p-type to give a hole density in the bulk of 2×10^{19} cm^{-3}. The depth dependence of the CBM and VBM with respect to the Fermi level, E_F, is also shown.

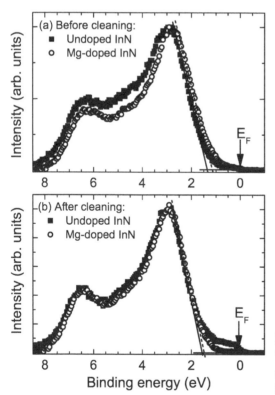

FIGURE 12.21

Valence band photoemission spectra from undoped and Mg-doped InN (a) before and (b) after surface cleaning.

samples, respectively. This indicates that the surface Fermi level is slightly lower for the Mg-doped sample than for the undoped sample before cleaning. More investigations are required to determine whether this is generally the case. However, after atomic hydrogen cleaning, both undoped and Mg-doped InN samples exhibit very similar surface Fermi levels, with 1.54 ± 0.10 eV and 1.50 ± 0.10 eV being found for the undoped and Mg-doped samples, respectively.

The presence of an electron-rich layer at the surface of Mg-doped InN is the main reason for the difficulties in characterizing suspected p-type InN films and the resulting lack of a large number of reports of successful p-type doping. In the inversion layer at the surface of p-type InN, the electron-rich surface layer is separated from the holes in the bulk by a depletion region. This is shown in Fig. 12.20, where for the particular bulk doping level, the minimum carrier density occurs at ~ 40 Å below the surface, where the Fermi level crosses the middle of the fundamental band gap. The inversion layer means that conventional Hall effect measurements of p-type InN thin films are dominated by conduction through the surface electron-rich layer.

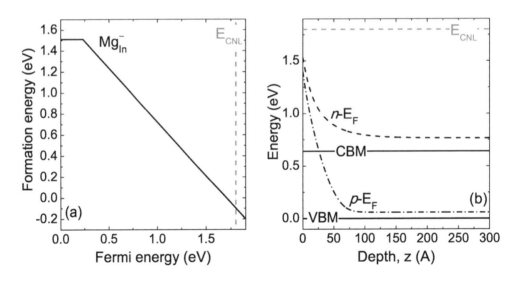

FIGURE 12.22

(a) The formation energy of Mg_{In} acceptors in InN under In-rich conditions as a function of Fermi level, E_F, calculated by Stampfl *et al.* [12.67]. $E_F = 0$ corresponds to the valence band maximum. (b) The variation of the Fermi level in the surface space-charge region for 'typical' n- and p-type InN. The Fermi level increases close to the surface, corresponding to a reduction of the Mg_{In} formation energy.

However, somewhat paradoxically, the same property that makes p-type doping of InN hard to quantify may very well be responsible for making p-type doping with Mg relatively efficient. In Mg-doped InN there are already indications of hole densities in excess of 10^{19} cm^{-3}, in contrast to Mg-doped GaN where the maximum achievable free-hole density is typically $2\text{--}3 \times 10^{18}$ cm^{-3} [12.68, 12.69]. The relative ease of *p*-type doping of InN is perhaps surprising considering the proclivity of InN toward *n*-type conductivity. However, one factor that may assist efficient *p*-type doping of InN is the influence of the Fermi level on the formation energy of Mg acceptors. The formation energy of Mg$_{In}$ acceptors in zinc-blende InN under In-rich conditions as a function of Fermi level calculated by Stampfl *et al.* [12.67] is plotted in Fig. 12.22(a). Formation energies of defects and impurities in the wurtzite phase are similar to the zinc-blende phase, except for interstitial configurations, where the geometry of the wurtzite phase leads to higher formation energies [12.67]. These calculations indicate that substitional Mg acceptors are more favorable with increasing Fermi level. The increase of the Fermi level toward the surface of *n*- and *p*-type InN is shown in Fig. 12.22(b). This suggests that the formation of Mg$_{In}$ acceptors at the surface, where Mg is initially incorporated during growth, is more favorable due to the electron accumulation than it would be otherwise.

12.4 Structural properties of InN surfaces

In spite of the extensive studies of the electronic properties of InN surfaces described above, the structural properties of InN surfaces have received very little attention. A few theoretical studies have been undertaken [12.70, 12.71], as discussed in detail in Chapter 13. Experimentally, most information concerning the structural and morphological nature of the surface of InN has been obtained from epitaxial growth studies, as discussed in Chapters 2 and 3. Here, the preliminary results are presented of some of the first experimental investigations of the surface structure of InN. A combination of core-level XPS, low energy electron diffraction, and co-axial impact collision ion scattering spectroscopy, has been performed on *ex situ*-prepared In-polarity InN(0001) surfaces. Data from these techniques has been combined with information from reflection high energy electron diffraction patterns recorded immediately after growth to enable a model of one of the reconstructions of the InN(0001) surface to be proposed. XPS has also been used to probe the surface stoichiometry of *ex situ*-prepared N-polarity InN(000$\bar{1}$) and non-polar InN(11$\bar{2}$0) surfaces.

12.4.1 Microscopic origins of surface electron accumulation?

Despite the low Γ-point CBM being indentified as the overriding band structure property responsible for the electron accumulation at InN surfaces, a microscopic origin of the phenomenon has only very recently been proposed. Based on first principles calculations, it has been suggested by Segev and Van de Walle that the donor-type surface states responsible for the electron accumulation are associated with In-In bonds in a surface In adlayer [12.72]. The experimental observation of In adlayers at InN surfaces exhibiting electron accumulation is discussed in section 12.4.2 below. The calculations, discussed in detail in Chapter 13, also indicated that electron accumulation would be absent from non-polar InN surfaces in the absence of In adlayers. This has been confirmed experimentally by Wu *et al.* by scanning photoemission microscopy and spectroscopy of *in situ* cleaved *a*-plane InN surfaces [12.73]. While the as-grown *c*-plane and *a*-plane InN surfaces exhibited surface Fermi levels of 1.54 eV and 1.65 eV above the VBM, respectively, the cleaved *a*-plane surface Fermi level was significantly lower, suggesting the absence of band bending. However, the reported Fermi level of 0.50 ± 0.10 eV above the VBM, if true, would correspond to non-degenerately doped InN, but contradicts the reported electron concentration of 1.2×10^{18} cm^{-3} in the 2.4-μm-thick InN film. This bulk electron density would correspond to a VBM to Fermi level separation of 0.7–0.8 eV, which seems to be consistent with the reported photoemission data. The existence of flat bands at *in situ* cleaved non-polar surfaces has long been known to be a common feature of all III-V semiconductors [12.74, 12.75].

In addition to electron accumulation being observed in the presence of In adlayers and its absence on an *in situ* cleaved non-polar surface, it is also apparent that In adlayers on the surface may be a sufficient, but not always necessary, condition for the existence of electron accumulation at InN surfaces. While In-In bonds in In adlayers may be the microscopic cause of electron accumulation at pristine InN surfaces in ultra-high vacuum, the donor surface states on *ex situ* oxidized samples, which also exhibit electron accumulation, must have an alternative microscopic origin, as the adlayers are disrupted by the formation of the native oxide. Possible candidates include steps, point defects and impurities such as oxygen and hydrogen.

12.4.2   In adlayers at InN surfaces

Relatively little attention has been paid to the chemical and structural nature of InN surfaces. In contrast, extensive studies of the GaN surface structure have revealed a strong tendency for Ga atoms to stabilize in the surface layer. This is the dominant force driving GaN surface reconstruction, overriding the usual principal mechanisms for conventional III-Vs of obeying electron counting and reducing the number of dangling bonds. Consequently, GaN grown under Ga-rich conditions has surface reconstructions involving Ga-adlayers, the coverage of which has been found to depend on the polarity of the GaN, with 1.1 monolayers (ML) of Ga terminating GaN(000$\bar{1}$) and 2.4 ML on GaN(0001) [12.76], in agreement with the predictions of first-principles theory [12.77, 12.78].

The propensity for GaN surfaces to be stabilized by Ga-adlayers is due to the small lattice constant and high anion-anion bond strength compared to those of conventional III-Vs [12.79]. The expectation from theoretical studies has long been that all polar III-nitride surfaces (including those of InN) will resemble those of GaN [12.79]. This idea that group III-atom stabilization of the surface is a general property of III-nitrides is supported by evidence of Al-adlayers on AlN(0001) surfaces [12.80] and In adlayers on InGaN(0001) surfaces [12.81].

As already mentioned, first-principles calculations have shown that In adlayers are energetically favorable at c-plane InN surfaces [12.70, 12.72]. In order to investigate this phenomenon experimentally, the surface stoichiometry of InN films prepared by atomic hydrogen cleaning has been investigated by x-ray photoemission spectroscopy [12.33]. All the cleaned InN surfaces studied exhibited surface Fermi level to VBM separations of 1.53 eV, indicating the presence of electron accumulation. The In:N XPS intensity ratio was calculated from the core-level peak areas divided by the atomic sensitivity factors for each core level for the Scienta ESCA300 spectrometer as a function of polarity for two different emission angles. The In:N ratio is greater in each case for the more surface sensitive 30° emission than for 90° emission and the In:N ratio also increases on going from N-polarity to In-polarity. These results indicate that all the samples studied have In-rich surfaces and that the surface of the In-polarity InN is more In-rich than the surface of the N-polarity InN.

To obtain quantitative values of the In-coverage on the different samples from the XPS core-level results, the In:N intensity ratios have been compared with model calculations for different surface reconstructions of In- and N-polarity InN, including different numbers of In adlayers and also, for the In-polarity InN, with and without a laterally contracted topmost layer containing 4/3 of a monolayer (ML) of In. This layer corresponds to the $(\sqrt{3} \times \sqrt{3})$R30° reconstruction observed by reflection high-energy electron diffraction (RHEED) after MBE growth at Ritsumeikan University [12.33] and also recently at Ilmeanau [12.82].

The analysis of the XPS core-level data revealed that the In-polarity InN has ~3.4 ML of In above the In-terminated InN bulk and the N-polarity InN has ~2.0 ML of In above the In-terminated InN bulk, as shown in Fig. 12.23. An In adlayer coverage of ~2–3 ML was also determined for the non-polar a-plane InN surface after cleaning. The In-coverages on c-plane InN are both 1 ML greater than the Ga-coverages for the respective GaN surfaces grown under Ga-rich conditions [12.76]. Additionally, these In-coverages are greater than the highest coverages for which theoretical calculations have been performed for wurtzite InN [12.70]. However, for InN(111), which is very similar to InN(0001) in total-energy terms, 3 ML of In on top of In-terminated InN(111) has been found to be more energetically favorable than lower InN coverages [12.83]. The theoretical studies of both the polar wurtzite InN surfaces and InN(111) indicate that InN has a stronger tendency for metallic overlayers to form than GaN.

Two properties unique to group-III nitrides are responsible for the unusual stability of group-III rich reconstructions previously observed on GaN surfaces, compared with on conventional III-V surfaces. First, Ga atoms on GaN form stronger Ga-Ga bonds than on, for example, GaAs. The origin of the stronger bonding lies in the sizeable mismatch of the covalent radii of Ga and N. Because of the small radius of the N atoms, the Ga atoms in GaN have approximately the same distance between them as in bulk Ga. The Ga atoms on the surface can form metallic

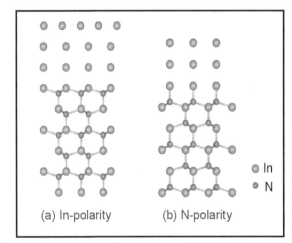

(a) In-polarity   (b) N-polarity

○ In
● N

FIGURE 12.23

The In- and N-polarity surfaces terminated by $\sim$3.4 ML and $\sim$2.0 ML of In. The topmost 1.4 ML of In for In-polarity InN is in a top laterally contracted and rotated layer, consistent with the $(\sqrt{3} \times \sqrt{3})R30°$ reconstruction observed by RHEED.

bonds similar to those in Ga metal even without any relaxation, thus stabilizing the Ga-terminated surface [12.84]. As there is an even greater mismatch between the covalent radii of In and N than for Ga and N, the In-In bonds are expected to be strong on InN, enabling the stabilization of In-terminated surfaces. The second mechanism stabilizing metal surfaces on group-III nitrides is the very different cohesive energies of the bulk phases of the group-III metal and N: The cohesive energy of bulk Ga is 2.81 eV/atom while that of the N_2 molecule is 5.0 eV/atom. In contrast, the cohesive energy of bulk As (2.96 eV/atom) is only slightly larger than that of bulk Ga. Because of this asymmetry between the Ga and N 'reservoir', more energy is needed to transfer N atoms from its reservoir to the surface than to transfer Ga atoms to the surface [12.84]. Moreover, because the cohesive energy of bulk In is 2.5 eV/atom [12.85] (even further from that of the N_2 molecule than bulk Ga), stabilization of a group-III-terminated surface may be even more favorable for InN than for GaN.

12.4.3 Structure of In-polarity InN(0001) surface under In-rich conditions

The aforementioned RHEED and the XPS In:N intensity ratio are insensitive to the registry of the surface overlayer with respect to the bulk InN, making them unsuitable for a full surface structure determination; the RHEED is sensitive only to the lateral periodicity and the XPS In:N ratio to the composition of the layers along the growth direction. Therefore, ion scattering spectroscopy has been used to determine the structure of the In-polarity InN(0001) surface prepared under In-rich conditions [12.86].

The c-plane InN film was grown at Cornell University by MBE to a thickness of 1500 nm on c-plane sapphire substrates, incorporating a GaN/AlN buffer layer. It was found to be In-polarity (that is, (0001) orientation) from scanning electron microscopy after KOH etching [12.87]. Hall effect measurements indicated an average free electron density and a mobility of 1.8×10^{18} cm^{-3} and 1200 cm^{2}V^{-1}s^{-1}.

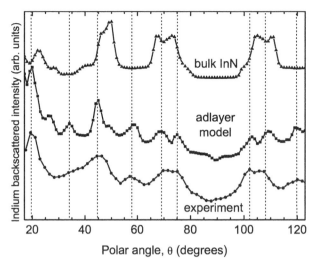

FIGURE 12.24

Experimental CAICISS In-back-scattered intensity as a function of polar angle for the [1000] azimuth. Simulated CAICISS spectra are also shown for bulk-like In-terminated In-polarity InN(0001) and InN(0001) terminated with In adlayers, as depicted in Fig. 12.25. Reprinted from T. D. Veal, P. D. C. King, M. Walker, C. F. McConville, H. Lu, and W. J. Schaff, In adlayers on non-polar and polar InN surfaces: Ion scattering and photoemission studies, *Physica B*, 401–402 (2007) 351–354. Copyright 2007, with permission from Elsevier.

The CAICISS data recorded from an In-polarity *c*-plane InN sample and corresponding simulations are shown in Fig. 12.24. From the variation of the energy and flux of the scattered ions as a function of incidence or azimuthal angle, it is possible to analyze the surface structure of the epilayer. To interpret the data, trial structures are simulated using the FAN simulation program developed by Niehus [12.88] until agreement is achieved between the experimental data and the simulated intensity as a function of polar angle.

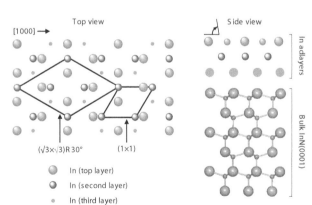

FIGURE 12.25

Structural models showing the top and side views of the In adlayer surface reconstruction for the In-polarity *c*-plane InN surface. There are three layers of In on top of the In-terminated bulk InN. The topmost layer consists of 4/3 ML of In with $(\sqrt{3} \times \sqrt{3})R30°$ periodicity. Reprinted from T. D. Veal, P. D. C. King, M. Walker, C. F. McConville, H. Lu, and W. J. Schaff, In adlayers on non-polar and polar InN surfaces: Ion scattering and photoemission studies, *Physica B*, 401–402 (2007) 351–354, Copyright 2007, with permission from Elsevier.

The simulated polar angle dependence of the CAICISS signal is shown in Fig. 12.24 for In-terminated bulk In-polarity *c*-plane InN. Comparison of this calculated spectrum with the experimental data reveals many differences. The peaks at ~33°, ~58° and ~120° in the experimental data are not reproduced in the calculated spectrum for the bulk-like termination of InN. Additionally, features at $\sim 20°$, $\sim 45°$, ~ 69-$75°$ and ~ 102-$108°$ in the experimental spectrum appear a few degrees away from similar features in the calculated spectrum for bulk termination.

Due to the lack of agreement between the measured ion scattering spectrum and that calculated for bulk-terminated InN, the ion scattering spectra for other trial structures were calculated. These consisted of various models containing In adlayers, the structure of which was based on analogies with Ga-adlayers on Ga-polarity GaN and previous photoemission and reflection high energy electron diffraction (RHEED) results from InN [12.33]. The best agreement between the experimental data and a simulated spectrum was obtained for the structure shown in Fig. 12.25. This structure consists of three layers of In on top of the bulk-terminated In-polarity InN, with 4/3 ML of In in the topmost layer with $(\sqrt{3} \times \sqrt{3})R30°$ periodicity, as shown in Fig. 12.25. The registry shown is required to produce, for example, the $58°$ peak in the spectrum which results from shadowing of In in the topmost layer onto atoms in the second layer. The best fit structure is consistent with both the observation of ~ 3.4 ML of In adlayers determined from the In:N ratio in core-level photoemission spectra and the $(\sqrt{3} \times \sqrt{3})R30°$ reconstruction previously observed by RHEED [12.33, 12.87]. Only integer order diffraction spots with relatively high background intensity were observed by LEED in the CAICISS chamber, indicating a relatively disordered surface but not ruling out the presence of regions of $(\sqrt{3} \times \sqrt{3})R30°$ periodicity.

The overall agreement between the simulated CAICISS intensity and the experimental spectrum is good, apart from the fact that the spectral features in the calculated spectrum are angularly narrower than those in the experimental data. The additional broadness in the experimental data compared with the simulation may be related to the In atoms in the topmost layer being mobile. This could result in the presence of rapidly moving domain boundaries separating regions of the surface having different registries. This is consistent with the observed (1×1) LEED pattern with high background intensity. Evidence for this phenomenon has previously been found for Ga-adlayers at GaN surfaces which exhibit a pseudo-(1×1) surface reconstruction [12.78]. Such effects would broaden the spectral features in the CAICISS spectra.

12.5   Surface electronic properties of InGaN and InAlN alloys

As described in sections 12.2 and 12.3, InN is known to exhibit extreme electron accumulation at the surface of *n*-type material and an inversion layer exists at the surface of samples which exhibit bulk *p*-type conductivity. Conversely, *n*- and *p*-type GaN display electron and hole depletion regions at the surface, respectively. This leads to the expectation that there will be a transition from one type of surface space-charge region to the other as the InGaN alloy composition is varied. This has been extensively investigated using x-ray photoemission spectroscopy for both undoped (*n*-type) and Mg-doped (*p*-type) InGaN alloys [12.89–12.91]. Additionally, the electronic properties of the surfaces of undoped and Mg-doped InAlN alloys have also been studied [12.92, 12.93]. The results of this work are presented below.

12.5.1 Undoped and Mg-doped InGaN alloys

Many years before the observation of surface electron accumulation in InN, the Fermi level at InAs surfaces and interfaces was known to lie above the conduction band edge [12.94], resulting in *n*-InAs/metal contacts exhibiting Ohmic behavior with electron accumulation layers rather than Schottky barriers with depletion space-charge regions. Moreover, it has been shown that this is also true for In-rich $In_xGa_{1-x}As$ alloys, with the Au/InGaAs barrier height varying as $\Phi_B = 0.95\text{-}1.90x + 0.90x^2$. This implies that the transition from positive to negative 'barrier height' and the corresponding change from interface electron depletion to accumulation in $In_xGa_{1-x}As$ occurs at $x \sim 0.8$ [12.95, 12.96]. Here, investigations of the transition in InGaN alloys from electron accumulation at InN surfaces to electron depletion at GaN surfaces are described. Similarly, the transition from an inversion layer at the surface of *p*-type InN to a hole depletion layer at the surface of *p*-type GaN is studied in Mg-doped InGaN samples.

The composition dependence of the Fermi-level position with respect to the band-edges at the surface of $In_xGa_{1-x}N$ alloys has been investigated using high-resolution XPS [12.89–12.91]. While clean surfaces of both GaN and InN have previously been obtained outside the growth chamber [12.97, 12.98], surface preparation remains very difficult and would require a different method to be optimized for each $In_xGa_{1-x}N$ composition. Consequently, the composition dependence of the surface Fermi-level pinning has been studied in the presence of the native oxide on the $In_xGa_{1-x}N(0001)$ surfaces. Knowledge of the surface Fermi level is important for surface sensitive devices, such as chemical and biological sensors, where an InGaN active layer is exposed to the environment. Indeed, the potential of InN's surface for use as a sensor has already been demonstrated by studies of its electrical response to chemical exposures [12.44–12.46].

The position of the surface Fermi level as a function of alloy composition determined using valence band XPS measurements is shown in Fig. 12.26. For the undoped $In_xGa_{1-x}N$ samples, an increase in VBM to surface Fermi level separation is observed with increasing Ga-fraction, although no such simple trend is observed for the Mg-doped alloys. In order to elucidate this differing behavior, it is useful to consider the surface Fermi level variation on an absolute energy scale, relative to the charge neutrality level (CNL). This is shown in Fig. 12.27. The pinning of the surface Fermi level for the Mg-doped and undoped materials diverges with increasing Ga-content due to the differing nature of the surface space-charge regions for the different alloy and doping configurations. This different behavior across the composition range indicates that all the Mg-doped alloys have *p*-type bulk conductivities.

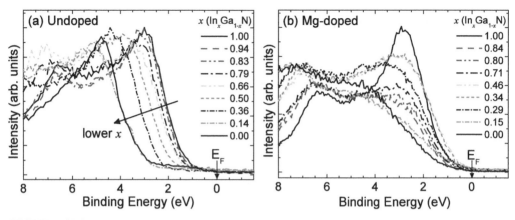

FIGURE 12.26

Valence band XPS spectra for (a) undoped and (b) Mg-doped $In_xGa_{1-x}N$ alloys with respect to the Fermi level, E_F. Reprinted with permission from P. D. C. King, T. D. Veal, H. Lu, P. H. Jefferson, S. A. Hatfield, W. J. Schaff, and C. F. McConville, physica status solidi (b), 245, 881 (2008). Copyright 2008 by Wiley-VCH Publishers, Inc.

The observed variation of surface Fermi level position is consistent with the alloy composition dependence of the band edges with respect to the charge neutrality level or branch-point energy (E_{CNL}), where the surface states change their charging character, from donor to acceptor type [12.47]. The positions of the band edges relative to the CNL in Fig. 12.27 are derived from our experimentally determined CNL position in InN of 1.83 eV above the VBM [12.43], our XPS-measured InN/GaN valence band offset of 0.58 eV [12.99] and the assumption of a linear variation of the valence band offset as a function of alloy composition, as proposed by Mönch [12.47].

For the In-rich undoped alloys, the surface Fermi level is below E_{CNL}, allowing donor-type surface states in the vicinity of E_{CNL} to be empty and donate electrons into the accumulation layer. For the Ga-rich alloys, the surface Fermi level is above E_{CNL} such that acceptor-type surface states close to E_{CNL} are occupied, negatively charged and contribute to the electron depletion layer. The electron accumulation layer charge profile and band bending for undoped In-rich alloys is illustrated schematically in the top right of Fig. 12.27. The electron depletion layer for undoped Ga-rich alloys is shown in the top left. The $In_xGa_{1-x}N$ composition marking the transition from surface electron accumulation (In-rich alloys) to electron depletion (Ga-rich alloys) can be estimated as the crossing of the surface Fermi level pinning position and the CNL. From Fig. 12.27, this is estimated to occur at $x \sim 0.41$.

The behavior of the Mg-doped InGaN alloys is different. The surface Fermi level is pinned below the CNL across the composition range, indicating the existence of unoccupied and hence negatively charged surface states, leading to downward band bending for all alloy compositions. For the In-rich alloys, the surface Fermi level pins above the mid-gap level, and above the CBM for the highest In-content

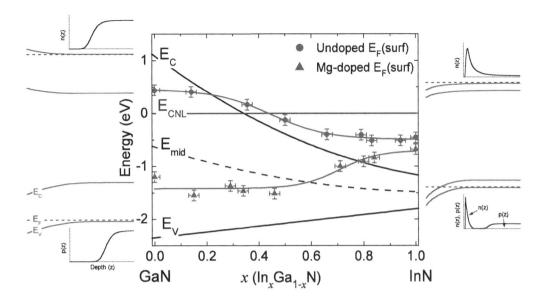

FIGURE 12.27

The CBM (E_C), VBM (E_V) and mid-gap position (E_{mid}) with respect to the universal charge neutrality level (E_{CNL}) as a function of $In_xGa_{1-x}N$ alloy composition. The position of the Fermi level at the surface [E_F(surf)] as determined from the photoemission results is shown for both undoped and Mg-doped alloys. The error bars represent the accuracy of the linear extrapolation method. The fits to the data are to guide the eye. The pinning of the surface Fermi level for the Mg-doped and undoped materials diverges with increasing Ga-content. Schematic representations of the band bending and carrier concentration profiles as a function of depth below the surface are also shown for the undoped and Mg-doped InN and GaN end points.

alloys, leading to an electron-rich surface region of *n*-type conductivity and hence surface inversion layers as discussed above for InN. This inversion layer charge profile and band bending is shown schematically in the bottom right of Fig. 12.27. With increasing Ga content, the VBM and hence the Fermi level in the *p*-type bulk move further away from the CNL, requiring the surface Fermi level to pin at lower energies to maintain charge neutrality. The large difference between the *n*- and *p*-type surface Fermi level pinning positions for Ga-rich alloys can be attributed to the Fermi level approaching the CNL from above for *n*-type alloys and below for *p*-type alloys. The measured differences in GaN are consistent with previous results [12.100]. When the surface Fermi level becomes pinned below mid-gap, electrons no longer accumulate at the surface and the downward band bending just produces a depletion of holes in the near surface region, as represented for *p*-type Ga-rich alloys in the bottom left of Fig. 12.27. From the main part of Fig. 12.27, this transition from inversion to hole depletion surface space-charge regions occurs at an $In_xGa_{1-x}N$ alloy composition of $x \sim 0.59$. This *p*-type transition is, however, different from that of the *n*-type alloys. For the *p*-type alloys, downward band bending occurs across the composition range, with the transition simply marking a change in the magnitude of the band bending, leading to a change in the dominant surface carrier type. Conversely, for the *n*-type alloys, the transition marks a change in the direction of band bending.

12.5.2   Undoped and Mg-doped InAlN alloys

The transition from electron accumulation (inversion) to electron (hole) depletion for *n*-type (*p*-type) material described above for *n*- and *p*-InGaN was explained by the CNL lying significantly above the CBM for In-rich alloys, but below the CBM for Ga-rich alloys. The band lineup relative to the CNL of the III-N semiconductors is shown in Chapter 4 in Fig. 4.4(a). The CNL lies close to the middle of the direct band gap in AlN, similar to the situation in GaN, although much further below the CBM. Similar transitions from electron accumulation (inversion) to electron (hole) depletion might therefore be expected on moving from In-rich to Al-rich *n*-type (*p*-type) InAlN alloys. In this section, XPS studies of the surface Fermi level pinning position for oxidized undoped and Mg-doped InAlN alloys across the composition range are described [12.92, 12.93].

The XPS measurements were performed in the same way as for the InN and In-GaN experiments described above, apart from the requirement to use an electron flood gun for charge compensation when measuring the most Al-rich insulating In-AlN alloys. For these samples, the binding energy scale was calibrated from the binding energy of the C 1*s* core-level XPS peak from physisorbed carbon, using the average C 1*s* binding energy from the samples that did not require charge compensation as a reference, which varied by only 40 meV between the samples.

Valence band photoemission spectra from undoped and Mg-doped InAlN samples are shown in Fig. 12.28. For the undoped InAlN samples, the VBM to surface Fermi level separation increases with decreasing In content. For the Mg-doped samples, no such monotonic change with alloy composition is apparent. It is instructive to consider the variation of the surface Fermi level pinning positions on an absolute energy scale, relative to the CNL, as shown in Fig. 12.29. The positions of the band edges relative to the CNL in Fig. 12.27 are derived from our experimentally determined CNL position in InN of 1.83 eV above the VBM [12.43], our XPS-measured InN/AlN valence band offset of 1.52 eV [12.101] and the assumption of a linear variation of the valence band offset as a function of alloy composition, as proposed by Mönch [12.47]. In contrast to its location high above the CBM in InN, the CNL is located at approximately the center of the fundamental band gap in AlN, causing the energy separation of the CNL and the CBM to decrease upon increasing the Al-content in In-rich InAlN alloys. This causes a reduction in the density of charged surface states and, for *p*-type alloys, for example, where the bulk Fermi level is located close to the VBM, the surface Fermi level to become pinned further below the CNL, as observed for the In-rich Mg-doped InAlN alloys in Fig. 12.29. When the surface Fermi level moves below the mid-gap energy, estimated from Fig. 12.29 to occur for an alloy composition of $x \sim 0.7$, electrons will no longer be the dominant carrier species at the surface. As the surface Fermi level is still pinned below the CNL, positively charged surface states will again be present, leading to downward band bending, although this will now result in a depletion of holes at the surface, as observed for Ga-rich InGaN:Mg alloys. This transition from inversion to hole depletion occurs at a higher In-content than for InGaN alloys; this is not surprising, as the CNL lies further from the band edges in AlN than in GaN, causing the changes in surface electronic properties with alloy composition to occur more rapidly for InAlN than for InGaN. Similarly, the transition from electron accumulation to depletion in undoped InAlN films can be seen from Fig. 12.29 to occur at $x = 0.59$, again at a higher In content than for the n-InGaN films.

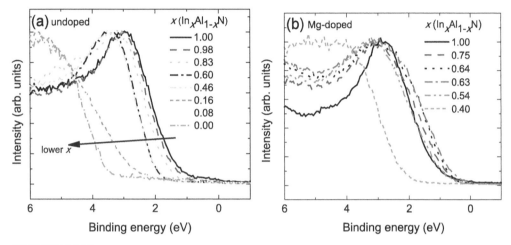

FIGURE 12.28

Valence band XPS spectra for (a) undoped and (b) Mg-doped $In_xAl_{1-x}N$ alloys with respect to the Fermi level, E_F.

For the more Al-rich alloy compositions ($x < 0.55$), the surface Fermi level pinning is in approximately the same place for undoped and Mg-doped alloys. This is in contrast to the behavior of InGaN alloys, and suggests that the Mg-doping is not causing p-type bulk conductivity in the most Al-rich alloys investigated here. Theoretical calculations for Mg acting as an acceptor in AlN show a pronounced increase in formation energy as the Fermi level moves toward the valence band [12.102]. Furthermore, as the VBM moves lower with respect to the charge neutrality level, the amphoteric defect model predicts that native point defects acting as compensating centers, attempting to drive the Fermi level toward the CNL, will become more favorable [12.103]. This is supported by first-principles calculations for AlN [12.102], which show, for example, that the formation energy for the nitrogen vacancy, acting as a donor, moves below that of the Mg-acceptor under p-type conditions. Consequently, for AlN and Al-rich InAlN alloys, compensating native defects make the realization of either n-type or p-type conductivity very difficult. The Fermi level in the bulk would therefore be expected to be located close to the middle of the direct band gap in both the undoped and the Mg-doped Al-rich InAlN samples investigated here, due to native defects also acting to drive the Fermi level

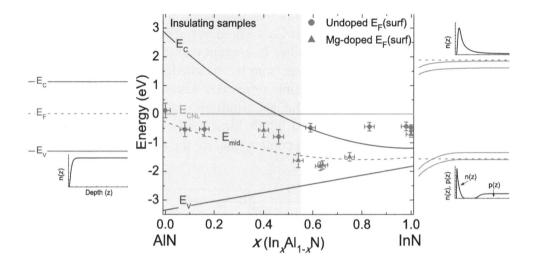

FIGURE 12.29

The CBM (E_C), VBM (E_V) and mid-gap position (E_{mid}) with respect to the universal charge neutrality level (E_{CNL}) as a function of In$_x$Al$_{1-x}$N alloy composition. The position of the Fermi level at the surface [E_F(surf)] as determined from XPS is shown for both undoped and Mg-doped samples. Schematic representations of the band bending and carrier concentration profiles as a function of depth below the surface are also shown for the InN and AlN end points.

toward the mid-gap position. The surface Fermi level pins close to the mid-gap position for both undoped and Mg-doped Al-rich alloys, leading to approximately flat band conditions, as represented schematically on the left of Fig. 12.29.

12.6   Conclusions

The electronic and structural properties of InN surfaces have been investigated. Intrinsic electron accumulation at the surface of wurtzite InN has been observed using HREELS and valence band XPS, indicating a surface Fermi level 1.53 eV and 1.40 eV above the VBM for clean and oxidized surfaces, respectively, and a surface sheet density of $\sim 2 \times 10^{13}$ cm^{-2}. The surface space-charge properties were described by solutions of the Poisson equation within the modified Thomas-Fermi approximation, incorporating conduction band nonparabolicity. The surface Fermi level position associated with the electron accumulation was found to be independent of surface orientation for a-plane and In- and N-polarity c-plane wurtzite InN. Zinc-blende InN also exhibited surface electron accumulation. The quantized sub-band electron states associated with the extreme downward band bending at the InN surface have been investigated using electron tunneling spectroscopy and angle-resolved photoemission spectroscopy. The one-electron potential, carrier concentration profile, quantized subband state energies, and parallel dispersion relations were calculated for the surface electron accumulation layer by numerically solving the Schrödinger equation for the potential well obtained from solving the Poisson equation. The inversion space-charge layer at the surface of p-type InN was also described. The overriding origin of the surface electron accumulation was identified as the particularly low Γ-point conduction band minimum in the bulk band structure that enables the existence of positively charged donor-type surface states. Possible microscopic origins of the donor surface states were identified as being In-In bonds in In adlayers for pristine surfaces and defects or impurities for air-exposed surfaces. The structure of InN surfaces, including the presence of indium adlayers, was investigated using core-level XPS, electron diffraction and ion scattering spectroscopy. The transition from surface electron accumulation to surface electron depletion with increasing Ga-content in n-type InGaN alloys was probed with valence band XPS, along with the transition from a surface inversion layer to a surface hole depletion layer in p-type InGaN alloys. The surface electronic properties of undoped and Mg-doped InAlN thin films as a function of alloy composition were also determined across the entire composition range.

Acknowledgments

We are grateful to former and current colleagues at Warwick, Imran Mahboob, Louis Piper, Marc Walker, Louise Bailey, Paul Jefferson, and Rob Johnston, for their contributions to this work. Bill Schaff, Steve Durbin, Yasushi Nanishi, Alex Georgakilas, Joan Redwing, Ted Moustakas, Donat As, Miguel Sánchez-García, Paul Chalker, Peter Parbrook, Tom Foxon, and Jim Speck and their colleagues are thanked for supplying samples. Frank Fuchs, Jurgen Furthmüller, and Friedhelm Bechstedt are thanked for providing the results of their first-principles calculations. Matt Phillips, Kevin Smith, Rüdiger Goldhahn, and Tom Myers and their colleagues are gratefully acknowledged for sharing the results of their tunneling spectroscopy, angle-resolved photoemission, ellipsometry and multiple-field Hall effect experiments, respectively. All of the above are thanked for useful discussions.

The Engineering and Physical Sciences Research Council (EPSRC), UK, funded both the work at the University of Warwick under grants GR/S56030/01, EP/C535-553/1 and EP/G004625/1 and the National Centre for Electron Spectroscopy and Surface Analysis under grants GR/S14252/01 and EP/E025722/1. Jennifer Houghton of EPSRC is particularly thanked for supporting our initial application for funding on indium nitride. TDV acknowledges financial support from the University of Warwick's study leave scheme and an EPSRC Career Acceleration Fellowship EP/G004447/1.

References

[12.1] I. Mahboob, T. D. Veal, C. F. McConville, H. Lu, and W. J. Schaff, Intrinsic electron accumulation at clean InN surfaces, *Physical Review Letters*, 92 (2004) 036804:1–4.

[12.2] W. J. Schaff, H. Lu, J. Hwang, and H. Wu, Growth of InN for heterojunction field effect transistor applications by plasma enhanced MBE, *Proceedings / IEEE / Cornell Conference on High Performance Devices* 225–231, Ithaca, New York, USA, August 2000.

[12.3] H. Lu, W. J. Schaff, L. F. Eastman, and C. Wood, Study of interface properties of InN and InN-based heterostructures by molecular beam epitaxy, *Material Research Society Symposium Proceedings*, 693 (2002) I1.5.1–I1.5.6.

[12.4] H. Lu, W. J. Schaff, L. F. Eastman, and C. E. Stutz, Surface charge accumulation of InN films grown by molecular-beam epitaxy, *Applied Physics Letters*, 82 (2003) 1736–1738.

[12.5] K. A. Rickert, A. B. Ellis, F. J. Himpsel, H. Lu, W. Schaff, J. M. Redwing, F. Dwikusuma, and T. F. Kuech, X-ray photoemission spectroscopic investigation of surface treatments, metal deposition, and electron accumulation on InN, *Applied Physics Letters*, 82 (2003) 3254–3256.

[12.6] V.-T. Rangel-Kuoppa, S. Suihkonen, M. Sopanen, and H. Lipsanen, Metal contacts on InN: Proposal for Schottky contact, *Japanese Journal of Applied Physics*, 45 (2006) 36–39.

[12.7] R. Khanna, B. P. Gila, L. Stafford, S. J. Pearton, F. Ren, I. I. Kravchenko, A. Dabiran, and A. Osinsky, Thermal stability of Ohmic contacts to InN, *Applied Physics Letters*, 90 (2007) 162107:1–3.

[12.8] S. X. Li, K. M. Yu, J. Wu, R. E. Jones, W. Walukiewicz, J. W. Ager III, W. Shan, E. E. Haller, Hai Lu, and W. J. Schaff, Fermi-level stabilization energy in group III–nitrides, *Physical Review B*, 71 (2005) 161201:1–4.

[12.9] J. W. Ager III, N. Miller, R. E. Jones, K. M. Yu, J. Wu, W. J. Schaff, W. Walukiewicz, Mg-doped InN and InGaN – Photoluminescence, capacitance-voltage and thermopower measurements, *physica status solidi (b)*, 245 (2008) 873-877.

[12.10] J. W. L. Yim, R. E. Jones, K. M. Yu, J. W. Ager, W. Walukiewicz, W. J. Schaff, and J. Wu, Effects of surface states on electrical characteristics of InN and $In_{1-x}Ga_xN$, *Physical Review B*, 76 (2007) 041303(R):1–4.

[12.11] X. Wang, S.-B. Che, Y. Ishitani, and A. Yoshikawa, Systematic study on p-type doping control of InN with different Mg concentrations in both In and N polarities, *Applied Physics Letters*, 91 (2007) 242111:1–3.

[12.12] G. Koblmüller, C. S. Gallinat, S. Bernardis, J. S. Speck, G. D. Chern, E. D. Readinger, H. Shen, and M. Wraback, Optimization of the surface and structural quality of N-face InN grown by molecular beam epitaxy, *Applied Physics Letters*, 89 (2006) 071902:1–3.

[12.13] C. Gallinat, G. Koblmüller, J. S. Brown, S. Bernardis, J. S. Speck, G. D. Chern, E. D. Readinger, H. Shen, and M. Wraback, In-polar InN grown by plasma-assisted molecular beam epitaxy, *Applied Physics Letters*, 89 (2006) 032109:1–3.

[12.14] C. H. Swartz, R. P. Tompkins, N. C. Giles, T. H. Myers, Hai Lu, W. J. Schaff, and L. F. Eastman, Investigation of multiple carrier effects in InN epilayers using variable magnetic field Hall measurements, *Journal of Crystal Growth*, 269 (2004) 29–34.

[12.15] C. H. Swartz, R. P. Tomkins, T. H. Myers, Hai Lu, W. J. Schaff, Demonstration of nearly non-degenerate electron conduction in InN grown by molecular beam epitaxy, *physica status solidi (c)*, 2 (2005) 2250–2253.

[12.16] T. B. Fehlberg, G. A. Umana-Membreno, B. D. Nener, G. Parish, C. S. Gallinat, G. Koblmüller, S. Rajan, S. Bernardis, and J. S. Speck, Characterisation of multiple carrier transport in indium nitride grown by molecular beam epitaxy, *Japanese Journal of Applied Physics*, 45 (2006) L1090–L1092.

[12.17] T. B. Fehlberg, C. S. Gallinat, G. A. Umana-Membreno, G. Koblmüller, B. D. Nener, J. S. Speck, and G. Parish, Effect of MBE growth conditions on multiple electron transport in InN, *Journal of Electron Materials*, 37 (2008) 593–596.

[12.18] I. Mahboob, T. D. Veal, L. F. J. Piper, C. F. McConville, H. Lu, W. J. Schaff, J. Furthmüller, and F. Bechstedt, Origin of electron accumulation at wurtzite InN surfaces, *Physical Review B*, 69 (2004) 201307(R):1–4.

[12.19] L. F. J. Piper, T. D. Veal, M. Walker, I. Mahboob, C. F. McConville, H. Lu, and W. J. Schaff, Clean wurtzite InN surfaces prepared with atomic hydrogen, *Journal of Vacuum Science and Technology A*, 23 (2005) 617–620.

[12.20] R. Fuchs and K. L. Kliewer, Optical modes of vibration in an ionic crystal slab, *Physical Review*, 140 (1965) A2076–A2088.

[12.21] T. D. Veal and C. F. McConville, Profiling of electron accumulation layers in the near-surface region of InAs(110), *Physical Review B*, 64 (2001) 085311:1–8.

[12.22] Ph. Lambin, J. P. Vigneron, and A. A. Lucas, Electron-energy-loss spectroscopy of multilayered materials: Theoretical aspects and study of interface optical phonons in semiconductor superlattices, *Physical Review B*, 32 (1985) 8203–8215.

[12.23] Ph. Lambin, J.-P. Vigneron, A. A. Lucas, Computation of the surface electron-energy-loss spectrum in specular geometry for an arbitrary plane-stratified medium, *Computer Physics Communications*, 60 (1990) 351–364.

[12.24] J.-P. Zöllner, H. Übensee, G. Paasch, T. Fiedler, and G. Gobsch, A novel self-consistent theory of the electronic structure of inversion layers in InSb MIS structures, *physica status solidi (b)*, 134 (1986) 837–845.

[12.25] E. O. Kane, Band structure of indium antimonide, *Journal of Physics and Chemistry of Solids*, 1 (1957) 249–261.

[12.26] P. D. C. King, T. D. Veal, and C. F. McConville, Nonparabolic coupled Poisson-Schrödinger solutions for quantized electron accumulation layers: Band bending, charge profile, and subbands at InN surfaces, *Physical Review B*, 77 (2008) 125305:1–7.

[12.27] S. A. Chambers, T. Droubay, T. C. Kaspar, and M. Gutowski, Experimental determination of valence band maxima for $SrTiO_3$, TiO_2, and SrO and the associated valence band offsets with Si(001), *Journal of Vacuum Science and Technology B*, 22 (2004) 2205–2215.

[12.28] H. Übensee, G. Paasch, and J.-P. Zöllner, Modified Thomas-Fermi theory for depletion and accumulation layers in n-type GaAs, *Physical Review B*, 39 (1989) 1955–1957.

[12.29] R. Goldhahn, P. Schley, A. T. Winzer, G. Gobsch, V. Cimalla, O. Ambacher, M. Rakel, C. Cobet, N. Esser, H. Lu, and W. J. Schaff, Detailed analysis of the dielectric function for wurtzite InN and In-rich InAlN alloys, *physica status solidi (a)*, 203 (2006) 42–49.

[12.30] T. Hofmann, T. Chavdarov, V. Darakchieva, H. Lu, W. J. Schaff, and M. Schubert, Anisotropy of the Γ-point effective mass and mobility in hexagonal InN, *physica status solidi (c)*, 3 (2006) 1854–1857.

[12.31] P. D. C. King, T. D. Veal, C. F. McConville, F. Fuchs, J. Furthmüller, F. Bechstedt, P. Schley, R. Goldhahn, J. Schörmann, D. J. As, K. Lischka, D. Muto, H. Naoi, Y. Nanishi, H. Lu, and W. J. Schaff, Universality of electron accumulation at wurtzite *c*- and *a*-plane and zincblende InN surfaces, *Applied Physics Letters*, 91 (2007) 092101:1–3.

[12.32] J. Schörmann, D. J. As, K. Lischka, P. Schley, R. Goldhahn, S. F. Li, W. Löffler, M. Hetterich, and H. Kalt, Molecular beam epitaxy of phase pure cubic InN, *Applied Physics Letters*, 89 (2006) 261903:1–3.

[12.33] T. D. Veal, P. D. C. King, P. H. Jefferson, L. F. J. Piper, C. F. McConville, H. Lu, W. J. Schaff, P. A. Anderson, S. M. Durbin, D. Muto, H. Naoi, and Y. Nanishi, In adlayers on *c*-plane InN surfaces: A polarity-dependent study by x-ray photoemission spectroscopy, *Physical Review B*, 76 (2007) 075313:1–8.

[12.34] P. D. C. King, T. D. Veal, C. S. Gallinat, G. Koblmüller, L. R. Bailey,

J. S. Speck, and C. F. McConville, Influence of growth conditions and polarity on interface-related electron density in InN, *Journal of Applied Physics*, 104 (2008) 103703:1–5.

[12.35] P. D. C. King, T. D. Veal, C. F. McConville, F. Fuchs, J. Furthmüller, F. Bechstedt, J. Schörmann, D. J. As, K. Lischka, H. Lu, and W. J. Schaff, Valence band density of states of zinc-blende and wurtzite InN from x-ray photoemission spectroscopy and first-principles calculations, *Physical Review B*, 77 (2008) 115213:1–7.

[12.36] B. R. Nag, On the band gap of indium nitride, *physica status solidi (b)*, 237 (2003) R1–R2.

[12.37] T. D. Veal, L. F. J. Piper, M. R. Phillips, M. H. Zareie, H. Lu, W. J. Schaff, and C. F. McConville, Scanning tunnelling spectroscopy of quantized electron accumulation at In$_x$Ga$_{1-x}$N surfaces, *physica status solidi (a)*, 203 (2006) 85–92.

[12.38] T. D. Veal, L. F. J. Piper, M. R. Phillips, M. H. Zareie, H. Lu, W. J. Schaff, and C. F. McConville, Doping-dependence of subband energies in quantized electron accumulation at InN surfaces, *physica status solidi (a)*, 204 (2007) 536–542.

[12.39] L. Colakerol, T. D. Veal, H. K. Jeong, L. Plucinski, A. DeMasi, T. Learmonth, P. A. Glans, S. Wang, Y. Zhang, L. F. J. Piper, P. H. Jefferson, A. Fedorov, T. C. Chen, T. D. Moustakas, C. F. McConville, and K. E. Smith, Quantized electron accumulation states in indium nitride studied by angle-resolved photoemission spectroscopy, *Physical Review Letters*, 97 (2006) 237601:1–4.

[12.40] S. R. Streight and D. L. Mills, Influence of surface charge on free-carrier density profiles in GaAs films: Application to second-harmonic generation by free carriers, *Physical Review B*, 37 (1988) 965–973.

[12.41] S. Abe, T. Inaoka, and M. Hasegawa, Evolution of electron states at a narrow-gap semiconductor surface in an accumulation-layer formation process, *Physical Review B*, 66 (2002) 205309:1–8.

[12.42] J. M. Ziman, *Principles of the Theory of Solids*, Cambridge University Press, Cambridge, 1972.

[12.43] P. D. C. King, T. D. Veal, P. H. Jefferson, S. A. Hatfield, L. F. J. Piper, C. F. McConville, F. Fuchs, J. Furthmüller, F. Bechstedt, H. Lu, and W. J. Schaff, Determination of the branch-point energy of InN: Chemical trends in common-cation and common-anion semiconductors, *Physical Review B*, 77 (2008) 045316:1–6.

[12.44] H. Lu, W. J. Schaff, and L. F. Eastman, Surface chemical modification of InN for sensor applications, *Journal of Applied Physics*, 96 (2004) 3577–

3579.

[12.45] Y. S. Lu, C. C. Huang, J. A. Yeh, C. F. Chen, and S. Gwo, InN-based anion selective sensors in aqueous solutions, *Applied Physics Letters*, 91 (2007) 202109:1–3.

[12.46] Y. S. Lu, C. L. Ho, J. A. Yeh, H. W. Lin, and S. Gwo, Anion detection using ultrathin InN ion selective field effect transistors, *Applied Physics Letters*, 92 (2008) 212102:1–3.

[12.47] W. Mönch, *Electronic Properties of Semiconductor Interfaces* (Springer, Berlin, 2004).

[12.48] R. M. Feenstra, Tunneling spectroscopy of the (110) surface of direct-gap III-V semiconductors, *Physical Review B*, 50 (1994) 4561–4570.

[12.49] S. Krischok, V. Yanev, O. Balykov, M. Himmerlich, J. A. Schaefer, R. Kosiba, G. Ecke, I. Cimalla, V. Cimalla, O. Ambacher, H. Lu, W. J. Schaff, and L. F. Eastman, Investigations of MBE grown InN and the influence of sputtering on the surface composition, *Surface Science*, 566–568 (2004) 849–855.

[12.50] L. F. J. Piper, T. D. Veal, C. F. McConville, H. Lu, and W. J. Schaff, InN: Fermi level stabilization by low-energy ion bombardment, *physica status solidi (c)*, 3 (2006) 1841–1845.

[12.51] H. Lüth, *Surfaces and Interfaces of Solid Materials* (Springer, Berlin, 1995).

[12.52] W. Mönch, *Semiconductor Surfaces and Interfaces* (Springer, Berlin, 2001).

[12.53] J. Tersoff, Schottky barriers and semiconductor band structures, *Physical Review B*, 32 (1985) 6968–6971.

[12.54] M. Noguchi, K. Hirakawa, and T. Ikoma, Intrinsic electron accumulation layers on reconstructed clean InAs(100) surfaces, *Physical Review Letters*, 66 (1991) 2243–2246.

[12.55] P. D. C. King, T. D. Veal, D. J. Payne, A. Bourlange, R. G. Egdell, and C. F. McConville, Surface electron accumulation and the charge neutrality level in In_2O_3, *Physical Review Letters*, 101 (2008) 116808:1–4.

[12.56] E. Burstein, Anomalous optical absorption limit in InSb, *Physical Review*, 93 (1954) 632–633.

[12.57] T. S. Moss, The Interpretation of the properties of indium antimonide, *Proceedings of the Physical Society B*, 67 (1954) 775–782.

[12.58] S. X. Li, E. E. Haller, K. M. Yu, W. Walukiewicz, J. W. Ager III, J. Wu, W. Shan, H. Lu, and W. J. Schaff, Effect of native defects on optical properties of $In_xGa_{1-x}N$ alloys, *Applied Physics Letters*, 87 (2005) 161905:1–3.

[12.59] C. G. Van de Walle and J. Neugebauer, Universal alignment of hydrogen levels in semiconductors, insulators, and solutions, *Nature*, 423 (2003) 626–628.

[12.60] J. Robertson and B. Falabretti, Band offsets of high K gate oxides on III-V semiconductors, *Journal of Applied Physics*, 100 (2006) 014111:1–8.

[12.61] S. H. Wei and A. Zunger, Calculated natural band offsets of all II-VI and III-V semiconductors: chemical trends and the role of cation d orbitals, *Applied Physics Letters*, 72 (1998) 2011–2013.

[12.62] W. Mönch, Empirical tight-binding calculation of the branch-point energy of the continuum of interface-induced gap states, *Journal of Applied Physics*, 80 (1996) 5076–5082.

[12.63] T. D. Veal, L. F. J. Piper, W. J. Schaff, and C. F. McConville, Inversion and accumulation layers at InN surfaces, *Journal of Crystal Growth*, 288 (2006) 268–272.

[12.64] R. E. Jones, K. M. Yu, S. X. Li, W. Walukiewicz, J. W. Ager, E. E. Haller, H. Lu, and W. J. Schaff, Evidence for p-Type doping of InN, *Physical Review Letters*, 96 (2006) 125505:1–4.

[12.65] P. A. Anderson, C. H. Swartz, D. Carder, R. J. Reeves, and S. M. Durbin, S. Chandril, and T. H. Myers, Buried p-type layers in Mg-doped InN, *Applied Physics Letters*, 89 (2006) 184104:1–3.

[12.66] X. Wang, S. B. Che, Y. Ishitani, and A. Yoshikawa, Systematic study on p-type doping control of InN with different Mg concentrations in both In and N polarities, *Applied Physics Letters*, 94 (2007) 242111:1–3.

[12.67] C. Stampfl, C. G. Van de Walle, D. Vogel, P. Krüger, and J. Pollmann, Native defects and impurities in InN: First-principles studies using the local-density approximation and self-interaction and relaxation-corrected pseudopotentials, *Physical Review B*, 61 (2000) R7846-R7849.

[12.68] J. W. Orton and C. T. Foxon, Group III nitride semiconductors for short wavelength light-emitting devices, *Reports on Progress in Physics*, 61 (1998) 1–75.

[12.69] Z. Liliental-Weber, T. Tomaszewicz, D. Zakharov, J. Jasinski, and M. A. O'Keefe, Atomic structure of defects in GaN:Mg grown with Ga polarity, *Physical Review Letters*, 93 (2004) 206102:1–4.

[12.70] D. Segev and C. G. Van de Walle, Surface reconstructions on InN and GaN polar and nonpolar surfaces, *Surface Science*, 601 (2007) L15–L18.

[12.71] C. K. Gan and D. J. Srolovitz, First-principles study of wurtzite InN (0001) and (000$\bar{1}$) surfaces, *Physical Review B*, 74 (2006) 115319:1–5.

[12.72] D. Segev and C. G. Van de Walle, Origins of Fermi-level pinning on GaN and InN polar and nonpolar surfaces, *Europhysics Letters*, 76 (2006) 305–311.

[12.73] C. L. Wu, H. M. Lee, C. T. Kuo, C. H. Chen, and S. Gwo, Absence of Fermi-level Pinning at Cleaved Nonpolar InN Surfaces, *Physical Review Letters*, 101 (2008) 106803:1–4.

[12.74] A. Huijser, J. van Laar, and T. L. van Rooy, Electronic surface properties of UHV-cleaved III-V compounds, *Surface Science*, 62 (1977) 472–486.

[12.75] W. E. Spicer, P. W. Chye, P. E. Gregory, T. Sukegawa, and I. A. Babalola, Photoemission studies of surface and interface states on III-V compounds, *Journal of Vacuum Science and Technology*, 13 (1976) 233–240.

[12.76] G. Koblmüller, R. Averbeck, H. Riechert, and P. Pongratz, Direct observation of different equilibrium Ga adlayer coverages and their desorption kinetics on GaN (0001) and (000$\bar{1}$) surfaces, *Physical Review B*, 69 (2004) 035325:1–9.

[12.77] A. R. Smith, R. M. Feenstra, D. W. Greve, J. Neugebauer, and J. E. Northrup, Reconstructions of the GaN(000$\bar{1}$) surface, *Physical Review Letters*, 79 (1997) 3934–3937.

[12.78] J. E. Northrup, J. Neugebauer, R. M. Feenstra, and A. R. Smith, Structure of GaN(0001): The laterally contracted Ga bilayer model, *Physical Review B*, 61 (2000) 9932–9935.

[12.79] T. K. Zywietz, J. Neugebauer, M. Scheffler, and J. E. Northrup, Novel reconstruction mechanisms: A comparison between group-III-nitrides and "traditional" III-V-semiconductors, HCM Newsletter (Psi k Network) 29 (1998) 1–13; arXiv:physics/9810003.

[12.80] C. D. Lee, Y. Dong, R. M. Feenstra, J. E. Northrup, and J. Neugebauer, Reconstructions of the AlN(0001) surface, *Physical Review B*, 68 (2003) 205317:1–11.

[12.81] H. Chen, R. M. Feenstra, J. E. Northrup, T. Zywietz, J. Neugebauer, and D. W. Greve, Surface structures and growth kinetics of InGaN(0001) grown by molecular beam epitaxy, *Journal of Vacuum Science and Technology B*, 18 (2000) 2284–2289.

[12.82] M. Himmerlich, A. Eisenhardt, J. A. Schaefer, and S. Krischok, PAMBE growth and in-situ characterisation of clean (2×2) and ($\sqrt{3} \times \sqrt{3}$)R30° reconstructed InN(0001) thin films, unpublished.

[12.83] U. Grossner, J. Furthmüller, and F. Bechstedt, Initial stages of III-nitride growth, *Applied Physics Letters*, 74 (1999) 3851–3853.

[12.84] J. Neugebauer, Ab initio analysis of surface structure and adatom kinetics of group-III nitrides, *physica status solidi (b)*, 227 (2001) 93–114.

[12.85] H. Chen, R. M. Feenstra, J. Northrup, J. Neugebauer, and D. W. Greve, Indium incorporation and surface segregation during InGaN growth by molecular beam epitaxy: experiment and theory, *MRS Internet Journal of Nitride Semiconductor Research*, 6 (2001) 11.

[12.86] T. D. Veal, P. D. C. King, M. Walker, C. F. McConville, H. Lu, and W. J. Schaff, In-adlayers on non-polar and polar InN surfaces: Ion scattering and photoemission studies, *Physica B*, 401–402 (2007) 351–354.

[12.87] D. Muto, T. Araki, H. Naoi, F. Matsuda, and Y. Nanishi, Polarity determination of InN by wet etching, *physica status solidi (a)*, 202 (2005) 773–776.

[12.88] H. Niehus, W. Heiland, and E. Taglauer, Low-energy ion scattering at surfaces, *Surface Science Reports*, 17 (1993) 213-303; H. Niehus, FAN Simulation Software, Humboldt-Universität zu Berlin, Institut für Physik, Berlin (http://asp2.physik.hu-berlin.de/).

[12.89] T. D. Veal, P. H. Jefferson, L. F. J. Piper, C. F. McConville, T. B. Joyce, P. R. Chalker, L. Considine, H. Lu, and W. J. Schaff, Transition from electron accumulation to depletion at InGaN surfaces, *Applied Physics Letters*, 89 (2006) 202110:1–3.

[12.90] P. D. C. King, T. D. Veal, P. H. Jefferson, C. F. McConville, H. Lu, and W. J. Schaff, Variation of band bending at the surface of Mg-doped InGaN: Evidence of *p*-type conductivity across the composition range, *Physical Review B*, 75 (2007) 115312:1–7.

[12.91] P. D. C. King, T. D. Veal, H. Lu, P. H. Jefferson, S. A. Hatfield, W. J. Schaff, C. F. McConville, Surface electronic properties of n- and p-type InGaN alloys, *physica status solidi (b)*, 245 (2008) 881–883.

[12.92] P. D. C. King, T. D. Veal, A. Adikimenakis, H. Lu, L. R. Bailey, E. Iliopoulos, A. Georgakilas, W. J. Schaff, and C. F. McConville, Surface electronic properties of undoped InAlN alloys, *Applied Physics Letters*, 92 (2008) 172105:1–3.

[12.93] P. D. C. King, T. D. Veal, W. J. Schaff, and C. F. McConville, Surface electronic properties of Mg-doped InAlN alloys, *physica status solidi (b)* (2009) DOI:10.1002/pssb.200880766.

[12.94] C. A. Mead and W. G. Spitzer, Fermi level position at metal-semiconductor interfaces, *Physical Review*, 134 (1964) A713–A716.

[12.95] K. Kajiyama, Y. Mizushima, and S. Sakata, Schottky barrier height of *n*-In$_x$Ga$_{1-x}$As diodes, *Applied Physics Letters*, 23 (1973) 458–460.

[12.96] H. Lüth, Research on III-V semiconductor interfaces: Its impact on technology and devices, *physica status solidi (a)*, 187 (2001) 33–44.

[12.97] V. M. Bermudez, Study of oxygen chemisorption on the GaN(0001)-(1×1) surface, *Journal of Applied Physics*, 80 (1996) 1190–1200.

[12.98] L. F. J. Piper, T. D. Veal, M. Walker, I. Mahboob, C. F. McConville, H. Lu, and W. J. Schaff, Clean wurtzite InN surfaces prepared with atomic hydrogen, *Journal of Vacuum Science and Technology A*, 23 (2005) 617–620.

[12.99] P. D. C. King, T. D. Veal, C. E. Kendrick, L. R. Bailey, S. M. Durbin, and C. F. McConville, InN/GaN valence band offset: High-resolution x-ray photoemission spectroscopy measurements, *Physical Review B*, 78 (2008) 033308:1–4.

[12.100] K. M. Tracy, W. J. Mecouch, R. F. Davis, and R. J. Nemanich, Preparation and characterization of atomically clean, stoichiometric surfaces of *n*- and *p*-type GaN(0001) *Journal of Applied Physics*, 94 (2003) 3163–3172.

[12.101] P. D. C. King, T. D. Veal, P. H. Jefferson, C. F. McConville, T. Wang, P. J. Parbrook, H. Lu, and W. J. Schaff, Valence band offset of InN/AlN heterojunctions measured by x-ray photoelectron spectroscopy, *Applied Physics Letters*, 90 (2007) 132105:1–3.

[12.102] C. Stampfl and C. G. Van de Walle, Theoretical investigation of native defects, impurities, and complexes in aluminum nitride, *Physical Review B*, 65 (2002) 155212:1–10.

[12.103] W. Walukiewicz, Mechanism of Schottky barrier formation: The role of amphoteric native defects, *Journal of Vacuum Science and Technology B*, 5 (1987) 1062–1067.

13

Theory of InN surfaces

C. G. Van de Walle

Materials Department, University of California, Santa Barbara, CA 93106-5050, United States of America

13.1   Introduction

InN is a fascinating material that exhibits a host of unusual properties. Even though it is a member of the III-nitride family, its band gap (0.7 eV) is significantly smaller than that of AlN (6.2 eV) or GaN (3.4 eV). This offers, of course, great opportunities for band-gap engineering. It turns out that most of the band-gap difference between InN and GaN is accommodated in the conduction band (see Fig. 13.1, based on band lineups calculated in Ref. [13.1]). This places the conduction band of InN at an unusually low position on an absolute energy scale (for example, with respect to the vacuum level [13.2]), and one might expect this would affect the properties of InN. Such a low conduction-band position renders it more likely for a semiconductor to exhibit *n*-type conductivity, and indeed, InN has a strong tendency to be highly *n*-type doped. We will see that this conduction-band position also affects the properties of surfaces, and indeed is responsible for the high electron concentrations that have been observed at InN surfaces [13.3].

The recently acquired ability to grow high-quality InN, covered extensively in other chapters of this volume, has caused a surge in experimental studies of InN since the year 2000. Very few of the experimental studies have explicitly addressed structural aspects of InN surfaces, but it has been universally observed that high electron concentrations are present on the surface [13.3–13.6]. In this chapter, an overview of theoretical information that has been obtained on InN surfaces is presented. Most of this information has been derived from first-principles calculations based on density functional theory. Such calculations have become established as a powerful tool for studying surface reconstructions of semiconductors. A number of such calculations have indeed been performed for GaN surfaces by various groups; some of these will be cited at appropriate points in the text below. For InN, to our knowledge the only studies that have been published are by Großner *et al.* [13.7], Gan and Srolovitz [13.8] and by Segev and Van de Walle [13.9, 13.10]. The study of Großner *et al.* [13.7] was limited to investigations of In adlayers on a zinc-blende (111) 1×1 surface and concluded that formation of such adlayers is indeed favorable under In-rich conditions. Gan and Srolovitz [13.8] investigated wurtzite

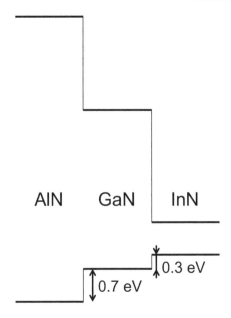

FIGURE 13.1
Natural band lineups between AlN, GaN, and InN, obtained from first-principles calculations [13.1].

(0001) and (000$\bar{1}$) surfaces. Segev and Van de Walle covered a broader range of surface reconstructions on various surface orientations of wurtzite InN, and not only addressed atomic structure but also the electronic structure of the surface, a feature that is extremely relevant in the case of InN. The results will be discussed in detail below.

In Section 13.2 the computational techniques are briefly reviewed. Section 13.3 presents results for surface reconstructions, while Section 13.4 focuses on the electronic structure. Section 13.5 provides a summary and outlook.

13.2 Computational methodology

13.2.1 Density functional theory

The computational approach is based on density functional theory (DFT), which has a long and strong track record of producing reliable results for surface reconstructions of semiconductors. The calculations use either the local density approximation (LDA) for the exchange-correlation potential, with the Ceperley-Alder data [13.11] as parameterized by Perdew and Zunger [13.12], or the generalized gradient approximation (GGA) [13.13]. Norm-conserving pseudopotentials [13.14] are used, as well as the projected augmented wave approach (PAW) [13.15], as implemented in the VASP code [13.16].

The surfaces are simulated in a repeated slab geometry, using slabs of up to ten double layers of GaN or InN, with a vacuum region of up to 25 Å. The bottom surface of the slab is passivated with fractionally charged hydrogen atoms [13.17], and the lower four layers are kept fixed during the atomic relaxations. Convergence with respect to **k**-point sampling, supercell size, slab and vacuum thickness is always explicitly checked.

13.2.2 Surface energies

The formalism for surface energy calculations is well established [13.17, 13.18]. The energy $\Delta\gamma(\mu_{Ga})$ of a given reconstructed surface with respect to the bare, relaxed 1×1 unreconstructed GaN surface, is defined as

$$\Delta\gamma(\mu_{Ga}) = E_S^{tot} - E_{bare}^{tot} - (n_{Ga} - n_N)\mu_{Ga} - n_N\mu_{GaN}^{bulk}. \tag{13.1}$$

Note there is no attempt here to define an "absolute" surface energy. In fact, for the polar surface of wurtzite InN it is not possible to define such an absolute energy; only the *sum* of the energies of the (0001) and (000$\bar{1}$) surfaces can be determined. E_S^{tot} and E_{bare}^{tot} are the total energies of the surface under consideration and of the reference surface, respectively. The chemical potential terms μ_{Ga} and μ_N appear because the surface can exhibit a stoichiometry that differs from the bulk, and atoms are exchanged with gallium and nitrogen reservoirs. n_{Ga} (n_N) is the number of Ga (N) atoms added to (positive) or substracted from (negative) the surface. In equilibrium μ_{Ga} and μ_N are related to the energy μ_{GaN}^{bulk} per formula unit of bulk GaN through the expression

$$\mu_{Ga} + \mu_N = \mu_{GaN}^{bulk} = \mu_{Ga}^{bulk} + \mu_{N_2} + \Delta H_{GaN}^f. \tag{13.2}$$

The Ga chemical potential μ_{Ga} is the free energy of the Ga reservoir; it varies between the Ga-rich limit, for which $\mu_{Ga} = \mu_{Ga}^{bulk}$, and the N-rich limit ($\mu_N = \mu_{N_2}$), for which $\mu_{Ga} = \mu_{Ga}^{bulk} + \Delta H_{GaN}^f$, where ΔH_{GaN}^f is the formation enthalpy of GaN (negative for a stable compound). Pressure and vibrational contributions to the free energy are neglected, as justified in Ref. [13.17].

Using this formalism, results for surface reconstructions and surface energies for GaN were obtained [13.19] that were in good agreement with published results [13.17–13.21]. For the (0001) and (000$\bar{1}$) surfaces, 2×2 reconstructions were found to be particularly stable, especially those that satisfy the "electron counting rule". The essence of this rule is that energetically favorable arrangements of atoms occur when (1) all bonds between atoms are filled with two electrons; (2) any cation dangling bonds (which have energies high in the gap, near the conduction band) are unoccupied; and (3) any anion dangling bonds (which have energies low in the gap, close to the valence bands) are fully occupied. The calculated formation energies as a function of chemical potential are shown in the left panels of Fig. 13.2. As the Ga chemical potential increases, the (0001) surface undergoes a transition from a 2×2 N-adatom structure to a 2×2 Ga-adatom reconstruction (2×2 Ga$_{T4}$)

for moderate Ga/N ratios, and then to a laterally contracted Ga-bilayer structure at highly Ga-rich conditions. The $(000\bar{1})$ surface evolves from a 2×2 Ga-adatom reconstruction (2×2 Ga$_{H3}$) to a 1×1 Ga-adlayer reconstruction (1×1 Ga$_{atop}$, with Ga atoms sitting directly on top of N) with increasing Ga/N ratio.

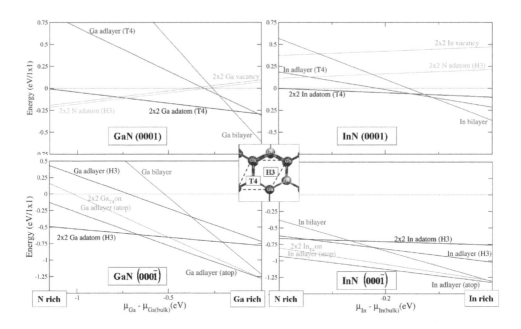

FIGURE 13.2

Energies of surface reconstructions on GaN (0001) and $(000\bar{1})$ (left panels) and InN (0001) and $(000\bar{1})$ (right panels) surfaces as a function of Ga and In chemical potentials. The inset shows a ball-and-stick representation of a top view of the (0001) surface, indicating the $T4$ and $H3$ sites. The small balls are N atoms. Reprinted from Surface Science 601, D. Segev and C. G. Van de Walle, "Surface reconstructions on InN and GaN polar and non-polar surfaces", L15–L18, Copyright 2007, with permission from Elsevier.

13.2.3   The band-gap problem

Both LDA and GGA produce a negative band gap for InN. This is obviously a severe problem when trying to address the electronic structure of the surface. The underestimate of the band gap is a well-known problem of DFT; it is related to the fact that DFT was constructed as a ground-state theory aimed at obtaining accurate charge densities and energies, while the electronic states that appear in the associated band structure are only auxiliary quantities that are not guaranteed to have physical meaning. Still, it has become well established that the calculated LDA or GGA band structures generally compare well with experiment, except for the underestimate of the band gap. Clearly, to be able to perform studies of surface electronic structure, an approach must be adopted that would produce a finite band gap for InN.

One approach that has been used to address band-gap problems in semiconductors is to perform quasiparticle calculations in the GW formalism [13.22–13.24]. This method is much more computationally intensive than regular DFT calculations, and particularly its application to surfaces is far from straightforward [13.24, 13.25]. An additional problem in the case of InN is that, in order to properly describe the band structure of InN, GW has to be combined with the exact exchange method [13.26, 13.27]. Exact exchange is even more computationally expensive than GW, and therefore impractical for systematic studies of surfaces of InN. Moreover, current implementations of GW do not allow for calculation of total energies, and therefore one could not self-consistently obtain surface structures that are consistent with the electronic structure.

In Chapter 11, several other methods are reviewed that can be used to overcome the band-gap problem. Here we focus on the so-called "modified pseudopotential" (MPP) approach, which has been successfully used to study surfaces of both GaN and InN. In this approach [13.28], an atom-centered repulsive potential of Gaussian shape is added at the all-electron stage of the pseudopotential generation, within LDA and the norm-conserving scheme. The approach is similar to a scheme developed by Christensen [13.29] in the context of the linearized muffin-tin-orbital method. The repulsive potential acts primarily on the lowest-energy $1s$ state, and affects higher-lying states through orthogonality of the wave functions. As a consequence, the semicore In $4d$ states are less screened and become more localized. Judicious choice of the Gaussian potentials can result in band structures in close agreement with experiment, while preserving the ability of the pseudopotentials to produce structures and energies that are as accurate as those obtained with regular pseudopotentials. When applied to surfaces, the approach allows for performing fully self-consistent calculations of total energies and atomic relaxations, while simultaneously obtaining band structures in which surface states are calculated within a band gap that is very close to the experimental value.

The modified pseudopotentials are generated to yield a direct band gap of 0.75 eV for InN, using a plane-wave cutoff of 60 Ry and theoretical lattice parameters of a=3.58 Å, c/a=1.617, and u=0.379, within 1% of the experimental values. The band structures produced with the modified pseudopotentials agree remarkably well with experiment (better than expected based on fitting just the direct gaps to experimental values). This was tested in particular for GaN, for which more reliable band structures are available than for InN. Features other than the direct band gap, such as the energetic positions of semi-core d states and of higher-lying conduction-band minima, exhibited marked improvements as well. Details about these comparisons can be found in Ref. [13.28].

Before applying them to InN the modified pseudopotentials were extensively tested for surface reconstructions on GaN, for which a certain amount of experimental information is available. Reassuringly, the tests for both GaN and InN show that the modified pseudopotentials result in very similar surface structures as well as energies compared to LDA and GGA. This justifies the systematic and consistent use of these pseudopotentials for all aspects of the calculations, structural as well as electronic.

13.2.4   Electronic structure

To evaluate the reliability of modified pseudopotentials to calculate surface electronic structure, tests were performed [13.28] for a case where experimental information is available, namely, for the GaN (0001) surface. The 2×2 Ga_{T4} reconstruction on the GaN (0001) surface consists of one Ga adatom added to a 2×2 surface unit cell. This adatom binds to three Ga atoms in the outermost surface layer, while one surface Ga remains threefold coordinated. Both this surface atom and the Ga adatom therefore have one dangling bond. Electron counting shows that these dangling bonds are unoccupied. The calculated surface band structures indeed show the presence of two nearly degenerate surface states in the upper part of the band gap, and plotting of the associated wave functions clearly shows they are localized on the Ga dangling bonds.

These surface states play an important role for the electronic structure of the surface. They have a very high areal density of $\sim6\times10^{14}$ cm^{-2}, and on n-type GaN they can accommodate the electronic charge that is removed from a depletion layer in the near-surface region. Upward band bending occurs, and the Fermi level becomes pinned near the lower edge of these unoccupied surface states. In fact, an experimental determination of this pinning position has been reported: Kočan *et al.* [13.30] used *in situ* x-ray photoemission spectroscopy (XPS) and found that under experimental conditions (moderate Ga/N ratios) consistent with the presence of the 2×2 Ga_{T4} reconstruction, the Fermi level was pinned at 2.9 eV (that is, 0.5 eV below the conduction-band minimum (CBM)). The results also agree well with the experimental data of Long *et al.* [13.31] (E_F-E_v=2.7 eV, in the limit of low doping concentration), as well as with the results of Wu *et al.* [13.32] (2.6 eV).

In LDA calculations, the unoccupied surface states occur at about 2.0 eV above the valence-band maximum (VBM) in a band gap of 2.1 eV; that is, they are located 0.1 eV below the (theoretical) CBM. This situation illustrates the uncertainties faced when using typical DFT-LDA results to try and compare with experiment. Should we focus on the relative position with respect to the VBM, or with respect to the CBM? The situation becomes even more confused when also inspecting GGA results, where the unoccupied surface states occur at about 1.9 eV above the VBM, with a GGA band gap of 1.7 eV; that is, here the surface states are actually located *above* the (theoretical) CBM. None of these agrees with the experimental result.

Using modified pseudopotentials, the unoccupied surface states are found to lie at 2.8 eV above the VBM, that is, 0.6 eV below the CBM, a result that agrees to within 0.1 eV with the experimental Fermi-level pinning position. This result provides confidence that the MPP approach is capable of predicting electronic structure.

13.3 Results for surface reconstructions

13.3.1 Polar surfaces

Results for the (0001) and (000$\bar{1}$) surfaces of InN are shown in the right panels of Fig. 13.2. Note that the scale of the chemical potential axis is quite different for the two materials. The 2×2 In$_{T4}$ structure is stable on InN(0001) at moderate In/N ratios and a contracted In-bilayer structure is stable under highly In-rich conditions. It is interesting that, in contrast with GaN, no reconstructions involving N adatoms are found under N-rich conditions. The 2×2 In$_{T4}$(0001) structure thus remains stable even under In-poor conditions. Northrup *et al.* [13.33] have reported that on the AlN(0001) surface the 2×2 N-adatom structure is stable over almost the entire range of Al chemical potentials. Therefore, the finding for InN seems consistent with a trend of destabilization of the N adatom structure with increasing cation atomic number in the III-nitride compounds. This may be attributed to a decrease in bond strength of the cation-nitrogen bond when going from Al to Ga to In.

Reconstructions on the InN(000$\bar{1}$) surface are also very different from the GaN case: 1×1 In$_{atop}$ is stable over the entire range of In chemical potentials (see Fig. 13.2). Since this is a structure in which one monolayer of In is added to the surface, it results in a metallic surface. This structure is lower in energy than other metallic structures, including an In adlayer with the atoms in $H3$ positions, an In bilayer, and a reconstruction with additional In adatoms [2×2 In$_{T4}$ on In adlayer (atop)]. The 2×2 In$_{H3}$ (000$\bar{1}$) structure is found not to be stable for any In chemical potential, in contrast to the situation on the GaN(000$\bar{1}$) surface. The results for InN (0001) and (000$\bar{1}$) surfaces presented in Fig. 13.2 are quite similar to those of Gan and Srolovitz [13.8]. Again, an interesting trend can be observed: on the AlN(000$\bar{1}$) surface the 2×2 Al$_{H3}$ reconstruction is stable over almost the entire range of Al chemical potentials [13.33]. The trend seems to be that the 2×2 cation$_{H3}$ structure becomes less stable on the nitride (000$\bar{1}$) surface with increasing cation atomic number.

13.3.2 Non-polar surfaces

Energies of surface reconstructions on non-polar *m*- and *a*-planes of InN are shown in Fig. 13.3. Again, results for GaN are included for comparison. For both materials, the unreconstructed (but relaxed) surface consisting of Ga(In)-N dimers is stable over a large range of chemical potentials. For GaN, this result agrees with previous calculations [13.18]. In these dimers, the N atom relaxes outward (toward an s^2p^3 configuration) while the Ga atom relaxes inward (toward sp^2), accompanied by a charge transfer from the Ga dangling bond to the N dangling bond. This is typical behavior for non-polar surfaces of compound semiconductors.

Both GaN and InN also exhibit other surface reconstructions under cation-rich conditions, that is, when the cation chemical potential is within 0.2 eV of its max-

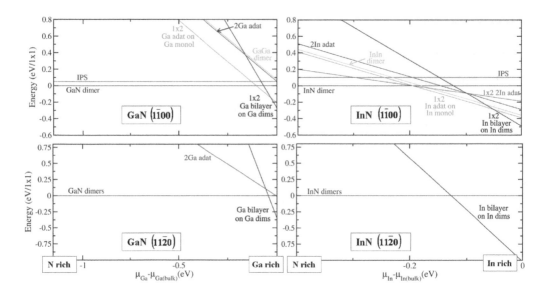

FIGURE 13.3

Energies of surface reconstructions on GaN m ($1\bar{1}00$) and a ($11\bar{2}0$) planes (left panels) and InN m ($1\bar{1}00$) and a ($11\bar{2}0$) planes (right panels) as a function of Ga and In chemical potentials. Reprinted from Surface Science 601, D. Segev and C. G. Van de Walle, "Surface reconstructions on InN and GaN polar and nonpolar surfaces", L15–L18, Copyright 2007, with permission from Elsevier.

imum value. The reconstructions for GaN were discussed in Ref. [13.19]. For the m-plane of InN at high In chemical potentials a 1×2 reconstruction appears which consists of an In adatom on top of a (pseudo-)monolayer of In comprising three In atoms per unit cell. The most stable structure at the highest In chemical potentials consists of an In bilayer on top of a surface in which two N atoms in the outermost GaN surface layer are replaced by two Ga atoms. For an In chemical potential equal to that of bulk In, the energy of this reconstruction is about 0.5 eV per 1×1 unit cell lower than the relaxed, bare InN m-plane.

On the a-plane of InN, a stable 1×1 reconstruction (labeled "In bilayer on In dims") appears at high In chemical potentials. It is formed by replacing two N atoms by two In atoms in the outermost surface layer, and adding an In pseudo-monolayer or bilayer consisting of four In atoms. Its energy is more than 1 eV lower than the relaxed, bare InN a-plane, for an In chemical potential equal to that of bulk In.

13.4 Results for electronic structure

13.4.1 Polar surfaces

Figure 13.4 shows band structures for selected reconstructions of the (0001) and (000$\bar{1}$) surfaces; plots for GaN are included for comparison. As discussed in Section 13.2.4, the 2×2 Ga$_{T4}$ (0001) is stable at moderate Ga/N ratios (see Fig. 13.2) and contains a Ga adatom that binds to three surface Ga atoms, with one surface Ga remaining threefold coordinated. The corresponding band structure is shown in Fig. 13.4(a). The 2×2 Ga$_{T4}$ (0001) structure results in a semiconducting surface, with two sets of surface states appearing within the band gap. By inspecting the electronic charge densities associated with the surface states, it is possible to identify their microscopic origins. The upper, empty states (at ∼0.6 eV below the CBM) arise from the dangling bond on the Ga adatom, with some charge density also on nearby sublayer N atoms, and from the dangling bond on the threefold-coordinated Ga surface atom. The lower states at ∼1.7 eV above the VBM are fully occupied and are well localized on bonds between the Ga adatom and three Ga surface atoms.

For the GaN (000$\bar{1}$) surface, a 2×2 Ga$_{H3}$ reconstruction is stable at moderate Ga/N ratios (see Fig. 13.2), in which a Ga adatom binds to three surface N atoms, and one surface N atom remains threefold coordinated. The upper, empty state in the corresponding band structure, shown in Fig. 13.4(b), is mainly derived from the Ga-adatom dangling bond, while the lower occupied state corresponds to a nitrogen dangling bond and is close in energy to the VBM. We note that these surface reconstructions give rise to states with rather small dispersion, due to relatively small interactions between the states that give rise to the bands. These surface states therefore result in fairly sharp peaks in the density of states (DOS) [13.10].

As noted in Section 13.2.4, these unoccupied and occupied states have an areal density of ∼6×10^{14} cm^{-2} each. The unoccupied surface states can therefore readily accept electrons taken from a near-surface depletion layer, effectively acting to pin the Fermi level at the surface of *n*-type GaN at ∼0.6 eV below the CBM for the (0001) plane; as noted in Section 13.2.4, this is in agreement with experiment, giving confidence that similar calculations for InN are also reliable.

The electronic band structures of the 2×2 In$_{T4}$ (0001) and 1×1 In$_{atop}$ (000$\bar{1}$) reconstructed surfaces of InN at moderate In/N ratios are shown in Figs. 13.4(c) and (d). Two sets of surface states appear for the 2×2 In$_{T4}$ (0001) structure, with similar character as in the case of GaN. However, the small band gap and large electron affinity of InN results in a low energetic position of the CBM at the Γ point. As a consequence, both sets of surface states occur at energies *above* the CBM.

The presence of such occupied surface states above the CBM provides an immediate explanation for the observed electron accumulation on InN polar surfaces.

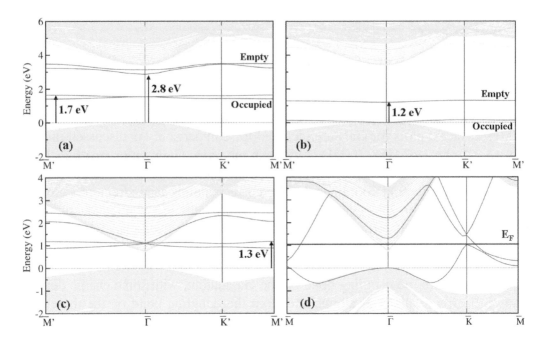

FIGURE 13.4

Electronic band structures of the (a) 2×2 Ga_{T4} (0001); (b) 2×2 Ga_{H3} (000$\bar{1}$); (c) 2×2 In_{T4} (0001); (d) 1×1 In_{atop} (000$\bar{1}$) reconstructed surfaces. The gray background indicates the projected bulk band structure. The zero of energy is set at the bulk VBM. Relevant energy differences between the VBM and surface states (bold lines) are indicated by arrows. For the 1×1 In_{atop} (000$\bar{1}$) structure, with highly dispersive surface states, the Fermi level E_F is also indicated. Reprinted with permission from C. G. Van de Walle and D. Segev, Journal of Applied Physics 101, 081704 (2007). Copyright 2007, American Institute of Physics.

Because the number of surface states is much larger than the number of available bulk states in the near-surface accumulation layer, the surface Fermi-level position is approximately determined by the position of the upper portion of the occupied surface states, which is $\sim$0.6 eV above the CBM. This result compares well with the experimental value of $\sim$0.8 eV [13.5, 13.34].

We can also make a simple order-of-magnitude estimate of the resulting sheet carrier density. Wu *et al.* [13.35] reported that a Fermi-level position of 0.6 eV above the CBM corresponds to a bulk electron concentration of $\sim 2\times10^{20}$ cm^{-3}. At that concentration, and using the InN dielectric constant and a band bending of 0.6 eV, the space-charge-layer width can be estimated to be $\sim$2 nm. Assuming a constant electron density over that width, we would find an sheet carrier density of 4×10^{13} cm^{-2} confined to a 2-nm thin layer near the surface. Our estimate is in reasonable agreement with the experimental estimates of the surface sheet density reported in Chapter 12.

Under highly Ga-rich conditions, the (0001) polar surface is covered by a laterally contracted double layer of Ga [13.21]. At this high coverage, the Ga-Ga bonding and dangling-bond states that were distinct at moderate Ga/N ratios now strongly interact, leading to a large energy dispersion within the band gap. This results in a metallic surface; the cations in the bilayer form relatively strong metallic bonds within the plane and between the two adlayers, and the Fermi level is located at ~1.8 eV above the VBM, close to the experimental value of 1.7 eV reported in Ref. [13.30]. An analogous situation occurs on the $(000\bar{1})$ structure under highly Ga-rich conditions.

Similar physics applies in the case of the In bilayer on InN (0001) under highly In-rich conditions. Since the electronic structure is again dominated by In-In bonds, it is not surprising that the Fermi level is again at ~0.7 eV above the CBM, very close to the 0.6 eV found for moderate In/N ratios.

For the $(000\bar{1})$ surface, Fig. 13.2 shows that the 1×1 In$_{atop}$ reconstruction is stable over the entire range of chemical potentials. It results in relatively dispersive states due to the interaction between neighboring In adatoms, and the surface is metallic, with a Fermi level at about 0.3 eV above the CBM, as shown in Fig. 13.4(d).

The occurrence of electron accumulation on InN surfaces has been attributed to the unusual position of the "charge neutrality level" (CNL) in InN [13.3]. This level occurs at approximately the average midgap energy, when valence and conduction bands are averaged over the entire Brillouin zone [13.36]. Because of the unusually low position of the CBM at Γ, the CNL in InN lies above the CBM. The CNL can be interpreted as the energy at which localized states change their character from donor-like to acceptor-like. Surface states below the CNL have donor character, and if the CNL is above the CBM such states can donate electrons to near-surface bulk conduction-band states, leading to electron accumulation [13.3, 13.37]. This picture is appealing in its simplicity and offers nice physical insight; however, it does *not* provide for a microscopic identification of the surface states, nor indeed does it guarantee that such surface states will be present. All it tells us is that there will be a *tendency* for such states with donor-like character to form. We will see in the next section that it is actually possible to obtain stable surface reconstructions for which such states are not present.

13.4.2 Non-polar surfaces

For the non-polar *m* and *a* surfaces, we noted that at moderate Ga(In)/N ratios the unreconstructed dimer structure is most stable. Consequently, the electronic structure shows an unoccupied Ga-dangling-bond state at ~0.7 eV below the CBM for the GaN *m*-plane (Fig. 13.5(a)), and two unoccupied Ga-dangling-bond states at ~0.5 eV below the CBM for the *a*-plane (Fig. 13.5(b)) (the unit cell of the *a*-plane is twice as large as that of the *m*-plane). The Ga-derived dangling bond energy level is nearly the same as for the polar (0001) surface, resulting in similar Fermi-level pinning on *n*-type GaN. In contrast, the *occupied* surface states behave differently on non-polar *versus* polar surfaces: on the *m*- and *a*-planes, these states are associated with dangling bonds on the N atom and overlap with the valence band; that is, they do *not* create levels within the band gap (Figs. 13.5(a) and (b)).

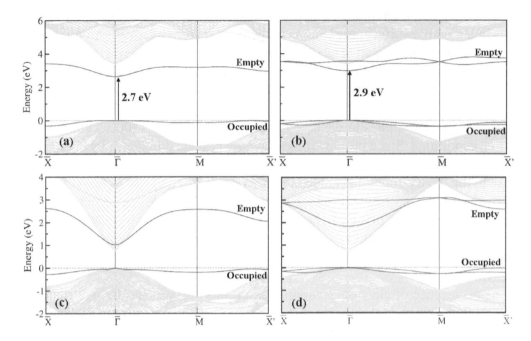

FIGURE 13.5

Electronic band structures of the (a) GaN dimer $(1\bar{1}00)$; (b) GaN dimers $(11\bar{2}0)$; (c) InN dimer $(1\bar{1}00)$; and (d) InN dimer $(11\bar{2}0)$ reconstructed surfaces. The gray background indicates the projected bulk band structure. The zero of energy is set at the bulk VBM. Relevant energy differences between the VBM and surface states (bold lines) are indicated by arrows. Reprinted with permission from C. G. Van de Walle and D. Segev, Journal of Applied Physics 101, 081704 (2007). Copyright 2007, American Institute of Physics.

Once we understand this electronic structure, it is easy to understand why non-polar surfaces of InN behave very differently from the polar surfaces. On polar surfaces, In-In bonds are present that give rise to occupied surface states above the CBM, and hence to generation of an electron accumulation layer. For moderate In/N ratios, such In-In bonds are absent from the non-polar surfaces. On the InN *m*- and *a*-planes the occupied surface states are close to or below the VBM, as shown in Figs. 13.5(c) and (d). Therefore, for moderate In/N ratios these results predict an absence of electron accumulation on the non-polar InN $(1\bar{1}00)$ and $(11\bar{2}0)$ surfaces, in contrast with the polar surfaces. The unoccupied surface states associated with In dangling bonds are well above the CBM, with no discernible effect on the electronic properties of the surface.

For highly Ga(In)-rich conditions, the electronic structure of the non-polar surfaces is qualitatively similar to that of the polar surfaces. Indeed, all of these surface reconstructions involve metallic adlayers, and therefore the electronic structure is again characterized by highly dispersive bands crossing the Fermi level. For GaN this pins the Fermi level at $\sim$1.8 eV above the VBM. In the case of InN, we find that the Fermi level is located about 0.6 eV above the CBM, as for the In bilayer on the (0001) polar surface. Non-polar surfaces of InN therefore exhibit electron accumulation when In adlayers are present on the surface. However, it may be possible to remove such adlayers in post-growth processing, resulting in a surface consisting purely of In-N dimers, for which electron accumulation is absent.

Some experimental studies of non-polar surfaces have already been performed. Non-polar epitaxial films have already been grown [13.38]. King *et al.* [13.37] used x-ray photoemission spectroscopy and found an accumulation layer not only on *c*-plane but also on *a*-plane InN, as well as on the zinc blende (001) surface. Other experiments have investigated nanocolumns that have non-polar side facets [13.39]; electrical measurements also showed the presence of an accumulation layer. In all these cases, the presence of an accumulation layer could of course be due to the presence of impurities such as oxygen on the surface, or due to the presence of metallic surface reconstructions. Further studies are called for, both to examine the effects of different growth conditions and enhanced impurity control, and to investigate the effects of postgrowth manipulation.

13.5 Summary and outlook

In summary, we have presented an overview of computational studies of recon-structed InN surfaces in various orientations, including $(1\bar{1}00)$ (*m*-plane), $(11\bar{2}0)$ (*a*-plane), as well as the polar (0001) $(+c)$ and $(000\bar{1})$ $(-c)$ planes. Similarities but also important differences with GaN have been discussed.

Reliable self-consistent calculations of surface electronic structure were possible thanks to the use of modified pseudopotentials. The resulting band structures allow the surface states to be indentified with specific atomistic features of the surface. For GaN, it is found that Ga dangling bonds give rise to empty states at above 2.7 eV above the VBM, N dangling bonds produce occupied states near the VBM, and Ga-Ga bonds give rise to occupied states around 1.7 eV above the VBM. The same qualitative features occur for InN, but with the important difference that the band gap is much smaller and the CBM occurs at a much lower energy than in GaN. As a result, the occupied surface states associated with In-In bonds occur above the CBM. Electrons in these states can lower their energy by transferring to near-surface bulk states, thus giving rise to the observed electron accumulation layer. We note that the observed high surface conductivity is really due to the presence of electrons in extended bulk-like states (even though these electrons are confined to the near-surface layer due to band bending). Electrons in such states can exhibit high mobility; the electrons residing in the surface states themselves, on the other hand, are expected to have very limited mobility since they can only move through a hopping process.

Interestingly, the electronic structure of non-polar InN is found to be very differ-ent from that of polar surfaces, since no occupied surface states are present above the CBM (at least when metallic adlayers on the surface are avoided). Further experimental studies of non-polar surfaces should be informative. Electrical measurements performed on nanocolumns that have non-polar side facets showed the presence of an accumulation layer [13.39]. As the authors acknowledged, this could of course be due to the presence of impurities such as oxygen on the surface, or due to the presence of metallic surface reconstructions. Further studies are called for, both to examine the effects of postgrowth manipulation and to investigate epitaxial films rather than nanostructures.

Acknowledgments

I gratefully acknowledge collaborations with D. Segev and A. Janotti, and helpful discussons with C. Gallinat, G. Koblmüller, and J. S. Speck. The work was supported in part by ONR under Award #N00014-02-C-0433, monitored by C. Baatar, through a subcontract from the Palo Alto Research Center Incorporated, and by ONR under Award # N00014-08-1-0095, monitored by P. Maki. It also made use of the CNSI Computing Facility under NSF grant No. CHE-0321368.

References

[13.1] C. G. Van de Walle and J. Neugebauer, Small valence-band offsets at GaN/InGaN heterojunctions, *Applied Physics Letters*, 70 (1997) 2577–2579.

[13.2] C. G. Van de Walle and J. Neugebauer, Universal alignment of hydrogen levels in semiconductors, insulators, and solutions, *Nature*, 423 (2003) 626–628.

[13.3] I. Mahboob, T. D. Veal, L. F. J. Piper, C. F. McConville, H. Lu, W. J. Schaff, J. Furthmüller, and F. Bechstedt, Origin of electron accumulation at wurtzite InN surfaces, *Physical Review B*, 69 (2004) 201307(R):1–4.

[13.4] H. Lu, W. J. Schaff, L. F. Eastman, and C. E. Stutz, Surface charge accumulation of InN films grown by molecular-beam epitaxy, *Applied Physics Letters*, 82 (2003) 1736–1738.

[13.5] I. Mahboob, T. D. Veal, and C. F. McConville, H. Lu, and W. J. Schaff, Intrinsic electron accumulation at clean InN surfaces, *Physical Review Letters*, 92 (2004) 036804:1–4.

[13.6] T. D. Veal, I. Mahboob, L. F. J. Piper, C. F. McConville, H. Lu, and

W. J. Schaff, Indium nitride: Evidence of electron accumulation, *Journal of Vacuum Science and Technology B*, 22 (2004) 2175–2178.

[13.7] U. Großner, J. Furthmüller, and F. Bechstedt, Initial stages of III-nitride growth, *Applied Physics Letters*, 74 (1999) 3851–3853.

[13.8] C. K. Gan and D. J. Srolovitz, First-principles studies of wurtzite InN (0001) and (000$\bar{1}$) surfaces, *Physical Review B*, 74 (2006) 115319:1–5.

[13.9] D. Segev and C. G. Van de Walle, Origins of Fermi-level pinning on GaN and InN polar and nonpolar surfaces, *Europhysics Letters*, 76 (2006) 305–311.

[13.10] C. G. Van de Walle and D. Segev, Microscopic origins of surface states on nitride surfaces, *Journal of Applied Physics*, 101 (2007) 081704:1–6.

[13.11] D. M. Ceperley and B. J. Alder, Ground state of the electron gas by a stochastic method, *Physical Review Letters*, 45 (1980) 566–569.

[13.12] J. P. Perdew and A. Zunger, Self-interaction correction to density-functional approximations for many-electron systems, *Physical Review B*, 23 (1981) 5048–5079.

[13.13] J. P. Perdew, K. Burke, and M. Ernzerhoff, Generalized gradient approximation made simple, *Physical Review Letters*, 77 (1996) 3865–3868.

[13.14] N. Troullier and J. L. Martins, Efficient pseudopotentials for plane-wave calculations, *Physical Review B*, 43 (1991) 1993–2006.

[13.15] P. E. Blöchl, Projector augmented-wave method, *Physical Review B*, 50 (1994) 17953–17979.

[13.16] G. Kresse and D. Joubert, From ultrasoft pseudopotentials to the projector augmented-wave method, *Physical Review B*, 59 (1999) 1758–1775.

[13.17] C. G. Van de Walle and J. Neugebauer, First-principles surface phase diagram for hydrogen on GaN surfaces, *Physical Review Letters*, 88 (2002) 066103:1–4.

[13.18] J. E. Northrup and J. Neugebauer, Theory of GaN($10\bar{1}0$) and ($11\bar{2}0$) surfaces, *Physical Review B*, 53 (1996) R10477–R10480.

[13.19] D. Segev and C. G. Van de Walle, Surface reconstructions on InN and GaN polar and nonpolar surfaces, *Surface Science*, 601 (2007) L15–L18.

[13.20] A. R. Smith, R. M. Feenstra, D. W. Greve, J. Neugebauer, and

J. E. Northrup, Reconstructions of the GaN(0001$\bar{1}$) surface, *Physical Review Letters*, 79 (1997) 3934–3937.

[13.21] J. E. Northrup, J. Neugebauer, R. M. Feenstra, and A. R. Smith, Structure of GaN(0001): The laterally contracted Ga bilayer model, *Physical Review B*, 61 (2000) 9932–9935.

[13.22] L. Hedin, New method for calculating the one-particle Green's function with application to the electron-gas problem, *Physical Review*, 139 (1965) A796–A823.

[13.23] M. S. Hybertsen and S. G. Louie, Electron correlation in semiconductors and insulators: Band gaps and quasiparticle energies, *Physical Review B*, 34 (1986) 5390–5413.

[13.24] M. Hedström, A. Schindlmayr, G. Schwarz, and M. Scheffler, Quasiparticle corrections to the electronic properties of anion vacancies at GaAs(110) and InP(110), *Physical Review Letters*, 97 (2006) 226401:1–4.

[13.25] C. Freysoldt, P. Eggert, P. Rinke, A. Schindlmayr, and M. Scheffler, Screening in 2D: GW calculations for surfaces and thin films using the repeated-slab approach, *Physical Review B*, 77 (2008) 235428:1–11.

[13.26] P. Rinke, A. Qteish, J. Neugebauer, C. Freysoldt, and M. Scheffler, Combining GW calculations with exact-exchange density-functional theory: an analysis of valence-band photoemission for compound semiconductors, *New Journal of Physics*, 7 (2005) 126:1–35.

[13.27] P. Rinke, M. Scheffler, A. Qteish, M. Winkelnkemper, D. Bimberg, and J. Neugebauer, Band gap and band parameters of InN and GaN from quasiparticle energy calculations based on exact-exchange density-functional theory, *Applied Physics Letters*, 89 (2006) 161919:1–3.

[13.28] D. Segev, A. Janotti, and C. G. Van de Walle, Self-consistent band-gap corrections in density functional theory using modified pseudopotentials, *Physical Review B*, 75 (2007) 035201:1–9.

[13.29] N. E. Christensen, Electronic structure of GaAs under strain, *Physical Review B*, 30 (1984) 5753–5765.

[13.30] M. Kočan, A. Rizzi, H. Lüth, S. Keller, and U. K. Mishra, Surface potential at as-grown GaN(0001) MBE layers, *Physica Status Solidi (b)*, 234 (2002) 773–777.

[13.31] J. P. Long and V. M. Bermudez, Band bending and photoemission-induced surface photovoltages on clean n-and p-GaN (0001) surfaces, *Physical Review B*, 66 (2002) 121308:1–4.

[13.32] C. I. Wu, A. Kahn, N. Taskar, D. Dorman and D. Gallagher, GaN (0001)-(1×1) surfaces: Composition and electronic properties, *Journal of Applied Physics* 83 (1998) 4249–4251.

[13.33] J. E. Northrup, R. Di Felice, and J. Neugebauer, Atomic structure and stability of AlN(0001) and (000$\bar{1}$) surfaces, *Physical Review B*, 55 (1997) 13878–13883.

[13.34] T. D. Veal, L. F. J. Piper, I. Mahboob, H. Lu, W. J. Schaff, and C. F. McConville, Electron accumulation at InN/AlN and InN/GaN interfaces, *Physica Status Solidi (c)*, 2 (2005) 2246–2249.

[13.35] J. Wu, W. Walukiewicz, S. X. Li, R. Armitage, J. C. Ho, E. R. Weber, E. E. Haller, H. Lu, and W. J. Schaff, Effects of electron concentration on the optical absorption edge of InN, *Applied Physics Letters*, 84 (2004) 2805–2807.

[13.36] M. Cardona and N. E. Christensen, Acoustic deformation potentials and heterostructure band offsets in semiconductors, *Physical Review B*, 35 (1987) 6182–6194.

[13.37] P. D. C. King, T. D. Veal, C. F. McConville, F. Fuchs, J. Furthmüller, F. Bechstedt, P. Schley, and R. Goldhahn, Universality of electron accumulation at wurtzite *c*- and *a*-plane and zinc-blende InN surfaces, *Applied Physics Letters*, 91 (2007) 092101:1–3.

[13.38] H. Lu, W. J. Schaff, L. F. Eastman, J. Wu, W. Walukiewicz, V. Cimalla, and O. Ambacher, Growth of a-plane InN on r-plane sapphire with a GaN buffer by molecular-beam epitaxy, *Applied Physics Letters*, 83 (2005) 1136–1138.

[13.39] E. Calleja, J. Grandal, and M. A. Sánchez-García, M. Niebelschütz, V. Cimalla, and O. Ambacher, Evidence of electron accumulation at nonpolar surfaces of InN nanocolumns, *Applied Physics Letters*, 90 (2007) 262110:1–3.

14

Structure of InN and InGaN: Transmission electron microscopy studies

Z. Liliental-Weber

Materials Sciences Division, Lawrence Berkeley National Laboratory, Berkeley, California 94720, United States of America

14.1 Introduction

Indium nitride is a key element in InGaN-based light emitting diodes (LEDs) [14.1, 14.2]. It is also an important III-V compound semiconductor with potential applications in microelectronics, optoelectronics and solar cells because of its low electron effective mass and high mobility [14.3, 14.4]. The controversy associated with the discovery of the narrow band gap of ~ 0.7 eV [14.3, 14.4] has led to intense research efforts aimed at understanding the electronic structure of InN and In-rich group III-nitride alloys. In 2002, Davydov *et al.* [14.3] and Wu *et al.* [14.4] suggested that the band gap of InN is close to 0.7 eV based on photoluminescence (PL), optical absorption, and modulated reflectance measurements. This value was different from the initial estimate of 1.9 eV in polycrystalline InN deposited by sputtering [14.5]. Since then many other reports have confirmed the narrow band gap of intrinsic InN [14.4, 14.6]. The previously reported band gap near 1.9 eV was ascribed to the contribution from indium oxide related phase in the sample [14.3]. Other possible origins of the absorption edge of InN being in the visible part of the spectrum are quantum size effects at high electron concentrations [14.7], the Burstein-Moss shift at high electron concentration [14.8–14.10], and excess nitrogen in InN [14.11].

InN has the smallest effective electron mass of all the group-III nitrides, which leads to a potentially high mobility, saturation velocity and a large drift velocity at room temperature. Due to its narrow band gap [14.4], alloying with GaN and AlN ensures light emission from the ultraviolet to the infrared, covering in this way the whole solar spectrum. Among the III-nitrides, InN is the least studied compound, largely because obtaining material of high structural quality is difficult, largely due to the low dissociation temperature and high equilibrium vapor pressure of nitrogen. Growth of InN is also more difficult than that of AlN and GaN due to the weak bonding of In and N [14.3].

In recent years, however, structural properties of InN grown by metal organic vapor phase epitaxy (MOVPE) and molecular beam epitaxy (MBE) have been greatly improved. Particularly, InN films grown by MBE have shown Hall mobilities above $1000 \text{ cm}^2/\text{Vs}$ and carrier concentrations below 10^{18} cm^{-3} [14.12–14.15]. However, device quality material has still not been produced since most of the films have threading dislocation densities higher than 10^{10} cm^{-2} and high residual carrier concentrations of the order of 10^{18} cm^{-3} [14.16]. It turned out that a proper buffer layer plays an important role in obtaining a reasonable structural quality of InN. Some researchers used AlN [14.13, 14.14], low-temperature (LT) InN [14.17] or LT-InN/LT-GaN [14.18] buffer layers. It was also reported that crystal quality strongly depends on the thickness of a LT-AlN buffer layer [14.19]. For layers grown by MOCVD, an improvement of surface roughness was observed for decreased buffer layer thickness (down to 30 Å) [14.20]. On the contrary, for layers grown by MBE, only increased buffer layer thicknesses (about 200 nm) substantially improved the InN layer quality [14.13]. Since the structural quality of InN governs the optical and electrical properties of InN, this chapter will be devoted to structural characterization of InN grown by MBE.

14.2 Structural defects in as-grown InN

Lack of large diameter wafers of native substrates for growth of InN layers and its alloys forces the use of heteroepitaxial growth using foreign substrates. The most commonly used substrates are (0001) Al_2O_3. The lattice mismatch between InN and α-Al_2O_3 is about 25%; therefore, the density of structural defects is very high. It has been predicted that dislocations may act as non-radiative recombination centers [14.20, 14.21] and that scattering at charged dislocations could be a limiting factor of carrier mobility in a two-dimensional electron gas [14.22].

As mentioned above, in order to overcome these difficulties, a buffer layer is usually introduced. In this chapter, samples have a thin AlN nucleation layer grown on a sapphire substrate, followed by a thicker layer of GaN, as a binary buffer layer, grown at 550–600°C, on which the InN was grown. All layers were grown by molecular beam epitaxy (MBE). Their structural properties will be described based on transmission electron microscopy (TEM) studies. In order to obtain structural information about these samples, cross-section TEM samples were prepared by gluing face-to-face two thin slabs of the samples (Al_2O_3/AlN/GaN/InN) which were perpendicular to each other. This was done in order to be able to observe InN layers in $[11\bar{2}0]$ and $[1\bar{1}00]$ orientations. These glued samples were embedded into brass rings and polished with a diamond paste to obtain a thickness below 80 μm, which was followed by 'dimpling down' to 10 μm. After careful washing in methanol (or ethanol), samples were placed in an Ar ion mill (Precision Ion Mill) to obtain sample perforation of which the perimeters were transparent for electrons. Cooling of these samples was necessary to avoid In droplet formation.

These samples were studied by conventional transmission electron microscopy using a JEOL 3010 microscope (JEOL Ltd, Tokyo, Japan) and high-resolution electron microscopy (HREM) was performed using a Philips CM 300 sub-Angstrom microscope. Both microscopes were operated at 300 kV.

We observed that the microstructure and morphology of undoped InN varies from sample to sample, and also for different areas in the same sample. High-resolution images of the interface between Al_2O_3 and AlN/GaN are shown in Figs. 14.1 and 14.2. Figure 14.1 shows an abrupt interface where the AlN layer is about 2 unit cells thick. However, Fig. 14.2 shows that some three-dimensional islands of AlN might be occasionally formed. Since these islands are rather small, their formation does not lead to additional defects. The GaN buffer layer, with a thickness close to 100 nm, was grown on top of the AlN. The orientation relationship between the sapphire and AlN/GaN was determined to be $(0001)_{AlN,GaN} \parallel (0001)_{Al_2O_3}$ and $[11\bar{2}0]_{AlN,GaN} \parallel [1\bar{1}00]_{Al_2O_3}$. The orientation between the AlN/GaN does not require a 30° rotation and remains $(0001)_{AlN,GaN} \parallel (0001)_{InN}$ and $[11\bar{2}0]_{AlN,GaN} \parallel [11\bar{2}0]_{InN}$.

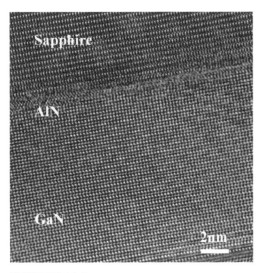

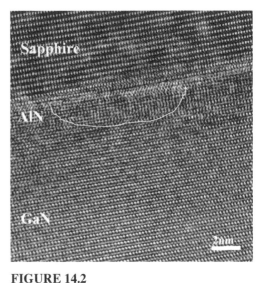

FIGURE 14.1

An abrupt interface between sapphire and the AlN/GaN buffer layer.

FIGURE 14.2

A small island (outlined) of AlN formed at the interface with GaN.

Figure 14.3 shows that InN crystallizes in the wurtzite structure with lattice parameters $a = 3.52$ Å and $c = 5.72$ Å. In the samples studied, the thicknesses of the InN layers are in the range of 0.4–2 μm. Figure 14.4(a) shows a micrograph of an InN layer with a thickness of 500 nm. It is observed that some additional dislocations are formed at the buffer/InN interface, shown in Fig. 14.4(b), since a higher density of dislocations is observed in the InN layer above this interface ($\sim 2 \times 10^{10}$ cm^{-2}) than in the underlying buffer layer ($\sim 7 \times 10^9$ cm^{-2}). The majority (almost two-thirds) of dislocations have a screw/mixed character and only one-third are of an edge character. Some dislocations are straight lines, but a large majority of them are very curved, often interacting with planar defects.

FIGURE 14.3
A high perfection wurtzite InN layer grown on top of a GaN layer.

The bright field TEM image of as-grown InN films (grown on c-plane Al_2O_3) shows that the InN layer makes an abrupt interface with an underlying GaN buffer layer (Fig. 14.4(b)). In high-resolution images, this interface appears as a straight line (Fig. 14.5), similar to the interface between sapphire and GaN. However, there are some areas where this interface is slightly undulated (Fig. 14.6) showing moiré fringes from two overlapped materials.

There are InN layers with flat top surfaces where surface undulation does not exceed a few nanometers, but there are also samples where a columnar growth is taking place. In these layers the surface corrugation can be as large as 20 nm (Fig. 14.7).

The distribution of defects is not uniform in various samples grown under similar growth conditions. There are areas (sometimes in the same sample) where many stacking faults are formed on basal and prismatic planes in addition to dislocations (Fig. 14.8). The basal stacking faults are delineated either by prismatic stacking faults (as found earlier in a-plane GaN [14.23, 14.24] grown on a-plane SiC or r-plane Al_2O_3) or by partial dislocations. Sometimes it is found that large thicknesses of cubic layers remain without converting back to the hexagonal structure, especially when planar defects are formed near the top surface (Fig. 14.9).

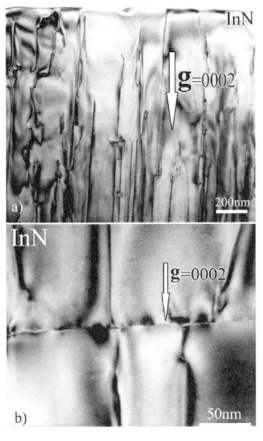

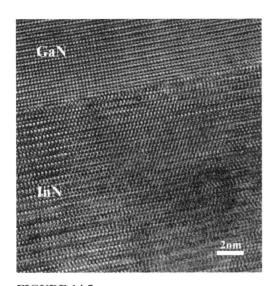

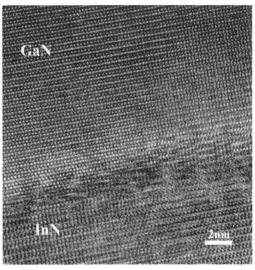

FIGURE 14.4

(a) Cross-section TEM micrograph showing distribution of screw and mixed dislocations in the InN layer. (b) The interface between the GaN buffer layer and InN. Note that locally the interface is slightly undulated and the density of dislocations in the InN layer is only slightly higher than in the GaN buffer layer.

FIGURE 14.5

High-resolution micrograph of the abrupt part of the interface between GaN and InN.

FIGURE 14.6

High-resolution micrograph of the undulated part of the interface between GaN and InN. Note moiré fringes at the interface.

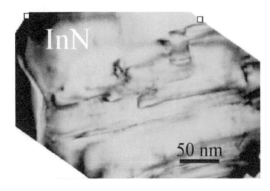

FIGURE 14.7
Defects (dislocations and planar) in the as-grown InN. Note the presence of surface corrugation.

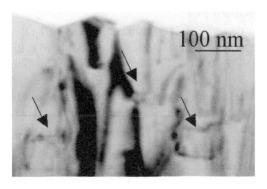

FIGURE 14.8
Stacking faults (indicated by the arrows) in the InN layer.

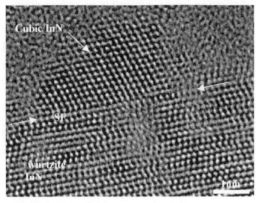

FIGURE 14.9
Formation of a cubic InN region as a result of stacking fault formation (indicated by the arrows) in wurtzite InN.

14.3 Growth polarity of the InN epi-layers

Convergent Beam Electron Diffraction (CBED) was applied to study the growth polarity of InN layers grown on c-plane sapphire using a JEOL 3010 TEM. Patterns were taken for different sample thicknesses, and are shown in Fig. 14.10. CBED patterns for the same zone axis and sample thickness, simulated for the accelerating voltage (300 keV) used in the experiment, are also shown. Good agreement between experimental and calculated patterns was obtained. Based on these experiments and taking into account the rotation angle between the image and a diffraction pattern in our microscope, it was determined that the layers were grown with In-polarity, as schematically indicated in the figure. Earlier studies of InN grown on a GaN template using two-step metal-organic vapor-phase-epitaxy (MOVPE) on a (0001) sapphire substrate show N-polarity [14.25] using the same CBED method for polarity determination. The GaN for this case was grown at 1010°C followed by InN grown at 500°C [14.26]. It was shown that radio frequency plasma assisted molecular beam epitaxy (RF-MBE) growth of InN at 550°C on low-temperature (LT) InN deposited at 300°C leads to N-polarity growth, but samples grown at 450°C on LT-GaN also grown at 300°C are In-polarity. This shows that InN can grow with both polarities.

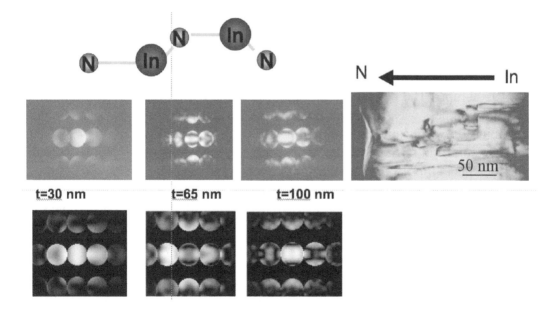

FIGURE 14.10

Experimental and calculated convergent beam electron diffraction (CBED) patterns for different thicknesses of InN. The schematic atomic arrangement for which these patterns were calculated is also shown. A micrograph of InN for which these CBED patterns were taken is shown on the right side, indicating the sample surface. Based on CBED patters, In-polarity growth was determined.

14.4 Mg-doped layers

In order to use InN in devices, one needs to be sure that p-type doping of this material is possible. While undoped InN films are always n-type, p-type doping is still not well understood due to the unusual band structure of InN (see Chapters 4, 10 and 12). Early studies [14.27, 14.28] did not confirm that p-doping was possible due to very high background donor concentrations ($\sim 10^{20}$ cm^{-3}). It was argued that, because of the presence of an electron accumulation layer in InN, measurement of p-type InN by the Hall effect method is challenging. Recently, determination of a p-type bulk beneath the electron accumulation layer was achieved using electrolyte-based capacitance-voltage (ECV) measurements [14.29] and a combination of multiple-field Hall effect, photoluminescence and ECV measurements [14.30]. Mg-doped InN thin films were therefore confirmed to be p-type.

Samples with different Mg concentrations grown by MBE have been studied using cross-sectional TEM. The InN:Mg layers were grown on top of an undoped InN layer, which was grown on c-plane sapphire with an AlN nucleation layer followed by a GaN buffer layer. The InN:Mg samples were grown using three Mg cell temperatures: 275°C, 300°C and 400°C. SIMS analysis of two of them is shown in Fig. 14.11. These results show that a higher Mg cell temperature results in a higher Mg concentration. The sample grown with a 300°C Mg cell temperature has a Mg concentration [Mg] of $\sim 2 \times 10^{20}$ cm^{-3} and that grown with a 400°C Mg cell temperature has [Mg] $= 2 \times 10^{21}$ cm^{-3}. The sample with the lowest Mg cell temperature (275°C) should have the lowest Mg content ($\sim 2 \times 10^{19}$ cm^{-3}). The latest sample was more than twice as thick (1.25 μm) as the other samples (Fig. 14.12). The interface between the GaN buffer and the InN layer in this sample was abrupt. The density of dislocations was about 2.5×10^{10} cm^{-2} but the surface roughness was substantially increased by up to 50 nm in comparison with the thinner InN:Mg samples with higher Mg content. The distribution of dislocations was uniform throughout the entire area of the TEM sample. There were no other defects in this sample besides the dislocations.

The sample with [Mg] $= 2 \times 10^{20}$ cm^{-3} (GS1810 on Fig. 14.11) had a thickness of about 400 nm and dislocation density of about 3.6×10^{11} cm^{-2}. The sample surface was reasonably flat with the roughness not exceeding 5–10 nm (Fig. 14.13). An abrupt interface between GaN buffer layer and InN was also observed. As expected, all dislocations from the AlN/GaN buffer layer propagated into the InN layer, but some new dislocations were also formed at the interface with GaN, similarly as observed for the undoped samples (Fig. 14.4(b)). One notices the formation of an amorphous material (possible oxide) on the entire InN sample (Fig. 14.14). The oxide layer on the sample surface (marked by an arrow on the figure) has a different contrast from the glue used for cross-section sample preparation. It was also noticed that a thin amorphous layer covered all InN cross-section samples since the contrast on the InN layer is not as crisp as on the GaN, on which InN

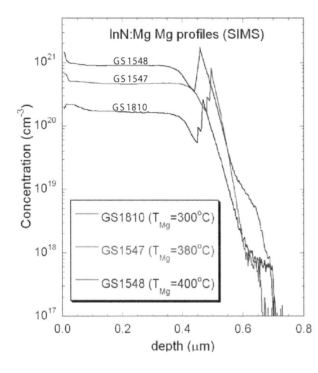

FIGURE 14.11

SIMS data for three different InN:Mg samples showing Mg concentration. The sample GS1810, grown at 300°C Mg cell temperature, and GS1548, grown at 400°C, are described in the text. The sample with the lowest Mg cell temperature, 275°C (GS1650) is not shown here but it is described in the text.

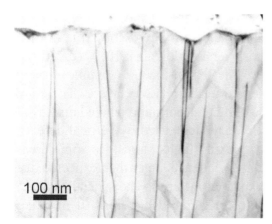

FIGURE 14.12

Cross-section of InN:Mg (GS1650 – expected to have the lowest Mg concentration of $\sim 10^{19}$ cm^{-3}). The sample thickness is $t = 1.25\ \mu m$ and the surface roughness ~ 50 nm. The density of dislocations in this sample is $d = 2.5 \times 10^{10}$ cm^{-2}.

was grown. This might be an indication of the formation of a surface electron accumulation layer with a high carrier concentration and described in the earlier literature [14.29].

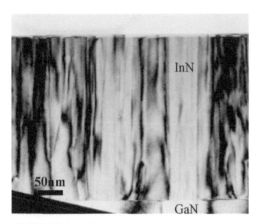

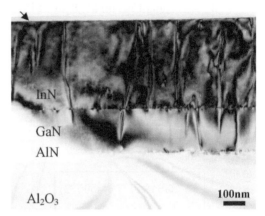

FIGURE 14.13

Cross-section micrograph of InN:Mg (GS1810; intermediate Mg concentration of 2×10^{20} cm^{-3}; sample thickness $t = 400$ nm). The surface of this sample is reasonably flat with a roughness $\sim$5–10 nm. The density of dislocations in this sample is $d = 3.6 \times 10^{11}$ cm^{-2}.

FIGURE 14.14

The InN layer (GS1810) grown on GaN. Note a different contrast on these two layers suggesting formation of additional amorphous material (possibly oxide) on the InN layer.

The InN:Mg sample with $[\text{Mg}] = 2 \times 10^{21}$ cm^{-3} (GS1548) was not as uniform as the former sample. This sample has a thickness of $\sim$500 nm (Fig. 14.15). There are areas of the sample where the dislocation density is only 1.4×10^{10} cm^{-2} and the surface roughness is about 30 nm. Convergent beam electron diffraction shows that this sample is also grown with In polarity, as for the undoped samples. However, there are inversion domains, as shown in Fig. 14.16, embedded in the matrix layer. The inversion was confirmed by high-resolution electron microscopy (Fig. 14.17) and also by CBED studies. Such embedded inversion domains (IDs) were previously observed in GaN:Mg [14.31, 14.32]. The IDs in the GaN samples have a specific pyramidal shape, which is not observed in the InN:Mg layers. There are also many areas in the sample with the highest Mg concentration where lamellas of stacking faults are formed (Fig. 14.18) in domains delineated by vertical defects. Within each domain, the distance between stacking faults varies. In some areas there appear to be regions with a cubic stacking sequence larger than expected due to the regular three types of basal SFs in hexagonal material. The special arrangements of these planar defects and presence of the large cubic areas (if in larger concentration) might substantially influence the optical and electrical properties of such layers.

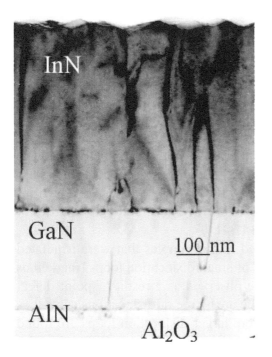

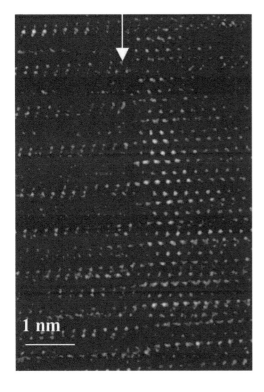

◄ **FIGURE 14.15**

High perfection area of InN:Mg (GS1548); the surface roughness is ~ 30 nm. The thickness of this sample is $t \sim 500$ nm and dislocation density $d = 1.4 \times 10^{10}\ cm^{-2}$.

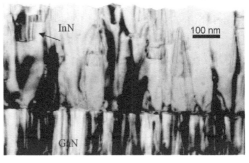

▲ **FIGURE 14.16**

Embedded inversion domain (indicated by the arrow) in the InN:Mg (sample GS1548).

◄ **FIGURE 14.17**

High-resolution micrograph from the embedded inversion domain boundary.

▲ **FIGURE 14.18**

High density of planar defects (stacking fault lamellas) in the InN:Mg layer.

14.5   Irradiated InN

Our earlier studies of the effect of 2 MeV ^{4}He$^+$ irradiation followed by thermal annealing showed that this process generates point defects and creates films with high electron concentrations and high electron mobilities. There is a linear relationship between electron concentration and applied ^{4}He$^+$ ion fluence, which changes further upon annealing [14.33]. It was suggested that the increase in mobility in the irradiated and annealed InN samples could be the result of defect ordering. The electrical and optical properties of these samples are discussed in detail in Chapter 10. The irradiation fluence ranged from 1.1×10^{15} to 8.9×10^{15} cm^{-2}. The films were annealed in the temperature range of 375–500°C. The structural properties of these layers have been studied using TEM [14.34].

Only small structural changes are observed in the samples that were irradiated with lower fluences (Fig. 14.19). Formation of small dislocation loops (marked as 'l') and nanopipes (marked as 'n') could be observed as a result of point defect agglomeration. These loops were found in the areas free of dislocations as well as along dislocation lines. There are areas where larger dislocation loops are found arranged on top of each other along the growth direction (Fig. 14.19(b)). One can assume that the high density of point defects pinched off dislocations present in the as-grown samples leading to the formation of these chains of elongated loops. The density of small dislocation loops in the areas free of linear or planar defects was estimated to be 8.6×10^9 cm^{-2}. For higher fluences, the size of the dislocation loops and their density increase slightly and in some areas their density can be measured as 2.2×10^{10} cm^{-2}.

High densities of planar defects have been found in this particular sample. They were present in the entire layer, close to the interface (Fig. 14.20) as well as in the center part of the layer and closer to the sample surface (Fig. 14.19(a)). These planar defects are grouped together and form kind of domains (Fig. 14.19(a) and Fig. 14.20). The distances between particular planar defects are, however, larger than those in the samples that were Mg-doped, described in the previous section. As was discussed above, such defects can also occasionally be seen in the as-grown samples. Therefore, it is believed that the presence of these defects is rather characteristic of as-grown material which is subsequently irradiated, and not a result of the irradiation itself. However, one can emphasize that the density of planar defects in this particular sample was much higher than in any other sample; therefore more statistics would be needed to come to an unambiguous conclusion.

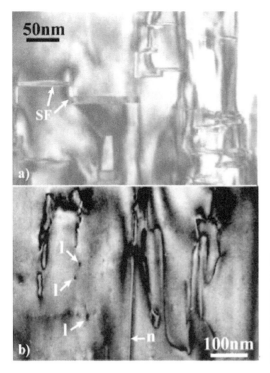

FIGURE 14.19

TEM micrographs showing ^{4}He$^+$ irradiated samples: (a) with a fluence of 1.1×10^{15} cm^{-2}, and (b) with a fluence of 8.9×10^{15} cm^{-2}. Note the high density of planar defects (SF) in (a). Examples of dislocation loops (l) and nanopipes (n) are shown.

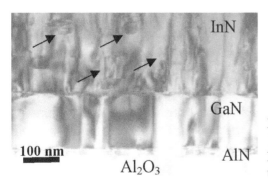

FIGURE 14.20

Presence of planar defects and dislocations in the InN layer irradiated with 50 μC of ^{4}He$^+$. This is the area close to the interface with GaN.

14.6   Annealed samples

The samples studied were also subjected to rapid thermal annealing (RTA). This was performed on as-grown and also on previously irradiated samples. A Heatpulse 10T-02 Rapid Thermal Pulsing System with flowing N_2 gas was used. The samples were annealed at 375°C or 425°C for different times varying from 10 to 300 sec. Some samples were also annealed at 475°C and 500°C. There were no observable structural changes in the as-grown samples annealed at 375°C. However, samples annealed at 475°C for 10 sec showed some changes (Fig. 14.21). The main defects were small voids distributed along screw and edge dislocations. Their size did not exceed more than 5 nm and the distance between voids was about 20 nm. These voids were not observed in the as-grown samples. Some surface deterioration is also found (marked by an arrow on Fig. 14.21).

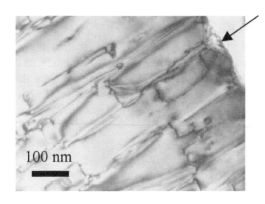

100 nm

FIGURE 14.21
Annealed (475°C for 10 sec) as-grown InN sample. Note formation of voids along screw dislocation lines, some surface deterioration (marked by the arrow), and the high density of dislocations.

Annealing at 475°C for 300 seconds leads to a larger surface deterioration (Fig. 14.22). The areas with lighter contrast on this micrograph indicate smaller thickness in comparison to the surrounding areas, for example, due to the formation of grooves. Based on their shape one can conclude that formation of these grooves occurs either on grain boundaries or on a group of threading dislocations intersecting the sample surface. In cross-section samples, these grooves will have a triangle shape (like the one shown by an arrow in Fig.14.21). Similar to the samples annealed for a shorter time, voids are also found in samples annealed for 300 sec, but with larger diameters. For samples annealed for 300 sec, one could occasionally find inclusions where moiré fringes indicate the presence of In precipitates. Their density, however, was rather low. Only three precipitates were found in the TEM sample. Figure 14.23 also shows an example of an elongated precipitate with a length of 90 nm and a width of 25 nm. This precipitate did not show any moiré fringes, which might suggest that this precipitate is located close to the sample surface. The precipitate of a size of 15×20 nm^2 was connected with a void formed at one end. This might indicate a specific mechanism of void/precipitate formation.

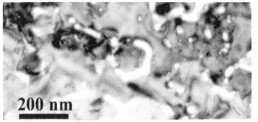

FIGURE 14.22

Plan-view image of the as-grown InN sample after annealing at 475°C for 300 sec. The white areas show formation of deep grooves.

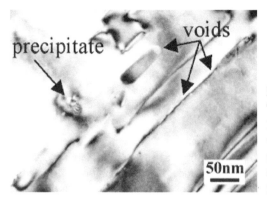

FIGURE 14.23

TEM micrograph of the annealed (475° for 300 sec) as-grown sample revealing formation of voids (three arrows) along the dislocation lines and voids associated with a precipitate. Another arrow indicates an In precipitate with moiré fringes.

Samples that were irradiated with a low fluence (1.1×10^{15} cm^{-2}) and subsequently annealed show the formation of similar defects as annealed as-grown samples, but the funnels formed at the surface enter deeper into the sample. Many nanopipes were also formed along the dislocations. Similar to annealed samples, lamellas of stacking faults were often present in this sample.

Samples that were irradiated with a higher fluence (8.9×10^{15} cm^{-2}) and subsequently annealed show substantial structural changes (Figs. 14.24 and 14.25). High densities of dislocation loops (2.8×10^{11} cm^{-2}) are formed through the samples. They can be found in the areas close to the interface (Fig. 14.24) and closer to the sample surface (Fig. 14.25). Their sizes vary from 5–20 nm. An interaction of the loops with dislocation lines leads to some undulation of dislocations. The formation of larger loops and an increase of their density are most probably due to the agglomeration of individual point defects (probably vacancies) created by the irradiation. Large voids are also formed at the interface with GaN. Annealing at 500°C led to the formation of In clusters and delamination of the film from the substrates.

14.7 Structure of non-polar InN

It was shown theoretically [14.35] and proven experimentally [14.36] that spontaneous and piezoelectric induced polarization is present within GaN-based active layer when this material is grown on (0001) Al_2O_3 or 6H-SiC. The same applies

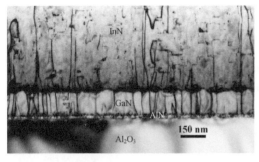

FIGURE 14.24
Formation of a high density of dislocation loops $(8.9 \times 10^{15} \text{ cm}^{-2})$ in the sample irradiated and later annealed at 475°C. The area is close to the substrate.

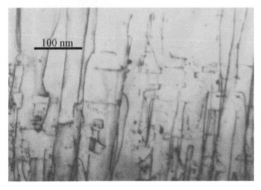

FIGURE 14.25
The same sample as in Fig. 14.24 but closer to the sample surface. Note the same type of defects as in the area close to the interface with the GaN buffer layer.

to any wurtzite III-nitrides grown along the c-axis, including InN. The total polarization along the [0001] direction of InN (GaN) leads to high interface charge densities and spatial separation of the electron and hole wave function in quantum well structures, which alters the optoelectronic properties of devices. One possible solution to eliminate these effects is to grow III-nitride based layers in non-polar orientations, such as $(1\bar{1}00)$ (m-plane) or $(11\bar{2}0)$ (a-plane). Studies of such layers demonstrate polarization free behavior along the growth direction. One can notice that, for growth on polar surfaces, the c-plane is occupied by one kind of atom: In or N. Atom distribution on the $(11\bar{2}0)$ a-plane is different: An equal number of In and N must be present on these planes. In these circumstances, growth of the layers proceeds along the $[11\bar{2}0]$ direction and c-planes are along this direction. Since the energy of stacking faults formed on the c-plane is the lowest, these defects are expected to propagate to the sample surface.

InN samples grown in the non-polar direction followed a similar layer sequence arrangement as those grown on c-plane sapphire, with an AlN nucleation layer followed by a GaN buffer layer. TEM studies show that different defects are formed in the InN layers grown on polar and non-polar substrates. Columnar growth is observed for the layers grown in non-polar directions. The column diameters are in the range of 100–250 nm. Some separation between these columns is visible (Fig. 14.26). The faceting of the columns is observed at the sample surface, making these surfaces much rougher in comparison with layers grown along the c-axis. The measured height difference between the tip and valley of the column can be as large as 100 nm, while the roughness of layers with the same sample thickness grown along the *c*-axis does not exceed 10–20 nm. Much larger tilt/twist between these columns is observed for growth along the $[11\bar{2}0]$ direction in comparison with the deviation from grain to grain for growth along the *c*-axis. As described earlier, for growth in the polar direction, the main defects are threading dislocations, which propagate along the growth direction with their density in the range of 8×10^9 cm^{-2} to 2×10^{10} cm^{-2}, which is only slightly larger than the density of dislocations in the underlying GaN buffer layer. For the growth in non-polar directions, the density of dislocations is at least $\sim 10^{11}$ cm^{-2}, but in such layers basal stacking faults (BSFs) formed on *c*-planes propagate to the sample surface, which is deleterious for device application. For the samples grown in the non-polar direction, prismatic stacking faults (PSFs) are observed in addition to dislocations and BSFs (Fig. 14.27). The $\frac{1}{2}[0\bar{1}11]$ displacement fault vector was determined using the contrast invisibility criteria. PSFs are always found to intersect BSFs. As the fault vectors of a PSF and a BSF are not equal, stair-rod dislocations are present at such intersections, resulting in slightly higher dislocation densities than in the layers grown in the polar direction. PSFs have also been observed previously in GaN grown in non-polar orientations [14.23]. The non-polar InN samples were also irradiated with a low fluence of $\sim 1 \times 10^{15}$ cm^{-2} ^{4}He$^+$ ions, as for those grown along *c*-axis. These samples were also annealed. TEM studies show some change of contrast in the areas free of structural defects, therefore suggesting that conglomeration of point defects is taking place. This is similar to the irradiated and annealed samples grown in the polar orientation. For annealing at 425°C for 300 sec, large voids are formed at the interface with the AlN/GaN buffer layer (Fig. 14.28). This might lead to layer delamination.

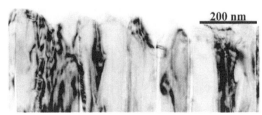

200 nm

FIGURE 14.26
Columnar growth of *a*-plane InN samples on *r*-plane Al$_2$O$_3$.

14.8 Compositional modulation of MBE-grown InGaN alloys with high In content

In$_x$Ga$_{1-x}$N semiconductor materials are technologically important constituents in microelectronic and optoelectronic devices. For example, they are used as the active layer in short-wavelength light-emitting diodes and lasers [14.37, 14.38]. Use

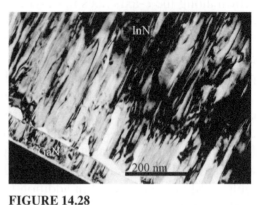

FIGURE 14.28

Micrograph of the ^{4}He$^+$ irradiated (50 μC) a-plane InN layer grown on r-plane Al$_2$O$_3$ and annealed at 425°C for 10 s followed by 300 s annealing. Note the formation of large voids at the interface with the GaN layer.

FIGURE 14.27

Formation of prismatic stacking faults (shown by arrows) in an a-plane InN layer grown on r-plane Al$_2$O$_3$.

of these alloys together with GaN and AlN makes it possible to produce nitride-based light-emitting diodes operating from the ultraviolet well into the red. However, the large size difference between Ga and In atoms makes growth of In$_x$Ga$_{1-x}$N challenging, particularly for $x > 0.2$. A miscibility gap for this material was predicted theoretically [14.39, 14.40] and phase separation has been observed experimentally in films grown by MOVPE and MBE for $x > 0.25$ [14.41–14.43].

Compositional fluctuations have been observed in several ternary alloys such as InGaP [14.44, 14.45] and were associated with spinodal decomposition driven by asymmetry in the elastic coefficients. There are other reports suggesting that compositional modulation is developed on the surface during epitaxy and leads to the formation of a columnar structure, as reported by Chu *et al.* [14.46]. There is also a comprehensive review on phase separation and ordering written by Zunger and Mahajan [14.40], where the authors discussed lateral and vertical phase separation from both a theoretical and experimental point of view.

We previously investigated [14.47] the compositional modulation observed in In$_x$Ga$_{1-x}$N films (with $0.3 < x < 0.8$) grown by MBE and the results of these studies will be reviewed here. The samples were grown in the temperature range 530–600°C. To obtain samples with higher In content, lower growth temperatures were used. The samples were grown either on a GaN or AlN buffer layer grown on c-plane sapphire. According to Ho and Stringfellow, a strong thermodynamic driving force for spinodal decomposition is expected for all of these compositions [14.39].

TEM studies show that edge dislocations dominate in the sample with $x = 0.34$ with a density of about 2×10^{10} cm^{-2}. No phase separation was observed in these samples (Fig. 14.29). Some increase in dislocation density (10^{11} cm^{-2}) was observed for a sample with higher In content ($x = 0.5$ or 0.55) grown on an AlN buffer layer. However, for a much higher In concentration of $x = 0.78$ (grown on

AlN), a decrease of the dislocation density to 6×10^{10} cm^{-2} was observed. The In-GaN/AlN interface was rough, with roughnesses up to 150 Å measured along the c-axis. In all three samples with In concentrations $x = 0.5$, 0.55 and 0.78 white/black "tweed"-like fringes were observed (Fig. 14.30). These black/white fringes are parallel to each other within the column, and were observed in both $[1\bar{1}00]$ and $[11\bar{2}0]$ projections through the entire sample thickness. From selected area diffraction, it was noticed that each diffraction spot from the InGaN layer was surrounded by extra diffraction "spots" suggesting a structure with possible periodic compositional modulation (Fig. 14.31). These extra "spots" were in the form of small arcs due to some deviation between particular grains of InGaN (Fig. 14.31(b)). Based on the measured distances from the main InGaN reflections and the center of the arc (equivalent to the extra spot along the c-axis) the period of this modulation along the c-axis was determined to be $\Delta = 45$ Å for the sample with $x = 0.5$, $\Delta = 47$ Å for the sample with $x = 0.55$, and $\Delta = 66$ Å for the sample with $x = 0.78$. These results show that a larger modulation period is observed in samples with higher In content. Growth of the layers with a similar In content on GaN buffer layers did not show compositional modulation. The interface in the latter case was much smoother than the one formed between InGaN and AlN. Therefore, it is concluded that two factors lead to these modulations: larger strain between AlN and InGaN and surface roughness. It is most probable that the roughness of the AlN growth surface and large lattice mismatch with the underlying AlN promotes strong In segregation on particular crystallographic planes and leads to compositional modulation.

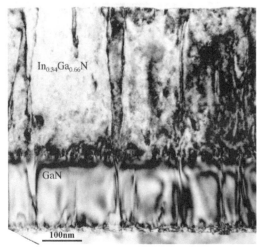

FIGURE 14.29
The micrograph showing a high perfection of the InGaN layer with 34% of In grown on GaN.

The presence of compositional modulation was confirmed by x-ray diffraction (XRD) in the $\theta - 2\theta$ coupled geometry using Cu-K$_\alpha$ x-rays. The XRD pattern reveals satellite reflections around the 0002 and 0004 InGaN peaks (Fig. 14.32). The period values determined by XRD are in a good agreement with those determined by TEM. Energy dispersive x-ray spectroscopy measurements across white/black

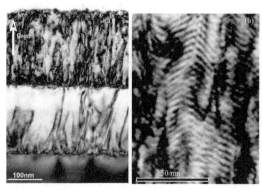

FIGURE 14.30

TEM micrographs showing compositional modulation observed in the $In_xGa_{1-x}N$ layers with (a) $x = 0.5$ and (b) $x = 0.78$.

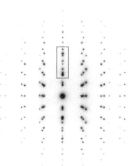 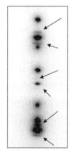

FIGURE 14.31

(a) Experimental diffraction pattern from the sample with $x = 0.5$. (b) Magnified view of the outlined area is also shown. The arrows indicate extra maxima (arcs) related to the corrugated nature of the black/white fringes.

fringes show systematically that the intensity of the In-K_α peak decreases when the Ga-K_α peak intensity increases, confirming compositional modulation within the film [14.47]. The average estimated concentration on the black fringe was about 40% In and about 60% Ga. Based on these results, we can postulate that alternative layers of GaN(+In)/InN(+Ga) were formed.

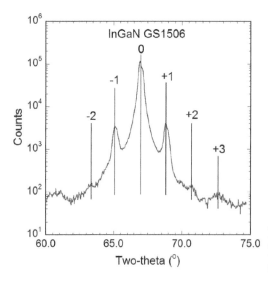

FIGURE 14.32

XRD data showing extra reflections from the ordered sample ($x = 0.78$).

This study shows that it is possible to grow InGaN layers that are uniform in composition for high In content up to $x = 0.34$ using MBE, and also shows the possibility to grow self-assembled thin layers on a nanometer scale separated from each other at equal distances (ordered-like structure) throughout the entire layer. It was also observed that modulated films tend to have large "Stokes shifts" between their absorption edge and photoluminescence peak [14.48, 14.49].

14.9 Summary

In this chapter, detailed structural characterization of MBE-grown undoped and Mg-doped InN layers was reported. It was shown that the density of defects ($\sim$few times 10^{10}cm^{-2}) in the as-grown InN layers is at least one order of magnitude higher than in the GaN buffer layer deposited on sapphire with a thin AlN nucleation layer. This density of defects is much higher than in GaN grown by MOVPE or HVPE. Further development in growth procedure is needed for device application. Besides dislocations, planar defects are also observed. The distribution of planar defects is not uniform in the studied samples. It is clear that the density of planar defects increases with Mg concentration for Mg-doped samples. Sometimes they are distributed with equal distances, but often they form lamellas with a different thickness of cubic layers. All InN layers studied were grown with In growth polarity, although the layers with high Mg concentration have inversion domains embedded in the matrix.

Contrast on TEM micrographs from the InN and GaN layer differs even for freshly prepared samples, showing formation of a thin amorphous layer. Sometimes a thin amorphous layer can be noticed also on the InN sample surface in cross-section samples. This might be related to the formation of the accumulation layer discussed in the literature.

Sample surfaces of as-grown InN show a typical roughness of 10–20 nm. The surface roughness tends to increase with sample thickness and/or Mg doping. Some columnar growth is observed and the height of such columns varies. The largest height to valley distance was observed in the InN samples grown in non-polar orientations such as the $[11\bar{2}0]$ growth direction. In these samples, basal stacking faults propagated to the sample surface. Besides the basal stacking faults, prismatic stacking faults are also observed in the samples grown in non-polar orientations.

Samples irradiated with 2 MeV ^{4}He$^+$ and annealed were also studied. Structural changes in these samples depend drastically on the irradiation fluence, annealing temperature and duration. Small voids often distributed along dislocation lines are observed in annealed samples. Longer annealing times lead also to the formation of In precipitates, sometimes directly attached to the voids. Irradiated samples show the presence of small dislocation loops as a result of conglomeration of point defects. These dislocation loops interact with dislocation lines leading to their undulation. Delamination of the InN layers from the GaN buffer layer was observed for the irradiated samples annealed at 475°C for 300 sec.

The structure of $In_xGa_{1-x}N$ films ($0.3 < x < 0.8$) grown by MBE was also described. There is no phase separation for compositions up to 34% In. For higher In contents, compositional modulation through the entire sample (grown on an AlN buffer layer) was observed. The modulation period increased with the In content. These samples showed large Stokes shifts between their absorption edge and photoluminescence peak.

Acknowledgments

This work is supported by the Director, Office of Science, Office of Basic Energy Sciences, Division of Materials Sciences and Engineering, of the U.S. Department of Energy under Contract No. DE-AC02-05CH11231. The use of the facilities at the National Center for Electron Microscopy at Lawrence Berkeley National Laboratory is greatly appreciated. Many thanks to M. Hawkridge for his help in formatting the micrographs for this chapter.

References

[14.1] S. Chichibu, T. Azuhata, T. Sota, and S. Nakamura, Spontaneous emission of localized excitons in InGaN single and multiquantum well structures, *Applied Physics Letters*, 69 (1996) 4188–4190.

[14.2] K. P. O'Donell, R. W. Martin, and P. G. Middleton, Origin of luminescence from InGaN diodes, *Physical Review Letters*, 82 (1999) 237–240.

[14.3] V. Y. Davydov, A. A. Klochikhin, R. P. Seisyan, V. V. Emtsev, S. V. Ivanov, F. Bechstedt, J. Furthmüller, H. Harima, A. V. Mudryi, J. Aderhold, O. Semchinova, and J. Graul, Absorption and emission of hexagonal InN. Evidence of narrow fundamental band gap, *physica status solidi (b)*, 229 (2002) R1–R3.

[14.4] J. Wu, W. Walukiewicz, K. M. Yu, J. W. Ager III, E. E. Haller, H. Lu, W. J. Schaff, Y. Saito, and Y. Nanishi, Unusual properties of the fundamental band gap of InN, *Applied Physics Letters*, 80 (2002) 3967–3969.

[14.5] T. L. Tansley and C. P. Foley, Optical band gap of indium nitride,

Journal of Applied Physics, 59 (1986) 3241–3244.

[14.6] T. Matsuoka, H. Okamoto, M. Nakao, H. Harima, and E. Kurimoto, Optical bandgap energy of wurtzite InN, *Applied Physics Letters*, 81 (2002) 1246–1248.

[14.7] B. R. Nag, On the band gap of indium nitride, *physica status solidi (b)*, 237 (2003) R1–R2.

[14.8] J. Wu, W. Walukiewicz, W. Shan, K. M. Yu, J. W. Ager III, E. E. Haller, H. Lu, W. J. Schaff, Effects of the narrow band gap on the properties of InN, *Physical Review B*, 66 (2002) 201403:1–4.

[14.9] J. Wu and W. Walukiewicz, Band gaps of InN and group III nitride alloys, *Superlattices and Microstructures*, 34 (2003) 63–75.

[14.10] J. Wu, W. Walukiewicz, S. X. Li, R. Armitage, J. C. Ho, E. R. Weber, E. E. Haller, H. Lu, W. J. Schaff, A. Barcz, and R. Jakiela, Effects of electron concentration on the optical absorption edge of InN, *Applied Physics Letters*, 84 (2004) 2805–2807.

[14.11] T. V. Shubina, S. V. Ivanov, V. N. Jmerik, M. M. Glazov, A. P. Kalvarskii, M. G. Tkachman, A. Vasson, J. Leymarie, A. Kavokin, H. Amano, I. Akasaki, K. S. A. Butcher, Q. Guo, B. Monemar, and P. S. Kop'ev, Optical properties of InN with stoichoimetry violation and indium clustering, *physica status solidi (a)*, 202 (2005) 377–382.

[14.12] Y. Nanishi, Y. Saito, T. Yamaguchi, F. Matsuda, T. Araki, H. Naoi, A. Suzuki, H. Harima, T. Miyajima, Band-Gap energy and physical properties of InN grown by RF-molecular beam epitaxy, *Materials Research Society Symposium Proceedings*, 798 (2004) Y12.1.1.

[14.13] H. Lu. W. J. Schaff, J. Hwang, H. Wu, G. Koley, and L. F. Eastman, Effect of an AlN buffer layer on the epitaxial growth of InN by molecular-beam epitaxy, *Applied Physics Letters*, 79 (2001) 1489–1491.

[14.14] H. Lu, W. J. Schaff, L. F. Eastmann, J. Wu, W. Walukiewicz, D. C. Look, and R. J. Molnar, Growth of thick InN by molecular beam epitaxy, *Materials Research Society Symposium Proceedings*, 743 (2003) L4.10.1–L4.10.6.

[14.15] Y. Nanishi, Y. Saito, and T. Yamaguchi, RF-molecular beam epitaxy growth and properties of InN and related alloys, *Japanese Journal of Applied Physics – Part 1*, 42 (2003) 2549–2559.

[14.16] T. Araki and Y. Nanishi, Structural characterization of low-temperature InN buffer layer grown by RF-MBE, *Materials Research Society Symposium Proceedings*, 798 (2004) Y10.68.1.

[14.17] Y. Saito, N. Teraguchi, A. Suzuki, T. Araki, and Y. Nanishi, Growth of high-electron-mobility InN by RF molecular beam epitaxy, *Japanese Journal of Applied Physics – Part 2*, 40 (2001) L91–L93.

[14.18] M. Higashiwaki and T. Matsui, High-quality InN film grown on a low-temperature-grown GaN intermediate layer by plasma-assisted molecular-beam epitaxy, *Japanese Journal of Applied Physics – Part 2*, 41 (2002) L540–L542.

[14.19] A. Jain and J. M. Redwing, Study of the growth mechanism and properties of InN films grown by MOCVD, *Materials Research Society Symposium Proceedings*, 798 (2004) Y12.8.1.

[14.20] J. E. Northrup, Screw dislocations in GaN: The Ga-filled core model, *Applied Physics Letters*, 78 (2001) 2288–2290.

[14.21] J. E. Nothrup, Theory of intrinsic and H-passivated screw dislocations in GaN, *Physical Review B*, 66 (2002) 045204:1–5.

[14.22] D. Jena, A. C. Gossard, and U. K. Mishra, Dislocation scattering in a two-dimensional electron gas, *Applied Physics Letters*, 76 (2000) 1707–1709.

[14.23] D. Zakharov, Z. Liliental-Weber, B. Wagner, Z. J. Reitmeier, E. A. Preble, and R. F. Davis, Structural TEM study of nonpolar *a*-plane gallium nitride grown on $(11\bar{2}0)$ 4H-SiC by organometallic vapor phase epitaxy, *Physical Review B*, 71 (2005) 235334:1–9.

[14.24] Z. Liliental-Weber, D. Zakharov, B. Wagner, and R. F. Davis, TEM studies of laterally overgrown GaN layers grown in polar and nonpolar directions, *Proceedings of the SPIE – The International Society for Optical Engineering*, 6121 (2006) 612101.

[14.25] T. Mitate, S. Mizuno, H. Takahata, R. Kakegawa, T. Matsuoka, and N. Kuwano, InN polarity determination by convergent-beam electron diffraction, *Applied Physics Letters*, 86 (2005) 134103:1–3.

[14.26] T. Matsuoka, H. Okamoto, H. Takahata, T. Mitate, S. Mizuno, Y. Uchiyama, and N. T. Makimoto, MOVPE growth and photoluminescence of wurtzite InN, *Journal of Crystal Growth*, 269 (2004) 139–144.

[14.27] A. V. Blant, T. S. Cheng, N. J. Jeffs, L. B. Flannery, I. Harrison, J. F. W. Mosselmans, A. D. Smith, and C. T. Foxon, EXAFS studies of Mg doped InN grown on Al_2O_3, *Materials Science and Engineering: B*, 59 (1999) 218–221.

[14.28] V. V. Mamutin, V. A. Vekshin, V. Y. Davydov, V. V. Ratnikov, Y. A. Kudriavtsev, B. Y. Ber, V. V. Emtsev, and S. V. Ivanov, Mg-doped hexagonal InN/Al_2O_3 films grown by MBE, *physica status solidi (a)*,

176 (1999) 373–378.

[14.29] R. E. Jones, K. M. Yu, S. X. Li, W. Walukiewicz, J. W. Ager III, E. E. Haller, H. Lu, and W. J. Schaff, Evidence for *p*-type doping of InN, *Physical Review Letters*, 96 (2006) 125505:1–4.

[14.30] P. A. Anderson, C. H. Swartz, D. Carder, R. J. Reeves, S. M. Durbin, S. Chandril, and T. H. Myers, Buried *p*-type layers in Mg-doped InN, *Applied Physics Letters*, 89 (2006) 184104:1–3.

[14.31] Z. Liliental-Weber, M. Benamara, J. Washburn, I. Grzegory, and S. Porowski, Spontaneous ordering in bulk GaN:Mg samples, *Physical Review Letters*, 83 (1999) 2370–2373.

[14.32] Z. Liliental-Weber, T. Tomaszewicz, D. Zakharov, J. Jasinski, and M. A. O'Keefe, Atomic structure of defects in GaN:Mg grown with Ga polarity, *Physical Review Letters*, 93 (2004) 206102:1–4.

[14.33] R. E. Jones, S. X. Li, E. E. Haller, H. C. M. van Genuchten, K. M. Yu, J. W. Ager III, Z. Liliental-Weber, W. Walukiewicz, H. Lu, and W. J. Schaff, High electron mobility InN, *Applied Physics Letters*, 90 (2007) 162103:1–3.

[14.34] Z. Liliental-Weber, R. E. Jones, H. C. M. van Genuchten, K. M. Yu, W. Walukiewicz, J. W. Ager III, E. E. Haller, H. Lu, and W. J. Schaff, TEM studies of as-grown, irradiated and annealed InN films, *Physica B*, 401–402 (2001) 646–649.

[14.35] F. Bernandini and V. Fiorentini, Macroscopic polarization and band offsets at nitride heterojunctions, *Physical Review B*, 57 (1998) R9427–R9430.

[14.36] T. Nishida and N. Kobayashi, Ten-milliwatt operation of an AlGaN-based light emitting diode grown on GaN Substrate, *physica status solidi (a)*, 188 (2001) 113–116.

[14.37] S. Nakamura, T. Mukai, and M. Senoh, Candela-class high-brightness InGaN/AlGaN double-heterostructure blue-light-emitting diodes, *Applied Physics Letters*, 64 (1994) 1687–1689.

[14.38] I. Akasaki, S. Sota, H. Sakai, T. Tanaka, M. Koike, and H. Amano, Shortest wavelength semiconductor laser diode, *Electronics Letters*, 32 (1996) 1105-1106.

[14.39] H. Ho and G. B. Stringfellow, Solid phase immiscibility in GaInN, *Applied Physics Letters*, 69, (1996) 2701–2703.

[14.40] A. Zunger and S. Mahajan, S. Mahajan (editor), Handbook on Semiconductors (North Holland, Amsterdam, (1994), Vol. 3. pp. 1399–1514.

[14.41] A. Wakahara, T. Tokuda, X. Dang, S. Noda, and A. Sasaki, Compositional inhomogeneity and immiscibility of a GaInN ternary alloy, *Applied Physics Letters*, 71 (1997) 906–908.

[14.42] N. A. El-Masry, E. L. Piner, S. X. Liu, and S. M. Bedair, Phase separation in InGaN grown by metalorganic chemical vapor deposition, *Applied Physics Letters*, 72 (1998) 40–42.

[14.43] D. Doppalapudi, S. N. Basu, K. F. Ludwig, and T. D. Moustakas, Phase separation and ordering in InGaN alloys grown by molecular beam epitaxy, *Journal of Applied Physics*, 84 (1998) 1389–1395.

[14.44] O. Ueda, S. Isozumi, and S. Komiya, Composition-modulated structures in InGaAsP and InGaP liquid phase epitaxial layers grown on (001) GaAs substrates, *Japanese Journal of Applied Physics*, 23 (1984) L241–L243.

[14.45] P. Henoc, A. Izrael, M. Quillec, and H. Launois, Composition modulation in liquid phase epitaxial $In_xGa_{1-x}As_yP_{1-y}$ layers lattice matched to InP substrates, *Applied Physics Letters*, 40 (1982) 963–965.

[14.46] S. N. G. Chu, S. Nakahara, K. E. Strege, and W. D. Johnston Jr., Surface layer spinodal decomposition in $In_{1-x}Ga_xAs_yP_{1-y}$ and $In_{1-x}Ga_xAs$ grown by hydride transport vapor-phase epitaxy, *Journal of Applied Physics*, 57 (1985) 4610-4615.

[14.47] Z. Liliental-Weber, D. N. Zakharov, K. M. Yu, J. W. Ager III, W. Walukiewicz, E. E. Haller, H. Lu, and W. J. Schaff, Compositional modulation in $In_xGa_{1x}N$: TEM and X-ray studies, *Journal of Electron Microscopy*, 54 (2005) 243–250.

[14.48] Z. Liliental-Weber, D. N. Zakharov, K. M. Yu, J. W. Ager III, W. Walukiewicz, E. E. Haller, H. Lu, and W. J. Schaff, Compositional modulation in $In_xGa_{1-x}N$, *Physica B*, 376–377 (2006) 468–472.

[14.49] Z. Liliental-Weber, J. Jasinski, K. M. Yu, J. Wu, S. X. Li, J. W. Ager III, W. Walukiewicz, E. E. Haller, H. Lu, and W. J. Schaff, Relation between structural and optical properties of InN and $In_xGa_{1-x}N$ thin films, Proceedings of the 27th International Conference on the Physics of Semiconductors, Flagstaff, AZ, July 26–30, 2004; *AIP Conference Proceedings*, 772 (2005) 209–210.

15

InN-based dilute magnetic semiconductors

S. M. Durbin

Department of Electrical and Computer Engineering, The MacDiarmid Institute for Advanced Materials and Nanotechnology, University of Canterbury, Te Whare Wānanga o Waitaha, Christchurch 8140, New Zealand

15.1 Introduction

Over 1300 journal publications from research groups around the world [15.1] reflect the considerable effort put into InN growth and characterization following the 2002 announcement of Davydov and co-workers that the band gap of this material is at least 1 eV narrower than originally proposed [15.2]. Consequently, new applications are actively being sought for this long neglected material in light of its more accurately determined properties. With a theoretical electron Hall mobility [15.3] of $\sim$4000 cm^2/Vs and experimental evidence [15.4] of values approaching 3500 cm^2/Vs, transistor applications are among those devices now being pursued. The timing of the band gap announcement led to the natural consideration of InN for a specific type of transistor structure based on segregating carriers not so much by their electronic charge, but rather on their spin — a concept termed in popular writing as *spintronics*. For this to be feasible, a nonmagnetic host semiconductor must be made ferromagnetic, with the obvious route being doping with transition metal (TM) elements as has been done with other semiconductors. In this chapter, we examine both theoretical predictions and results of several experimental studies of InN doped with magnetic ions, with a view toward their suitability for a new class of spin filtering based electronic devices.

Interest in the class of materials commonly referred to as dilute magnetic semiconductors (DMS) can be traced to the late 1970s, when unexpectedly large magneto-optical effects were observed at low temperatures in HgMnTe and CdMnTe alloys with $\sim$2–50 atomic per cent Mn [15.5–15.7]. In particular, a significant Zeeman splitting resulted in a large Faraday rotation, leading to investigation of CdMnTe for optical isolators [15.8]. These early studies were based on bulk crystals grown using the Bridgman technique [15.9], but advances in molecular beam epitaxy (MBE) growth of CdMnTe and related DMS materials allowed greater control over Mn incorporation, higher quality crystals, and more complex structures such as superlattices [15.10, 15.11]. Since MnTe forms in the NiAs structure under equilibrium conditions and low Mn content CdMnTe was determined to be purely zinc-blende, it was also realized that the observed increase in band gap energy with increasing Mn content pointed to a significantly wider zinc-blende phase band gap for MnTe [15.12, 15.13]. Subsequent studies determined that zinc-blende MnTe, representing the lesser characterized binary component of CdMnTe alloys, is anti-ferromagnetic with a Néel temperature of approximately 65 K [15.14]. Later work included investigations of other nonmagnetic II-VI semiconductor host matrices doped with varying amounts of Mn, including ZnMnSe — which is reasonably well lattice-matched to GaAs, a common substrate and light emitting diode (LED) material. None of these semiconductors, however, exhibited ferromagnetic properties, and little was done in terms of device structures designed to exploit their magnetic characteristics.

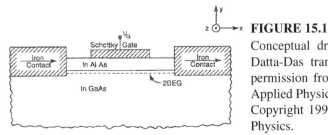

FIGURE 15.1

Conceptual drawing of the spin-filter Datta-Das transistor. Reprinted with permission from S. Datta and B. Das, Applied Physics Letters, 56 (1990) 665. Copyright 1990, American Institute of Physics.

Then, in 1990 Supriyo Datta of Purdue University and his student Biswajit Das proposed a radical new device concept designed around the proposed ability to filter electron spin in a two dimensional electron gas (Fig. 15.1). Interestingly, in their words, this was an "electronic analog of [an] electro-optic modulator" [15.15] where just a short time before Datta had contributed to the development of what could be termed a magneto-optic modulator based on CdMnTe [15.8]. Not even two decades later, this landmark paper has already been directly cited by over 1300 other publications [15.16], and arguably launched the drive toward development of spin-based transistors within the new field popularly known as "spintronics". In the original Datta-Das design, iron contacts were proposed as the spin filtering/aligning layers. Although several recent studies have suggested that this might be a viable approach [15.17], reports of poor spin selectivity across the Fe-semiconductor interface [15.18, 15.19] led many to return to hybrid devices which incorporated a DMS layer such as ZnMnSe for spin filtering [15.20, 15.21]. Unfortunately, such devices only work in the presence of a rather large (for example, 2–5 Tesla) external magnetic field, relegating them more to simply proof-of-concept devices.

15.2 Ferromagnetism in nitride semiconductors

A significant step forward was made in 1992 when Ohno and co-workers at the IBM T. J. Watson Research Center reported partial ferromagnetic ordering below 7.5 K in MBE-grown samples of *p*-type InMnAs [15.22]. Although not a practical temperature for the type of commercial device applications envisioned for spin-based transistors, this result suggested that it was worthwhile considering transition-metal doped mainstream compound semiconductors as candidates for ferromagnetic layers. Then, four years later, Ohno along with a number of colleagues in Japan reported clear evidence of ferromagnetic ordering in MBE-grown GaMnAs thin films at 110 K [15.23]. A significant aspect of the InMnAs and GaMnAs work, in contrast with Mn-doped II-VI DMS materials, was that Mn acts as an acceptor in these semiconductors. Even more significant, however, was the realization that the magnetization of a sample increased with the concentration of holes — in other words, that the observed ferromagnetism was somehow dependent upon the free carriers, and not merely the magnetic ions. Since then, other III-Mn-V compounds such as InMnSb [15.24] have been reported to exhibit ferromagnetism, but their Curie temperatures are well below room temperature.

In 2001, Dietl *et al.* put forth a prediction of a number of ferromagnetic Mn-doped *p*-type semiconductors having Curie temperatures above the 110 K value of GaMnAs [15.25]. Of particular interest was the prediction of a Curie temperature in excess of 400 K in the case of GaMnN, and ~300 K in the case of InMnN (Fig. 15.2). A key aspect of the theory was the role of hole-mediated exchange interactions, and the predictions in Fig. 15.2 are in fact for rather large ($\sim 10^{20}$ cm^{-3}) hole concentrations. The underlying approach was described as being based on the Zener model for ferromagnetism [15.26]. Whether this is the best explanation for the observed dependence of magnetization on hole concentration in GaMnN, however, is a matter of ongoing debate, with one of the issues being whether the holes in fact reside in a separate impurity band [15.27]. Regardless, the theoretical prediction of room temperature ferromagnetism in nitride semiconductors spurred a flurry of experimental activity encompassing a wide range of thin film deposition techniques from sputtering to various epitaxial methods.

Further support for investigating III-Mn-N compounds came from theoretical predictions of Katayama-Yoshida and Sato [15.28], who studied a range of III-V compounds doped with Mn — including AlN, GaN and InN — with the intention of determining under what conditions a stable ferromagnetic state could be obtained. The work was later extended to investigate the concentration dependence of Curie temperatures in more detail [15.29, 15.30]. This first-principles local spin density approach, based on the Korringa-Kohn-Rostoker method, also predicts ferromagnetic states in Mn-doped III-V semiconductors. However, in contrast to the work of Dietl *et al.* [15.25], Katayama-Yoshida and Sato assert that InN with 5 atomic percent Mn substituted for In would have a more stable ferromagnetic state than

AlN or GaN (or any of the III-V semiconductors considered, for that matter). They also predicted a spin glass state in GaMnN, the primary focus of their study, for Mn concentrations exceeding approximately 20 atomic percent.

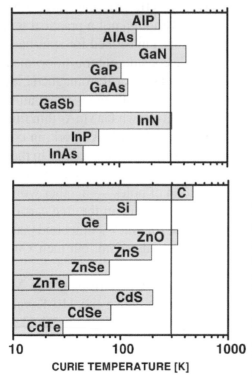

FIGURE 15.2

Calculated Curie temperatures for various semiconductors doped with 5% Mn and a hole concentration of 3.5×10^{20} cm^{-3}. Reprinted with permission from T. Dietl, H. Ohno, and F. Matsukura, Hole-mediated ferromagnetism in tetrahedrally coordinated semiconductors, Physical Review B, 63 (2001) 195205. Copyright 2001, American Physical Society.

15.3　Overview of transition metal doping experiments in GaN and AlN

Given that InN is a rather difficult material to grow due to its low decomposition temperature, as well as the fact that it is less understood than other nitrides, the majority of TM-doped nitride studies to date have been focussed on AlN [15.31–15.36] and GaN [15.32, 15.37, 15.38]. Frazier *et al.* have reported successful growth of AlMnN using gas-source MBE [15.31], although apparently the normal two-dimensional growth experienced for AlN became three-dimensional in character for even weakly Mn-doped AlMnN films. A precise Mn content is not reported, but estimates from Auger electron spectroscopy suggest a rather low value, on the order of 1 atomic percent. At high Mn fluxes, secondary phases formed in the films and AlMn was detected by post-growth x-ray diffraction. Hysteresis is reported in the range of 10 to 300 K, and a coercive field of ~70 Oe.

Studies of AlCrN report ferromagnetic behavior as a general rule, with possible indications that it can persist to as high as 900 K [15.32]. The same study reports a saturation magnetic moment of 0.6 μ_B per chromium atom for a film having 7 atomic percent Cr, suggesting 20% of the Cr in the sample is contributing to the observed magnetization based on an expected 3 μ_B per chromium atom. These values are slightly larger than those of Zhang *et al.*, who examined 50 nm thick films with 15% Cr content grown by sputtering and reported a saturation moment of 0.23 μ_B per chromium atom and a coercive field of 50 Oe [15.35]. It is worth noting that Liu *et al.* report a linear reduction in *c*-axis lattice constant with increasing Cr content up to 7.4 atomic percent Cr [15.32], and Polyakov *et al.* note a small (100 meV) reduction in apparent band gap energy in the case of a film doped to 2 atomic percent Cr [15.34]. Polyakov *et al.* also report detection of multiple phases in material grown under "nonoptimized conditions" [15.34].

Considerably more effort has been applied to growth and characterization of GaMnN. Since it has been fabricated by a large number of bulk and thin film techniques, perhaps it is not surprising that a wide range of magnetic behavior has been reported to date [15.39]. This material system appears to be subject to mixed-phase growth, as Reed *et al.* point out in the context of structural compatibility between cubic Mn_3GaN and wurtzite GaN [15.37]. Even very early reports of GaMnN present hysteresis above room temperature, and there are indications that the magnetization can be strongly dependent on the actual Mn content of a film [15.40]. In contrast, reports of GaCrN are less common, likely due in part to the low vapor pressure of Cr and the need for a high-temperature effusion cell for MBE growth. Still, Singh *et al.* report a saturation magnetic moment of 1.8 Bohr magnetons per chromium atom, while indicating 0.35 μ_B/Cr atom is the highest stable value observed [15.38]. Worth noting is that films grown below 775°C show uniform Cr distribution by high-resolution transmission electron microscopy (HRTEM), but energy-filtered TEM revealed "significant" Cr clustering in films grown at 825°C. This analysis was supported by electron energy loss spectroscopy and indicates that spatial uniformity of any Cr-doped film is an issue that needs to be investigated.

15.4 Manganese doping of indium nitride

In 2004, Chen *et al.* at Tohoku University reported the growth and magnetic characterization of InMnN having Mn mole fractions in the range of 0.04 to 0.1 [15.41–15.44]. These films were grown using a plasma-assisted molecular beam epitaxy system with In and Mn provided using standard effusion cells, and active nitrogen species provided by an RF inductively coupled plasma source operated at 250 W and a chamber pressure of 2×10^{-5} Torr. Films were grown on (0001) sapphire substrates chemically cleaned and outgassed at 800°C for 20 minutes. Following thermal cleaning, substrates were "nitrided" by exposure to the nitrogen plasma while heated to 700°C for 50 minutes. The typical structure is shown schematically in Fig. 15.3; both capped and uncapped films were characterized.

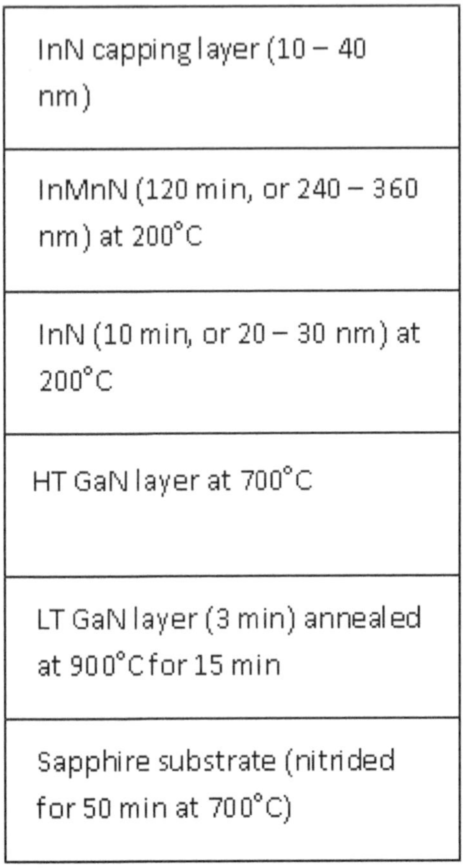

InN capping layer (10 − 40 nm)
InMnN (120 min, or 240 − 360 nm) at 200°C
InN (10 min, or 20 − 30 nm) at 200°C
HT GaN layer at 700°C
LT GaN layer (3 min) annealed at 900°C for 15 min
Sapphire substrate (nitrided for 50 min at 700°C)

FIGURE 15.3

Schematic of typical InMnN structure grown by the Tohoku University group, as described in Chen *et al.* [15.41] The low temperature (LT) and high temperature (HT) GaN layers were grown at 500°C and 700°C, respectively.

The InN layers were grown at the unusually low temperature of 200°C (as opposed to ~500°C) in an effort to improve the solubility of Mn by moving the growth kinetics away from equilibrium. This enabled Mn contents up to 10% to be achieved, although spotty reflection high-energy electron diffraction (RHEED) patterns and post-growth atomic force microscopy images confirmed somewhat rough films compared to high temperature growth. Interestingly, the authors report that addition of Mn led to a small but noticeable improvement in the RHEED pattern which is reminiscent of GaMnN growth, where the existence of a Mn "floating layer" has been hypothesized despite uniform Mn incorporation as determined by secondary ion mass spectrometry (SIMS) [15.45]. Films with Mn content greater than 10% showed clear signs of polycrystalline composition and hence were not included in the study. For Mn content less than 10%, *c*-axis lattice constants exhibited a linear dependence on Mn mole fraction, consistent with Vegard's law and hence substitutional incorporation.

In the case of Ney *et al.*, a collaboration between IBM Almaden Research Center, Stanford University and Lawrence Berkeley National Laboratory, very sensitive x-ray diffraction revealed the presence of a secondary phase in the form of Mn_3N_2 (evident through the (103) reflection) even for a moderate Mn concentration of 1% [15.46–15.48]. These films were also grown by a standard plasma-assisted MBE process using sapphire substrates and GaN buffer layers. It is worth noting that these InMnN films are only 20–50 nm thick, on top of 50–100 nm thick undoped InN buffer layers, in contrast to Chen *et al.* who grew essentially the opposite structure in terms of layer thickness [15.41]. No mention is made of any surfactant-like effect, although the authors report streaky RHEED patterns during growth. Based on the observation of secondary phases, Ney *et al.* did not pursue InMnN further, and reported that magnetization versus temperature measurements of the 1% Mn film exhibited purely paramagnetic behavior which was attributed to Mn_3N_2 clusters [15.47].

The Tohoku University group reported the magnetic properties of two InMnN films, one with a Mn mole fraction of 0.04 and the other with a mole fraction of 0.1. The lower doped film exhibited unambiguous paramagnetic behavior from 300 K down to 1.7 K, as evidenced by the abrupt decrease ($\sim 1/T$) in magnetisation (emu/cm^3) with increasing temperature (Fig. 15.4) [15.43, 15.49]. Somewhat disappointingly, the Curie-Weiss plot shown as the inset to Fig. 15.4 indicates a negative Curie temperature (θ) of -4.5 K, which means the interaction between Mn ions is antiferromagnetic. The higher doped film, in contrast, showed behavior that was initially attributed to ferromagnetism [15.42] but subsequently identified as a paramagnetic to spin-glass transition [15.41]. This conclusion is based on the behavior of the magnetization at low temperature, which exhibits a cusp at approximately 3 K (Fig. 15.5), and also an essentially temperature-independent magnetization in the field-cooled (500 Oe) case which is subsequently independent of sample history. Interestingly, uncapped films (or films capped with less than 2 nm of undoped InN) exposed to air for approximately 1 month exhibited what was interpreted as a "clear ferromagnetic response" [15.41].

The same authors also reported absorption measurements of several Mn-doped films [15.41]. Based on a linear fit to α^2 as a function of energy, which assumes the absorption coefficient α in the range of fitting is proportional to $\sqrt{E - E_{g0}}$, the intercept E_{g0} is taken to be the effective optical band gap. It is well known that this quantity differs significantly from the actual band gap energy of InN however, due to band filling from high background electron concentrations (Moss-Burstein effect). It is not clear that the authors were able to fully account for this effect despite proposing that Mn might be a deep acceptor in InN [15.41]. Although somewhat different fitting can be envisioned for the same data as the α^2 versus energy plots are not well-behaved for all films, it was suggested that Mn doping leads to a monotonic decrease in the actual band gap energy of InN, which is consistent with what Song *et al.* later reported in the case of AlMnN [15.50].

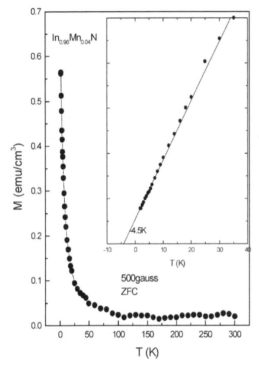

FIGURE 15.4

Magnetization as a function of temperature for an $In_{0.96}Mn_{0.04}N$ film. Inset: Curie-Weiss plot showing a negative Curie temperature consistent with antiferromagnetic coupling between Mn ions. Reprinted from P. P. Chen, H. Makino, and T. Yao, InMnN: a nitride-based diluted magnetic semiconductor, Solid State Communications, 130 (2004) 25–29, with permission from Elsevier.

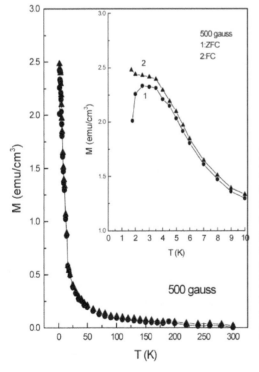

FIGURE 15.5

Magnetization of a 10% InMnN film as a function of temperature showing paramagnetic behavior across the measured temperature range. The inset shows the so-called "cusp" feature at low temperature indicative of a spin glass. Reprinted from P. P. Chen, H. Makino, and T. Yao, InMnN: a nitride-based diluted magnetic semiconductor, Solid State Communications, 130 (2004) 25–29, with permission from Elsevier.

There is a wide variety of MnN compounds reported in the literature, including MnN, Mn_4N, and Mn_3N_2, although their energy gaps are not well characterized [15.51–15.53]. Perhaps particularly relevant to understanding the magnetic behavior of Mn-doped nitrides, Marques *et al.* studied magnetic ground states of zinc-blende and wurtzite MnN using the Vienna ab initio simulation package (commonly known as VASP) [15.51]. Their results suggested a strong dependence of magnetic ground state on lattice constant, with a transition from antiferromagnetic to ferromagnetic possible with a strong hydrostatic strain in the case of a zinc-blende crystal structure. In contrast, they predict a ferromagnetic ground state in unstrained wurtzite MnN. Thus, interpretation of ferromagnetism in InMnN where manganese nitride clusters are possible must be done carefully.

15.5   Chromium doping of indium nitride

Subsequent to their Mn-doped InN study, the Tohoku University group also grew a number of Cr-doped InN samples as well [15.42, 15.43]. The same structure was used for InCrN as they had used for InMnN, including a 10–40 nm thick InN capping layer. As in the case of their InMnN films, they again employed a low substrate temperature (300°C) in an effort to inhibit multiple phase formation. Uniform Cr incorporation up to 4 atomic percent was reported as determined by electron probe microanalysis [15.42]. They report a "slight" decrease in lattice constant with increasing Cr content up to 4 atomic percent, consistent with what they observed for similar concentrations of Mn [15.42]. The magnetic properties of the Cr-doped films were substantially different, however, in that evidence of ferromagnetism was unambiguous. Hysteresis was observed at both 5 and 300 K for an $In_{0.98}Cr_{0.02}N$ film, with a room temperature (zero field cooled) magnetization of approximately 2.5 emu/cm^3. The saturation moment at 300 K was not quoted, but from the magnetization (M-H) curves can be estimated as approximately 7.5 emu/cm^3 [15.42].

In 2005, we reported the results of a similar study, where approximately 700 nm thick InCrN films were grown by a plasma-assisted molecular beam epitaxy technique on a 150 nm InN layer also grown at 450°C, on top of a 150 nm GaN buffer layer grown at 650°C on a (0001) sapphire substrate which had been nitrided at 650°C for 15 minutes. The cryopumped system utilized an Oxford Applied Research model HD25 radio frequency inductively coupled plasma source, a 60 cc conical crucible effusion cell for In evaporation, and a 10 cc high-temperature effusion cell for Cr evaporation. Fluxes were measured using a quartz crystal microbalance placed in the substrate position, and ranged from approximately 1.2×10^{14} atoms/cm^2s in the case of In, to between 2×10^{11} and 5×10^{12} atoms/cm^2s for Cr, as estimated from theoretical flux curves calibrated with a high Cr flux measurement. The resulting Cr compositions ranged from 0.05 to 4 atomic percent as determined by a combination of Rutherford backscattering spectrometry (RBS), nuclear reaction analysis, and particle-induced x-ray emission [15.54].

Unlike what Cheng *et al.* reported for the case of InMnN, increased Cr content led to rapid deterioration of the overall film quality, with weak arcs superimposed on the nominally streaky RHEED pattern by the end of a three-hour growth at the higher Cr concentrations. However, it should be noted that the InCrN thickness in the two studies differs by over an order of magnitude. Regardless of the initial quality of the InCrN layer upon nucleation, internal stress eventually led to formation of large voids on the order of several hundred nanometers throughout the film (Fig. 15.6) [15.55]. Subsequent analysis using electron induced x-ray emission in conjunction with scanning transmission electron microscopy revealed that the Cr in these films is rather mobile, with almost the same concentration detected in the undoped InN and GaN buffer layers [15.55]. X-ray diffraction analysis suggested the presence of both interstitial as well as substitutional Cr [15.54].

FIGURE 15.6

Scanning electron microscope image of an InCrN film containing 2 atomic percent Cr. Inset: RHEED pattern observed at end of growth showing overall the sample is single crystalline, despite the presence of large voids. Reprinted from R. J. Kinsey, P. A. Anderson, Z. Liu, S. Ringer, and S. M. Durbin, Evidence for room temperature ferromagnetism in the $In_{1-x}Cr_xN$ system, Current Applied Physics, 6 (2006) 579–582, with permission from Elsevier.

The magnetic behavior of four films was studied in detail using a SQUID magnetometer; specifically 0.2, 1.0, 2.5 and 4.0 atomic percent Cr. Temperature-dependent magnetization measurements showed essentially constant separation between ZFC and FC curves, with no indication of the rapid increase at low temperatures expected from a paramagnetic component (Fig. 15.7). In all cases the films displayed ferromagnetic behavior, with clear hysteresis observable. For the $In_{0.975}Cr_{0.025}N$ film, a saturation magnetic moment of 7 emu/cm^3 corresponding to approximately 1 μB per Cr atom agrees well with the estimated value of 7.5 emu/cm^3 measured by Cheng *et al.* for a film of essentially the same Cr content (but much smaller thickness). It is unclear at present why these values are higher than what has been reported for AlCrN and GaCrN, but it is worth noting that the samples prepared at Canterbury have apparently retained their magnetic characteristics over a period of several years after storage in atmosphere [15.56]. Separate vibrating sample magnetometer measurements of two films having 2.5 and 4.0 atomic percent Cr, respectively, indicated the ferromagnetic behavior persisted to 470 K, at which the signal-to-noise ratio prevented further measurements [15.54].

The IBM-Stanford collaboration also looked at InCrN during the same period,

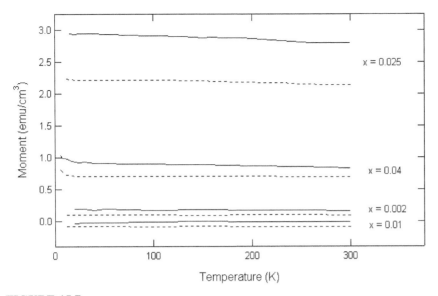

FIGURE 15.7

Magnetization as a function of temperature for four different Cr concentrations. Dashed curves correspond to ZFC measurements and solid curves to FC (1 Tesla) measurements. Reproduced with permission from P. A. Anderson, R. J. Kinsey, S. M. Durbin, A. Markwitz, V. J. Kennedy, A. Asadov, W. Gao, and R. J. Reeves, Journal of Applied Physics, 98 (2005) 043903. Copyright 2005, American Institute of Physics.

growing significantly thinner films (only 10 to 20 nm) at a lower substrate temperature (350°C) [15.57, 15.58]. They employed an approximately 50 nm thick InN buffer layer grown at the more typical substrate temperature of 450°C, on top of an MOVPE GaN layer on (0001) sapphire. The RHEED patterns of these very thin films are virtually identical to those of the much thicker University of Canterbury samples 30 minutes into growth. Cr content was determined to be approximately 3 atomic percent using RBS, but SIMS indicated the Cr was confined to the top 16 nm of the InN layer. Due to uncertainty in the precise thickness of each layer, it is unclear whether any Cr segregation occurred, but they do not report any contamination of the buffer layers as seen in the films we grew at a higher temperature. In terms of magnetic properties of the layers, Rajaram *et al.* report a clear paramagnetic component present in the film, but overall it exhibits hysteresis and remanent magnetisation, consistent with ferromagnetism [15.48]. Further studies suggested an "optimal" chromium concentration of approximately 1–2 atomic percent [15.58], consistent with our observation of a larger magnetization for a film having 2.5 atomic percent Cr compared to one with 4 atomic percent Cr [15.54].

Photoluminescence (PL) studies of the series of films grown at the University of Canterbury revealed the presence of the typically observed feature peaked slightly below 0.7 eV in the case of an undoped film. This peak shifted to slightly greater than 0.8 eV for a Cr mole fraction of 0.0005, consistent with band filling effects expected for the $\sim 10^{20}$ cm^{-3} electron concentration as measured by Hall effect

[15.54]. The peak did not shift further with increasing Cr content, although an additional high energy feature in the vicinity of 1 to 1.2 eV was also present for mole fractions of 0.002 (0.2 atomic per cent) and above. A (111) oriented rocksalt structure CrN film, grown directly on sapphire at 650°C (resulting in a background electron concentration of approximately 9×10^{18} cm^{-3}), did not exhibit any detectable PL, which is consistent with reports of CrN having an indirect band gap. Whether the additional high energy feature observed in PL of InCrN can be attributed to a non-rocksalt phase of CrN, or transitions involving a Γ-valley conduction band minimum more than 0.7 eV above the valence band is yet to be resolved, but it is interesting to note that neither is inconsistent with reported band gap reduction in AlCrN.

The CrN film was also studied using SQUID magnetometry, with initial measurements suggesting ferromagnetism but all subsequent measurements consistently resulting in antiferromagnetic behavior; a Néel temperature slightly below 300 K was also evident in temperature-dependent resistivity. Ney and co-workers, however, examined CrN grown on both (001) MgO and (0001) sapphire, and observed evidence of ferromagnetism in the films on sapphire [15.59]. The fact that films grown on different substrates yet exhibiting ostensibly the same lattice structure (but different crystallographic orientation) and simultaneously exhibiting different magnetic properties is a very interesting result and worth pursuing further, as it clearly has the potential to significantly affect interpretation of magnetization measurements of Cr-doped nitrides. It should also be noted here that Ney *et al.* have recently reported observation of metastable magnetic properties in their InCrN films [15.60], with a stability measured in hours but decaying over several days, although this was not observed in any of the much thicker films grown at the University of Canterbury [15.56].

Although GaMnAs showed clear evidence of a direct role played by holes in determining the magnetic properties of samples, other materials, in particular those highlighted by Dietl *et al.*, have been more controversial. One reason for this is that the vast majority of semiconductors has an inherent tendency to be *n*-type through material-specific defects, so that the addition of an impurity to the host lattice can often lead to an increase in free electron concentration. Thus, simply seeing an increase in magnetic moment correlated to an increase in free electron concentration is really only circumstantial evidence at best; the standard for causality is much higher. Since the nitride semiconductors, and InN in particular, suffer from large background electron concentrations in undoped material, this is particularly problematic. All three TM-doped InN studies described in this chapter concluded that the samples were *n*-type, and in the case of Cr-doping, extremely *n*-type, although in the latter case this appears consistent with evidence of Cr occupying interstitial sites where it would be expected to act as a donor. However, it has only been in the past several years that a true appreciation for the surface electron accumulation layer of InN has developed [15.61], to the extent that standard techniques such as single magnetic field Hall effect measurements and even capacitance-voltage profiling cannot be relied upon to detect p-type conductivity in the "bulk" of a film [15.62–15.64].

15.6 The issue of clustering

Quite a few TM-doped nitride studies allude to the possibility of clusters of various phases which might precipitate out of the otherwise homogeneous alloy structure, regardless of whether they detect any using the experimental techniques at hand. This is a nontrivial issue since some of those phases could at least theoretically exhibit comparable ferromagnetic behavior. Also relevant are theoretical predictions of ferromagnetic behavior attributable to TM clusters of only a few atoms [15.65–15.67], for the simple reason that experimentally, such sized clusters are extremely difficult to detect. Thus, as mentioned previously, interpretation of magnetic behavior — or at least ferromagnetic behavior — in TM-doped nitrides can prove rather problematic. It may be that only once a device whose operation depends upon spin injection based on such a ferromagnetic layer is constructed will the evidence be incontrovertible.

15.7 Conclusions

The investigation of DMS materials beyond the original II-VI:Mn alloys and the unquestionably successful GaMnAs has largely been prompted by predictions of Curie temperatures of approximately 300 K (or higher), which are predicated on the assumption of carrier-mediated ferromagnetism. As discussed in Chapters 4 and 5, InN appears to suffer from two major issues regarding carrier concentration. First, the lack of homoepitaxial bulk substrates means that epitaxial samples are grown on foreign substrates, all of which have a substantially different lattice constant. As a result of the finite amount of strain an epilayer can withstand, threading dislocations form near the interface and propagate through the film. Although the dislocation density can fall off rapidly with increasing film thickness, their density remains high (10^{10} cm^{-2} is not uncommon), and nitrogen vacancies along edge dislocations appear to be effective shallow donors. Therefore, in thin epilayers of 1 μm or less, the background electron concentration in *undoped* samples is typically 10^{18} cm^{-3}, making control of the carrier type and concentration somewhat difficult.

It is of course worth noting that the original drive to study TM-doped nitride semiconductors was based at least in part by theoretical predictions of high Curie temperatures in these materials as a result of hole-mediated exchange interactions with p-type material ($\sim 10^{20}$ cm^{-3}) being required. Yet, the TM-doped InN films discussed in this chapter are all likely n-type. There have been recent suggestions that in fact transition metal ions will bind holes in these materials, thereby making it practically impossible to achieve p-type conducting layers [15.68]. To investigate this in more detail, it will be necessary to determine under what circumstances (if any) Mn, Cr or other transition metals can act as acceptors in InN, and whether hole doping, for example, through co-doping with Mg, leads to any observable changes in magnetization.

Still, regardless of its fundamental origin, three separate groups have successfully prepared single crystal samples of Cr-doped InN which exhibit similar ferromagnetic characteristics. Such a result is perhaps surprising but nevertheless encouraging, as the overall quality of these initial structures is rather poor, due to low temperature growth, interstitial Cr, and large strain. Clearly there remains more work to be done in this area, including growth on better lattice-matched substrates, and further into the indium-rich flux regime. Co-doping with Mg would also be interesting, as it has since been established as a viable p-type dopant for InN. It is less clear as to whether Mn doping holds as much promise, although now that it has been established that p-type layers are difficult to detect, it might be worth reinvestigating InN:Mn at least, with the intent to establish whether Mn acts as an acceptor in the material.

Acknowledgments

I would like to acknowledge the efforts of Drs. Phillip Anderson, Robert Kinsey, and Roger Reeves in the InCrN and CrN growth, SQUID magnetometry and photoluminescence. Additional assistance from Chito Kendrick, Tse En Daniel Lee, Lyndon Williams, Dr. John Kennedy and Gary Turner is also gratefully acknowledged. I am also grateful to Dr. Andreas Ney for remeasuring several of the InCrN films. Financial support for the research performed at the University of Canterbury was provided by the MacDiarmid Institute for Advanced Materials and Nanotechnology as well as a University Research Committee internal grant.

References

[15.1] Scopus physical sciences database search on "InN", 2002 onwards. Amsterdam: Elsevier B.V. Accessed 15 February 2008.

[15.2] V. Y. Davydov, A. A. Klochikhin, R. P. Seisyan, V. V. Emtsev, S. V. Ivanov, F. Bechstedt, J. Furthmüller, H. Harima, A. V. Mudryi, J. Aderhold, O. Semchinova, and J. Graul, Absorption and emission of hexagonal InN. Evidence of narrow fundamental bandgap, *physica status solidi (b)*, 229 (2002) R1–R3.

[15.3] B. R. Nag, Electron mobility in indium nitride, *Journal of Crystal Growth*, 269 (2004) 35–40.

[15.4] C. H. Swartz, R. P. Tompkins, N. C. Giles, T. H. Myers, H. Lu, W. J. Schaff, and L. F. Eastman, Investigation of multiple carrier effects in InN epilayers using variable magnetic field Hall measurements, *Journal of Crystal Growth*, 269 (2004) 29–34.

[15.5] G. Bastard, C. Rigaux, and A. Mycielski, Giant spin splitting induced by exchange interactions in $Hg_{1-k}Mn_k Te$ mixed crystals, *physica status solidi (b)*, 79 (1977) 585–593.

[15.6] J. A. Gaj, R. R. Gatazka, and M. Nawrocki, Giant exciton Faraday rotation in $Cd_{1-x}Mn_x Te$ mixed crystals, *Solid State Communications*, 25 (1978) 193–195.

[15.7] J. A. Gaj, J. Ginter, and R. R. Galazka, Exchange interaction of manganese 3d5 states with band electrons in $Cd_{1-x}Mn_x Te$, *physica status solidi (b)*, 89 (1978) 655–662.

[15.8] A. E. Turner, R. L. Gunshor, and S. Datta, New class of materials for optical isolators, *Applied Optics*, 22 (1983) 3152–3154.

[15.9] J. Stankiewicz and A. Aray, Electrical properties of p-type $Mn_x Cd_{1-x} Te$ crystals, *Journal of Applied Physics*, 53 (1982) 3117–3120.

[15.10] L. A. Kolodziejski, T. Sakamoto, R. L. Gunshor, and S. Datta, Molecular beam epitaxy of $Cd_{1-x}Mn_x Te$, *Applied Physics Letters*, 44 (1984) 799–801.

[15.11] L. A. Kolodziejski, T. C. Bonsett, R. L. Gunshor, S. Datta, R. B. Bylsma, W. M. Becker, and N. Otsuka, Molecular beam epitaxy of diluted magnetic semiconductor ($Cd_{1-x}Mn_x Te$) superlattices, *Applied Physics Letters*, 45 (1984) 440–442.

[15.12] S. H. Wei and A. Zunger, Alloy-stabilized semiconducting and magnetic zinc-blende phase of MnTe, *Physical Review Letters*, 56 (1986) 2391–2394.

[15.13] S. M. Durbin, J. Han, O. Sungki, M. Kobayashi, D. R. Menke, R. L. Gunshor, Q. Fu, N. Pelekanos, A. V. Nurmikko, D. Li, J. Gonsalves, and N. Otsuka, Zinc-blende MnTe: Epilayers and quantum well structures, *Applied Physics Letters*, 55 (1989) 2087–2089.

[15.14] W. Szuszkiewicz, E. Dynowska, J. Z. Domagala, E. Janik, E. Lusakowska, M. Jouanne, J. F. Morhange, M. Kanehisa, K. Ortner, and C. R. Becker, Néel temperature of zinc-blende, MBE-grown MnTe layers: modification by the crystal growth conditions, *physica status solidi (c)*, 1 (2004) 953–956.

[15.15] S. Datta and B. Das, Electronic analog of the electro-optic modulator, *Applied Physics Letters*, 56 (1989) 665–667.

[15.16] Scopus physical sciences database cited reference search. Amsterdam: Elsevier B.V. Accessed 15 February 2008.

[15.17] See, for example, O. M. J. van't Erve, A. T. Hanbicki, M. Holub, C. H. Li, C. Awo-Affouda, P. E. Thompson, and B. T. Jonker, Electrical injection and detection of spin-polarized carriers in silicon in a lateral transport geometry, *Applied Physics Letters*, 91 (2007) 212109:1–3.

[15.18] P. R. Hammar, B. R. Bennett, M. J. Yang, and M. Johnson, Observation of spin injection at a ferromagnet-semiconductor interface, *Physical Review Letters*, 83 (1999) 203–206.

[15.19] H. X. Tang, F. G. Monzon, R. Lifshitz, M. C. Cross, and M. L. Roukes, Ballistic spin transport in a two-dimensional electron gas, *Physical Review B*, 61 (2000) 4437–4440, and references therein.

[15.20] M. Oestreich, J. Hübner, D. Hägele, P. J. Klar, W. Heimbrodt, W. W. Rühle, D. E. Ashenford, and B. Lunn, Spin injection into semiconductors, *Applied Physics Letters*, 74 (1999) 1251–1253.

[15.21] B. T. Jonker, A. T. Hanbicki, Y. D. Park, G. Itskos, M. Furis, G. Koseoglou, A. Petrou, X. Wei, Quantifying electrical spin injection: Component-resolved electroluminescence from spin-polarized light-emitting diodes, *Applied Physics Letters*, 79 (2001) 3098–3100.

[15.22] H. Ohno, H. Munekata, T. Penney, S. von Molnár, and L. L. Chang, Magnetotransport properties of p-type (In,Mn)As diluted magnetic III-V semiconductors, *Physical Review Letters*, 68 (1992) 2664–2667.

[15.23] H. Ohno, A. Shen, F. Matsukura, A. Oiwa, A. Endo, S. Katsumoto, and Y. Iye, (Ga,Mn)As: A new diluted magnetic semiconductor based on GaAs, *Applied Physics Letters*, 69 (1996) 363–365.

[15.24] M. Frazier, R. N. Kini, K. Nontapot, G. A. Khodaparast, T. Wojtowicz, X. Liu, and J. K. Furdyna, Time resolved magneto-optical studies of ferromagnetic InMnSb films, *Applied Physics Letters*, 92 (2008) 061911:1–3.

[15.25] T. Dietl, H. Ohno, and F. Matsukura, Hole-mediated ferromagnetism in tetrahedrally coordinated semiconductors, *Physical Review B*, 63 (2001) 195205:1–21.

[15.26] T. Dietl, H. Ohno, F. Matsukura, J. Cibert, and D. Ferrand, Zener model description of ferromagnetism in zinc-blende magnetic semiconductors, *Science*, 287 (2000) 1019–1022.

[15.27] T. Dietl, Origin and control of ferromagnetism in dilute magnetic semiconductors and oxides, *Journal of Applied Physics*, 103 (2008) 07D111:1–6.

[15.28] H. Katayama-Yoshida and K. Sato, Materials design for semiconductor spintronics by ab-initio electronic-structure calculation, *Physica B*, 327 (2003) 337–343.

[15.29] K. Sato, P. H. Dederichs, H. Katayama-Yoshida, and J. Kudrnovsky, Magnetic impurities and materials design for semiconductor spintronics, *Physica B*, 340–342 (2003) 863–869.

[15.30] K. Sato, P. H. Dederichs, and H. Katayama-Yoshida, Curie temperatures of dilute magnetic semiconductors from LDA + U electronic structure calculations, *Physica B*, 376–377 (2006) 639–642.

[15.31] R. Frazier, G. Thaler, M. Overberg, B. Gila, C. R. Abernathy, and S. J. Pearton, Indication of hysteresis in AlMnN, *Applied Physics Letters*, 83 (2003) 1758–1760.

[15.32] H. X. Liu, S. Y. Wu, R. K. Singh, L. Gu, D. J. Smith, N. Newman, N. R. Dilley, L. Montes, and M. B. Simmonds, Observation of ferromagnetism above 900 K in Cr-GaN and Cr-AlN, *Applied Physics Letters*, 85 (2004) 4076–4078.

[15.33] A. Y. Polyakov, N. B. Smirnov, A. V. Govorkov, R. M. Frazier, J. Y. Liefer, G. T. Thaler, C. R. Abernathy, S. J. Pearton, and J. M. Zavada, Properties of highly Cr-doped AlN, *Applied Physics Letters*, 85 (2004) 4067–4069.

[15.34] A. Y. Polyakov, N. B. Smirnov, A. V. Govorkov, R. M. Frazier, J. Y. Liefer, G. T. Thaler, C. R. Abernathy, S. J. Pearton, and J. M. Zavada, Optical and electrical properties of AlCrN films grown by molecular beam epitaxy, *Journal of Vacuum Science and Technology B*, 22 (2004) 2758–2763.

[15.35] J. Zhang, X. Z. Li, B. Xu, and D. J. Sellmyer, Influence of nitrogen growth pressure on the ferromagnetic properties of Cr-doped ALN thin films, *Applied Physics Letters*, 86 (2005) 212504:1–3.

[15.36] Y. Endo, T. Sato, A. Takita, Y. Kawamura, and M. Yamamoto, Magnetic, electrical properties, and structure of Cr-ALN and Mn-AlN thin films grown on Si substrates, *IEEE Transactions on Magnetics*, 41 (2005) 2718–2720.

[15.37] M. L. Reed, N. A. El-Masry, H. H. Stadelmaier, M. K. Ritums,

M. J. Reed, C. A. Parker, J. C. Roberts, and S. M. Bedair, Room temperature ferromagnetic properties of (Ga,Mn)N, *Applied Physics Letters*, 79 (2001) 3473–3475.

[15.38] R. K. Singh, S. Y. Wu, H. X. Liu, L. Gu, D. J. Smith, and N. Newman, The role of Cr substitution on the ferromagnetic properties of $Ga_{1-x}Cr_xN$, *Applied Physics Letters*, 86 (2005) 012504:1–3.

[15.39] A. Bonanni, Ferromagnetic nitride-based semiconductors doped with transition metals and rare earths, *Semiconductor Science and Technology*, 22 (2007) R41-R56.

[15.40] S. J. Pearton, C. R. Abernathy, M. E. Overberg, G. T. Thaler, D. P. Norton, N. Theodoropoulou, A. F. Hebard, Y. D. Park, F. Ren, J. Kim, and L. A. Boatner, Wide band gap ferromagnetic semiconductors and oxides, *Journal of Applied Physics*, 91 (2003) 1–13.

[15.41] P. P. Chen, H. Makino, and T. Yao, InMnN: a nitride-based diluted magnetic semiconductor, *Solid State Communications*, 130 (2004) 25–29.

[15.42] P. P. Chen, H. Makino, and T. Yao, MBE growth and magnetic properties of InMnN diluted magnetic semiconductor, *Physica E*, 21 (2004) 983–986.

[15.43] H. Makino, J. J. Kin, P. P. Chen, M. W. Cho, and T. Yao, Making ferromagnetic semiconductors out of III-V nitride semiconductors, *Proc. SPIE*, 5774 (2004) 11–16.

[15.44] P. P. Chen, H. Makino, and T. Yao, MBE growth and properties of InN-based dilute magnetic semiconductors, *Journal of Crystal Growth*, 269 (2004) 66–71.

[15.45] S. Dhar, O. Brandt, A. Trampert, K. J. Friedland, Y. J. Sun, and K. H. Ploog, Observation of spin-glass behavior in homogeneous (Ga,Mn)N layers grown by reactive molecular beam epitaxy, *Physical Review B*, 67 (2003) 165205:1–7.

[15.46] A. Ney, R. Rajaram, E. Arenholz, J. S. Harris Jr., M. Samant, R. F. C. Farrow, and S. S. P. Parkin, Structural and magnetic properties of Cr and Mn doped InN, *Journal of Magnetism and Magnetic Materials*, 300 (2006) 7–11.

[15.47] A. Ney, R. Rajaram, R. F. C. Farrow, J. S. Harris Jr., and S. S. P. Parkin, Mn- and Cr-doped InN: A promising diluted magnetic semiconductor material, *Journal of Superconductivity*, 18 (2005) 41–46.

[15.48] R. Rajaram, A. Ney, R. F. C. Farrow, S. S. P. Parkin, G. S. Solomon, and J. S. Harris Jr., Structural and magnetic behavior of transition metal doped InN grown by molecular beam epitaxy, *Journal of Vac-*

uum Science and Technology B, 24 (2006) 1644–1648.

[15.49] J. K. Furdyna, Diluted magnetic semiconductors, *Journal of Applied Physics*, 64 (1988) R29–R64.

[15.50] Y.-Y. Song, P. H. Quang, V.-T. Pham, K. W. Lee, and S.-C. Yu, Change of optical band gap and magnetization with Mn concentration in Mn-doped AlN films, *Journal of Magnetism and Magnetic Materials*, 290–291 (2005) 1375–1378.

[15.51] M. Marques, L. K. Teles, L. M. R. Scolfaro, J. Furthmüller, F. Bechstedt, and L. G. Ferreira, Magnetic properties of MnN: Influence of strain and crystal structure, *Applied Physics Letters*, 86 (2005) 164105:1–3.

[15.52] A. Leineweber, R. Niewa, H. Jacobs, and W. Kockelmann, The manganese nitrides η-Mn_3N_2 and θ-Mn_6N_{5+x}: nuclear and magnetic structures, *Journal of Materials Chemistry*, 10 (2000) 2827–2834.

[15.53] H. Yang, H. Al-Brithen, E. Trifan, D. C. Ingram, and A. R. Smith, Crystalline phase and orientation control of manganese nitride grown on MgO(001) by molecular beam epitaxy, *Journal of Applied Physics*, 91 (2002) 1053–1059.

[15.54] P. A. Anderson, R. J. Kinsey, S. M. Durbin, A. Markwitz, V. J. Kennedy, A. Asadov, W. Gao, and R. J. Reeves, Magnetic and optical properties of the InCrN system, *Journal of Applied Physics*, 98 (2005) 043903:1–5.

[15.55] R. J. Kinsey, P. A. Anderson, Z. Liu, S. Ringer, and S. M. Durbin, Evidence for room temperature ferromagnetism in the $In_{1-x}Cr_xN$ system, *Current Applied Physics*, 6 (2006) 579–582.

[15.56] A. Ney, private communication, 2007.

[15.57] R. Rajaram, A. Ney, G. Solomon, J. S. Harris Jr., R. F. C. Farrow, and S. S. P. Parkin, Growth and magnetism of Cr-doped InN, *Applied Physics Letters*, 87 (2005) 172511:1–3.

[15.58] A. Ney, R. Rajaram, J. S. Harris Jr., and S. S. P. Parkin, Dilute magnetic semiconductors based on InN, *Phase Transitions*, 79 (2006) 785–791.

[15.59] A. Ney, R. Rajaram, S. S. P. Parkin, T. Kammermeier, and S. Dhar, Magnetic properties of epitaxial CrN films, *Applied Physics Letters*, 89 (2006) 112504:1–3.

[15.60] A. Ney, R. Rajaram, S. S. P. Parkin, T. Kammermeier, and S. Dhar, Experimental investigation of the metastable magnetic properties of Cr-doped InN, *Physical Review B*, 76 (2007) 035205:1–7.

[15.61] I. Mahboob, T. D. Veal, C. F. McConville, H. Lu, and W. J. Schaff, Intrinsic electron accumulation at clean InN surfaces, *Physical Review Letters*, 92 (2004) 036804:1–4.

[15.62] R. E. Jones, K. M. Yu, S. X. Li, W. Walukiewicz, J. W. Ager, E. E. Haller, H. Lu, and W. J. Schaff, Evidence for p-type doping of InN, *Physical Review Letters*, 96 (2006) 125505:1–4.

[15.63] P. A. Anderson, C. H. Swartz, D. Carder, R. J. Reeves, S. M. Durbin, S. Chandril, and T. H. Myers, Buried p-type layers in Mg-doped InN, *Applied Physics Letters*, 89 (2006) 184104:1–4.

[15.64] J. W. L. Yim, R. E. Jones, K. M. Yu, J. W. Ager III, W. Walukiewicz, W. J. Schaff, and J. Wu, Effects of surface states on electrical characteristics of InN and $In_{1-x}Ga_xN$, *Physical Review B*, 76 (2007) 041303(R):1–4.

[15.65] B. K. Rao and P. Jena, Giant magnetic moments of nitrogen-doped Mn clusters and their relevance to ferromagnetism in Mn-doped GaN, *Physical Review Letters*, 89 (2002) 185504:1–4.

[15.66] X. Y. Cui, B. Delley, A. J. Freeman, and C. Stampfl, Neutral and charged embedded clusters of Mn in doped GaN from first principles, *Physical Review B*, 76 (2007) 045201:1–12.

[15.67] H. Raebiger, S. Lany, and A. Zunger, Impurity clustering and ferromagnetic interactions that are not carrier induced in dilute magnetic semiconductors, *Physical Review Letters*, 99 (2007) 167203:1–4.

[15.68] W. Pacuski, P. Kossacki, D. Ferrand, A. Golnik, J. Cibert, M. Wegscheider, A. Navarro-Quezada, A. Bonanni, M. Kiecana, M. Sawicki, and T. Dietl, Observation of strong-coupling effects in a diluted magnetic semiconductor $Ga_{1-x}Fe_xN$, *Physical Review Letters*, 100 (2008) 037204:1–4.

16

InN-based low dimensional structures

S. B. Che and A. Yoshikawa

Graduate School of Electrical and Electronics Engineering, Chiba University, 1-33 Yayoi-cho, Inage-ku, Chiba 263-8522, Japan

16.1 Introduction

After finding that the energy band gap of InN is as small as 0.63 eV at room temperature, it was realized that the III-nitride family, consisting of AlN, GaN and InN, can cover a very wide optical wavelength range from the deep UV of ~ 200 nm to the near IR of ~ 2 μm. We can therefore expect much functionality of InN-based photonic devices, not only for light emitters operating over wide wavelength ranges, but also for photodetectors and III-nitride based tandem solar cells being able to cover almost the whole solar spectrum. Further, since the conduction band offset at the heterointerface between InN and GaN or AlN is extremely large (2 eV or more), it is possible to fabricate novel photonic devices, such as laser diodes able to work at high temperatures due to the strong confinement of electrons in quantum wells, and very high-speed optical switches using intersubband electron transitions. When considering fabrication of those photonic devices, it is almost always necessary to fabricate nano-heterostructures of AlN, GaN, InN and their ternary and quaternary alloys.

Growth of nitride-based multiple quantum wells (MQWs) or double heterostructures including InN is, however, generally difficult due to large differences in (1) lattice constants and (2) optimal epitaxial growth temperatures between the composite materials of QWs. For the first point, since the lattice mismatch between InN and GaN or AlN is larger than 10%, as shown in Fig. 16.1, a high density of misfit dislocations would be induced at the hetero-interface between InN and GaN or AlN. For the second point, it is known that the optimum growth temperature of InN is about 500–600°C, which is much lower than that of GaN (about 800°C) in molecular beam epitaxy (MBE) [16.1]. Therefore, InN-based quantum well structures should be generally grown at temperatures lower than 600°C, which causes big difficulties for getting high crystalline quality heteroepitaxial layers of GaN, AlN and their ternary alloys.

Therefore, in order to fabricate high-crystalline quality quantum well structures consisting of InN and other nitrides, a detailed understanding of the effects of the

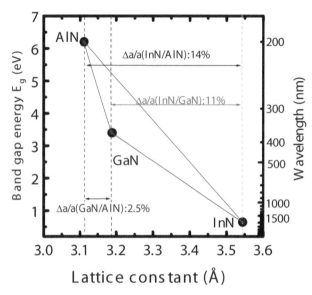

FIGURE 16.1

Lattice constant vs. band gap energy of nitride semiconductors.

two issues mentioned above on structural properties is necessary, as is careful epitaxy control of the hetero-interface.

In this chapter, we discuss systematic studies of MBE growth of InN-based low dimensional structures taking into account the effect of these two issues. First, fabrication of simple InN/GaN single quantum well (SQW) structures is reported and their structural qualities and optical properties in the near infrared region are discussed. Although the lattice mismatch between InN and GaN is as large as 11%, we first examined/tried to fabricate InN/GaN heterostructures. It is shown that the large lattice mismatch seriously affects the growth and properties of InN/GaN heterostructures as expected. Then, as a next step to reduce the effects arising from lattice mismatch, we examined the use of InGaN ternary alloys as barriers instead of GaN. We show that drastic improvement in crystalline quality of InN-based QWs can be achieved by using $In_{0.7}Ga_{0.3}N$ barriers, where it is possible to drastically decrease the lattice mismatch between well and barriers compared with the InN/GaN QWs while keeping the energy band gap difference between the well and barrier at about 0.3 eV. As for the InN/GaN QW structures, we also discuss precise estimations of the critical thickness of an InN well on an $In_{0.7}Ga_{0.3}N$ barrier and the quantum confined Stark effect (QCSE) in this QW structure.

Furthermore, as one of the methods to markedly reduce the problems/effects arising from the large lattice mismatch in the InN-GaN system, we have been paying strong attention to the "InN-based nanostructure family" shown in Fig. 16.2. The fundamental member of the "InN-based nanostructure family" is a novel InN-based nanostructure where extremely thin InN QWs were embedded in a GaN matrix. We have proposed and achieved novel InN-based nanostructures consisting of one monolayer (ML) thick InN wells inserted in a GaN matrix/barriers. Since the InN well layer thickness is ultimately as thin as 1 ML, which is thinner than the critical thickness for InN epitaxy on GaN, a coherent heterostructure is constructed

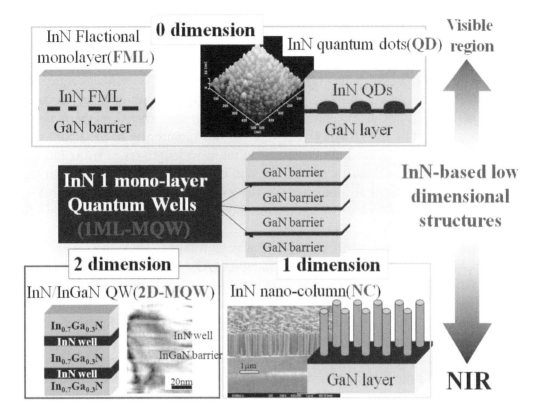

FIGURE 16.2
InN nanostructure family.

in this InN/GaN MQW system, and problems arising from the large lattice mismatch can be avoided in principle. Finally, we describe the growth of other InN based nanostructures, such as zero-dimensional quantum dots and one-dimensional nano-columns, where a remarkable reduction in defect density and enhancement of quantum confinement effects are expected. For these nanostructures, with careful modification of the growth conditions, it is possible to extend the fundamental 1 ML-thick InN/GaN matrix QW structure for fabricating fractional monolayer (FML) MQWs, quantum dots (QDs), nanocolumns (NCs) and also conventional two-dimensional MQWs.

16.2 InN-based quantum well structures

16.2.1 InN/GaN single quantum well

There are few reports on the growth of InN/GaN-based QWs until recently [16.2, 16.3]. This is mainly due to the difficulty in hetero-epitaxy of InN on GaN and vice versa owing to the large differences in both lattice mismatch and optimum growth temperature between them. Therefore, it is important to investigate how these two issues affect the epitaxy and crystalline quality of InN/GaN QW structures.

Samples were grown by conventional RF-MBE. This MBE system was equipped with *in situ* spectroscopic ellipsometry (SE) and reflection high-energy electron diffraction (RHEED). RHEED patterns and diffraction spot intensities were monitored and analyzed by the kSA-400 software. *In situ* monitoring of SE signal was carried out using M2000FI (J. A. Woollam) in the wavelength range from 250 nm to 1696 nm. These enabled us to observe and control epitaxy behaviors on the growth-front surface in real time, such as the monitoring and precise control of lattice relaxation behaviors, growth rate, surface roughness and III/V ratio during growth.

The growth sequence for InN/GaN SQW was as follows. Prior to growth, a *c*-plane sapphire substrate was thermally cleaned in UHV at 830°C for 30 min, then nitrided at 500°C for 20 min. First, a GaN buffer layer was grown at 610°C and a 500 nm GaN epilayer was grown at 790°C. Second, the substrate temperature was decreased to 570–590°C and then the InN well layer and GaN cap layer, of which the thickness was varied from 5 to 40 nm and 50 to 100 nm, respectively, were grown continuously. For the InN/GaN SQW samples with an InN well layer thickness of 5 nm and 7 nm, GaN cap layers were grown by migration-enhanced epitaxy (MEE) for the first 10 nm. For the other SQW samples with well layer thickness of 20 and 40 nm, the GaN cap layers were grown by a conventional MBE process. The polarity of these samples was nitrogen polar, which was confirmed by coaxial impact collision ion scattering spectroscopy (CAICISS) measurements [16.1].

Figure 16.3 shows typical SE-signal traces of the real part of the pseudo-dielectric function $\langle \varepsilon_1 \rangle$ of the InN(5 nm)/GaN SQW measured at $\lambda = 450$ nm and 700 nm. During the growth of the first GaN layer and the GaN cap layer, steady oscillations due to the interference effect can be observed for each wavelength. This means that these GaN layers are grown with well-controlled surface stoichiometry or surface III/V ratio resulting in an optically flat surface [16.1, 16.4]. Generally, the steady-oscillation growth window becomes narrower with decreasing growth temperature [16.5]. Therefore, the precise control of surface stoichiometry was necessary for the growth of the GaN cap layer at such a low temperature of 570–590°C. The *in situ* SE monitoring was quite helpful to control the stoichiometry in real time because, when the surface III/V ratio exceeds unity stoichiometry, an abrupt change in SE signals takes place, especially in the imaginary part of the pseudo-dielectric function $\langle \varepsilon_2 \rangle$ [16.6].

Figure 16.4(a) shows how the RHEED and SE signal traces change during the deposition of InN/GaN SQW, that is, two signal traces during the InN well (5 nm) and MEE GaN growth are shown for the in-plane lattice constant obtained from the *d*-spacing between the (01) and $(0\bar{1})$ RHEED patterns in the $[1\bar{1}00]$ azimuth and the $\langle \varepsilon_1 \rangle$ signal at $\lambda = 450$ nm. Figure 16.4(b) shows streaky RHEED patterns for both layers during growth. The change in $\langle \varepsilon_1 \rangle$ during InN growth indicates the in-

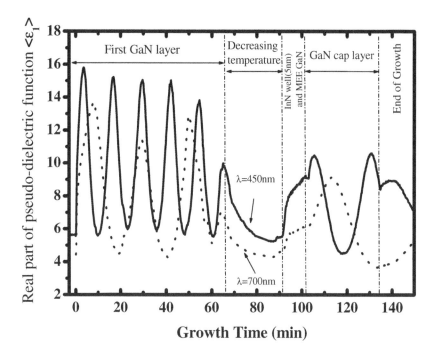

FIGURE 16.3

Signal traces of the real part of the pseudo-dielectric function $\langle \varepsilon_1 \rangle$ of the InN(5 nm)/GaN SQW measured at $\lambda = 450$ nm and 700 nm.

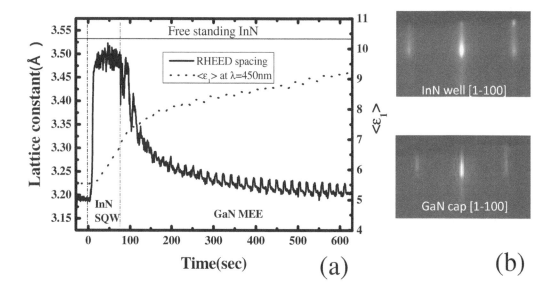

FIGURE 16.4

(a) *In situ* RHEED and SE monitoring during the growth of an InN well (5 nm) and MEE-GaN cap layer. The solid and dashed lines indicate the *d*-spacing and the real part of the pseudo-dielectric function $\langle \varepsilon_1 \rangle$, respectively. (b) RHEED patterns for each layer.

crease of InN thickness under a relaxed InN lattice. Moreover, the d-spacing shows the relaxation behavior of the InN well and GaN cap layers. The InN well layer is almost relaxed after 1–2 monolayers of growth [16.7], and the lattice relaxation in the InN/GaN interface causes the generation of a high density of threading dislocations. In this way, the growth processes of InN well and GaN cap layers could be monitored and analyzed.

Figure 16.5 shows a transmission electron microscopy (TEM) bright field image of an InN(7 nm)/GaN SQW. The InN well and GaN underlayer and cap layer are clearly identifiable. The InN well thickness is estimated to be about 8 nm, which is almost the same as the intended value (7 nm). Note that the threading dislocation density (TDD) suddenly increases at the InN well and GaN cap regions: the TDD in the GaN cap is finally as high as about 10^{11} cm^{-2}, which is more than 10 times higher than that in the GaN underlayer. The high TDD originates from the lattice relaxation discussed above and this seriously affects the optical properties of the InN/GaN SQW samples.

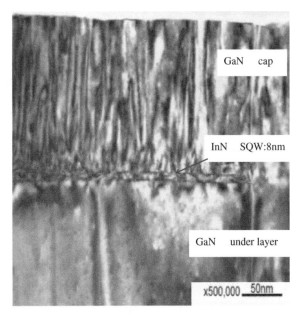

FIGURE 16.5

Cross-sectional TEM micrograph of InN(7 nm)/GaN SQW (bright field).

Figure 16.6 shows photoluminescence (PL) spectra at 13 K for several InN/GaN SQWs with different InN well thicknesses. Here, we also show the PL spectrum of a 2.2-μm-thick InN film for comparison. The properties of the InN films are as follows: XRD full width at half maximum (FWHM) for (002): 460 arcsec; (102): 850 arcsec; carrier concentration: 2×10^{18} cm^{-3}; and electron mobility: ~ 1400 cm^2/Vs at RT. Single peak emissions from the InN SQWs are observed in the range from 1.55 μm to 1.79 μm and these emission peaks red shift with increasing InN well thickness. The PL peak shift is attributed to the quantum size effect of InN/GaN SQWs. For the InN SQW sample with a 7 nm well, the emission peak shifts to around 1.55 μm, indicating the blue shift due to the quantum size effect in the well. In the SQW samples with 20 and 40 nm well layer thickness, emission peaks are also observed at around 1.7 μm. These peak wavelengths are still much shorter than that of thick InN at 1.79 μm. The blue shifts in the SQW samples might also be attributed to the effect of residual carrier concentration in the InN well [16.8]. Moreover, emission intensities of InN SQWs are much weaker than that of the thick InN film. This is caused by a high TDD in the InN wells due to the lattice relaxation, as shown in Fig. 16.5. In order to improve their optical properties, a remarkable decrease in the density of defects such as misfit dislocations is essential.

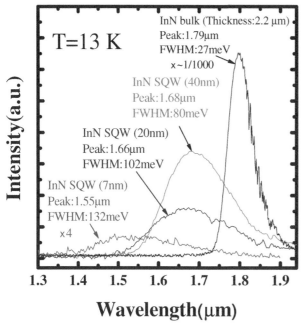

FIGURE 16.6

13 K PL spectra of the InN/GaN SQW samples and InN bulk.

Although the observed quality of InN/GaN SQWs was very poor, this study was the first report of the fabrication of InN/GaN SQW structures with different well thicknesses and the observation of their single PL peaks clearly depending on the well layer thickness. It was very difficult to fabricate high-quality InN/GaN QW structures due to their large lattice mismatch and also due to a big difference in optimum growth temperatures for InN below 600°C and for GaN of approximately 800°C in MBE. In particular, the large lattice mismatch between the well and barrier layers introduced a high density of misfit dislocations and induced a 3-dimensional growth mode, resulting in poor structural and optical properties.

16.2.2   InN-based multiquantum wells with In-rich InGaN barriers

In order to solve the problems stated above, an In-rich InGaN ternary alloy was used as the barrier layer instead of a GaN layer in the InN-based QWs. The In content of the InGaN barrier layer was set at approximately 0.7 to reduce the lattice mismatch with the InN well. The lattice mismatch in the InN/In$_{0.7}$Ga$_{0.3}$N heterostructure was remarkably reduced, but approximately 3.1% mismatch still remained. The energy band gap difference between the InN well and the In$_{0.7}$Ga$_{0.3}$N barrier was estimated to be approximately 0.3 eV [16.9]. As compared with the InN/GaN QWs, a significant improvement was observed in the structural and optical properties of InN-based MQWs when using the In$_{0.7}$Ga$_{0.3}$N barrier.

Two N-polarity InN/In$_{0.7}$Ga$_{0.3}$N MQW samples (samples 1A and 1B) were grown by RF-MBE. Samples 1A and 1B consisted of 10-period MQWs with different InN well layer thicknesses of ~ 16 nm and ~ 3 nm, respectively, but having the same InGaN barrier thickness of ~ 10 nm and GaN cap layer thickness of ~ 100 nm. The In content of the InGaN barriers estimated from the flux ratio of In and Ga sources and was approximately 0.68 for both samples 1A and 1B. Furthermore, in order to eliminate excess In and/or Ga droplets on the surface/interface during the growth of MQWs, the surface was irradiated with RF-plasma-excited nitrogen for 10 sec at each growth interval between the well and barrier. Moreover, a 7 period InN/GaN MQW structure (sample 1C) consisting of ~ 5 nm well and ~ 10 nm GaN barrier layers was fabricated for comparison. Both the well and barrier layers were grown at 570°C.

Figure 16.7 shows cross-sectional TEM images of the samples 1A, 1B and 1C. These images clearly show that the structural quality of the InN/In$_{0.7}$Ga$_{0.3}$N MQWs is much better than that of the InN/GaN MQWs (sample 1C). In sample 1C, a higher density of dislocations is generated in the MQWs region, and it is difficult to identify the interface between the InN wells and GaN barriers. The significant increase in dislocation density is primarily due to the large lattice mismatch between InN and GaN which causes a high density of misfit dislocations, as observed in the InN/GaN SQW structure (Fig. 16.5). These results suggest that the InN/GaN heterointerface induces a large number of dislocations, and the fabrication of fine structure InN/GaN SQW and MQWs is extremely difficult.

On the other hand, for samples 1A and 1B of InN/InGaN MQWs, fine periodic MQW structures and sharp/clear interfaces are observed, as shown in Fig. 16.7(a) and (b). The well and barrier thicknesses were easily estimated to be 16.1 nm and 9.2 nm for sample 1A, and 3.6 nm and 9.4 nm for sample 1B, respectively. The successful growth of these fine structure MQWs was due to the reduction in the lattice mismatch by employing InGaN barriers instead of GaN. These results indicate that a reduction in the lattice mismatch and control of strain in QW structures are essential for achieving the fabrication of high quality InN-based MQWs.

Figure 16.8 shows $2\theta - \omega$ XRD scans for the (002) reflections of the InN-based MQW samples. Solid and dashed lines indicate the experimental and simulated results, respectively. For sample 1C, no satellite peaks are observed, indicating

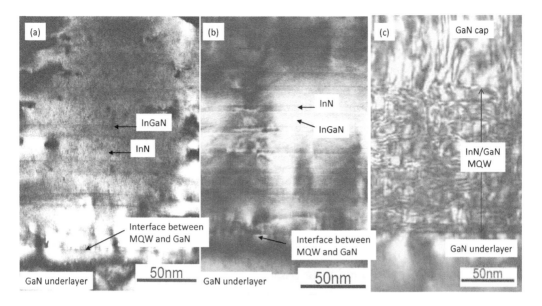

FIGURE 16.7

Cross-sectional TEM microgaphs of MQWs (dark field, $g = 002$): (a) Sample 1A: InN(16.1 nm)/InGaN(9.2 nm) 10-QWs, (b) sample 1B: InN(3.6 nm)/InGaN(9.4 nm) 10-QWs and (c) sample 1C: InN(5 nm)/GaN(10 nm) 7-QWs.

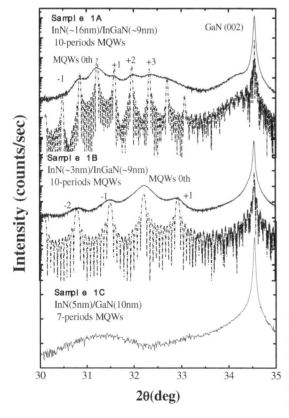

FIGURE 16.8

$2\theta - \omega$ XRD scans of InN/InGaN MQWs (samples 1A and 1B) and InN/GaN MQW (sample 1C). The solid and dashed lines indicate the experimental and simulated results, respectively.

that no periodic structure is formed. This is caused by the very high density of dislocations as shown in the TEM image in Fig. 16.7(c). Compared with sample 1C, clear satellite peaks up to the second order are observed for the InN/InGaN MQW samples 1A and 1B. This result indicates that the InN/InGaN MQWs have fine periodic structures, which is consistent with the TEM results. From the observed satellite peak positions and the simulated data, the thicknesses of each well and barrier in the MQWs and the In contents of the InGaN barriers were estimated. The layer thicknesses of the combinations of InN/InGaN were consistent with the observed TEM images (sample 1A: InN(16.0 nm)/InGaN(9.0 nm), sample 1B: InN(3.5 nm)/InGaN(9.5 nm)). The In contents of the InGaN barriers were estimated to be 0.67 in sample 1A and 0.56 in sample 1B. In the case of sample 1B, the value slightly deviated from the beam flux ratio of In and Ga sources, which was 0.68. It might be caused by fluctuation of the flux during the growth.

The PL spectrum of the InN(16 nm)/In$_{0.67}$Ga$_{0.33}$N(9 nm) MQWs (sample 1A) at 13 K is shown by a solid line in Fig. 16.9. A PL spectrum of one of the best quality 2.2-μm-thick InN epilayers with carrier concentration of 2×10^{18} cm^{-3} is also shown as a dashed line for comparison. In the InN/InGaN MQWs, a single peak emission at approximately 1.75 μm is observed. The PL peak wavelength of the MQWs is slightly shorter than that of the thick InN epilayer. The difference in the peak wavelengths may be due to the differences in the residual carrier concentrations in the InN well layers. In comparison to the spectrum (FWHM = 18 meV) of the InN epilayer, the PL from the MQWs is weak and broad (FWHM = 96 meV). This is attributed to poor crystalline quality of the MQWs, primarily due to the lattice relaxation which occurred at the interface between the first-InGaN and GaN underlayer. Misfit dislocations induced by the lattice relaxation strongly affect the crystalline quality and optical properties of MQWs. Actually, no PL emission was observed in the InN/GaN MQWs (sample 1C) with such a high density of dislocations and it was difficult to observe PL emission at RT even for sample 1A. Therefore, in order to achieve good optical properties of the InN-based QWs comparable to the InN bulk, further improvement in their crystalline quality, especially the first InGaN layer grown on the GaN underlayer, is necessary.

16.2.3   In polarity growth of InN/InGaN MQWs for improved crystal quality

It is well known that the polarity control is an important issue in III-nitride epitaxy due to its great influence on structural quality and surface properties. In the case of InN epitaxy, the highest possible growth temperature of InN is about 600°C for N-polarity, which is about 100°C higher than that for In-polarity [16.1]. The higher growth temperature would be a big advantage for high-quality epitaxial growth and nano-heterostructure fabrication with other III-nitrides. In fact, N-polarity InN/InGaN MQWs have been successfully fabricated as discussed in the previous section [16.10]. On the other hand, study of In-polarity InN growth was reported by a few research groups, and they confirmed that comparable crystalline quality with that for the N-polarity growth could be obtained in the In-polarity growth in spite

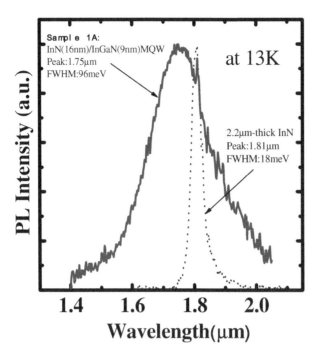

FIGURE 16.9

13 K PL spectra from a InN(16.1 nm)/In$_{0.67}$Ga$_{0.33}$N(9.2 nm) MQW structure (solid line) and from a 2.2-μm-thickness InN epilayer (dashed line).

of its lower growable temperature [16.11–16.13]. Moreover, an atomically flat surface with single monolayer height steps was obtained in In-polarity InN epitaxy, but not in the N-polarity case, where a spiral-mode growth morphology with bilayer or multiple-layer height steps was generally observed [16.14]. The achievement of an atomically flat surface/interface is very important in the fabrication of high-quality and high-performance quantum structure photonic/electronic devices. It has been reported that for GaN and InGaN with smaller In composition ($X_{In} < 0.36$) [16.15], high-quality growth and smooth surface morphology were obtained in the group III polarity growth compared with the N-polarity growth. In addition, superior crystalline quality of In-rich InGaN ($X_{In} \sim 0.7$) has been reported for group III polarity growth rather than N-polarity growth [16.16]. These results indicate that further improvement in structural quality and optical properties of the InN-based MQWs should be possible by growing the QWs in the group III polarity growth regime.

Figure 16.10 shows $2\theta - \omega$ scans for (002) diffraction of both polarity 20-period InN/InGaN MQWs. Here, the In composition in the InGaN barriers was fixed to 0.7 and the well/barrier thicknesses in both samples were about 5 nm/6 nm. The solid and dashed lines indicate the experimental and simulated results, respectively. The diffraction peak positions in simulated curves fit well to the experimental ones. The layer thickness combinations of well/barrier and In compositions in the In-GaN were estimated by the simulations. The estimated results are also indicated in Fig. 16.10. Clear satellite peaks up to second order are observed for both samples. However, it is obvious by comparing the two curves of both polarity MQWs that the experimental curve of the In-polarity MQWs agrees better with the simulated curve than the case of the N-polarity MQWs. The FWHM values of the 0th and

-1st order peaks of the MQWs were estimated to be 222 arcsec and 330 arcsec for the In-polarity MQWs and 440 arcsec and 680 arcsec for the N-polarity MQWs, respectively. These results indicate that superior structural quality MQWs can be obtained in In-polarity growth, which is mainly due to the quality of the In-rich InGaN barrier layers.

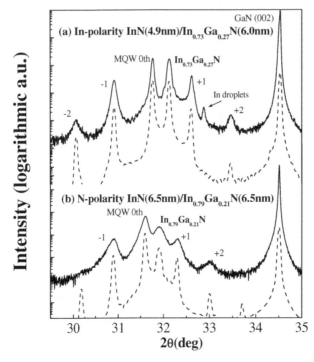

FIGURE 16.10

$2\theta - \omega$ XRD scans of the InN/InGaN MQWs. (a) In-polarity and (b) N-polarity. The solid and dashed lines indicate the experimental and simulated results.

PL spectra at room temperature (RT) and 15 K of the In-polarity MQWs (50 periods) are shown in Fig. 16.11. The well/barrier thicknesses of the MQWs were InN(1.1 nm)/$In_{0.70}Ga_{0.30}N$(4.0 nm). A single peak emission at around 1.62 μm at RT and 1.58 μm at 15 K is clearly observed from this MQW structure. This result indicates that the In-polarity InN/$In_{0.7}Ga_{0.3}N$ MQWs can provide much stronger PL emission compared with the N-polarity MQWs. On the basis of these results, we conclude that the In-polarity growth regime is preferable to fabricate better structural and interface quality InN/InGaN QWs.

16.2.4   Critical thickness of InN well in InN/$In_{0.7}Ga_{0.3}N$ MQW

In the fabrication of MQW structures, precise estimation of the critical thickness between the well and the barrier layers is very important to realize high structural quality MQWs with a reduced density of defects arising from the lattice relaxation. For the InN-based MQWs, a few studies on the critical thickness are found [16.10, 16.17, 16.18], but investigations for estimating the critical thickness and/or lattice relaxation and their effects on the structural quality have not been performed in

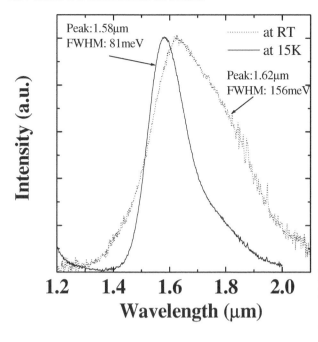

FIGURE 16.11
PL spectrum of the In-polarity InN (1.1 nm)/In$_{0.7}$Ga$_{0.3}$N(4.0 nm) MQWs (50 periods) at RT and LT.

detail yet.

Here, the critical thickness and lattice relaxation behavior for InN wells were investigated by fabricating 20 periods of In-polarity InN/In$_{0.7}$Ga$_{0.3}$N MQWs. The MQWs were grown on MOCVD-grown 2–3-μm-thick Ga-polarity GaN/c-plane sapphire to grow the samples under a group III polarity regime. Prior to the MQW growth, a 100–200-nm-thick free standing In$_{0.7}$Ga$_{0.3}$N interlayer was deposited so that the lattice constants of the barrier in MQWs and the underlying interlayer would be the same. Then, 20 period InN/In$_{0.7}$Ga$_{0.3}$N MQWs were fabricated at 450–470°C. Finally, the surface was capped by a 20 nm In$_{0.7}$Ga$_{0.3}$N layer. The InN well layer thickness was varied from 0.7 to 4 nm, while the InGaN barrier layer thickness was kept unchanged at about 4 nm. Table 16.1 summarizes experimentally estimated structural parameters for all samples examined in this study. These values were estimated by simulated results of XRD scans and reciprocal space mapping (RSM) measurements. Here, the strain in the InN wells was evaluated by using the lattice relaxation parameter R,

$$R = \frac{(a_{well} - a_{barrier})}{(a_0 - a_{barrier})} \times 100\% \tag{16.1}$$

where a_{well}, a_0, and $a_{barrier}$ are the in-plane lattice constants for InN wells, free standing InN (that is, 3.5 Å), and In$_x$Ga$_{1-x}$N barriers, respectively.

Figure 16.12 shows $2\theta - \omega$ scans for the (002) diffraction peak of samples 2B, 2D, and 2F, where the thicknesses of the InN wells were designed to be 1, 2, and 4 nm, respectively. The solid and dotted lines indicate experimental and corresponding best-fit simulated diffraction curves, respectively. The peaks labelled "In$_x$Ga$_{1-x}$N" and "GaN" indicate (002) diffraction peaks from the interlayer and

TABLE 16.1

Summary of In-polarity InN/In$_{0.7}$Ga$_{0.3}$N MQW samples examined in Sections 16.2.4 and 16.2.5. InN well thickness (t_{well}), InGaN barrier thickness ($t_{barrier}$), In content in InGaN layer (X_{In}) and relaxation parameter (R) were estimated by simulated diffraction curves and RSM measurement.

Sample	Sample number	t_{well} (nm)	$t_{barrier}$ (nm)	X_{In} in InGaN barrier/interlayer	R in InN wells (%)
2A	U391	0.7	4.1	0.71	0
2B	U374	1.0	4.1	0.72	0
2C	U378	1.3	3.4	0.70	75
2D	U364	2.0	4.2	0.74	80
2E	U404	2.5	3.6	0.71	80
2F	U365	4.2	4.1	0.73	100

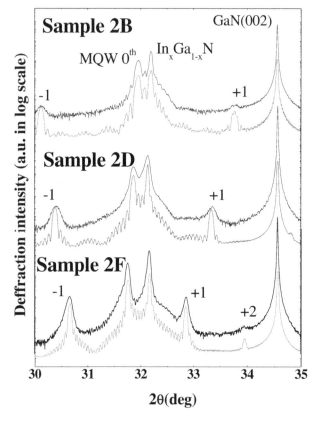

FIGURE 16.12

$2\theta - \omega$ XRD scans (solid lines) of In-polarity InN/In$_{0.7}$Ga$_{0.3}$N MQWs (samples 2B, 2D, and 2F), and their simulated diffraction curves (dotted lines).

substrate, respectively.

It is shown in Fig. 16.12 that clear and sharp satellite peaks are observed from the MQWs and simulated diffraction curves agree well with the experimental ones, indicating that fine periodic structures are formed. Compared with the previous results for N-polarity MQWs (Fig. 16.8 and Fig. 16.10(b)), much sharper diffraction peaks were observed, which means that In-polarity MQWs with similar structure were remarkably improved compared to the N-polarity ones. This is attributed to the surface/interface flatness in the In-polarity MQWs, which can be more easily improved than in the case of N-polarity.

Figure 16.13 shows the RSM profiles around the asymmetric (104) reflection of MQWs for four different InN well thicknesses of 0.7, 1, 1.3 and 2 nm (samples 2A, 2B, 2C and 2D, respectively, as indicated in Table 16.1), where Q_x and Q_y coordinates indicate the reciprocals of in-plane lattice spacing of 100 and 004 planes, respectively. It is shown in Fig. 16.13 that clear satellite peaks of 0th and $\pm 1^{st}$ order are observed from the MQWs in all samples, but their position in Q_x with respect to that from each $In_{0.7}Ga_{0.3}N$ interlayer is different depending on the InN well layer thickness. For the samples 2A and 2B, which have relatively thinner InN well thicknesses among the examined samples, the 0th order main diffraction peak from the MQWs is located at almost the same position in Q_x as that from the $In_{0.7}Ga_{0.3}N$ interlayer, that is, ΔQ_x is negligibly small. The estimated differences in the in-plane lattice constant (Δa) between them for both samples were less than 0.15% (5×10^{-3} Å). This means that the InN wells in the MQWs were coherently grown on the $In_{0.7}Ga_{0.3}N$ barrier/interlayer and almost no lattice relaxation took place when the InN well thickness was less than 1 nm. On the other hand, the corresponding difference $\Delta Q_x s$ for samples 2C and 2D are remarkably large, indicating that lattice relaxation took place when the InN well thickness was larger than 1.3 nm. This means that the experimentally obtained critical thickness is much thinner than the theoretically estimated value of 4 nm [16.10].

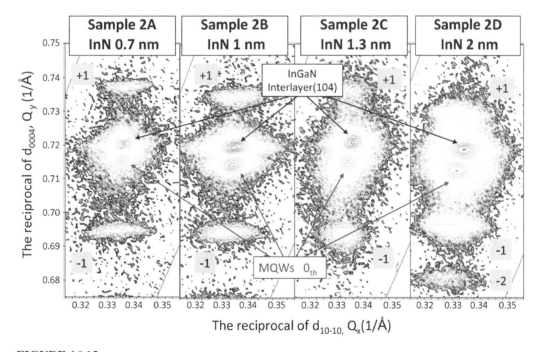

FIGURE 16.13

XRD-reciprocal space mapping for asymmetric (104) diffractions for In-polarity $InN/In_{0.7}Ga_{0.3}N$ MQWs (samples 2A–2D). Q_x and Q_y indicate the reciprocal of d_{100} and d_{004}, respectively.

The dependence of the structural quality of the MQWs on InN well thickness was also investigated by atomic force microscopy (AFM) measurements. Figure 16.14(a)–(c) shows 3×3 μm^2 AFM images for samples 2B, 2D, and 2F, respectively. It is obvious that the thickness of the InN wells greatly affects the surface morphology. As shown in the figures, fairly deep "V-shape" pits were observed (see Fig. 16.14(d)) and their density was increased markedly with increasing InN well thicknesses: 7×10^7 cm^{-2}, 7×10^8 cm^{-2} and 1×10^9 cm^{-2} for samples 2B, 2D, and 2F, respectively. The relationship between the pit density as well as its depth and InN well thickness is shown in Fig. 16.15. Here, the difference, Δa, in the in-plane lattice constant between the MQWs and InGaN interlayer as a function of InN well thickness is also shown for comparison. It is shown that both the pit density and Δa suddenly increase when the InN well thickness exceeds 1 nm. It is also shown in Fig. 16.15 that the depth of the pits tends to increase with the InN well thickness. It was found that the pit depth in sample 2F with 4.2-nm-thick InN wells was deeper than the thickness of the InGaN cap layer (~ 20 nm). This means that these pits were formed in the MQWs regions following the lattice relaxation at the hetero-interface during MQW growth, and we have concluded that the critical thickness of the InN well in the MQW structure is 1.0 nm.

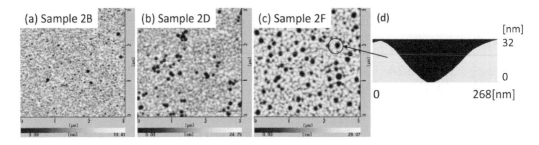

FIGURE 16.14

(a)–(c) 3×3 μm^2 AFM images of samples 2B, 2D, and 2F. (d) A line profile of one of the pits in the surface morphology of sample 2F.

16.2.5 Quantum confined Stark effect in InN/In$_{0.7}$Ga$_{0.3}$N MQW

The dependence of RT-PL spectra of the In-polarity InN/In$_{0.7}$Ga$_{0.3}$N MQWs (listed in Table 16.1) on InN well thickness was investigated. The results for samples 2A, 2B, 2C, 2D and 2F are shown in Fig. 16.16. The solid and dotted lines indicate PL spectra and their Gaussian-fit curves correcting the effect of optical interference in the samples, respectively. Near infrared emission around optical communication wavelengths was observed at RT for these samples and it was found that the structural and optical properties of the MQW samples are superior, from the viewpoint of observing PL at RT, to the N-polarity MQWs samples previously discussed in Sections 16.2.1 and 16.2.2. When the InN well thickness was varied in the range

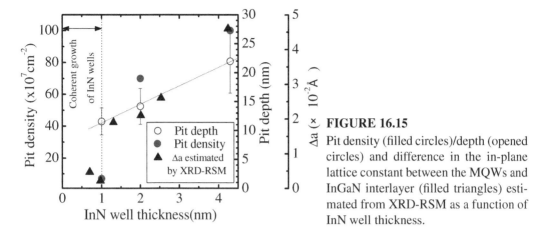

FIGURE 16.15
Pit density (filled circles)/depth (opened circles) and difference in the in-plane lattice constant between the MQWs and InGaN interlayer (filled triangles) estimated from XRD-RSM as a function of InN well thickness.

from 0.7 to 2.0 nm, a peak shift in PL emission wavelength from 1.40 to 1.93 μm was observed. Their PL intensities were remarkably decreased, however, with increasing InN well thickness. Finally, no emission was detected at RT in the sample 2F having 4.2-nm-thick InN wells.

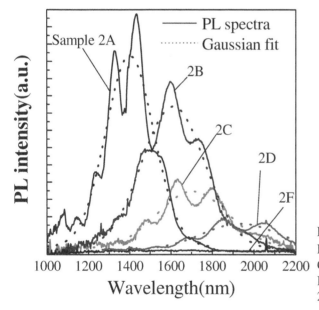

FIGURE 16.16
RT-PL spectra and their fitting with Gaussian function of In-polarity InN/In$_{0.7}$Ga$_{0.3}$N MQWs (samples 2A, 2B, 2C, 2D, and 2F).

One of the possible reasons for the decrease in PL intensity with increasing InN well thickness is poorer crystalline quality for thicker InN well samples due to the effect of lattice relaxation as confirmed in Figs. 16.14 and 16.15. The other important reason is the decrease of optical emission efficiency due to the QCSE in the InN/In$_{0.7}$Ga$_{0.3}$N MQW system.

In order to investigate the effect of the InN well thickness on PL emission properties, we theoretically calculated the optical transition energies in the InN/In$_{0.7}$Ga$_{0.3}$N

MQWs for different InN well thicknesses, paying attention to the QCSE in particular. The calculation was carried out by using a commercially available device simulator (LASTIP, Crosslight Software Inc.), where the Schrödinger and Poisson equations were solved self-consistently and the screening effect of the electric field in the InN wells by free carriers was taken into account. The following device/material parameters were assumed for the numerical calculation: (1) 7 InN quantum wells; (2) band gap energies for InN and $In_{0.7}Ga_{0.3}N$ of 0.63 and 0.93 eV, respectively (that is, the bowing parameter of 2.5 eV for InGaN alloys was assumed [16.9]); (3) a conduction and valence band offset ratio of 0.62:0.38 [16.19]; (4) carrier concentrations for InN and $In_{0.7}Ga_{0.3}N$ layers of 5 and 1×10^{18} cm^{-3}, respectively [16.13, 16.16]; (5) a static dielectric constant for InN of 9.3 [16.20–16.22]; (6) the MQWs structures were coherently grown, that is, the effect of lattice relaxation on the transition energies was not taken into account; and (7) other parameters, such as effective masses of electrons and holes [16.23] and deformation potentials [16.17] were taken from those reported in other references.

Comparison between experimental PL peak energies and calculated transition energies as a function of InN well thickness is illustrated in Fig. 16.17. Solid triangles and circles are experimentally obtained PL peak energies for the coherently grown and relaxed MQWs samples, respectively, and one solid- and the other three dashed- theoretical-lines correspond to an internal electric field of 0, 1, 2, and 3 MV/cm, respectively. It is shown that experimental PL peak energies are smaller than those for the no internal electric field case (the upper solid line), which indicates the internal electric field remarkably affects PL properties. When the InN well thickness is less than its critical thickness, a fairly close agreement is confirmed between experimental and calculated results for the case of 3 MV/cm. This value for the internal electric field is larger than that of 1.8 MV/cm reported by Ohashi *et al.* for $InN/In_{0.75}Ga_{0.25}N$ MQWs [16.17] and this would be attributed to the larger strain because of using a higher In content InGaN barrier here.

When the InN well thickness is increased above the critical thickness, however, it seems that the internal electric fields tend to be smaller. As shown in Table 16.1, lattice relaxation took place for these samples, resulting in smaller strain or smaller piezoelectric field in these samples with thicker InN wells. Actually, for the samples with InN wells thicker than 1.5 nm, the internal electric fields are expected to be 1–2 MV/cm.

Thus, it was confirmed that large piezoelectric fields of 3 MV/cm are induced in the In-polarity $InN/In_{0.7}Ga_{0.3}N$ MQWs and this results in the remarkable effect of the QCSE causing the red shift of PL peak energies from the unstrained case as well as a reduction in their intensities. It should be noted, however, that we did not take into account the effects of fluctuations in well/barrier thicknesses nor interface flatness in this study. In order to directly evaluate the QCSE for PL emission energy and intensity as well in this MQW system, additional experiments, such as pumping power dependent measurements or carrier dynamics measurements using a high-speed IR-detection system, are helpful.

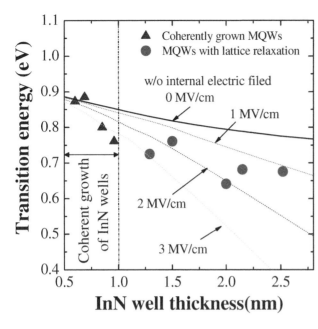

FIGURE 16.17
Comparison between PL peak energies and calculated transition energies of InN/In$_{0.7}$Ga$_{0.3}$N MQWs with different InN well thickness. Solid triangles and circles indicate experimentally obtained PL peak energies for the coherently grown MQWs and the MQWs with lattice relaxation. In the calculation, the internal electric fields in the MQWs were assumed to be 0 (solid line), 1.6, 3 and 4 MV/cm (dashed lines).

16.3 Novel nanostructure one monolayer InN quantum wells in GaN matrix

If the well layer thickness in the MQW structure is less than the critical thickness, each layer of the MQW structure is just strained and problems arising from introducing dislocations due to the large lattice mismatch can be avoided or remarkably reduced. We have succeeded in extending this idea to the InN/GaN-based MQW system, that is, ultimately thin InN layer insertion into a GaN matrix, on the basis of our understanding of InN epitaxy.

We have reported the first successful achievement of a novel structure and new functionality of InN/GaN-based III-nitride MQWs consisting of ultimately thin InN wells, that is, one monolayer (ML) and/or fractional monolayer (FML) InN wells coherently embedded in the GaN matrix [16.24, 16.25, 16.26]. Since the critical thickness for the epitaxy of InN on *c*-plane GaN is about 1 ML, these MQWs are in principle free of newly generated misfit dislocations at the hetero-interface. Furthermore, the growth temperature for 1 ML InN epitaxy on GaN can be intentionally remarkably higher than those for normal InN layer epitaxy, resulting in higher quality epitaxy of each part of the MQW structure thanks to much more enhanced surface migration during epitaxy under elevated growth temperatures. This higher growth temperature can be achieved because of an increased effective bonding strength between In and N in this case due to the effect of the GaN matrix [16.27], that is, each N atom bonded to In still has three other stronger bonds with surrounding Ga atoms. Therefore this novel InN/GaN-based MQW structure possesses a potentially high quality nature.

Some new functionality of the 1 ML InN/GaN-MQW system is expected from its similarity to ultra-thin InAs well insertion in a GaAs matrix, that is, a ~ 1 ML InAs/GaAs-MQW system. Both systems are common-anion systems, where narrow band gap ultra-thin wells are inserted in wide band gap matrices, where the energy band gaps of InN/GaN and InAs/GaAs are (0.63 eV)/(3.39 eV) and (0.35 eV)/ (1.42 eV) at room temperature, respectively. In the InAs/GaAs MQW structure, excitons as well as holes in GaAs can be effectively localized at the InAs well due to the effects of the large confinement potential at the well and smaller electronegativity of In than Ga, resulting in much stronger oscillator strength of the excitons and/or much brighter luminescence [16.26–16.28]. Since the Bohr radius of free excitons in GaN is much smaller than that of GaAs, and also the energy band gap ratio between the well and barrier is smaller in InN/GaN than that in InAs/GaAs, excitons in GaN would be more effectively localized at the inserted type-I InN well, resulting in two dimensional excitons.

One very important difference between the InN/GaN and InAs/GaAs systems is that the former is an immiscible alloy system but the latter is a miscible one [16.31]. Depending on this nature, as shown later, we have found that structurally very highquality InN layer insertion into a GaN matrix, that is, atomically flat and sharp interface between them, can be expected by the self atomic-ordering effect [16.32, 16.33].

Furthermore, it should be noted here that this newly proposed 1 ML InN/GaN MQW structure may have significant practical application, leading to stable roomtemperature operation of GaN-based excitonic devices. Because the free exciton binding energy in GaN is 25 meV and almost the same as the room temperature thermal energy, 26 meV, even a small increase of binding energy up to four times the original value would result in a much more effective localization of free excitons at room temperature, and correspondingly a remarkable increase of their oscillator strength. We expect that this MQW structure will afford the possibility for developing room temperature operating GaN-based excitonic photonic devices. Furthermore, the optical transition energy of those excitons localized there would be remarkably red shifted compared with that for GaN itself due to the additional effective confinement effect of electrons and holes in very narrow but very deep InN wells in the GaN matrix. Therefore, we hopefully expect the development of highly efficient new structure GaN-based light emitters on the basis of the proposed fundamental MQW structure which are working in much longer wavelengths than those available at present, that is, GaN-based laser diodes in the green and/or light emitting diodes in the red.

On the basis of our idea explained above, higher temperature growth of InN/GaN MQW structures consisting of 1 ML and FML InN well insertion into a GaN matrix under the In-polarity growth regime [16.13] was attempted by conventional RF-MBE. Ga-polarity GaN substrates (~ 2 μm thick) grown by MOCVD were used as templates in this study. Prior to the MQW growth, the GaN template was thermally cleaned at 860°C for 15 min and re-growth of 100-nm-thick GaN was carried out at

845°C. One basic period of the MQW studied here consisted of a $\sim$1 ML InN well and $\sim$15 nm GaN barrier. In order to reduce the effect of accumulated strain in the MQW structure, we inserted a $\sim$150-nm-thick GaN spacer after depositing 5–6 periods of this basic structure. Then this process was repeated 8 times, resulting in 40–48 InN well insertions in the GaN matrix. Finally the top surface was capped by a 100 nm GaN layer.

Figure 16.18 shows XRD $2\theta - \omega$ patterns around the GaN (002) reflection for samples grown at temperatures of (a) 500°C, (b) 600°C and (c) 650°C with a GaN spacer layer and 1 ML InN supplied using a deposition rate of 0.5 Å/sec (total supply of InN is 1 ML). From the observation of satellite peaks, it was found that MQW structures could be fabricated at up to 600°C, which is 100°C higher than the highest epitaxy temperature of In-polarity InN layers by MBE (about 500°C [16.1]), and also, from the number and sharpness of those satellite peaks, it was found that the structural properties of these MQWs were remarkably improved with increasing growth temperature. In particular, the sample grown at 600°C indicated sharp satellite peaks up to 5th order. Improvement in the surface/interface flatness with increasing growth temperature was also confirmed by *in situ* observation of RHEED patterns, which changed from spotty to streaky with increasing temperature. These results basically originate from an improvement in both the interface sharpness and the quality of the GaN matrix by the effect of enhanced migration under elevated temperatures. Anyway, we could confirm here that 1 ML InN insertion in GaN matrix was possible at temperatures as high as 600°C. When the temperature was raised to 650°C, however, no such satellite peak was observed, indicating that the InN layers did not stick during source supply and/or decomposed/re-evaporated during growth. Then, in order to achieve InN layer deposition at 650°C, we examined the fabrication of a MQW structure under conditions without a GaN spacer and using 3 MLs of InN supplied using a higher InN deposition rate of 1.5 Å/sec. Figure 16.18(d) shows the XRD pattern of this sample, where very clear and sharp satellite peaks up to 6th order are observed. It was confirmed here that the deposition of InN was possible at as high a temperature as 650°C and the structural quality of the obtained MQWs was significantly improved with increasing growth temperature as expected.

To confirm the structural quality of these MQW samples and also to precisely determine the inserted InN well layer thickness, high resolution XTEM characterization together with careful comparison of the experimental XRD data with simulated diffraction patterns was carried out. Figure 16.19 shows the XTEM dark field images for the sample grown at 600°C, of which the XRD pattern was shown in Fig. 16.18(b), where the image shown in (a) is taken for the condition under $g = 002$ and (b) is a magnified image for the same sample. In these XTEM images, it was confirmed that surprisingly sharp and atomically flat InN well layer insertion into the GaN matrix over a wide area was achieved. Furthermore, it is shown in Fig. 16.19(b) that the InN layer is almost 1 ML and the GaN layer is about 14 nm. Another important point is that we do not observe any serious generation of misfit

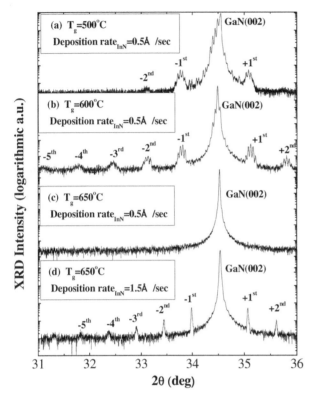

FIGURE 16.18

$2\theta - \omega$ XRD scans of InN/GaN MQW samples grown at different temperatures of (a) 500°C, (b) 600°C, and (c) and (d) 650°C. Samples for (a) to (c) were deposited with GaN spacer layers under 1 ML InN supply with a deposition rate of 0.5 Å/sec, and the sample for (d) was grown without a GaN spacer layer under 3 MLs InN supply, with a deposition rate of 1.5 Å/sec.

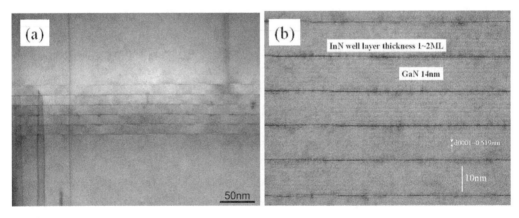

FIGURE 16.19

(a) Cross-sectional TEM dark field image ($g = 002$) for the sample grown at 600°C, of which the XRD pattern is shown in Fig. 16.18(b). The magnified image is shown in (b).

dislocations at the heterointerfaces, at least in the TEM images shown here.

Figure 16.20 shows a dark field XTEM image for the sample grown at 650°C, of which the XRD pattern was shown in Fig. 16.18(d). It is shown that a very clear and sharp MQW structure is fabricated in this sample also, but InN is not a continuous layer any more because of the lowered sticking coefficient for In and N species at such high temperatures as 650°C and/or dissociation and reevaporation of InN at this high temperature. As clearly shown in this figure, this resulted in the

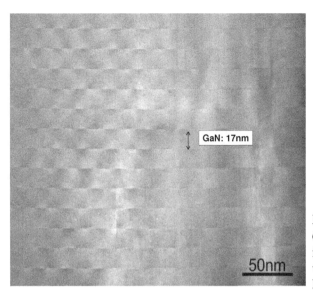

FIGURE 16.20
Cross-sectional TEM dark field image for the sample grown at 650°C, of which the XRD pattern is shown in Fig. 16.18(d).

fabrication of a fractional ML InN insertion in a GaN matrix. Furthermore, we can observe that the strain induced by the insertion of FML InN affected the following deposition of the InN layer and finally this resulted in fabrication of a coherently stacked and/or self-aligned structure of FML InN/GaN MQWs. As explained already, 3 MLs of InN was supplied under 3 times increased deposition rate in this case. It was found that this effectively increased InN surface coverage is necessary to assure its deposition at higher temperatures above 650°C.

According to our experience of fabrication and characterization of conventional InN-based MQWs including InN/InGaN as discussed in Section 16.2, the structural quality of the MQWs shown in Figs. 16.18, 16.19, and 16.20 is surprisingly better than what we expected, that is, the heterointerface between InN and GaN is very sharp and atomically flat over a very wide/long range. To the best of our knowledge, fabrication of this kind of very fine structure 1 ML and FML InN well insertion in a GaN matrix is reported for here the first time. We now believe that this kind of fine heterointerface is formed by the self-ordering mechanism during growth owing to the immiscibility between InN and GaN [16.33].

Next we investigated how the growth temperature affects the dislocation density of the MQW structures and the results are shown in Fig. 16.21, where dislocation densities with both screw and edge components were estimated from XTEM images for a series of samples grown at different temperatures from 500°C to 650°C. It is shown that the density of dislocations remarkably decreases with increasing growth temperature as expected. For screw components, it is about 1×10^{10} cm^{-2} at 500°C and decreases down to about 8×10^{8} cm^{-2} above 600°C. For edge components, however, samples grown at above 600°C show only a slightly smaller dislocation density of about 9×10^{9} cm^{-2} compared with the sample grown at 550°C (about 5×10^{10} cm^{-2}). The observed dependence on temperature of dislocation densities suggests that higher temperature growth is preferable to improve the crys-

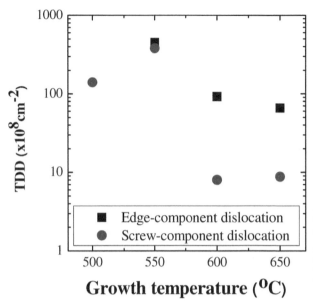

FIGURE 16.21

Growth temperature dependence of dislocation densities in the InN/GaN MQW structure. The squares are for edge components and the circles for screw components.

talline quality of the MQWs.

Figure 16.22 shows PL spectra of samples grown at 650°C for three different InN deposition rates of (a) 0.5 Å/sec, (b) 1.5 Å/sec and (c) 2.5 Å/sec, corresponding to a total InN supply of 1 ML, 3 MLs, and 5 MLs, respectively. The sample for (a) is the same one for which the XRD pattern is shown in Fig. 16.18(c), and no InN layer is inserted in this sample. The sample for (b) is the same one indicating FML InN insertion in a GaN matrix, for which the XRD pattern and its TEM image are shown in Fig. 16.18(d) and Fig. 16.20, respectively. The sample for (c) was newly grown to confirm thicker InN insertion into a GaN matrix and it was found from the XRD pattern analysis that the InN layers were more than 1 ML and resulted in poorer structural quality than that of the sample for (b). The PL emission peak for (a) is located at 363 nm and originated from GaN. The peaks for (b) and (c) are located at 398 nm and 437 nm, respectively. Because of the increased InN content in the GaN matrix, a much larger red shift was observed in the sample (c). It was found that the PL intensity of the sample for (b) was much stronger, probably due to the strong localization of excitons at FML InN wells compared to sample (c). This is because the crystalline quality of the latter sample was deteriorated by the effect of thicker InN well insertion than 1 ML. We consider that skillful design of the MQW structures and properties, such as the spacing of the InN wells, in-plane coverage of FML, thickness of InN wells, and the composition of well and matrix, will give us a new development of exciton-based functionality of photonic devices.

Finally, we have fabricated a preliminary LED structure using the novel structure InN/GaN MQWs grown by RF-MBE and have succeeded in getting fairly bright electroluminescence from the samples [16.25, 16.26]. Figure 16.23 shows the EL spectrum for the sample grown under the same conditions as that shown in Fig. 16.18(b). It is shown that the luminescence peak energy is almost the same

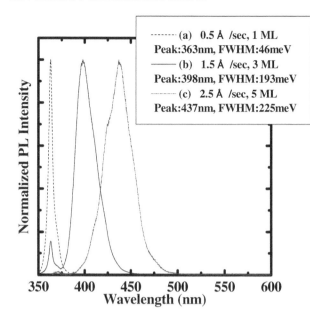

FIGURE 16.22

Room temperature PL spectra of InN/GaN MQWs grown at 650°C under different InN deposition rates: (a) 0.5 Å/sec; (b) 1.5 Å/sec; and (c) 2.5 Å/sec, corresponding to a total InN supply of 1 ML, 3 MLs, and 5 MLs, respectively.

between them. Further, we have extended our understanding of fabricating the novel structure InN/GaN MQWs by MBE to growth by MOCVD, and have successfully developed similar MOCVD processes fabricating the novel structure InN/GaN MQWs consisting of ultimately thin InN wells, approximately 1 ML thick, in a GaN matrix.

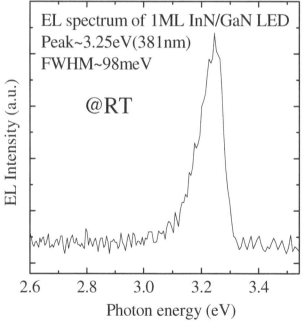

FIGURE 16.23

Electroluminescence spectrum for the preliminary LED sample made of novel structure InN/GaN MQWs with 1 ML InN wells in a GaN matrix.

16.4   InN-based quantum dots

Since the lattice mismatch between InN and GaN or AlN is larger than 10%, the density of misfit dislocations at the heterointerface between InN and GaN or AlN or their ternary/quaternary alloys such as InGaN, InAlN, and/or InAlGaN is very high. These defects would drastically deteriorate device performance. One possible way to avoid this problem is to utilize quantum dots (QDs) in the active region for photonic devices, as the defect density in the QDs can be remarkably reduced and the quantum confinement effects for electrons in the QDs are effective to improve the device performances.

The formation of InN QDs can be achieved by the Stranski-Krastanov (S-K) growth mode and their density is as large as 10^{11} cm^{-2}. Figure 16.24 shows AFM images of typical InN QDs grown on N-polarity GaN by RF-MBE at (a) 450°C, (b) 500°C, and (c) 550°C, respectively. The effective InN supply (hereafter, this is expressed as the nominal surface coverage) was 3.7 MLs for all samples. In order to avoid the deposition of In droplets and also to decrease the surface migration of adsorbed In, InN dots were grown under N-rich conditions (the V/III ratio was about 3 for the In-beam pressure normally used for this experiment). Successful formation of InN QDs is clearly indicated in these figures. The temperature dependence of the average size of the QDs, that is, the diameter and the height, is shown in Fig. 16.25. The average diameter and the height of the QDs grown at 450 and 500°C are about 15–20 nm and 1–1.5 nm, respectively, and they slightly increased with increasing the temperature to 550°C. The density of QDs grown at 450 and 500°C was about 3×10^{11} cm^{-2} and it slightly decreased at 550°C to be about one third of that at lower temperatures, resulting in a slight increase in the diameter of the QDs. Furthermore, it was found that the QDs tend to be formed at the step edges first, followed by uniform formation on the terrace area: this behavior is also indicated in the AFM images shown in Fig. 16.24(c). Note that the growth temperature of the InN QDs could be remarkably increased here compared with other studies on MBE growth of In polarity InN QDs [16.34, 16.35].

In order to study the formation mechanism of the InN dots, we studied how the lateral lattice constant of the deposited InN varies with the nominal surface coverage. Since the critical layer thickness of InN on GaN was expected to be extremely thin, the lattice relaxation process was carefully investigated for much slower than normal InN supply rates, varied from 0.02 to 0.22 ML/s. Furthermore, the change of pseudo-dielectric constants during the formation of QDs was simultaneously observed using *in situ* SE together with the change of lattice constant from RHEED. Figure 16.26(a) summarizes the evolution of the lateral lattice constant a of InN against the nominal surface coverage for four different deposition rates of InN. Here, the growth temperature was kept constant at 500°C. It is shown that the lattice constant is the same as that of GaN in the beginning and it suddenly changes and approaches that of free-standing InN after about 1 ML coverage. Figure 16.26(b)

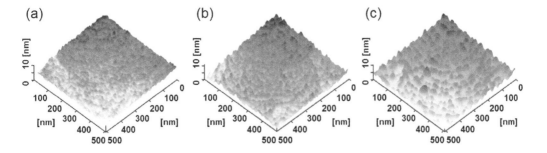

FIGURE 16.24

AFM images of InN QDs grown at (a) 450°C, (b) 500°C and (c) 550°C. (Scanning area: 500 × 500 nm^2).

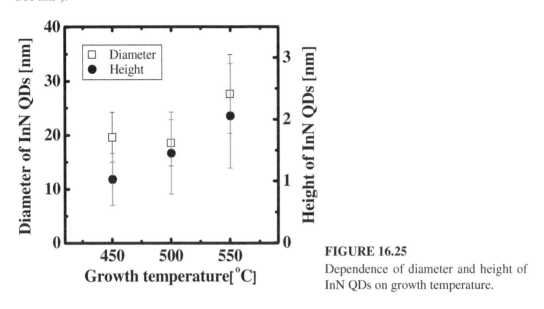

FIGURE 16.25

Dependence of diameter and height of InN QDs on growth temperature.

shows the corresponding change in the real part of the pseudo-dielectric constants observed at a wavelength of 445 nm. It is shown that the pseudo-dielectric constant immediately begins to increase just after starting the growth of InN dots. The increase in the pseudo-dielectric constant with increasing surface coverage corresponds to the increase of the effective thickness of InN. Therefore, the deposition of the InN layer is confirmed by the observation of a SE signal change beginning just after the InN supply, though the lattice constant is kept the same as that of GaN during the initial 1 ML surface coverage. Therefore we can conclude that the InN QDs are formed exactly by the S-K growth mode with the wetting layer thickness of about 1 ML at 500°C and almost full lattice relaxation takes place during the formation of QDs. The lattice relaxation process corresponding to the results shown in Fig. 16.26(a) was also confirmed by the observation of the evolution of RHEED patterns from very streaky at first with a gradual change to spotty.

Optical properties of the InN QDs grown by MBE are very poor and it is difficult to observe PL emission from the QDs. Maleyre *et al.* and Ke *et al.* reported

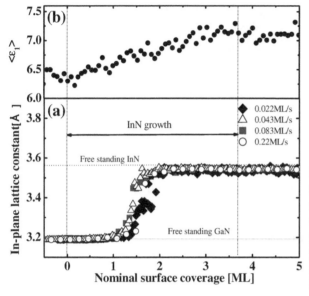

FIGURE 16.26

(a) Evolution of the lateral lattice constant *a* of InN against the nominal InN surface coverage for four different deposition rates. (b) Change of the real part of pseudo-dielectric constants at $\lambda = 445$ nm as a function of nominal InN surface coverage.

near-infrared PL emission from InN QDs grown by MOCVD [16.36, 16.37]. In Ref. [16.37], a blue shift in the emission energy from 0.78 eV to 1.07 eV was observed as the average dot height was reduced from 32.4 nm to 6.5 nm. However, the size of dots was much larger than that of the QDs grown by MBE. The diameter of the dots grown by MOCVD ranged from 70 nm to 255 nm, and the density of the dots was less than 10^{10} cm^{-2}. Therefore, optical emission from InN QDs has only been observed in relatively large-size dots grown by MOCVD at present.

16.5   InN-based nanocolumns

It is known that GaN nanocolumns (NCs) are almost dislocation free [16.38]. Therefore, it can be expected that in nanocolumn-based optical devices such effects as quantum confinement of electrons/holes lead to high-efficiency light emission as well as its tunability [16.39, 16.40]. Fabrication of InN NCs has been reported by several research groups. In those studies, InN NCs were synthesized by several methods such as vapor-liquid-solid, solvothermal method, and halide chemical vapor deposition. However, these growth methods were often used for the growth of InN NCs under spontaneous nucleation [16.39–16.42] and are not suitable for fabricating photonic or electronic devices except for special applications such as electron emitters, because the growth direction of the InN NCs was randomly distributed. MBE growth of InN NCs was also reported, and self-assembled and structurally uniform NCs were grown along the *c*-axis direction, perpendicular to the substrate surface [16.45]. We examined the growth of InN NCs in both In and N-polarity growth regimes on GaN templates by MBE [16.44–16.46]. It was found that the shape control of NCs was easier in the In-polarity case. We explain here the growth of *c*-axis oriented InN NCs in In-polar growth regime studied by *in situ* investigation of SE and RHEED.

Figure 16.27 shows scanning electron microscopy (SEM) images of the (a) top surface, (b) 30°-tilted-surface, (c) cross-section, and (d) 5°-tilted cross-section of typical In-polarity InN NCs. The InN NCs were grown by RF-MBE on an MOVPE-grown 0.85-μm-thick Ga-polarity GaN template. The N/In flux ratio for the InN NC epitaxy was about 7, which was a very N-rich condition compared with the InN-QD growth of about 3 (see Section 16.4). The InN NC epitaxy was *c*-oriented with an epitaxial relationship of $(0001)_{InN} \parallel (0001)_{GaN}$. All NCs were in a hexagonal-column shape with faceted top surface and smooth side-walls. The diameters of the InN NCs were Gaussian distributed with a mean size of about 220 nm. The standard deviation (σ) obtained from a Gaussian fit was about 60 nm. This Gaussian-like distribution of the diameter is typical for self-organized growth of NCs. The density of InN NCs was about 4.4×10^8 cm^{-2}. Here, we would like to point out that the density and diameter of the NCs could be controlled by changing the growth temperature and In/N flux ratio. From the cross-sectional image, we found that the diameter of the InN NCs hardly changes with the growth time under this growth condition. It indicates that the diameter of the NCs is almost determined by the initial growth stage of InN. Furthermore, the InN NCs have nearly the same height, with a value of about 1.2 μm which is independent of their diameters.

In order to study the growth mechanism of InN NCs in the initial growth stage, we investigated the *in situ* in-plane lattice constant *a* by RHEED *d*-spacing (Fig. 16.28(a)), *in situ* SE signal (the real part of pseudo dielectric constants $\langle \varepsilon_1 \rangle$) and *in situ* intensity of RHEED diffraction beam (Fig. 16.28(b)) as a function of growth time. The $\langle \varepsilon_1 \rangle$ was recorded for $\lambda = 1600$ nm. At first, we simply describe the evolution of growth by the change of RHEED intensity. In Fig. 16.28, several points are marked as A–E during the growth procedure. The growth started from point A and only In was supplied during A–B with a coverage corresponding to 1 ML InN. The RHEED intensity was immediately decreased indicating the deposition of In. Then, In supply was stopped and only the N beam was supplied from point B to point C for 30 seconds. One ML of InN was grown in this region. From point C, both In and N beams were supplied and the RHEED intensity was decreased again, indicating InN deposition. Point D was marked at the lowest intensity of RHEED. The InN coverage at point D was about 5 ML. Point E was marked at the InN coverage of about 10 ML.

From the *in situ* signal of $\langle \varepsilon_1 \rangle$ shown in Fig. 16.28(b), $\langle \varepsilon_1 \rangle$ is slightly increased first due to the deposition of In and then slightly decreased because of the formation of 1 ML of InN under the irradiation by the N-beam in region A–C. In region C–E, $\langle \varepsilon_1 \rangle$ is first increased to that at point D and then is decreased to that at point E. Here, we have arranged the GaN layer thickness so that the increase of thickness corresponds to the increase of $\langle \varepsilon_1 \rangle$ at the initial growth of InN. Furthermore, we have previously reported that the $\langle \varepsilon_1 \rangle$ tends to be smaller when the surface becomes

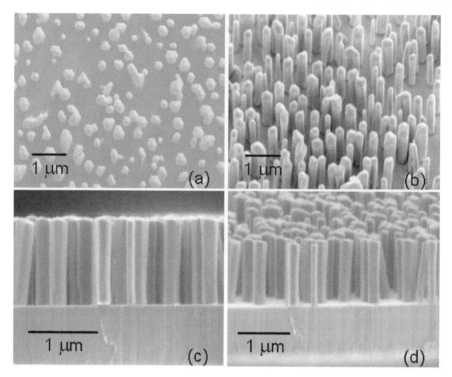

FIGURE 16.27

SEM images of typical InN nanocolumns: (a) Surface (top view); (b) 30°-tilted surface; (c) cross-sectional image; (d) 5°-tilted cross-sectional image.

N-rich [16.6]. Therefore, in this region, the variation of $\langle \varepsilon_1 \rangle$ depends on the competition between the increase of InN thickness and surface roughening. In region C–D, the increase of $\langle \varepsilon_1 \rangle$ indicates that the increase of InN thickness is dominant while the surface is not rough. In region D–E, the quick decrease of $\langle \varepsilon_1 \rangle$ indicates that the surface quickly becomes rough. Here, we notice that point D is located at the same position as the lowest intensity of the RHEED. From the *in situ* investigation of in-plane lattice constant (Fig. 16.28(a)), InN has the same lattice constant as GaN in region A–C and is not relaxed. The relaxation of InN starts from point C and is almost completed at point D. This coincides with the typical behavior of self-organized growth of QDs in the S-K mode [16.7, 16.35, 16.46]. The thickness of coherently grown InN (1 ML) in region A–C corresponds well with the study on N-polar InN QDs growth, where the wetting layer thickness of InN on GaN is about 1 ML (see Section 16.4). The strain due to the large lattice mismatch between InN and GaN is relaxed by the formation of QDs. The evolution of $\langle \varepsilon_1 \rangle$ is similar to the previous report of the growth of GaN islands in the S-K mode [16.46]. In region D–E, as we discussed above, the quick decrease of $\langle \varepsilon_1 \rangle$ indicates that the surface becomes rough quickly. It indicates that embryonic InN NCs are formed in this region leading to a rough surface. This procedure is completed at point E. The formations of InN QDs at point D and embryonic InN NCs at point E are con-

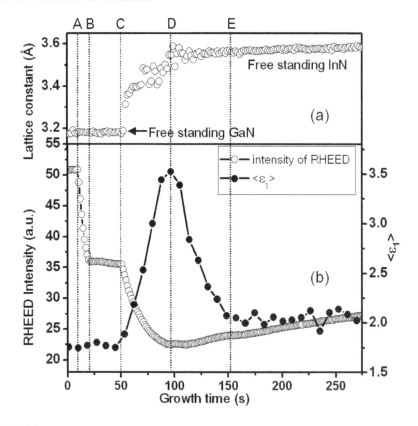

FIGURE 16.28

In situ investigation of the initial growth of InN nanocolumns by RHEED and SE. (a) Evolution of the in-plane lattice constant of InN as a function of growth time. The lattice constant is calculated from the spacing d between (01) and $(0\bar{1})$ RHEED diffraction spots. (b) *In situ* change of RHEED intensity and the real part of the pseudo-dielectric constants ($\langle\varepsilon_1\rangle$) at $\lambda = 1600$ nm as a function of growth time. A–E are marked as different growth points.

firmed by AFM observation, as shown in Fig. 16.29. It is clear that InN QDs are grown at point D with high density. The mean size of InN QDs is about 70 nm. Some large islands with relatively high height are also observed, which might be due to the coalescence of small dots. The density of these large islands becomes higher while their height becomes larger with further growth as shown in the AFM image at point E. These islands are the embryonic InN NCs. The surface roughness (root mean square value) of the InN surface at point D and point E are 1.2 nm and 7.7 nm, respectively, in $10 \times 10~\mu m^2$ scanned areas. This coincides well with the quick decrease of $\langle\varepsilon_1\rangle$. The mean diameter of the embryonic InN NCs at point E is about 210 nm, which is close to the diameter of InN NCs shown in Fig. 16.27. Actually, the density of these islands is about 7.5×10^8 cm^{-2}, which is slightly larger than that of InN NCs. This indicates that the density and diameter of InN NCs are almost determined at point E with InN coverage of about 10 ML. Furthermore, as mentioned above, the growth rate of InN NCs is about 3 times higher than that of InN films at the same In beam flux. This indicates that In adatoms at other regions migrate to the side-walls of the InN NCs and most of them move to the top faces. This can also be confirmed by the surface morphologies at points D and E shown in Fig. 16.29. In the area without the embryonic InN NCs at point E, the surface is still dominated by the InN QDs that show similar morphology to that at point D. The embryonic InN NCs originate from the coalescence of self-organized InN QDs, indicating that the growth of InN NCs is also self-organized. After point E, the growth is along the *c*-direction and finally results in the InN NCs as shown in Fig. 16.27.

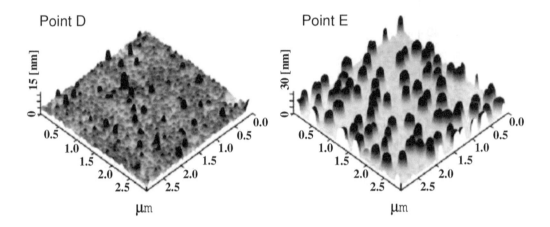

FIGURE 16.29
AFM surface morphologies of InN at point D and point E.

The crystalline quality of the InN NCs was investigated by HR-XRD, where diffraction peaks from (002) InN, (002) GaN and (006) sapphire were observed, indicating a *c*-axis oriented epitaxy of the InN NCs. It was found that the FWHM of the ω-scan for (002) InN NCs showed a comparable value to that of the GaN template, in the range of 500–600 arcsec, indicating that tilt of the InN NCs was very small. InN NCs showed strong PL with an emission peak at about 0.70–0.71 eV at 14 K. The PL intensity of InN NCs was typically several times stronger than that of InN film. Strong PL of InN NCs was also observed with an emission peak energy at about 0.67 eV at room temperature. These results indicate that the InN NCs would be a good candidate for high-efficiency photonic nanodevices.

16.6 Conclusions

Systematic studies of MBE growth of InN-based low dimensional structures have been presented. First, fabrication of simple InN/GaN single quantum well structures was reported and their structural qualities and optical properties in the near infrared region were discussed. The large lattice mismatch was shown to seriously affect the growth and properties of InN/GaN heterostructures. Therefore, the use of InGaN ternary alloys as barriers instead of GaN was explored. A dramatic improvement in crystalline quality of InN-based QWs was achieved by using $In_{0.7}Ga_{0.3}N$ barriers, where the lattice mismatch between well and barriers is significantly reduced compared with the InN/GaN QWs while keeping the band gap difference between the well and barrier at about 0.3 eV. The critical thickness was estimated for an InN well on both GaN and $In_{0.7}Ga_{0.3}N$ barriers and the influence of the quantum confined Stark effect was evaluated.

Novel InN-based nanostructures consisting of one monolayer thick InN wells inserted in a GaN matrix have been proposed and achieved. Since the InN well layer thickness is as thin as 1 ML, that is, below the critical thickness for InN epitaxy on GaN, a coherent heterostructure could be constructed in this InN/GaN MQW system, avoiding the problems arising from the large lattice mismatch. Finally, the growth of other InN based nanostructures was described, such as zero-dimensional quantum dots and one-dimensional nanocolumns. These structures offer both a remarkable reduction in defect density and enhancement of quantum confinement effects. It is possible to encompass the entire InN-based nanostructure family by growing not only the fundamental 1-ML-thick InN/GaN matrix QW structure, but also fractional monolayer MQWs, quantum dots, nanocolumns and also conventional two-dimensional MQWs.

References

[16.1] K. Xu and A. Yoshikawa, Effects of film polarities on InN growth by molecular-beam epitaxy, *Applied Physics Letters*, 83 (2003) 251–253.

[16.2] T. Ohashi, T. Kouno, M. Kawai, A. Kikuchi, and K. Kishino, Growth and characterization of InGaN double heterostructures for optical devices at 1.5-1.7 μm communication wavelengths, *physica status solidi (a)*, 201 (2003) 2850–2854.

[16.3] S. B. Che, W. Terashima, T. Ohkubo, M. Yoshitani, N. Ashimoto, K. Akasaka, Y. Ishitani, and A. Yoshikawa, InN/GaN SQW and DH structures grown by radio frequency plasma-assisted MBE, *physica status solidi (c)*, 2 (2005) 2258–2262.

[16.4] K. Xu, N. Yano, A. W. Jia, A. Yoshikawa, and K. Takahashi, In-situ real-time analysis on strain relaxation process in GaN growth on sapphire by RF-MBE, *Journal of Crystal Growth*, 237–239 (2002) 998–1002.

[16.5] B. Heying, R. Averbeck, L. F. Chen, E. Haus, H. Riechert, and J. S. Speck, Control of GaN surface morphologies using plasma-assisted molecular beam epitaxy, *Journal of Applied Physics*, 88 (2000) 1855–1860.

[16.6] M. Yoshitani, K. Akasaka, X. Wang, S. B. Che, Y. Ishitani, and A. Yoshikawa, In situ spectroscopic ellipsometry in plasma-assisted molecular beam epitaxy of InN under different surface stoichiometries, *Journal of Applied Physics*, 99 (2006) 044913:1–6.

[16.7] A. Yoshikawa, N. Hashimoto, N. Kikukawa, S. B. Che, and Y. Ishitani, Growth of InN quantum dots on N-polarity GaN by molecular-beam epitaxy, *Applied Physics Letters*, 86 (2005) 153115:1–3.

[16.8] Y. Ishitani, H. Masuyama, W. Terasima, M. Yoshitani, S. B. Che, and A. Yoshikawa, Band gap energy of InN and its temperature dependence, *physica status solidi (c)*, 2 (2005) 2276–2280.

[16.9] V. Y. Davydov, A. A. Klochikhin, V. V. Emtsev, D. A. Kudyukov, S. V. Ivanov, V. V. Vekshin, F. Bechstedt, J. Furthmüller, J. Aderhold, J. Graul, A. V. Mudryi, H. Harima, A. Hashimoto, A. Yamamoto, and E. E. Haller, Band gap of hexagonal InN and InGaN alloys, *physica status solidi (b)*, 234 (2002) 787–795.

[16.10] S. B. Che, W. Terashima, Y. Ishitani, T. Matsuda, H. Ishii, S. Yoshida, and A. Yoshikawa, Fine-structure N-polarity InN/InGaN multiple quantum wells grown on GaN underlayer by molecular-beam epitaxy, *Applied Physics Letters*, 86 (2005) 261903:1–3.

[16.11] E. Dimakis, E. Iliopoulos, K. Tsagaraki, and A. Georgakilas, Physical model of InN growth on Ga-face GaN (0001) by molecular-beam epitaxy, *Applied Physics Letters*, 86 (2005) 133104:1–3.

[16.12] C. S. Gallinat, G. Koblmüller, J. S. Brown, S. Bernardis, J. S. Speck, G. D. Chern, E. D. Readinger, H. Shen, and M. Wraback, In-polar InN grown by plasma-assisted molecular beam epitaxy, *Applied Physics Letters*, 89 (2006) 032109:1–3.

[16.13] X. Wang, S. B. Che, Y. Ishitani, and A. Yoshikawa, Step-flow growth of In-polar InN by molecular beam epitaxy, *Japanese Journal of Applied Physics*, 45 (2006) L730–L733.

[16.14] X. Wang, S. B. Che, Y. Ishitani, and A. Yoshikawa, Effect of epitaxial temperature on N-polar InN films grown by molecular beam epitaxy, *Journal of Applied Physics*, 99 (2006) 073512:1–5.

[16.15] X. Q. Shen, T. Ide, M. Shimizu, and H. Okumura, Growth and characterizations of InGaN on N- and Ga-polarity GaN grown by plasma-assisted molecular-beam epitaxy, *Journal of Crystal Growth*, 237–239

(2002) 1148–1152.

[16.16] S. B. Che, T. Shinada, T. Mizuno, X. Wang, Y. Ishitani, and A. Yoshikawa, Effect of precise control of V/III ratio on In-rich In-GaN epitaxial growth, *Japanese Journal of Applied Physics*, 45 (2006) L1259–L1262.

[16.17] T. Ohashi, P. Holmstrm, A. Kikuchi, and K. Kishino, High structural quality InN/In$_{0.75}$Ga$_{0.25}$N multiple quantum wells grown by molecular beam epitaxy, *Applied Physics Letters*, 89 (2006) 041907:1–3.

[16.18] S. B. Che, Y. Ishitani, and A. Yoshikawa, Fabrication and characterization of 20 periods InN/InGaN MQWs, *physica status solidi (c)*, 3 (2006) 1953–1957.

[16.19] G. Martin, A. Botchkarev, A. Rockett, and H. Morkoç, Valence-band discontinuities of wurtzite GaN, AlN, and InN heterojunctions measured by x-ray photoemission spectroscopy, *Applied Physics Letters*, 68 (1996) 2541–2543.

[16.20] B. Arnaudov, T. Paskova, P. P. Paskov, B. Magnusson, E. Valcheva, B. Monemar, H. Lu, W. J. Schaff, H. Amano, and I. Akasaki, Energy position of near-band-edge emission spectra of InN epitaxial layers with different doping levels, *Physical Review B*, 69 (2004) 115216:1–5.

[16.21] A. A. Klochikhin, V. Y. Davydov, V. V. Emtsev, A. V. Sakharov, V. A. Kapitonov, B. A. Andreev, H. Lu, and W. J. Schaff, Acceptor states in the photoluminescence spectra of n-InN, *Physical Review B*, 71 (2005) 195207:1–16.

[16.22] R. E. Jones, H. C. M. Van Genuchten, S. X. Li, L. Hsu, K. M. Yu, W. Walukiewicz, J. W. Ager III, E. E. Haller, H. Lu, W. J. Schaff, Electron transport properties of InN, *Materials Research Society Symposium Proceedings*, 892 (2006) 105.

[16.23] Y. Ishitani, W. Terashima, S. B. Che, and A. Yoshikawa, Conduction and valence band edge properties of hexagonal InN characterized by optical measurements, *physica status solidi (c)*, 3 (2006) 1850–1853.

[16.24] A. Yoshikawa, S. B. Che, W. Yamaguchi, H. Saito, X. Q. Wang, Y. Ishitani, and E. S. Hwang, Proposal and achievement of novel structure InN/GaN multiple quantum wells consisting of 1 ML and fractional monolayer InN wells inserted in GaN matrix, *Applied Physics Letters*, 90 (2007) 073101:1–3.

[16.25] A. Yoshikawa, S. B. Che, N. Hashimoto, H. Saito, Y. Ishitani, and X. Q. Wang, Fabrication and characterization of novel monolayer InN quantum wells in a GaN matrix, *Journal of Vacuum Science and Tech-*

nology B, 26 (2008) 1551–1559.

[16.26] A. Yoshikawa and S. B. Che, Proposal of novel structure light emitting devices consisting of InN/GaN MQWs with ultrathin InN wells in GaN matrix, *International Journal of High Speed Electronics and Systems*, 18 (2008) 993–1003.

[16.27] J. E. Northrup, J. Neugebauer, R. M. Feenstra, and A. R. Smith, Structure of GaN(0001): The laterally contracted Ga bilayer model, *Physical Review B*, 61 (2000) 9932–9935.

[16.28] O. Brandt, L. Tapfer, R. Cingolani, K. Ploog, M. Hohenstein, and F. Phillipp, Structural and optical properties of (100) InAs single-monolayer quantum wells in bulk like GaAs grown by molecular-beam epitaxy, *Physical Review B*, 41 (1990) 12599–12606.

[16.29] P. D. Wang, N. N. Ledentsov, C. M. Sotomayor Torres, I. N. Yassievich, A. Pakhomov, A. Yu. Egovov, P. S. Kop'ev, V. M. Ustinov, Magneto-optical properties in ultrathin InAs-GaAs quantum wells, *Physical Review B*, 50 (1994) 1604–1610.

[16.30] M. V. Belousov, N. N. Ledentsov, M. V. Maximov, P. D. Wang, I. N. Yasievich, N. N. Faleev, I. A. Kozin, V. M. Ustinov, P. S. Kop'ev, and C. M. Sotomayor, Energy levels and exciton oscillator strength in submonolayer InAs-GaAs heterostructures, *Physical Review B*, 51 (1995) 14346–14351.

[16.31] I. Ho and G. B. Stringfellow, Solid phase immiscibility in GaInN, *Applied Physics Letters*, 69 (1996) 2701–2703.

[16.32] P. Ruterana, G. Nouet, W. Van der Stricht, I. Moerman, and L. Considine, Chemical ordering in wurtzite $In_xGa_{1-x}N$ layers grown on (0001) sapphire by metalorganic vapor phase epitaxy, *Applied Physics Letters*, 72 (1998) 1742–1744.

[16.33] D. Doppalapudi, S. N. Basu, K. F. Ludwig Jr., and T. D. Moustakas, Phase separation and ordering in InGaN alloys grown by molecular beam epitaxy, *Journal of Applied Physics*, 84 (1998) 1389–1395.

[16.34] C. Nörenberg, R. A. Oliver, M. G. Martin, L. Allers, M. R. Castell, and G. A. D. Briggs, Stranski-Krastanov growth of InN nanostructures on GaN studied by RHEED, STM and AFM, *physica status solidi (a)* 194 (2002) 536–540.

[16.35] Y. F. Ng, Y. G. Cao, M. H. Xie, X. L. Wang, and S. Y. Tong, Growth mode and strain evolution during InN growth on GaN(0001) by molecular-beam epitaxy, *Applied Physics Letters*, 81 (2002) 3960–3962.

[16.36] B. Maleyre, O. Briot, and S. Ruffenach, MOVPE growth of InN films and quantum dots, *Journal of Crystal Growth*, 269 (2004) 15–21.

[16.37] W. C. Ke, C. P. Fu, C. Y. Chen, L. Lee, C. S. Ku, W. C. Chou, W.-H. Chang, M. C. Lee, W. K. Chen, W. J. Lin, and Y. C. Cheng, Photoluminescence properties of self-assembled InN dots embedded in GaN grown by metal organic vapor phase epitaxy, *Applied Physics Letters*, 88 (2006) 191913:1–3.

[16.38] V. V. Mamutin, N. A. Cherkashin, V. A. Vekshin, V. N. Zhnerik, and S. V. Ivanov, Transmission electron microscopy of GaN columnar nanostructures grown by molecular beam epitaxy, *Physics of the Solid State*, 43 (2001) 151.

[16.39] A. P. Alivisatos, Semiconductor clusters, nanocrystals, and quantum dots, *Science*, 271 (1996) 933–937.

[16.40] K. Yamano, A. Kikuchi, and K. Kishino, Self-organized GaN/AlN superlattice nanocolumn crystals grown by RF-MBE, *Materials Research Society Symposium Proceedings*, 831 (2004) E8.39.1.

[16.41] J. Zhang, L. Zhang, X. Peng, and X. Wang, Vapor-solid growth route to single-crystalline indium nitride nanowires, *Journal of Materials Chemistry*, 12 (2002) 802–804.

[16.42] Z. H. Lan, W. M. Wang, C. L. Sun, S. C. Shi, C. W. Hsu, T. T. Chen, K. H. Chen, C. C. Chen, Y. F. Chen, and L. C. Chen, Growth mechanism, structure and IR photoluminescence studies of indium nitride nanorods, *Journal of Crystal Growth*, 269 (2004) 87–94.

[16.43] Y. J. Bai, Z. G. Liu, X. G. Xu, D. L. Cui, X. P. Hao, X. Feng, and Q. L. Wang, Preparation of InN nanocrystals by solvothermal method, *Journal of Crystal Growth*, 241 (2002) 189–192.

[16.44] N. Takahashi, A. Niwa, T. Takahashi, T. Nakamura, M. Yoshioka, and Y. Momose, Growth of InN pillar crystal films by means of atmospheric pressure halide chemical vapor deposition, *Journal of Materials Chemistry*, 12 (2002) 1573–1576.

[16.45] M. Yoshizawa, A. Kikuchi, M. Mori, N. Fujita, and K. Kishino, Growth of self-organized GaN nanostructures on Al_2O_3 (0001) by RF-radical source molecular beam epitaxy, *Japanese Journal of Applied Physics – Part 2*, 36 (1997) L459–L462.

[16.46] A. Yoshikawa, K. Xu, Y. Taniyasu, and K. Takahashi, Spectroscopic ellipsometry in-situ monitoring/control of GaN epitaxial growth in MBE and MOVPE, *physica status solidi (a)*, 190 (2002) 33–41.

[16.47] X. Wang, S. B. Che, Y. Ishitani, and A. Yoshikawa, Growth of In-polar and N-polar InN nanocolumns on GaN templates by molecular beam epitaxy, *physica status solidi (c)*, 3 (2006) 1561–1565.

[16.48] X. Wang, S. B. Che, Y. Ishitani, and A. Yoshikawa, In situ spectroscopic ellipsometry and RHEED monitored growth of InN nanocolumns by molecular beam epitaxy, *Journal of Crystal Growth*, 301–302 (2007) 496–499.

17

InN nanocolumns

J. Grandal, M. A. Sánchez-García, and E. Calleja

ISOM-Departamento Ingeniería Electrónica, ETSI Telecomunicación, Universidad Politécnica de Madrid, Ciudad Universitaria, 28040 Madrid, Spain

S. Lazić, E. Gallardo and J. M. Calleja

Department Física de Materiales, Universidad Autónoma de Madrid, E-28049 Madrid, Spain

E. Luna and A. Trampert

Paul-Drude-Institut für Festkörperelektronik, Hausvogteiplatz 5-7, 10117 Berlin, Germany

M. Niebelschütz

Institut für Mikro- und Nanotechnologien, Technische Universität Ilmenau, Gustav-Kirchhoff Str. 7, 98693 Ilmenau, Germany

V. Cimalla and O. Ambacher

Fraunhofer Institute for Applied Solid State Physics, Tullastr. 72, 79108 Freiburg, Germany

17.1   Introduction

GaN nanocolumns grown by molecular-beam epitaxy have been proven to be defect and strain free, exhibiting very high crystal quality. This is also the case for InN nanocolumns, offering an ideal opportunity to study fundamental properties of the least studied III-nitride binary. In addition, the peculiar optoelectronic properties of InN open up the possibility to use this material system for applications in photovoltaics or high power/high frequency electronic devices. However, the fabrication of devices based on nanocolumnar heterostructures is still a challenge. These aspects are reviewed in this chapter.

InN has attracted much attention in the last few years, especially since its band gap value was determined to be close to 0.7 eV [17.1, 17.2], extending the operation range of III-nitrides into the visible and infrared (IR) spectral regions. This fact allows one to envisage applications in long wavelength optoelectronic devices. InN is also a good candidate for high power/high frequency devices due to a small electron effective mass and a high electron drift velocity [17.3, 17.4].

Good crystal quality InN layers are difficult to grow because of two main factors: the low dissociation temperature of InN and the lack of an appropriate substrate. Thermal dissociation can be avoided by growing at low enough temperatures when using molecular-beam epitaxy (MBE). On the other hand, the lack of a lattice matched substrate leads to the generation of extended defects during heteroepitaxial growth. Though this problem can be partially eased using different types of buffer layers to accommodate the lattice mismatch, a rather high dislocation density is almost unavoidable. However, as will be shown later, a careful choice of the growth conditions by MBE yields very high crystal quality InN nanocolumns (NCs) oriented along the *c*-axis, having a hexagonal crystal structure [17.5].

Nanostructures based on III-nitrides are the object of increasing attention for basic physics and novel applications, such as strong light-matter coupling in microcavities [17.6, 17.7] or high temperature single photon emitters in the UV range [17.8]. However, reports on the growth and characterization of InN-based nanostructures are scarce because of their critical growth conditions [17.9] and the limited knowledge of the fundamental parameters of InN, as compared to GaN.

Transmission electron microscopy (TEM) and Raman spectroscopy measurements carried out in III-nitride NCs prove that they are defect-free (containing no dislocations or stacking faults) and strain-free [17.5, 17.10, 17.11], thus, allowing for a proper and reliable study of the basic material properties. The use of buffer layers is not critical for either InN or GaN NCs, as it is for continuous (compact) films. This is due to the fact that the NC aspect ratio and free lateral surface can accommodate both lattice and thermal mismatches [17.12]. Indeed, III-nitrides NCs have been successfully grown on different substrates such as Si(111), sapphire, SiC and also recently on Si(100) surfaces [17.13]. The growth of very high quality wurtzite nitrides on Si(100) substrates represents a crucial step to achieving a reliable integration of optoelectronic nitride-based devices with complementary metal oxide semiconductor (CMOS) technology in integrated circuits.

InN is currently the subject of controversy and discussion to explain the natural this material to have strong *n*-type conductivity. The origin of this high electron concentration is still unclear and it may relate to different sources depending on the growth technique and conditions (material quality, unintentional doping, interface effects). This tendency is also responsible for the strong surface electron accumulation already observed at both polar and non-polar InN surfaces [17.14–17.16]. In fact, the Burstein-Moss shift produced by the high electron densities hindered, until recently, the determination of the true band gap of InN [17.1, 17.2] and it may also be partly responsible for the difficulties, so far, in achieving efficient *p*-type doping of InN.

17.2 Growth of InN nanocolumns on silicon substrates

In order to determine the optimal conditions to grow InN NCs, it is worth starting from the optimal growth of compact (2D) InN layers, which in general proceeds under stoichiometry or slightly metal-rich conditions. When the effective III/V molecular flux ratio is decreased, the layer morphology changes dramatically until a full columnar regime is reached [17.17].

It is found that the substrate temperature is the most critical parameter in the growth of InN, and it needs to be determined before adjusting the III/V ratio by means of varying the In and/or active nitrogen fluxes [17.9]. The growth of In-polar layers performed at substrate temperatures above 500°C (in N-polar films, this limit is about 600°C [17.18]) leads to the dissociation of the InN layer and the formation of metallic In droplets on the surface which are difficult to remove by thermal treatment during the growth (Fig. 17.1(a)). This indicates that the dissociation temperature of InN is lower than that needed to desorb the excess In, as has previously been reported by other authors [17.18–17.21]. However, setting the temperature below the InN dissociation onset is not sufficient to achieve uniform NC growth. At high temperature (but below the dissociation limit) In adatoms may diffuse rapidly from the NC base along their sidewalls up to their top, enlarging the NC diameter and typically yielding a conical shape (Fig. 17.1(b)), also observed by Wang *et al.* [17.10], Stoica *et al.* [17.22] and Ristić *et al.* [17.23]. Once the growth temperature is set to the optimal value, in our case 475°C (well below the dissociation temperature), the III/V ratio becomes the fundamental parameter controlling the morphology of the films. N-rich conditions lead to columnar InN samples (Fig. 17.1(c)) with very intense photoluminescence (PL) emission, while moving conditions toward In-rich yields compact layers (Fig. 17.1(d)) with a PL emission intensity two orders of magnitude lower.

Figure 17.2 shows a typical low temperature (16 K) PL spectrum (obtained using a PbS detector) of a NC sample (Fig. 17.1(b)) characterized by an emission peak at 0.75 eV. A more thorough analysis of the PL in this type of sample has been performed varying the temperature and the excitation power [17.17]. The emission intensity decreases as the temperature increases, while the peak energy position roughly follows the theoretical band gap dependence given by Varshni [17.24]. Additionally, there is no change of the peak position with excitation power over two orders of magnitude. While the first behavior points to an excellent crystal quality (defect-free as observed by TEM techniques) of the InN NCs [17.5], the second one suggests an excitonic nature of this emission.

A detailed description of the growth mechanism has been published recently by Calleja *et al.* [17.25] and by Ristić *et al.* [17.23], from which it seems clear that III-nitride NCs grown by MBE under highly N-rich conditions (without catalysts) do not follow the vapor-liquid-solid mechanism. Instead, it suggests a Volmer-Weber spontaneous nanocolumnar nucleation driven by a strong lattice mismatch, while

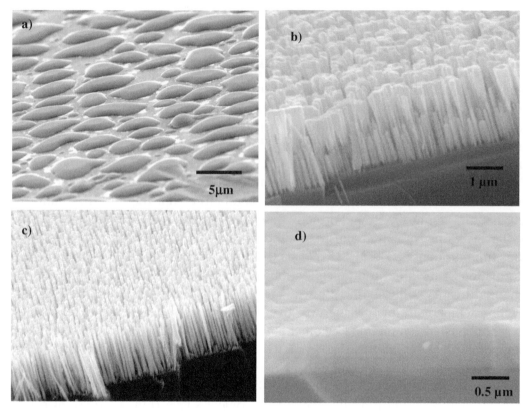

FIGURE 17.1

Scanning electron microscope micrographs: (a) InN sample grown above the dissociation temperature (550°C); (b) InN nanocolumns grown above the optimal substrate temperature yielding a conical shape (500°C); (c) InN nanocolumns grown under N-rich conditions (475°C); and (d) compact
InN layer grown close to stoichiometry (475°C).

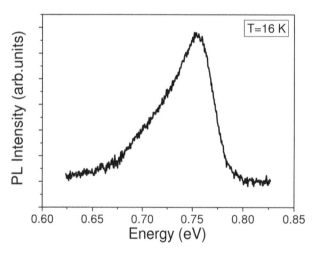

FIGURE 17.2
Low temperature PL spectrum of an
InN nanocolumn sample.

the excess of N prevents coalescence of the nucleation sites. Further nanocolumnar growth proceeds by direct In incorporation on the NC's top and by In diffusion
along the NC's sidewalls up to their apex. InN NCs preserve a constant diameter

if growth conditions are not modified because of a strong metal adatom diffusion length along their sidewalls. This strong diffusion is most probably at the origin of the observed undulation of the NCs sidewalls (Fig. 17.3(a)). No signs of metal droplets have ever been observed at the top of the NCs (Fig. 17.3(b)). It has therefore been proven that III-nitride NCs grow on very different types of substrates, with or without buffer layers, the only requirement being to have a nitrogen flux high enough to significantly reduce metal adatom surface diffusion.

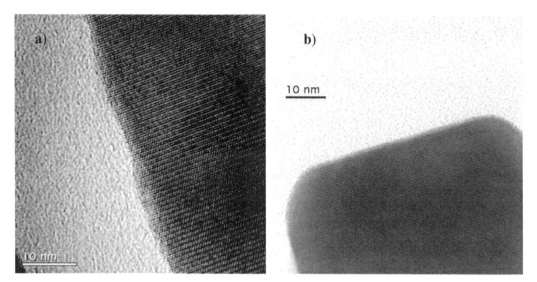

FIGURE 17.3

Transmission electron microscope micrographs of an InN NC showing (a) slightly undulated sidewalls, and (b) no metal droplets at the atomically flat top. Reprinted with permission from E. Calleja, J. Risticacute, S. Fernández-Garrido, L. Cerutti, M. A. Sánchez-García, J. Grandal, A. Trampert, U. Jahn, G. Snchez, A. Griol, B. Sánchez, physica status solidi (b), 244, 2816 (2007). Copyright 2007 by Wiley-VCH Publishers, Inc.

17.3 Strain Relaxation in InN nanocolumns grown on Si(111)

Another important issue is the lattice mismatch accommodation mechanism in InN NCs (as in GaN NCs) that make them grow strain- and defect-free. InN nanocolumnar samples were grown either directly on Si(111) substrates or using a high temperature (HT) AlN buffer following the optimal procedure [17.5, 17.26] in order to clarify the role of the different interfaces on the lattice mismatch accommodation. Once the buffer layer has been grown, the substrate temperature is decreased down to 475°C, the optimal value for In-polar InN [17.9]. At the beginning of the InN NC growth, reflection high energy electron diffraction (RHEED) patterns show clearly

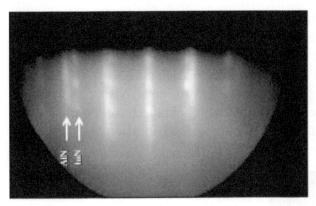

FIGURE 17.4
RHEED pattern of InN NCs super-imposed on that of the AlN buffer. Reprinted with permission from J. Grandal, M. A. Sánchez-García, E. Calleja, E. Luna, and A. Trampert, Applied Physics Letters, 91, 021902 (2007). Copyright 2007, American Institute of Physics.

the superposition of both the InN NCs (spotty features) and the AlN buffer layer (streaks) (Fig. 17.4).

For InN NCs grown without buffer layers, directly on bare Si(111), the substrate temperature was also kept to 475°C. In this case, the RHEED pattern at the beginning of the growth showed concentric rings that were gradually segmenting. After 30 minutes, the rings turned into well defined and bright thin segments lying on the position of the initial concentric rings, which remained until the end of the growth.

Characterization by scanning electron microscopy (SEM) and PL in both series of samples showed no significant differences in terms of morphology and optical properties. The growth on bare Si(111) yields a thin (~ 3 nm) amorphous Si_xN_y layer, observed by TEM between the substrate and the InN NCs, most probably because in this case, no Al metal was previously deposited [17.26]. This Si_xN_y film is frequently found when growing III-nitrides on bare Si surfaces due to the strong affinity of N to bond Si atoms. Nevertheless, NCs are well aligned (perpendicular) to the substrate if the Si_xN_y layer is thin enough (< 3 nm) and homogeneous. It is worth mentioning that, even in a given sample, NCs may have random orientations with respect to the substrate if the Si_xN_y film is not homogeneous or is too thick, causing the NCs to lose their "connection" with the substrate. A similar behavior is observed when NCs are grown on thick SiO_2 [17.27]. In all cases, the crystal orientation of NCs is unaffected, being along the c-direction.

When growing on AlN buffered Si(111), HRTEM images reveal that the interface between the InN NCs and the AlN buffer is clean and abrupt (Fig. 17.5(a)). All investigations by TEM indicate a perfect epitaxial alignment between the Si(111) substrate, the AlN buffer layer, and the InN NCs. TEM also reveals that both the AlN buffer and the InN NCs are strain-free [17.5], which is also corroborated by Raman spectroscopy measurements [17.28]. In this case, the strain relaxation of InN NCs is accommodated by an array of misfit dislocations periodically spaced (Fig. 17.5(b)), as was also observed in InN NCs grown on GaN/Al_2O_3(0001) templates [17.29]. These dislocations are only present at the InN/AlN interface and the NC volume is free of extended defects (Fig. 17.5(a)). For the case of the NCs grown directly on bare Si(111), the epitaxial alignment with the substrate is not perfect. This lack of perfect epitaxial relationship may be associated with the presence of steps on the surface of the Si(111) substrate, which are not present at the InN/AlN interface (Fig. 17.5).

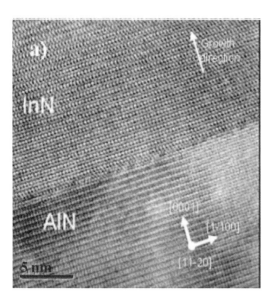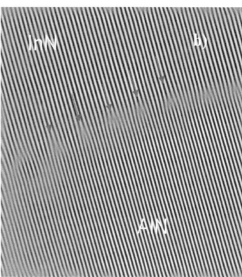

FIGURE 17.5

(a) Abrupt interface between an InN nanocolumn and the AlN buffer layer. The NC volume is free of extended defects. (b) Array of misfit dislocations periodically spaced at the InN-AlN interface. Reprinted with permission from J. Grandal, M. A. Sánchez-García, E. Calleja, E. Luna, and A. Trampert, Applied Physics Letters, 91, 021902 (2007). Copyright 2007, American Institute of Physics.

17.4 InN nanocolumns grown on Si(100)

Wurtzite (c-axis) InN NCs have been successfully grown on Si(111) with AlN layers, on a thin amorphous Si_xN_y film [17.5], and on GaN/Al_2O_3 templates [17.29]. The work by Cerutti *et al.* [17.13] demonstrated that high crystal quality wurtzite GaN NCs can also be grown on Si(100) substrates. These facts suggest the possibility to grow wurtzite InN NCs on Si(100) substrates as well. Using the same growth parameters as in the case of Si(111) substrates, a set of InN nanocolumnar samples was grown directly on Si(100). The optical and structural characterization carried out on these samples by PL, Raman spectroscopy and SEM revealed some interesting features:

- Low temperature (16 K) PL spectrum showed an emission at 0.7 eV with no other emissions at higher energies (Fig. 17.6(a)).

- The peak positions in the Raman spectrum point to strain-free, wurtzite NCs. No differences were observed between spectra from InN NCs grown on Si(100) and those grown on Si(111) (Fig. 17.6(b)).

- The SEM (Fig. 17.6(c)) and TEM (Fig. 17.6(d)) images show that the NCs grow isolated, with slightly random orientations.

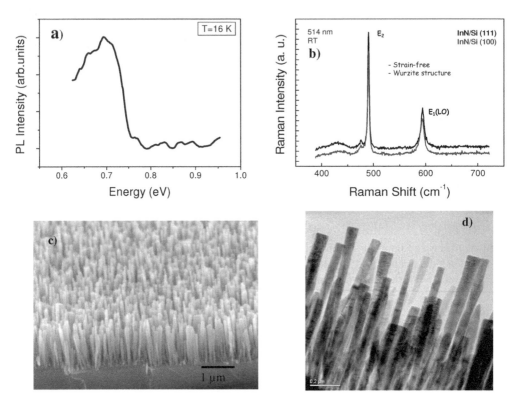

FIGURE 17.6

(a) Low temperature (16 K) PL spectrum of InN NCs grown on Si(100); (b) Raman spectra of InN NCs grown on Si(100) and Si(111) substrates; (c) SEM image of InN NCs grown on Si(100); and (d) magnified image by TEM of the same sample.

A more detailed TEM analysis of the NCs interface with the Si(100) substrate shows an epitaxial alignment that is not as perfect as for growth on AlN-buffered Si(111). As for the case of growth directly on Si(111), the steps present on the surface, clearly observed in Fig. 17.7, may be responsible for this misalignment.

Due to this misalignment, the selected area electron diffraction (SAED) patterns, associated with the InN NCs grown directly either on Si(100) or Si(111) do not present a single spot (Fig. 17.8(a)) as is the case for those grown on AlN-buffered Si(111) (Fig. 17.8(b)).

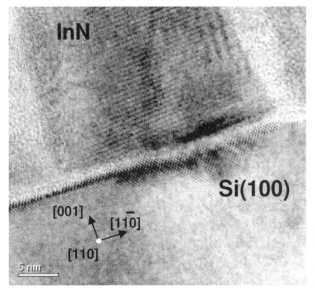

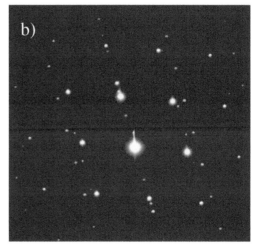

FIGURE 17.7
TEM image of the InN nanocolumn misalignment with the Si(100) substrate.

FIGURE 17.8
Selected area electron diffraction patterns of (a) InN nanocolumns grown directly on Si(100) (similar for the case of Si(111) and (b) InN nanocolumns grown on AlN/Si(111).

17.5 Electron accumulation in InN nanocolumns

Non-intentionally doped InN layers, grown by different epitaxial techniques, exhibit a high propensity to show n-type conductivity (electron concentrations between 10^{17} and 10^{21} cm^{-3}). A strong surface conductivity has also been observed [17.14]. In fact, the surface electron accumulation is one of the the hottest topics under debate.

Segev and Van de Walle [17.30, 17.31] calculated the Fermi level pinning position at the surface of GaN and InN in polar (c-plane) and non-polar (m-plane, *a*-plane) orientations, based on band structure and total energy methods (see Chapter 13). Their calculations predict electron accumulation in polar and in non-polar InN provided that In adlayers are present at the surface. Experimental evidence of surface electron accumulation was obtained by different characterization techniques in polar InN, such as Capacitance-Voltage (*C-V*) [17.14, 17.32], Hall effect [17.14] and inelastic electron scattering [17.33] measurements (see Chapter 12).

In order to determine the presence of electron accumulation on non-polar InN surfaces, a very high-quality material is needed, like the InN NCs presented in the previous sections. InN NCs grown along the *c*-axis are single crystalline [17.5], defect free and have non-polar lateral surfaces (sidewalls).

In Section 17.2, it was shown how the NC growth mechanism involves In adatom diffusion along the NC sidewalls. A high temperature enhances the diffusion process giving way to an increasing diameter toward the NC top, resulting in a conical shape (Fig. 17.1(b)) [17.23]. Although less pronounced, this effect also occurs at lower temperature, causing an undulation of the nanocolumn sidewall (Fig. 17.3(a)). In both cases, In-rich sidewalls are expected, providing the necessary conditions for metal-rich reconstructions of the non-polar InN sidewalls. This could seem to be in contradiction with the NC growth conditions, always claimed to proceed under a N-rich regime. However, the latter refers to the nucleation process on the substrate surface, while the former applies to the NC growth process, so there is no contradiction.

Calleja *et al.* [17.15] reported direct evidence of an electron accumulation layer present at the non-polar surfaces of InN nanocolumns, by measuring current-voltage (*I-V*) characteristics with an atomic force microscope (AFM) in individual InN nanocolumns with different diameters. A work by Niebelschütz *et al.* [17.34] explains in detail the experimental setup. For this purpose, a set of nanocolumnar InN samples with diameters varying from 40 to 120 nm were grown on bare *n*-type Si(111).

Figure 17.9 shows a plot of the InN NCs conductivity as a function of their reciprocal diameter ($1/d$). If an electron accumulation layer is present at the InN NCs non-polar surface (sidewalls), the conductivity will have two contributions: that from the background volume carrier density (n_{bulk}), and that from the surface sheet carrier density (N_S). If the different NCs are grown in a similar way, just changing the diameter, it is assumed that the above mentioned carrier densities would not change significantly between them. In addition, due to the high quality of the InN NCs (low contamination and defect free), n_{bulk} values are assumed to be much smaller than the contribution of N_S. The "two-channel" conductivity may be expressed as

$$\sigma = e\mu \left(n_{bulk} + \frac{4N_s}{d} \right), \qquad (17.1)$$

and it should increase with the reciprocal diameter since the $4N_S/d$ term for surface conduction becomes higher and dominant over n_{bulk} for smaller NC diameters.

For high enough NC diameters the conductivity should become independent of the reciprocal diameter, since the current contribution through the NC volume (n_{bulk}) becomes dominant. In between these two regimes, a smooth transition should be expected. This is exactly what is experimentally observed, as shown in Fig. 17.9.

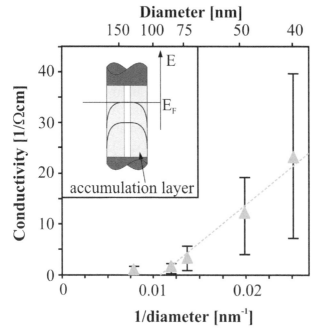

FIGURE 17.9
Conductivity dependence on the reciprocal diameter of InN nanocolumns. The conductivity trend is roughly linear with $1/d$. The inset shows a schematic representation of electron accumulation at the surface of an InN nanowire.

17.6 Phonon-plasmon coupling in InN nanocolumns

Inelastic light scattering (ILS) measurements provide an excellent tool to investigate strain and crystal quality, as well as the presence of free electrons in the InN nanocolumns. Free electrons also affect the phonon spectrum of InN since the longitudinal optical (LO) phonons couple with electrons, producing coupled phonon-plasmon excitations [17.35]. Coupled modes in InN films have been studied by several groups using Raman scattering [17.3, 17.36, 17.37], and other optical characterization techniques such as IR reflectance and IR ellipsometry [17.3].

An electron density-dependent low energy coupled phonon-plasmon mode (L−) at around 400–450 cm^{-1} has been reported in InN films [17.36], coexisting with the uncoupled A$_1$(LO) mode at 586 cm^{-1}. Surface or interface depletion layers, where the uncoupled A$_1$(LO) phonons could exist [17.37], have been claimed to explain this coexistence. However, infrared ellipsometry measurements [17.3] showed that this hypothetical depletion layer would be too thin to be responsible for the strong

A_1(LO) modes observed. Alternatively, wavevector non-conservation has been proposed [17.3] but this assumption is difficult to understand in defect-free samples with high crystalline quality, which is the case for InN nanocolumns.

The weak and broad feature detected in the NC samples at 429 cm^{-1} (Fig. 17.10(a)) is assigned to the lower coupled phonon-plasmon mode (L_-) arising from the interaction of the E_1(LO) mode with the collective oscillation of free electrons through associated electric fields. This peak is not observed in InN compact layers, probably due to the comparatively low specific surface as compared to NCs. The presence of both the uncoupled E_1(LO) phonon and a coupled L_- mode in the NCs can be explained by intrinsic surface electron accumulation at the NC sidewalls [17.33]. Figure 17.10(b) shows micro-Raman measurements performed in a backscattering geometry perpendicular to the growth axis. According to the polarization selection rules for the wurtzite structure, the intensity of the peak at 489 cm^{-1} is significantly lower for the $x(z,-)\bar{x}$ than for the $x(y,-)\bar{x}$ configuration, indicating its E_2 symmetry. The observation of the symmetry-allowed light scattering by the E_1(LO) phonon from the lateral surface (cross-section) of a thick (2 μm) compact sample confirms the E_1(LO) assignment (inset in Fig. 17.10(b)). Phonon-plasmon coupled modes originate from the thin electron accumulation layer present at the NCs sidewalls, while the uncoupled E_1(LO) mode should arise from the underlying scattering volume within the penetration depth of the incident light. There are no significant differences between Raman spectra of InN NCs grown on Si(111) and those grown on Si(100), as expected [17.38] (Fig. 17.6(b)).

Lazić et al. [17.38] also studied the correlation between the electron concentration, estimated from the L^- frequency, and the growth temperature (450–480°C), observing a significant electron concentration decrease for increasing temperature for both Si(111) and Si(100) substrates. As stated before, the NCs growth mechanism involves In adatom diffusion along the NC sidewalls up to its top. It has been observed that the higher the growth temperature (480°C) the higher the NC height for a given growth time, interpreted as due to a strong (exponential) increase of the In adatom diffusion. For a fixed In flux during the growth, the instantaneous number of In atoms "climbing" up the NC sidewall is smaller as their speed increases. Consequently, at higher growth temperatures, the "instantaneous In coverage" of the NC sidewalls decreases. Upon growth termination, this coverage "freezes out", giving way to a specific surface reconstruction. If we consider that a lower coverage represents less In-rich conditions at the NC sidewall, this picture would agree with a lower surface electron concentration [17.30, 17.31]. This assumption needs more experimental evidence, but so far is a qualitative argument.

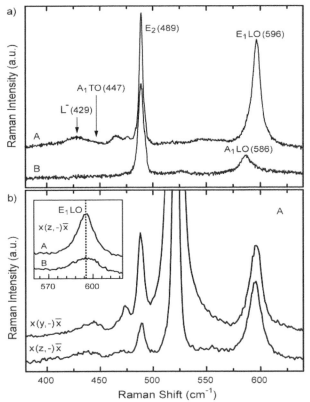

FIGURE 17.10

Room-temperature Raman spectra of (a) an InN NC (Sample A) and compact layer (Sample B). (b) Micro-Raman measurements performed in a backscattering geometry perpendicular to the growth axis.

17.7 Conclusions

In summary, high crystal quality InN NCs can be grown by MBE without any previous substrate treatment, catalyst, or surfactant. The most critical parameter in the growth is the substrate temperature, to avoid thermal dissociation or excessive In mobility. High crystal quality InN NCs have been grown on different substrate orientations, either AlN-buffered or bare Si(111) and Si(100). In the first case (AlN buffer) the InN/AlN interfaces were abrupt, showing an array of misfit dislocations that accommodate the NC strain. TEM analysis of the InN NCs grown on bare Si (both orientations (111) and (100)) revealed a thin Si_xN_y layer at the InN/Si(111) and InN/Si(100) interfaces due to nitridation of the Si prior to the NC growth. The thickness and flatness of the Si_xN_y layer may affect the relative alignment of the NCs with the substrate. In all cases, InN NCs have wurtzite structure, and are defect-free and strain-free.

The electrical characterization of individual InN NCs reveals the presence of electron accumulation at the non-polar InN surfaces (NC sidewalls). This fact follows the theoretical prediction when NCs are grown under In-rich conditions. These results were also corroborated by Raman spectroscopy, showing that this electron layer couples with the $E_1(LO)$ phonon, giving rise to the coupled L_- mode.

References

[17.1] V. Y. Davydov, A. A. Klochikhin, R. P. Seisyan, V. V. Emtsev, S. V. Ivanov, F. Bechstedt, J. Furthmüller, H. Harima, A. V. Mudryi, J. Aderhold, O. Semchinova, and J. Graul, Absorption and emission of hexagonal InN. Evidence of narrow fundamental band gap, *physica status solidi (b)*, 229 (2002) R1–R3.

[17.2] J. Wu, W. Walukiewicz, K. M. Yu, J. W. Ager III, E. E. Haller, H. Lu, W. J. Schaff, Y. Saito, and Y. Nanishi, Unusual properties of the fundamental band gap of InN, *Applied Physics Letters*, 80 (2002) 3967–3969.

[17.3] A. Kasic, M. Schubert, Y. Saito, Y. Nanishi, and G. Wagner, Effective electron mass and phonon modes in n-type hexagonal InN, *Physical Review B*, 65 (2002) 115206:1–7.

[17.4] S. P. Fu and Y. F. Chen, Effective mass of InN epilayers, *Applied Physics Letters*, 85 (2004) 1523–1525.

[17.5] J. Grandal, M. A. Sánchez-García, E. Calleja, E. Luna, and A. Trampert, Accommodation mechanism of InN nanocolumns grown on Si(111) substrates by molecular beam epitaxy, *Applied Physics Letters*, 91 (2007) 021902:1–3.

[17.6] T. Butté, G. Christmann, E. Feltin, J.-F. Carlin, M. Mosca, M. Ilegems, and N. Grandjean, Room-temperature polariton luminescence from a bulk GaN microcavity, *Physical Review B*, 73 (2006) 033315:1–4.

[17.7] S. Christopoulos, G. Baldassarri, P. G. Lagoudakis, A. Grundy, A. V. Kavokin, J. J. Baumberg, G. Christmann, R. Butté, E. Feltin, J.-F. Carlin, and N. Grandjean, Room-temperature polariton lasing in semiconductor microcavities, *Physical Review Letters*, 98 (2007) 126405:1–4.

[17.8] C. Santori, S. Götzinger, T. Yamamoto, S. Kako, K. Hoshino, and Y. Arakawa, Photon correlation studies of single GaN quantum dots, *Applied Physics Letters*, 87 (2005) 051916:1–3.

[17.9] J. Grandal and M. A. Sánchez-García, InN layers grown on silicon substrates: effect of substrate temperature and buffer layers, *Journal of Crystal Growth*, 278 (2005) 373–377.

[17.10] X. Wang, S.-B. Che, Y. Ishitani, and A. Yoshikawa, Growth of In-polar and N-polar InN nanocolumns on GaN templates by molecular beam epitaxy, *physica status solidi (c)*, 3 (2006) 1561–1565.

[17.11] M. A. Sánchez-García, J. Grandal, E. Calleja, S. Lazic, J. M. Calleja, and A. Trampert, Epitaxial growth and characterization of InN nanorods and compact layers on silicon substrates, *physica status solidi (b)*, 243 (2006) 1490–1493.

[17.12] J. Ristić, M. A. Sánchez-García, J. M. Ulloa, E. Calleja, J. Sánchez-Páramo, J. M. Calleja, U. Jahn, A. Trampert, and K. Ploog, Al-GaN Nanocolumns and AlGaN/GaN/AlGaN nanostructures grown by molecular beam epitaxy, *physica status solidi (b)*, 234 (2002) 717–721.

[17.13] L. Cerutti, J. Ristić, S. Fernández-Garrido, E. Calleja, A. Trampert, K. H. Ploog, S. Lazić, and J. M. Calleja, Wurtzite GaN nanocolumns grown on Si(001) by molecular beam epitaxy, *Applied Physics Letters*, 88 (2006) 213114:1–3.

[17.14] H. Lu, W. J. Schaff, L. F. Eastman, and C. E. Stutz, Surface charge accumulation of InN films grown by molecular-beam epitaxy, *Applied Physics Letters*, 82 (2003) 1736-1738.

[17.15] E. Calleja, J. Grandal, M. A. Sánchez-García, M. Niebelschütz, V. Cimalla, and O. Ambacher, Evidence of electron accumulation at nonpolar surfaces of InN nanocolumns, *Applied Physics Letters*, 90 (2007) 262110:1–3.

[17.16] P. D. C. King, T. D. Veal, C. F. McConville, F. Fuchs, J. Furthmüller, F. Bechstedt, P. Schley, R. Goldhahn, J. Schörmann, D. J. As, K. Lischka, D. Muto, H. Naoi, Y. Nanishi, H. Lu, and W. J. Schaff, Universality of electron accumulation at wurtzite *c*- and *a*-plane and zincblende InN surfaces, *Applied Physics Letters*, 91 (2007) 092101:1–3.

[17.17] J. Grandal, M. A. Sánchez-García, F. Calle, and E. Calleja, Morphology and optical properties of InN layers grown by molecular beam epitaxy on silicon substrates, *physica status solidi (c)*, 2 (2005) 2289–2292.

[17.18] K. Xu and A. Yoshikawa, Effects of film polarities on InN growth by molecular-beam epitaxy, *Applied Physics Letters*, 83 (2003) 251–253.

[17.19] M. Higashiwaki and T. J. Matsui, Plasma-assisted MBE growth of InN films and InAlN/InN heterostructures, *Journal of Crystal Growth*, 251 (2003) 494–498.

[17.20] S. Gwo, C.-L. Wu, C.-H. Shen, W.-H. Chang, T. M. Hsu, J.-S. Wang, and J.-T. Hsu, Heteroepitaxial growth of wurtzite InN films on Si(111) exhibiting strong near-infrared photoluminescence at room temperature, *Applied Physics Letters*, 84 (2004) 3765–3767.

[17.21] Y. Nanishi, Y. Saito, T. Yamaguchi, F. Matsuda, T. Araki, H. Naoi,

A. Suzuki, H. Harima, and T. Miyajima, Band-gap energy and physical properties of InN grown by RF-molecular beam epitaxy, *Materials Research Society Symposium Proceedings*, 798 (2004) Y12.1.1–Y12.1.12.

[17.22] T. Stoica, R. Meijers, R. Calarco, T. Richter, and H. Lüth, MBE growth optimization of InN nanowires, *Journal of Crystal Growth*, 290 (2006) 241–247.

[17.23] J. Ristić, E. Calleja, S. Fernández-Garrido, L. Cerutti, A. Trampert, U. Jahn, and K. H. Ploog, On the mechanisms of spontaneous growth of III-nitride nanocolumns by plasma-assisted molecular beam epitaxy, *Journal of Crystal Growth*, 310 (2008) 4035–4045.

[17.24] Y. P. Varshni, Temperature dependence of the energy gap in semiconductors, *Physica*, 34 (1967) 149–154.

[17.25] E. Calleja, J. Ristić, S. Fernández-Garrido, L. Cerutti, M. A. Sánchez-García, J. Grandal, A. Trampert, U. Jahn, G. Sánchez, A. Griol, and B. Sánchez, Growth, morphology, and structural properties of group-III-nitride nanocolumns and nanodisks, *physica status solidi (b)*, 244 (2007) 2816-2837.

[17.26] E. Calleja, M. A. Sánchez-García, F. J. Sánchez, F. Calle, F. B. Naranjo, E. Muñoz, S. I. Molina, A. M. Sánchez, F. J. Pacheco, and R. García, Growth of III-nitrides on Si(111) by molecular beam epitaxy: Doping, optical, and electrical properties, *Journal of Crystal Growth*, 201/202 (1999) 296–317.

[17.27] S. Guha, N. A. Bojarczuk, M. A. L. Johnson, and J. F. Schetzina, Selective area metalorganic molecular-beam epitaxy of GaN and the growth of luminescent microcolumns on Si/SiO_2, *Applied Physics Letters*, 75 (1999) 463–465.

[17.28] J. M. Calleja, S. Lazić, J. Sanchez-Páramo, F. Agulló-Rueda, L. Cerutti, J. Ristić, S. Fernández-Garrido, M. A. Sánchez-García, J. Grandal, E. Calleja, A. Trampert, and U. Jahn, Inelastic light scattering spectroscopy of semiconductor nitride nanocolumns, *physica status solidi (b)*, 244 (2007) 2838–2846.

[17.29] Th. Kehagias, A. Delimitis, Ph. Komninou, E. Iliopoulos, E. Dimakis, A. Georgakilas, and G. Nouet, Misfit accommodation of compact and columnar InN epilayers grown on Ga-face GaN(0001) by molecular-beam epitaxy, *Applied Physics Letters*, 86 (2005) 151905:1–3.

[17.30] D. Segev and C. G. Van de Walle, Origins of Fermi level pinning on GaN and InN polar and nonpolar surfaces, *Europhysics Letters*, 76 (2006) 305–311.

[17.31] D. Segev and C. G. Van de Walle, Surface reconstructions on InN and GaN polar and nonpolar surfaces, *Surface Science*, 601 (2007) L15–L18.

[17.32] S. X. Li, K. M. Yu, J. Wu, R. E. Jones, W. Walukiewicz, J. W. Ager III, W. Shan, E. E. Haller, H. Lu, and W. J. Schaff, Fermi-level stabilization energy in group III nitrides, *Physical Review B*, 71 (2005) 161201(R):1–4.

[17.33] I. Mahboob, T. D. Veal, C. F. McConville, H. Lu, and W. J. Schaff, Intrinsic electron accumulation at clean InN surfaces, *Physical Review Letters*, 92 (2004) 036804:1–4.

[17.34] M. Niebelschütz, V. Cimalla, O. Ambacher, T. Machleidt, J. Ristic, and E. Calleja, Electrical performance of gallium nitride nanocolumns, *Physica E*, 37 (2007) 200–203.

[17.35] G. Abstreiter, M. Cardona, and A. Pinczuk, *Light Scattering in Solids IV*, edited by M. Cardona and G. Güntherodt (Springer-Verlag, Berlin, 1984), p. 5.

[17.36] V. Y. Davydov, V. V. Emtsev, I. N. Goncharuk, A. N. Smirnov, V. D. Petrikov, V. V. Mamutin, V. A. Vekshin, S. V. Ivanov, M. B. Smirnov, and T. Inushima, Experimental and theoretical studies of phonons in hexagonal InN, *Applied Physics Letters*, 75 (1999) 3297–3299.

[17.37] T. Inushima, M. Higashiwaki, and T. Matsui, Optical properties of Si-doped InN grown on sapphire (0001), *Physical Review B*, 68 (2003) 235204:1–7.

[17.38] S. Lazić, E. Gallardo, J. M. Calleja, F. Agulló-Rueda, J. Grandal, M. A. Sánchez-García, E. Calleja, E. Luna, and A. Trampert, Phonon-plasmon coupling in electron surface accumulation layers in InN nanocolumns, *Physical Review B*, 76 (2007) 205319:1–6.

Index